The Science of
Forensic Entomology

The Science of Forensic Entomology

Second Edition

David B. Rivers
Department of Biology
Loyola University Maryland
Baltimore, MD, USA

Gregory A. Dahlem
Department of Biological Sciences
Northern Kentucky University
Highland Heights, KY, USA

Registered Office
John Wiley & Sons Ltd, The Atrium, Southern Gate, Chichester, West Sussex, PO19 8SQ, UK

Editorial Office
9600 Garsington Road, Oxford, OX4 2DQ, UK

For details of our global editorial offices, customer services, and more information about Wiley products visit us at www.wiley.com.

Wiley also publishes its books in a variety of electronic formats and by print-on-demand. Some content that appears in standard print versions of this book may not be available in other formats.

Library of Congress Cataloging-in-Publication Data

Names: Rivers, David, 1966– author. | Dahlem, Gregory, author.
Title: The science of forensic entomology / David B. Rivers, Gregory A.
   Dahlem.
Description: Second edition. | Hoboken, NJ : John Wiley & Sons, Ltd, 2023.
   | Includes index.
Identifiers: LCCN 2022027271 (print) | LCCN 2022027272 (ebook) | ISBN
   9781119640660 (paperback) | ISBN 9781119640684 (adobe pdf) | ISBN
   9781119640615 (epub)
Subjects: MESH: Forensic Entomology | Arthropods–physiology
Classification: LCC RA1063.45 (print) | LCC RA1063.45 (ebook) | NLM W 750
   | DDC 614/.17–dc23/eng/20220810
LC record available at https://lccn.loc.gov/2022027271
LC ebook record available at https://lccn.loc.gov/2022027272

Cover Design: Wiley
Front Cover Image: Courtesy of David Cappaert, Bugwood.org
Back Cover Images: Courtesy of Krzysztof Szpila, Nicolaus Copernicus University, Toruń

Set in 10/12pt MinionPro by Straive, Pondicherry, India
Printed and bound by CPI Group (UK) Ltd, Croydon, CR0 4YY

C9781119640660_220922

Contents

Preface

Welcome to the second edition of *The Science of Forensic Entomology*. Much has changed in forensic entomology since the first edition was published. An incredible amount of research has occurred in carrion ecology, providing new observations on insect associations with human and other animal remains as well as helping develop baseline data on insect succession in several geographical regions of the world. New fields have emerged (i.e., forensic microbiomes, insect pattern analysis), at least one has experienced a rejuvenation (forensic entomotoxicology) and others (molecular entomology and carrion ecology) continue to pave new ground into our understanding of and use of forensic insects in legal investigations. All are signs that not only is forensic entomology growing in interest, but also that the discipline continues to gain a foothold in judicial systems around the world. As you can imagine, trying to capture all that is new from almost a decade of work is nearly impossible. However, we have tried our best. Each chapter has been revised to include recent and updated information, images, and tables, all designed to improve student learning. The book has also been expanded to include new topics, more in-depth coverage on familiar subjects and increased resources for student learning. The latter is reflected in additional chapter end questions, expanded supplementary readings and an increased number of links to additional resources. New to this edition are chapters focused on aquatic insects relevant to forensic investigations, microbiomes of flies and carrion, forensic entomotoxicology, and veterinary and wildlife forensics. Instructors using the first edition also asked for information to be included on professional standards in forensic entomology, as well as to integrate case studies into the text. The latter was accomplished by developing a chapter focused on case studies that cover a range of conditions and insects associated with medico-legal entomology. References are provided for additional case examples for students to expand their understanding of the practical applications of forensic entomology. Along with the 24 chapters, a glossary of common terms is included, as well as four appendices detailing information about methods for collection and preservation of necrophagous flies (Appendix I), how to request help in getting specimens identified (Appendix II), a comprehensive list of research articles with developmental data for forensically important insects (Appendix III), and a new appendix (IV) that provides taxonomic information regarding adult Sarcophagidae. The result is a comprehensive textbook useful to undergraduates, graduate students, and practitioners.

Staying true to the goals of the first edition, our intention here was to create a resource that sheds light on the underlying concepts and principles of ecology and entomology that are essential to the use of insects as evidence in legal investigations. As such, the text is not meant to be a "how to" for death investigations or collection and preservation of insect evidence. Other authors have covered those topics and have provided excellent resources for practitioners. This text is designed to do what the title suggests; to unveil the science that serves as the foundation for forensic entomology and that provides the background necessary to analyze, interpret, and draw conclusions about insects and their activity, primarily in the context of violent crimes. The latter represents a haven for necrophagous and necrophilous insects. Correspondingly, *The Science of Forensic Entomology* is clearly focused on topics relevant to medico-legal entomology and particular attention is given to the biology, ecology, and behavior of necrophagous Diptera, without question the most important type of ecological evidence that can be discovered at crime scenes. What follows is an introduction to a macabre world filled with fascinating and incredibly efficient beasts that are sources of a wealth of information if you understand what to look for.

How to use this textbook

Organization

The Science of Forensic Entomology is organized into 24 chapters that can be explored in any order. The only prerequisite is that for students lacking any kind of entomological background, it is recommended that you first review Chapter 4 before diving into the more advanced entomological topics covered in Chapters 6–24. The early chapters function as an introduction to forensic science (Chapter 1), the history of forensic entomology (Chapter 2), and the role of insects in legal investigations (Chapter 3). Each chapter is organized into a brief overview of the contents, followed by "The big picture" – a list of key concepts or ideas to be presented in the book – followed by in-depth discussion of the concepts and ideas. At the end of each chapter is a representation of the "big picture" concepts and ideas along with at least two to three key points. Questions, references cited, a supplemental reading list, and a list of other useful resources (i.e., websites, organizations) are also included at the end of each chapter.

Pedagogy

With any textbook you have likely encountered, the reading is not meant to work like a novel which you are compelled to read cover to cover. There are several points of entry to this text, so to aid you in finding the meat of each chapter, key concepts are listed at the front as "The big picture." Each of these concepts or ideas serves as the subheadings throughout the text, with the content under a subheading designed to develop the conceptual idea. At the end of each chapter, the "big picture" is presented again, only as a more developed outline with key facts and ideas presented along with the key concepts. In the truest since, these outlines can serve as study guides for the chapters, although the level of detail to be expected for an exam or quiz will likely be less than expected by an instructor. Key terms are set in **bold** when mentioned for the first time in the book. Students should learn these terms as they represent part of the working vocabulary associated with forensic entomology. The emphasis of this book will not be on the terms, but instead on the concepts and application of ideas to solving biological questions.

Study questions are included at the end of each chapter so that students can monitor their progress with learning the material presented. In most chapters, a series of questions following Bloom's taxonomy will be given so that assessment of rote memorization, conceptual understanding, application of ideas and concepts, and synthesis can be self-assessed. Typically, questions will progressively become more challenging, with the latter ones more consistent with higher-order learning.

Study materials

Additional information can be found at the end of each chapter to aid in student learning and allow for further exploration of topics. A supplemental reading list is found after the references cited section and provides more in-depth coverage of topics that oftentimes are only superficially dealt with in a given chapter. URL addresses for websites that provide information on additional readings, topics, or organizations related to chapter topics will follow the supplemental reading list. Because a single resource like this text is not sufficient to do full justice to forensic entomology, students are encouraged to review this information as a means to fully engage in the fascinating world of insects and the field of forensic entomology.

About the companion website

This book is accompanied by a companion website:

www.wiley.com/go/forensicentomology2

The website includes:
- Figures
- Tables.

Chapter 1

Role of forensic science in criminal investigations

For decades, the forensic science disciplines have produced valuable evidence that has contributed to the successful prosecution and conviction of criminals as well as to the exoneration of innocent people.

Committee on Identifying the Needs of the Forensic Sciences Community,
National Research Council[1]

Overview

Before an in-depth discussion of forensic entomology can really begin, there is a need to define the relationship between this discipline and the broader field of forensic science. As the name implies, science is the core of forensic analyses. It is only fitting, then, that Chapter 1 begins with an exploration of the application of science to legal matters, which also serves as a simple working definition of **forensic science**. Throughout the chapter, emphasis will be placed on the use of the scientific method in all forms of forensic analyses, from the process of analyzing physical and trace evidence to understanding the types of outcomes associated with forensic analyses. When the scientific method is applied properly to forensic investigations, integrity is maintained, and "junk" science should not encroach on the pursuit of justice. The different specialty areas of forensic science will be discussed to allow a perspective of the broad impact of science on criminal and civil investigations.

The big picture

- What is forensic science?
- Application of science to criminal investigations.
- Ensuring "good" science in forensic analysis.
- Recognized specialty disciplines in forensic science.

1.1   What is forensic science?

Science is used to solve crimes. In fact, it is instrumental in resolving cases involving both civil and criminal issues, particularly those of a violent nature. Not surprisingly, crime too has become more sophisticated, with today's criminals relying on aspects of science to "hide" their crimes or even to commit an offense. One has to look no further than cybersecurity attacks to see a clear linkage between scientific understanding and criminal activity. This chapter is devoted to understanding the relationship

---

The Science of Forensic Entomology, Second Edition. David B. Rivers and Gregory A. Dahlem.
© 2023 John Wiley & Sons Ltd. Published 2023 by John Wiley & Sons Ltd.
Companian website: www.wiley.com/go/forensicentomology2

between science and legal investigations. Particular attention is given to understanding the **scientific method**, a defined way of doing science, as it serves as the core principle for studying natural phenomena and in forensic analyses.

Forensic science has become a broad term, departing somewhat from the simple definition given earlier in which it was stated to be the application of science to law. The term "forensic" is defined as pertaining to or connected with the law, while "science" is the study of the physical and natural world through systematically arranged facts and principles that are rigorously tested by experimentation. When used together the two terms yield a discipline that addresses issues pertaining to or connected to the law through the application of tested facts and principles and by use of rigorous experimentation. As mentioned previously, the definition of forensic science has become more encompassing, now representing a vast array of medical, scientific (natural and applied), and social scientific disciplines (Table 1.1). So now, we may revise our definition of forensic science to reflect modern, broader approaches: "the use of scientific knowledge and technologies in civil and criminal matters, including case resolution, enforcement of laws, and national security." The term criminalistics refers to the application of scientific techniques or tests in connection with detecting a crime or in processing a crime scene (Harris & Lee, 2020). Most aspects of

applying science to the law, including those associated with forensic entomology, fall under the umbrella of criminalistics.

Use of the term "forensics" as a substitute for "forensic" has created confusion about the terminology to some degree. The former term originally meant the study or art of debate or argumentation. Hence, a school debate team practices forensics or debating. Although "debate" between attorneys has a defined role in the courtroom, it does mean pertaining to the law. However, within the court of public or popular opinion, "forensics" has come to imply forensic science. In fact, a word search on the internet or in some dictionaries yields results which indicate that "forensics" can also be defined as referring to the law. In today's society, practice tends to set policy or norms and, as such, "forensics" is quickly becoming an accepted term for forensic science. No doubt this expanding definition has its origins with the popular television crime shows.

Yet another impact of the rising popularity of forensic science through television programming is the phenomenon known as the crime scene investigation **(CSI) effect** (Saferstein, 2017). The name is derived from the very popular television series *CSI: Crime Scene Investigation* (which was aired on the CBS television network from 2000 to 2015). In general terms, the increased public attention to forensic science is usually linked to this TV series. However, there are numerous other influences that have contributed to the soaring popularity. Regardless of the source of influence, the public's perception of what science can do for a criminal investigation has become distorted. Many individuals, including those who potentially serve as jurors, have become convinced from various media sources that when the experts (i.e., forensic scientists) are called in to investigate a crime, they will always find **physical and/or trace evidence** and that detailed analyses in the crime lab, using real and imaginary technologies[2], will ultimately solve the crime by identifying the perpetrator (Figure 1.1). When delays occur during an investigation or when there is simply little or no evidence to go on, the victim(s), families, and even jurors become frustrated and believe the problem is the incompetence of the investigative team. After all, it only takes one hour for the CSI team to examine the crime scene, find evidence, analyze it, identify suspects, interview the suspects, and seal a full confession! This impressive effort is usually achieved by only one or two people, who perform all the functions that in real life would normally require a team of

Table 1.1 Specialized areas of forensic science recognized by the American Academy of Forensic Sciences (AAFS).

Section	Membership totals*
Anthropology	567
Criminalistics	2720
Digital & Multimedia Sciences	130
Engineering & Applied Sciences	121
General	767
Jurisprudence	192
Odontology	342
Pathology/Biology	951
Psychiatry & Behavioral Science	125
Questioned Documents	167
Toxicology	552
AAFS Total Membership	6634

*Membership data as of December 20, 2019 at http://www.aafs.org/ sections. Forensic entomologists typically belong to the Pathology/Biology section of AAFS.

Figure 1.1 A knife found at a crime scene is an example of physical evidence. *Source*: Ricce/Wikimedia Commons/Public domain.

individuals. Of course, in reality the process is much more time-consuming, requiring many individuals working together, and often a crime goes unsolved. When television fantasy is not separated from reality, the result is that unrealistic expectations are placed on law enforcement officials based on the public's belief that television reflects the real world of forensic science and criminal investigations. It is important to note that some studies (Shelton *et al.*, 2006; Holmgren & Fordham, 2011) investigating the legitimacy of the CSI effect have concluded that negative juror bias does not result from watching crime shows. In fact, potential jurors may be better informed about the appropriate types of evidence to expect for different types of crimes from their faithful viewing of *CSI* and similar sources of entertainment (Shelton *et al.*, 2006).

The reality is that the application of science to legal matters can profoundly influence the resolution of a crime. However, there are limitations to what can and cannot be done, some of which will be addressed later in this chapter. The real value of science in legal matters is that it relies on validation via scientific inquiry using the scientific method. The scientific method is the key, as its use requires adherence to defined, unbiased approaches to designing, conducting, and interpreting experiments. Human emotions or desires, as well as error, are minimized so that the facts, or truths in the case of law, can come to light. A more detailed discussion of the scientific method can be found in Section 1.2.3.

1.2 Application of science to criminal investigations

What can forensic science do to help in civil and criminal cases? Or more to the point, what do forensic scientists do? Forensic investigation is used to address numerous issues associated with criminal, civil, and administrative matters. Indeed, most forensic scientists actually work on cases of a

civil or administrative nature, or deal with issues related to national security such as those under the umbrella of the Department of Homeland Security in the United States (Harris & Lee, 2020). The focus of the majority of this book is **medicocriminal entomology**, so the emphasis in this chapter is placed on criminal matters.

So, how do forensic scientists contribute to criminal investigations? In Section 1.1, we spent some time discussing what they cannot do: solve crimes as on *CSI*. In real cases, forensic scientists spend the majority of their time applying the principles and methodologies of their discipline to the elements of the crime. In other words, a great deal of time is devoted to using the scientific method. Interestingly, training in scientific inquiry is not a universal feature of the curricular pedagogy of all the disciplines contributing to forensic science. Graduates in traditional science subjects such as biology, chemistry, and physics (collectively referred to as the natural sciences), and even geology, are trained in rigorous use of the scientific method. Other disciplines may incorporate aspects of scientific inquiry into their curricula but the approaches are not the *core* of the training as is common in the natural sciences. Thus, our attention will be directed to what forensic scientists do when trained in the natural sciences.

The major functions performed by a forensic scientist include analysis of physical and trace evidence, providing expert testimony to the court and, in some cases, collection of evidence at a crime scene. Details of evidence collection go beyond the scope of this textbook and the reader should consult such excellent works as Saferstein (2017) and Swanson *et al.* (2018) for a general discussion of crime scene techniques, and Haskell and Williams (2008) and Byrd and Tomberlin (2020) for information specific to the collection of insect and arthropod evidence. The majority of this section will focus on analyses of physical and trace evidence. However, before discussing the means of forensic analyses, we need to spend some time determining what is physical evidence.

1.2.1 Physical evidence

Physical evidence is any part or all of a material object used to establish a fact in a criminal case. Items as diverse as bullet casings, bone fragments, a dental crown, matches, or fly maggots can serve as physical evidence. Each is a physical object that may be directly related to a violent act that has been committed or that results from a criminal deed. It is this physical evidence that a prosecutor must use to "prove" the elements of a case, or **corpus delicti**, to a jury beyond reasonable doubt. Proving something true is contrary to the training of a scientist well versed in the scientific method, and thus a forensic scientist faces an ethical challenge to stay focused on facts or data and not to make absolute statements more inclined to come from an attorney. Some of the work of the forensic scientist is to help establish the elements of the case. For example, in a scenario in which a police officer confiscates a brown powder from a suspect or alleged criminal, it is the job of a forensic chemist or toxicologist to determine whether the powder is a narcotic like heroin, in which case a crime has been committed, or whether it is some other substance. In most instances, however, forensic analyses are performed on an object or material collected from what has already been determined to be a crime scene.

In contrast, some evidence is the result of the interaction that occurs between individuals, presumably the victim and assailant. According to **Locard's exchange principle**, every contact between individuals leaves a trace; that is, physical contact between two individuals (or physical objects) will inevitably lead to transference of materials that can serve as trace amounts of physical evidence (Harris & Lee, 2020). These minute amounts of materials are referred to as **trace evidence** and can include such items as hairs, clothing or fabric fibers, gunshot residue, **bloodstains**, and other types of body fluids (**Figure 1.2**). Entomological trace evidence is commonly encountered as insect stains (bloodstains resulting from insect activity), artifacts (similar to insect stains), and frass produced by **necrophagous** or **necrophilous** insects (Figure 1.3), as bite marks on the corpse created when feeding, or trails generated when fly maggots disperse from human remains. In reality, there are a multitude of small pieces of evidence that may be found at a crime scene that are useful to make linkages, identification of persons or objects, or which aid in reconstruction of events associated with the crime.

1.2.2 Collection of evidence

Details of proper methods of evidence recovery will not be covered in this textbook. However, it is important to emphasize that before any evidence is sent to a forensic laboratory for further analysis, the physical

Figure 1.2 Human hair is a common form of trace evidence found at a crime scene or on a victim. *Source*: Edward Dowlman/Wikimedia Commons / Public domain.

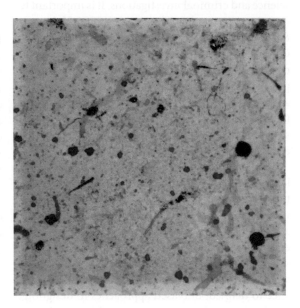

Figure 1.3 Fly artifacts are forms of trace evidence that can be virtually indistinguishable from bloodstains. *Source*: Photo by D.B. Rivers.

evidence collected at a crime scene must be properly preserved. In this respect, evidence from a crime scene must be accounted for during the entire process of investigation, from the time the physical or trace evidence is recovered at a crime scene and analyzed at a forensic laboratory until the evidence is presented in the courtroom by expert witnesses, many of whom are the forensic scientists conducting the analyses. The "accounting" is in the form of a record (i.e., paper, digital, etc.) that provides a complete flowchart showing with whom and where the evidential object has been at all times. This is referred to as **continuity or chain of evidence**. The

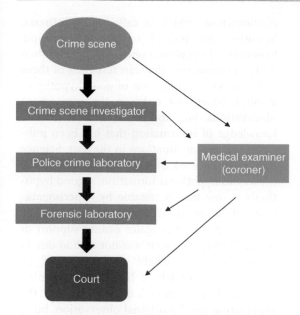

Figure 1.4 Chain of custody of physical evidence collected at a crime scene. *Source*: Modified from Jackson & Jackson (2016).

term also refers to processing the evidence, for once collected, any form of physical and trace evidence must be "preserved" in the condition discovered at the crime scene. In other words, the evidentiary object should not change, be modified, or be allowed to deteriorate as it changes hands during examination and storage. If it does, the evidence may be deemed inadmissible in a court of law. Improper storage of physical or trace evidence is one reason contributing to why DNA testing cannot be performed on evidentiary objects associated with many cold cases dating before the mid-1990s, when forensic DNA became a recognizable force in criminal investigations. The group of individuals responsible for maintaining continuity of evidence during a criminal investigation is termed **chain of custody** and a typical evidence progression is illustrated in Figure 1.4.

1.2.3   The scientific method is the key to forensic analyses

Scientific inquiry using the scientific method is the foundation for the natural sciences as well as forensic science. More generally, science is a process of asking questions about natural phenomena and then seeking the answers to those questions. Scientists as a whole are inquisitive in nature and it is this core make-up that leads individuals to study a particular

scientific discipline. Asking questions is one feature of scientific inquiry. Asking the right questions in the right way and then designing means (experiments) to test those questions is what scientists do (Barnard *et al.*, 1993). Anyone can ask questions and try to find the answers. However, testing questions using an approach centered on formulating hypotheses, making observations from carefully designed experiments, refining questions, and narrowing possible explanations is a skill that is learned or acquired from rigorous training. The process outlined is referred to as the scientific method, a systematic approach or procedure for investigating natural phenomena. Simply stated, it is a defined way of doing science. Not everyone is trained in the scientific method, including many who engage in forensic analyses. Such training is typically associated with education in the natural sciences. The inherent value of the scientific method to scientists is that it provides a roadmap for conducting scientific investigations and also serves as a means for peers in the scientific community to scrutinize research in their respective fields. It can be viewed, then, as a means of validation of results (observations), methodology, and explanations. In the applied world of forensic science, the scientific method provides not only validation but also a systematic approach for distinguishing between alternate hypotheses for elements of a crime.

The scientific method is our way to really understand cause-and-effect relationships in the world around us. Carefully crafted, controlled experiments allow scientists to move beyond observed correlations between one variable and another to a real understanding of the underlying causal relationships (or a realization that two events, while correlated, are not intimately linked to each other).

Research, especially published peer-reviewed research, provides the background for quality scientific testimony in the courtroom. The landmark ruling by the US Supreme Court in *Daubert v. Merrell Dow Pharmaceuticals, Inc.* provides a framework for judges to assess the scientific opinion. The decision provides guidance by establishing four criteria to assess scientific testimony (Faigman, 2002):

1. Is the information testable and has it been tested?
2. What is the error rate and is that error rate acceptable?
3. Has the information been peer reviewed and published?
4. What is its general acceptance by other scientists in the same field?

The process of scientific inquiry is straightforward and use of the scientific method is not meant to be intimidating. That said, asking the right questions in the right way generally requires a fundamental understanding of the phenomena or organisms to be observed as well as practice in developing the skill of scientific inquiry. If you are not convinced that scientific investigation is a skill that is developed, read through any research journal and compare the "quality" of the experiments detailed.

The scientific method can be broken down into specific steps that can be followed much like a flowchart (Figure 1.5):

1. *Make observations that lead to questions.* Observing a natural phenomenon generally leads to formulation of questions: Why did this happen? How did this happen? Will it happen again? The last two questions can lead to the development of explanations that can be tested by the scientific method, but not necessarily the first. "Why" questions imply evolutionary meaning and, while fun for speculation, often cannot be framed, at least not simply, in a manner that allows experimental testing.

2. *Formulation of hypotheses.* The "right" or "good" question is one that lends itself to being tested. What is tested is not the question directly, rather the explanation that has been formulated to account for the initial observation. In other words, an educated guess or explanation of the initial observation or phenomenon, which is called a **hypothesis**. Scientists are trained to develop multiple hypotheses to explain a phenomenon and then design experiments that can test each of these explanations. Formulation of good hypotheses is often based not only on the researcher's observations but also on a comprehensive knowledge of information that has been published on similar situations in the past. Science builds on the work of past scientists. Here lies the key to hypothesis formation: a good hypothesis is one that is testable by experimentation. If after conducting a well-designed experiment the investigator cannot support or refute the hypothesis, it was not a good one to begin with. Conversely, the results of an experiment can falsify a hypothesis, meaning that the data generated do not support the explanation for the original observation, but a hypothesis can never be proven true (Morgan & Carter, 2011). As should be obvious, a conflict potentially exists between the outcomes available to a scientist using the scientific method and those desired by officials working in a judicial system.

3. *Testing hypotheses.* Hypotheses are tested through carefully controlled experiments. In this case, "controlled" means that all possible variables or factors that could influence the outcome of the experiment must be taken into consideration so that only one of them – the one being tested in the hypothesis – is allowed to vary during the study. The others are held as

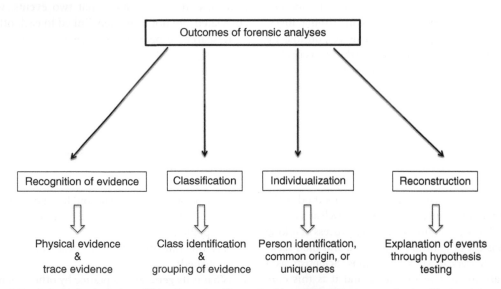

Figure 1.5 Schematic depiction of how the scientific method is used in the process of scientific inquiry.

constant or static as possible for the experimental conditions. For example, in a study interested in testing the influence of temperature on the rate of development of necrophagous fly larvae, the hypothesis will be based on the idea that temperature is the most important factor influencing development. In this example, temperature is the **independent variable**. Aspects of fly development that can be measured as impacted by temperature are called **dependent variables** and might include overall length of development or the duration of each stage of larval development. Other factors (e.g., food, humidity, species of fly, and size of maggot mass) which could potentially be independent variables that influence fly development must be maintained at a constant value (as much as is possible) and are referred to as **control variables**. Only carefully controlled experiments allow an investigator to examine the impact of one variable alone on the condition of being examined.

4. *Evaluating observations or data.* Once the observations have been made, it is imperative that the scientist or investigator thoroughly interprets the data collected. In many cases, this involves a series of comparative evaluations (addressed in more detail in Section 1.2.4) with other data in the scientific literature, databanks test, or **voucher specimens** (a specimen archived in a permanent collection and serving as a reference for a taxon), or other resources. The data are also evaluated by statistical analyses to determine if what has been observed differs significantly from what was predicted or from other treatments. Statistical analyses require that the investigator understands the type of data to be collected prior to initiating the tests so that the experimental design is appropriate for the data and statistical test to be used. Experiments need to be replicated several times to increase the "power" of statistical analysis. For example, a single experiment may be thought of as an anecdotal observation. However, data derived from experiments performed multiple times result in less statistical variation. The interpretations from such data are also considered more reliable.

5. *Refining hypotheses.* After careful evaluation of the data, the original hypotheses are reevaluated or refined so that they can be retested, repeatedly, by the original researchers or by others. The process of experimental testing, if done well, should lead to new, more narrowly focused explanations of the original phenomenon that can be retested over and over. The idea is that with each subsequent round of experimentation, the scientist is moving closer and closer to the real explanation, or in the case of criminal investigation, a step closer to the truth.

It is important to understand that the scientific method is not simply a cookbook approach to scientific inquiry, meaning that once the final step is complete, the answer is known. Rather, doing science is a process whereby each observation leads to new questions with new hypotheses. The journey may seem endless, and in the sense of new observations stimulating new ideas and new questions, it is. However, this systematic approach to addressing questions also guides us closer to the real answers and away from incorrect or false explanations.

It is also important to understand that the scientific method is not the sole source of information used by forensic entomologists and other scientists. Much of our understanding of the relationships between insects and decomposing carrion has come from publications relying on careful systematic observations rather than controlled experiments where a single variable is manipulated to establish cause and effect. For example, we might be interested in knowing what insects are associated with the early decay process of several species of wildlife (Watson & Carlton, 2005) or what species are attracted to human remains in a geographical area (Carvalho *et al.*, 2000). These studies generate new data for forensic inquiry through systematic observations rather than by testing a clearly stated hypothesis. The basic research that we compare against to establish an estimate of the postmortem interval (PMI) is based on detailed observations of a species life stage and size at different times and temperatures (Byrd & Butler, 1998).

1.2.4 Analysis of physical evidence

Physical or trace evidence collected from a crime scene is typically delivered to a crime laboratory or sent directly to a forensic expert for further analyses. The initial characterization may be simply a quantitative or qualitative assessment of the evidence (Jackson & Jackson, 2016). In other words, the analyses may be needed to determine the identity of the evidential sample (**qualitative analysis**) in order to determine if a crime has even been committed, such as drug identification. In other instances, **quantitative analysis** is performed to determine the amount of a particular substance that has been discovered in order to affirm whether legal limits have been exceeded. Both forms of analysis are relevant to entomological evidence.

In the majority of cases, the forensic scientist evaluates physical evidence through comparison testing: the evidential object may be compared to known objects in databases or validated reference collections such as voucher specimens, or compared with the outcomes of controlled experiments. There are several ways in which comparison testing is utilized in forensic science and the most common are discussed briefly here.

1.2.4.1 Recognition of evidence

Recognizing whether a physical object is actually an evidence is the first step in forensic analysis (Harris & Lee, 2020). Such determinations often rely on the experience of the forensic investigator or scientist, whereby the object has been previously observed during training, prior cases, or experimentation.

When entomological evidence is present at a crime scene, care must be taken to collect as much material as possible, even if much of the material will never be used. In general, entomological evidence needs to be collected at the time the crime is discovered; with a few exceptions we cannot go back a month later to get additional specimens to use as evidence. Empty puparia associated with a shallow grave at the time of a body's discovery are much more relevant than empty puparia collected two months later at the old crime scene. In order to estimate PMI on a fairly fresh corpse, the entomologist needs to see the biggest maggots (presumed to be the oldest) of as many different species as possible to make the determination. Just collecting a few maggots from the first larval mass encountered could seriously compromise the ability of an entomologist to accurately estimate the time of death.

1.2.4.2 Classification

Once an item has been recognized as physical evidence, the object is processed in an attempt to identify it (Figure 1.6). Broadly, this means

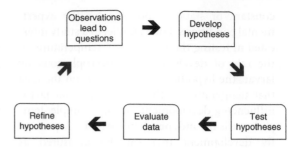

Figure 1.6 Possible outcomes that can result from forensic analysis relying on various forms of comparison testing.

classifying the evidential object into groups or categories. Hairs, bloodstains, and other body fluids must be identified to determine if in fact they are human and, if not, to classify what animal or organism the samples may have been derived from. For example, fly artifacts created by adult flies regurgitating undigested food (such as blood) onto a surface appear almost indistinguishable from some types of blood spatter at a crime scene (Parker *et al.*, 2010). An initial analysis of the artifacts is needed to classify them as either a bloodstain or some other object like an insect stain. Likewise, paint chips, powders, and other materials must be identified so that objects can be grouped with similar items. Comparisons are generally performed between the object of interest and known databases, reference collections, or other validated resources. In the case of insects, voucher specimens are used to confirm relationships with an insect group (usually a family or genus).

The process of classification can also lead to the exclusion of objects from a grouping. Such is the case with material such as paint chips, fibers, or glass fragments, which may be broadly classified or grouped but determined to be not the same as those found on a victim, perpetrator, and/or at some other location of interest. The object would be excluded or considered dissociated from the crime scene since it does not belong to the same class or group of interest.

1.2.4.3 Individualization

The process of classification leads to the identification of the object so that it can be grouped or classified. Classification is not intended to identify specifically say the individual or, in the case of insects, the species of fly or beetle found on a corpse. Individualization is a further level of identification, or narrowing of classification, that involves comparison testing to distinguish an object as being unlike or unique from others in a grouping, or to determine that the physical evidence has the same source or origin when compared with another item in the same class (Harris & Lee, 2020). The latter is commonly done when impressions of shoe prints or tire tread marks from a crime scene (**scene impressions**) are compared with **test impressions** (Figure 1.7). The impressions can then be identified further by making comparisons with specific brand characteristics using databases supplied with information from the manufacturers.

The identity of a victim or an attacker like a rapist can be determined or individualized through the use of DNA profiling on blood, semen, or other body fluid evidence, or from fingerprint analyses. In the case of an individual with a prior criminal record, the DNA (STR) sequences or profiles can then be compared in databases such as CODIS or IAFIS, respectively. Insect evidence can

be identified in a similar fashion by using a series of dichotomous identification keys, voucher specimens, and even DNA analysis to determine which species was collected from a crime scene. These testing procedures can lead to **positive identification** of a victim, criminal, and/or specimen. As with classification, individualization comparisons can also yield negative identifications or exclusions, for example an alleged suspect may be exonerated because the fingerprints or DNA do not match those found at the crime scene or on the victim.

1.2.4.4 Reconstruction

The investigative efforts of a forensic scientist that perhaps most closely showcases use of the scientific method is reconstruction. Because the process of reconstruction involves using the physical evidence and results from analyzing evidential objects to try and piece together the events of the crime, it can be thought of as analogous to hypothesis testing. Crime scene reconstruction requires formulation of explanations to account for the evidence collected, testing the explanations, and then, based on the test results, refining the initial hypotheses so that further testing can be performed. The results of reconstruction can shed light on the events that occurred before, during, or immediately after the crime was committed. This information is useful in corroborating or refuting statements made by a victim, suspect, or eyewitness. As with scientific inquiry, this form of forensic analysis yields information that is mostly speculative theory based predominantly on physical evidence. Reconstruction does not "prove" anything is true.

1.2.4.5 Intelligence information

Ordinarily, gathering of intelligence information related to the activities of criminals falls outside the realm of the natural sciences, and is more consistent with disciplines focused on profiling. However, changes in the global interactions between different groups of people, namely the widespread acts of terrorism, have broadened the scope of forensic science. Terrorist acts have become very sophisticated as has the "war on terror" employed by some nations. In the United States, several forensic scientists work to gather information on terrorist groups or cells by analyzing the weapons or components used to make the devices so that material suppliers or locations can be determined. Even insects have

Figure 1.7 Tire treads pressed into mud, dirt, sand, or snow can represent a scene impression found at crime scene. *Source*: HobsonRoad / Wikimedia Commons/CC BY-SA 4.0.

been used, as adult flies and larvae can be tested for bomb or other explosives residues in the hope that those responsible can be identified and/or the location of explosives assembly, and hence the terrorist group, can be identified. This approach relies on the assumption that a terrorist cell or group has an established **modus operandi**, i.e., that they use characteristic explosive materials.

The preceding simply provides an overview of some of the activities performed by forensic scientists. A more in-depth presentation of the roles of forensic scientists in forensic science and criminalistics can be found in National Research Council Committee on Identifying the Needs of the Forensic Science Community (2009) and Daeid (2010).

1.3   Ensuring "good" science in forensic analysis

Forensic science evidence presented in court is presumed to be unbiased and to have been rigorously tested. Indeed, the very definition of science is based on the premise that the underlying facts and principles of a given scientific discipline have been systematically and rigorously tested through experimentation. For years, the public and legal system has had an almost blind faith in forensic science. This belief at its peak elevated to almost religious proportions, for if science demonstrated it to be "true" then it must be so. No better place to put this into practice than in a courtroom to proclaim guilt or innocence of the accused based on science. The seemingly flawless tools of forensic DNA nearly single-handedly pushed forensic science onto this unrealistic pedestal. Unfortunately (or fortunately depending on the view), that bond of trust has been shattered. High-profile scandals in forensic laboratories and a number of successful defense challenges to forensic science evidence, especially those related to individualization, have shown an unfavorable light on the forensic sciences (Giannelli, 2002). The once impenetrable façade of forensic science showed cracks – large ones – and most importantly, the public viewed for the first time what had always been true, that science of any type is not infallible. A new term emerged in the media, "junk" science, to account for the problems with forensic science. Junk science essentially lacks the features we discussed earlier in describing the scientific method. It refers to forensic science applications that are not based or derived from the basic sciences, have not

been validated through systematic and extensive hypothesis testing via experimentation, and generally do not rely on quantifiable measures of uncertainty. Good science is obviously the polar opposite of this definition, although as stated previously in Section 1.2.3, some forms of scientific investigation do not rely on the scientific method.

The harshest critics indicate that disciplines built on junk science provide unsupported assumptions that lack statistical validation as testimony (Saks & Faigman, 2008). Within forensic science disciplines, the comparative sciences have come under the heaviest scrutiny as being purveyors of junk science. The techniques identified by the National Academy of Science (2009) in their scathing report of forensic science in the United States included bite mark analysis, hair and fiber evidence, bloodstain pattern analysis, bullet matching, and even fingerprinting. Why were these disciplines and techniques singled out by the NAS report? For two major reasons. First, the committee tasked with reviewing forensic science in the United States asked representatives from each forensic science discipline to provide evidence in the form of scientific studies demonstrating validation and reliability of the techniques and methods used by each. Shockingly, only materials on nuclear and mitochondrial DNA and drug analysis were provided (Edwards, 2009). The absence of such scientific research for all other disciplines immediately called into question the validity of forensic evidence across the forensic sciences. Which brings us to the second issue: When unvalidated methods are used in forensic investigation and analyses, especially to make a person's identification through individualization, the result can be a wrongful conviction or the guilty remaining free. In other words, junk science undermines the entire judicial process.

So, what can be done to expunge junk science from the forensic science disciplines? The first step is to recognize where the problems lie in all areas of forensic science. A large portion of that task has already been done, with the following major problems identified (Edwards, 2009; NAS Report, 2009):

1. For many disciplines, a scientific foundation has never been established. Most importantly, scientific research is lacking or altogether absent with respect to confirmation of the validity.
2. No quantifiable measures of uncertainty have been established. Consequently, error rates, which are a fundamental feature of the *Daubert*

standard for admissibility of evidence in US federal courts and for many state courts (Daubert v Merrill Dow Pharmaceuticals, 1993), do not exist, nor are methods of statistical analysis used;

3. An absence of ongoing or new research focused on new technology and innovation is occurring in many disciplines;

4. No standard operating procedures for most disciplines in terms of techniques, analyses, interpretations, terminology, and establishing limits of methodology.

5. No oversight of the disciplines or individuals functioning as practitioners. In other words, *mandatory* certification is not required to be a practitioner for many forensic science disciplines;

6. Forensic or crime laboratories in the public sector are not autonomous from law enforcement;

7. Human nature. This problem was not specifically mentioned in the NAS report (2009) but it has been common knowledge that some individuals stray from the path of good science due to an array of biases (i.e., cognitive bias), adoption of a vigilante mentality, or even to become an "expert for hire" whereby testimony is catered to the desires of the employer.

There were and are other issues that plague the forensic sciences, and the problems are not restricted to the United States. The list above suffices to provide a picture of the major issues that need to be addressed. However unpleasant it was to unmask these deficiencies, it has fueled sweeping changes and for the better (Kaye, 2010). The NAS report outlined a blueprint for structural and cultural reform throughout the forensic science community. Organizations such as the American Academy of Forensic Sciences and others in cooperation with the Department of Justice have established working groups to develop discipline-specific standards, certification processes, and codes of ethics. The American Board of Forensic Entomology (https://forensicentomologist.org/certification/), the North American Forensic Entomology Association (https://www.nafea.net/about/objectives/), and European Association of Forensic Entomology (http://eafe.org/EAFE_Constitution.html) have followed suit, developing codes of ethics for practitioners and members alike, and laying the foundations for standard operating procedures within the discipline. While these represent positive advances in forensic science, there is also a need for reform in other areas of the judicial system to ensure good science in the courtroom. For example, despite the deficiencies identified with several forensic disciplines, safeguards were in place to prevent junk science from ever making its way to the jury. The *Daubert* standard for federal courts and the earlier *Frye* standard still used in many US states established specific criteria for the admissibility of scientific evidence in court (Gabel & Wilkinson, 2008). When exercised properly by the judges who serve as the gatekeepers, improperly analyzed evidence should not be admissible. Yet it did occur and with regular frequency. This example serves as a testament that all issues with forensic science have not originated from bad science. Rather, it reinforces the powerful influence that human nature can have on the judicial system.

While there are many layers associated with the reform in forensic sciences, the key to a solid foundation for any of the disciplines is good scientific research. The NAS committee placed a clear premium on scientific research and rightfully so. When reviewing the tenants of the scientific method, if followed as prescribed, good science should be the result. By extension, good science is essential to ensuring unbiased and meticulously performed forensic analysis in forensic entomology and other disciplines in forensic science.

1.4   Recognized specialty disciplines in forensic science

Forensic investigation is used to examine issues common to criminal, civil, and administrative matters. To address such a vast array of topics, experts from many disciplines are needed to perform forensic analyses. The remaining chapters focus exclusively on forensic entomology. Eleven major subdivisions of forensic science are recognized by the American Academy of Forensic Sciences (http://aafs.org, see Table 1.1), considered to be the largest organization of forensic scientists in the world, with as many as 31 subdisciplines contributing expert analyses in legal cases. Here, we provide a brief snapshot of some of the related fields of forensic science, and forensic entomologists may collaborate with workers in these fields while working on a case.

1.4.1   Forensic pathology

Synonymous with forensic medicine, forensic pathology is the discipline concerned with determining the cause and manner of death through examination of a corpse. A forensic pathologist determines the medical reason for the person's death and also attempts to decipher the circumstances surrounding the death. The pathologist performs an autopsy at the request of the medical examiner (a physician by training) or coroner (an elected official who may or may not be trained in medicine).

1.4.2   Forensic anthropology

Forensic anthropology uses applied principles of physical anthropology, the study of human form via the skeleton system and osteology (study of bones) to examine human remains in a legal context. Reconstructing a body from skeletal remains or making individual or gender identifications are some of the main functions of a forensic anthropologist. The "body farm" at the University of Tennessee in Knoxville is a major research facility maintained by the Department of Anthropology that essentially thrust forensic anthropology into the limelight. The facility has been instrumental in conducting basic and applied research using human corpses to shed light on factors influencing decomposition and altering remains **postmortem**. Today, 10 body farms or human taphonomy centers exist worldwide (8 in the United States, and 1 each in Australia and Amsterdam) and the first in the United Kingdom is scheduled to open in 2020.

1.4.3   Forensic odontology

Forensic odontology is the area of forensic science concerned with dentition or teeth as it pertains to legal matters. The discipline can essentially be divided into activities where dental patterns or individual teeth are used for identification of an individual or to whom an individual tooth belongs, and use of bite marks to individualize a potential attacker in which a victim has been bitten. Teeth and saliva can also be used as a source of DNA for subsequent identification as well. In recent years, bite mark analysis has become one of the most controversial areas of forensic science, frequently receiving the label of "junk" or "failed" science due to the lack of validated methodology or adherence to scientific principles during investigations.

1.4.4   Forensic psychology and psychiatry

Forensic psychologists and forensic psychiatrists have similar roles in the judicial system. In some instances, they determine if a defendant is competent to stand trial for the accused offence. In other instances, the expertise of a psychologist or psychiatrist is needed in intelligence gathering, i.e., trying to characterize the patterns or other features of the modus operandi of a criminal in an effort to apprehend the individual before they commits another crime. This role is generally referred to as **forensic or criminal profiling**.

1.4.5   Forensic toxicology

Forensic toxicology primarily functions to analyze samples associated with poisoning, drug use, or death. Forensic analyses performed by a forensic toxicologist include qualitative analysis and classification to determine what the substance is, as well as quantitative analysis to determine amounts of substance. The latter can be significant in determining if a crime has been committed, such as when alcohol has been consumed, or deciphering causation of death when poisoning or an overdose is suspected.

1.4.6   Computer forensic science/ computer forensics

This discipline is considered a branch of digital forensic science and is focused on the information or data found on computer devices and other forms of digital media. As one might expect, the recent explosion of electronic data devices such as smartphones smart appliances, smart home security systems (including smart doorbells), smart speaker, music players, readers, and all forms of laptop computers has created an upsurge in training associated with digital media. Computer forensic science focuses on the identification, retrieval, preservation, and storage of information found on digital media and associated devices as it pertains to civil and criminal matters. Digital and cybercrimes have exploded over the last decade, placing individuals trained in this form of forensic investigation in great demand.

1.4.7   Forensic botany

Forensic botany is the application of plant science to legal matters. Identification of plant species and application of plant development can be used to determine if a crime has been committed in a particular location, a body has been moved before or after death, and can help to calculate a portion of the PMI. In many ways, the roles of a forensic botanist are similar to those of a forensic entomologist.

Chapter review

What is forensic science?

- Forensic science can be defined broadly as the use of scientific knowledge and technologies in civil and criminal matters, including case resolution, enforcement of laws, and national security.
- Criminalistics is a term used to describe the functions of a crime or forensic laboratory and represents a narrower definition of forensic science.
- The real value of science in legal matters is that it relies on validation via scientific inquiry using the scientific method, an approach that requires adherence to defined unbiased approaches to designing, conducting, and interpreting experiments.
- Public opinion of what forensic science is and what it can do with regard to legal matters is riddled with unrealistic expectations, termed the CSI effect from popular television crime shows such as *CSI: Crime Scene Investigation*.

Application of science to criminal investigations

- Forensic science is used to investigate several issues associated with criminal, civil, and administrative matters. Most forensic scientists work on cases of a civil or administrative nature, or deal with issues related to national security such as those under the umbrella of the Department of Homeland Security in the United States.
- The major functions performed by a forensic scientist include analysis of physical and trace evidence, providing expert testimony to the court and, in some cases, collection of evidence at a crime scene.
- Forensic scientists spend the majority of their time applying the principles and methodologies of their discipline to the elements of the crime, principally analyzing the physical and trace evidence, which is any part or all of a material object used to establish a fact in a criminal case. Items as diverse as bullet casings, bone fragments, a dental crown, matches, or fly maggots can serve as physical evidence.

- Many aspects of forensic analyses utilize the scientific method for the examination of physical and trace evidence, as well as other elements of a crime. The scientific method is a systematic approach or procedure for investigating natural phenomena. It relies on testing questions using an approach centered on formulating hypotheses, making observations from carefully designed experiments, refining questions, and narrowing possible explanations, so that further testing and observation can occur. The method is a skill that is learned or acquired from rigorous training and extensive practice.

- Physical and trace evidence collected from a crime scene is delivered to a crime laboratory or sent directly to a forensic expert for quantitative or qualitative analyses. In the majority of cases, the forensic scientist evaluates physical evidence by comparison testing, which includes recognition of evidence, classification (classifying or grouping the object), individualization (identification of individual or determining if two similar objects have a common origin), reconstruction (hypothesis testing when reconstructing events of the crime), and intelligence information (making inferences about criminals based on modus operandi).

Ensuring "good" science in forensic analysis

- Contrary to public opinion, forensic science is not infallible. When forensic analysis is performed in the absence of standard operating procedures, independent of error or statistical analysis, and/or overstepping the boundaries of the scientific method, junk science will prevail. The result is the potential for wrongful convictions or the guilty remaining free, undermining the judicial process
- The National Academy of Sciences in the United States published a scathing report in 2009 detailing the current state of forensic sciences. A multitude of problems were identified with most forensic science disciplines, including a lack of oversight of "who" practices forensic science, no mandatory certification of practitioners, the absence of a scientific foundation in many disciplines but especially associated with the comparative sciences,

and failure to establish error rates or measures of uncertainty.

- Since 2009, sweeping changes began to occur throughout the forensic science community. Reform efforts were designed to overcome the deficiencies identified by the NAS report as well as to ensure that safeguards are in place to prevent junk science from ever gaining a foothold again.

Recognized specialty disciplines in forensic science

- Forensic investigation is used to examine issues common to criminal, civil, and administrative matters, and requires experts from many disciplines to perform the multiple forensic analyses.
- The American Academy of Forensic Sciences recognizes 11 major subdivisions of forensic science, with as many as 31 subdisciplines contributing expert analyses in legal cases.

Test your understanding

Level 1: knowledge/comprehension

1. Define the following terms:

   (a) modus operandi
   (b) physical evidence
   (c) scientific method
   (d) hypothesis
   (e) classification
   (f) trace evidence
   (g) junk science

2. Match the terms (i–vi) with the descriptions (a–f).

   (a) Identification of a victim based on DNA profiling
   (b) Study of poisons, drugs, and death
   (c) Shoeprints found at crime location
   (d) Validate identified insect in a museum collection
   (e) Identifying an object to class or group
   (f) Factor that is measured in an experiment

   (i) Scene impression
   (ii) Dependent variable
   (iii) (Individualization)
   (iv) Forensic toxicology
   (v) Qualitative analysis
   (vi) Voucher specimen

3. Explain how qualitative and quantitative analyses are used generally in forensic analyses.
4. Discuss how test and scene impressions are used in criminal investigations.

Level 2: application/analysis

1. Describe the process of classification and individualization for entomological evidence such as the presence of second- and third-stage larvae of the necrophagous blow fly *Lucilia sericata* collected from a corpse discovered in a wooded area in the south-eastern region of the United States.
2. In an experiment aimed at determining which odors emanating from a corpse are attractive to adult flesh flies, identify potential independent variables that must be controlled.
3. Explain why controlled experiments are an important feature of well-designed reconstruction analyses.
4. Discuss how the scientific method can be considered a major stopgap in preventing junk science from surfacing during forensic analysis.

Level 3: synthesis/evaluation

1. Design an experiment to test the hypothesis that blow fly larvae develop faster at 30 °C than 20 °C. In your answer, identify the independent, dependent, and control variables.
2. Discuss what steps that the discipline of Forensic Entomology can take to avoid the issues that have plagued the comparative sciences in terms of credibility in the judicial system.

Notes

1. From the National Academy of Sciences report "Strengthening Forensic Science in the United States: A Path Forward (2009)."
2. Several TV shows portray computer applications, databases, molecular biology techniques, and other technologies that do not yet exist, but the general public is unaware of this and expects similar approaches to be used today.

References cited

Barnard, C., Gilbert, F. & McGregor, P. (1993) *Asking Questions in Biology*. Addison Wesley Longman, Harlow, UK.

Byrd, J.H. & Butler, J.F. (1998) Effects of temperature on *Sarcophaga haemorrhoidalis* (Diptera: Sarcophagidae) development. *Journal of Medical Entomology* 35: 694–698.

Byrd, J.H. & Tomberlin, J.K. (eds) (2020) *Forensic Entomology: The Utility of Arthropods in Legal Investigations*, 3rd edn. CRC Press, Boca Raton, FL.

Carvalho, L.M.L., Thyssen, P.J., Linhares, A.X. & Palhares, F.A.B. (2000) A checklist of arthropods associated with pig carrion and human corpses in southeastern Brazil. *Memórias do Instituto Oswaldo Cruz* 95: 135–138.

Daeid, N.N. (2010) *Fifty Years of Forensic Science*. Wiley Blackwell, Oxford.

Daubert v. Merrill Dow Pharmaceuticals (1993) 509 US 579.

Edwards, H.T. (2009) Solving the problems that plague the forensic science community. *Jurimetrics* 50: 5–19.

Faigman, D.L. (2002) Is science different for lawyers? *Science* 297: 339–340.

Gabel, J.D. & Wilkinson, M.D. (2008) "Good" science gone bad: How the criminal justice system can redress the impact of flawed forensics. *Hastings Law Review* 59: 1001–1030.

Giannelli, P.C. (2002) Fabricated reports. *Criminal Justice* 16: 49–50.

Harris, H.A. & Lee, H.C. (2020) *Introduction to Forensic Science and Criminalistics*, 2nd edn. CRC Press, Boca Raton.

Haskell, N.H. & Williams, R.E. (2008) *Entomology and Death: A Procedural Guide*, 2nd edn. Forensic Entomology Partners, Clemson, SC.

Holmgren, J.A. & Fordham, J. (2011) The CSI effect and the Canadian and the Australian jury. *Journal of Forensic Sciences* 56: S63–S71.

Jackson, A.R.W. & Jackson, J.M. (2016) *Forensic Science*, 4th edn. Pearson, Harlow, UK.

Kaye, D.H. (2010) The good, the bad, the ugly: The NAS report on strengthening forensic science in America. *Science and Justice* 50: 8–11.

Morgan, J.G. & Carter, M.E.B. (2011) *Investigating Biology Laboratory Manual*, 7th edn. Benjamin Cummings, Boston.

National Research Council Committee on Identifying the Needs of the Forensic Science Community (2009) *Strengthening Forensic Science in the United States: A Pathway Forward*. National Academies Press, Washington, DC.

Parker, M.A., Benecke, M., Byrd, J.H., Hawkes, R. & Brown, R. (2010) Entomological alteration of bloodstain evidence. In: J.H. Byrd & J.L. Castner (eds) *Forensic Entomology: The Utility of Arthropods in Legal Investigations*, pp. 539–580. CRC Press, Boca Raton, FL.

Saferstein, R. (2017) *Criminalistics: An Introduction to Forensic Science*, 12th edn. Pearson, Boston.

Saks, M.J. & Faigman, D.L. (2008) Failed forensics: How forensic science lost its way and how it might yet find it. *Annual Review of Law and Social Science* 4: 149–171.

Shelton, D.E., Kim, Y.S. & Barak, G. (2006) A study of juror expectations and demands concerning scientific evidence: Does the CSI effect exist. *Vanderbilt Journal of Entertainment and Technical Law* 9: 331–368.

Swanson, C., Chamelin, N., Territo, L. & Taylor, R. (2018) *Criminal Investigation*, 12th edn. McGraw-Hill, New York.

Watson, E.J. & Carlton, C.E. (2005) Insect succession and decomposition of wildlife carcasses during fall and winter in Louisiana. *Journal of Medical Entomology* 42: 193–203.

Supplemental reading

Houck, M.M. & Siegel, J.A. (2010) *Fundamentals of Forensic Science*, 2nd edn. Academic Press, San Diego, CA.

Bell, S. (2017) *Measurements of Uncertainty in Forensic Science: A Practical Guide*. CRC Press, Baco Raton, FL.

Ogle, R.R. (2011) *Crime Scene Investigation and Reconstruction*. Prentice Hall, Upper Saddle River, NJ.

Pechenik, J.A. (2009) *A Short Guide to Writing about Biology*. Longman, New York.

Roberts, J. & Márquez-Grant, N. (2012) *Forensic Ecology: From Crime Scene to Court*. Wiley Blackwell, Oxford.

Smith, K.G.V. (1986) *A Manual of Forensic Entomology*. Cornell University Press, Ithaca, NY.

Tomberlin, J.K., Mohr, R., Benbow, M.E., Tarone, A.M. & VanLaerhoven, S. (2011) A roadmap for bridging basic and applied research in forensic entomology. *Annual Review of Entomology* 56: 401–421.

Additional resources

American Academy of Forensic Sciences: www.aafs.org

American Board of Forensic Anthropology: www.theabfa.org

American Board of Forensic Psychology: www.abfp.com

American Board of Forensic Toxicology: http://abft.org

American Board of Odontology: www.abfo.org

American Society of Forensic Odontology: http://asfo.org

Combined DNA Index System (CODIS): https://www.fbi.gov/services/laboratory/biometric-analysis/codis

Forensic medicine for medical students: www.forensicmed.co.uk

International Association of Computer Investigative Specialists: www.iacis.com

International Society of Forensic Computer Examiners: www.isfce.com

Next Generation Identification (NGI): https://www.fbi.gov/services/cjis/fingerprints-and-other-biometrics/ngi

Society of Forensic Toxicology: www.soft-tox.org

Chapter 2
History of forensic entomology

Overview

Entomology has its origins in applied biology, simply meaning that the founding of the discipline was fueled by humankind's desire to quench the negative consequences of insect–human interactions. This is the reality of entomology's formation, rather than an intellectual motivation driven by an undying appreciation of, and desire to learn more about, the endlessly fascinating world of insects. It is not surprising that a specialized branch of this discipline known as forensic entomology, more specifically medicocriminal entomology, is both applied in nature and focused on negative interactions. In these cases, it is generally predicated by humans inflicting devastation on other humans. Insects are merely the opportunistic bystanders. What follows in this chapter is an exploration of the origins of forensic entomology. Key historic events are examined that directly led to the formation of the discipline or indirectly contributed to current knowledge that serves as the foundation of forensic entomology.

The big picture

- Historical records of early human civilizations suggest understanding of insect biology and ecology.

- Early influences leading to forensic entomology.
- Foundation for discipline is laid through casework, research, war, and public policy.
- Turn of the twentieth century brings advances in understanding of necrophagous insects.
- Forensic entomology during the "great" wars.
- Growth of the discipline due to the pioneering efforts of modern forensic entomologists leads to acceptance by judicial systems and public.
- Forensic entomology in the twenty-first century is characterized by increased casework and development of professional standards.

2.1 Historical records of early human civilizations suggest understanding of insect biology and ecology

For as long as humankind has inhabited the planet, insects have made their presence known by chewing, biting, sucking, flying, annoying, and every other conceivable means of interaction with humans. The results are obvious: loss of food, spread of disease, destruction of dwellings and property, and attacks on people, livestock, and pets.

The Science of Forensic Entomology, Second Edition. David B. Rivers and Gregory A. Dahlem.
© 2023 John Wiley & Sons Ltd. Published 2023 by John Wiley & Sons Ltd.
Companian website: www.wiley.com/go/forensicentomology2

In fact, the rise and fall of many human civilizations has rested on the ability to successfully coexist with these six-legged beasts (McNeill, 1977). The negative impacts of insect activity have been recorded in some of the earliest forms of written history and can be traced to the writings and symbolism of ancient Greece, China, and Egypt and throughout Aztec and Mayan hieroglyphics (Berenbaum, 1995). Some of the earliest records date back to more than 2500 years ago. Even the Christian Bible details God's use of insects in three plagues ("lice,"[1] flies, and locusts), possibly a fourth if the plague of boils originated from biting flies, to inflict pain and suffering on Egypt as a means to prompt Pharaoh to release the Israelites (Exodus 8–10) (Figure 2.1). If the events are assumed to be historically correct, the most plausible "Pharaoh" referenced in Exodus is Rameses II of the 19th Dynasty, placing the timeline of the insect plagues to between 1290 and 1213 BC. These descriptions can be viewed as some of the first cases in forensic entomology since they represent acts that fall under today's definitions of criminal intent and terrorist activity! Take for instance the fourth plague of Egypt in which God sends swarms of flies (Exodus 8: 20–23). The Hebrew word referenced in the biblical text refers to biting flies[2], conceivably the stable fly *Stomoxys calcitrans*, suggesting that the "swarms" were not just an annoyance but also a source of pain and potentially disease. Regardless

Figure 2.1   *The Plague of Flies*, **c.**1896–**c.**1902, by James Jacques Joseph Tissot (1836–1902), depicting the fourth plague of Egypt. Image available in public domain at http://commons.wikimedia.org/wiki/File:Tissot_The_Plague_of_Flies.jpg.

of the accuracy of fly species identification, details provided in Exodus seem to describe a clear attack on the national security of Egypt along with a confession from God to Moses, the author of the book of Exodus. Though **statutes of limitation** generally do not apply to particularly heinous crimes, it is doubtful that any attorney would wish to tackle this case.

Perhaps to avoid further controversial speculation, it is time to turn our attention to more concrete examples of the early history of forensic entomology. The path we are about to begin would seem to lack the historical footing expected of a scientific text, yet when the focus is on chronology, the Christian Bible may serve as the first source book for forensic entomology. Either God or "His" prophets discuss knowledge of **insect succession** and postmortem decomposition. The prophet Isaiah states that "Thy pomp is brought down to the grave and the noise of thy vols, the worm is spread under thee and the worms cover thee" (Isaiah 14: 11). As Berenbaum (1995) points out, the verse provides reference to the activity of necrophagous fly larvae (the worm), presumably blow flies (Diptera: Calliphoridae), the most common and important group of insects feeding on a corpse (Smith, 1986). Isaiah also provides acknowledgment that large feeding aggregations or **maggot masses** form on the body (Figure 2.2), giving the appearance of covering the entire body. Similarly, Job (21: 26) indicates that "worms," meaning maggots, cover the bodies of the dead, "dead" being freshly dead as "they lie down alike in dust…" (Job 21: 26) (Figure 2.3). Unlike the earlier references to biblical literature, the descriptions of blow fly activity do not depend on "faith" to realize that the authors detail the relationship between necrophagous flies and dead animals (**carrion**) much earlier (eighth to fourth century BC; Blenkinsopp, 1996; Dell, 2003) than the first recorded case of forensic entomology, typically credited to a thirteenth-century murder investigation in China (discussed later in the chapter).

Ancient Egyptians, Chinese, Mayans, and Aztecs, and no doubt other civilizations, have associated death and reincarnation with insects. Butterflies, cicadas, and beetles are just some of the insects revered in part because of their metamorphic changes (Laufer, 1974; Beutelspacher, 1988). Of particular interest were those associated with complete or **holometabolous** metamorphosis, because the transformation that occurred with a **molt** from **pupa** to **imago** or adult represented a

Figure 2.2 Large feeding aggregation or maggot mass of necrophagous flies feeding on a piglet located in a rural woodlot during late summer in the eastern United States. Photo by D.B. Rivers.

Figure 2.3 The Christian Bible is one of the earliest written records describing human understanding of necrophagous insect activity. Photo courtesy of the Gerald R. Ford Presidential Museum. The image is in public domain and is available at https://commons. wikimedia.org/wiki/File:Bicentennial_Bible.jpg.

"rebirth" of the dead either on Earth or in the after-life. In ancient Egypt, dung beetles, a type of scarab (Family Scarabaeidae), were linked to the god Khepri (Figure 2.4), a sun or solar deity. The association stems from the dung beetle's behavior of forming large balls of dung that are rolled from one location to another, which to Egyptians of long ago represented the physical movements of the sun

Figure 2.4 The Egyptian sun god Khepri. Note the head in the form of a dung beetle. Photo by manna nadar (Gabana Studios, Cairo. The image is in public domain and available at https://commons.wikimedia.org/wiki/File:QV66_Khepri_Tomb_of_Nefertari_entrance.jpg.

across the sky. Egyptian mythology also links these scarabs to rebirth or reincarnation because adult females lay eggs in the dung balls where the larvae feed and develop and, in turn, the dung or dead matter gives rise to new beetles (Berenbaum, 1995).

Ancient Chinese associated immortality after death with cicadas due to the perceived rebirth that occurs with each molt of this **hemimetabolous** insect (Laufer, 1974). Aztecs identified insect life cycles, particularly those of butterflies, with numerous aspects of their lives, including death and reincarnation. In fact, two Aztec goddesses, Xochiquetzal and Izpapalotl (Figure 2.5), are depicted in some paintings as having wings of butterflies, and at least one (Izpapalotl) was claimed to have the ability to swallow darkness, perhaps meaning to control death (Beutelspacher, 1988). This latter idea may be the root of the superstition in Mexico that if a black butterfly (actually a moth, *Ascalapha odorata*) stops at your door, somebody will die (Ponce-Ulloa, 1997).

The shared beliefs of the Egyptians, Chinese, and Aztecs relating insects to death and then immortality stem from their understanding of key features of insect life cycles (Laufer, 1974; Beutelspacher, 1988) and, to some degree, recognition of the necrophilic activity of some species of insects. So though forensic entomology has evolved at its fastest over the last 30–40 years, the underpinnings of medicocriminal entomology seem to be derived from anecdotal observations dating back to times when the dead and their insects were worshipped rather than feared or despised.

2.2   Early influences leading to forensic entomology

2.2.1   Thirteenth-century China

Any discussion of the history of forensic entomology inevitably begins in thirteenth-century China. The first acknowledged case in the discipline is credited to the Death Investigator Sung Tz'u for his sleuthing that ultimately led to the confession by a murderer. As the story is told in his training manual for investigating death *His yüan chi lu* (*Washing Away of Wrongs*, from 1235), a stabbing victim was found near a rice field and the stab wound was consistent with a common sickle (Figure 2.6) used in the fields. On the day after the body was discovered, also presumed to be the next day after the murder, Tz'u arrived in the village, had all the workers report, bringing with them their working tools. As Tz'u examined each sickle resting in front of its respective owner, flies were found on and around just one of the sickles, and the owner reportedly confessed to the crime and "knocked his head on the floor" (as retold by Benecke, 2001). The interpretation of this description is that small traces of blood and tissue were apparently still present on the murderer's sickle, which drew the attention of blow flies (Diptera: Calliphoridae). Thus, the worker was so bereft with stress and guilt that it took very little questioning to garner a full confession. (It seems that Sung Tz'u's text was also the foundation for television crime shows in which the murderer always seems to confess with some prodding!)

What is never discussed is why the accused murderer did not recognize that flies were hovering around the sickle immediately after the assault. Tz'u states that confrontation with the villagers did not occur until the next day, plenty of time to clean the sickle or at least recognize the insect activity. Equally curious is the presumption that blow flies

Figure 2.5   Image of the Aztec goddess Izpapalotl at the Jorge R. Acosta Museum in Tula. Photo by Alejandro Linares Garcia. The image is available at https://commons.wikimedia.org/wiki/File:TulaSite25.JPG.

Figure 2.6   Workers using common sickle to harvest rice. Photo by F.H. King. Image available in public domain via https://commons.wikimedia.org/wiki/File:Farmers_of_forty_centuries_Japanese_farmers_harvesting_rice_with_sickles.jpg.

were the insect evidence that yielded the confession. While we know today that adult blow flies are typically the first colonizers on carrion (Smith, 1986), it is difficult to ascertain whether that aspect of insect succession on a corpse was understood in thirteenth-century China. Perhaps the lack of concern by the villager to clean his sickle argues it was not common knowledge. However, Cheng (1890, cited in Greenberg & Kunich, 2002) wrote that the Chinese understood as early as the tenth century that flies and other insects could reveal much at a crime scene. So it is quite likely that at least Tz'u was familiar with these earlier works and applied his understanding of insect behavior and biology to a murder investigation. Undoubtedly this also confirms that Tz'u was not the first to use entomological evidence to solve a crime, and may not even have been the first to document a forensic entomology case. Regardless, his detective work was sound and did serve as a major contribution that led to the modern discipline.

2.2.2 Seventeenth-century Europe

Entomology as a discipline was still a fledgling at the turn of the seventeenth century by comparison to other fields of life sciences. Almost nothing is

Figure 2.7   Necrophagous flies on fresh corpse of a South African porcupine (*Hystrix africaeaustralis*). Photo by Paul Venter. Image available at https://commons.wikimedia.org/wiki/File:Decomposition01.jpg.

recorded that relates to forensic entomology from Sung Tz'u's work in 1235 until the nineteenth century. However, landmark work in biology was occurring throughout Europe that had a profound impact on the applied use of insects in legal investigations. Among the most significant was the research conducted by the Italian physician and naturalist Francesco Redi (1626–1697). In 1667, Redi tested the theory of spontaneous generation or **abiogenesis**, the idea that maggots spontaneously formed on meat (Figure 2.7). His classic experiments demonstrated conclusively that flies and their juveniles do not form from meat but rather that the adults were attracted to the spoiling tissues and lay eggs on or near the food (Ross *et al.*, 1982). Redi was instrumental in reestablishing what the ancient Chinese had previously determined that certain species of flies were attracted to dead animals and that fly reproduction relied on utilization of the carcass.

2.2.3   Eighteenth-century Europe

By the middle of the eighteenth century, the Swedish naturalist Carolus Linnaeus (also known as Carl von Linné, 1707–1778) was establishing himself as an exceptional systematist (Figure 2.8). Linnaeus developed a system of nomenclature (binomial) for classifying plants and animals that unified and organized botanical and zoological specimens. His method so simplified earlier attempts that it was almost immediately adopted by researchers in all other related fields and earned him the title of the father of taxonomy. More importantly, the binomial system allowed botanists, zoologists, and naturalists across the world to communicate much more effectively about the

Figure 2.8   Portrait of the Swedish naturalist Carl von Linné (1707–1778) by Alexander Roslin (1718–1793). Image available in public domain at http://commons. wikimedia.org/wiki/File:Carl_von_Linn%C3%A9.jpg

same organisms because the names were unified (or stabilized) in approach.

Linnaeus was also instrumental in collecting and identifying thousands of animals; among them, about 2000 insects were initially described by his efforts. Several species are forensically important including the common house fly, *Musca domestica* L. (Diptera: Muscidae), which he described in 1758. From his field observations of blow flies, Linnaeus made the statement that three flies (adults) would destroy a horse as fast as a lion (Müller and von Linné, 1774, cited in Benecke, 2001). This would seem to be a clear reference to the reproductive capacity of calliphorids when using carrion and also to the ability of feeding larvae (maggots) to remove all soft tissue from animal remains in just a matter of days. The phrase "spineless vultures" aptly captures the observations of Linnaeus.

2.3   Foundation for discipline is laid through casework, research, war, and public policy

The literature is devoid of reference to direct casework in forensic entomology from the time of Sung Tz'u in 1235 until the mid-1800s. Several important events occurred in the nineteenth century that laid the foundation for modern forensic entomology.

2.3.1   Casework in Europe

The French physician Louis François Etienne Bergeret (1814–1893) is the author of the first modern case report in forensic entomology. Bergeret was apparently called in to investigate the discovery of mummified remains of a newborn baby. The child's body was discovered in March 1850 encased behind a chimney in a boarding house undergoing renovations. During his autopsy of the body, Bergeret found larvae of a fly identified as *Sarcophaga carnaria* (L.) and pupae of "butterflies of the night," which today are presumed to have been moths belonging to the family Tineidae (Lepidoptera) (Benecke, 2001). Since Bergeret apparently had some insight into necrophilic insect life cycles, he offered an interpretation of what the entomological evidence revealed about the case. It was his opinion that adult flies lay eggs in summer months, larvae metamorphose to pupae in the following spring, and the pupae then hatch (adult **eclosion**) during the ensuing summer. In essence, Bergeret argued that the flies were **monovoltine**, and thus to account for the approximately 8–10 months required for completion of fly development, from **oviposition** or egg laying to emergence of an adult, the child's body must have been available to the insects, and hence deceased, since the middle part of 1849. This time period corresponded with the previous occupants of the tenement and not the current residents. The interpretation was filled with mistakes about the life cycles of the flies but, as Benecke (2001) points out, Bergeret did not focus on the entomological evidence in his case report to the court – it was merely one aspect he used along with other approaches to determine the time of death. It is important to note that Bergeret's efforts are the first known recorded use of insect evidence to establish a **postmortem interval (PMI)**.

In 1878, another French physician (pathologist) made a mark on forensic entomology, but not due to his own efforts. Paul Camille Hippolyte Brouardel (1837–1906) described a child autopsy that he had worked on in which a newborn child was covered with numerous arthropods. Identification of the specimens was outside the realm of his expertise, so he enlisted the help of Monsieur Perier from the Museum of Natural History in Paris and an army veterinarian Dr Jean Pierre Mégnin (1828–1905). It was Mégnin's research work that helped to transform the use of insects and acari from anecdotal observations at crime scenes to bona fide physical evidence. Mégnin performed countless experiments throughout his career investigating insect succession on corpses. His seminal works *Fauna des Tombeaux* and *La Fauna des Cadares* are two of the most important texts that pertain to forensic entomology. In them, Mégnin describes waves of insect succession, in which he recognized eight distinct waves associated with insect succession on bodies located in terrestrial environments and two associated with buried bodies. His book *La Fauna des Cadares* provided extraordinary details of adult and larval morphology of several families of flies and Mégnin spent a great deal of time collecting and identifying species of necrophagous flies using the binomial system developed by Linnaeus. Much of this work still influences modern views on the use of cadavers by necrophagous insects. Certainly a strong case can be made for Mégnin being recognized as the father of forensic entomology.

Figure 2.9   An adult scuttle fly, *Megaselia scalaris* Loew (Family Phoridae). Photo by Charles Schurch Lewallen. Image available at https://commons.wikimedia.org/wiki/File:Megaselia_scalaris.jpg.

Overlapping Mégnin's works were the investigations of Hermann Reinhard (1816–1892). This German physician is credited with conducting the first systematic research in forensic entomology (Benecke, 2001), focusing his attention on insect succession of buried bodies. Reinhard's extensive work with exhumed corpses was instrumental in establishing which insects were truly necrophagous on buried cadavers and which were merely associated with the burial site (Reinhard, 1881). The work led to the identification of phorids (Diptera: Phoridae), a group of small flies commonly inhabiting buried or enclosed bodies, earning them the common name coffin or mausoleum flies (Figure 2.9).

2.3.2   Influences from the United States

While most early work in forensic entomology was occurring in Europe during the nineteenth century, a few key events had an indirect influence on developing the field in the United States. Among these were the research studies of Murray Galt Motter working for the US Bureau for Animal Industry. Motter's research group examined the exhumed remains of 150 individuals, providing information on the insect fauna, burial conditions (burial depth), and soil types associated with each corpse (Motter, 1898). Although the entomological descriptions were far more detailed in the studies of Mégnin and Reinhold, Motter's was perhaps the first to place faunal succession of buried bodies in context of the physical parameters of the burial itself.

Perhaps the most profound forensic entomological event in North America during this time was the United States Civil War (1861–1865). Insects of all sorts played havoc on the soldiers of both the North and South throughout the duration of the war (Miller, 1997), an all-too-common relationship between insects and humans (Berenbaum, 1995). Numerous eyewitness accounts have attested to the horrendous stench emanating from the battlefields, where thousands of soldiers, horses, and other animals lay dead (Stotelmyer, 1992) (Figure 2.10). Estimates of the total death toll approached 620,000, with more than 200,000 Union and Confederate soldiers lost on the battlefields. Adding to these staggering totals were the horses used by cavalry and artillery units. At the Battle of Gettysburg alone, an estimated 1500 artillery horses were left for dead after three days of intense fighting (Miller, 1997) (Figure 2.11). Most of the battles occurred during the hot temperatures of summer, promoting rapid decomposition of all manner of corpses (Miller, 1997). Odors indicative of tissue decay, comprising of hundreds of volatile organic compounds over the course of decomposition, served as cues to draw the attention of necrophagous flies and beetles. Black bloated bodies were

Figure 2.11   Dead and bloated artillery horses lying in the fields of Gettysburg, Pennsylvania, during the US Civil War. Photo courtesy of the Massachusetts Commandery Military Order of the Loyol Legion. Image available at https://commons.wikimedia.org/wiki/File:The_barn_at_Trostle%27s_farm_still_stands_where_a_Union_battery_held_its_ground_at_a_great_cost_to_men_and_horses._Photographer_Timothy_0%E2%80%99Sullivan_emphasized_the_dead_artillery_horses_when_he_reached_(9160161836).jpg.

Figure 2.10   One of many dead soldiers left on the battlefield at the Siege of Petersburg during the US Civil War. Photo by Thomas C. Roche. Image available in public domain at https://commons.wikimedia.org/wiki/File:Dead_soldier_(American_Civil_War_-_Siege_of_Petersburg,_April_1_1865).jpg.

noted to be covered with worms (maggots) or have the appearance of "maggoty bodies" (Broadhead, 1864; Stotelmyer, 1992). There was no reprieve from the sights and smells even when the soldiers were removed from the scene of battle as the hospitals contained wounded with maggots feeding on necrotic and living tissues (Miller, 1997). Piles of amputated body parts that had not been buried accumulated outside the hospitals and became infested with maggots (Strong, 1961). Physicians in the Union army apparently attempted to rid the wounds of maggots by rinsing the sites of infestation with chloroform. By contrast, Confederate physicians were rarely equipped with the necessary medical supplies that the Union possessed, and as a result observed that the "untreated maggots" could be beneficial (Brooks, 1966). Some physicians noted that the fly maggots were actually more effective in cleaning necrotic tissue from a wound than nitric acid or a scalpel, and at least one Confederate field surgeon, J.F. Zacharias, acknowledged using the maggots deliberately to treat battle wounds (Greenberg, 1973). This may be the first reported use of **maggot therapy**.

An interesting side note to the presence of flies in hospitals and on the battlefields was the observation that a tiny gnat-like insect appeared from the pupae of some fly species (Whiting, 1967). The insect was later identified as a parasitic wasp, *Nasonia* (*Mormoniella*) *vitripennis* (Hymenoptera: Pteromalidae), which utilizes puparial stages of flies from several families of forensic importance, and which itself has been used as entomological evidence collected from crime scenes (Turchetto & Vanin, 2004).

During the early years of the Civil War, the Federal government passed the Morrill Act of Congress in 1862. The Act established land grant universities that were to focus on education in agriculture, mechanical arts, and natural sciences. Entomology was included in biology curricula, but not as an official discipline or necessarily even as a separate course (Cloudsley-Thompson, 1976). Nonetheless, the foundation for the discipline in the United States was linked to an act of Congress, and arguably this is the single most important event that led to the eventual establishment of entomology and its subdisciplines in the United States. As a key feature of its inclusion with the Morrill Act, entomological subject matter was mostly applied in nature, focused on topics related to agriculture and insect control. This same applied emphasis is what drives modern forensic entomology.

2.4 Turn of the twentieth century brings advances in understanding of necrophagous insects

If the fourteenth through nineteenth centuries were characterized by the absence of forensic entomology casework, the beginnings of the twentieth century quickly filled the void. The German physician Klingelhöffer reported on a case in which a 9-month-old baby had died in May 1889 and the local police had arrested the child's father on suspicion of murder. A local doctor who had performed the initial examination of the dead baby believed the child had been poisoned, most likely with sulfuric acid, since "patches" had been detected on the nose and lips and in the throat (Benecke, 2001). Other features consistent with sulfuric acid poisoning had not

been observed, but this did not prevent the police from arresting the father. When Klingelhöffer performed an autopsy on the child three days after death, he reportedly found no evidence of poisoning, and concluded that the abrasions were most likely induced by cockroaches. In fact, the damage to the child's body probably occurred postmortem.

The Austrian pathologist Stefan von Horoszkiewicz reported on a case nearly identical to that of Klingelhöffer in that a child's body showed obvious abrasions but no signs of poisoning. During the autopsy in April 1899, lesions were found on the face, neck, left hand, fingers, genitals, and inner thighs, skin damage that could not be explained by sulfuric acid or any other type of poisoning (Benecke, 2001). When the child's mother was questioned, she revealed that after leaving the home to make funeral preparations, upon her return she noticed that the child's body was covered by a black shroud of cockroaches (Figure 2.12). Horoszkiewicz conducted a series of experiments with cockroaches and tissues collected from freshly dead corpses to determine if these insects could inflict the type of damage witnessed on the child's body (Benecke, 2001). He concluded not only that cockroach feeding inflicted nearly identical abrasions on the isolated tissues, but that the lesions were not readily apparent until after the skin began to dry.

The Austrian medical examiner Maschka wrote about multiple cases in which ants and other arthropods were likely responsible for skin lesions and abrasions detected on deceased children's bodies

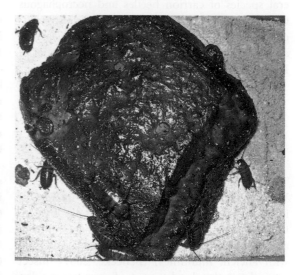

Figure 2.12 Lesions on fresh bovine liver after only four hours of feeding by the American cockroach, *Periplaneta americana*, at 25 °C. Photo by D.B. Rivers.

(Benecke, 2001). Like Horoszkiewicz, Maschka concluded that the damage was inflicted postmortem not **antemortem**. The cases reported by Klingelhöffer, Horoszkiewicz, and Maschka share the commonality of insect feeding or bite marks appearing postmortem. Why was this so prevalent? During this era, human bodies were not buried immediately after death. Typically, several days would pass before the funeral. The deceased body lay in waiting, either non-embalmed or **embalmed** but using relatively poor preservatives by today's standards, usually in the home of the immediate family. The net effect was a rich food source (the corpse) that was decomposing and emitting odors and which was available for days to an array of insects. The end result was that the insect feeding damage initially resembled lesions typical of sulfuric acid poisoning, a common method of murder or suicide during the nineteenth century. The careful investigative efforts of these physicians helped to change interpretations at autopsy.

The early twentieth century was also characterized by advances in general entomology, namely though the excellent descriptive works of Jean Henri Casimir Fabre (1823–1915). Fabre was renowned for excellence in teaching as well as in field research, where his approach of exquisite attention to detail and exactness were obvious in his writings. Over the course of his career, Fabre wrote a series of papers and texts on insects that collectively are referred to as *Souvenirs Entomologiques* (*Souvenirs of Insect Life*). In these writings, Fabre provided observations on numerous insects and arachnids, including several species of carrion beetles and necrophagous flies. His writing style and observations also influenced many scientists, including Charles Darwin, although interestingly Fabre was a skeptic of Darwin's theories (Fabre, 1921).

2.5 Forensic entomology during the "great" wars

The period of time encompassing World War I and World War II ushered in increased interest in insects associated with such topics as maggot therapy, pest control, and even the beginnings of archaeoentomology (specifically insects associated with ancient mummies) (Benecke, 2001). Insect control, via the development of new insecticides, tended to dominate entomological research from the 1920s until the late 1950s. From a forensic

entomology perspective, the observations of Karl Meixner and Hermann Merkel provided new insights into the study of insect succession on cadavers. Meixner noticed that the rate of decomposition of a corpse infested with maggots was influenced by the age of the deceased, with children decomposing far faster than adults while placed in storage (Meixner, 1922). In 1925, Merkel wrote about a murder investigation in which a son killed his parents by shooting. Blow flies had infested both the mother and father but the rate of decay was significantly different between the two. Merkel described the mother as obese and fully bloated when discovered, with fly larvae present in the eyeballs and feeding on brain tissue. The father, by contrast, was slim in build and his body was post-bloat. Maggots were found throughout the body cavity and pupae (**puparia**) were evident (Merkel, 1925). Merkel reported that other than weight differences, the only other major deviation between the parents was that the father had also been stabbed. This led Merkel to conclude that the method of death and/or wounding could accelerate decomposition of a body, since the flies would have direct access to the body cavity via injuries like stab wounds.

Several works during this period contributed to the understanding of **myiasis**, the disease or parasitic condition in which fly larvae infest live human tissue. In 1948, D.G. Hall published his treatise *The Blowflies of North America,* which included descriptions of myiasis occurring in the nasal sinuses. In the same year, M.T. James thorough work *The Flies that Cause Myiasis in Man* provided a classification system of myiasis based on ecological and anatomical criteria. Many clinicians and forensic entomologists still reference the text. Perhaps one of the most interesting (to an entomologist) and disgusting cases of myiasis was described by Hurd (1954), in which a fly infestation was reported in association with the use of an aspirator for insect collection. A modern examination of myiasis within forensic contexts is presented in Chapter 16.

The war era marked some of the first investigative work involving submerged bodies and aquatic insects. Josef Holzer (1937) and later Hubert Casper (1950s) showed the importance of caddisflies (Order Trichoptera) in investigations involving bodies submerged in freshwater environments (Figure 2.13). Holzer detailed the type of skin aberrations that resulted from the feeding activity of caddisfly larvae on a corpse, observations very much akin to those described for cockroaches and

Figure 2.13   A caddisfly larva (Order Trichoptera) in its case. Photo by Bob Hendricks. Image available at https://commons.wikimedia.org/wiki/File:Saddle-case_case-maker_caddisfly_larva_(7234729126).jpg.

ants in terrestrial scenarios. Casper added a layer of understanding to aquatic succession by documenting an investigation in which caddisfly larvae incorporated thread from socks still present on a submerged victim to build larval casings. Matched with details of how and when caddisflies typically form larval cases within the context of ambient temperatures, Casper was able to conclude that the corpse had been submerged for at least one week (Benecke, 2001).

The end of World War II marked a period of entomological research centered predominantly on insecticide development. The initial results of insecticide treatments yielded such impressive suppression of pest populations that many entomologists feared their jobs would become extinct (Sweetman, 1958). Of course, the "optimism" faded as deleterious effects started to be revealed from overuse of insecticides, including the development of widespread resistance in many medically and economically important species of insects. Rachel Carson's *Silent Spring* published in 1962 had a profound impact on the insecticide era in the United States and is often cited as the major force that shifted attention to broader views of insect life. The period also marked a time of excellence in research, with outcomes directly impacting forensic entomology. In 1916, J.M. Aldrich published Sarcophaga *and Allies* proceeded by DG Hall's treatise in 1948 both of which have served as comprehensive references for entomologists with diverse interests in these flies. Adel S. Kamal followed in 1958 with his extensive efforts characterizing the bionomics of thirteen species of necrophagous flies representing the families Calliphoridae and Sarcophagidae. The research yielded information on several key life-history features of the adult flies and offered developmental data on each stage of fly development at multiple rearing temperatures. Kamal's paper is still cited frequently in casework and research in which calculations of PMI, **accumulated degree hours** or **days**, and developmental comparisons are warranted. However, it should also be noted that many recent authors contend that the paper is filled with errors that compromise the utility of the development data. Adding to the mounting body of forensic entomological research was that of the famed ecologist Jerry Payne. In 1965, Payne's seminal paper on insect succession first appeared. In it, he described the insect fauna that colonized pigs located in the southwestern region of the United States, characterizing waves of succession through the different stages of decomposition. The paper also highlighted the value of using pigs for succession studies. Today, it has become common practice to use adults of the pig *Sus scrofa* L. as models for emulating decomposition of adult humans. That said this practice has recently been challenged by field studies that suggest that pigs and other nonhuman models do not capture the pattern, rate, and variability of human decomposition (Dautartas *et al.*, 2018). Continued investigation is warranted to determine the suitability of nonhuman models as surrogates for humans in decomposition and carrion ecology research, especially those focused on insect colonization and faunal succession.

2.6   Growth of the discipline due to the pioneering efforts of modern forensic entomologists leads to acceptance by judicial systems and public

Forensic entomology has undergone a growth spurt since sometime in the 1980s. As mentioned in Chapter 1, crime shows airing in the United States have contributed to the increasing popularity of all forms of forensic science, including forensic entomology. Yet this growth should really be attributed to the tireless efforts of recent pioneers in the field. As a result of the research and casework of such individuals as B. Greenberg, E.P. Catts, M.L. Goff, N.H. Haskell, P. Nuorteva, W. Lord, K. Kim, M. Leclercq, R. Merritt, Z. Erinçlioglu, and many others, forensic entomology has come to be recognized as a legitimate subdiscipline of entomology and forensic science, increasingly gaining acceptance in judicial systems around the world. The continued efforts of several very talented forensic scientists, including J. Amendt, M. Archer, G. Anderson, J. Byrd, M. Benecke, C. Campobasso, I. Dadour, M.J.R. Hall, R. Kimsey, K. Schoenly, A. Tarone, J. Tomberlin, S. VanLaerhoven, J. Wallace, J. Wallman, J. Wells, S. Vanin, M. Villet and others[3] are leading the discipline into the twenty-first century by ensuring that high-quality research is conducted in areas in need of understanding and by promoting professional standing for all individuals who wish to represent the discipline in a court of law. The latter aspect is absolutely essential to maintaining the integrity of the discipline and to ensure high-quality scientific analyses are provided in civil and criminal matters (Hall & Huntington, 2010; Michaud et al., 2012).

The growth of the field has led to the publication of manuals detailing procedures for collecting and using entomological evidence in criminal investigations (Smith, 1986; Haskell & Williams, 2008), a comprehensive text to train law enforcement officers and criminal investigators about forensic entomology (Byrd & Tomberlin, 2020), compilations of current research in the field (Amendt et al., 2010; Tomberlin & Benbow, 2015), an introductory book to fuel the growing interest of the lay public and those with a scientific background alike (Gennard, 2012), and this work as a textbook for training undergraduate and graduate students. Amendt et al. (2010) have documented the increased output of research articles focused on forensic entomological topics, which corresponds with an increase in the number of scientific journals that have at least a partial focus on forensic entomology (Table 2.1).

In the last two decades, organizations dedicated to forensic entomology have been formed in North America and Europe and include the North American Forensic Entomology Association (NAFEA), the American Board of Forensic Entomology (ABFE), and the European Association for Forensic Entomology (EAFE). Such organizations as the American Academy of Forensic Sciences (AAFS), the Canadian Society of Forensic Sciences

Table 2.1 Scientific journals typically reporting research results in forensic entomology*.

Journal	Publisher
Canadian Society of Forensic Sciences Journal	Canadian Society of Forensic Sciences
Forensic Science International	Elsevier Publishing
International Journal of Legal Medicine	Springer Publishing
Journal of Forensic Sciences	Wiley-Blackwell for AAFS[†]
Journal of Medical Entomology	Oxford University Press for Entomological Society of America

[†]American Academy of Forensic Sciences.

(CSFS), and the Entomological Society of America (ESA) also have subsections dedicated to topics relevant to forensic entomology. The membership in each of these specialized organizations (those directly associated with forensic entomology) is typically small in comparison with other scientific fields, though modest increases in membership numbers has occurred over the past 10 years. Likewise, undergraduate and graduate courses focused on forensic entomology have sprung up at colleges and universities around the world, providing training for the next generation of researchers and caseworkers (Butin et al., 2020). At present, only two institutions [Purdue University (undergraduate) and the University of Huddersfield (graduate) provide a degree in forensic entomology but that, too, may change as the discipline grows.

Part of the explanation for why forensic entomology is not offered as a degree program is that job opportunities are quite limited. Agencies involved in civil and criminal investigations generally do not hire individuals to serve solely as forensic entomologists. Of course, there are exceptions to this rule, but a career solely focused on forensic entomology has been uncommon worldwide. Typically, scientists trained in medical entomology, insect taxonomy, and/or insect ecology are employed by a college, university, government agency, or some other employer and offer their services as forensic entomology consultants. Students with a desire to work in this field should obtain a strong background in general and medical entomology and insect ecology and taxonomy as well as developing a proficiency in biological statistics and an understanding of insect physiology. For specific degree and coursework recommendations, consult the education section of the American Board of Forensic Entomology's by-laws (https://forensicentomologist.org/).

2.7   Forensic entomology in the twenty-first century is characterized by increased casework and development of professional standards

The new millennium has been unrivaled in terms of casework involving forensic entomology. This is a testament to the dedicated work of the early pioneers of the field to establish the utility of insect evidence in all manner of legal investigations. Since the early 2000s, some forensic entomologists have reported involvement in nearly 100 cases a year, though the average is closer to between two and five annually. Casework during this era has centered predominantly on insect colonization/activity on human remains associated with suspicious deaths and homicides. The highest profile among these investigations was that of Casey Anthony, a young Florida mother accused of first-degree murder of her two-year old daughter Caylee Anthony (Figure 2.14). Skeletal remains of the young girl were recovered in December 2008 in Orlando, Florida while earlier in the year the car of the accused was impounded and presumed to have contained decomposing remains based on odors emanating from the trunk. The case pitted two forensic entomologists against each other in the interpretation of entomological evidence discovered in the mother's car. Additional evidence was collected with the skeletal remains but was not a focus of the defense team since it could not be linked to their client (Casey). The murder trial brought forensic entomology into the daily lives of the public, as the media provided nearly nonstop coverage of each day's court proceedings. The case reports and trial testimony of each entomologist have been used as classroom examples, scrutinized by other practitioners, and even relived as part of an annual meeting of the North American Forensic Entomology Association. A strong argument can be made that the Casey Anthony Murder trial has

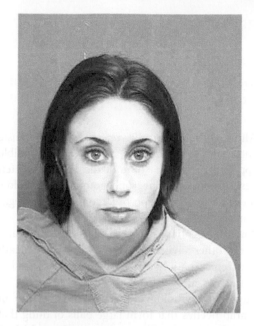

Figure 2.14   Casey Anthony at the time of her arrest in 2016. Photo by the Orange County Sheriff's Department. Image available at https://commons.wikimedia.org/wiki/File:Casey_Anthony_Mugshot.jpeg.

supplanted the work of Sung Tz'u as the most referenced case in forensic entomology.

Numerous non-death investigations have also typified forensic entomology casework, including matters involving medical malpractice, neglect and abuse, insect infestations of human habitation (urban entomology), and the presence of insects or their parts in food and stored products (stored product entomology). Each of these topics will be discussed more thoroughly in subsequent chapters and will include relevant case examples. The growing involvement of entomological evidence in legal matters has necessitated discussion, debate and eventual development of professional standards for practitioners. The process is ongoing but Amendt *et al.* (2007) and Sanford *et al.* (2020) outlined a blueprint for moving toward standard operating procedures that can be used universally by all forensic entomologists. Similarly, several important research and opinion articles (Michaud *et al.*, 2012; Tarone & Sanford, 2017; Wells & LaMotte, 2020) have laid the groundwork for improving the statistical measures and methods of validation used to assess insect development and activity associated with human remains. The next several years should see as much research dedicated to validation and methods of uncertainty in forensic entomology as those focused on insect biology and carrion ecology.

Chapter review

Historical records of early human civilizations suggest understanding of insect biology and ecology

- Although not typically referenced in forensic entomology discussions, the Christian Bible appears to provide some of the earliest written descriptions of the necrophagous activity of blow flies and the formation of large feeding aggregations or maggot masses on cadavers.
- Ancient Egyptians, Chinese, Mayans, Aztecs, and likely other civilizations have associated death and reincarnation with insects such as cicadas, butterflies, and beetles.
- Egyptian mythology describes the worship of scarab beetles that form large balls of dung that are rolled from one location to another, which to Egyptians of long ago represented the physical movements of the sun across the sky.
- The shared beliefs of the Egyptians, Chinese, and Aztecs relating insects to death and then immortality stem from their understanding of key features of insect life cycles and recognition of the necrophilic activity of some species of insects.

Early influences leading to forensic entomology

- The first acknowledged case in forensic entomology is credited to the Chinese Death Investigator Sung Tz'u for his sleuthing in 1235 that ultimately led to a murder confession based on the attraction of blow flies to the murder weapon.
- Francesco Redi's experiments in 1667 testing the theory of spontaneous generation demonstrated not only that flies and their juveniles do not form from meat but rather that the adults were attracted to the spoiling tissues and lay eggs on or near the food. These observations reestablished what the ancient Chinese had previously determined, that certain species of flies are attracted to dead animals and fly reproduction relies on utilization of the carcass.
- In the middle of the eighteenth century Carolus Linnaeus established the binomial nomenclature system for classifying zoological and botanical specimens. The approach won universal appeal as it provided a simplified approach for organizing and naming plants and animals, allowing scientists from different regions to be able to share information about the same organisms. Linnaeus was also prodigious as an entomologist, naming over 2000 insect species, including several of forensic importance.

Foundation for discipline is laid through casework, research, war, and public policy

- The literature is devoid of reference to direct casework in forensic entomology from the time of Sung Tz'u in 1235 until around the mid-1800s. Several important events occurred in the nineteenth century that laid the foundation for modern forensic entomology.
- The French physician Louis François Etienne Bergeret is regarded as the author of the first modern case report in forensic entomology. Bergeret applied expertise in pathology along with an apparent knowledge of necrophagous insects to his investigation of the mummified remains of a newborn baby.
- During the late 1800s, Jean Pierre Mégnin helped to transform the use of insects and acari from anecdotal observations at crime scenes to bona fide physical evidence. His seminal works *Fauna des Tombeaux* and *La Fauna des Cadares* are two of the most important texts that pertain to forensic entomology and describe waves of insect succession: eight distinct waves associated with insect succession on bodies located in terrestrial environments and two waves associated with buried bodies. Mégnin provided extraordinary details of adult and larval morphology of several families of flies, identifying numerous species of necrophagous flies using the binomial system developed by Linnaeus.
- The German physician Hermann Reinhard is credited with conducting the first systematic research in forensic entomology, focusing his attention on insect succession of buried bodies. Reinhard's work with exhumed remains led to the identification of flies in the family Phoridae.
- Research in the United States led by Murray Galt Motter focused on the exhumed remains of 150 individuals, providing information on the insect fauna, burial conditions (burial depth), and soil types associated with each corpse.
- The United States Civil War provided direct observations of the devastating influence of medically important insects on the human condition. Several eyewitness accounts documented the necrophagous and carnivorous activity of carrion flies, including perhaps the first reports on the use of maggot therapy for removal of necrotic tissues.

Turn of the twentieth century brings advances in understanding of necrophagous insects

- Case reports from the physicians Klingelhöffer, Horoszkiewicz, and Maschka document the distortions of cadavers that can result from insect feeding or bite marks by cockroaches, ants, and lesser necrophiles. Their observations were critical to the determination of antemortem versus postmortem injury.
- The early twentieth century was also characterized by the excellent descriptive works of Jean Henri Casimir Fabre. Fabre was renowned for excellence in teaching as well as in field research, the latter characterized by exquisite attention to detail and exactness. Many consider him the father of modern entomology.

Forensic entomology during the "great" wars

- The observations of Karl Meixner and Hermann Merkel provided new insight into insect succession on cadavers, detailing differences in human decomposition associated with age of the deceased and manner of death.
- Josef Holzer (1937) and later Hubert Casper (1950s) showed the importance of caddisflies (Order Trichoptera) in investigations involving bodies submerged in freshwater environments.
- Following World War II, forensic entomology progressed through the treatise of D.G. Hall (*Blowflies of North America*) and the seminal research on fly bionomics by A.S. Kamal and ecological succession conducted by J. Payne using baby pigs.

Growth of the discipline due to the pioneering efforts of modern forensic entomologists leads to acceptance by judicial systems and public

- Since the 1980s, forensic entomology has grown as a discipline and gained acceptance as a subfield of forensic science and as a valued tool used in civil and criminal matters.
- Evolution of the discipline has resulted through the tireless efforts in casework and research by numerous individuals located in the United States, Australia, Germany, the United Kingdom, Finland, Russia, and other countries.
- In the last two decades, organizations dedicated to forensic entomology have been formed in North America and Europe and include the North American Forensic Entomology Association (NAFEA), the American Board of Forensic Entomology (ABFE), and the European Association for Forensic Entomology (EAFE).

Forensic entomology in the twenty-first century is characterized by increased casework and development of professional standards

- Forensic entomology in the twenty-first century has been characterized by an unrivaled amount of casework involving insect evidence. This increased recognition of the value of entomological evidence in legal matters is a testament to the dedicated work of the early pioneers of the field.
- Casework during this era has centered predominantly on insect colonization/activity on human remains associated with suspicious deaths and homicides. The highest profile among these investigations was that of Casey Anthony, a young Florida mother accused of first-degree murder of her two-year-old daughter Caylee Anthony in 2008.
- Numerous non-death investigations have also typified forensic entomology casework, including matters involving medical malpractice, neglect and abuse, insect infestations of human habitation (urban entomology), and the presence of insects or their parts in food and stored products (stored product entomology).

Test your understanding

Level 1: knowledge/comprehension

1. Define the following terms:

   (a) abiogenesis
   (b) maggot mass
   (c) carrion
   (d) hemimetabolous
   (e) binomial classification
   (f) embalmed

2. Match the terms (i–v) with the descriptions (a–e).

(a) Act of laying eggs outside of body   (i) Puparium

(b) Events occurring prior to death   (ii) Monovoltine

(c) Sclerotized skin of last stage larva of dipteran species covering pupa   (iii) Antemortem

(d) Emergence of the imago from pupal covering   (iv) Oviposition

(e) Only one generation per year   (v) Eclosion

3. Describe the contributions of Mégnin that justify him receiving the title of father of modern forensic entomology.
4. What evidence exists that ancient civilizations had some understanding of the relationships between insects and death prior to Redi's experiments examining spontaneous generation?

Level 2: application/analysis

1. Explain the significance of Klingelhöffer's observations to the understanding of postmortem decomposition.
2. Detail how myiasis can have "positive" medical application in specific situations.

Level 3: synthesis/evaluation

1. The Chinese magistrate Sung Tz'u is often credited with having performed the first documented case in forensic entomology. Explain how an argument can be made that his work really did not serve as a foundation for the development of forensic entomology as it exists today.

Notes

1. Scholars generally agree that the term "lice" probably more correctly refers to one of three possible types of insects: gnats or midges, mosquitoes, or tsetse fly.
2. There is considerable debate as to whether the original Hebrew text refers to swarms of biting flies, wild animals, or even beetles.
3. Several other outstanding forensic entomologists practice worldwide but are too numerous to be listed here.

References cited

Aldrich, J.M. (1916) Sarcophaga and Allies. Murphy-Bivins Publishing Company, Lafayette, IN.

Amendt, J., Campobasso, C.P., Gaudry, E., Reiter, C., LeBlanc, H.N. & Hall, M.J. (2007) Best practice in forensic entomology—standards and guidelines. International Journal of Legal Medicine 121: 90–104.

Amendt, J., Campobasso, C.P., Goff, M.L. & Grassberger, M. (2010) Current Concepts in Forensic Entomology. Springer, London.

Benecke, M. (2001) A brief history of forensic entomology. Forensic Science International 120: 2–14.

Berenbaum, M. (1995) Bugs in the System: Insects and their Impact on Human Affairs. Helix Books, Berkeley, CA.

Beutelspacher, C.R. (1988) Las Mariposas Entre los Antiguos Meicanos. Fondo de Cultra Econonica, Avenida de la Universidad, Mexico.

Blenkinsopp, J. (1996) A History of Prophecy in Israel. Westminster John Knox Press, Louisville, KY.

Broadhead, S.M. (1864) The Diary of a Lady of Gettysburg, Pennsylvania. 1992 transcription, G.T. Hawbaker, Hershey, PA.

Brooks, S. (1966) Civil War Medicine. Charles C. Thomas Publishers, Springfield, IL.

Butin, E. & Rivers, D., & Wallace, J.R. (2020) Practical considerations for teaching forensic entomology. In: J.H. Byrd & J.K. Tomberlin (eds) Forensic Entomology: The Utility of Arthropods in Legal Investigations, pp. 577–594. CRC Press, Boca Raton, FL.

Byrd, J.H. & Tomberlin, J.K. (eds) (2020) Forensic Entomology: The Utility of Arthropods in Legal Investigations. CRC Press, Boca Raton, FL.

Carson, R. (1962) Silent Spring. Houghton Mifflin Publishers, Boston.

Cheng, K. (1890) Zhe yu gui jian. China Lu shih (no page numbers).

Cloudsley-Thompson, J.L. (1976) Insects and History. St Martin's Press, New York.

Dautartas, A., Kenyhercz, M.W., Vidoli, G.M., Jantz, L.M., Mundorff, A. & Steadman, D.W. (2018) Differential decomposition among pig, rabbit, and human remains. Journal of Forensic Sciences 63: 1673–1683.

Dell, K. (2003) Job. In: J.D.G. Dunn & J.W. Rogerson (eds) Eerdmans Bible Commentary, pp. 337–363. W.B. Eerdmans Publishing Co., Grand Rapids, MI.

Fabre, A. (1921) The Life of Jean Henri Fabre. Dodd, Mead and Co., New York.

Gennard, D.E. (2012) Forensic Entomology: An Introduction, 2nd edn. John Wiley & Sons Ltd., Chichester, UK.

Greenberg, B. (1973) Flies and Disease. Vol. 2. Biology and Disease Transmission. Princeton University Press, Princeton, NJ.

Greenberg, B. & Kunich, J.C. (2002) Entomology and the Law: Flies as Forensic Indicators. Cambridge University Press, New York.

Hall, D.G. (1948) *The Blowflies of North America*. Thomas Say Foundation, Baltimore, MD.

Hall, R.D. & Huntington, T.E. (2010) Introduction: Perceptions and status of forensic entomology. In: J.H. Byrd & J.L. Castner (eds) *Forensic Entomology: The Utility of Arthropods in Legal Investigations*, pp. 1–16. CRC Press, Boca Raton, FL.

Haskell, N.H. & Williams, R.W. (2008) *Entomology and Death: A Procedural Guide*, 2nd edn. East Park Printing, Clemson, SC.

Holy Bible (1892) *Translated Out of the Original Tongues*. National Bible Press, Philadelphia.

Hurd, P.D. (1954) "Myiasis" resulting from the use of the aspirator method in the collection of insects. *Science* 119: 814–815.

James, M. T. (1947) *The Flies That Cause Myiasis in Man (No. 631)*. US Department of Agriculture, Washington, DC.

Kamal, A.S. (1958) Comparative study of thirteen species of sarcosaprophagous Calliphorida and Sarcophagidae (Diptera). I. Bionomics. *Annals of the Entomological Society of America* 51: 261–270.

Laufer, B. (1974) *Jade. A Study in Chinese Archaeology and Religion*. Dover Publishers, New York.

McNeill, W.H. (1977) *Plagues and Peoples*. Doubleday, New York.

Mégnin, P. (1894) *La Faune des Cadavres. Application de l'Entomologie a la Medicine Legale*. Encyclopedie Scientifique des Aides-Memoire. G. Masson and Gauthier-Villars, Paris.

Meixner, K. (1922) Leichenzerstorung durch Fliegenmaden. *Zeitschrift fur Medizinalbeamte* 35: 407–413.

Merkel, H. (1925) Die Bedeutung der Art der Totung fur die Leichenzerstorung durch Madenfrass [in German]. *Deutsche Zeitschrift für die gesamte gerichtliche Medizin* 5: 34–44.

Michaud, J.-P., Schoenly, K.G. & Moreau, G. (2012) Sampling flies or sampling flaw? Experimental design and inference strength in Forensic Entomology. *Journal of Medical Entomology* 49: 1–10.

Miller, G.L. (1997) Historical natural history: insects and the Civil War. *American Entomologist* 43: 227–245.

Motter, M.G. (1898) A contribution to the study of the fauna of the grave. A study of one hundred and fifty disinterments, with some additional experimental observations. *Journal of the New York Entomological Society* 6: 201–233.

Müller, P.L.S. & von Linné, D.R.C. (1774) Vollständiges Natursystem nach der zwölften lateinischen Ausgabe (…). 5 Theil, I. Band, Von den Insecten. Raspe, Nürnberg [in German].

Payne, J.A. (1965) A summer carrion study of the baby pig *Sus scrofa Linnaeus*. *Ecology* 46: 592–602.

Ponce-Ulloa, H. (1997) Beutelspacher's butterflies of ancient Mexico. *Cultural Entomology Digest*, Issue 4. Available at http://www.insects.org/ced4/beutelspacher.html.

Reinhard, H. (1881) Beitrage zur Graberfauna. *Verhandlungen der k.-k. zoologisch-botanischen Gesellschaft in Wien* 31: 207–210.

Ross, H.H., Ross, C.A. & Ross, J.R.P. (1982) *A Textbook of Entomology*, 4th edn. John Wiley & Sons Ltd., New York.

Sanford, M.R., Byrd, J.H., Tomberlin, J.K. & Wallace, J.R. (2020) Entomological evidence collections methods: American Board of Forensic Entomology approved protocols. In: J.H. Byrd & J.K. Tomberlin (eds) *Forensic Entomology: The Utility of Arthropods in Legal Investigations*, pp. 68–89. CRC Press, Boca Raton, FL.

Smith, K.G.V. (1986) *A Manual of Forensic Entomology*. Cornell University Press, Ithaca, NY.

Stotelmyer, S.R. (1992) *The Bivouacs of the Dead*. Toomy Publishing, Baltimore, MD.

Strong, R.H. (1961) *A Yankee Private's Civil War*. Henry Regnery Press, Chicago.

Sweetman, H.L. (1958) *The Principles of Biological Control*. Wm C. Brown Company, Dubuque, IA.

Tarone, A.M. & Sanford, M.R. (2017) Is PMI the hypothesis or the null hypothesis? *Journal of Medical Entomology* 54: 1109–1115.

Tomberlin, J.K. & Benbow, M.E. (2015) *Forensic Entomology: International Dimensions and Frontiers*. CRC Press, Boca Raton, FL.

Turchetto, M. & Vanin, S. (2004) Forensic evaluations on a crime scene with monospecific necrophagous fly population infected by two parasitoid species. *Aggrawal's International Journal of Forensic Medicine and Toxicology* 5: 12–18.

Wells, J.D. & LaMotte, L.R. (2020) Estimating the postmortem interval. In: J.H. Byrd & J.K. Tomberlin (eds) *Forensic Entomology: The Utility of Arthropods in Legal Investigations*, pp. 229–240. CRC Press, Boca Raton, FL.

Whiting, A. (1967) The biology of the parasitic wasp *Mormoniella vitripennis*. *Quarterly Review of Biology* 42: 333–406.

Supplemental reading

Anderson, G.S. (2001) Forensic entomology in British Columbia: a brief history. *Journal of the Entomological Society of British Columbia* 98:127–136.

Catts, E.P. & Goff, M.L. (1992) Forensic entomology in criminal investigations. *Annual Review of Entomology* 37: 253–277.

Erzinçlioglu, Z. (2000) *Maggots, Murder and Men*. Thomas Dunne Books, New York.

Goff, M.L. (2000) *A Fly for the Prosecution: How Insect Evidence Helps Solve Crimes*. Harvard University Press, Cambridge, MA.

Klotzbach, H., Krettek, R., Bratzke, H., Püschel, K., Zehner, R. & Amendt, J. (2004) The history of forensic entomology in German-speaking countries. *Forensic Science International* 144: 259–263.

Leclercq, M. (1969) *Entomology and Legal Medicine*. Pergamon Press, Oxford.

Villet, M.H. & Williams, K.A. (2006) A history of southern Africa research relevant to forensic entomology: review article. *South African Journal of Science* 102: 59–65.

Additional resources

American Academy of Forensic Sciences: www.aafs.org

Entomological Society of America: www.entsoc.org

Forensic Entomology: Insects in Legal Investigations: http://forensic-entomology.com/

US Army Heritage and Education Center: www.usahec.org

US Civil War homepage: www.civil-war.net

Chapter 3

Role of insects and other arthropods in urban and stored product entomology

Overview

Forensic entomology is a multifaceted branch of forensic science that requires an understanding of insects reaching far beyond just those that display necrophilic and necrophagous activity on human cadavers. There are three subfields that comprise the discipline: stored product entomology, urban entomology, and medicocriminal entomology. Although the remainder of this book delves into topics centered on medicocriminal entomology, this chapter is devoted to urban and stored product entomology. The chapter will explore the defining features of each branch, both in terms of legal and nonlegal matters, and will also examine the life cycles of some important insects associated with civil and criminal issues. A discussion of civil, criminal, and administrative law also follows, with examples of how insects and other arthropods assist with such investigations.

The big picture

- Insects and other arthropods are used in civil, criminal, and administrative matters pertinent to the judicial system.

- Civil cases involve disputes over private issues.
- Criminal law involves more serious matters involving safety and welfare of people.
- Administrative law is concerned with rulemaking, adjudication, or enforcement of specific regulatory agendas.
- Stored product entomology addresses issues of insect infestations of food and stored products.
- Urban entomology is focused on insects invading human habitation.

3.1 Insects and other arthropods are used in civil, criminal, and administrative matters pertinent to the judicial system

Medicocriminal entomology is the focus of our textbook. Why? In part because the crimes being investigated are often more serious in nature (e.g., violent) and thus the outcomes of the criminal investigations have enormous impact on the lives of all involved. The 2011 trial of Casey Anthony in Orlando, Florida for the murder of her 2-year-old

The Science of Forensic Entomology, Second Edition. David B. Rivers and Gregory A. Dahlem.
© 2023 John Wiley & Sons Ltd. Published 2023 by John Wiley & Sons Ltd.
Companian website: www.wiley.com/go/forensicentomology2

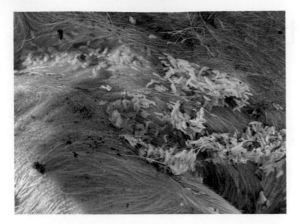

Figure 3.1 Medicocriminal or legal entomology receives more media attention than the other the two disciplines of forensic entomology. Correspondingly, the insects that feed on a human corpse, such as blow flies and flesh flies are Almost "household" names with the lay public. Photo by D.B. Rivers.

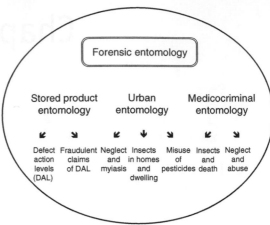

Figure 3.2 The three branches that comprise forensic entomology.

daughter is a testament to the devastating consequences of being confronted with homicide, regardless of the jury's decision (Ablow, 2011; Ashton & Pulitzer, 2011). There is also no denying the intrigue associated with such crimes. As discussed in Chapter 1, homicides and suspicious deaths draw tremendous interest from the public and news media. The fascination likely extends to the scientists engaged in the investigations, at least from the perspective of the intellectual stimulation in the deductive process of piecing together the insect clues and in presenting the findings to the court. The courtroom can be full of drama and energy as the environment is often adversarial and hostile to expert witnesses, yet at the same time may serve as an invigorating challenge to defend data generated via the scientific method. (Not everyone would agree with the latter statement, as appearing in court is among the most dreaded experiences for some forensic scientists.) Finally, most books and reviews written on forensic entomology cover only medicocriminal entomology, likely because this is where most interest lies for investigators. The result, however, is that a narrow focus generally leaves a gap in the resources available for students and practitioners interested in other aspects of forensic entomology (Figure 3.1).

Medicocriminal entomology is only a part of forensic entomology. In fact, a forensic entomologist is much more likely to be involved in cases involving urban or stored product entomology than those concerned with medicolegal questions

(Figure 3.2). Legal matters involving the presence of insects in food products or other foodstuffs, or that appear in homes can lead to litigation between individuals or corporations. Such issues fall under the jurisdiction of stored product and urban entomology, and the species in question are typically different from those relevant to medicocriminal entomology.

Since many of the legal cases involving these two branches of forensic entomology involve civil law, we will also briefly explore the differences between the three major areas of law common to forensic entomologists: civil, criminal, and administrative law. **Civil law** involves disputes between individuals or organizations in which compensation may be awarded to the victim. An entomological example is that of a pest control company failing to "properly" control or eradicate[1] a pest insect from a home, a matter that is commonly tried in civil court. By contrast, a matter of **criminal law** is more serious in stature in that it addresses conduct that threatens the safety and welfare of an individual(s) (Fletcher, 1998). Improper use of pesticides in the treatment of a home or business that threatens the health or causes harm to an individual, particularly if the violation was done deliberately, would constitute a criminal issue. **Administrative law** has become increasingly relevant to all forensic sciences as enforcement of laws and regulations has become a high priority in the United States and in many other countries facing threats to their national and individual security. Each of these branches of law will be examined in more detail in Sections 3.2–3.4 (Figure 3.3).

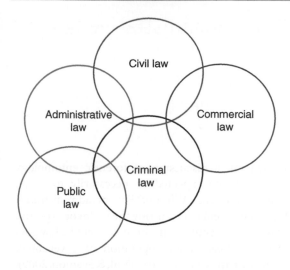

Figure 3.3 Simplified schematic depicting relationships between the major types of law in the United States.

3.2 Civil cases involve disputes over private issues

In the United States, civil law is the body of rules that define the rights and remedies of private citizens. It regulates disputes between two or more parties (between individuals, between individuals and organizations, and between organizations) in matters involving accidents, negligence or libel, collectively referred to as **torts** (Glannon, 2010). A tort is a breach of civil duty, or simply a wrong associated with non-public issues. Civil law also governs contract disputes, probate of wills, trusts, disputes over property, commercial law, administrative law, and any other private (as opposed to public) issues involving private parties and/or organizations (Bevans, 2007). An action addressed under criminal law does not necessarily preclude filing suit in civil court. In fact, in some instances victims of a crime may seek compensation from the defendant through civil action.

The latter statement reflects a fundamental difference between criminal and civil law: civil law attempts to right a wrong, enforce or honor a contract, or settle disputes between parties. If someone is wronged, a favorable decision in civil court means that they will be compensated, and the individual responsible for the wrong will be required to pay compensation. This can be thought of as legal revenge and may be the only recourse for victims of both civil and criminal offenses. For instance, in the case of O.J. Simpson,

though he was exonerated of the murder of his estranged-wife Nicole Brown Simpson in criminal court, Nicole's family did win a multi-million-dollar ($33.5 million) suit against him in civil court.

Another fundamental difference between civil and criminal law is the process. In a criminal case, the prosecuting attorney is responsible for "proving" the elements of the case to the jury beyond a reasonable doubt (Fletcher, 1998). However, in civil matters, the burden is placed on the plaintiff or victim, and the plaintiff must merely show that the defendant is liable by a preponderance of the evidence, meaning more than half. The plaintiff must file a complaint in civil court that describes the injury or dispute, how the defendant is responsible for the injury or dispute, and detail the remediation being requested. Remediation may be compensation for the injury, or simply asking the defendant to stop an action or honor a contract. Examples of civil matters relevant to urban and stored product entomology are discussed latter in the chapter.

3.3 Criminal law involves more serious matters involving safety and welfare of people

Criminal law is defined as the body of rules that deal with crime. This begs the question "What is a crime?" A **crime** is any act that violates rules (laws) established to protect public safety and welfare (Fletcher, 1998). Failure to act or an omission when action is required also constitutes a criminal violation (Kaplan *et al.*, 2008). This occurs in situations such as when a parent or legal guardian does not protect a child from harm as prescribed by law. In other words, a crime is failure to abide by public law (as opposed to civil law). Criminal law is generally considered more serious in nature than other forms of law. Correspondingly, it is distinctive in the severity of punishments or sanctions associated with commission of a crime or failure to comply with the law.

Crime is generally categorized on the severity of the offense. For example, a crime that is termed a **felony** is the most serious violation of public law and includes such acts as murder, rape, and armed robbery. These examples share the commonality of direct threats or actual attacks on individual safety and welfare. By contrast, a less severe form of crime is called a **misdemeanor.** Such offenses can range from traffic and parking violations to petty theft (stealing items valued at less than $500). Misdemeanor offenses generally do not involve any aspect of violence or threats toward individuals. It is also important to note that several aspects of civil law may be violated during a criminal act, and thus some offenses may be subject to both criminal and civil prosecution.

As alluded to during our discussion of civil law, the process of criminal prosecution is more challenging than with civil matters. The prosecuting attorney must prove all elements of the case to a jury beyond a reasonable doubt. In most criminal cases, two elements must be established: an **overt criminal act** and **criminal intent.** An overt criminal act is committed when a person knowingly, purposefully, and/or recklessly does something directly in violation of public law (Kaplan *et al.*, 2008). Such willful acts are done voluntarily as opposed to by accident or mistake. The physical evidence described in Chapter 1 is used to prove the elements of the case, or corpus delicti, with regard to criminal act.

Criminal intent deals more with a mental state. This does not refer to the mental competency of an individual, such as whether a person is deemed fit to stand trial. Rather, intent addresses the mental state of the individual accompanying the criminal act. Was it planned or did it occur by accident? The former is consistent with criminal intent in that the act was anticipated or developed prior to the crime occurring. The length of time before the act was committed is not really an issue, only that the intent to commit a crime preceded the act (Fletcher, 1998). The crime is thus deemed willful. An example can be quite simple, such as inferring intent to commit murder when an accused brings a weapon to a meeting with another individual, who is later found dead. Why carry a loaded gun or a knife to the encounter in the first place if the intent was not to harm the other person? Proof of the intent element is generally fulfilled if the defendant was aware that he or she was in fact violating the law by committing the act (or by not acting in the case of an omission) (Kaplan *et al.*, 2008). To achieve a conviction for the crime, the elements of criminal intent *and* the criminal act must be proven beyond a reasonable doubt.

3.4 Administrative law is concerned with rulemaking, adjudication, or enforcement of specific regulatory agendas

In general, a discussion of forensic entomology would not be expected to necessarily focus on administrative law. Why not? For one, administrative law would seem to be outside the realm or expertise of forensic scientists. This branch of public law governs the creation and operation of government agencies in the United States (Funk & Seamon, 2009). It is also the body of law that results from the activities of government administrative agencies. So, what possible connection is there to forensic science or any branch of forensic entomology? The connection becomes more obvious when examining some of the outcomes associated with government agencies. Broadly speaking, most federal and state agencies develop the policies and regulations that establish standards associated with manufacturing, trade, importation, the environment, transport, and several other facets of the national regulatory scheme (Davis, 1975). Science-based regulations are critical to many of these areas, including food safety, environmental issues, and workplace health and safety. From this vantage point, it is fairly easy to recognize the connection between food safety regulations, particularly defect action levels (DALs), and stored product entomology. **DALs** are defined by the Food and Drug Administration (FDA, 2018) as the amounts of naturally occurring or nonpreventable defects in foods that present no health hazards for humans if consumed (FDA Food Defect Levels Handbook). DALs will be discussed in more detail in Section 3.5.

Numerous government agencies practice rulemaking: the creation and implementation of rules and policies regulating various aspects of public law. FDA standards, guidelines concerning border protection, cybersecurity and terrorism detection and prevention by the Department of Homeland Security, and the policies set forth by the US Department of Agriculture are all outcomes of the rulemaking process of specific government agencies. Each of these examples also has implications for forensic science, with the need for forensic analyses, rule enforcement, and the development of new technologies. Forensic entomology expertise in stored product entomology, urban entomology, and medicocriminal entomology may also be warranted depending on the specific government regulations or issues. Specific examples of matters of administrative law relevant to forensic entomology are discussed in Sections 3.5 and 3.6 and also in Chapter 21.

3.5 Stored product entomology addresses issues of insect infestations of food and stored products

Now that you have a foundation in the forms of law typically associated with forensic science investigations, it is time to look at how they are specifically aligned with forensic entomology. We focus here on branches of forensic entomology that more commonly deal with issues of civil and administrative law. That is not to say that criminal cases cannot or do not result from stored product or urban entomology issues. Rather, criminal matters are simply less common than those found with medicocriminal entomology. Both branches also have a bias toward insect infestations of homes/dwellings or "stuff" in such structures. "Stuff" ranges from food products to clothing to building materials.

Our journey begins with **stored product entomology,** the area of entomology that deals with insect pests of raw and processed cereals, seeds, dried fruits, nuts, and other types of dry food commodities (Hagstrum & Subramanyam, 2009). This statement implies two points that need to be addressed: this discipline has a broader scope than just forensic entomology (i.e., all stored product issues do not have legal implications) and the associated insects generally are "pests" before becoming evidence. Before diving into this topic any further, we need to spend a moment discussing what is a pest.

3.5.1 Pest status

What is a pest? The answer to this question can vary depending on context. From an entomological perspective, pest status is achieved when the population density of a given insect exceeds some unacceptable

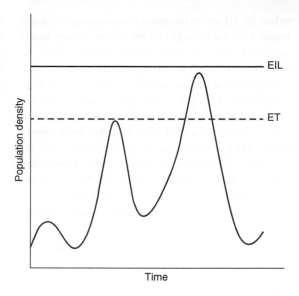

Figure 3.4 Relationship between an insect population density and achievement of pest status based on economic considerations. EIL, economic injury level; ET, economic threshold. Based on information from Pedigo & Rice (2006).

subjective threshold level, beyond which economic damage occurs (Horn, 1988). Economic damage implies a linkage to agricultural economics, the primary area of global concern for stored product entomology. In this context, the pest status of an insect can be modeled as in Figure 3.4. As the model illustrates, insect population densities are monitored over time and related to some threshold level of acceptable or unacceptable insect damage established for a cropping system or stored product. The threshold level is used as part of the decision-making process in pest management programs. Without establishing calculated or arbitrary levels of damage assessment, control efforts can be unwarranted and not cost-effective. The **economic injury level (EIL)** is an example of such a threshold. It represents the lowest number of insects that will evoke damage but is also defined as an arbitrary value at which the economic damage induced by insect activity is equal to the cost of managing the pest population (Pedigo & Rice, 2006). Population density levels above the EIL justify control measures, and those below generally do not. EILs are not static and can be influenced by several factors, including market values, agricultural practices, location, and season (Pedigo & Rice, 2006). Threshold levels are also influenced by the products in question as well as their use. For example, in urban locations, economic considerations typically mean little to individuals intolerant of any insect activity in their homes or food (Horn, 1988). In these scenarios, EILs are established by the spending habitats of the consumer (this is an oversimplification, but it does get to the core of the difference) and are often set at levels that are unfeasible or impossible to achieve (Figure 3.5). **Aesthetic injury level (AIL)** is sometimes used to describe thresholds that reflect consumer desire rather than actual injury or damage limits (Horn, 1988). Often the AIL is placed at zero because the homeowner or consumer is unwilling to accept any insect presence in their food or shelter. It is important to note that an AIL threshold is arbitrarily influenced by the consumer, independent of any real damage or health risk, and thus generally would not be suitable grounds for civil or criminal legal action.

3.5.2   Beyond pest status

To a degree, the considerations of EIL and AIL also go hand in hand with the food DALs associated with insect damage and/or insect parts in fresh and stored food products. Of course, DALs differ from AIL in that they represent limits taking into consideration human health, not discomfort, and the limits or thresholds are established through the regulatory activity (administrative law) of the FDA, not the consumer. As we have discussed earlier, the amounts of naturally occurring or non-preventable defects in foods that present no health hazards for humans if consumed are termed defect action levels or DALs. The FDA must set these limits because it is economically impractical, if not impossible, to grow, harvest, or process agricultural products that are completely free of nonhazardous, naturally occurring, unavoidable damage or defects (FDA Defect Action Level Handbook). Products known to be harmful to consumers are subject to regulatory action (as well as civil and/or criminal litigation) regardless of whether they exceed the action levels. Naturally occurring entomological defects include insect parts, secretions (e.g., saliva and digestive enzymes) and excretions **(frass)**, and evidence of damage due to insect injury. Table 3.1 provides some examples of DLAs

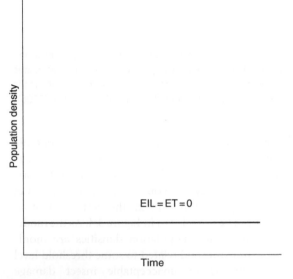

Figure 3.5   Insect status as a pest due to mere presence in habitat or locale. Control measures are applied at any population density.

Table 3.1   Examples of food defect action levels established by the US Food and Drug Administration with regard to insect parts and activity.

Food item	Defect limits	
Broccoli, frozen	Insects and mites	Average of 60 or more aphids and/or thrips and/or mites per 100 g
Corn meal	Insects	Average of 1 or more whole insects (or equivalent) per 50 g
	Insect filth	Average of 25 or more insect fragments per 25 g
Cinnamon, ground	Insect filth	Average of 400 or more insect fragments per 50 g
Citrus fruit juices, canned	Insects and insect eggs	Five or more *Drosophila* and other fly eggs per 250 mL or 1 or more maggots per 250 mL
Mushrooms, canned, and dried	Insects	Average of over 20 or more maggots of any size per 100 g of drained mushrooms and proportionate liquid or 15 g of dried mushrooms *or* average of 5 or more maggots 2 mm or longer per 100 g of drained mushrooms and proportionate liquid or 15 g of dried mushrooms
Peanut butter	Insect filth	Average of 30 or more insect fragments per 100 g

Information derived from the US Food and Drug Administration's Defect Action Level Handbook. Courtesy of USDA.

of foods and stored products with reference to whole insects, their parts, and products. When the insect-related defects and damage exceed the limits established by the FDA, the matter becomes an issue to be addressed by the forensic entomology side of stored product entomology.

Our earlier definition of a pest as that population density of an insect which exceeds a threshold level beyond which economic damage occurs is particularly relevant to agriculture, which in the United States is regulated by the Department of Agriculture (USDA). Since we have used terms of damage and defects in discussing DLAs, can the EIL concept be applied to the issues under the regulatory arm of the FDA? The short answer is no. EILs are difficult to use with insects that pose a human health risk because a market value or economic loss cannot be assigned to human life, regardless of whether an individual is "merely" injured or dies. These limitations also hinder the use of economic modeling of some urban pests, particularly those that are medically important to humans or pets (Pedigo *et al.*, 1986).

3.5.3 Forensic entomology considerations

The guidelines published by the FDA recognizes the fact that it is impossible to produce and distribute food that is completely defect-free, or for our purposes, insect-free. Thus, food manufacturers in the United States are permitted to sell food items and related foodstuffs with an acceptable level of insect-derived parts and products. When those limits are exceeded, the consumer may decide to take legal action (Catts & Goff, 1992). As discussed with civil law, the plaintiff (consumer) is required to establish how they were injured and that the defendant is responsible for that injury. A forensic entomologist trained in stored product entomology would need to assess the food or products in question to determine the species of insect responsible for contamination, what degree of contamination (whole insects, body parts, or evidence of activity such as feeding damage or webbing), and if possible, when the food item became contaminated. The latter is a critical piece of evidence as it establishes who is responsible for the insect contamination. For example, the identity of the insect may indicate that it is a pest during the growing phase, and thus the contamination occurred prior to processing or packaging. Alternatively, the insect

may still be alive and has been actively feeding on the product in question. Dependent on the different developmental stages found in the food product, it may be revealed that the insect must have been in the food prior to purchase. The responsibility then shifts to the grocery chain or possibly the food distributor. As you can see, the "when" aspect of contamination is critical to the plaintiff's case.

Such information is also important to dismantle a fraudulent claim. It is not uncommon for an individual to file suit claiming that an entire insect (or rodent) was found in a canned product or in food purchased at a restaurant. More than one restaurant chain has faced accusations of whole cockroaches (usually dead) being discovered in a sandwich or as an unwanted extra topping on pizza. (Interestingly, there never seems to be the discovery of a partially eaten cockroach in such food items!) Careful examination of the facility in question by a forensic entomologist can reveal whether that insect species is indeed an inhabitant. Further examination of any records of preventive or reactive (i.e., performed because a specific insect was detected) pest control can aid the case of either the defendant or plaintiff.

Stored product entomology cases can become issues for criminal courts if the insect-related defect leads to injury (long or short term) or potentially death of an individual. Cases of fraud will also be tried according to criminal law depending on the severity of monetary damage (potential or realized) associated with the claim and also for lying to officers of the court. Fraudulent matters can lead to civil suits by the defendant, particularly for large corporations, as the means to protect or attempt to restore their public image.

Chapter 5 provides details of several forensically important groups of insects, mostly with relevance to medicocriminal entomology. Few of those presented overlap in cases of stored product entomology. Five of the most common stored product insect pests in the United States are discussed in the remainder of this section.

3.5.4 Stored product insects

3.5.4.1 *Plodia interpunctella* Hübner (Order Lepidoptera: Family Pyralidae)

Indian meal moths, sometimes also called flour or pantry moths, are common household pests of such stored products as milled grains (flour, oats), breakfast cereal, energy bars, dried fruits, and related food items.

Figure 3.6 Larvae and adults of the Indian meal moth *Plodia interpunctella*. Photo courtesy of Clemson University-USDA Cooperative Extension Slide Series, www.bugwood.org

Figure 3.7 An adult confused flour beetle *Tribolium confusum*. Photo courtesy of Clemson University-USDA Cooperative Extension Slide Series, www.bugwood.org. References: Hagstrum & Subramanyam (2008); Rees (2008).

The insect may be detected in dry dog or cat food, as well as birdseed. Larvae are small whitish to tan caterpillars, with a brown head capsule including the mandibles. Prior to pupation, the larvae may reach 10–12 mm in length. Larvae leave distinctive silken webbing throughout the food, which tends to cause food items like cereals, oats, and seeds to stick together in loose clumps. The weblike material is diagnostic for this moth and closely related members of the Pyralidae. Pupation can occur either in the food material or outside the packaging, along shelves, walls and ceilings, typically in crevices. Adult moths are small to medium in size, typically about 8–10 mm in length with a wingspan approaching 12–15 mm in width (Figure 3.6). The adult has distinctive coloration: the tergum extending from the pronotum to the mesothorax is bronze, copper, or a deep gray while the abdomen appears to have dark bands. Forewings share similar patterns as the distal regions are copper, bronze, or dark gray; proximally the wings appear gray; and medially black lines intermixed with dark gray is evident. Adult moths do not feed and die shortly after mating and oviposition.

3.5.4.2 *Tribolium confusum* Jacquelin du Val (Order Coleoptera: Family Tenebrionidae)

The confused flour beetle and the closely related red flour beetle, *T. castaneum* (Herbst), are abundant pests of milled grain products such as flour,

cereal, corn meal, oats, rice, and crackers. Larvae and adults consume fine grain dust and broken kernels but are not capable of feeding on intact kernels. All developmental stages may be found in the same area. High infestations are characterized by a pungent odor. Larvae are elongate, reaching about 4–5 mm in length, yellow to tan in coloration, with the exception of the head which is usually dark brown, and possess three pairs of legs on the "thorax." Pupation typically occurs in or on the food source and pupae blend in quite well as the cocoon is white to tan in color. Adults of the confused flour beetle are typically about 3–4 mm in length (Figure 3.7). Bodies are flat and a shiny reddish-brown color throughout, and the antennae gradually increase in diameter from the scape to the last segment of the flagellum, yielding a club appearance. Adults of *T. castaneum* appear very similar with the exception that the terminal segments of the antennae enlarge profoundly not gradually.

3.5.4.3 *Tenebrio molitor* Linnaeus (Order Coleoptera: Family Tenebrionidae)

The yellow mealworm or darkling beetle frequents milled grain, cereals, cake mixes, meat scrapes in kitchens, dog and reptile food, birdseed, and even bedding materials for aquarium-based pets. Adults and larvae prefer food items that are damp and located in dark damp areas such as basements.

Figure 3.8 Larvae and adults of the yellow mealworm *Tenebrio molitor* feeding on oats, cereal, and potato wedges. Photo by D.B. Rivers. References: Penn State Extension (2017); Rees (2008).

Figure 3.9 An adult female vinegar fly *Drosophila melanogaster*. Photo courtesy of Pest and Diseases Image Library, www.bugwood.org. References: Hagstrum & Subramanyam (2008); Rees (2008).

Larvae are long (up to 30–33 mm) and cylindrical with a dark yellow to golden colored exoskeleton and dark brown head capsule as mature larvae, and lighter colors when younger. This beetle has an indeterminate number of larval stages. Pupation occurs in dry conditions around the food source, and pupae are naked and white to light yellow in color. Adult beetles are dark brown, reach 12–18 mm in length, and possess metathoracic wings hidden by dark brown elytra (Figure 3.8). Both males and females can be found walking slowly through and on food infested with eggs, juveniles, and pupae.

3.5.4.4 *Drosophila melanogaster* Meigen (Order Diptera: Family Drosophilidae)

The vinegar fly, universally referred to as the fruit fly, is a frequent pest of ripened and decaying produce like apples, bananas, peaches, tomatoes, grapes, melons, squash, and pumpkins. Adults will oviposit on items other than fruits and vegetables, including garbage disposals, trash containers, empty bottles, floor drains in food preparatory areas, wet mops, or anywhere a wet film of microorganisms is active. Eggs are generally deposited on or near ripened or rotting fruits and vegetables; following egg hatch, very small (>1 mm) translucent to white larvae begin to feed. Development is rapid, and within a few days (four days at 25 °C) last-stage larvae are white in appearance and may reach 2–3 mm in length. Pupariation followed by pupation usually occurs away from the food, in dryer but still damp locations. Pupation requires four days at 25 °C to complete. The puparium is translucent so that the events of adult

metamorphosis can be observed, including darkening of the exoskeleton and formation of deep red eye color. Adult flies are sexually dimorphic but both sexes have yellow to light brown body coloration with dark bands on the tergum of the abdomen (Figure 3.9). Males possess sex combs, a row of dark hairs, on the tarsus of the prothoracic legs and are only about three-quarters the length of the female. Females are quite prolific and may lay more than 500 eggs each during their lifespan.

References: Ashburner *et al.* (2005); Rees (2008).

3.5.4.5 *Oryzaephilus surinamensis* (Linnaeus) (Order Coleoptera: Family Silvanidae)

The sawtoothed grain beetle infests milled grains, cereals, bread, popcorn, dried fruits, macaroni products, crackers, and similar items found in kitchen pantries and cabinets. Along with the closely related merchant beetle, *O. mercator* (Fauvel), the two are considered the most common grain pests in the United States. Adults use mandibles to chew into unopened cardboard boxes or through cellophane windows of packaging. Once reaching the food source, population densities can increase rapidly, and individuals will then spread to other stored products. Larvae are small, whitish, elongate cylinders with a brown head and three pairs of legs on the thorax. As the larvae mature, yellow sclerites are evident on the thoracic terga. Last instar larvae attain a length near 4–5 mm. Pupation occurs in the food as the last-stage larva covers its body with food material to construct a protective capsule in which to pupate. Adults are

Figure 3.10 Adults of the sawtoothed grain beetle *Oryzaephilus surinamensis*. Photo courtesy of Clemson University-USDA Cooperative Extension Slide Series, www.bugwood.org. References: Hagstrum & Subramanyam (2008); Lyon (1997).

flattened dorsoventrally with a reddish-brown exoskeleton (Figure 3.10). Along the margins of the thorax are six saw-like projections that account for the common name. These beetles have a maximum life expectancy near four years.

3.6 Urban entomology is focused on insects invading human habitation

Urban entomology is the branch of entomology that deals with insects and other arthropods that are associated with human habitation or the human environment (Table 3.2) (Hall & Huntington, 2010). The terms "human habitation" and "human environment" are more inclusive than simply meaning human dwellings. Urban entomology does indeed address insects located in homes and other buildings, as well as those occurring in yards and neighborhoods (Catts & Goff, 1992), including those typically thought of as agricultural that invade human space. However, if we stop our definition here, then a comparison with stored product entomology can be made: both urban and stored product entomology have a primary scope outside of forensic entomology, and each discipline has a focus on insects that can be

Table 3.2 Common arthropods associated with urban entomology.

Classification	Common name	Habitat
Noninsect arthropods		
Arachnida	Spiders of all sorts	Indoors and outdoors
	Ticks, mites	Indoors and outdoors
Chilopoda	Centipedes	Warm, damp locations
Diplopoda	Millipedes	Warm, damp locations
Insecta		
Blattodea	Cockroaches	Warm, humid, dark indoors
Coleoptera	Wide range of beetles	Indoors and outdoors
Diptera	Flies	Indoors and outdoors
	Mosquitoes	Outdoors*
Hemiptera	Box elder bugs, bed bugs	Indoors and outdoors
Hymenoptera	Ants	Indoors and outdoors
	Bees, wasps, and hornets	Outdoors*
Isoptera	Termites	Subterranean, galleries
Siphonaptera	Fleas	Indoors
Thysanura[†]	Silverfish, firebrats	Homes, buildings

*Typically occur in outdoor environments but will move indoors for brief periods.
[†]Are now considered separate from the insects but closely related in the Class Hexapoda.

found in homes and buildings. Urban entomology does not have an agricultural focus as does stored product entomology, but agriculturally important insects do become relevant to this discipline. For example, several species of flies associated with livestock feedlots, dairy farms, poultry houses, or hog facilities may invade nearby homes and businesses, creating an urban entomology problem that may lead to litigation. This example can be viewed more commonly as a civil matter but can become a criminal issue with the potential for spread of disease through mechanical transfer or biting action of flies.

Insects that are structural pests of any type of building fall under the umbrella of urban entomology. Such insects include post beetles, termites, carpenter ants, and carpenter bees. One of the most common responsibilities of an urban entomologist is developing methods for detecting and controlling structural insects. However, when controversies arise, frequently between a homeowner and a pest control company, these become matters involving the forensic entomology side of urban entomology. Legal action may be pursued over whether adequate control measures were used, species identification was correct, or whether there was misuse of pesticides, which, if health issues arise, may be instrumental in changing a civil case into a criminal case.

Our discussion of structural pests is a fitting place to discuss economic considerations and thresholds. Can mathematical modeling and decision thresholds like an EIL be applied to urban entomology? As with stored product entomology, EILs are generally not applicable in an urban insect context (Horn, 1988). Why not? The answer is multi-tiered in that (i) the urban environment (dwellings, healthcare facilities) does not allow easy sampling or modeling of insect populations, (ii) the insects are not predictable pests in that their movement into an area is often ephemeral, and (iii) the market value considerations of human life mentioned earlier prevail in these scenarios as well. A structural pest would seemingly fall outside of such considerations, which is true in the sense that they do not impact human health. It is also true that economic damage can be measured in terms of destruction to a home or other building (Pedigo & Rice, 2006). However, discovery of a structural pest is akin to a medically important insect in that measures to control and eradicate will be taken at any population density of a structural pest. This is not the same as an AIL which is entirely arbitrary, yet these scenarios can be analogous to an EIL being equal to zero, which also means equal to an economic threshold level that warrants preventive or immediate action (Pedigo & Rice, 2006).

Cases of adequate control are always dealt with by urban entomology because consumers generally are intolerant of any insects or other arthropods in their domiciles and thus demand eradication. Certainly, it is understandable that a homeowner wants a structural pest totally eliminated to prevent further damage to such an enormous financial investment such as a home. However, in some cases the biology of the insect does not match up well with available technologies to achieve the desired outcome of total eradication. This situation is best illustrated with the major urban pest *Solenopsis invicta*, the red, imported, fire ant. In several regions of the southern United States, this ant has made its presence known by invading farmland, backyards, and even moving indoors. The ants bite and sting, producing painful welts and pustules on humans, pets and livestock, and in some cases may induce death. Although several measures have been attempted to manage local populations of *S. invicta*, true control is not easily achieved and eradication has proven nearly impossible. This has not stopped some homeowners from filing civil suits against pest control companies for unsatisfactory control.

Fire ants do not restrict their invasions to outdoor locations. Numerous reports detail attacks on patients in hospitals and nursing homes, in some cases leading to death. Similarly, fly infestations of patients in healthcare facilities (and domiciles) is not an uncommon phenomenon throughout the United States. Fly eggs or larvae may be found on necrotic tissues of a patient, associated with catheters or in or around diapers, conditions known as **myiasis** (See Chapter 16 for in-depth discussions of myiasis). Any of these examples generally imply some form of neglect. Such matters frequently become the subject of either civil or criminal litigation.

The United States has experienced a resurgence of some urban entomological pests that were thought to be problems of the past. Over the last decade, bed bug populations have soared in Australia, Europe, and the United States (Dogget *et al.*, Doggett *et al.*, 2004; Potter, 2005). The reasons for the resurgence have not been fully deciphered, although development of insecticide resistance (Romero *et al.*, 2007) and/or decreases in use of broad-spectrum pesticides mandated by the Environmental Protection Agency (EPA) in the United States (Potter, 2005) have been postulated as contributing factors. In fact, the latter explanation has also been cited for recent increases in a number of stored product and urban pests (Hagstrum & Subramanyam, 2008). Increased travel by students and other individuals, particularly as participants in study programs abroad associated with colleges and universities, is a working hypothesis that may partly explain elevated populations of the bed bug, *Cimex lectularius* Linnaeus (Hemiptera: Cimicidae)[2], but this would not account for the rise in other urban pests.

It is worth noting that recently there are few insects that evoke such emotional responses from the public as bed bugs. Fear, loathing, and panic are all commonly witnessed from individuals learning or believing that bed bugs are present in their luggage, home, dorm room, or hotel (Davies, 2004; Intini, 2008). The immediate response seems to be application of some sort of control measures to rid "the bugs" from the premises. No doubt the second thought is "Who is responsible"? Civil litigation soon follows. For the time being, urban entomologists in the United States will have a fair share of civil and possibly criminal cases to work involving *C. lectularius* (Sharkey, 2003; Goddard & deShazo, 2009).

Five of the most common urban insect pests found in the United States are discussed in the remainder of this section. The one exception is the house fly, *Musca domestica* Linnaeus (Diptera: Muscidae), a common household pest but also a species of medicolegal concern.

3.6.1   Urban insects

3.6.1.1 *Periplaneta americana* (Linnaeus) (Order Blattaria: Family Blattidae)

The American cockroach is one of the most common insect pests found in homes, commercial buildings, restaurants, and hotels in the United States. It is one of several common cockroach species that frequent dwellings and display overlapping food preferences. Adults and juveniles (nymphs) are found in the same habitats, typically dark moist areas such as basements, crawl spaces, crevices along walkways, laundry rooms, and in kitchen and bathrooms near water supplies. This insect is omnivorous and will feed on almost anything, including all kinds of human food, the fingernails of small children, tissue and fluids of a corpse, and each other (cannibalism). Adults are large, reaching nearly 4 cm in length, and display a reddish-brown coloration over the entire body (Figure 3.11). They also have two long antennae and large black compound eyes. Newly hatched nymphs are about 3–4 mm in length, lack wings, and have similar coloration as the adults. With each subsequent molt, nymphs will gradually become longer and wings will begin to develop as external buds in the notal region of mesothorax and metathorax. Nymphs and adults are generally nocturnal but may be observed in daylight hours if populations are high and/or food is abundant. High populations tend to produce a distinctive unpleasant odor that is usually not obvious with just a few individuals. Juveniles and adults will opportunistically feed on a corpse found indoors, creating artifacts in the form of lesions where feeding has occurred and depositing stains or modifying existing ones.

Figure 3.11 Nymphs and adults of the American cockroach *Periplaneta americana*. A recently molted adult is located in the center of the image. Photo by D.B. Rivers. References: Gold & Jones (2000); Robinson (2005).

3.6.1.2 *Solenopsis invicta* Buren (Order Hymenoptera: Family Formicidae)

The red imported fire ant is native to South America but has become firmly established in parts of the United States, ranging east to Florida, north to Maryland, and west to Texas. What do they eat? It may be easier to point out what they will not eat. In general, fire ants are carnivorous, feeding on a range of arthropods, including several that are agricultural pests. The ants have been reported to feed on vegetation as well, but this seems to occur when other forms of food are scarce. Damage evoked by this ant is often through mound building activities, where livestock or pets that disturb the mounds are viciously attacked by hundreds to thousands of individuals. Ant bites and stings leave painful welts on the skin. The only members of the ant colony encountered are workers or soldiers and these have a reddish-brown head and thorax, and a dark colored abdomen (Figure 3.12). The "waist" or region between the thorax and abdomen has two humps. Adults range in length between 3 and 6 mm.

3.6.1.3 *Reticulitermes flavipes* Kollar (Order Isoptera: Family Reticulitermitidae)

Subterranean termites in the genus *Reticulitermes* are the most destructive termites in the United States. The eastern subterranean termite is most commonly encountered in the eastern region of the United States. They are a species that requires a connection to moist ground, which the worker

Figure 3.12 Adults of the red imported fire ant *Solenopsis invicta*. Photo courtesy of the USDA APHIS PPQ Archive, USDA APHIS PPQ, www.bugwood.org. References: Vinson & Greenberg (1986); Robinson (2005)

Figure 3.13 Workers of the subterranean termite *Reticulitermes flavipes*. Photo courtesy of Clemson University-USDA Cooperative Extension Slide Series, www.bugwood. org. References: Osmun (1962); EPA; Robinson (2005).

termites maintain by forming tunnels to food sources. Almost anything containing cellulose is suitable as food and includes paper products, books, wood-based building materials, furniture, and materials composed of cotton. Termites are social insects and form an elaborate caste system. The castes include workers, soldiers, and reproductives, which usually is just a queen and king. Workers are cryptobiotic and confine their feeding to the inside of wood, so they are seldom observed. The workers are about 4–5 mm in length, wingless, and creamy white in appearance with an oval to round head, and have short antennae and lack eyes (Figure 3.13). The reproductive stage is the most likely to be observed, but generally only during early months of spring when swarming occurs. Swarmers are winged dark-bodied male and female adults that pair up, engage in mating, lose their wings and then form new colonies. Active healthy termite colonies may contain 10,000 to

more than 1 million individuals with a queen laying 5000–10,000 eggs per year.

3.6.1.4 *Cimex lectularius* Linnaeus (Order Hemiptera: Family Cimicidae)

The common bed bug has a cosmopolitan distribution worldwide and all members of this group are blood-feeding parasites on warm-blooded mammals. Both males and females are obligate blood feeders, and females require a relatively continuous blood supply to mature multiple clutches. Males display mating preference for recently fed females. Bed bugs typically spend the majority of their time in a refugia, waiting to feed once the host displays minimal activity. This typically coincides with the nocturnal sleep patterns of humans. Feeding by adults lasts 10–20 minutes, after which time the bed bugs return to their refugia to engage in mating. Mating is unique for *C. lectularius* in that traumatic insemination is used by adult males. Adults are virtually indistinguishable from each other: both sexes are reddish-brown in color, flattened dorso-ventrally, oval in shape, and wingless. Mesothoracic wings are vestigial pads located on the tergum. Adults reach 4–5 mm in length and 1–2 mm in width before expanding greatly following a blood meal (Figure 3.14). Nymphs are smaller and lighter in color than the adults and gradually darken with each molt. Bed bug activity may be evident in areas around bedding and furniture due to "spotting" from fluid excretion.

3.6.1.5 *Blattella germanica* Linnaeus (Order Blatodea: Family Blattellidae)

The German cockroach is one of the most gregarious household pests found throughout the world. Their habitat is nearly identical to that of

Figure 3.14 An adult bed bug *Cimex lectularius*. Photo courtesy of Clemson University-USDA Cooperative Extension Slide Series, www.bugwood.org. References: Harlan (2006); Reinhardt & Siva-Jothy (2007).

Figure 3.15 Adult of the German cockroach *Blattella germanica*. Photo courtesy of Clemson University-USDA Cooperative Extension Slide Series, www.bugwood.org. References: Osmun (1962); Robinson (2005).

P. americana with the exception that they are almost never found outdoors in temperate regions, consistent with the tropical origins of this insect. In commercial buildings, hotels, dormitories or apartment buildings, adults and nymphs can easily move from room to room or to other floors by following pipelines or other types of service connections. These are the cockroaches most likely encountered in restaurants, food processing facilities, nursing homes, or other types of commercial buildings. Juveniles and adults are found together in moist dark environs. The insects, at all feeding stages, are efficient scavengers, consuming almost any type of plant or animal material. Adults reach 10–12 mm in length, possess wings but almost never use them, and appear tan to light brown in color. Two diagnostic dark longitudinal bands are present on the pronotum (Figure 3.15). Females can be distinguished from the males since they are usually larger and carry an embryonic "sac" or ooetheca at the tip of the abdomen. Juveniles undergoing hemimetabolous development (nymphs), are light brown, and display wing buds on the thoracic terga.

3.6.2 Delusionary parasitosis

Any discussion of urban entomology in a forensic context is not complete without having spent some time on examining **delusionary parasitosis**. Delusional parasitosis or infestation is a clinical term referencing a broad range of psychopathological conditions characterized by individuals' fixation that they are infested with small pathogens or animals (e.g., insects, "worms") on the skin or that are internal. Despite evidence to the contrary from medical, psychiatric or academic professionals (e.g., entomologists, microbiologists), an "afflicted" individual is unwavering in believing an infestation has occurred and details tactile sensations in the form of pruritus, formication, and biting/stinging that have persisted over an extended period (Freudenmann & Lepping, 2009; Hinkle, 2011). An array of other symptoms can be associated with delusional infestation and have been detailed in the excellent reviews by Murray & Ash (2004), Freudenmann and Lepping (2009), and Hinkle (2011). The condition is more commonly known as delusional parasitosis, but also has been reported by many other synonyms including delusory parasitosis, entomophobia, aracophobia, parasitophobic neurodermatitis, Morgellons, and Ekbom syndrome (Wong & Koo, 2013). Regardless of the name applied to the symptoms, two main forms of delusional infestation are recognized: primary and secondary. The primary form is characterized by the delusion or conviction of being infested with no medical or physical evidence to support the contention and is accompanied by hallucinogenic tactile sensations of the skin. The latter frequently leads to intense and obsessive cleansing, scratching, and at times, severely intrusive means for removal of the perceived pathogens or parasites (Murray & Ash, 2004). Aside from the delusional condition, afflicted individuals generally display no other psychosis and engage in relatively normal conversations outside of the delusional theme. By contrast, the secondary form is believed to arise from some other disorder, trauma, or substance abuse (Freudenmann, 2003; Aw *et al.*, 2004). The exact causations are not known and thus a spectrum of symptoms is exhibited in addition to the two unifying features of the delusional belief and abnormal skin sensations.

Delusional infestation is considered an uncommon if not rare condition. However, the number of individuals reported with any form of delusional infestation is on the rise (Hinkle, 2011). In the case of the unclassified condition Morgellons, undoubtedly this can be attributed in part to use of the Internet for self-diagnosis, which in turn leads to an individual contacting experts directly for help, bypassing or rejecting medical or psychiatric diagnosis (Pearson *et al.*, 2012). Increases in instances of secondary delusional infestation are likely linked to elevations in the number of users of illicit or pre-scribed medications. Though the characteristics of either form of delusional infestation are distinct and readily apparent to anyone with familiarity with the spectrum of delusional disorders, diagnosis requires consultation with medical and psychiatric professionals. The willingness to provide specimens is common with either form of delusional infestation. Providing perceived physical proof of an infestation is referred to as the specimen or matchbox sign (Freudenmann & Lepping, 2009). In nearly all cases of delusionary parasitosis, the specimens provided are inanimate objects rather than biological creatures (Figure 3.16), supporting the diagnosis of a psychosomatic event. Individuals suffering from any form of delusionary infestation require psychiatric consultation, outside of any training of a forensic entomologist.

Afflicted individuals commonly contact forensic entomologists, as they believe infestations or attacks involve insects or related creatures, and the contact information for most forensic entomologists is readily accessible via a search on Google or other search engines. Consequently, it is important for any entomologists to be aware of physicians/psychiatrists with appropriate training, so that referrals can be made if contacted by an individual suspected of suffering from delusionary parasitosis.

Chapter review

Insects and other arthropods are used in civil, criminal, and administrative matters pertinent to the judicial systems

- The use of insects and related arthropods in legal issues is the remit of forensic entomology, the discipline being represented by three branches: stored product entomology, urban entomology, and medicocriminal entomology.
- The presence of insects in food products or other foodstuffs or that appear in homes (either as an annoyance, such as fly infestations due to nearby livestock facilities, or because of improper treatment by a pest control company to rid a home of termites or ants) often leads to litigation between individuals or corporations, falls under the jurisdiction of stored product and urban entomology, and the insects in question are typically different from those relevant to medicocriminal entomology.
- Usually a forensic entomologist specializes in one of the branches rather than attempting to be an expert in all matters relevant to forensic entomology.

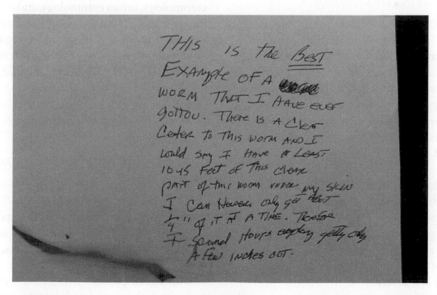

Figure 3.16 A letter from an individual suffering from delusionary parasitosis, in which a specimen or "worm" was provided as evidence of infestation. The "worm" in the lower left corner was identified as a piece of black thread. Photo by D.B. Rivers.

Civil cases involve disputes over private issues

- Civil law is the body of rules that governs disputes between two or more parties (between individuals, between individuals and organizations, and between organizations) over private matters and also governs contract disputes, probate of wills, trusts, disputes over property, commercial law, administrative law, and any other private issues.
- An action addressed under criminal law does not necessarily preclude filing suit in civil court.
- Civil law differs from criminal law in that the former attempts to right a wrong, enforce or honor a contract, or settle disputes between parties.
- In civil court, the process places the burden of proof on the plaintiff or victim, and the plaintiff must merely show that the defendant is liable by a preponderance of the evidence, meaning more than half. The plaintiff must file a complaint in civil court that describes the injury or dispute, how the defendant is responsible for the injury or dispute, and detail the remediation being requested.

Criminal law involves more serious matters involving safety and welfare of people

- Criminal law is defined as the body of rules that deals with crime. Generally, criminal law deals with more serious issues than civil law, and correspondingly the punishments and sanctions are more severe than those with civil cases.
- A crime is any act that violates rules (laws) established to protect public safety and welfare. Failure to act or an omission when action is required also constitutes a criminal violation. In other words, a crime is failure to abide by public law (as opposed to civil law).
- Crime is generally categorized based on the severity of the offense. A felony is a more serious violation of public law, whereas a misdemeanor is considered less severe.
- The process of criminal prosecution is more challenging than that of civil matters. A prosecuting attorney must prove all elements of the case to a jury beyond reasonable doubt. In most criminal cases, two elements must be established: an overt criminal act and criminal intent. To achieve a conviction for the crime, the elements of criminal intent and the criminal act must be proven beyond reasonable doubt.

Administrative law is concerned with rulemaking, adjudication, or enforcement of specific regulatory agendas

- Administrative law is the branch of public law that governs the creation and operation of government agencies in the United States, and is also the body of law that results from the activities of government administrative agencies.
- FDA standards, guidelines concerning border protection, cybersecurity, and terrorism detection and prevention by the Department of Homeland Security, and the policies set forth by the United States Department of Agriculture are all outcomes of the rulemaking process of specific government agencies. Each of these examples also has implications for forensic science with the need for forensic analyses, rule enforcement, and the development of new technologies.
- Forensic entomology expertise in stored product entomology, urban entomology, and medicocriminal entomology may also be warranted depending on the specific government regulations or issues, but often are most closely aligned to rules established by the USDA, FDA, and EPA.

Stored product entomology addresses issues of insect infestation of food and stored products

- Stored product entomology is the area of entomology that deals with insect pests of raw and processed cereals, seeds, dried fruits, nuts, and other types of dry food commodities.
- Not all stored product issues have legal implications and the associated insects are generally "pests" before becoming evidence.
- Pest status is achieved when the population density of a given insect exceeds a threshold level,

beyond which economic damage occurs. Economic damage implies a linkage to agricultural economics, the primary area of global concern for stored product entomology.

- The FDA recognizes the fact that it is impossible to produce and distribute food that is completely defect-free, or for our purposes insect-free. Thus food manufacturers in the United States are permitted to sell food items and related foodstuffs with an acceptable level of insect-derived parts and products. When those limits are exceeded, the consumer may want to take legal action.
- A forensic entomologist trained in stored product entomology would need to assess the food or products in question to determine the species of insect responsible for contamination, what degree of contamination, and if possible when the food item became contaminated.
- Stored product entomology cases can become issues for criminal courts if the insect-related defect leads to injury or potentially death of an individual. Cases of fraud will also be tried according to criminal law depending on the severity of monetary damage (potential or realized) associated with the claim and also for lying to officers of the court.

Urban entomology is focused on insects invading human habitation

- Urban entomology is the branch of entomology that deals with insects and other arthropods that are associated with human habitation or the human environment. The terms "human habitation" and "human environment" are more inclusive than simply indicating human dwellings, and mean to address insects located in homes and other buildings, as well as those occurring in yards and neighborhoods, including those typically thought of as agricultural that invade human space.
- Urban entomology does not have an agricultural focus as does stored product entomology, but agriculturally important insects do become relevant to this discipline, as are structural pests of any type of building material and potentially destructive or annoyance insects like nest builders that sting or bite.

- When controversies do arise, frequently between a homeowner and a pest control company, the issues become matters of the forensic entomology side of urban entomology. Legal action may be pursued over whether adequate control measures were used, species identification was correct, or due to misuse of pesticides. The latter issue may well move from a civil matter to a criminal case if health issues arise.
- Some urban insects have medical relevance such as fly myiasis, implying neglect, or as the result of a bite or sting from social Hymenoptera or bed bugs.
- The United States has experienced a resurgence of some urban entomological pests that is thought to be attributed to increased world travel by individuals, development of insecticide resistance in some insects (e.g., bed bugs), or a decrease in the use of broad-spectrum pesticides in urban environments.
- Delusional infestation or parasitosis is a clinical term referencing a broad range of psychopathological conditions characterized by individuals' fixation that they are infested with small pathogens or animals on the skin or that are internal. Despite evidence to the contrary from medical, psychiatric, or academic professionals, an "afflicted" individual is unwavering in believing an infestation has occurred and details tactile sensations in the form of pruritus, formication, and biting/stinging that have persisted over an extended period. Afflicted individuals commonly contact forensic entomologists, as they believe infestations or attacks involve insects or related creatures, and the contact information for most forensic entomologists is readily accessible via a search on Google or other search engines.

Test your understanding

Level 1: knowledge/comprehension

1. Define the following terms:
   (a) civil law
   (b) tort
   (c) criminal law
   (d) delusionary parasitosis
   (e) EIL
   (f) stored product entomology
   (g) urban entomology.

2. Match the terms (i–vii) with the descriptions (a–g).

(a) Type of crime that is generally considered a lesser offense

(i) Criminal intent

(b) Infestation of body tissues by fly larvae

(ii) AIL

(c) Arbitrary pest threshold based on zero tolerance for any level of insects

(iii) Felony

(d) Limit of acceptable insect parts or damage to a food item like fresh produce

(iv) Crime

(e) Failing to abide by public law

(v) Myiasis

(f) A violent act such as murder or rape

(vi) DAL

(g) Mental state of an individual when committing a violation of public law

(vii) Misdemeanor

3. Compare and contrast the differences in legal processes associated with a civil versus criminal litigation to establish guilt of a defendant.

4. Distinguish between an issue of interest to urban entomology that has legal versus nonlegal implications.

5. Describe the differences between economic thresholds, EILs, and AILs.

Level 2: application/analysis

1. Under what conditions would the treatment of a private residence by a pest control company for carpenter ants potentially lead to criminal charges filed by a district attorney.

2. Explain how matters of administrative law directly impact stored product and urban entomology.

3. Provide examples of the types of forensic analyses that a forensic entomologist would engage in if hired to investigate a case in which larvae of *Plodia interpunctella* were discovered in a box of biscuit mix recently purchased by the plaintiff.

Level 3: synthesis

1. During a recent trip to New York City, an insurance adjuster stays at a very nice hotel in the financial district of Manhattan. He discovers upon returning home that there are several small reddish, inflamed bumps on both legs. Puzzled by what caused the "rash," the man soon has his answer. He finds several nymphs and adults of the bed bug *Cimex lectularius* in his suitcase. The man concludes that the bed bugs and the bites had to have been incurred while staying at the hotel in New York. As he loathes insects and is allergic to bee stings, the man becomes very angry. He promptly calls his attorney to begin the process of suing the hotel. Discuss the potential strengths and weaknesses of the man's case should it evolve into civil litigation. In your answer, explain what type of evidence is needed for the plaintiff to have a strong case against the defendant.

Notes

1. Eradication of insects has been a desire of many people but generally is an impossible goal to attain. The exception can be local control, such as a home or business building, in which eradication may be achieved depending on the insect and degree of infestation.

2. *Cimex lectularius* is a flightless insect that is only able to migrate by walking or hitching a ride on clothing or in luggage.

References cited

Ablow, K. (2011) *Inside the Mind of Casey Anthony: A Psychological Portrait*. St Martin's Press, New York.

Ashburner, M., Golic K.G. & Hawle, R.S. (2005) *Drosophila: A Laboratory Handbook*, 2nd edn. Cold Spring Harbor Laboratory Press, Cold Spring Harbor, NY.

Ashton, J. & Pulitzer, L. (2011) *Imperfect Justice: Prosecuting Casey Anthony*. William Morrow Publishers, New York.

Aw, D.C.W., Thong, J.Y., & Chan, H.L. (2004) Delusional parasitosis: case series of 8 patients and review of the literature. *Annals of the Academy of Medicine Singapore* 33: 89–94.

Bevans, N. (2007) *Civil Law and Litigation for Paralegals*. McGraw-Hill, New York.

Catts, E.P. & Goff, M.L. (1992) Forensic entomology in criminal investigations. *Annual Review of Entomology* 37: 253–272.

Davies, E. (2004) Australian plague of bed bugs costs tourist industry millions. *The Independent*, November 6, p. 42.

Davis, K.C. (1975) *Administrative Law and Government*. West Publishing, St Paul, MN.

Doggett, S.L., Geary, M.J. & Russell, R.C. (2004) The resurgence of bed bugs in Australia: with notes on their ecology and control. *Environmental Health* 4: 30–38.

Environmental Protection Agency (n.d.) Termites: How to identify and control them. Available at https://www.epa.gov/safepestcontrol/termites-how-identify-and-control-them.

Fletcher, G.P. (1998) *Basic Concepts of Criminal Law*. Oxford University Press, London.

Food and Drug Administration (2018) *Defect Levels Handbook: Food Defect Action Levels*. Available at http://www.fda.gov/Food/GuidanceRegulation/GuidanceDocumentsRegulatoryInformation/SanitationTransportation/ucm056174.htm

Freudenmann, R.W. (2003) A case of delusional parasitosis in severe heart failure. Olanzapine within the framework of a multimodal therapy. *Nervenarzt* 74: 591–595.

Freudenmann, R.W. & Lepping, P. (2009) Delusional infestation. *Clinical Microbiology Reviews* 22(4): 690–732.

Funk, W.F. & Seamon, R.H. (2009) *Administrative Law: Examples and Explanations*, 3rd edn. Aspen Publishers, New York.

Glannon, J.W. (2010) *The Law of Torts: Examples and Explanations*, 4th edn. Aspen Publishers, New York.

Goddard, J. & deShazo, R. (2009) Bed bugs (*Cimex lectularius*) and clinical consequences of their bites. *Journal of the American Medical Association* 301: 1358–1366.

Gold, R.E. & Jones, S.C. (2000) *Handbook of Household and Structural Insect Pests*. Entomological Society of America Handbook Series, Lanham, MD.

Hagstrum, D.W. & Subramanyam, B. (2008) *Fundamentals of Stored-product Entomology*. American Association of Cereal Chemists, St Paul, MN.

Hagstrum, D.W. & Subramanyam, B. (2009) A review of stored-product entomology information resources. *American Entomologist* 55: 174–183.

Hall, R.D. & Huntington, T.E. (2010) Introduction: Perceptions and status of forensic entomology. In J.H. Byrd and J.L. Castner (eds) *Forensic Entomology: The Utility of Arthropods in Legal Investigations*, pp. 1–16. CRC Press, Boca Raton, FL.

Harlan, H.J. (2006) Bed bugs 101: the basics of *Cimex lectularius*. *American Entomologist* 52: 99–101.

Hinkle, N.C. (2011) Ekbom syndrome: A delusional condition of "bugs in the skin". *Current Psychiatry Reports* 13: 178–186.

Horn, D.J. (1988) *Ecological Approach to Pest Management*. Guilford Press, New York.

Intini, J. (2008) Sleeping with the enemy. Available at https://archive.macleans.ca/article/2008/1/14/sleeping-with-the-enemy.

Kaplan, J., Weisberg, R. & Binder, G. (2008) *Criminal Law: Cases and Materials*, 6th edn. Aspen Publishers, New York.

Lyon, W.F. (1997) Sawtoothed and merchant grain beetles. Ohio State University Extension Fact Sheet HYG-2086-97.

Available at https://www1.maine.gov/dacf/php/gotpests/bugs/factsheets/grain-beetles-wv.pdf.

Murray, W.J. & Ash, LR. (2004) Delusional parasitosis. *Clinical Microbiology News* 26(10): 73–77.

Osmun, J.V. (1962) Household insects. In: R.E. Pfadt (ed.) *Fundamentals of Applied Entomology*. Macmillan Company, New York.

Pearson, M.L., Selby, J.V., Katz, K.A., Cantrell, V., Braden, C.R., Parise, M.E., Paddock, C.D., Lewin-Smith, M.R., Kalasinsky, V.F., Goldstein, F.C., Hightower, A.W., Papier, A., Lewis, B., Motipara, S. & Eberhard, M.L. (2012) Clinical, epidemiologic, histopathologic and molecular features of an unexplained dermopathy. *PLoS One* 7(1): e29908.

Pedigo, L.P. & Rice, M.E. (2006) *Entomology and Pest Management*, 5th edn. Prentice Hall, Upper Saddle River, NJ.

Pedigo, L.P., Hutchins, S.H. & Higley, L.G. (1986) Economic injury levels in theory and practice. *Annual Review of Entomology* 31: 341–368.

Penn State Extension (2017) Confused flour beetle and red flour beetle. https://extension.psu.edu/confused-flour-beetle-and-red-flour-beetle.

Potter, M.F. (2005) A bed bug state of mind: emerging issues in bed bug management. *Pest Control Technology* 33: 82–85.

Rees, D. (2008) *Insects of Stored Products*. SBS Publishers, Dartford, UK.

Reinhardt, K. & Siva-Jothy, T. (2007) Biology of the bed bugs (Cimicidae). *Annual Review of Entomology* 52: 351–374.

Robinson, W.H. (2005) *Urban Insects and Arachnids: A Handbook of Urban Entomology*. Cambridge University Press, Cambridge, UK.

Romero, A., Potter, M.F., Potter, D.A. & Hayes, K.F. (2007) Insecticide resistance in the bed bug: a factor in the pest's sudden resurgence. *Journal of Medical Entomology* 44: 175–178.

Sharkey, C. (2003) Modern tort litigation trends. Punitive damages as societal damages. *Yale Law Journal* 113: 347–454.

Vinson, S.B. & Greenberg, L. (1986) The biology, physiology and ecology of imported fire ants. In: S.B. Vinson (ed.) *Economic Impact and Control of Social Insects*, pp. 193–226. Praeger Press, New York.

Wong, J.W. & Koo, J.Y.M. (2013) Delusions of parasitosis. *Indian Journal of Dermatology* 58(1): 49–52.

Supplemental reading

Augerland, R. & Strange, J. (2007) *The Bugman on Bugs: Understanding Household Pests and the Environment*. University of New Mexico Press, Albuquerque, NM.

Byrd, J.H. & Castner, J.L. (eds) (2010) *Forensic Entomology: The Utility of Arthropods in Legal Investigations*, 2nd edn. CRC Press, Boca Raton, FL.

Eisenberg, J. (2011) *The Bed Bug Survival Guide: The Only Book You Will Need to Eliminate or Avoid This Pest Now*. Grand Central Publishing, New York.

Gaensslen, R.E., Harris, H.A. & Lee, H.C. (2007) *Introduction to Forensic Science and Criminalistics*. McGraw-Hill, Boston.

Hames, J.B. & Ekern, Y. (2009) *Introduction to Law*, 4th edn. Prentice Hall, Upper Saddle River, NJ.

James, S. & Norby, J.J. (2009) *Forensic Science: An Introduction to Scientific and Investigative Techniques*, 3rd edn. CRC Press, Boca Raton, FL.

Singer, R.G. & La Fond, J.Q. (2007) *Criminal Law*, 4th edn. Aspen Publishers, New York.

Tschinkel, W. (2006) *The Fire Ants*. Belknap Press, Cambridge, MA.

Additional resources

Center for Urban and Structural Entomology at Texas A&M University: http://urbanentomology.tamu.edu/

Delusional parasitosis: https://www.health.state.mn.us/diseases/pests/dp.html

Stored product pests in the pantry: http://www.ca.uky.edu/entomology/entfacts/ef612.asp

United States Department of Agriculture: http://www.usda.gov/wps/portal/usda/usdahome

United States Department of Justice: www.justice.gov

Urban Entomology Program at University California-Riverside: http://urban.ucr.edu/

Chapter 4
Introduction to entomology

Overview

Entomology is the scientific study of insects, but the field is generally broadened to include study of other terrestrial arthropod groups. Here, we will look at the characteristics that define the insects and see what makes them different from other groups in their phylum, the Arthropoda.

When we try to explain why an insect is present in a particular forensic setting, it can be very important to "think like an insect" – to try to understand its life and motivations. In order to have a feeling for how insects perceive the world and why they do the things they do, we need to have an understanding of the body plan of these creatures, including the sensory structures that determine their perception of their environments. Knowledge of their morphology, especially their external morphology, is critical to enable us to identify insects and their relatives.

The big picture

- Insecta is the biggest class of the biggest phylum of living things, the Arthropoda.

- The typical adult insect has three body parts, six legs, two antennae, compound eyes, external mouthparts, and wings.
- Tagmosis has produced the three functional body segments of insects: the head, thorax, and abdomen.
- Sensory organs and their modifications allow insects to perceive and react to their environments.
- The structure and function of an insect's digestive system is intimately tied to the food that it prefers to eat.
- In insects, a tubular tracheal system transports oxygen to the body's cells while blood moves through the body without the aid of blood vessels.
- The nervous system of insects integrates sensory input and drives many aspects of behavior.
- In order to grow, insects need to shed their "skin."
- Many insects appear and behave in an entirely different way as a larva than as an adult – the magic of metamorphosis.
- The desire to reproduce is a driving force for unique reproductive behaviors and copulatory structures in insects.

The Science of Forensic Entomology, Second Edition. David B. Rivers and Gregory A. Dahlem.
© 2023 John Wiley & Sons Ltd. Published 2023 by John Wiley & Sons Ltd.
Companian website: www.wiley.com/go/forensicentomology2

4.1 Insecta is the biggest class of the biggest phylum of living organisms, the Arthropoda

The largest animal phylum (i.e., the one with the greatest number of living species) is undeniably the Arthropoda (see Boxes 4.1 and 4.2). Spiders, scorpions, insects, and crustaceans are some of the common members of this diverse group. Arthropods creep and crawl on all continents and swim in salt-water oceans and freshwater environments. They range in size from tiny wasps no bigger than a single-celled *Paramecium* to spider crabs with arm spans of over 3 m (10 feet). Insects and other arthropods are extremely numerous across the land, even if most go unnoticed due to their small size. A study of the microfauna of a terrestrial habitat in North Carolina (Pearse, 1946) yielded an estimate of approximately 124 million arthropods per acre.

Major databases dealing with the diversity of life (e.g., Catalogue of Life, World Register of Marine Species) list approximately 1.25 million described species, while another 700,000 additional species are estimated to have been described but their names have not been entered into the main data-bases yet. Realize that we are talking about *all* organisms, from bacteria and plants to birds and

House fly	

Scientific classification

Kingdom:	Animalia
Phylum	Arthropoda
Class	Insecta
Order	Diptera
Family	Muscidae
Genus	*Musca*
Species	*domestica* Linnaeus

Binomial name

Musca domestica Linnaeus, 1758

Figure 4.1   Classification of the house fly, *Musca domestica*. *Source*: James Lindsey/Wikimedia Commons/CC BY-SA 3.0.

Box 4.1   Classification categories

While we are introducing the phylum Arthropoda, it might be a good idea to review what a phylum is. A **phylum** is a major taxo-nomic category in the Linnaean hierarchy used for the classification of life. Phyla (pleural of phylum) have a rank below kingdom and above class. There are approximately 35 currently rec-ognized phyla in the animal kingdom. The Linnaean hierarchy classifies animals into seven major categories, ending in the species. These ranks are (from most inclusive to least): kingdom, phylum, class, order, family, genus, and species (Figure 4.1). A good way to remember these rank names in proper order is to make up a simple statement, using the first letter of each rank as the first letter of seven consecutive words (e.g., "King Phillip Came Over For Ginger Snaps" or "Keep Plates Clean Or Family Gets Sick").

monkeys. Over a million of these organisms, approximately two-thirds of the described species of life on planet Earth, are arthropods. And the vast majority of arthropods are insects (Figure 4.2).

One of the most basic questions that we might ask about insects is "How many different kinds are there?" Despite centuries of work ded-icated to the naming and describing of life on planet Earth, there is little consensus on the answer to this question. Most of the estimates of species diversity are little more than educated guesses without reliable empirical data to support them. One estimate that has received some notoriety is the 30 million proposed by Erwin (1982), based on the number of beetle species associated with individual tropical rain-forest tree species. Hamilton *et al.* (2010) provided new estimates based on a similar

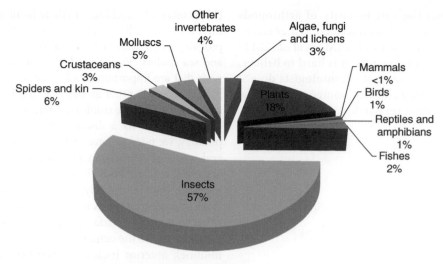

Figure 4.2   Relative numbers of described eukaryotic species.

Box 4.2   Importance of scientific names

Modern classifications of life are based on phylogenetic hypotheses that group together organisms based on common ancestry and evolutionary origin rather than basic morphological similarity. Familiarity with these strange-sounding Latin and Greek names is *essential* for the forensic entomologist. Most authors of scientific articles just use scientific names, expecting their audience to know what organisms they are talking about. In courtroom testimony, the proper use of scientific names is the mark of an expert witness, while their improper use or overreliance on common names can indicate amateur status (Figure 4.3).

Figure 4.3   Response differences of amateur versus professional forensic entomologist. *Source*: Courtesy of G. A. Dahlem.

assumption that the number of beetle species associated with tropical rainforest tree species can be used to estimate the species richness of tropical arthropods. The authors use two separate but related mathematical models to estimate arthropod diversity with medians of 3.7 million and 2.5 million tropical species, with 90% confidence intervals of 2.0–7.4 million and 1.1–5.4 million species, respectively. These estimates are considerably lower than Erwin's estimates, but still show that the vast majority of arthropods remain unnamed and undescribed. Other recent studies on the diversity of life (Mora *et al.*, 2011) have yielded similar results. It is hard to believe, in this day and time, that entomologists do not have a better answer when someone asks how many species exist on Earth.

4.2   The typical adult insect has three body parts, six legs, two antennae, compound eyes, external mouthparts, and wings

The defining characteristics of arthropods include a segmented body, jointed appendages, chitinous exoskeleton, tubular alimentary canal, open circulatory system, and a number of other internal and developmental features. The name Arthropoda literally means joint (arthro) foot (poda) and an easy way to picture an arthropod is as an armor-plated relatively small animal with jointed legs (Figure 4.4). Phyla are subdivided into distinctive groups called classes. The class Insecta includes arthropods that share a common ancestor and exhibit distinctive characteristics, including six legs, three body parts, external mouthparts, one pair of antennae, compound eyes, and (usually) four wings as adults.

Other classes in the Arthropoda that you are probably familiar with are the arachnids (Arachnida) such as spiders and scorpions, centipedes (Chilopoda), millipedes (Diplopoda), and crabs and their relatives

Figure 4.4 Typical arthropod showing "armor-plated" exterior. *Source*: Courtesy of G. A. Dahlem.

(Malacostraca). In addition to these familiar animals, there are many other distinctive groups that you have probably never seen or heard of, such as pauropods and sea spiders (Pycnogonida). While most arthropods that are important in a forensic context will be insects, there are other groups that may be of importance in specific case studies (Merritt *et al.*, 2007).

Going back to a discussion of external morphology, we need to have an understanding of general terms of orientation in order to find our way around the insect body. Insects are bilaterally symmetrical, and their body can be described along three axes: ventral refers to the lower surface; dorsal refers to the upper surface; lateral refers to the side; medial refers to the center (usually the longitudinal midline); anterior indicates toward the head end; posterior indicates toward the hind end; distal or apical refers toward the tip; proximal or basal refers toward the body or base (Figure 4.5). These terms are used repeatedly in just about any identification key that you may use. For example, Whitworth (2006) uses key characters such as "one or more accessory notopleural setae between the usual *anterior* and *posterior* notopleural setae . . ." or "*basal* section of stem vein setose . . ." in his key to the genera of North American blow flies.

An insect's body is covered with a "skin" termed exoskeleton, which is shed as the insect grows. It does not grow as the animal grows, like our skin. The supportive "skeletons" of insects are on the outside of the body and the muscles are attached inside. The exoskeleton is composed of a noncellular coating called the **cuticle** that is secreted by the outer living cell layer of the body, the epidermis. The combination of the epidermis and the cuticle is called the **integument** (Figure 4.6). The hard rigid plates of the exoskeleton are called **sclerites** and these are connected by soft flexible membranes. Internal invaginations of the exoskeleton are called **apodemes** and **sulci.** These serve to strengthen the exoskeleton and provide attachment sites for the internal musculature. Sclerites covering the dorsal surface of the insect are called **tergites**. Sclerites on the ventral surface are called **sternites** and those on the lateral surfaces are called **pleurites.**

The larvae of many insects with complete metamorphosis do not show this armor plate-like covering and the body appears to be partially to entirely enveloped with a continuous, flexible, and tough membranous cover. The chemical composition of this softer "skin" is very similar to that of the hardened exoskeleton of adults. The difference is that the cuticle has not undergone the chemical reactions associated with sclerotization that the adult cuticle undergoes.

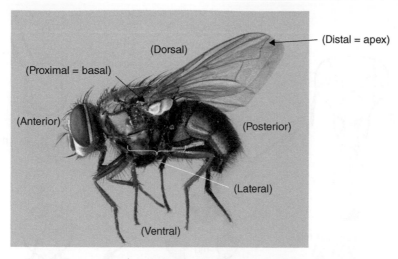

(Dorsal)

(Distal = apex)

(Proximal = basal)

(Anterior)

(Posterior)

(Lateral)

(Ventral)

Figure 4.5   Body orientation of an adult insect. *Source*: Courtesy of G. A. Dahlem.

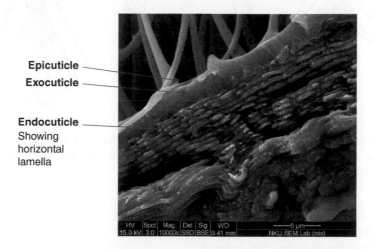

Epicuticle
Exocuticle

Endocuticle
Showing
horizontal
lamella

Figure 4.6   Scanning electron micrograph showing cross section of cuticle. *Source*: Courtesy of G. A. Dahlem.

The larvae of higher Diptera, commonly referred to as "maggots," have a membranous body covering but their mouthparts are composed of hard sclerotized plates.

4.3   Tagmosis has produced the three functional body segments of insects: the head, thorax, and abdomen

Arthropod body plans evolved from multiple repeating body plans of an ancestor much like the common earthworm, as the body segments fused into functional units. Arthropods like harvestmen ("daddy-long-legs") fused everything together to one functional body. Other groups, like Arachnids

(spiders, ticks, etc.), have bodies made up of two functional segments. The insects evolved a three-part body, comprising head, thorax, and abdomen. **Tagmosis** is the evolutionary process that involves the modification and fusion of body segments into functional units.

4.3.1   Head

The head of an insect is where the feeding appendages are located and most of the structures associated with sensory input. It is also the site where the brain is located. Insects have evolved a wide variety of different feeding strategies and mouthpart morphologies. Most adult insects that are of medicolegal importance (carrion feeders) will have biting/chewing mouthparts or lapping mouthparts (Figure 4.7). Biting and chewing mouthparts are characteristic of adult insects

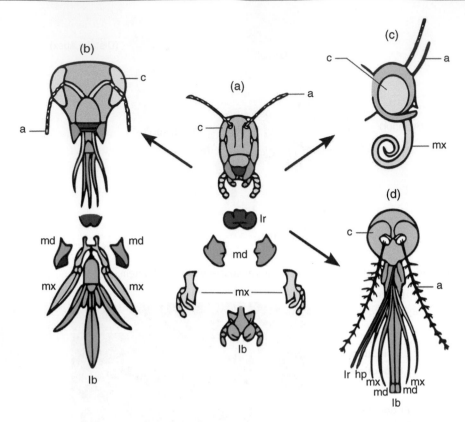

Figure 4.7   Types of insect mouthparts: (a) chewing/biting, (b) chewing/Lapping, (c) siphoning, and (d) piercing/sucking. a, antennae; c, compound eye; md, mandible; mx, maxillae; lr, Labrum; lb, labium hp, hypopharynx. *Source*: Xavier Vaszquez/Wikimedia/CC BY-SA 3.0.

like beetles, wasps, and ants and allow feeding on solid materials. The highly specialized lapping mouthparts of flies limit the adults to a mainly liquid diet. Flies can feed on solid food but must first liquefy it with their saliva before sucking up the resulting dissolved materials (see Box 4.3). Note that blow flies and flesh flies feed a little differently when ingesting fluid as opposed to solid: the liquid is imbibed, mixed with digestive enzymes from salivary glands, foregut and possibly midgut, and then regurgitated onto a solid surface for consumption of the latter. This aspect of fly feeding has huge significance for crime scene reconstruction in which blood spatter and fly spots are both present.

The senses of sight, taste, smell and touch, but not hearing, are perceived to a large extent by structures associated with the head. The characteristic adult insect has two large compound eyes and a pair of antennae. These structures are discussed further in Section 4.4 dealing with sensory organs. All these sensory inputs are processed by the insect brain, which is the initiation point for the majority of behaviors exhibited by insects.

Box 4.3   Fly feeding in the movies

This mode of eating by flies was sensationalized in the classic 1986 horror movie *The Fly* where a scientist named Seth Brundle (played by Jeff Goldblum) experiments with teleportation and accidentally fuses his DNA with that of a fly, causing him to mutate into a monstrous insect. During the transformation process we get the memorable quote from Seth: "How does Brundlefly eat? Well, he found out the hard and painful way that he eats very much the way a fly eats. His teeth are now useless, because although he can chew up solid food, he can't digest them. Solid food hurts. So like a fly, Brundlefly breaks down solids with a corrosive enzyme, playfully called "vomit drop." He regurgitates on his food, it liquefies, and then he sucks it back up. Ready for a demonstration, kids? Here goes. . ." (from http://www.imdb.com/title/tt0091064/quotes). The ensuing scene is one that needs to be seen, rather than read, for the full effect.

Source: Based on the Fly/20th Century Fox.

The insect head is usually composed of a very hardened cuticle reinforced with robust internal apodemes and sulci, which protect the brain and serve as attachment points for the strong muscles involved with the mouthparts. The beginning of the alimentary canal consists of a small diameter tube extending through an often narrow neck into the thorax.

The immature stages of insects that undergo complete metamorphosis appear totally different from the adult and their head is very different in construction. In most maggots, only the mandibles, or mouth hooks, are visible externally and the rest of the cephalopharyngeal skeleton (Figure 4.8) is found inside the head segments of the larvae. This cephalopharyngeal skeleton is usually the only hard and durable part of the fly larva and its shape may be used to separate different species as well as different life stages. Special techniques need to be employed to examine these distinctive larval features, which in some cases can be used as a means of identification (Sukontason *et al.*, 2004a). New techniques have been developed to allow DNA extractions from fly larvae without damaging the structure of the cephalopharyngeal skeleton for morphological conformation of molecular identifications (Martoni *et al.*, 2019).

4.3.2   Thorax

The thorax is the locomotion segment of the body. This is where the legs and wings are found. The thorax represents the fusion of three ancestral segments, each possessing a pair of legs but only the second and third segments normally possess wings. The three segments are named the prothorax (anterior), mesothorax (middle), and metathorax (posterior). Almost all insects have four wings, the one exception being the Diptera or true flies. The order name Diptera means two (di) wings (ptera), and the one pair of wings of these insects are present on the middle segment (mesothorax). The hind wings of flies have been reduced to two knob-like gyroscopic appendages called the **halteres** (Figure 4.9). They are found on the dorsolateral surface of the hindmost segment of the thorax. The halteres are complex mechanosensory organs that help steer the wings and stabilize the fly's gaze (see Agrawal *et al.*, 2017 for more about the halteres).

Each leg is segmented, with complex articulation points to facilitate the insect's ability to walk and/or manipulate its environment. A leg is divided into five basic parts: the coxa, trochanter, femur, tibia, and tarsi (from base to apex) (Figure 4.10). Note the similarity in terminology between the last three segments and the names of the bones in a human leg and foot.

Internally the thorax is packed with muscles to power the movement of the legs and wings. Large **spiracles** (external openings for the ventilatory system) are generally found on this segment to facilitate the movement of oxygen through the respiratory tracheal system to power the energy needs of the thoracic musculature. The pleural sclerites (lateral plates) of this segment are highly developed and each pleurite has a particular name, unlike the abdomen. The alimentary canal is basically a thin tube which passes through this segment on its way to the abdomen. The abdomen, not the thorax, is where we find the vast majority of internal organs.

Similar to the discussion of the head, the thorax of the immature forms of many insects can exhibit a

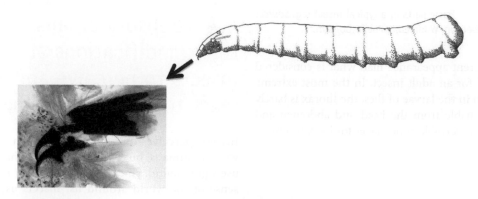

Figure 4.8   Cephalopharyngeal skeleton typical of a necrophagous fly larva. Illustration of a fly Larva by Art Cushman and provided courtesy of the Department of Entomology at the Smithsonian Institution (http://www. entomology.si.edu/IllustrationArchives.htm). *Source*: U.S. Department of Health & Human Services.

Figure 4.9 An adult crane fly with the pair of halteres clearly visible. *Source*: Pinzo/Wikimedia Commons/Public Domain and Courtesy of G. A. Dahlem.

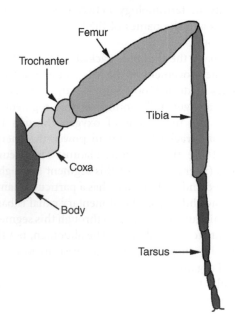

Figure 4.10 Basic parts of a typical insect leg. *Source*: Image Nwbeeson/Wikimedia/Public Domain.

very different appearance than what is considered "normal" for an adult insect. In the most extreme case, seen in the larvae of flies, the thorax is barely distinguishable from the head and abdomen and bears no discernible appendages for locomotion.

4.3.3 Abdomen

The largest body region is the last. The abdomen appears to be pretty simple in appearance on the outside, but the real complexity and importance can

be found on the inside. Here is where we find the majority of the organs of the digestive, excretory, and reproductive systems. Storage and digestion of food, osmoregulation (maintenance of water and ionic homeostasis), and production of sperm and egg cells are all the responsibility of organs in the abdomen. Externally, the structures associated with copulation and oviposition are located at the posterior tip, as well as the anus for the release of waste products. Unlike mammals, insects do not separate liquid from solid waste: their feces are more similar to bird feces than mammal feces in this regard. The **aedeagus** of the male (similar to the penis of male mammals, Figure 4.11) and vagina of the female are not involved with urination, only reproduction. Note that the external and internal features of the larval abdomen can be substantially different from the features of the adult abdomen.

4.4 Sensory organs and their modifications allow insects to perceive and react to their environments

Insects perceive their environment in similar ways to humans, but with different emphasis. We use sight, sound, smell, taste, and touch to make sense of the world around us. Insects use these same senses, but they are much more sensitive to the chemical world surrounding them than the visual world.

Boettcheria latisterna   *Boettcheria bisetosa*

Figure 4.11   Aedeagi of two flesh flies. Scanning electron micrographs. *Source*: Courtesy of G. A. Dahlem.

When we think of forensically important insects, it is important to have an understanding of how they find the food sources they will feed on and deposit their offspring on or near. Chemical cues dispersed in the air lead them to these resources. Chemicals important to insects are usually volatile (have low boiling points and vaporize easily) (see Chapter 7 for an in-depth discussion of insect olfaction). These chemicals dissipate as a complex molecular mixture, with much higher concentrations closer to the source. If you have been around a dead animal in the summer, you know that the smell is much stronger the closer you get and that it is better to stand upwind to lessen the impact of the rotting stench. To the carrion fly, that stench is as appealing as the smell of freshly baked cookies to a human, alerting them to a potential feast. Wind and concentration gradients direct the searching insect to the potential food or oviposition media.

The primary structures involved with the sense of smell in insects are the antennae. The antennae are usually packed with chemoreceptors (Sukontason *et al.*, 2004b; Yan et al., 2018) that possess a fantastic ability to perceive particular chemicals, in concentrations much too low for human perception. Studies have investigated the attraction of insects to the complex smells of potential food resources in the laboratory with the use of modified wind tunnels. The sense of taste is connected to the sense of smell, in that both are forms of **chemoreception.** Taste is a type of contact chemoreception involved with short-range discrimination.

While olfaction requires thousands of different types of receptor neurons, taste usually involves a relatively small group of receptors that distinguish broader groups of chemical constituents. Both flies and humans exhibit similar taste discrimination, especially regarding sweet and bitter compounds (Scott, 2005). While the human sense of taste is pretty much restricted to the tongue, insects can taste with receptors on a variety of different morphological structures. The gustatory receptors in flies can be found on bristles scattered all over the body, including the legs (especially the tarsal pads), the wings, the proboscis, and the ovipositor. When a fly walks across your picnic lunch, it literally tastes where it steps. When it steps in something "good," it will lower its proboscis to feed.

Vision is another important sense for insects, but an insect's ability to see is generally not as important as its ability to smell. Insects see with compound eyes. Each facet picks up a separate picture of the immediate surroundings and these many pictures are fused into a useful visual input by the insect's brain. The number of separate ommatidia (facets of the compound eye, Figure 4.12) varies by species and sex (see Box 4.4). Decisions based on visual cues come into play for short-range decisions when insects are hunting for food materials. Despite being a short-range input, visual input is transmitted at a very rapid pace. This allows an insect like a fly to react quickly to a potential predator (or the slap of a human hand). To get a feeling for this quicker input and reaction to visual stimuli, as compared to humans, realize that the typical fluorescent

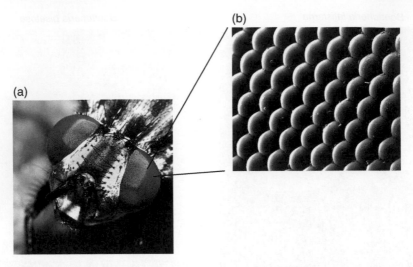

(b)

(a)

Figure 4.12 Anterior view of the compound eyes of a fly (a) and a surface view using scanning electron microscopy (b). *Source*: Rude/Wikimedia Commons/GNU Free Documentation License. Scanning electron micrograph of flesh fly ommatidia. *Source*: Courtesy of G. A. Dahlem.

Box 4.4   Ommatidia in a fly eye

In a recent study by Sukontason *et al.* (2008) comparing the number of ommatidia in several species of flies, a significant difference was found. For example, they found that a male flesh fly, *Sarcophaga dux*, had 6032 ommatidia per eye while the house fly, *Musca domestica*, had only 3484.

Figure 4.13 Setae in sockets. Scanning electron micrograph. *Source*: Courtesy of G. A. Dahlem.

light bulb flickers on and off 120 times per second (120 Hz). Humans can notice lights flicking on and off up to about 50 Hz but a house fly notices over 200 Hz (Ruck, 1961). To a house fly flying through a building with fluorescent bulbs, the environment would appear to be illuminated with strobe lights, constantly flashing on and off. The extremely quick visual perception of blow flies is currently being investigated for potential use in robotics (Anonymous, 2009).

Most of the insects of medicolegal importance are effectively deaf, with no known structures involved in the detection of sound. Certainly, there are insects with refined abilities to discriminate sound signals. In most cases, these are insects that use sound for reproductive behaviors, especially finding mates. The "ears" of insects that can hear can be found on a variety of body parts, depending on the species involved. Many crickets have their audio receivers on their front legs. Many grasshoppers have sound-receiving organs on the sides of their abdomen. Some flies have sound receptors on their neck, but these are usually associated with

host location behaviors of insect parasitoids that use the mating songs of other insects to locate their prey (Schniederkötter & Lakes-Harlan, 2004).

Mechanoreception, or the sense of touch, is accomplished differently in insects than in humans. We usually sense touch by a change in pressure on the skin. Insects are covered by a hardened cuticle that usually does not allow pressure detection. Insects use body setae (hairs) and their deflection or movement for mechanoreception. The setae involved with touch are located in a socket beneath the cuticle (Figure 4.13). When the hair is deflected, there is a pressure change in the corresponding socket and this causes a neuron to fire. It is through this mechanism that an insect perceives that it has touched an object or that something has touched it.

4.5 The structure and function of an insect's digestive system is intimately tied to the food that it prefers to eat

Moving from the external morphology of insects to the internal, we start with a discussion of the insect digestive system. This includes the alimentary canal, through which the insect's food and water pass, and associated feeding structures like the salivary glands. The alimentary canal can be divided into three functional parts, the foregut, midgut, and hindgut (Figure 4.14). The foregut and hindgut are formed from invaginations at the head and posterior tip of the abdomen and are lined with a relatively impermeable layer of cuticle. The midgut is not coated in cuticle, is permeable, and is where digestion and nutrient absorption occurs. The midgut of both fly larvae and adults is the longest part of the alimentary canal, comprising approximately two-thirds of the length in an adult blow fly (Boonsriwong *et al.* 2011).

The foregut (stomodaeum) is subdivided into a pharynx, an esophagus, and a crop (food storage pouch). In some insects a proventriculus (similar to the gizzard of birds) is found associated with the crop and is used for grinding up solid food into smaller particles. The stomodael valve marks the end of the foregut and serves as the control mechanism for movement of food into the midgut.

Since no digestion occurs in the crop, food material retained in this organ may be used to obtain quality DNA from food, the insect has been eating. Analysis of gut contents may allow the association of a fly or maggot with a corpse through DNA analysis in a murder investigation (Wells *et al.*, 2001; Mohammad *et al.*, 2021).

The major morphological divisions of the midgut are the gastric caeca and **ventriculus.** The midgut is the portion of the alimentary canal where most of the digestion of food takes place. Digestive enzymes are produced by cells that line the midgut to break down the ingested food materials. Absorption of water, ions, glucose, amino acids, and other nutrients occurs across the midgut wall. The gastric caeca often appear as a number of pouch-like extensions near the anterior end of the midgut and these provide extra surface area for secretion and absorption. As noted earlier, the midgut is not lined with cuticle like the foregut and hindgut. Most insects produce a relatively tough membrane that surrounds the bolus of food. This peritrophic membrane serves to protect the delicate cells that line the interior surface from possible abrasion.

The midgut is the location where many microbial endosymbionts will reside that an insect often requires for its particular nutritional requirements. It is also the location where most parasites and pathogens ingested by the insect will leave the

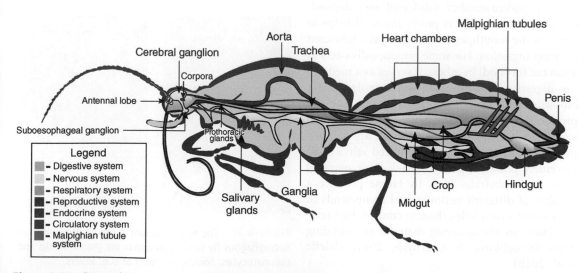

Figure 4.14 Internal anatomy of an adult insect. *Source*: Bugboy52.40/Wikimedia/CC BY-SA 3.0.

alimentary canal and move into the hemolymph to potentially infect the insect's body.

The hindgut starts at the pyloric valve, which controls the movement of food out of the midgut. This part of the alimentary canal is divided into three morphological parts, the ileum, colon, and rectum. Excretion and ionic balance are controlled by specialized excretory organs called Malpighian tubules, which empty into the anterior end of the hindgut. These excretory organs are involved in the removal of metabolic waste products from the body and in osmoregulation. The intestine is often modified in structure to house and support important mutualistic microorganisms. The rectum is particularly important for water and ion reabsorption from the digested remains of the food before it is ejected from the body through the anus.

Excess food is stored in a white or yellowish tissue called the fat body. These amorphous energy-storage organs may occupy a large amount of the available space within the abdomen of an insect adult or larva. They serve as a storage place for glycogen, fat, and protein. In addition to a storage function, they serve as a site for metabolism of a wide range of nutrient molecules. While separate from the alimentary canal, the fat body is considered a part of the insect digestive system.

The salivary glands are also an important part of the insect digestive system that are not directly connected to the alimentary canal. Insects have a pair of salivary glands that lie ventral to the foregut in the head and thorax, and occasionally extend posteriorly into the abdomen. These glands vary in size, shape, and the type of secretion produced.

Saliva plays a number of different roles, depending on the insect that is producing it. It helps to moisten the mouthparts and serves as a lubricant for food ingestion. For some insects, saliva acts as a solvent for food. In others, it serves as a medium for digestive enzymes and anticoagulants (for blood feeders) or as a source of toxins. The silk produced by butterfly and moth caterpillars and a variety of bee, wasp, and ant larvae are salivary products. Certain flies use specialized saliva as an extremely powerful glue to attach their puparial cases to a substrate. Blow fly larvae produce a number of different antimicrobial compounds in their saliva which helps them to control the bacterial fauna in the decaying material surrounding them (Kruglikova & Chernysh, 2011; Caleffe et al., 2018).

4.6 A tubular tracheal system transports oxygen to the body's cells while blood moves through the body without the aid of a vascular system

The basic purpose of the respiratory and circulatory systems of animals is to provide each cell of the body with the oxygen and food it needs to undergo cellular respiration. The way that food and oxygen is transported to cells, and waste products removed, is very different in insects compared with humans and other vertebrates. Insects do not "breathe" with their heads and they do not have arteries and veins to transport blood around their body.

The major components of the insect circulatory system are the hemocytes (blood cells, Figure 4.15), hemolymph (blood), and the dorsal blood vessel, which helps to circulate the hemolymph through the body. The dorsal blood vessel has two functional parts, the valved heart in the abdomen and the unvalved aorta which extends through the thorax. The valves of the heart are important in directing the posterior to anterior flow of hemolymph. Insects have an open circulatory system – there are no blood vessels extending through their bodies. Even without veins and arteries, the hemolymph does move along predictable pathways through the body and appendages, transporting food to cells and removing metabolic waste products. In some insects where movement of hemolymph is restricted

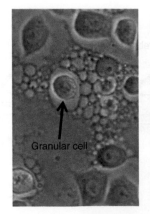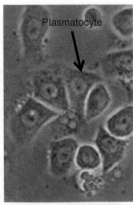

Figure 4.15 The two dominant types of hemocytes in necrophagous fly larvae and pupae are granular cells and plasmatocytes. *Source*: Courtesy of D.B. Rivers.

to the dorsal vessel between the abdomen and thorax, there are periodic heartbeat reversals where the flow changes to anterior to posterior (Wasserthal, 2012). The hemolymph does *not* transport oxygen, except for larvae of a few species that live in oxygen-poor aquatic environments.

Transport of oxygen, and disposal of gaseous waste products like carbon dioxide, is accomplished through a complex network of hollow tubes that make up an insect's ventilatory system. This organ system has three functional parts: (i) the spiracles, or holes through the body to the outside atmosphere (Figure 4.16), (ii) the cuticle-lined tracheae, which branch out from the spiracles throughout the insect's body, and (iii) the tiny, branching tracheoles, which form the terminal endings of the tracheal system. Transfer of oxygen to individual cells, and removal of carbon dioxide, occurs in the tracheoles. Most insects have eight or nine pairs of spiracles located on the lateral sides of their bodies. There are no spiracles on the head, one or two on the thorax, and the rest are often found associated near or within the membrane that connects the dorsal tergites with the ventral sternites in the abdomen.

Their number can be much reduced from this normal state, especially in immature stages. The larvae of flies, for example, have only two pairs of spiracles, one located on the lateral sides of the thorax and the other pair at the posterior end. Maggots submerse themselves in a decaying liquid "soup" and are able to breathe through the posterior spiracles. The posterior spiracular openings are usually distinctive in fly larvae and serve as one of the diagnostic features for identification to species or to life stage (instar) (e.g. Szpila *et al.*, 2014).

4.7 The nervous system of insects integrates sensory input and drives many aspects of behavior

The basic components of the insect nervous system are the neurons, or nerve cells. Each neuron has three basic parts: (i) the dendrite, which receives input, (ii) the cell body, where the nucleus and most organelles are found, and (iii) the axon, which transmits information to another neuron or to an effector organ (e.g., muscle tissue). One neuron does not directly touch another neuron, but the cells are very close to one another. The small gap between the dendrite of one neuron and the axon of another is called the synapse and signals are passed from one cell to the next by a group of chemical messengers called neurotransmitters.

These neurotransmitters can either stimulate or inhibit the neuron or other tissue at the tip of the axon branches. Many of these chemical messengers are very similar to those in mammalian

Prothoracic spiracle of *Boettcheria bisetosa*

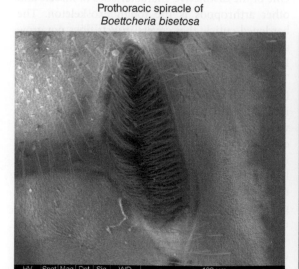

Abdominal spiracle of *Boettcheria bisetosa*

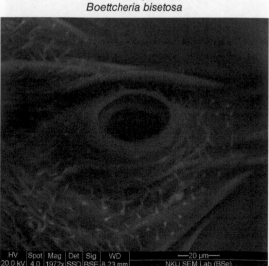

Figure 4.16 Spiracles of adult flesh flies. Scanning electron micrographs. *Source*: Courtesy of G. A. Dahlem.

nervous systems. Two common examples are ace-tylcholine and dopamine. Many common insecticides are neurotoxins that affect the chemicals necessary for transmission of signals between neurons.

There are four functional types of neurons. Sensory neurons receive information from the environment and transmit signals to the central nervous system (CNS). Interneurons receive information and transmit it to other neurons along a neural pathway. Motor neurons receive information from interneurons and transmit appropriate signals to muscle fibers to initiate or halt contraction, thus controlling movement of the insect's appendages. The final type of neuron is called the neuroendocrine cells and these are involved in hormone production.

The CNS (Figure 4.17) consists of a series of ganglia joined by paired longitudinal nerve cords called connectives. Ganglia are nerve centers where the cell bodies of interneurons and motor neurons are aggregated. In the most primitive condition, we would expect to find one pair of ganglia per body segment. In most insects of forensic interest, the two ganglia of each thoracic

and abdominal segment are fused into a single structure. The ganglia of the head are fused to form two centers known as the brain (or supraesophageal ganglion) and the subesophageal ganglion. The chain of thoracic and abdominal ganglia is called the ventral nerve cord. All the ganglia are connected together in a chain by the connectives, from the tip of the abdomen to the brain along the ventral midline of the body. The brain is composed of three pairs of fused ganglia, each with a special primary function. The anterior portion is associated with the eyes and is particularly associated with processing visual signals. The middle portion directly connects to the antennae and processes most of the sense of smell. The posterior portion is concerned with handling the signals that arrive from most of the rest of the body. The fused ganglia of the mouthpart-bearing segments form the subesophageal ganglion, which innervates and controls the movement of mouthparts. The brain is the only portion of the nerve cord that is dorsal to the alimentary canal. Some reflex actions of insects are processed in the thoracic or abdominal ganglia, without interpretation by the brain, but most complex behaviors are initiated by neurons contained in the brain.

4.8   In order to grow, insects need to shed their "skin"

One of the distinctive characteristics of insects and other arthropods is the tough exoskeleton. The hardened cuticle that covers insects does not grow as the insect grows and it does not stretch much. Insects need to shed their exoskeleton as they grow. This process is called **ecdysis** or molting (Figure 4.18). This is a much more complicated and involved process than what a snake goes through when it sheds its skin. When an insect undergoes ecdysis all of the cuticular structures are shed, including inner parts of the exoskeleton (e.g., the linings of the foregut, hindgut, and basal sections of the tracheae) and an often radically different body plan has been constructed under the old skin before molting occurs.

As an insect prepares for ecdysis, a series of chemical and physiological changes occur under the hardened skin. The exoskeleton will undergo **apolysis** (separation of the old exoskeleton from the underlying epidermal cells). During apolysis there

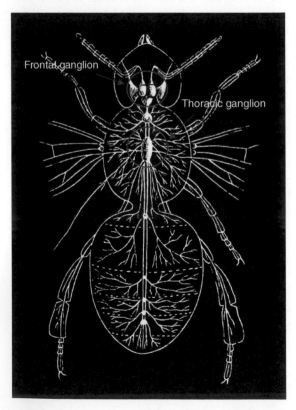

Frontal ganglion

Thoracic ganglion

Figure 4.17 Nervous system of an adult bee. *Source*: Unknown author/Wikimedia Commons/Public Domain.

Figure 4.18   Cicada in the act of molting. *Source*: Brian1442/ Wikimedia Commons/Public Domain.

is secretion of fluid from the molting glands of the epidermal layer and a loosening of the bottom of the cuticle. Once the old cuticle has separated from the epidermis, a new cuticle begins to form and a digesting fluid is secreted into the space between the new and old cuticle. Digestion of the non-sclerotized portion of the old cuticle allows some recycling of the biological molecules that comprise the exoskeleton. This process is under hormonal control, discussed in Section 4.9 dealing with metamorphosis.

When the insect sheds its old skin, we usually see a characteristic splitting down the dorsal midline that allows the new instar to emerge. This initial crack is often initiated by an increase in body fluid pressure accompanied by distinctive combinations of body movements. The new insect wiggles out of the old skin, and it appears soft and vulnerable. Actually, this is one of the most vulnerable times in an insect's life, where it is especially susceptible to predation. If you have ever eaten a soft-shelled crab, you know that you can eat the entire animal, shell and all. This is because it was killed and cooked just after it had molted, before the new cuticle had a chance to sclerotize or harden. The new instar must spend some time after it molts undergoing expansion of the body and appendages (including new wings if it is molting to the adult stage) and hardening and darkening of the new cuticle. Newly emerged instars are often white or very pale in color until sclerotization occurs and are referred to as "teneral" until their cuticle hardens. Members of the general public often find these teneral insects and mistakenly think they have found an albino form because of their pale coloration.

4.9   Many insects appear and behave in an entirely different way as a larva than as an adult – the magic of metamorphosis

The postembryonic life of an insect is divided into life stages called stadia (singular, stadium). Within a stadium, different stages or forms are termed instars. For example, the larval stage of development of many necrophagous fly species is characterized by three molts (from egg hatch to pupation), with each period between molts designated as an instar (first, second, and third). Metamorphosis describes the changes in form and size of the body during an insect's life as it molts and progresses from one instar to the next. There are three broad groupings of metamorphic life histories:

1. The most primitive orders of insects, which never develop wings (e.g., silverfish), change little in form as they molt from one instar to the next. These insects exhibit ametabolous metamorphosis.
2. The second type of life history is known as hemimetabolous metamorphosis (or "incomplete" metamorphosis). The larva is similar to the adult in appearance, food habits and habitat, but only the adult has wings and functional reproductive organs. Some common examples of insects showing hemimetabolous metamorphosis include grasshoppers, cockroaches, stink bugs, and termites.
3. The most extreme form of metamorphosis is seen in the holometabolous insects. The larvae and adult differ greatly in body form and habits. A pupal stage occurs between larva and adult. Holometabolous metamorphosis is often referred to as "complete" metamorphosis. The primary responsibility of the larval stage is feeding, while the adult stage mainly involves mating and dispersal. This type of metamorphosis is characteristic of the "advanced" orders of insects such as Diptera (flies), Coleoptera (beetles), Lepidoptera (butterflies and moths), and Hymenoptera (wasps, bees, and ants).

Insect metamorphosis and development is under hormonal control. Hormones are biological molecules produced in one part of the body which cause an effect in a different part of the body. The three major endocrine organs are the neurosecretory cells in the brain, the prothoracic gland, and the **corpora allata,** which produce hormones involved in molting and metamorphosis. The corpora cardiaca does not produce molting hormones but does serve as a storage and release site for important hormones. Three major hormones are involved in controlling ecdysis and metamorphosis: prothoracicotropic hormone (PTTH), ecdysone, and juvenile hormone.

PTTH is produced by neurosecretory cells in the brain. Axons from these cells transport PTTH to the corpora cardiaca. These structures are storage-release organs for PTTH. When released, PTTH acts on the prothoracic gland in the insect's thorax.

Ecdysone is often referred to as the molting hormone. It is a steroid hormone, related to cholesterol and vertebrate sex hormones. Ecdysone is manufactured and released by the prothoracic gland when stimulated by PTTH. Release of this hormone triggers the molting process. Note that we use the term "ecdysone" to refer to a complex of ecdysteroid hormones produced by insects to help simplify this discussion.

A molt can lead to a larva, pupa, or adult instar. The choice between these alternatives is largely determined by juvenile hormone. Juvenile hormone is produced by the corpora allata. Again, we use the term "juvenile hormone" to refer to multiple molecular forms of this hormone in order to simplify the discussion. This hormone maintains the juvenile stage and prevents metamorphosis. In holometabolous insects, release of a high concentration of juvenile hormone and normal release of ecdysone produce a larval molt. A low level of juvenile hormone combined with ecdysone produces a pupal molt. Ecdysone alone leads to an adult molt. Low levels or no juvenile hormone activates the **imaginal discs,** clusters of undifferentiated cells that activate during the metamorphosis of holometabolous insects to help form the pupal and adult body structure and form.

4.10 The desire to reproduce is a driving force for unique reproductive behaviors and copulatory structures in insects

One of the prerequisites for life on land is the provision of some means for internal fertilization of the egg for sexually reproducing species. The sexes are separate in

Figure 4.19 Mating dung flies, *Scatophaga stercoraria*. *Source*: CopyrightFreePhotos/Wikimedia Commons/Public Domian.

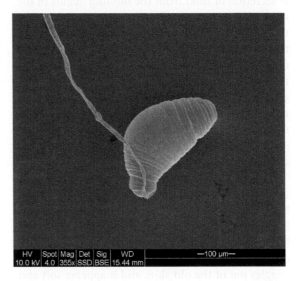

Figure 4.20 Spermatheca of flesh fly (one of three). Scanning electron micrograph. *Source*: Courtesy of G. A. Dahlem.

insects and reproduction is usually a sexual event, involving fusion of an egg and sperm cell nucleus. This implies a form of copulation, during which semen is transferred to the female (Figure 4.19).

In most insects, the semen must be transported to a specialized sperm-storage organ in the female called the **spermatheca** (Figure 4.20). Sperm are stored and nourished here until they are used to fertilize the eggs. Copulation usually involves the insertion of the male phallus, or aedeagus, into the female's genital opening. The aedeagus is basically a single, modified hollow tube for sperm and semen transmission. One exception to this basic plan is found in mayflies which have a pair of aedeagi.

Mating position varies by taxonomic group. Blow flies and carrion beetles mate with the male on top. Grasshoppers and praying mantises have the male on top, but the genitalia attach from the side. Some tree crickets mate with the female on top. Many moths and primitive flies mate end to end, with males and females facing opposite directions. In some insects (e.g., damselflies, bed bugs) the mating position can be much more stylized and complex.

The most common way that male insects deliver sperm to the female is within a protective sac called the **spermatophore**, which is produced from excretions of the male reproductive system. In some Orthoptera (grasshoppers and relatives), part of the spermatophore may extend externally from the female's genital opening. It may serve a nutritional role for the female which often eats much of the exposed spermatophore. In some Lepidoptera (butterflies and moths) and Trichoptera (caddisflies), the spermatophore may have a hard coating that serves as a "genital plug" to prevent other males from mating with that female.

Other insects transfer sperm directly into the spermathecae. Many of these animals have an extremely long aedeagus that is coiled up until erection (expansion) occurs during copulation. Others have aedeagi that are exceptionally complex in morphological structure and which vary by species. In many insects, the shape of the aedeagus is the most striking and useful morphological feature for identification to species level (e.g. Vairo *et al.*, 2011). Surprisingly, the female genital opening does not usually show the same degree of specialization, indicating that this is not a "lock and key" type of reproductive isolation. Some species with direct sperm delivery ejaculate into the female genital tube but stay in copulation for hours. It is thought that the long copulation time ensures that the sperm transfers to the spermathecae properly.

Courtship and reproductive behaviors are often the most complex, and interesting, behaviors exhibited by an insect during its lifespan (see Box 4.5). There are many books and research articles devoted to this facet of entomology, and an understanding of these instincts can be critical for development of control strategies against insect pests.

Insect eggs are covered by a tough shell called the **chorion.** The chorion is often species-specific in sculpturing pattern and shape and may include one or more micropiles or openings for sperm transmission (Peterson & Newman, 1991). During fertilization several sperm may penetrate the micropile(s) but only one sperm nucleus fuses with

---

Box 4.5   Bed bug reproduction

One particularly strange variation in insect mating behavior is exhibited by bed bugs (Figure 4.21). Sperm transfer in bed bugs is accomplished by traumatic insemination, where the male punctures the female abdomen with his aedeagus, ejaculates into the hemolymph, and the sperm swim through the female's body and hemolymph to the spermatheca.

Figure 4.21   Bed bug. *Source*: Piotr Naskrecki/ Wikimedia Commons/Public Domian.

---

the egg nucleus. It may be possible to estimate the age of blow fly eggs with microscopic examination (Martin-Vega & Hall, 2016; Pais & Archer, 2018)

Chapter review

Distinguishing features of arthropods and insects

- Arthropods share a variety of structural and physiological characteristics, including a segmented body, jointed appendages, chitinous exoskeleton, and an open circulatory system.
- Insects are the largest class of Arthropoda, which also includes animals such as spiders, scorpions, centipedes, millipedes, crabs, and their relatives.
- The outer body covering of insects is called an exoskeleton. This tough integument is composed of a non-cellular layer called the cuticle which is secreted by the outer living cell layer of the body, the epidermis.
- The exoskeleton is divided into plates connected by membranes, much like a suit of armor. Each plate and group of plates have different names based on their location on the insect body.

External morphology

- Tagmosis involves the modification and fusion of body segments into functional units. Insects have three functional body parts, called the head, thorax, and abdomen.
- The head is where the feeding appendages are located and where most of the structures involved with sensory input are found. The brain is located within the head.
- The thorax is the body part associated with locomotion. Most adult insects have six legs and four wings. There are notable exceptions in certain groups of insects. Flies, for example, are the only insects which possess two wings.
- The abdomen is the location for the majority of the organs involved with digestion, excretion, and reproduction.

Sensory structures

- Insects perceive their world with similar senses as humans, but with different emphasis. While sight is important, smell is the most important sense for most insects.
- The primary structures involved with the sense of smell are the single pair of antennae which extend from the head. Chemoreceptors on these organs are extremely sensitive to a range of chemical cues, responding to scents at much lower concentrations than a human could ever notice.
- Taste receptors are a different type of chemoreceptors that rely on direct contact with a substance. They can be found on a variety of different surfaces of the body beyond the mouthparts.
- Vision is accomplished with a pair of compound eyes made up of thousands of individual eyes packed together. Each facet of the eye picks up a slightly different visual input and these multiple images are processed into a usable signal by the brain.
- Hearing is mainly used for courtship in some insects. Insects that do not use sound to find a mate are usually deaf.
- Touch is sensed by the movement of bristles on the body.

Internal morphology

- The digestive system is composed of the alimentary canal, salivary glands, and fat bodies. The alimentary canal is divided into three functional regions: the foregut, midgut, and hindgut. The foregut is involved with food intake and storage. The midgut is where the majority of digestion and absorption of nutrients occurs, while the hindgut is mainly involved with water and ion homeostasis. The fat bodies serve as storage centers for glycogen, fat, and protein. Salivary glands play important and varied roles, depending on the species of insects involved, from lubrication of mouthparts to silk production.
- Insects have an open circulatory system. There are no networks of vessels inside the insect body to move the hemolymph. Insects do have a muscular dorsal vessel that serves to keep the hemolymph moving throughout the body. The hemocytes do not normally contain hemoglobin and are not used for transport of oxygen.
- The ventilatory system of insects is composed of a complex branching set of tubes, called tracheae. This internal network of tracheae opens to the outside of the body at holes in the integument called spiracles. Oxygen is supplied to the cells and carbon dioxide is removed by direct diffusion across the tracheole surface.
- The nervous system of insects consists of a network of peripheral and sensory neurons connected to a ventral nerve cord composed of nerve centers called ganglia connected by paired nerve cords called connectives. The primary ganglia of insects are found in the head and consist of the brain and subesophageal ganglion. The brain is where sensory information is processed and where behaviors originate.

Insect development and life cycles

- Insects grow by shedding their skin in a process called ecdysis. This is a complicated, multistep process which is under hormonal control. Some parts of the newly formed cuticle harden into plates by a chemical process called sclerotization, while other parts stay soft. The most notable of these unsclerotized regions are the flexible membranes between moveable sclerites.
- Insects change in body form as they grow and mature. This process is called metamorphosis. Ametabolous insects are primitively wingless and they change little as they molt from instar to the next. Hemimetabolous insects have larvae that appear similar and eat similar food as the adults,

but only the adults have wings. This type of development is often referred to as "incomplete" metamorphosis. Holometabolous insects show "complete" metamorphosis. Their larvae are very different in appearance and food preference as compared to the adults, and there is a transformation stage, called the pupa, between the larval instars and the adult stage. The comparative concentration of hormones signals the developing cells to produce the appropriate developmental body form. Ecdysone is the hormone that triggers molting, but it is the concentration of juvenile hormones that determines the life stage that the molt will result in.

- Most insects reproduce sexually and have internal fertilization. Males transmit their sperm to the females with an aedeagus. Aedeagi can vary in form from a rather simple tube to wildly elaborate structures, depending on the species involved. Some insects pass their sperm in a package called a spermatophore, while others transfer their sperm directly into the female genital tract. The females store the sperm until the time of fertilization in a specialized internal organ called the spermatheca. Courtship and reproductive behaviors are often the most complex behaviors exhibited by insects.

Test your understanding

Level 1: knowledge/comprehension

1. Define the following terms:

  (a) integument
  (b) tagmosis
  (c) spermathecae
  (d) open circulatory system
  (e) ecdysis
  (f) teneral.

2. Match the terms (i–v) with the descriptions (a–e).

(a) Incomplete metamorphosis	(i) Crop
(b) Molting hormone	(ii) Ecdysone
(c) Food storage pouch of foregut	(iii) Spiracles
(d) Respiratory holes in body to outside atmosphere	(iv) Chorion
(e) Egg shell	(v) Hemimetabolous

3. Draw a typical insect leg. Label the tibia, coxa, trochanter, tarsi, and femur.
4. Describe the pathway that food would take through an insect's alimentary canal, including parts of the foregut, midgut, and hindgut and the valves separating these regions.

Level 2: application/analysis

1. Explain the role of PTTH, ecdysone, and juvenile hormone in the control of molting and metamorphosis.
2. Describe the cues and senses used by a blow fly to locate a dead squirrel that has been recently killed by a car.
3. "Keep Plates Clean Or Family Gets Sick" is one possible mnemonic device for remembering the order of the major classification categories (kingdom, phylum, class, order, family, genus, and species). Make up your own phrase for remembering these groups in proper order.

References cited

Agrawal, S., Grimaldi, D. & Fox, J.L. (2017) Haltere morphology and campaniform sensilla arrangement across Diptera. *Arthropod Structure & Development* 46: 215–229.

Anonymous (2009) Flies' vision system aids robots. InTech online technical news magazine. Available at http://www.isa.org/Content/ContentGroups/News/2009/August42/Flies_vision_system_aid:robots.htm.

Boonsriwong, W., Sukontason, K., Vogtsberger, R.C. & Sukontason, K.L. (2011) Alimentary canal of the blow fly *Chrysomya megacephala* (F.) (Diptera: Calliphoridae): an emphasis on dissection and morphometry. *Journal of Vector Ecology* 36: 2–10.

Caleffe, R.R.T., de Oliveira, S.R., Gigliolli, A.A.S., Rufolo-Takasusuki, M.C.C. & Conte, H. (2018) Bioprospection of immature salivary glands of *Chrysomya megacephala* (Fabricius, 1794) (Diptera: Calliphoridae). *Micron* 112: 55–62.

Erwin, T.L. (1982) Tropical forests: their richness in Coleoptera and other arthropod species. *The Coleopterists Bulletin* 36: 74–75.

Hamilton, A.J., Basset, Y., Benke, K.K., Grimbacher, P.S., Miller, S.E., Novotny, V., Samuelson, G.A., Stork, N.E., Weiblen, G.D. & Yen, J.D.L. (2010) Quantifying uncertainty in estimation of tropical arthropod species richness. *American Naturalist* 176: 90–95.

Kruglikova, A.A. & Chernysh, S.I. (2011) Antimicrobial compounds from the excretions of surgical maggots, *Lucilia sericata* (Meigen) (Diptera, Calliphoridae). *Entomological Review* 91: 813–819.

Martin-Vega, D. & Hall, M.R.J. (2016) Estimating the age of *Calliphora vicina* eggs (Diptera: Calliphoridae): determination of embryonic morphological landmarks and preservation of egg samples. *International Journal of Legal Medicine* 130: 845–854.

Martoni, F., Valenzuela, I. & Blacket, M.J. (2019) Non-destructive DNA extractions from fly larvae (Diptera: Muscidae) enable molecular identification of species and enhance morphological features. *Austral Entomology* 58: 848–856.

Merritt, R.W., Snider, R., de Jong, J.L., Benbow, M.E., Kimbirauskas, R.K. & Kolar, R.E. (2007) Collembola of the grave: a cold case history involving arthropods 28 years after death. *Journal of Forensic Science* 52: 1359–1361.

Mohammad, Z., Alajmi, R., Alkuriji, M., Metwally, D., Kaakeh, W. & Almeaiweed, N. 2021. Role of *Chrysomya albiceps* (Diptera: Calliphoridae) and *Musca domestica* (Diptera: Muscidae) maggot crop contents in identifying unknown cadavers. *Journal of Medical Entomology* 58: 93–98.

Mora, C., Tittensor, D.P., Adl, S., Simpson, A.G.B. & Worm, B. (2011) How many species are there on Earth and in the ocean? *PLoS Biology* 9: e1001127. https://doi.org/10.1371/journal.pbio.1001127.

Pais, M. & Archer, M.S. (2018) Histological age estimation of the eggs of *Calliphora vicina* Robineau Desvoidy (Diptera: Calliphoridae). *Forensic Sciences Research* 3: 40–51.

Pearse, A.S. (1946) Observations on the microfauna of the Duke Forest. *Ecological Monographs* 16: 127–150.

Peterson, R.D. II & Newman, S.M. Jr (1991) Chorionic structure of the egg of the screwworm, *Cochliomyia hominivorax* (Diptera: Calliphoridae). *Journal of Medical Entomology* 28: 152–160.

Ruck, P. (1961) Photoreceptor cell response and flicker fusion frequency in the compound eye of the fly, *Lucilia sericata* (Meigen). *Biological Bulletin* 120: 375–383.

Schniederkötter, K. & Lakes-Harlan, R. (2004) Infection behavior of a parasitoid fly, *Emblemasoma auditrix*, and its host cicada *Okanagana rimosa*. *Journal of Insect Science* 4: 36. Available at http://insectscience.org/4.36.

Scott, K. (2005) Taste recognition: food for thought. *Neuron* 48: 455–464.

Sukontason, K., Methanitikorn, R., Sukontason, K.L., Piangjai, S. & Olson, J.K. (2004a) Clearing technique to examine the cephalopharyngeal skeletons of blow fly larvae. *Journal of Vector Ecology* 29: 192–195.

Sukontason, K., Sukontason, K.L., Piangjai, S., Boonchu, N., Chaiwong, T., Ngern-klun, R., Sripakdee, D., Bogtsberger, R.C. & Olson, J.K. (2004b) Antennal sensilla of some forensically important flies in families Calliphoridae, Sarcophagidae and Muscidae. *Micron* 35: 671–679.

Sukontason, K., Chiwong, T., Piangjai, S., Upakut, S., Moophayak, K. & Sukontason, K. (2008) Ommatidia of blow fly, house fly, and flesh fly: implication of their vision efficiency. *Parasitology Research* 103: 123–131.

Szpila, K., Pape, T., Hall, M.J.R. & Madra, A. 2014. Morphology and identification of first instars of European and Mediterranean blowflies of forensic importance. Part III: Calliphorinae. *Medical and Veterinary Entomology* 28: 133–142.

Vairo, K.P., de Mello-Patiu, C.A. & de Carvalho, C.J.B. (2011) Pictorial identification key for species of Sarcophagidae (Diptera) of potential forensic importance in southern Brazil. *Revista Brasileira de Entomologia* 55: 333–347.

Wasserthal, L.T. (2012) Influence of periodic heartbeat reversal and abdominal movements on hemocoelic and tracheal pressure in resting blowflies *Calliphora vicina*. *Journal of Experimental Biology* 215: 362–373.

Wells, J.D., Introna, F. Jr, Di Vella, G., Campobasso, C.P., Hayes, J. & Sperling, F.A.H. (2001) Human and insect mitochondrial DNA analysis from maggots. *Journal of Forensic Science* 46: 685–687.

Whitworth, T. (2006) Keys to the genera and species of blow flies (Diptera: Calliphoridae) of America north of Mexico. *Proceedings of the Entomological Society of Washington* 108: 689–725.

Yan, G., Liu, S., Schlink, A.C., Flematti, G.R., Brodie, B.S., Bohman, B., Greeff, J.C., Vercoe, P.E., Hu, J. & Martin, G.B. (2018) Behavior and electrophysiological response of gravid and non-gravid *Lucilia cuprina* (Diptera: Calliphoridae) to carrion-associated compounds. *Journal of Economic Entomology* 111: 1958–1965.

Supplemental reading

Catalogue of Life: www.catalogueoflife.org.

Chapman, R.F. (1998) *The Insects: Structure and Function*, 4th edn. Cambridge University Press, New York.

Eberhard, W.G. (1985) *Sexual Selection and Animal Genitalia*. Harvard University Press, Cambridge, MA.

Evans, H.E. (1993) *Life on a Little Known Planet*. The Lyons Press, Guilford, CT.

Gullan, P.J. & Cranston, P.S. (2014) *The Insects: An Outline of Entomology*, 5th edn. Wiley Blackwell, Malden, MA.

Richards, A.G. & Richards, P.A. (1979) The cuticular protuberances of insects. *International Journal of Insect Morphology and Embryology* 8: 143–157.

Stewart, A. (2011) *Wicked Bugs: The Louse That Conquered Napoleon's Army and Other Diabolical Insects*. Algonquin Books, Chapel Hill, NC.

Wheeler, Q.D. (1990) Insect diversity and cladistics constraints. *Annals of the Entomological Society of America* 83: 1031–1047.

World Register of Marine Species: www.marinespecies.org.

Additional resources

University of Florida Book of Insect Records (UFBIR) (names insect champions and documents their achievements): http://entnemdept.ufl.edu/walker/ufbir/

Department of Entomology at the Smithsonian Institute: http://entomology.si.edu/ Insect evolution: http:// www.fossilmuseum.net/Evolution/evolution-segues/ insect_evolution.htm

Insects explained: external morphology: http://www.in sectsexplained.com/03external.htm

Understanding evolution: the arthropods: https://evolution. berkeley.edu/evolibrary/article/arthropodstory

Chapter 5

Biology, taxonomy, and natural history of forensically important insects

Overview

Insects at a potential crime scene involving a corpse can be assigned to one of four biological relationships: necrophagous insects, parasites and predators, omnivorous species, and adventive species. While most forensic entomology investigations focus on the necrophagous species, the other three groups can be very important to understanding what has actually occurred to a victim, as well as when, how, and where the tragedy happened. While it is impossible to cover all the possible insects that may be found in association with a corpse, we will take a look at a variety of species that commonly show up in forensic investigations. We will focus on the most important necrophagous species but will also look at the biology and forensic importance of a variety of other entomological associates. Since each species has its own distinctive life cycle and natural history, accurate identification of the species involved is absolutely essential for forensic reconstruction of events at a possible crime scene (see Boxes 5.1 and 5.2).

The big picture

- A variety of different insects and terrestrial arthropods are attracted to a dead body.

- The fauna of insects feeding on a body is determined by location, time, and associated organisms.
- Necrophagous insects include the taxa feeding on the corpse itself.
- Parasitoids and predators are the second most significant group of carrion-frequenting taxa.
- Omnivorous species include taxa which feed on both the corpse and associated arthropods.
- Adventitious species include taxa that use the corpse as an extension of their own natural habitat.
- The proper use of insect names is important for reporting observations and for information retrieval.

5.1 A variety of different insects and terrestrial arthropods are attracted to a dead body

When a person dies, the body immediately becomes a potential resource for insect colonization, but it takes some time before the first insects to arrive. Gravity will generally lower the body and, much like a tree falling to the ground, the body becomes part of the ground-level environment, offering

The Science of Forensic Entomology, Second Edition. David B. Rivers and Gregory A. Dahlem.
© 2023 John Wiley & Sons Ltd. Published 2023 by John Wiley & Sons Ltd.
Companian website: www.wiley.com/go/forensicentomology2

Box 5.1 Order and family names

Entomological publications use order and family names to indicate groups of insects. These are rankings within the Linnaean hierarchy, as discussed in Chapter 4. Journal article titles will often emphasize the order and family names for taxa and these are capitalized, separated by a colon (or comma), and surrounded with parentheses, for example "Keys to the genera and species of blow flies (Diptera: Calliphoridae) of America north of Mexico" by Whitworth (2006). Order and family names are also seen within tables that list organisms collected in a particular study. Authors often assume that their reading audience knows the organisms that are being referred to and do not use common names for the groups.

Here are some of the order and family names you should know as a student of forensic entomology. The most common forensically important orders include Coleoptera (beetles), Diptera (true flies), and Hymenoptera (bees, wasps, and ants). Some of the most commonly encountered families are Calliphoridae (blow flies), Sarcophagidae (flesh flies), Silphidae (carrion beetles), Staphylinidae (rove beetles), and Dermestidae (carpet beetles). These insect family names have widely used common names, as indicated in parentheses in the previous sentence. Many families do not have English common names. An alternative to a common name is the "shortcut" use of the family name without the "ae" on the end to refer to the included organisms. So, parasitic wasps in the family Braconidae are commonly referred to as braconids (note that this "common" usage does not start with a capital letter). This may also be seen where common names are available (e.g., the use of sarcophagids as a synonym of "flesh flies" for members of the family Sarcophagidae).

Box 5.2 Understanding scientific names of species

In the scientific literature (and this book), you will find that species of insects are usually referred to by their scientific name, rather than by a common name. There are several reasons for this. First, while all described species have a scientific name, very few have a recognized common name, for example there are over 54 recognized species of blow flies (Calliphoridae) known to occur in North America but only five of these species have an English common name approved for use by the Entomological Society of America. Second, many of the common necrophagous species occur in many countries throughout the world. Common names change from one language to the next, for example fly (English), la mouche (French), la mosca (Spanish), or ذبابة (Arabic). Scientific names are always written the same, regardless of the language involved, for example the scientific name of the common house fly, *Musca domestica*, is spelled and typed the same irrespective of whether the journal article is written in English or Deutsch (German) or русский язык (Russian) or 中國語言 (Chinese).

You may have learned that a scientific name is made up of two words, a binomial, consisting of the genus name first and the specific epitaph second. The name is italicized in print (or underlined if handwritten) and the genus name always begins with an upper-case letter while the species name always begins with a lower-case letter (e.g., *Cynoma mortuorum*, a species of blow fly found in the far north near the Arctic Circle). While that is true, there are more details to consider and understand when reading and using entomological literature. Here are some scientific name variations that you may encounter and an explanation of their meaning.

In entomological literature, when an author first uses a scientific name in an article the name must be followed by the name of the author that first described that species. Some journals also require that the date of the initial description be included. So *Cynomya cadaverina* should be written as *Cynomya cadaverina* Robineau-Desvoidy,

protection from the open environment to nearby arthropods. We refer to the first phase of the postmortem interval (PMI) as the exposure phase of the **precolonization interval** (Tomberlin *et al.*, 2011). It extends from the time of death until the body is detected by arthropods. While the exposure phase can be very important in the

1830 (with or without the "1830" depending on the journal involved). This tells us that the species *C. cadaverina* was described by Robineau-Desvoidy in 1830. The full spelling of the genus name and inclusion of author and date usually only occurs the first time the species is mentioned. After that the name will usually appear with the genus abbreviated, often to a single letter, followed by the specific epitaph. Authors' names are never italicized.

Sometimes, you will notice that one species has the author's name enclosed in parentheses, while other species have the author's name just listed after the specific epitaph. There is a reason for this. If the author's name is not in parentheses, this indicates that the author described that species in the currently recognized genus. If the author's name is in parentheses, then the species was described with a different genus name than we currently recognize. So, looking at the name *Lucilia cuprina* Wiedemann tells us that Wiedemann described the species "*cuprina*" and included it (at the time of description) in the genus *Lucilia*. The name *Lucilia sericata* (Meigen) tells us that Meigen described the species "*sericata*" but did not include it in the genus *Lucilia* at the time of description (the original name he used in 1826 for this species was *Musca sericata*).

You may also notice that organisms are sometimes named with the genus then with "sp." or "spp." after the genus name. The ending "sp." indicates that the author was able to identify the particular organism to the generic level but was unable to identify which species it represents. Thus, *Sarcophaga* sp. indicates that we are talking about one species in the genus *Sarcophaga*, but we do not know what species it is. When you see "spp." (the plural of "sp."), it indicates that multiple species within that genus are probably involved or that the author cannot tell if he or she is dealing with one or several species. Species level identifications of many insects, including those commonly seen in forensic investigations, can be very difficult, even for a systematic specialist in the group involved! A systematist is a scientist who specializes in the identification and evolutionary relationships of a particular taxonomic group of organisms.

determination of PMI, especially when a body has been purposely manipulated to stop insect detection, it is not a focus of this chapter. In this chapter, we are looking at the types of insects and other arthropods that detect, feed on, and/or colonize the corpse. These actions take place during the detection phase of the precolonization interval and the acceptance, consumption, and dispersal phases of the **postcolonization interval** of vertebrate decomposition (Tomberlin *et al.*, 2011).

Fairly rapidly, depending to a large part on the ambient temperature and other environmental conditions, the dead body takes on a much different appearance to carrion flies and other necrophagous arthropods. Insects that would never think of even landing on a person as they slept, recognize the body as something else, something new, a potential food source to investigate further. Subtle chemical cues associated with decomposition begin to be released from the corpse within the first few hours after death. At the same time that lice and other **hematophagous** (blood-feeding) insects begin to abandon their home to search for new sustenance, a new community of insects begins to move in.

To most insects, the world is not understood or recognized based on visual cues – they "see" the world based on chemicals that they detect with their refined olfactory organs. You can think of a dead body as a light with a dimmer switch. After death, the chemical signals show the body as a dim light in a dark environment. As time progresses, the chemical signals intensify, making the body "shine" brighter and brighter to the chemically driven insects. After the body passes its prime, the chemicals given off by the process of decay change in intensity and composition, resulting in the body becoming "dimmer" and less attractive to the searching insects.

5.2   The fauna of insects feeding on a body is determined by location, time, and associated organisms

Location can refer to the geographical region or immediate surroundings of the carrion. Some insects are cosmopolitan, or found in locations all around the world. Others are restricted to a particular terrestrial ecozone. These eight major biogeographic

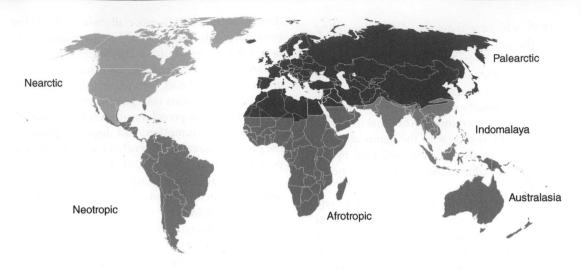

Figure 5.1   Major terrestrial ecozones of the world. *Source*: Carol/Wikimedia/CC BY-SA 3.0.

regions are the Nearctic, Neotropic, Palearctic, Afrotropic, Australasia, Indomalaya, Oceana, and Antarctic (Figure 5.1; note that Oceana and Antarctic zones are not labeled). The United States, Canada, and northern Mexico are included in the Nearctic ecozone. Within a particular ecozone, we find some species restricted to geographical areas within the ecozone (e.g., within the Nearctic ecozone we expect to see different insects in southern Florida compared with northern Ontario). Different assemblages of species should be expected when comparing rural woodlands with an urban parking lot. There are usually major differences between the insects found outside versus inside of a given suburban house. Location on all different scales can make major differences in the community of life attracted to a carrion resource.

Just as location helps to determine the insects that could potentially be at a given crime scene, timing also determines the cast of characters that show up at the dead body. Time may refer to the time of year, weather or immediate environmental conditions, or stage of corpse decomposition. A body left in a field or house in January in New Jersey will have a very different entomological fauna feeding on it than a body left in the same location in the heat of mid-July. Weather conditions beyond temperature can have a profound effect on how attractive the body is to certain groups of insects. We would expect to see a different group of species on a body left on a wet riverbank in spring than one left in an exposed gully in the dry weather of late summer. And timing comes into play as we look at the waves of different species that are attracted to a corpse based on its stage of

decomposition. A body left in a field may be very attractive to blow flies during the first week after death but the dried-up remains 2 months later will attract an entirely different assemblage of species.

Many of the organisms found in association with a dead body are determined by the presence (or absence) of other organisms associated with the decomposition process. Obviously, insects that are predators or parasites of the necrophagous species need those species to be present to make the corpse environment attractive. Each species that feeds on the corpse changes that food resource for the species coming later. This process is often referred to as "waves" of faunal succession. If certain key, early successional species are occluded, like blow flies and flesh flies, then an entirely different pattern of succession will emerge that may include insect species that might not be seen otherwise.

5.3   Necrophagous insects include the taxa feeding on the corpse itself

5.3.1   Insects that feed on but do not breed in carrion

The most important insects for determination of PMI are the species that use the corpse as a breeding site. These insects are more interested in the dead body as a place for their young to feed than as food for themselves. We will emphasize the biology of a variety of the most common carrion breeders in the United States in Section 5.3.2.

Before we look at the carrion breeders, let us take a quick look at insects that come to feed on the corpse, but which do not necessarily reproduce there. Many species of flies and other insects are attracted to the chemical cues emitted by a dead body but have no inclination to lay eggs there. Most carrion breeding insects do not mate at their oviposition site. Adult male insects collected on or near a body will not be there to lay eggs, but many need a good protein-rich meal after they emerge from their puparium for proper seminal fluid production, which is necessary for successful mating and fertilization. Blood and other bodily fluids available on the surface of a corpse will serve as a protein meal for these males, and for females that often require a protein meal for egg development after copulation and fertilization. Non-gravid females of fly species that breed in carrion and which are collected at a corpse are probably there to fulfill this reproductive requirement.

A wide variety of flies that do not breed in carrion are often collected feeding on a body. Some common examples are cluster flies and their relatives in the genus *Pollenia* (Figure 5.2), which are known to be parasitoids of earthworms (e.g., Thomson & Davies, 1973; Gisondi *et al.*, 2020). Many species of Sarcophagidae and Tachinidae that live as parasitoids of other insects and arthropods, and which do not develop on mammalian carrion, may show up to get a protein meal. Adults of a variety of dung-breeding flesh flies in the genera *Ravinia* and *Oxysarcodexia* are often collected on corpses during early stages of decomposition. While their usual association is

simply to obtain a protein meal, these species may become important to a forensic investigation if they breed in exposed feces associated with a corpse.

Other species may be using the body as a congregation site for mating purposes. Males of the flesh fly *Sarcophaga utilis* Aldrich are often found perching on or very close to carrion as they wait for a female to come by. This species is a parasitoid of scarab beetles (Dodge, 1966). Another flesh fly, *Ravinia pusiola* (Wulp), commonly uses weeds or bushes growing downwind from a body (within a couple of meters) as a male congregation or station site for mating purposes. This species usually breeds in omnivore or carnivore dung (e.g., Poorbaugh & Linsdale, 1971) but has been collected in association with dog carcasses by Reed (1958) and with baby pig carcasses by Payne and King (1972).

5.3.2 Insects that breed in carrion

The insects that are most useful for determination of PMI are those that use the decaying corpse as a larval food resource. Their immature stages are tightly associated with the body, do not fly or run away when disturbed, and their growth and development occur in a predictable pattern that can be used to establish the minimum period of time that they have been present at the body. Adults can come and go, but the larvae stay with the carrion throughout their maturation. The two groups of insects of highest importance to a forensic investigation are species in the orders of true flies (Diptera) and the beetles (Coleoptera).

The Diptera are distinguished from all other groups of insects by the single pair of membranous wings on the adults, located on the middle segment of the thorax (or mesothorax). The hind wings are reduced to two small gyroscopic pegs called halteres, one on each side of the posterior thoracic segment (or metathorax). Coleoptera is the largest order of insects, with approximately 40% of all described species. Adult beetles are also distinguished by their wings. They have four wings, with the front pair usually hardened into hard or leathery plates that cover the membranous hind wings. They use these larger hind wings for flight. At rest, the hind wings are usually folded up under the front wings.

The flies of primary interest to forensic investigations are members of more recently evolved families, often referred to as "higher Diptera." Lower Diptera refers to older evolutionary lineages

Figure 5.2 The cluster fly, *Pollenia rudis* (Fabricius). *Source*: Courtesy of G. A. Dahlem.

Figure 5.3   Fly larvae (maggots). *Source*: Courtesy of G. A. Dahlem.

and families that are considered to be more primitive in form (e.g., mosquitoes, crane flies, midges, and gnats). All flies undergo complete metamorphosis, but the higher flies are the ones with larvae known as maggots. Maggots (Figure 5.3) are generally legless, wormlike, and the head is not sclerotized (except for the mouthparts). The pupal stage is passed inside the last larval skin, which is called a puparium. The two families of flies used most often for determination of PMI, Calliphoridae and Sarcophagidae, are in a group called the **calyptrate** Diptera. Calyptrates also include the families Anthomyiidae, Muscidae, and Tachinidae. Defining characteristics of this group include wings with calypteres (lobe-like extensions of the basal posterior portion of the wing that give this group their name; Figure 5.4), a **suture** on the second antennal segment (Figure 5.5), and a transverse suture on the dorsal surface of the thorax (Figure 5.6). These families of flies (along with several other smaller families) form a well-recognized phylogenetic, or evolutionary, group within the Diptera.

5.3.2.1   The blow flies (Diptera: Calliphoridae)

The Calliphoridae includes many species that are well known to the general public. The adults are not secretive and often land in sunny locations that are

(a)

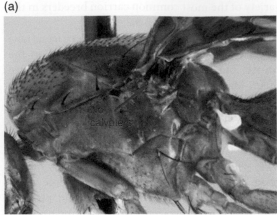

(b)

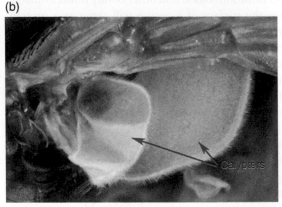

Figure 5.4   Distinguishing calyptrate from non-calyptrate Diptera: (a) without calypteres; (b) with calypteres. *Source*: Courtesy of G. A. Dahlem and R. Edwards.

conspicuous to the human eye. Many common blow flies are bright metallic green or blue and are considered as emerald and sapphire gemstones of the insect world. You may have heard of these insects referred to as "greenbottle" or "bluebottle" flies. But the beauty of these flies is tempered by our knowledge of their secret childhoods spent wriggling and feeding in decomposing flesh (Greenberg, 1991). Not all Calliphoridae are necrophagous, but many are.

Lucilia spp. and *Phormia regina* (Meigen)

The most commonly encountered necrophagous blow flies in most of North America are species in the genus *Lucilia* and the black blow fly, *Phormia regina* (the only species in the genus *Phormia* that occurs in North America). These flies have both the thorax and abdomen colored metallic green and adults are commonly encountered perching on garbage cans and sitting on piles of dog dung. Adults of *P. regina* have a dark, metallic, blue-green coloration

(a)   (b)

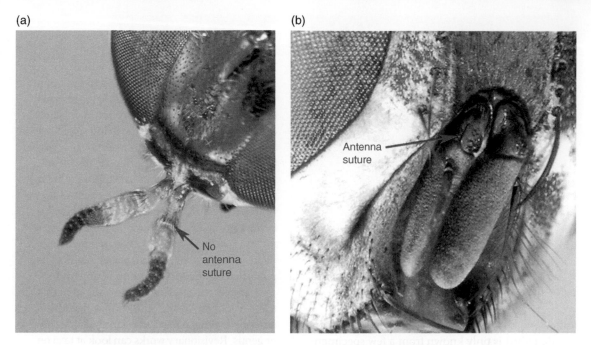

Figure 5.5   Distinguishing calyptrate from non-calyptrate Diptera: (a) no antenna suture; (b) with antenna suture. *Source*: Courtesy of G. A. Dahlem and R. Edwards.

(a)   (b)

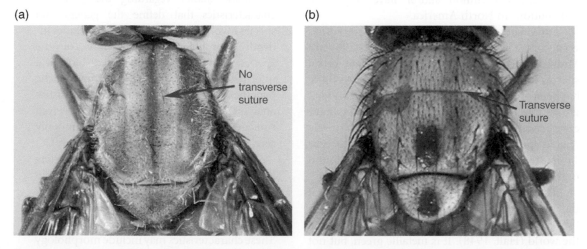

Figure 5.6   Distinguishing calyptrate from non-calyptrate Diptera: (a) no transverse suture; (b) with transverse suture. *Source*: Courtesy of G. A. Dahlem and R. Edwards.

but are not black in color, as their common name suggests.

Phormia regina (Figure 5.7) is the dominant necrophagous species in the northern United States and southern Canada during the summer months and the dominant species in the southern United States in the winter (Byrd & Allen, 2001). This species reproduces, often in huge numbers, in vertebrate carrion but also has economic importance because of the secondary myiasis it is responsible for, sometimes resulting from cattle castration and

dehorning or soiled wool of sheep (Hall, 1948). A large amount of literature has been devoted to the biology of this species. The references cited above should serve as a good starting point for the student interested in learning more about this common necrophagous species.

The genus *Lucilia* includes 12 species in North America (Jones, et al., 2019) (see Box 5.3). Some of these species are very uncommon (*L. elongata* Shannon is very rarely collected and only from localities in the Pacific Northwest) or have very

Figure 5.7 Black blow fly, *Phormia regina* (Meigen). *Source*: Courtesy of G. A. Dahlem and R. Edwards.

limited North American distributions *(L. eximia* (Wiedemann) is only known from a few specimens collected in Texas and Florida but is common in urban areas of Central America). Other species can be extremely common and/or have very wide distributions in North America.

Lucilia coeruleiviridis Macquart (Figure 5.8) is very common in the eastern United States and Midwest and is one of the most brilliant green species of forensic importance. Unfortunately, little has been documented on the life cycle of *L. coeruleiviridis* due to the great difficulty researchers have experienced with rearing of its larvae. A closely related species, *L. mexicana* Macquart, looks very similar to *L. coeruleiviridis* but it is primarily found in the Southwest.

Lucilia sericata (Meigen) (Figure 5.9) is one of the most common and widely distributed species in this genus, not only in North America but around the world (Hall, 1948). It is metallic green, but not as bright green as *L. coeruleiviridis* and *L. mexicana*, often exhibiting a bit of a coppery sheen. This is a species of high forensic importance and much has been written on its biology, including detailed work on larval growth and development (e.g., Tarone & Foran, 2008). This is the species that has been used in human and veterinary medicine in the United States since the Food and Drug Administration approved the technique of maggot therapy in 2004. Maggot therapy is the medical use of fly larvae to debride or clean problematic wounds that do not respond to other medical treatments (e.g., Sherman *et al.*, 2000; Zubir *et al.*, 2020). The closely related *L. cuprina* Wiedemann is an important veterinary

Box 5.3   *Lucilia* or *Phaenicia*? Name changes of taxonomic categories

When you read American literature about *Lucilia sericata* that was written 10 or more years ago, you may see the species being referred to as *Phaenicia sericata*. Why is this? Name changes occur throughout the living world but are especially common with insects. As scientists learn more about a group of organisms, they try to improve the understanding of their relationships to one another. The result of forward progress in our understanding of evolutionary groupings is reflected in name changes for the species involved. Major name changes often occur when a systematic revision is published on a particular group of organisms.

A systematic revision is an investigation that tries to look at all the species that have been described within a particular family, subfamily, or genus. Revisionary works can look at taxa on a worldwide basis or from a particular region of the world. Determinations are made by the systematic author regarding the particular characteristics that define the genera and species involved. A revision usually includes in-depth morphological descriptions of the species, summaries of published biological information for the species, and phylogenetic hypotheses based on results of cladistics analyses (for modern research) or perceived similarities and differences (more commonly seen in revisions from 30 years ago or more).

The systematist may use new characters that have not been used in the past to help define the particular taxa that he or she is investigating or may have a different point of view on the relative importance of characters. These characteristics may include morphology of adults, morphology of immature stages, physiology, behavior, and/or molecular data (especially DNA). New data may result in new interpretations of relationships, which require name changes. Many changes in our understanding of species groupings are occurring today because of the consideration of new molecular data (especially from mitochondrial and nuclear DNA sequencing).

Species once placed in one genus are found to actually share a closer evolutionary relationship with another genus and its name will change to reflect this. Another common source of change involves the systematist trying to make sure that organisms in a particular genus are all derived from a single common ancestor, as groupings are based on the organisms' evolutionary history. Some systematists prefer to split groupings into lots of small genera, others prefer to include more species in larger genera. You will often hear the first group referred to as "splitters" and the other as "lumpers." When there is more than one recognized systematic expert on a group in the world, you may see discussions or arguments about what generic classification should be used.

In some cases, different regions or countries in the world will use a broad generic concept while in a different place a narrower generic concept is accepted and used. For example, in the Sarcophagidae we see broad use of the genus *Sarcophaga*, often split into subgroupings called subgenera, in North America, Western Europe, and Australia. Authors from South America and Asia often consider these subgenera groupings as full genus names. So, the species we call *Sarcophaga (Neobellieria) bullata* in the United States may show up in Asian journals as *Neobellieria bullata*. (Note that subgeneric names, when used, are in italics and surrounded by parentheses between the genus name and the specific epitaph.) Same species, but using two different generic concepts. In general, you should be aware of, and use, the nomenclature preferred by the region of the world you are located in.

So, why do most people use *Lucilia* for the flies that used to be split into the genera *Phaenicia*, *Bufolucilia*, *Lucilia*, and *Francilia*? We see this change happening after the publication of "Blowflies (Diptera, Calliphoridae) of Fennoscandia and Denmark" by Knut Rognes in 1991. The change did not happen immediately, but as more and more systematists came to agree with the arguments Rognes presented for a larger generic concept of *Lucilia*, the genus is now usually used for these flies in modern literature. New information and new revisions may change usage in the future, but right now *Lucilia* in the broad sense or *sensu lato* (abbreviated as *s.l.*) is the generic name used by most world scientists. The qualifier *sensu stricto* (or *s.s.*) means "in the strict sense" when given after a generic name and would apply to the less inclusive generic classifications used by other systematists (Hall, 1948).

Figure 5.8 *Lucilia coeruleiviridis* Macquart. *Source*: Courtesy of G. A. Dahlem and R. Edwards.

Figure 5.9 *Lucilia sericata* (Meigen). *Source*: Courtesy of G. A. Dahlem and R. Edwards.

species due to its role in myiasis with sheep in Australia and Africa and it may be a very important forensic species in many areas of the world but is only locally significant in the southern United States (Whitworth, 2006).

Calliphora spp.

Adults of *Calliphora* are usually significantly larger than those of *Lucilia* and often have bright, metallic blue abdomens. Thirteen species are recognized in North America (Whitworth, 2006; Jones *et al.*, 2019). Most of these species are rare or have very restricted distributions where they may or may not be locally abundant, but there are several species that are very common throughout much of the United States or major portions of the continent. Two species in

Figure 5.10 *Calliphora vicina* Robineau-Desvoidy. *Source*: Courtesy of G. A. Dahlem and R. Edwards.

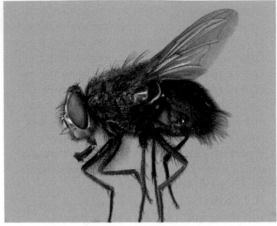

Figure 5.11 *Calliphora vomitoria* (Linnaeus). *Source*: Courtesy of G. A. Dahlem and R. Edwards.

particular are common in both the United States and many other parts of the world and are considered to be of high importance for forensic investigations: *Calliphora vicina* Robineau-Desvoidy and *C. vomitoria* (Linnaeus).

Calliphora vicina (Figure 5.10) is a very common necrophagous species around the world and has been used in a variety of criminal investigations (for example, see the "Case histories" chapter in Smith, 1986). This species is often seen around suburban and urban homes on warm and sunny spring mornings sitting in patches of sunlight. It is known to enter houses, which is one reason it may seem more common in forensic case studies than its abundance in the environment would seemingly predict. Much has been written about the biology of this species, including rearing rates of immature forms at various constant temperatures (e.g., Anderson, 2000; Marchenko, 2001), which make this a very useful species for determination of PMI.

Calliphora vomitoria (Figure 5.11) is superficially similar in appearance to *C. vicina*, but trends to a larger body size in healthy individuals. Ecologically, *C. vomitoria* is more common in rural settings while *C. vicina* is more common in urban settings, but overlap does occur. Apparently, it is much more common today than it was in the past, as Hall (1948) states "It is not common anywhere in North America…", but Whitworth (2006) states that "This is a common species throughout North America." This may be explained by the hypothesis mentioned by Hall (1948) that this species was native to the Palearctic region and was introduced to and spread across North America by commerce. So, it may be a rapidly expanding introduced species rather than one originally occurring in the New World. Regardless of

its origins, *C. vomitoria* is considered to be a species of high forensic importance and information has been published on the rearing rates of its immature forms (Marchenko, 2001).

Cochliomyia spp. and *Chrysomya* spp.

Flies in the genera *Cochliomyia* and *Chrysomya* tend to stand out with their brilliant metallic green coloration and bright yellow or silvery heads. The genus *Cochliomyia* contains four species known from North America. Two of these, *C. aldrichi* Del Ponte and *C. minima* Shannon, have a North American distribution that is limited to southern Florida and the Florida Keys. The screwworm, *C. hominivorax* (Coquerel), was once a devastating pest to livestock across the United States but has been eradicated in North America by the **sterile-male technique**, a type of biological control where massive numbers of flies are reared by control agencies and the puparia are exposed to radiation that leaves the flies healthy but sterile. The sterile males are released in huge numbers into the environment where they have a huge numeric advantage over wild-type and fertile males in finding mates. Females from the rearing process are destroyed. Males will mate as many times as they can, but females generally mate only once. Using this technique, the USDA has been able to eradicate this species in North America (eliminated from the United States in 1966) and control efforts continue to limit its distribution to South America and the West Indies (eliminated from Mexico in 1991 and Costa Rica by 2000). The fourth species is the secondary screwworm, *C. macellaria* (Fabricius) (Figure 5.12), which is a common species of high

Figure 5.12 *Cochliomyia macellarla* (Fabricius). *Source*: Courtesy of G. A. Dahlem and R. Edwards.

Figure 5.13 *Chrysomya rufifacies* (Macquart). *Source*: James Niland/Wikimedia Comons/CC BY 2.0.

forensic importance, especially in the southern parts of the United States. Information has been published on rearing rates at different temperatures for this species (e.g., Greenberg & Kunich, 2002).

The genus *Chrysomya* is of Old-World origin but several species have been accidentally introduced into the New World in recent years and two species are becoming increasingly important to forensic investigations, especially in the southeastern United States. The hairy maggot blow fly, *C. rufifacies* (Macquart) (Figure 5.13), and its relative *C. megacephala* (Fabricius) are rapidly expanding their ranges since their initial introduction into the United States. There is some controversy about the overwintering ability of this species, but *C. rufifacies* is known to disperse northward during the warm summer months and has been collected as far north as southern Canada (Rosati & Vanlaerhoven, 2007). The common name of "hairy maggot blow fly" derives from the distinctive appearance of the larvae of *C. rufifacies*, which has distinctive tubercles and patches of setae (Figure 5.14). These contrast with the normal smooth appearance of most other blow fly species, making them one of the only easy-to-identify calliphorid larvae in the field. Much has been written on growth rates of these two species of *Chrysomya* (e.g., Byrd & Butler, 1997), making them very valuable for the determination of PMI.

Other calliphorid species
Other species of Calliphoridae may be of local significance in North America, but we have discussed the most important and widespread species in the previous paragraphs. There are

Figure 5.14 *Chrysomya rufifacies* larva. *Source*: Austinh37/Wikimedia Comons/CC BY-SA 3.0.

many more forensically important species if you go to other parts of the world, particularly in the tropics (e.g., Carvalho & Mello-Patiu, 2008).

5.3.2.2 The flesh flies (Diptera: Sarcophagidae)

Despite their common name as "flesh flies," most species included in the family Sarcophagidae are not normally necrophagous, at least on vertebrate flesh. The necrophagous sarcophagids are gray flies with red eyes, dark stripes (or vittae) on the dorsum of their thorax, and gray tessellated (checkerboard patterned) abdomens. Many species have parasitoid, parasitic, or predatory relationships with a wide variety of other insects and arthropods, terrestrial mollusks (snails and slugs), and reptiles and amphibians. A variety of species are **coprophagous** (dung breeding) and many others are scavengers of dead insects and other invertebrates. Only a relatively small number of species specialize on colonizing larger, vertebrate carrion, but those that do can be very important for forensic analysis. While a relatively small number of species naturally colonize large carrion, many species can successfully develop on a ground meat (e.g., hamburger) or liver diet, if forced to (e.g., Sanjean, 1957). A wide variety of sarcophagids are attracted to carrion for a protein meal or as a mating site, even if they do not actively breed there. Unfortunately, identification of sarcophagids to species can be very difficult, especially for the larval stages. This has led to problems with references to these species in a forensic context in the literature due to incorrect identifications, for example Cherix *et al.* (2012) discuss several case studies where *Sarcophaga carnaria* (Linnaeus) is indicated as necrophagous on human bodies even though this species appears to be an obligate parasitoid of earthworms! *Sarcophaga* and allies (Aldrich, 1916) is one of the main references for identification of North American Sarcophagidae. Many names have changed, but look at Appendix 4 for a list associating the old Aldrich names with currently used scientific names for his included species.

An interesting biological note regarding all Sarcophagidae is that these flies do not lay eggs. The flies incubate their eggs in a bi-pouched uterus and almost always deposit active first instar larvae. Feeding begins almost immediately after deposition.

Sarcophaga spp.

Within the genus *Sarcophaga* (and here we are using *Sarcophaga sensu lato*) there are several species of fairly high forensic interest in North America. Two species found in both the New and Old World that commonly breed in carrion are *S. argyrostoma* (Robineau-Desvoidy) and *S. crassipalpis* Macquart (Figure 5.15). It is hypothesized that these species are not native to North America, but their introduction would have taken place many years ago, as both species appear in Aldrich's landmark publication on the North American fauna in 1916 (but under different species names than what we use now, due to subsequent synonymy; see also Box 5.4). These species are fairly common in suburban and urban areas, with extensive human disturbance to the environment, but are rarely found in more rural, or undisturbed, areas. A quick search for information on *S. crassipalpis* will yield a multitude of results, as this species has been one of the basic laboratory research animals for physiological research on insects. However, little focused research on forensic aspects of its biology has been published. Several recent papers on *S. argyrostoma* have increased the forensic usefulness of this species (e.g., Grassberger & Reiter, 2002; Draber-Monko *et al.*, 2009).

One of the most important species for forensic investigations in North America is the native species *S. bullata* Parker (Figure 5.16). *Sarcophaga bullata* has served as a basic laboratory animal for investigations of insect physiology and hundreds of articles related to this species, yet no focused

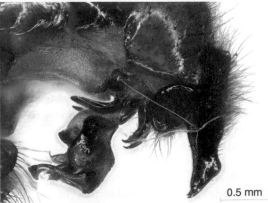

Figure 5.15 *Sarcophaga (Liopygia) crassipalpis* Macquart with close-up of male genitalia. *Source*: Courtesy of G. A. Dahlem and T. Pape.

While it does not happen extremely often, species name changes do occur and this can cause confusion. One of the most common species of Sarcophagidae mentioned in literature over the last several centuries went by the name *Sarcophaga haemorrhoidalis* (Fallén) until the 1980s. It was then discovered that the species name *"haemorrhoidalis"* was preoccupied (someone else had already used this name for a different species in the genus that the species was placed in at the time of description), and the species name was changed to the next valid name known for this species, *S. cruentata* Meigen (Verves, 1986). From the late 1980s to the late 1990s, you will see literature pertaining to this common scavenger species under this species epitaph. Then, in 1996 when Pape published his world catalog of sarcophagid names, the name was changed again. The new name *S. africa* (Wiedemann) (Figure 5.17) was applied when an investigation of Wiedemann's type specimens revealed that he had named this species before Meigen published the name *"cruentata."* We are still in the process of seeing this new name come into common usage in modern literature.

Why did the name change after so many years of use? It turns out that there are fairly strict rules adopted by the international community for handling species names. These rules are laid out in a book called the *International Code of Zoological Nomenclature* (International Commission on Zoological Nomenclature, 1999). The reasoning behind this Code is "to promote stability and universality in the scientific names of animals and to ensure that the name of each taxon is unique and distinct." The Principle of Priority is one of the main precepts and it states, "the valid name of a taxon is the oldest available name applied to it . . . ." There are many cases where species, especially common species, have been described and named more than once. *Sarcophaga africa* has had over 30 different names (Pape, 1996). Only the earliest name, that is uniquely applied, has official status.

research on larval rearing times at different temperatures is currently available.

The best reference available for *S. bullata*, and two other native necrophagous species (*S. cooleyi* Parker and *S. shermani* Parker), was written in 1958 by Kamal where he reared larvae at approximately 27 °C (80 °F) and monitored their growth and development.

Blaesoxipha spp.

The most important flesh fly not in the genus *Sarcophaga* in North America is *Blaesoxipha plinthopyga* (Wiedemann) (Figure 5.18), a necrophagous species that is commonly collected throughout the southern states of the United States. Very little information is available on the life history of this species, with no published information on rearing times. This could be a valuable species for forensic investigations but additional studies on its biology and larval development are sorely needed.

Other sarcophagid species

A variety of other sarcophagid species have been associated with carrion in the literature. Some of these may be opportunistic scavengers, others may be mistakenly associated with the carrion or misidentified. *Helicobia rapax* (Walker) is an example of a common scavenger species associated with a wide variety of small carrion, including invertebrates and vertebrates, which may be reared in a forensic investigation. In other parts of the world, a variety of species in the Sarcophagidae may be important forensic indicators (e.g., Vairo *et al.*, 2011) and they can show up as the dominant species in a particular situation or case study, even in North America.

5.3.2.3   The scuttle flies (Diptera: Phoridae)

The scuttle flies, or phorids, are small flies with a distinctive humpback appearance. These flies tend to show up at a corpse during later stages of decomposition and are very adept at colonizing bodies that are buried or enclosed in some way that stops the larger blow fly and flesh fly adults from reaching the body. This is a large family of flies and most species are not attracted to large vertebrate carrion, but the scavenger species that are present can be very important to a forensic investigation as they may represent the only entomological material collected at a scene.

Two species that have been used in forensic investigations in the United States are *Megaselia*

Figure 5.16   *Sarcophaga (Neobellieria) bullata* Parker with close-up of male genitalia. *Source*: Courtesy of G. A. Dahlem.

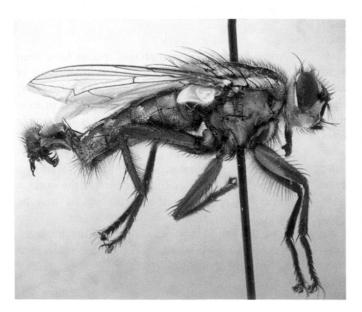

Figure 5.17   *Sarcophaga (Bercaea) africa* Wiedemann with close-up of male genitalia. *Source*: Courtesy of G. A. Dahlem.

abdita Schmitz and *M. scalaris* (Loew) (Greenberg & Wells, 1998). *Megaselia scalaris* (Figure 5.19) is often referred to as the "coffin fly," although this is not an officially recognized common name for this species (the Entomological Society of America, or ESA, is the source for "official" common names of insect species in the United States). Several other phorid species are occasionally referred to as the coffin fly, so it is probably best to use this common name loosely for any phorids present at a decomposing body

rather than using it to indicate a particular species.

Scuttle flies can complete multiple generations on a protected corpse, even when buried in a coffin. A recent case report found an actively breeding population of *Conicera tibialis* Schmitz on a body buried for 18 years in Spain (Martin-Vega *et al.*, 2011). Several species have had developmental data published on them, making these species very useful for PMI determination (e.g., Greenberg & Wells, 1998; Disney, 2006, 2008).

Figure 5.18 *Blaesoxipha (Gigantotheca) plinthopyga* (Wiedemann). *Source*: Courtesy of G. A. Dahlem and T. Pape.

Figure 5.20 *Piophila casei* (Linnaeus). *Source*: John Curtis/Wikimedia Commons/Public Domain.

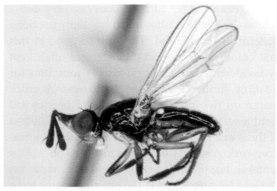

Figure 5.19 *Megaselia scalaris* (Loew). *Source*: AfroBrazilian/Wikimedia Commons/CC-BY-SA-3.0.

Figure 5.21 *Prochyliza xanthostoma* Walker (Diptera: Piophilidae). *Source*: Courtesy of G. A. Dahlem.

5.3.2.4 The skipper flies (Diptera: Piophilidae)

The skipper flies in the family Piophilidae are often associated with corpses in late stages of decomposition or skeletal remains. This is a relatively small family of flies, with less than 100 described species. The cheese skipper, *Piophila casei* (Linnaeus), is the best-known member of this family that includes many species with a scavenger life history. *Piophila casei* (Figure 5.20) is a pest of stored foods, especially processed meats and cheeses. The larva of this species, along with several other species in this family, has a remarkable ability to leap into the air (view this extraordinary behavior at http://www.youtube.com/watch?v=XCLCAqedEeY). A common place to find the larvae of these flies is on the interior of major bones, feeding on the bone marrow. Their

jumping (or "skipping") usually comes into play when they leave their larval food source and move away from the body for pupariation. While the cheese skipper is the best-known species, several others are more frequently associated with carrion in North America, including the distinctive looking *Prochyliza xanthostoma* Walker (Figure 5.21), with its elongated head and antennae (Martin-Vega, 2011).

5.3.2.5 The carpet beetles (Coleoptera: Dermestidae)

The carpet beetles are usually associated with late stages of decomposition, after most of the body has been consumed. These are relatively small beetles that are covered with short scale-like setae

Figure 5.22 Dermestid beetle larva. *Source*: Aiwok/ Wikimedia Commons/CC BY-SA 3.0.

as adults and with bands of long hair-like setae as larvae. Other common names associated with this family of beetles are lard beetles, hide beetles, and skin beetles. These are the clean-up animals that eat parts of a corpse that nothing else seems interested in. Dried bits of leftover flesh, skin, hair, cartilage, and other such remains are very desirable to the beetle adults and larvae. Only the bony skeleton is left behind. They are so good at this that colonies of dermestid beetles are often kept by taxidermists and natural history museums for cleaning flesh off vertebrate skeletons. A quick search on the internet for dermestid beetles will yield a variety of sources for purchasing colonies of these insects for business or hobby skeletonization work. A variety of species may feed on human remains, but the hide beetle, *Dermestes maculatus* DeGeer (Figure 5.22), is the one that seems to be frequently involved (e.g., Kulshrestha & Satpathy, 2001; Schroeder *et al.*, 2002).

5.3.2.6 Other necrophagous insects

A wide variety of other insects may be found feeding on human remains, depending on the location of the body and its relative availability to gravid females when compared with other, preferred larval food resources. In the Diptera, species of a variety of different families may develop on carrion alone, especially in situations where the dominant blow flies are excluded (e.g., Drosophilidae, Ephydridae, Muscidae, Sepsidae, Sphaeroceridae, Stratiomyidae) (Smith, 1986; Byrd & Castner, 2009). Other necrophagous Coleoptera include certain members of the families Cleridae, Nitidulidae, and Tenebrionidae. In the Lepidoptera (butterflies and moths), caterpillars of species in the family Tineidae (clothes moths) are found in late stages of decay. Many other insects that live as general scavengers

will take advantage of carrion food sources, if the situation is right.

5.4 Parasitoids and predators are the second most significant group of carrion-frequenting taxa

A variety of species are **necrophilous,** or specifically attracted to carrion while not feeding on the carrion itself. Many of these come to feast on the insects that rapidly develop on this ephemeral resource, especially the young and juicy maggots that can be present in huge numbers during the early stages of decomposition.

5.4.1 The rove beetles (Coleóptera: Staphylinidae)

Rove beetles have an atypical body form for a beetle. They are usually much longer than wide, run rapidly, and have small scale-like front wings that do not cover their abdomen but do cover a functional pair of complexly folded hind wings. When disturbed, many of these beetles arch the tip of their abdomen up and over their body, giving the impression to potential predators that they can deliver a scorpion-like sting (this is all show, they have no ability to "sting" anything). Only a relatively few species of this very diverse and speciose family are necrophilous, but those that are can be fairly large, colorful, and obvious near and on carrion. Some species associated with fly larvae have complex parasitoid lifestyles that are very unusual in the Coleoptera.

The hairy rove beetle, *Creophilus maxillosus* (Linnaeus) (Figure 5.23), is fairly common around carrion during early decomposition and this is the only staphylinid that currently has an ESA approved common name. Both adults and larvae of this beetle feed on maggots, but they may also feed on the body itself (which would put them into the omnivore category rather than a pure predator). A variety of staphylinids may show up to feed at a corpse, with as many as 50–60 species collected at small carrion over the course of a year (Smith, 1986).

Staphylinids in the genus *Aleochara* (Figure 5.24) may show up to take advantage of necrophagous Diptera puparia as a larval food source

Figure 5.23   Hairy rove beetle, *Creophilus maxillosus* (Linnaeus). *Source*: Martin Andersson/Wikimedia Commons/CC BY-SA 3.0.

Figure 5.24   *Aleochara lanuginosa* Gravenhorst. *Source*: Reginald Webster et al./Wikimedia Commons/CC BY 3.0.

(Klimaszewski, 1984). Beetles in this genus have a parasitoid lifestyle; the larvae feed on a single host during their development and kill the host during the process. The newly hatched beetle larvae find and enter a fly puparium. The tiny beetle larva feeds on the fly pupal stage slowly at first, without damaging vital organs or tissues, in order to keep its host alive until the very end. At the end of the grub's developmental process, it begins to grow rapidly, finally feeding freely and killing its host in the

process. After killing the developing fly, the beetle larva will pupate in the soil or in the puparium itself. After a short period of time, a new adult will emerge.

5.4.2   The parasitoid wasps (Hymenoptera: Braconidae and Pteromalidae)

A variety of small wasps come to carrion to reproduce in the insects feeding on the decaying flesh. Some of the most important wasps are parasitoids of the large blow fly and flesh fly larvae and puparia. As described above with the staphylinids in the genus *Aleochara*, these wasps infect a single Diptera larva with their offspring. The larva grows slowly at first, feeding on the living maggot. Only at the end of its development do the wasp larvae actually eat essential organs and tissues, causing the death of its host. These wasps can cause great frustration for the forensic entomologist who is trying to rear larvae from an experiment or crime scene in order to identify the fly species involved, only to find wasps emerging from the fly puparia!

One of the most common and important hymenopteran parasites is a small wasp in the family Pteromalidae, *Nasonia vitripennis* (Walker) (Figure 5.25). This is one of the most intensively studied species of all insects, often referred to as the "lab rat" of the Hymenoptera. It is one of the few species of insects (at present) that has had its entire genome decoded and it is a common subject for investigations on basic inheritance patterns (see https://fire.biol.wwu.edu//young/322/old_data_files/nasonia_1.pdf for life-history information and a genetic lab exercise). Part of its usefulness in

Figure 5.25   *Nasonia vitripennis* (Walker). *Source*: D.B. Rivers.

genetic studies comes from the peculiar sex determination found in all members of the Hymenoptera. In all wasps, bees, and ants, males come from unfertilized eggs (they are haploid), while females develop from fertilized eggs (diploid).

The intensive interest and research on *N. vitripennis* have uncovered a particularly fascinating life history of these tiny parasitoid wasps. The adult female wasps seek out puparia of calyptrate Diptera and "sting" the puparium, injecting venom that allows the developing wasp larvae to evade the immune system of the developing fly, redirect the metabolism of the host, and arrest fly development to favor parasitoid development (Danneels *et al.*, 2010). Female wasps are **gregarious,** laying anywhere from 10 eggs per host to as many as 200 eggs within a single fly puparium. Wasp larvae consume the fly relatively rapidly (5–7 days at 25 °C), and then pupate inside the puparium. Consistent with their designation as a **parasitoid,** *N. vitripennis* always kills the fly host, either directly by the action of the venom or through the feeding activity of the developing offspring. The association with the host is complete once adults emerge by chewing small holes in the puparium; males usually create the exit holes since they typically emerge 24 hours before their sisters. This parasitic relationship with necrophagous flies can be a problem within a forensic investigation where fly larvae are reared to adult to determine the species, and to be able to relate temperature growth charts for PMI determination. Generally, the presence of *N. vitripennis* is only a nuisance if parasitized puparia are collected at a crime scene, or if the wasp invades the forensic laboratory (which occurs all too frequently) to attack unprotected puparia reared from eggs and larvae for species identification.

Luckily, all is not lost if *N. vitripennis* emerges from a puparium rather than a fly. This wasp offers advantages over many other parasitic species in that fly hosts cannot be parasitized until after pupation is complete but prior to the onset of eclosion behavior, providing a window into the minimum length of host development on a corpse. Wasp development rates from egg to adult emergence at any given temperature show low variation among siblings (Whiting, 1967; Rivers & Denlinger, 1995), and since little outbreeding occurs with *N. vitripennis* in patchy environments like carrion (Werren, 1980), developing wasp progeny in multiple puparia collected at a crime scene are likely related.

Work has also been done on rearing times at different temperatures for this wasp, allowing *N. vitripennis* to serve as a forensic indicator species

Figure 5.26   A species of *Alysia*. *Source:* Photo by James Lindsey at Ecology of Commanster. James Lindsey/ Wikimedia Commons/CC BY-SA 3.0.

(Grassberger & Frank, 2003). However, caution must be exercised in using developmental data for *N. vitripennis* in which the frequency of parasitism and venom injection have not been controlled, as larval development is longer when multiple venom injections have occurred during conditions of multi- or super-parasitism (Rivers, 1996).

Unlike *N. vitripennis*, which develops multiple offspring within a single puparium, many other hymenopteran parasitoids develop just a single offspring within a single fly. One of the largest parasitoid families in the Hymenoptera is the Braconidae, and a variety of species have been reared from necrophagous Diptera. One of these braconids, *Alysia manducator* Panz (Figure 5.26), is known to attack blow fly larvae in the wandering stage before pupariation (Reznik *et al.*, 1992). Little is known about the growth rates of these parasitoids at different temperatures and they can be considered to be of little forensic importance (other than blocking identification of reared flies) until more research is completed on their life cycles.

5.4.3   Other parasitoids and predators

A variety of hymenopteran parasitoids have been reared from Calliphoridae and Sarcophagidae larvae, including members of the Encyrtidae and Chalcididae and additional species of Braconidae and Pteromalidae. The species composition of these parasitoid complexes will vary, based on the geographical location and microhabitat involved (Shaumar *et al.*, 1990; Geden, 2002; Oliva, 2008). Parasitoids and predators could, in the future, be valuable indicators for forensic studies but much more needs to be known about these associates before their potential can be realistically utilized.

5.5 Omnivorous species include taxa which feed on both the corpse and associated arthropods

Omnivorous insects include carrion beetles, ants, wasps, and a variety of other species. Large populations of these omnivores may retard the rate of carcass decomposition by depleting populations of necrophagous species. In this case, "omnivorous" does not refer to eating plant and animal tissue but to feeding on both the corpse and the insects feeding on the corpse. Some of these species may be useful for PMI determination but others have an unpredictable association with carrion which makes them unsuitable for use. However, omnivorous species must be taken into account when found, as they may affect the normal decay sequence if they have a sufficiently significant impact on the normal necrophagous communities.

5.5.1 The carrion beetles (Coleoptera: Silphidae)

Silphids are large and often charismatic members of the carrion community. Many are brightly colored, with oranges, yellows, and reds in contrast against black backgrounds. There seems to be some controversy as to whether the beetles are attracted to carrion for the carrion itself or the maggots it contains and the actual answer may be related to the particular species involved, but contradictory accounts are present in the literature (Smith, 1986). Evidently many species feed on both. Carrion beetles are most commonly encountered in rural or forested environments and are rarely noted in urban situations or inside houses. Carrion beetles also show species-specific preferences for the height above ground when locating a carcass to feed upon (Ikeda *et al.*, 2011). They are highly attuned to the odor of decay and are often part of the first wave of insects to arrive at a dead body.

Some species are known to specialize on small carrion, which they bury in an underground chamber and the male and female exhibit complex biparental care of their offspring. While these species have

Figure 5.27 A common carrion beetle, *Necrophila americana* (Linnaeus). *Source*: Ryan Hodnett/Wikimedia Commons/CC BY-SA 4.0.

fascinating life histories, they do not usually appear at larger corpses and rarely come into play in forensic investigations. There are a variety of other species that come to feed on, and breed in, larger corpses. A common example from forensic literature is *Necrophila americana* (Linnaeus) (Figure 5.27). This species shows little or no parental care for its offspring. An interesting publication by Brett Ratcliffe (1996) on the carrion beetles of Nebraska covers identification and includes a behavioral review of many North American species.

5.5.2 The ants (Hymenoptera: Formicidae)

A variety of different ant species have been associated with carrion. Many of these species will feed on both the flesh of the corpse and on the associated Diptera larvae. One example involves the red, imported fire ant, *Solenopsis invicta* Buren (Figure 5.28). In carrion-baited traps in Texas, fire ants caused changes in the daily occurrence of mature and wandering

Figure 5.28   Fire ants, *Solenopsis invicta* Buren. *Source*: MOs810/Wikimedia Commons/CC BY-SA 4.0.

larvae of secondary screwworms (Wells & Greenberg, 1994). A novel use of ants in a forensic setting can be seen in a case involving the time required for establishment of a colony to the point where winged adults were produced (Goff & Win, 1997).

5.5.3   The yellowjackets (Hymenoptera: Vespidae)

Several species of vespids have been associated with vertebrate carrion. They are important because of their feeding habits, which can include both the corpse and the adult and larval flies associated with the corpse (e.g., Moretti *et al.*, 2011). A common vespid in the United States and elsewhere in the world is the German yellowjacket, *Vespula german-ica* (Fabricius) (Figure 5.29). It can often be found foraging at carrion but the species exhibits different feeding behaviors in closed habitats as compared to more open habitats (d'Adamo & Lozada, 2007). These wasps may not be directly important for PMI determination, but their feeding behavior on the corpse at a crime scene should be considered when dealing with their Diptera prey.

5.5.4   Other omnivorous species

A wide variety of other insects, particularly flies and beetles, may come to a corpse to scavenge on carrion and necrophagous larvae. These often show up in faunal succession research (e.g., Michaud *et al.*, 2010) and are occasionally crucial indicator species in forensic entomology investigations. For example, *Synthesiomyia nudiseta* Wulp and *Hydrotaea* spp. (Diptera: Muscidae) are discussed

Figure 5.29   *Vespula germanica* (Fabricius). *Source*: User:Fir0002/Wikimedia Commons/GFDL 1.2

in the forensic entomology literature (e.g., Lord *et al.*, 1992; Dadour *et al.*, 2001) with the indication that they are purely necrophagous, but these mus-cids are known to eat both carrion and other fly larvae. Many of the omnivorous species recorded at a corpse are facultative predators who do not have a fixed relationship with the necrophagous insects on a corpse, but will take advantage of such a food resource when they happen across it.

5.6   Adventitious species include taxa that use the corpse as an extension of their own natural habitat

There are two times when a corpse becomes just a physical structure in the landscape for insects to hide under: (i) when the body first hits the ground until decay really starts; and (ii) after the remains dry and mummify during late decomposition. During these times we would be likely to find the same sorts of insects hiding under the body that we might find under a rock or log outside, or a under a couch or stack of books and magazines inside. A wide variety of insects might be found hiding under a corpse during the very early and late decomposition periods in an outdoor setting, from terrestrial isopods (Figure 5.30) to field crickets (Figure 5.31). Inside a house or apartment there may be cockroaches, silverfish, or house centipedes hiding under a body or in the cracks and crevices of clothing materials. While such insects may not provide good evidence for PMI determination, they may be useful indicators of potential movement of a body from one location to another. The community of arthropods will vary based on location and weather conditions that the body has been exposed to (e.g., Wolff *et al.*, 2001).

Figure 5.30 Terrestrial isopod. *Source*: Dat doris/ Wikimedia Commons/CC BY-SA 4.0

Figure 5.31 Field cricket. *Source*: Donald Hobern/ Wikimedia Commons/CC BY 2.0.

Chapter review

A variety of different insects and terrestrial arthropods are attracted to a dead body

- Death almost immediately changes a body's chemical signature, providing cues for detection and colonization by insects. Arthropods associated with the living organism (e.g., lice, ticks) move away from the body as the insects attracted to carrion arrive at the body.
- The attractiveness of a corpse changes over time, depending on the molecular composition of the volatile chemicals associated with the decay process. Different odors given off during the progress of decomposition will attract different kinds of insects.

The fauna of insects feeding on a body is determined by location, time, and associated organisms

- Some necrophagous insects have wide distributions across the planet while others have very restricted geographical ranges. The terrestrial environments are divided into eight major biogeographic regions, or ecozones. The United States, Canada, and northern Mexico are included in the Nearctic region. Species distributions are usually restricted to particular parts of particular ecozones based on climate. Within any particular geographical location, different species are found inside human structures compared with outside. Location on all different scales can make major differences in the community of life attracted to a carrion resource.
- Just as location determines the arthropod community at a given crime scene, timing also determines the species composition at a dead body. Time may refer to the time of year, weather or immediate environmental conditions, or stage of corpse decomposition.
- Many of the organisms found in association with a dead body are determined by the presence (or absence) of other organisms associated with the decomposition process. Predators and parasitoids will not be found unless their prey or hosts are present. Each species that feeds on the corpse changes that food resource for the species coming later.

Necrophagous insects include the taxa feeding on the corpse itself

- The most important insects for determination of PMI are the species that use the corpse for reproduction.
- Many insects come to feed on the corpse but do not use it as a breeding site. A wide variety of adult flies may be found feeding on a corpse to obtain a protein meal necessary for normal reproductive behavior. Many of these insects have reproductive strategies that do not involve carrion. Some species use the corpse as an ecological marker for mating aggregations.
- The most important necrophagous insects for PMI determination are the carrion flies (e.g., blow flies and flesh flies), especially during early stages of decomposition. These flies use carrion as their breeding site and their larvae feed on the decaying flesh.
- The Sarcophagidae (flesh flies) and Calliphoridae (blow flies) are the most important families of the order Diptera (true flies) for forensic investigations. They are classified in an evolutionary group of "higher Diptera" known as the calyptrate Diptera.
- The blow flies contain numerous familiar species, many of which have bright metallic green and blue coloration. They are commonly referred to as greenbottle and bluebottle flies. Each species has its own distinctive developmental natural history and distribution. Many of these species have been spread from their original geographic origin to places around the world via accidental introduction by human activity.
- The flesh flies also contain many familiar species, but their rather bland black and gray coloration often leads people to clump them into a mental grouping of common "flies," a false grouping of a wide variety of taxa that include house flies. While sarcophagids can be very important in forensic investigations, most of the species in this family do not breed in large vertebrate carrion.
- Some smaller species of flies can be important in later stages of decomposition or with bodies that are in places where larger flies cannot reach them (e.g., buried in a coffin). Two important groups are the scuttle flies in the family Phoridae and the skipper flies in the family Piophilidae. Small phorids can find bodies that are well hidden, covered, wrapped, or even buried. Piophilids are often associated with bones and many species have larvae that exhibit rather spectacular jumping ability.
- Carpet or hide beetles in the family Dermestidae are important necrophagous Coleoptera, especially in the late stages of decomposition. These small beetles and their larvae feed on many tissues that other species cannot utilize, like skin and hair. These are the insects that are used by taxidermists and museum curators to clean tissue off skulls and skeletons.

Parasitoids and predators are the second most significant group of carrion-frequenting taxa

- Many of the predators of necrophagous insects scavenge on both the corpse and the insects they find there, thus putting them in the realm of omnivorous species rather than pure predators. With that said, there are some insects that do not eat carrion and show up during decomposition to specifically feed on the insects that are there. Some of the most obvious at a body are beetles in the family Staphylinidae. These rove beetles are effective predators, especially of blow fly and flesh fly larvae during early stages of decay.
- Parasitoids are different from parasites in that parasitoids eventually kill their host while parasites tend to feed on and benefit from the living host and do not (in general) cause death. The parasitoid lifestyle is uncommon in the Coleoptera, but one genus of rove beetles exhibits this life history in association with the larvae of calyptrate Diptera. Many families and species of Hymenoptera are obligate parasitoids of other insects and some of these can be very important in a forensic setting.
- One of the most important parasitoids of necrophagous Diptera puparia is a pteromalid wasp, *Nasonia vitripennis*. This species is one of the most thoroughly studied species of insects, serving as a "lab rat" in the Hymenoptera for a wide variety of physiological and genetic studies. This species (and other parasitoids) can cause problems in forensic investigations when they destroy developing fly larvae that are being reared for identification and use in PMI determination.

Omnivorous species include taxa which feed on both the corpse and associated arthropods

- In this case we are using the term "omnivorous" to describe insects that feed on both carrion and its insect inhabitants. Some groups are obviously necrophilous (attracted to carrion) and represent common members of the carrion community. They include insects like carrion beetles, ants, and yellowjacket wasps.
- Large populations of these may retard the rate of carcass removal by depleting populations of necrophagous species. Many are not useful for PMI determination but their possible effects must be taken into account when assessing the potential crime scene.

Adventitious species include taxa that use the corpse as an extension of their own natural habitat

- Invertebrates such as springtails, spiders, centipedes, isopods, and insects commonly found under rocks and/or wood may use the corpse as a shelter during very early (or late) stages of decomposition.
- Adventitious species are usually not useful for PMI determination but may provide very useful information regarding the movement of a body from one environment to another.

Test your understanding

Level 1: knowledge/comprehension

1. Define the following terms:

   (a) coprophagous
   (b) systematist
   (c) puparium
   (d) maggot therapy
   (e) parasitoid.

2. Place the following insects in their proper taxonomic order: Coleoptera, Diptera, or Hymenoptera.

   (a) Black blow fly
   (b) Hairy rove beetle
   (c) Yellowjacket wasp

3. Explain how an insect could be necrophilous but not necrophagous. Give an example of a species that exhibits this difference.

Level 2: application/analysis

1. In 1956, a systematic revision of a group of flesh flies was published in the *Annals of the Entomological Society of America*. Harold R. Dodge described a new genus and several new species in his paper "A new sarcophagid genus with descriptions of fifteen new species." One of the species was given the name IDONEAMIMA SABROSKYI. As a result of subsequent synonymy, by William L. Downes Jr, this species is now known as SARCOPHAGA SABROSKYI. Write the current name in proper form for first mention in a journal article.

2. Many forensic entomology case studies that involve bodies found inside buildings rely on unusual species that do not show up in significant numbers (if at all) in successional results from outdoor research projects. Briefly explain why we see this striking difference in carrion communities.

Level 3: synthesis/evaluation

1. Forensic investigations take place around the globe. When looking at a forensic entomology publication that reports on species from a different part of the world, we can still get useful information for local investigations. What types of information can be used from investigations in a place like Nagasaki, Japan to answer questions about a forensic case in Ames, Iowa?

References cited

Aldrich, J.M. (1916) *Sarcophaga and Allies in North America*. Vol. 1 of the Thomas Say Foundation of the Entomological Society of America, LaFayette, IN.

Anderson, G.S. (2000) Minimum and maximum development rates of some forensically important Calliphoridae (Diptera). *Journal of Forensic Science* 45: 824–832.

Byrd, J.H. & Allen, J.C. (2001) The development of the black blow fly, *Phormia regina* (Meigen). *Forensic Science International* 120: 79–88.

Byrd, J.H. & Butler, J.F. (1997) Effects of temperature on *Chrysomya rufifacies* (Diptera: Calliphoridae) development. *Journal of Medical Entomology* 35: 353–358.

Byrd, J.H. & Castner, J.L. (eds) (2009) *Forensic Entomology: The Utility of Arthropods in Legal Investigations.* CRC Press, Boca Raton, FL.

de Carvalho, C.J.B. & de Mello-Patiu, C.A. (2008) Key to the adults of the most common forensic species of Diptera in South America. *Revista Brasileira de Entomologia* 52: 390–406.

Cherix, D., Wyss, C. & Pape, T. (2012) Occurrences of flesh flies (Diptera: Sarcophagidae) on human cadavers in Switzerland, and their importance as forensic indicators. *Forensic Science International* 220: 158–163.

d'Adamo, P. & Lozada, M. (2007) Foraging behavior related to habitat characteristics in the invasive wasp *Vespula germanica. Insect Science* 14: 383–388.

Dadour, I.R., Cook, D.F. & Wirth, N. (2001) Rate of development of *Hydrotaea rostrata* under summer and winter (cyclic and constant) temperature regimes. *Medical and Veterinary Entomology* 15: 177–182.

Danneels, E.L., Rivers, D.B. & de Graaf, D.C. (2010) Venom proteins of the parasitoid wasp *Nasonia vitripennis:* recent discovery of an untapped pharmacopee. *Toxins* 2: 494–516.

Disney, R.H.L. (2006) Duration of development of some Phoridae (Dipt.) of forensic significance. *Entomologists Monthly Magazine* 142: 129–138.

Disney, R.H.L. (2008) Natural history of the scuttle fly, *Megaselia scalaris. Annual Review of Entomology* 53: 39–60.

Dodge, H.R. (1966) *Sarcophaga utilis* Aldrich and allies (Diptera: Sarcophagidae). *Entomological News* 77: 85–97.

Draber-Monko, A., Malewski, T., Pomorski, J., Los, M. & Slipinski, P. (2009) On the morphology and mitochondrial DNA barcoding of the flesh fly *Sarcophaga (Liopygia) argyrostoma* (Robineau-Desvoidy, 1830) (Diptera: Sarcophagidae): an important species in forensic entomology. *Annales Zoologici (Warszawa)* 59: 465–493.

Geden, C.J. (2002) Effect of habitat depth on host location by five species of parasitoids (Hymenoptera: Pteromalidae, Chalcididae) of house flies (Diptera: Muscidae) in three types of substrates. *Environmental Entomology* 31: 411–417.

Gisondi, S, Rognes, K, Badano, D, Pape, T & Cerretti, P. (2020) The world Polleniidae (Diptera, Oestroidea): key to genera and checklist of species. *ZooKeys* 971: 105–155. https://doi.org/10.3897/zookeys.971.51283.

Goff, M.L. & Win, B.H. (1997) Estimation of postmortem interval based on colony development time for *Anoplolepis longipes* (Hymenoptera: Formicidae). *Journal of Forensic Science* 42: 1176–1179.

Grassberger, M. & Frank, C. (2003) Temperature-related development of the parasitoid wasp *Nasonia vitripennis* as forensic indicator. *Medical and Veterinary Entomology* 17: 257–262.

Grassberger, M. & Reiter, C. (2002) Effect of temperature on development of *Liopygia (= Sarcophaga) argyrostoma* (Robineau-Desvoidy) (Diptera: Sarcophagidae) and its forensic implications. *Journal of Forensic Sciences* 47: 1–5.

Greenberg, B. (1991) Flies as forensic indicators. *Journal of Medical Entomology* 28: 565–577.

Greenberg, B. & Kunich, J. (eds) (2002) *Entomology and the Law: Flies as Forensic Indicators.* Cambridge University Press, Cambridge, UK.

Greenberg, B. & Wells, J.D. (1998) Forensic use of *Megaselia abdita* and *M. scalaris* (Phoridae: Diptera): case studies, development rates, and egg structure. *Journal of Medical Entomology* 35: 205–209.

Hall, D.G. (1948) *The Blowflies of North America.* Vol. IV of the Thomas Say Foundation of the Entomological Society of America, LaFayette, IN.

Ikeda, H., Shimano, S. & Yamagami, A. (2011) Differentiation in searching behavior for carcasses based on flight height differences in carrion beetles (Coleoptera: Silphidae). *Journal of Insect Behavior* 24: 167–174.

International Commission on Zoological Nomenclature (ICZN) (1999) *International Code of Zoological Nomenclature,* 4th edn. The International Trust for Zoological Nomenclature, London.

Jones, N., Whitworth, T. & Marshall, S.A. (2019) Blow flies of North America: keys to the subfamilies and genera of Calliphoridae, and to the species of the subfamilies Calliphorinae, Luciliinae and Chrysomyinae. *Canadian Journal of Arthropod Identification* 39: 191 pp.

Kamal, A.S. (1958) Comparative study of thirteen species of sarcosaprophagous Calliphoridae and Sarcophagidae (Diptera). I. Bionomics. *Annals of the Entomological Society of America* 51: 261–271.

Klimaszewski, J. (1984) A revision of the genus *Aleochara* Gravenhorst of America north of Mexico. *Memoirs of the Entomological Society of Canada* 129: 1–211.

Kulshrestha, P. & Satpathy, D.K. (2001) Use of beetles in forensic entomology. *Forensic Science International* 120: 15–17.

Lord, W.D., Adkins, T.R. & Catts, E.P. (1992) The use of *Synthesiomyia nudesita* (van der Wulp) (Diptera: Muscidae) and *Calliphora vicina* (Robineau-Desvoidy) (Diptera: Calliphoridae) to estimate the time of death of a body buried under a house. *Journal of Agricultural Entomology* 9: 227–235.

Marchenko, M.I. (2001) Medicolegal relevance of cadaver entomofauna for the determination of the time of death. *Forensic Science International* 120: 89–109.

Martin-Vega, D. (2011) Skipping clues: forensic importance of the family Piophilidae (Diptera). *Forensic Science International* 212: 1–5.

Martin-Vega, D., Gomez-Gomez, A. & Baz, A. (2011) The "coffin fly" *Conicera tibialis* (Diptera: Phoridae) breeding on buried human remains after a postmortem interval of 18 years. *Journal of Forensic Science* 56:1654–1656.

Michaud, J.-P., Majka, C.G., Privé, J.-P. & Moreau, G. (2010) Natural and anthropogenic changes in the insect fauna associated with carcasses in the North American Maritime lowlands. *Forensic Science International* 202: 64–70.

de Carvalho Moretti, T., Giannotti, E., Thyssen, P.J., Solis, D.R. & Godoy, W.A.C. (2011) Bait and habitat preferences, and temporal variability of social wasps (Hymenoptera: Vespidae) attracted to vertebrate carrion. *Journal of Medical Entomology* 48: 1069–1075.

Oliva, A. (2008) Parasitoid wasps (Hymenoptera) from puparia of sarcosaprophagous flies (Diptera: Calliphoridae; Sarcophagidae) in Buenos Aires, Argentina. *Revista de la Sociedad Entomológica Argentina* 67: 139–141.

Pape, T. (1996) *Catalogue of the Sarcophagidae of the World (Insecta: Diptera)*. Memoirs on Entomology International, Vol. 8. Associated Publishers, Gainsville, FL.

Payne, J.A. & King, E.W. (1972) Insect succession and decomposition of pig carcasses in water. *Journal of the Georgia Entomological Society* 7: 153–162.

Poorbaugh, J.H. & Linsdale, D.D. (1971) Flies emerging from dog feces in California. *California Vector Views* 18: 51–56.

Ratcliffe, B. (1996) *The Carrion Beetles (Coleoptera: Silphidae) of Nebraska*, Vol. 13. Bulletin of the University of Nebraska State Museum. Available at https://museum.unl.edu/collections/publications/museum-bulletins/volume-13.html.

Reed, H.G. Jr (1958) A study of dog carcass communities in Tennessee, with special reference to the insects. *American Midland Naturalist* 59: 213–245.

Reznik, S.Y., Chernoguz, D.G. & Zinovjeva, K.B. (1992) Host searching, oviposition preferences and optimal synchronization in *Alysia manducator* (Hymenoptera: Braconidae), a parasitoid of the blowfly, *Calliphora vicina*. *Oikos* 65: 81–88.

Rivers, D.B. (1996) Changes in the oviposition behavior of the ectoparasitoids *Nasonia vitripennis* and *Muscidifurax zaraptor* (Hymenoptera: Pteromalidae) when using different species of fly hosts, prior oviposition experience, and allospecific competition. *Annals of the Entomological Society of America* 89: 466–474.

Rivers, D.B. & Denlinger, D.L. (1995) Fecundity and development of the ectoparasitoid *Nasonia vitripennis* are dependent upon the host nutritional and physiological condition. *Entomologia Experimentalis et Applicata* 76: 15–24.

Rognes, K. (1991) *Blowflies (Diptera, Calliphoridae) of Fennoscandia and Denmark*. E.J. Brill/Scandinavian Science Ltd, Leiden.

Rosati, J.Y. & VanLaerhoven, S.L. (2007) New record of *Chrysomya rufifacies* (Diptera: Calliphoridae) in Canada: predicted range expansion and potential effects on native species. *The Canadian Entomologist* 139: 670–677.

Sanjean, J. (1957) Taxonomic studies of *Sarcophaga* larvae of New York, with notes on the adults. *Memoirs of the Cornell University Agricultural Experiment Station* 349: 1–115.

Schroeder, H., Klotzbach, H., Oesterhelweg, L. & Püschel, K. (2002) Larder beetles (Coleoptera, Dermestidae) as an accelerating factor for decomposition of a human corpse. *Forensic Science International* 127: 231–236.

Shaumar, N.F., El-Agoze, M.M. & Mohammed, S.K. (1990) Parasites and predators associated with blow flies and flesh flies in Cairo region. *Journal of the Egyptian Society of Parasitology* 20: 123–132.

Sherman, R.A., Hall, M.J.R. & Thomas, S. (2000) Medicinal maggots: an ancient remedy for some contemporary afflictions. *Annual Review of Entomology* 45: 55–81.

Smith, K.G.V. (1986) *A Manual of Forensic Entomology*. British Museum (Natural History), London.

Tarone, A.M. & Foran, D.R. (2008) Generalized additive models and *Lucilia sericata* growth: assessing confidence intervals and error rates in forensic entomology. *Journal of Forensic Science* 53: 942–948.

Thomson, A.J. & Davies, D.M. (1973) The biology of *Pollenia rudis*, the cluster fly (Diptera: Calliphoridae). II. Larval feeding behaviour and host specificity. *The Canadian Entomologist* 105: 985–990.

Tomberlin, J.K., Benbow, M.E., Tarone, A.M. & Mohr, R.M. (2011) Basic research in evolution and ecology enhances forensics. *Trends in Ecology and Evolution* 26: 53–55.

Vairo, K.P., de Mello-Patiu, C.A. & de Carvalho, C.J.B. (2011) Pictorial identification key for species of Sarcophagidae (Diptera) of potential forensic importance in southern Brazil. *Revista Brasileira de Entomologia* 55: 333–347.

Verves, Y.G. (1986) Family Sarcophagidae. In: Á. Soós & L. Papp (eds) *Catalogue of Palaearctic Diptera*, Vol. 12, pp. 58–193. Akadémiai Kiadó, Budapest and Elsevier, Amsterdam.

Wells, J.D. & Greenberg, B. (1994) Effect of the red imported fire ant (Hymenoptera: Formicidae) and carcass type on the daily occurrence of postfeeding carrion-fly larvae (Diptera: Calliphoridae, Sarcophagidae). *Journal of Medical Entomology* 31: 171–174.

Werren, J.H. (1980) Sex ratio adaptations to local mate competition in a parasitic wasp. *Science* 208: 1157–1159.

Whiting, A. (1967) The biology of the parasitic wasp *Mormoniella vitripennis*. *Quarterly Review of Biology* 42: 333–406.

Whitworth, T. (2006) Keys to the genera and species of blow flies (Diptera: Calliphoridae) of America north of Mexico. *Proceedings of the Entomological Society of Washington* 108: 689–725.

Wolff, M., Uribe, A., Ortiz, A. & Duque, P. (2001) A preliminary study of forensic entomology in Medellín, Columbia. *Forensic Science International* 120: 53–59.

Zubir, M.Z.M., Holloway, S. & Noor, N.M. (2020) Maggot therapy in wound healing: a systematic review. *International Journal of Environmental Research and Public Health* 17: 6103.

Supplemental reading

Anderson, R.S. & Peck, S.B. (1985) *The Carrion Beetles of Canada and Alaska (Coleoptera: Silphidae and Agyrtidae). The Insects and Arachnids of Canada, Part 13.* Publication 1778, Research Branch Agriculture Canada, Ottawa.

Disney, R.H.L. (1994) *Scuttle Flies: The Phoridae.* Chapman & Hall, London.

Hanley, G.A. & Cuthrell, D. (2008) *Carrion Beetles of North Dakota: An Atlas and Identification Guide.* MSU Science Monograph No. 4. Minot State University, Minot, ND.

Pape, T. & Dahlem, G.A. (2010) Sarcophagidae (flesh flies). In: B.V. Brown, A. Borkent, J.M. Cumming, D.M. Wood, N.E. Woodley & M.A. Zumbado (eds) *Manual of Central American Diptera*, Vol. 2, pp. 1313–1335. National Research Council Canada, Ottawa.

Triplehorn, C.A., Johnson, N.F. & Borror, D.J. (2005) *Introduction to the Study of Insects*, 7th edn. Thompson Brooks/Cole, Belmont, CA.

Vargas, J. & Wood, D.M. (2010) Calliphoridae (blow flies). In: B.V. Brown, A. Borkent, J.M. Cumming, D.M. Wood, N.E. Woodley & M.A. Zumbado (eds) *Manual of Central American Diptera*, Vol. 2, pp. 1297–1304. National Research Council Canada, Ottawa.

Chapter 6
Reproductive strategies of necrophagous flies

Overview

Insect reproduction warrants a book unto itself. The topic draws the attention of an array of biologists, many with no real fondness for insects at all. However, when discussions turn to insects that cannot produce eggs, oviposit, or even have sex until eating something dead, well how could anyone resist! Obviously space constraints prevent an in-depth exploration of the multitude of processes, mechanisms, and morphologies that have evolved independently in several insect groups. Thus, this chapter will focus on the reproductive biology of the most important insect colonizers of carrion, necrophagous flies, with particular attention given to members of the families Calliphoridae and Sarcophagidae. Emphasis is directed toward aspects of reproductive fitness, including egg production, oviposition strategies, larval adaptations, interspecific and intraspecific resource partitioning, and reproductive considerations for adult females versus offspring. Coverage is intended to provide a framework for topics in forensic entomology, as well as advanced topics in behavioral ecology and insect physiology.

The big picture

- The need to feed: anautogeny and income breeders are common among necrophagous Diptera.
- Size matters in egg production.
- Progeny deposition is a matter of competition.
- Larvae are adapted for feeding and competing on carrion.
- Feeding aggregations maximize utilization of food source.
- Mother versus offspring: fitness conflicts.
- Resource partitioning is path to reproductive success.

6.1 The need to feed: anautogeny and income breeders are common among necrophagous Diptera

Reproduction in calyptrate Diptera is highly evolved and complex, with as many "exceptions" and modifications as general rules to the process. The common ground regarding the insects of

The Science of Forensic Entomology, Second Edition. David B. Rivers and Gregory A. Dahlem.
© 2023 John Wiley & Sons Ltd. Published 2023 by John Wiley & Sons Ltd.
Companion website: www.wiley.com/go/forensicentomology2

interest to this textbook is the need for carrion. Animal remains (including blood, other fluids, and feces) serve as the ultimate location for oviposition or larviposition as well as for subsequent larval feeding that follows either egg hatch or migration of larvae over the body. For many species of calliphorids and sarcophagids, prior to the events of progeny deposition, a corpse can serve as a food source for adult feeding, provide chemical cues to promote adult aggregation, and also function as the site for courting and copulation (Figure 6.1). In fact some species require multiple visits to carrion or extended periods of feeding to acquire the necessary nutriment to develop eggs (Barton Browne *et al.*, 1976; Wall *et al.*, 2002). Insects that require food during the adult stage to produce mature **oocytes** are known as income breeders, a topic that is discussed in more detail later in this section, but are commonly referred to as **anautogenous**.

Autogeny describes the condition in animals, often in the context of entomology, in which the adult female is capable of producing eggs upon emergence. The insects referenced are usually holometabolous since there is a period of non-feeding – the pupal stage – between the immature stages and the imago. In this case, the female insect does not have a requirement to obtain nutriment, specifically protein, prior to development of mature oocytes, regardless of whether the insect produces eggs singly or in clutches (Chapman, 1998). Subsequent clutch formation, especially in the case of necrophagous fly species, may require that the adult females feed

on food sources high in protein and possibly other nutrients to successfully provision additional eggs (Pappas & Fraenkel, 1977; Hammack, 1999). Autogeny is more common with calliphorids than sarcophagids.

Many of the forensically important fly species that frequent carrion are anautogenous rather than autogenous (Table 6.1) and thus do not have the necessary nutrient reserves stored from the larval stages to provision eggs. Rather, anautogenous insects must obtain protein through feeding as adults to produce mature oocytes. Typically the protein meal is necessary for the events of **oogenesis** and **vitellogenesis** to proceed (Webber, 1958; Stoffolano *et al.*, 1995; Chapman, 1998; Hahn *et al.*, 2008a) and, at least for some species, it does not matter whether the protein source is fresh or decomposed (Huntington & Higley, 2010). This means that carrion, human remains, body fluids, and feces are all potential sources of protein for adult flies. A critical or minimum amount of protein must be ingested to initiate the hormonal cascade that regulates ovarian development and the subsequent events of oogenesis and vitellogenesis (Webber, 1958; Wall *et al.*, 2002). This lower limit to protein intake generally reflects the key facet for necrophagous flies that they must surpass a nutritional or minimum threshold for reproduction (Yin *et al.*, 1999). For example, the Holarctic blow fly *Protophormia terraenovae* requires a minimum of 0.4 mg protein per adult female to provision eggs, while females of *Lucilia cuprina* need at least 3.6 mg of liver exudate (high in protein) for egg maturation to occur (Harlow, 1956; Williams *et al.*, 1979). For the calliphorid *Lucilia sericata*, two protein thresholds appear to regulate egg maturation (Figure 6.2). After an initial protein meal (first threshold), yolk deposition occurs in all available oocytes (Wall *et al.*, 2002). A second protein threshold needs to be surpassed to allow extensive yolk deposition followed by oosorption and maturation of small egg clutches (Wall *et al.*, 2002). However, if a critical amount of protein is not available, arrestment of oocyte development occurs early in yolk deposition. Flies that enter arrested development may resume egg provisioning after subsequent protein meals, while those that display no halt in oocyte development can oviposit sooner in the adult female's life. This partitioning in development may represent a two-part reproductive strategy designed to maximize progeny production when using a temporal resource (Hayes *et al.*, 1999).

Figure 6.1 Bottle flies in copula. *Source*: gailhampshire/ Wikimedia Commons.

Table 6.1   Linkage of adult feeding to egg production and "oviposition" in necrophagous flies.

Species	Family	Protein requirement	Progeny deposition
Calliphora vicina	Calliphoridae	Anautogenous	Oviparity
Calliphora vomitoria	Calliphoridae	Anautogenous	Oviparity
Cochliomyia hominivorax	Calliphoridae	Autogeny	Oviparity
Lucilia cuprina	Calliphoridae	Anautogenous	Oviparity
Lucilia sericata	Calliphoridae	Anautogenous	Oviparity
Phormia regina	Calliphoridae	Anautogenous	Oviparity
Protophormia terraenovae	Calliphoridae	Anautogenous	Oviparity
Sarcophaga africa	Sarcophagidae	Anautogenous	Ovoviviparity
Sarcophaga argryostoma	Sarcophagidae	Autogeny	Ovovivaparity
Sarcophaga bullata	Sarcophagidae	Anautogenous*	Ovovivaparity
Sarcophaga crassipalpis	Sarcophagidae	Anautogenous	Ovovivaparity

* Baxter *et al.* (1973) report that *S. bullata* is both autogenous and anautogenous.

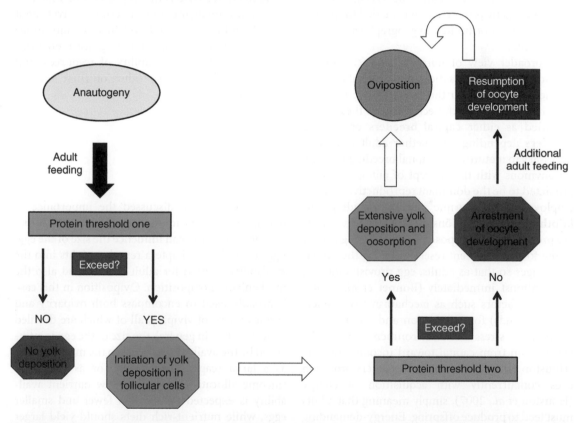

Figure 6.2   Critical protein thresholds associated with egg provisioning in the blow fly *Lucilia sericata*. *Source*: Adapted from Wall *et al.* (2002).

Protein can be acquired through a single feeding provided the meal is high in protein. Alternatively, protein intake may be required over an extended number of days: a single meal is not sufficient for any egg production or maximum egg output is only achieved if protein is available for a specific number of days (Hahn *et al.*, 2008b). For example, the length of time adult females of the flesh fly *Sarcophaga crassipalpis* feed on protein influences the rate of egg provisioning (slower when protein is limited), fecundity (clutch size), and the size of individual eggs (Hahn *et al.*, 2008a). Similarly, sexual receptivity as well as sex pheromone production are enhanced in protein-fed females of the Australian sheep blow fly, *Lucilia cuprina*, by comparison to those deprived of protein as adults (Barton Browne *et al.*, 1976).

Switching between autogeny and anautogeny apparently occurs across broods depending on the nutritional conditions faced by the larvae. In the case of the flesh fly *Sarcophaga bullata*, larvae developing in environments in which food is limited diminishes larval growth and results in small anautogenous adults (Baxter *et al.*, 1973). By contrast, when larval development is extended (in terms of time in instar) due to high food quality, the adults that are produced are large in size and autogenous (Baxter *et al.*, 1973). Autogeny is not commonly observed with this species, even when protein is not limited and adults are large, suggesting that this form of progeny production may be tied to specific protein sources or limited to geographical variants of *S. bullata*.

A broader view of nutrient requirements for reproduction examines flies based on total nutrient or energy needs rather than an exclusive focus on protein. In this context, necrophagous flies can be classified as either **capital breeders** or **income breeders** depending on whether adult feeding is necessary to mature eggs. Capital breeding is nearly synonymous with the concept of autogeny and is predicted to be the dominant reproductive strategy employed by ectothermic animals, including **poikilothermic** insects (Jönsson, 1997). Such insects are presumably predisposed toward storing energy in the form of nutrient reserves during the immature stages so that as adults, egg provisioning can begin almost immediately (Bonnet *et al.*, 1998). However, factors such as mechanisms (i.e., energetic demands) for locomotion and the degree of temperature stress in the environment can shift the expectation from capital toward income breeding (Houston *et al.*, 2007). Income breeders provision eggs concurrently with acquisition of energy (Houston *et al.*, 2007), simply meaning that adults must feed to produce offspring. Energy-demanding flight used by necrophagous flies, potentially over long distances since the food resource is **ephemeral** and non-predictable, may well position calliphorids and sarcophagids toward income breeding (Wessels *et al.*, 2010). Yet, at the same time, if the food source is nutrient-poor, exclusive income investment may not be sufficient to deliver adequate energy to offspring (Houston *et al.*, 2007). This latter example supports the contention that a life-history trait such as breeding strategy (i.e., energy allocation) lies on a continuum rather than being static.

Temperature stress is a particularly unique problem associated with the larval environment (maggot masses), rather than the adult. Although food assimilation is enhanced with elevated temperatures (below an upper temperature threshold) (Rivers *et al.*, 2011), long-term nutrient storage is likely limited due to competing physiological mechanisms like proteotaxic stress responses (Rivers *et al.*, 2010) from the so-called **larval-mass effect**[1] (Charabidze *et al.*, 2011). Consequently, exclusive capital breeding in these flies seems to be restricted to a few species, many of which are either facultative or obligate larval parasites (Webber, 1958; Thomas & Mangan, 1989). The more likely scenario is that reproductive strategies rely on an amalgamation of resource investment (e.g., blending of capital and income allocation) (Wessels *et al.*, 2010), changing with environmental conditions, availability of resources, and the frequency of clutch production (first, second, or third brood, etc).

6.2 Size matters in egg production

In Section 6.1, we discussed the importance of nutriment to provisioning eggs. Here, we will discuss how nutrition can influence the size of the egg. Egg size in higher Diptera reveals insights into the availability of food for adult females and also the mechanism of oviposition. Oviposition in this case is broadly used to encompass both oviparity and various forms of vivipary, all of which are detailed in Section 6.3. In general, the size of the egg is influenced by the available nutrients to the mother either as a larva (capital investment) or as an imago (income allocation). Certainly low nutrient availability is expected to result in fewer and smaller eggs, while nutrient-rich diets should yield larger clutches, possibly with large individual eggs.

For example, Hahn *et al.* (2008a) observed that with the anautogenous flesh fly *S. crassipalpis*, length and overall size of eggs are influenced by the availability of protein to adult females. Females remained responsive to allocating resources to eggs up to 6 days after reaching the reproductive threshold, while shorter periods of protein feeding yielded smaller eggs (Figure 6.3).

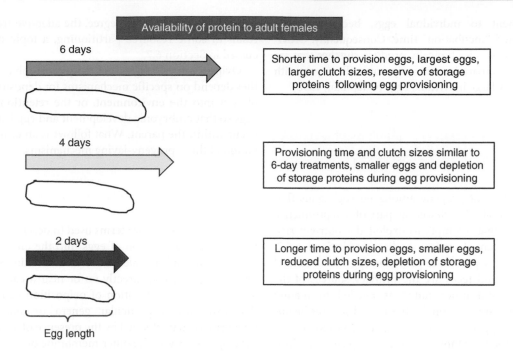

Figure 6.3 Relationship between reproductive output and feeding in adult females of *Sarcophaga crassipalpis*. *Source*: Adapted from Hahn *et al.* (2008a).

The influence of diet on egg size is expected to be greatest with flies that lay large clutches all at once like *S. crassipalpis*, as opposed to species that deposit eggs singly or asynchronously. Clutch layers (as opposed to solitary laying) make a large investment in resource allocation in a short period of time (Jervis *et al.*, 2007), so limitations in nourishment in general would seemingly result in smaller eggs. Tanaka *et al.* (1990) found that maternal size of three tropical sarcophagids correlated with egg size. Small adults produced small eggs, and large adults provisioned larger eggs. This does not always occur, however, with many species. Overcrowding on carrion frequently generates small larvae, which in turn yields stunted pupae and adults. In the case of *Calliphora vicina*, small adults provision fewer eggs than larger flies, but egg size is not compromised (Saunders & Bee, 1995). This trend is expected to occur with other closely related calliphorids.

The method of progeny deposition seems to yield characteristic egg sizes as well. What this means is that for those species that lay eggs so that embryonic development mostly occurs outside of the mother, a condition termed **oviparity** (Figure 6.4), females typically lay large numbers of relatively small eggs (Chapman, 1998). Resource allocation is generally less per individual egg than with flies that deposit

Figure 6.4 Oviparity, the laying of eggs into the environment, is the typical mechanism of progeny deposition used by necrophagous calliphorids. *Source*: D.B. Rivers.

progeny via some form of **vivipary**, a broad term conveying the idea of "live" birth. One of the most commonly used mechanisms of vivipary by calyptrate Diptera is **ovoviviparity** (examined in detail in Section 6.3), a condition in which embryos are retained in gravid females for an extended period. This association allows mothers to invest more nutriment to individual eggs, because of the increased "incubation" time. Consequently, ovoviviparous species typically produce clutches that are small comparative to oviparous flies, yet with relatively large individual eggs (Chapman, 1998).

6.3   Progeny deposition is a matter of competition

Deposition of progeny among necrophagous flies can generally be viewed as part of a reproductive strategy that attempts to exploit the nutrient-rich environment of carrion. As discussed in Section 6.2, flies require protein to provision eggs. Since animal remains and feces are the primary sources of the protein and other nutrients needed to mature oocytes, competition for these food sources begins prior to egg hatch. The timing of oviposition or larviposition by calliphorids and sarcophagids is absolutely critical to attain the necessary protein from the resource, both for adult females to provision eggs and for neonate larvae that begin to feed on the corpse (Levot et al., 1979). However, the corpse is a patchy nutrient island and, as such, is subject to intense intraspecific and interspecific competition (Hanski & Kuusela, 1977). Denno and Cothran (1975) contend that calliphorids and sarcophagids have evolved life-history strategies that allow coexistence on the same resource (i.e., carrion). The strategies are centered on differences in oviposition strategies. By oviposition, we are again more broadly defining the term to include oviparity and various forms of vivipary utilized by necrophagous flies. The differences are best illustrated by examining sarcophagids. In general terms, sarcophagids produce small clutches with large eggs that hatch inside an **ovisac** or **common oviduct**, resulting in the mother depositing first-stage larvae (neonates) on carrion (Denno & Cothran, 1975; Shewell, 1987a). In contrast, most calliphorids lay eggs in large clutches and the individual eggs are often smaller than the neonate larvae of sarcophagids (Shewell, 1987b). Larval development is considered slower for sarcophagids than calliphorids, contributing to the idea of coexistence or **exploitive competition** (Park, 1962). However, slower rates of development are characteristic of only some sarcophagids and can also be greatly influenced by the conditions of the microhabitat associated with the maggot mass itself (see Chapter 8 for more on the biology of maggot masses) (Kamal, 1958;

Rivers et al., 2011). To a degree, the adaptive traits result in some resource partitioning, a topic discussed in Section 6.7.

Oviposition strategies used by necrophagous flies depend on specific mechanisms for deposition of eggs into the environment, or the retention of eggs so that embryonic development and egg hatch occur within the parent. What follows is an examination of these progeny-laying mechanisms.

6.3.1   Ovipary or oviparity

Oviparity or ovipary are terms used to describe egg laying or the deposition of eggs into the environment by the adult female. In the case of calliphorid adults, eggs are laid directly on or near a carcass, and frequently the location of oviposition can be diagnostic for a particular genera or species. Oviparity is characterized by the presence of a well-developed chorion, the outer membrane or shell of the egg lying outside the vitelline envelope (Figure 6.5). The egg is laid prior to significant embryonic development. This means that most aspects of embryonic growth and development, including hatch, occur independently from the mother. The implication is that eggs receive a finite amount of nutriment (in the form of **yolk**) from the parent. Consequently, egg size is often smaller with

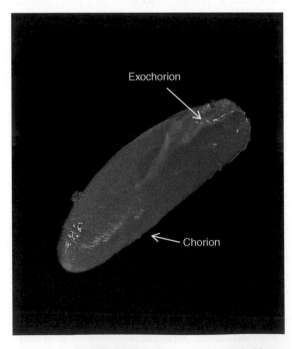

Figure 6.5   A typical calliphorid egg. *Source*: D.B. Rivers.

oviparous species in comparison with those that rely on some form of vivipary (Chapman, 1998). Among necrophagous flies, oviparity is predominantly used by calliphorids and is relatively rare among forensically important sarcophagids (Shewell, 1987a,b).

6.3.2 Vivipary

Vivipary describes the retention of eggs in the reproductive tract until hatching or the eggs hatch as they are passed out of the mother. The net effect is deposition of larvae rather than eggs, and hence these insects are said to have live birth. Significant embryonic development occurs while the eggs are maintained in the parent as opposed to oviparity, in which most aspects of embryonic development occur outside the mother's body. With some forms of vivipary (i.e., **viviparity**), the mother continues to provide nourishment to the developing embryo and possibly larvae beyond the initial contribution of yolk (Willmer *et al.*, 2000). Among necrophagous flies, the mother generally does not provide additional nutriment to the embryos and thus displays ovoviviparity (Chapman, 1998).

6.3.2.1 Viviparity

Viviparity among necrophagous insects is relatively rare. Shewell (1987b) states that sarcophagids are mostly viviparous and sometimes ovoviviparous. However, it seems that the opposite is probably correct. Truly viviparous insects, namely those that display viviparity, do not possess a chorion around the egg and use specialized structures akin to a uterus, milk gland, or similar adaptations that allow the parent (usually the mother) to provide nourishment to the embryo (Chapman, 1998). A reduction in the number of **ovarioles** typifies viviparous insects, resulting in fewer offspring being produced than with other forms of reproduction (Figure 6.6). In some cases, only one larva is produced at a time. In the case of the tsetse fly, *Glossina morsitans*, a single larva develops within the uterus of the mother, feeding on milk derived from an accessory gland, and at the time of **parturition**, the mother gives "birth" to a larva nearly three-quarters the length of her own body (Denlinger & Ma, 1974).

Three forms of viviparity are recognized in insects: pseudoplacental, adenotrophic, and hemocoelous. However, none occur with necrophagous

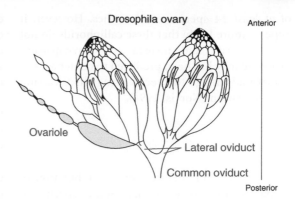

Figure 6.6 Cartoon of the paired ovaries of *Drosophila* spp., which is representative of most calliphorid and sarcophagid females. The number of ovarioles is reduced in viviparous species by comparison to flies displaying oviparity. Egg hatch occurs in the lateral or common oviduct of viviparous females. *Source*: El Mayimbe/ Wikimedia Commons/Public domain.

flies, although viviparity does occur in a few medically important insects, such as *Glossina* spp. (adenotrophic) and some cockroaches (pseudoplacental) (Chapman, 1998).

6.3.2.2 Ovoviviparity

Ovoviviparity involves the mother retaining the eggs within the reproductive tract, usually in an ovisac or common oviduct, until egg hatch. The egg hatches within the parent either immediately before deposition or at some prescribed time earlier that requires the mother to find a suitable substrate for deposition or else face being consumed by her offspring. Ovoviviparity is essentially oviparity (eggs with a chorion) with the exception that eggs hatch inside the mother. All nutriment for the embryo is derived from the egg (i.e., yolk) and no special anatomical structures are associated with nutrient transfer from mother to progeny following egg provisioning (Chapman, 1998). Flies that rely on ovoviviparity are referred to as larviparous, viviparous, and sometimes ovolarviparous. The deposition of larvae by the mother is commonly called larviposition.

Ovoviviparity appears to be the predominant strategy used by sarcophagids and is less commonly associated with calliphorids. With that said, several species of ovolarviparous calliphorids have been identified in Australia and include *Calliphora dubia*, *C. varifrons*, and *C. maritime* (Cook & Dadour, 2011). Shewell (1987a) used the term "viviparity" to describe the reproductive strategy of at least 14 species of blow flies. However, it appears more likely that these calliphorids do not use viviparity and instead are ovoviviparous. Ovoviviparity has been predicted to confer a competitive advantage to species using small carrion for oviposition (Norris, 1994).

6.3.3 Mixed strategies

A few species of flies rely on a combination of oviparity and ovoviviparity. Some species of *Musca* (Family Muscidae) that are normally oviparous can retain eggs until hatch to deposit live larvae (Chapman, 1998). Similarly, some oviparous calliphorids and sarcophagids retain fertilized eggs in the common oviduct while searching for an appropriate oviposition site. In these instances, the eggs remain in the adult female for one or more days, allowing significant embryonic development to occur (Smith, 1986; Wells & King, 2001). Villet *et al.* (2010) describes this situation as **precocious egg development**, in which an oviparous species displays ovovivipar-ity with a portion or an entire clutch (Table 6.2). In adults of *Calliphora vicina* collected in Germany, over 50% of gravid females contained either a precocious egg or neonate larvae in the genital tract (Lutz & Amendt, 2020). However, despite this high occurrence, no female was observed to contain more than one precocious egg or larva at a time. In contrast, Prado e Castro *et al.* (2016) reported ovoviviparity in the normally oviparous species *Calliphora loewi* in females collected from Madeira Island Portugal. In contrast to *C. vicina*, the females of *C. loewi*

displayed **oligolaraviparity** in that more than one precocious egg and larvae were present in each female. The capacity for oviparity with large clutches was not evident for this species on the island (Prado e Castro *et al.*, 2016).

Precocious egg development in Calliphoridae has not been well characterized but interestingly it has only been reported for the subfamily Calliphorinae (Lutz & Amendt, 2020). This would seemingly suggest an adaptive benefit critical for these flies. If true, the reason is not immediately obvious. Lutz & Amendt (2020) speculated that precocious egg development may be an adaptive reproductive strategy favored during harsh envi-ronmental conditions, such as the lack of food resources during low temperatures when this fly is active. While this could be true, more than one pre-cocious egg/larva per female would be expected under varying conditions. This may indeed be the case with *C. loewi*.

A slightly different mixed approach was observed with the Australian blow fly *C. dubia*. This species is ordinarily ovoviviparous but can lay eggs with larvae in the same clutch. Cook and Dadour (2011) found that under field conditions, the laid eggs of *C. dubia* were non-viable. However, they suggested that this mixed oviposition could conceivably reflect a strategy to distribute offspring across mul-tiple carcasses. Mixed oviposition, if both eggs and larvae are viable, has the potential to confound determinations of postmortem interval relying on fly development times (Villet *et al.*, 2010; Cook & Dadour, 2011). Thus, this is an area in need of further investigation to clarify our understanding of which fly species utilize mixed strategies and under what conditions it is most likely to occur.

Table 6.2 Necrophagous flies reported to display precocious egg development.

Species	Family	Frequency	Source of flies
Aldrichina grahami	Calliphoridae	1 egg/female	Laboratory
Calliphora loewi	Calliphoridae	1–3 larvae/female	Field
Calliphora nigribarbis	Calliphoridae	1 egg/female	Laboratory
Calliphoria stygia	Calliphoridae	Not determined	Laboratory
Calliphora terraenova	Calliphoridae	1 egg/female	Field
Calliphora vicina	Calliphoridae	1 egg/female	Field/laboratory
Calliphora vomitoria	Calliphoridae	1 egg/female	Field
Eucalliphora latrifons	Calliphoridae	1 egg/female	Laboratory
Lucilia sericata	Calliphorida	1 egg/female	Field

Source: Adapted from VanLaerhoven & Anderson (2001), Wells & King (2001), Prado e Castro *et al.* (2016), and Lutz & Amendt (2020).

6.3.4   Mating and oviposition

It is also worth noting that mating influences reproductive output. With some species, mated females produce more eggs than unmated (Adams & Hintz, 1969; Crystal, 1983; Hahn *et al.*, 2008b). At least with the sarcophagid *S. crassipalpis*, mating status (mated versus virgin) of the females influences other aspects of egg development such as onset of oogenesis and egg length or mass (Hahn *et al.*, 2008b). In other species, the stage of ovarian development, and hence oocyte maturation, affects either the receptivity of females to mating or the attractiveness of females to males (Strangways-Dixon, 1961; Crystal, 1983). The extent of mating status influence on oviposition and larviposition of calliphorids and sarcophagids has not been explored in many species, and thus remains largely unknown for the majority of these flies.

6.4   Larvae are adapted for feeding and competing on carrion

Competition to use human and animal corpses is perhaps most intense for necrophagous fly larvae. All nutriment for the larvae is derived from carrion, usually in the form of soft tissues. The carrion itself is a finite resource that is exploited by many different organisms other than just necrophagous flies. It is thus imperative that neonate fly larvae be equipped to compete effectively in the struggle for food in the harsh environment of animal remains. Several adaptations seem to facilitate larval utilization of the food resource (Figure 6.7). Adult flies can arrive within minutes after death and deposit eggs or larvae in locations that favor neonate larval feeding on tissues and liquids rich in protein (Rivers *et al.*, 2011). Following egg hatch (for oviparous species) or deposition of larvae (ovoviviparous), the larvae begin a period of intense feeding on soft tissues that is generally characterized by rapid and efficient food assimilation that contributes to accelerated growth rates (Ullyett, 1950; Roback, 1951). Coupled with a high metabolic rate and modified or unique digestive enzymes, released collectively for cooperative exodigestion (Scanvion *et al.*, 2018), the larvae can completely consume all soft tissues of a corpse in one generation (Greenberg & Kunich, 2002). Of course this depends on the size of the feeding aggregations that form on the corpse, ambient weather conditions, and other abiotic and biotic factors (Charabidze *et al.*, 2011; Rivers *et al.*, 2011). Chapter 8 examines larval feeding aggregations, providing a more detailed look into larval feeding.

Ullyett (1950) has argued that larval competition on carrion is the most important challenge faced by necrophagous flies in terms of reproductive success. As such, ovoviviparous or larviparous species would seemingly have a reproductive advantage over oviparous species, at least when protein is limited (Norris, 1965), since the deposited larvae are typically larger and can reach the most rapid phase of growth sooner than oviparous species

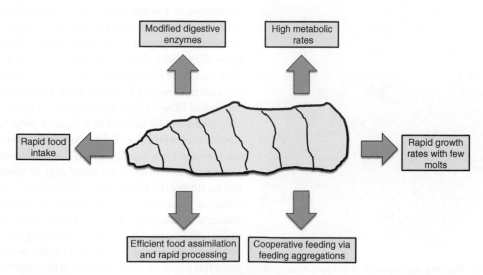

Figure 6.7   The larvae of necrophagous flies are adapted for competitive feeding battles.

(Levot *et al.*, 1979; Prinkkila & Hanski, 1995). These predictions do not always hold, as in many situations (e.g., larger carcasses) oviparous species outcompete ovolarviparous flies (Hanski, 1977; Levot *et al.*, 1979).

By comparison to other insects, necrophagous flies undergo few molts (two) during the larval stages. In general, this results in the larvae spending less time on the food, exposed to potential predators and parasites (Cianci & Sheldon, 1990), so that post-feeding development can begin earlier in the life cycle than observed with insects that require longer feeding windows (Zdárek & Sláma, 1972). The events of pupariation and pupation for most species occur while post-feeding larvae are buried in soil, usually several meters away from the corpse, conferring additional protection from predators and parasites.

6.5 Feeding aggregations maximize utilization of food source

With the exception of parasitic species, larvae of blow flies and flesh flies do not feed singly. In fact, many if not most cannot acquire sufficient nutriment when feeding alone to complete development (Rivers *et al.*, 2010). Rather, necrophagous flies tend to form large feeding aggregations or maggot masses on a corpse (Figure 6.8). The aggregations

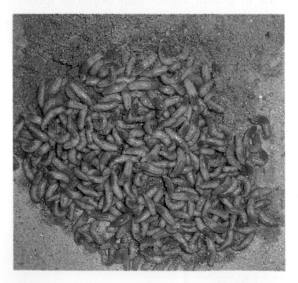

Figure 6.8 Feeding aggregation or maggot mass commonly found on carrion. *Source*: D.B. Rivers.

can be composed of mixed species representing several families of Diptera, but typically are dominated by calliphorids (Campobasso *et al.*, 2001). These maggot masses seem to facilitate maximum exploitation of the carcass as a food source, as the larvae are thought to participate in cooperative feeding (Greenberg & Kunich, 2002; Rivers *et al.*, 2011; Scanvion *et al.*, 2018). Cooperative feeding relies on manipulation of carcass tissues via the mouth hooks of hundreds to thousands of larvae in the aggregations and mass release of digestive enzymes (Greenberg & Kunich, 2002). Accelerated food acquisition and processing rely on heterothermic heat production associated with the internal environment of the feeding aggregations (Williams & Richardson, 1984; Slone & Gruner, 2007). The result is rapid rates of growth during the feeding stages (Hanski, 1977). In contrast, growth rates for the sarcophagid *Wohlfahrtia nuba* actually slow down with increases in larval density (Al-Misned, 2002). *Wohlfahrtia nuba* appears to be facultatively necrophagous; members of the genus *Wohlfahrtia* are predominantly parasitic (Lewis, 1955). This suggests that formation of larval feeding aggregations is an adaptive feature of necrophagous but not parasitic feeding among calliphorids and sarcophagids. An in-depth discussion of maggot masses can be found in Chapter 8.

6.6 Mother versus offspring: fitness conflicts

Utilization of a nutrient-rich yet ephemeral resource undoubtedly leads to oviposition decisions by gravid females that compromise progeny **fitness**. In fact, it is generally expected that gregarious exploitation of a finite resource like carrion will inevitably lead to an increase in **conspecific** competition and a decrease in individual fitness (Begon *et al.*, 1996). Fitness in this case is generically defined to indicate a propensity to survive and reproduce (Sober, 2001). For example, to compensate for a rapidly diminishing resource, flies are expected to oviposit far more eggs or larvae than can be supported by the carcass (Kneidel, 1984). The situation described is one in which a female fly attempts to maximize maternal fitness at the expense of her offspring's. The result is clustering of eggs by conspecifics during natural **faunal succession**, potentially leading to the formation of maggot masses composed of hundreds to thousands of individuals

Figure 6.9   Gravid females frequently cluster eggs with conspecifics and allospecifics during oviposition. *Source*: Courtesy of Susan Ellis.

(Figure 6.9) (Denno & Cothran, 1976; Hanski, 1977). It should be apparent that overcrowding would occur in the feeding aggregations.

Overcrowding means decreased food availability per individual, which in turn is expected to increase the length of larval development since it takes longer to acquire the critical nutrients associated with the next molt (Ullyett, 1950; Williams & Richardson, 1984) or to initiate pupariation (Zdárek & Sláma, 1972). If the feeding aggregations exceed a critical threshold density, larval development will be compromised due to depletion of nutrients and an accumulation of larval waste products (Levot *et al.*, 1979; Greenberg & Kunich, 2002; Rivers *et al.*, 2010). Some of the effects of overcrowding include slower growth rates, which may contribute to an increased chance of predation; reduced-size larvae that will yield smaller puparia and subsequent adults, which in turn may be less fecund; and potentially elevated mortality rates in all feeding and post-feeding stages (Figure 6.10) (Kamal, 1958; Cianci & Sheldon, 1990; Saunders & Bee, 1995; Erzinçlioglu, 1996).

The larvae may counter the mother's oviposition decisions through resource partitioning. Even if overcrowding is unavoidable, Kamal (1958) has shown that for several necrophagous fly species, extreme reductions in puparial size and in subsequent adults only modestly altered fecundity. Predation and parasitism on larvae aids in lowering larval density of maggot masses and thus may reduce the deleterious consequences of overcrowding. This latter example reflects an increase in larval fitness for those individuals not attacked and assuming the microhabitat of the feeding aggregation is not drastically altered, yet at the same time the mother's fitness is lowered as offspring are killed. Considering the potential pitfalls

outlined with overcrowding, why would the mother contribute to grouping or aggregations? One explanation is that formation of large aggregations, particularly when composed of conspecifics, reduces the likelihood of predator (or parasite) attack toward any single individual, thereby functioning as a predator avoidance strategy (Rohlfs & Hoffmeister, 2004). Under this view, though fitness for certain individuals (such as those located along the periphery of the maggot mass) is at risk, the overall fitness of the progeny, and hence the adult females contributing to the aggregations through oviposition, increases.

6.7   Resource partitioning is the path to reproductive success

As has been alluded to several times in this book, carrion is a nutrient-rich island, but as a patchy temporal resource its availability is neither predictable nor long-lasting. Consequently, intense intraspecific and interspecific competition exists among necrophagous organisms attempting to exploit a carcass. Necrophagous flies represent the dominant members of carrion communities in terms of species and individual abundance, and typically consume the largest portion of animal tissues (Hanski, 1987). Once oviposition occurs, whether it be by ovipary or vivipary, necrophagous fly larvae seemingly compete for the limited food resources in an identical manner. Rarely does coexistence occur between species competing identically for the same limited resource (Park, 1962). Rather, the species that is better adapted to use the resource will outcompete inferior competitors (Griffin & Silliman, 2011). Resource competition among calliphorid and sarcophagid larvae on carrion has been the subject of numerous investigations (Denno & Cothran, 1976; Levot *et al.*, 1979; Kneidel, 1984; Hanski, 1987; Ives, 1991). Indeed, many species of blow flies and flesh flies outcompete allospecifics and conspecifics, relying on such mechanisms as predation, cannibalism, and modified food acquisition and/or assimilation rates. Some species are even thought to alter the internal maggot mass temperatures to outcompete allospecifics (Richards *et al.*, 2009).

Despite the intense competition, coexistence does occur. Denno and Cothran (1976) have argued that the reproductive strategies employed

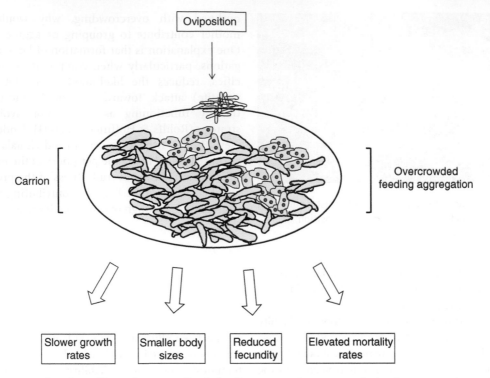

Figure 6.10   Deleterious effects of overcrowding in feeding aggregations on growth, development, and reproduction.

by calliphorids and sarcophagids results in **resource partitioning**. Resource partitioning is the idea that when at least two species are competing for the same limited resources, coexistence can occur by using the resource differently to suppress direct competition (Griffin & Silliman, 2011). From discussions earlier in this chapter, it should be apparent that differences in oviposition mechanisms (ovipary versus vivipary) contribute to resource partitioning, at least in terms of timing of progeny deposition. Spatial differences in terms of oviposition preferences (i.e., tissue specificity or body openings for egg laying) are common among calliphorid species and contribute to resource partitioning. Within heterogeneous larval aggregations (i.e., mixed species), spatial partitioning may also result depending on thermal tolerances of the fly larvae at a particular developmental stage (Richards *et al.*, 2009). Spatial and temporal patterns of oviposition (e.g., seasonal, geographic location) appear to also contribute to resource partitioning, as well as oviposition preferences based on carcass size (Denno & Cothran, 1975; Hanski & Kuusela, 1980).

Chapter review

The need to feed: anautogeny and income breeders are common among necrophagous Diptera

- Reproduction in calyptrate Diptera is highly evolved and complex, with as many "exceptions" and modifications as general rules to the process. The common ground with regard to forensically important flies is the need for carrion to reproduce.
- Several species of calliphorids are autogenous, simply meaning that the adult female is capable of producing eggs upon emergence. This condition is common among holometabolous insects since there is a period of non-feeding, the pupal stage, between the immature stages and the imago.
- Many sarcophagid species and some calliphorids that frequent carrion are anautogenous, meaning that they do not have the necessary nutrient

reserves stored from the larval stages to provision eggs. Rather, anautogenous insects must obtain protein through feeding as adults to produce mature oocytes. Typically the protein meal is necessary for the events of oogenesis and vitellogenesis to proceed.

- Switching between autogeny and anautogeny apparently occurs across broods depending on the nutritional conditions faced by the larvae of some fly species.
- A broader view of nutrient requirements is based on total nutrient or energy needs rather than a focus on protein. In this context, necrophagous flies can be classified as either capital or income breeders depending on whether adult feeding is necessary to mature eggs. Capital breeding is predisposed toward storing energy in the form of nutrient reserves during the immature stages so that as adults, egg provisioning can begin almost immediately, whereas income breeders must feed as adults to produce eggs.

Size matters in egg production

- The size of the egg is influenced by the available nutrients to the mother either as a larva (capital investment) or as an imago (income allocation). Low nutrient availability is expected to result in fewer smaller eggs, while nutrient-rich diets should yield larger clutches, possibly with large individual eggs.
- The influence of diet on egg size is expected to be greatest with flies that lay large clutches all at once as occuring in ovoviviparous sarcophagids, as opposed to species that deposit eggs singly or asynchronously. Clutch layers make a large investment in resource allocation in a short period of time, so limitations in nourishment in general would seemingly result in smaller eggs.
- The method of progeny deposition seems to yield characteristic egg sizes as well. What this means is that for those species that lay eggs so that embryonic development mostly occurs outside the mother, a condition termed oviparity, females typically lay large numbers of relatively small eggs. Resource allocation is generally less per individual egg than with flies that deposit progeny via some form of vivipary.

Progeny deposition is a matter of competition

- Deposition of progeny among necrophagous flies can generally be viewed as part of a reproductive strategy that attempts to exploit the nutrient-rich environment of carrion. Since animal remains are the primary sources of the protein and other nutrients needed to mature oocytes, the timing of oviposition or larviposition by calliphorids and sarcophagids is absolutely critical to attain the necessary protein from the resource.
- Oviposition strategies used by necrophagous flies depend on specific mechanisms for deposition of eggs into the environment, or retention of eggs so that embryonic development and egg hatch occur within the parent.
- Vivipary describes the retention of eggs in the reproductive tract until hatching or the eggs hatch as they are passed out of the mother. The net effect is deposition of larvae rather than eggs, and hence these insects are said to have live birth. Significant embryonic development occurs while the eggs are maintained in the parent as opposed to oviparity, in which most aspects of embryonic development occur outside the mother's body. Two forms of vivipary (viviparity and ovoviviparity) generally occur with insects, although only ovoviviparity is commonly encountered with necrophagous flies.
- Ovoviviparity involves the mother retaining the eggs within the reproductive tract, usually in an ovisac or common oviduct, until egg hatch. The egg hatches within the parent either immediately before deposition or at some prescribed time earlier that requires the mother to find a suitable substrate for deposition or else face being consumed by her offspring. Ovoviviparity is essentially oviparity (eggs with a chorion) with the exception that eggs hatch inside the mother.
- A few species of flies rely on a combination of oviparity and ovoviviparity. Some species of *Musca* (Family Muscidae) that are normally oviparous can retain eggs until hatch to deposit live larvae. Similarly, some oviparous calliphorids and sarcophagids retain fertilized eggs in the common oviduct while searching for an appropriate oviposition site.
- With some species, mated females produce more eggs than unmated. Mating status (mated versus

virgin) of the females may also influence other aspects of egg development, such as onset of oogenesis and egg length or mass. In other species, the stage of ovarian development, and hence oocyte maturation, affects either the receptivity of females to mating or the attractiveness of females to males.

Larvae are adapted for feeding and competing on carrion

- Competition to use human and animal corpses is perhaps most intense for necrophagous fly larvae. All nutriment for the larvae is derived from carrion, usually in the form of soft tissues.
- Several adaptations seem to facilitate larval utilization of the food resource and include gravid females ovipositing in areas that favor larval feeding, rapid and efficient food assimilation that contributes to rapid growth rates, high metabolic rates, modified or unique digestive enzymes, and few larval molts.

Feeding aggregations maximize utilization of food source

- Larvae of blow flies and flesh flies do not feed singly. In fact, many if not most cannot acquire sufficient nutriment when feeding alone to complete development. Rather, necrophagous flies tend to form large feeding aggregations or maggot masses on a corpse that can be composed of species from several families of Diptera.
- Maggot masses seem to facilitate maximum exploitation of the carcass as a food source, as the larvae are thought to participate in cooperative feeding. Cooperative feeding relies on manipulation of carcass tissues via the mouth hooks of hundreds to thousands of larvae in the aggregations and mass release of digestive enzymes.
- Accelerated food acquisition and processing rely on heterothermic heat production associated with the internal environment of the feeding aggregations. The result is rapid rates of growth during the feeding stages.

Mother versus offspring: fitness conflicts

- Utilization of a nutrient-rich yet ephemeral resource undoubtedly leads to oviposition decisions by gravid females that compromise

progeny fitness. In fact it is generally expected that gregarious exploitation of a finite resource such as carrion will inevitably lead to an increase in conspecific competition and a decrease in individual fitness.
- To compensate for a rapidly diminishing resource, flies are expected to oviposit far more eggs or larvae than can be supported by the carcass. The situation described is one in which a female fly attempts to maximize maternal fitness at the expense of her offspring's. In practice what occurs is clustering of eggs by conspecifics and allospecifics, frequently leading to overcrowding in feeding aggregations.
- Overcrowding means decreased food availability per individual, which in turn is expected to increase the length of larval development since it takes a longer time to acquire the critical nutrients associated with the next molt or to initiate pupariation. Some of the effects of overcrowding include slower growth rates, which may contribute to an increased chance of predation, reduced-size larvae that will yield smaller puparia and subsequent adults, which in turn may be less fecund, and potentially elevated mortality rates in all feeding and postfeeding stages.
- The larvae may counter the mother's developmental decisions through resource partitioning. Even if overcrowding is unavoidable, extreme reductions in puparial size and in subsequent adults only modestly alter fecundity with many fly species. Predation and parasitism on larvae aids in lowering individual density of maggot masses and thus may reduce the deleterious consequences of overcrowding.

Resource partitioning is path to reproductive success

- Carrion is a nutrient-rich island but is a patchy temporal resource, and consequently intense intraspecific and interspecific competition exists among necrophagous organisms attempting to exploit a carcass. Once oviposition occurs, whether it be by ovipary or vivipary, necrophagous fly larvae seemingly compete for the limited food resources in an identical manner. Rarely does coexistence occur between species competing identically for the same limited resource. Rather, the species that is better adapted to use the resource will outcompete inferior competitors.

- Many species of blow flies and flesh flies outcompete allospecifics and conspecifics relying on such mechanisms as predation, cannibalism, and modified food acquisition and/or assimilation rates. Some species are even thought to alter the internal maggot mass temperatures to outcompete allospecifics.
- Despite intense competition, coexistence does occur among carrion-breeding flies. Calliphorids and sarcophagids employ reproductive strategies that promote resource partitioning. Differences in oviposition mechanisms (ovipary versus vivipary), spatial differences in terms of oviposition preferences on a carcass, spatial partitioning among larvae, spatial and temporal patterns of oviposition (e.g., seasonal, geographic location), and oviposition preferences based on carcass size are part of reproductive strategies that contribute to resource partitioning.

Test your understanding

Level 1: knowledge/comprehension

1. Define the following terms:

   (a) autogeny
   (b) anautogeny
   (c) fitness
   (d) ovoviviparity
   (e) oviparity
   (f) ephemeral
   (g) capital breeder

2. Match the terms (i–vi) with the descriptions (a–f).

   (a) Individuals that belong to the same species   (i) Chorion
   (b) Condition in which a protein meal as an adult is not required to provision eggs   (ii) Income breeder
   (c) Progeny deposition in which egg hatch occurs in parent and mother provides nutriment to neonate larvae   (iii) Vitellogenesis
   (d) Process of yolk formation   (iv) Conspecific
   (e) Outer membrane of insect egg   (v) Viviparity
   (f) Adult fly that requires nutrient acquisition to produce oocytes   (vi) Autogeny

3. Explain how the terms autogeny, anautogeny, income and capital breeders have similar meanings but describe different aspects of egg provisioning.
4. Under what conditions would the fitness of a gravid female be favored over her progeny's fitness? Describe a scenario in which the offspring's fitness is favored over the mother's.
5. Explain the importance of protein thresholds to egg provisioning.
6. Compare and contrast oviparity and ovoviviparity as reproductive strategies used by necrophagous flies.

Level 2: application/analysis

1. Some fly species display mixed strategies in terms of progeny deposition, using a combination of oviparity and vivipary. For those species that produce viable eggs and larvae, would a conflict be expected between mother and offspring in terms of fitness? Explain why or why not.
2. Describe some potential physiological adaptations that larvae from oviparous species might possess that would allow them to compete with ovolarviparous species for nutrients on a small carcass.
3. Explain what the implications of precocious egg development might be in terms of estimation of the postmortem interval based on insect development.

Level 3: synthesis/evaluation

1. True viviparity does not occur with necrophagous calliphorids and sarcophagids. Speculate as to why this form of vivipary is not advantageous to necrophagous flies as opposed to ovoviviparity or oviparity.

Note

1. The larval-mass effect is the idea that depending on the number of individual fly larvae in a feeding aggregation and environmental temperatures, larvae release heat that has the potential to increase the local temperature of the mass and surrounding habitat.

References cited

Adams, T.S. & Hintz, A.M. (1969) Relationship of age, ovarian development, and the corpus allatum to mating in the housefly, *Musca domestica. Journal of Insect Physiology* 15: 201–215.

Al-Misned, F.A.M. (2002) Effects of larval population density on the life cycle of flesh fly, *Wohlfahrtia nuba* (Wiedemann) (Diptera: Sarcophagidae). *Saudi Journal of Biological Science* 9: 140–147.

Barton Browne, L., Bartell, R.J., van Gerwen, A.C.M. & Lawrence, L.A. (1976) Relationship between protein ingestion and sexual receptivity in females of the Australian sheep blowfly *Lucilia cuprina*. *Physiological Entomology* 1: 235–240.

Baxter, J.A., Mjeni, A.M. & Morrison, P.E. (1973) Expression of autogeny in relation to larval population density of *Sarcophaga bullata* Parker (Diptera: Sarcophagidae). *Canadian Journal of Zoology* 51: 1189–1193.

Begon, M.E., Harper, J.L. & Townsend, C.R. (1996) *Ecology: Individuals, Populations, and Communities.* Blackwell Publishing Ltd., Oxford.

Bonnet, X., Bradshaw, D. & Shrine, R. (1998) Capital versus income breeding: an ectothermic perspective. *Oikos* 83: 333–342.

Campobasso, C.P., Di Vella, G. & Introna, F. (2001) Factors affecting decomposition and Diptera colonization. *Forensic Science International* 120: 18–27.

Chapman, R.F. (1998) *The Insects: Structure and Function,* 4th edn. Cambridge University Press, Cambridge, UK.

Charabidze, D., Bourel, B. & Gosset, D. (2011) Larval-mass effect: characterization of heat emission by necrophagous blowflies (Diptera: Calliphoridae) larval aggregates. *Forensic Science International* 211: 61–66.

Cianci, T.J. & Sheldon, J.K. (1990) Endothermic generation by blowfly larvae *Phormia regina* developing in pig carcasses. *Bulletin of the Society of Vector Ecology* 15: 33–40.

Cook, D.F. & Dadour, I.R. (2011) Larviposition in the ovoviviparous blowfly *Calliphora dubia*. *Medical and Veterinary Entomology* 25: 53–57.

Crystal, M.M. (1983) Effect of age and ovarian development on mating in the black blow fly (Diptera: Calliphoridae). *Journal of Medical Entomology* 20: 220–221.

Denlinger, D.L. & Ma, W.-C. (1974) Dynamics of the pregnancy cycle in the tsetse *Glossina morsitans*. *Journal of Insect Physiology* 20: 1015–1026.

Denno, R.F. & Cothran, W.R. (1975) Niche relationships of a guild of necrophagous flies. *Annals of the Entomological Society of America* 68: 741–754.

Denno, R.F. & Cothran, W.R. (1976) Competitive interactions and ecological strategies of sarcophagid and calliphorid flies inhabiting rabbit carrion. *Annals of the Entomological Society of America* 69: 109–113.

Erzinçlioglu, Z. (1996) *Blowflies*. Richmond Publishing Co., Ltd, Slough, UK.

Greenberg, B. & Kunich, J.C. (2002) *Entomology and the Law*. Cambridge University Press, Cambridge, UK.

Griffin, J.N. & Silliman, B.R. (2011) Resource partitioning and why it matters. *Nature Education Knowledge* 2(1): 8.

Hahn, D.A., James, L.N., Milne, K.R. & Hatle, J.D. (2008a) Life-history plasticity after attaining a dietary threshold for reproduction is associated with protein storage in flesh flies. *Functional Ecology* 22: 1081–1090.

Hahn, D.A., Rourke, M.N. & Milne, K.R. (2008b) Mating affects reproductive investment into eggs, but not the timing of oogenesis in the flesh fly *Sarcophaga crassipalpis*. *Journal of Comparative Physiology B* 178: 225–233.

Hammack, L. (1999) Stimulation of oogenesis by proteinaceous adult diets for screwworm, *Cochliomyia hominivorax* (Diptera: Calliphoridae). *Bulletin of Entomological Research* 89: 433–440.

Hanski, I. (1977) An interpolation model of assimilation by larvae of the blowfly, *Lucilia illustris* (Calliphoridae) in changing temperatures. *Oikos* 28: 187–195.

Hanski, I. (1987) Carrion fly community dynamics: patchiness, seasonality and coexistence. *Ecological Entomology* 12: 257–266.

Hanski, I. & Kuusela, S. (1977) An experiment on competition and diversity in the carrion fly community. *Annals of the Entomological Society of Finland* 43: 108–115.

Hanski, I. & Kuusela, S. (1980) The structure of carrion fly communities: differences in breeding seasons. *Annales Zoologici Fennici* 17: 185–190.

Harlow, P.M. (1956) A study of ovarian development and its relation to adult nutrition in the blowfly *Protophormia terraenovae* (R.D.). *Journal of Experimental Biology* 33: 777–797.

Hayes, E.J., Wall, R. & Smith, K.E. (1999) Mortality rate, reproductive output, and trap response bias in populations of the blowfly *Lucilia sericata*. *Ecological Entomology* 24: 300–307.

Houston, A.I., Stephens, P.A., Boyd, I.L., Harding, K.C. & McNamara, J.M. (2007) Capital or income breeding? A theoretical model of female reproductive strategies. *Behavioral Ecology* 18: 241–250.

Huntington, T.E. & Higley, L.G. (2010) Decomposed flesh as a vitellogenic protein source for the forensically important *Lucilia sericata* (Diptera: Calliphoridae). *Journal of Medical Entomology* 47: 482–486.

Ives, A.R. (1991) Aggregation and coexistence in a carrion fly community. *Ecological Monographs* 61: 75–94.

Jervis, M.A., Boggs, C.L. & Ferns, P.N. (2007) Egg maturation strategy and survival trade-offs in holomemtabolous insects: a comparative approach. *Biological Journal of the Linnean Society* 90: 293–302.

Jönsson, K.I. (1997) Capital and income breeding as alternative tactics of resource use in reproduction. *Oikos* 78: 57–66.

Kamal, A.S. (1958) Comparative study of thirteen species of sarcosaprophagous Calliphorida and Sarcophagidae (Diptera). I. Bionomics. *Annals of the Entomological Society of America* 51: 261–270.

Kneidel, K.A. (1984) Competition and disturbance in communities of carrion-breeding Diptera. *Journal of Animal Ecology* 53: 849–865.

Levot, G.W., Brown, K.R. & Shipp, E. (1979) Larval growth of some calliphorid and sarcophagid Diptera. *Bulletin of Entomological Research* 69: 469–475.

Lewis, D.J. (1955) Calliphoridae of medical interest in the Sudan (Diptera). *Bulletin of the Entomological Society of Egypt* 39: 275–296.

Lutz, L. & Amendt, J. (2020) Precocious egg development in wild *Calliphora vicina* (Diptera: Calliphoridae) – an issue of relevance in forensic entomology? *Forensic Science International* 306: 110075.

Norris, K.R. (1965) The bionomics of blowflies. *Annual Review of Entomology* 10: 47–68.

Norris, K.R. (1994) Three new species of Australian "Golden blowflies" (Diptera: Calliphoridae: Calliphora), with a key to described species. *Invertebrate Taxonomy* 8: 1343–1366.

Pappas, C. & Fraenkel, G. (1977) Nutritional aspects of oogenesis in the flies *Phormia regina* and *Sarcophaga bullata*. *Physiological Zoology* 50: 237–246.

Park, T.B. (1962) Competition and populations. *Science* 138: 1369–1375.

Prado e Castro, C., Szpila, K., Rego, C., Boierro, M. & Serrano, A.R.M. (2016) First finding of larviposition in *Calliphora loewi* from an island relict forest. *Entomological Science.* https://doi.org/10.1111/ens.12163.

Prinkkila, M.-L. & Hanski, I. (1995) Complex competitive interactions in four species of *Lucilia* blowflies. *Ecological Entomology* 20: 261–272.

Richards, C.S., Price, B.W. & Villet, M.H. (2009) Thermal ecophysiology of seven carrion-feeding blowflies in Southern Africa. *Entomologia Experimentalis et Applicata* 131: 11–19.

Rivers, D.B., Ciarlo, T., Spelman, M. & Brogan, R. (2010) Changes in development and heat shock response in two species of flies (*Sarcophaga bullata* [Diptera: Sarcophagidae] and *Protophormia terraenovae* [Diptera: Calliphoridae]) reared in different sized maggot masses. *Journal of Medical Entomology* 47: 677–689.

Rivers, D.B., Thompson, C. & Brogan, R. (2011) Physiological trade-offs of forming maggot masses by necrophagous flies on vertebrate carrion. *Bulletin of Entomological Research* 101: 599–611.

Roback, S.S. (1951) A classification of the muscoid calyptrate Diptera. *Annals of the Entomological Society of America* 44: 327–361.

Rohlfs, M. & Hoffmeister, T.S. (2004) Spatial aggregation across ephemeral resource patches in insect communities: an adaptive response to natural enemies? *Oecologia* 140: 654–661.

Saunders, D. & Bee, A. (1995) Effects of larval crowding on size and fecundity of the blowfly *Calliphora vicina* (Diptera: Calliphoridae). *European Journal of Entomology* 92: 615–622.

Scanvion, Q., Hédouin, V. & Charabidzé, D. (2018) Collective exodigestion favours blow fly colonization and development on fresh carcasses. *Animal Behaviour* 141: 221–232.

Shewell, G.E. (1987a) Calliphoridae. In: J.F. McAlpine (ed) *Manual of Nearctic Diptera*, Vol. 2, pp. 1133–1146. Ariculture Canada, Ottawa.

Shewell, G.E. (1987b) Sarcophagidae. In: J.F. McAlpine (ed) *Manual of Nearctic Diptera*, Vol. 2, pp. 1159–1186. Ariculture Canada, Ottawa.

Slone, D.H. & Gruner, S.V. (2007) Thermoregulation in larval aggregations of carrion-feeding blow flies (Diptera; Calliphoridae). *Journal of Medical Entomology* 44: 516–523.

Smith, K.G.V. (1986) *A Manual of Forensic Entomology*. British Museum (Natural History), London.

Sober, E. (2001) The two faces of fitness. In: R. Singh, D. Paul, C. Krimbas & J. Beatty (eds) *Thinking About Evolution: Historical, Philosophical, and Political Perspectives*, pp. 309–321. Cambridge University Press, Cambridge, UK.

Stoffolano J.G., Mei-Fang, L.I., Sutton, J.A. & Yin, C.-M. (1995) Faeces feeding by adult *Phormia regina* (Diptera: Calliphoridae): impact on reproduction. *Medical and Veterinary Entomology* 9: 388–392.

Strangways-Dixon, J. (1961) The relationships between nutrition, hormones and reproduction in the blowfly *Calliphora erythrocepala* (Meig.). *Journal of Experimental Biology* 38: 637–646.

Tanaka, S., Guardia, M., Denlinger, D.L. & Wolda, H. (1990) Relationships between body size, reproductive traits, and food resources in 3 species of tropical flesh flies. *Research in Population Ecology* 32: 303–317.

Thomas, D.B. & Mangan, R.L. (1989) Oviposition and wound-visiting behavior of the screwworm fly, *Cochliomyia hominivorax* (Diptera: Calliphoridae). *Annals of the Entomological Society of America* 82: 526–534.

Ullyett, G.C. (1950) Competition for food and allied phenomena in sheep blowfly populations. *Philosophical Transactions of the Royal Society of London, Series B* 234: 77–175.

VanLaerhoven, S.L. & Anderson, G.S. (2001) Implications of using developmental rates of blow fly (Diptera: Calliphoridae) eggs to determine postmortem interval. *Journal of the Entomological Society of British Columbia* 98: 189–194.

Villet, M.H., Richards, C.S. & Midgley, J.M. (2010) Contemporary precision, bias, and accuracy of minimum post-mortem intervals estimated using development of carrion-feeding insects. In: J. Amendt, C.P. Campobasso, M.L. Goff & M. Grassberger (eds) *Current Concepts in Forensic Entomology*, pp. 109–137. Springer, London.

Wall, R., Wearmouth, V.J. & Smith, K.E. (2002) Reproduction allocation by the blow fly *Lucilia sericata* in response to protein limitation. *Physiological Entomology* 27: 267–274.

Webber, L.G. (1958) Nutrition and reproduction in the Australian sheep blowfly *Lucilia cuprina*. *Australian Journal of Zoology* 6: 139–144.

Wells, J.D. & King, J. (2001) Incidence of precocious egg development in flies of forensic importance (Calliphoridae). *Pan-Pacific Entomologist* 77: 235–239.

Wessels, F.J., Jordan, D.C. & Hahn, D.A. (2010) Allocation from capital and income sources to reproduction shift from first to second clutch in the flesh fly, *Sarcophaga crassipalpis*. *Journal of Insect Physiology* 56: 1269–1274.

Williams, H. & Richardson, A.M.M. (1984) Growth energetics in relation to temperature for larvae of four species of necrophagous flies (Diptera: Calliphoridae). *Australian Journal of Ecology* 9: 141–152.

Williams, K.L., Barton-Browne, L. & van Gerwen, A.C.M. (1979) Quantitative relationships between ingestion of protein rich material and ovarian development in the Australian sheep blowfly, *L. cuprina*. *International Journal of Invertebrate Reproduction* 1: 75–88.

Willmer, P., Stone, G. & Johnston, I. (2000) *Environmental Physiology of Animals*. Blackwell Publishing Ltd., Oxford.

Yin, C.-M., Qin, W.-H. & Stoffolano, J.G. (1999) Regulation of mating behavior by nutrition and the corpus allatum in both male and female *Phormia regina* (Meigen). *Journal of Insect Physiology* 45: 815–822.

Zdárek, J. & Sláma, K. (1972) Supernumerary larval instars in cyclorrhaphous Diptera. *Biological Bulletin* 142: 350–357.

dos Reis, S.F., von Zuben, C.J. & Godoy, W.A.C. (1999) Larval aggregation and competition for food in experimental populations of *Chrysomya putoria* (Wied.) and *Cochliomyia macellaria* (F.) (Dipt., Calliphoridae). *Journal of Applied Entomology* 123: 485–489.

Greenberg, B. (1990) Behavior of postfeeding larvae of some Calliphoridae and a muscid (Diptera). *Annals of the Entomological Society of America* 83: 1210–1214.

Ireland, S. & Turner, B. (2006) The effects of larval crowding and food type on the size and development of the blowfly, *Calliphora vomitoria*. *Forensic Science International* 159: 175–181.

Kneidel, K.A. (1985) Patchiness, aggregation and the coexistence of competitors for ephemeral resources. *Ecological Entomology* 10: 441–448.

Parrish, J.K. & Edelstein-Keshet, L. (1999) Complexity, pattern, and evolutionary trade-offs in animal aggregation. *Science* 284: 99–101.

VanLaerhoven, S.L. (2010) Ecological theory and its application to forensic entomology. In: J.H. Byrd & J.L. Castner (eds) *Forensic Entomology: The Utility of Arthropods in Legal Investigations*, 2nd edn, pp. 493–518. CRC Press, Boca Raton, FL.

Supplemental reading

Barton Browne, L. (1993) Physiologically induced changes in resource-oriented behavior. *Annual Review of Entomology* 38: 1–25.

Clark, K., Evans, L. & Wall, R. (2006) Growth rates of the blowfly, *Lucilia sericata*, on different body tissues. *Forensic Science International* 156: 145–149.

Additional resources

Bloomington *Drosophila* Stock Center: http://flystocks.bio.indiana.edu/

Dipterists Forum: https://dipterists.org.uk/home

International Society of Behavioral Ecology: www.behavecol.com

North American Dipterists Society: https://dipterists.org/

Chapter 7
Chemical attraction and communication

Overview

The system of chemical detection and signaling used by insects for intraspecific and interspecific communication is perhaps the most refined among all animal groups. This impressive system is manifested through the insect's ability to detect minute amounts of specific signals, despite a milieu of "environmental noise" (i.e., array of chemical compounds), and then orient themselves to find the source of emission. For some species, the chemical signal may originate in relatively close proximity, yet for others detection happens only after the chemical has traveled an incredibly long distance. The mechanisms for emission of chemical signals are equally impressive to that of detection, as several species have the capacity to communicate not only across taxa but participate in interkingdom signaling. Most features of this highly efficient system are on display during the formation of carrion communities: insects can detect various chemicals emanating directly from carrion that reflect stages of decay, pheromonal signaling is used for intraspecific communication, and allelochemicals are employed to modify the behavior and physiology of non-related insects and even

other organisms. This chapter will lay a foundation for understanding concepts of chemical ecology and the chemicals (semiochemicals) used for communication between and among organisms, particularly those that frequent carrion. A discussion of what is known about the attraction of necrophagous insects, particularly flies, to a corpse will be presented.

The big picture

- Insects rely on chemicals in intraspecific and interspecific communication.
- Chemical communication requires efficient chemoreception.
- Semiochemicals modify the behavior of the receiver.
- Pheromones are used to communicate with members of the same species.
- Allelochemicals promote communication across taxa and kingdoms.
- Chemical attraction to carrion by initial colonizers.
- Chemical attraction to carrion by subsequent fauna.
- Chemical attraction to bacteria on carrion.

The Science of Forensic Entomology, Second Edition. David B. Rivers and Gregory A. Dahlem.
© 2023 John Wiley & Sons Ltd. Published 2023 by John Wiley & Sons Ltd.
Companian website: www.wiley.com/go/forensicentomology2

7.1 Insects rely on chemicals in intraspecific and interspecific communication

The ability to sense and respond to chemicals present in the environment is inherent to almost all living organisms, regardless of the level of complexity (Vickers, 2000). By comparison to other animals, insects as a whole are more dependent on chemical stimuli for **interspecific** and **intraspecific** communication (Gullan & Cranston, 2014). Carrion-inhabiting insects, like all arthropods, can perceive their environment using a variety of visual, auditory, and chemical stimuli. However, the microcosm of a dead animal is particularly suited for chemical communication. This is true in terms of long-distance detection of chemicals emanating from a decomposing body by female flies desperate to obtain a protein meal or to release a clutch of eggs, or in the use of chemicals to signal conspecifics to mass oviposit or to attract a mate to the patchy resource that will serve as food for the insect couple's offspring (Tomberlin et al., 2011; Brodie et al., 2015). The odor profile may also reveal that unwanted "guests" are present, such as allospecific fly species that can reduce female fitness through competition, predation, or by other means (Brundage et al., 2017; Spindola et al., 2017). Predatory and parasitic insects can intercept intraspecific signals to locate necrophagous species or some simply zero in on the same odors of decomposition used by **saprophagous** insects to find the animal carcass that is home to thousands of individuals that may well become their next meal or host (Gião & Godoy, 2007). Several species of microorganisms seem to essentially function akin to a telemarketer[1] in that they relay information regarding a corpse to "interested" flies, thereby attracting them to the remains for mating, feeding, and oviposition (Ma et al., 2012; Tomberlin et al., 2012). No doubt this leads to payment for the services via transport (mechanical dispersal or inoculation) of the microorganisms to other ephemeral resources. The fact is carrion insects rely on a wide range of chemical signals to detect and relay information about the food resource and its occupants to members of the same species (intraspecific) or to those of another insect taxa (interspecific) or even an entirely different group of organisms, which could be in another kingdom (interkingdom) or domain.

The chemicals used for communication are broadly referred to as **semiochemicals**. Semiochemicals can be simply defined as chemicals that modify the behavior and/or physiology of the recipient. A wide range of semiochemicals are used to convey messages among and between insects and other organisms, ranging from highly volatile, complex blends of compounds to those with limited volatility that can only be perceived through direct contact between the emitter and receiver. These chemical signals are discussed in more detail in the remainder of this chapter.

7.2 Chemical communication requires efficient chemoreception

As Section 7.1 pointed out, insects depend on chemicals more than most other animal groups to communicate. One way this is apparent is simply in the wide range of semiochemicals that they produce and the elaborate network of glands and ducts used to synthesize and distribute these chemicals to the outside environment. A second piece of evidence revealing the importance of chemicals to communication is the presence of many varied receptors. Stated another way, efficient chemoreception (i.e., detection of chemical stimuli in the environment) is absolutely essential to any organism dependent on chemical signals for communicating messages (Vickers, 2000). Chemoreception is often broadly classified based on environment. For example, the ability of terrestrial insects to detect chemical signals is referred to as **olfaction** or the sense of smell. Alternatively, aquatic insects rely more on a sense of taste, otherwise referred to as contact detection or **gustation** (Gullan & Cranston, 2014). As with most classification schemes involving insects, static definitions are not satisfactory for describing all situations, and chemoreception is no different. Olfactory receptors do occur with some aquatic insects, and taste or gustatory reception is prevalent with most terrestrial species.

An in-depth discussion of how chemoreceptors operate in response to so many different stimuli lies beyond the scope of this textbook. However, it is important to have a basic understanding of chemoreception to place aspects of carrion ecology like chemical attraction to a corpse and oviposition and feeding stimulants into proper context. Thus, we will examine the basic ideas associated with chemical detection in insects to provide a framework for later discussions related to semiochemical functions. Our starting point is with the chemical receptors themselves, typically referred to as **sensilla** (sensillum in the singular). Sensilla generally have the appearance of fine thin hairs (setae) with one or more pore openings to allow chemical penetration (Chapman, 1998). Alternatively, some sensilla are short, squat, and broad yielding a peg-like morphology and still others have the shape of plates or depressions (pits) along the cuticular surface (Gullan & Cranston, 2014).

Regardless of the physical appearance of the receptor, one or more sensory neurons are associated with each sensillum. Dendrites extend the length of the exposed sensillum, either passing through the pore openings or lying in close proximity to the slits along the walls of the receptor (Figure 7.1). The number of pores as well as the shape of the sensillum reflects the type of chemoreception. In the case of olfaction, sensilla are either hair-like or peg-like in appearance and each sensillum has numerous pore openings along the thin walls of the receptor. Olfactory chemoreceptors are referred to as multiporous sensilla (Gullan & Cranston, 2014). Gustatory receptors more commonly represent the entire range of sensillum morphologies (hair, peg, plates, and pits) but possess only a single opening for chemical penetration. Hence, receptors associated with taste are termed uniporous sensilla (Gullan & Cranston, 2014). Cell bodies of the sensory neurons as well as the axons are buried deeper in the cuticle (Chapman, 1998).

Functionally, all sensilla detect chemical stimuli in a very similar fashion. Perception of chemical stimuli often begins with the trapping of chemical substances just inside the pore openings. The chemical substance is then moved to a recognition site along the dendritic membrane. In the family Calliphoridae (e.g., blow flies, bottle flies, and screwworms), the adult antennae possess three types of receptor protein gene families that bind odor molecules. These include odorant, gustatory, and ionotropic receptors (Leitch *et al.*, 2015). Binding at the recognition site does not guarantee further action: a threshold must be surpassed that often requires binding of multiple chemical molecules (Chapman, 2003). The next steps should come as no surprise to anyone who has studied the basics of excitable membranes: depolarization along the dendritic membrane occurs, leading to the generation of an action potential, which then travels the surface of the neuron cell body and axons to the presynaptic membranes at the axonal terminals. The subsequent events are not really relevant to our study of chemical communication. What is important to consider is that the location of the sensilla on the insect's body has a profound influence on sensitivity of detection (Gullan & Cranston, 2014), and hence the ability to achieve threshold binding of chemical stimuli to initiate the biochemical events associated with excitation of sensory neurons. Accordingly, sensilla associated with olfaction are usually located in a forward position on the body in relation to locomotion. Thus, olfactory receptors are commonly found in anterior regions such as antennae, head, and portions of the prothorax. In the case of adult blow flies, olfactory receptor neurons are primarily located on the antennae and maxillary palps (Hallem *et al.*, 2004) (Figure 7.2). Those used for taste can be found on mouthparts, legs, posterior abdominal segments, and even the ovipositor (Gullan & Cranston, 2014).

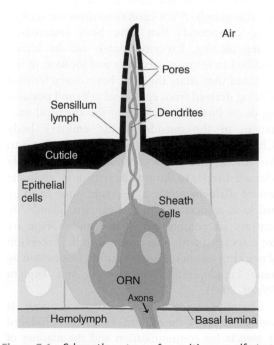

Figure 7.1 Schematic cartoon of a multiporous olfactory (basiconic) sensillum. *Source*: AlphaPsi/Wikimedia Commons/ CC BY-SA 4.0.

Figure 7.2 (a) Olfactory receptor neurons are located primarily on the antennae and maxillary palps of adult flies. *Source*: Richard Bartz/Wikimedia Commons/CC BY-SA 2.5. (b) Electronic micrograph of a fly antenna revealing a forest of chemosensory hairs. *Source*: CSIRO/Wikimedia Commons / CC BY 3.0.

7.3 Semiochemicals modify the behavior of the receiver

Olfactory chemoreceptors are used to detect the chemical signals found in the environment. As mentioned earlier in the chapter, these chemicals are called semiochemicals (from the Greek word *semion* or *semeon*, which literally means "signal"). Semiochemicals are used for communication in the external environment, as opposed to internally, where hormones are the major chemical messengers, to convey messages in intraspecific and interspecific communication. Chemical signals used to communicate with conspecifics are called **pheromones**, while those utilized in communication with allospecifics are broadly referred to as **allelochemicals**. As discussed in Sections 7.4 and 7.5, both groups of semiochemicals can be further subdivided by the behavioral and/or physiological changes imposed on the receiver of the message.

Almost all semiochemicals are produced in exocrine glands. A few exceptions do occur such as those compounds that have been sequestered during feeding. Exocrine glands can be highly modified in terms of structure and location in the body, but they share the same basic characteristics of being derived from epidermal cells and possessing ducts that allow release of the chemical substances to the outside of the emitter's body (Chapman, 1998). Salivary glands do not adhere to this rule as salivary secretions are not typically released directly to the outside environment. In the case of allelochemicals, a reservoir, often lined by cuticle, is associated with the exocrine glands. By contrast, pheromone production and storage are not associated with a reservoir, with the exception of marking compounds used during oviposition by some Lepidoptera, Hymenoptera, and Diptera (Wyatt, 2003).

Chemical composition of semiochemicals is highly varied, as is the amount of substance produced by insects for communication and the manner of distribution into the environment (i.e., airborne release of volatiles, physical contact, injection).

The physical characteristics of the chemical messages largely reflect how the signal is used or the message is to be conveyed. For example, pheromones tend to be volatile and structurally stable so that they can travel airborne over long distances. Their chemical stability as well as specificity means that only minute quantities need to be released. In contrast, some allelochemicals are volatile while others are not released unless the emitter is in direct contact with the target. Much larger quantities of allelochemicals are used in comparison with pheromones, in part reflecting more unstable molecules (i.e., proteins and peptides), and is an indication of the lower specificity of the chemical signals, particularly with defensive compounds like venoms or sprays (Blum, 1996).

In Sections 7.4 and 7.5, we will examine specific physical characteristics and functionality of pheromones and allelochemicals.

7.4 Pheromones are used to communicate with members of the same species

7.4.1 Basic characteristics of pheromones

Semiochemicals used in intraspecific communication are pheromones. When transferred from one individual to another, pheromones bind to olfactory receptors to modulate behavior and/or physiology of the receiver. Pheromones are produced in glandular (exocrine) epidermal cells that may be concentrated in specific regions beneath the cuticle or scattered throughout the body. In higher Diptera, for example, the sex pheromone-producing glands are frequently concentrated in the abdomen (Chapman, 1998). Likewise, in the American cockroach *Periplaneta americana* (Blattodea: Blattidae), sex pheromones are synthesized and released from exocrine (atrial) glands found in the genital atrium (Abed *et al.*, 1993). In both examples, the insects rely on contact for release and detection of sex pheromones, reflecting how the mechanism of distribution influences gland location. With that said, many lepidopterans possess **scent glands** (sex pheromone-producing exocrine glands) concentrated in the abdominal

region (Gullan & Cranston, 2014) but the pheromones are volatile and depend on the **chemotaxic** and **anemotaxic**[2] abilities of the receiver.

Pheromones are highly varied in terms of chemical composition. To a degree, this variability allows specificity for communication between members of the same species. However, this is not always the case. In many instances, multiple species use the same or very similar compounds or blends of chemicals as pheromones, which conceivably may be detected by non-related species. In such scenarios, the pheromone may actually permit interspecies communication and the signal functions as a type of allelochemical. When the message being conveyed requires an immediate response, particularly in cases of danger, the chemical structure may lack species specificity to accomplish the task of communicating alarm or need for quick aggregation (Chapman, 1998). Such compounds should be highly volatile to facilitate rapid dispersal. Volatility is constrained by the size of the molecule: smaller molecules are more volatile than larger, yet larger compounds provide more structural variation and thus a greater degree of specificity. Insect pheromones represent a chemical balance or compromise between volatility and specificity (Chapman, 1998). This is evident with the contact sex pheromones used by some **cyclorrhaphous** flies that have no requirement for volatility, which permits production of long-chain alkanes or alkenes to confer a high degree of specificity (Chapman, 1998).

7.4.2 Types of pheromones

Pheromones are typically classified by function or the effect the chemical signal has on the receiver of the message. These intraspecific messages are broadly divided into primers and releasers. Releaser pheromones elicit an immediate behavioral response in the recipient, binding directly to sensory neurons in an olfactory sensillum. Such pheromones convey alarm, aggregation, sex attraction, mate recognition, and oviposition stimulation. By contrast, primer pheromones do not cause an immediate reaction by the receiver. Rather, they operate, as the name implies, to "prime" the recipient for a behavioral or physiological action later in time. The time frame may be days, weeks, or even months later. With the longer time period in mind, events associated with growth, development, and reproduction are most likely influenced by primer pheromones (Chapman, 1998). A relationship between pheromone status (as primer or releaser) and volatility does not really exist. For example, sex attractants can be either volatile and airborne or work via direct contact, but in either case the chemical signal functions as a releaser pheromone.

An exhaustive discussion of the multitude of pheromone types will not be presented in this textbook but can be reviewed in more detail in Howse *et al.* (1998). Figure 7.3 provides an overview of many types of pheromones used by insects, including those frequenting carrion. Several of these pheromones are employed by necrophagous insects, and include sex pheromones used by the necrophagous flies that breed on site; aggregation signals to draw conspecifics to the decomposing body to feed, mate, or oviposit, and even to stimulate maggot mass formation (Boulay *et al.*, 2013); pheromones that allow mate recognition such as those used by sarcophagids; and marking pheromones applied to the egg chorion that are presumed to stimulate mass or cluster oviposition (Barton Browne *et al.*, 1969). However, as will be discussed later in this chapter, aggregated oviposition appears to be mediated by microbially-produced semiochemicals and not by oviposition pheromones, at least for some species (Brodie *et al.*, 2015; Uriel *et al.*, 2018).

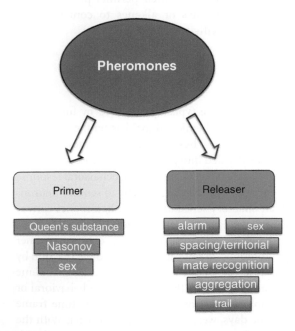

Figure 7.3 Types of pheromones used by insects in intraspecific communication.

7.4.3 Mode of action

Pheromones produced by some Lepidoptera and Coleoptera, especially those of agricultural significance, and social insects have been studied in greatest detail (Blum, 1996; Howse *et al.*, 1998; Vander Meer *et al.*, 2019). For other insects, the details are generally less well understood. This is a common theme in entomology: unless the insect has economic or medical value, it often goes unnoticed. The result is poor knowledge of key aspects of ecologically important species, or generalizations are made from seemingly distant species. In terms of pheromone mode of action, explicit details are not well known for most necrophagous insects. Certainly, the first steps involve binding of the chemical signal to an olfactory or gustatory receptor, followed by transmission of the message to the central nervous system (CNS) (Leal, 2005). However, even this feature does not always occur as some pheromones are ingested or swallowed, in which case alternative pathways may result. In one, the chemical is absorbed by the cells of the digestive tract, most commonly in the midgut or ventriculus, and then travels through the hemolymph to bind to a region in the CNS. The second option is for the pheromone to pass through the hemolymph to a target tissue or an organ lying outside the CNS. In this case, the chemical signal is functioning as both the sensory stimulant and effecter (Randall *et al.*, 2002). This latter pathway is more typical of primer pheromones than releasers.

Following threshold binding, pheromones are generally expected to activate common signal transduction pathways in target tissues (Gomperts *et al.*, 2002). In some cases, this changes the activity of the target tissue, as occuring in the corpora allata[3], in which lowered or suppressed cellular activity leads to gonad maturation in individuals and possibly developmental synchronization of the local population (Chapman, 1998). Pheromone-induced gonad maturation is manifested as accelerated rates of oogenesis in some insects such as *Tenebrio molitor* (Happ *et al.*, 1970). For several species of insects, the mechanism of action depends on multiple components in the pheromone to evoke a response in the receiver. A single compound may trigger one aspect of the "desired" behavior or is not capable by itself of eliciting any response. In such instances, the entire pheromonal blend is required to initiate the behavioral or physiological response in the recipient (Wyatt, 2003). Many of the sex pheromones function in this manner and the composition of the blend is what confers species specificity rather than unique compounds (Greenfield, 2002). The homestatic state is restored in the sensory (olfactory receptor) neuron once the odorant (pheromones) molecules are removed or degraded. For most insects, this is accomplished through the enzymatic action of various cytochrome P450s and esterases (Vogt, 2003; Leitch et al., 2015). Details of the mode of action of most primer pheromones is sparse and is an area in need of further investigation.

7.5   Allelochemicals promote communication across taxa and kingdoms

7.5.1   Basic characteristics of allelochemicals

Allelochemicals are the chemical signals used in interspecific communication. Mostly, every property of these chemicals is more varied in comparison with pheromones, owing to the multitude of signals that fall under the umbrella of allelochemicals. Broadly defined, any chemical that elicits a behavioral response in the receiver, favorable or not, is considered some sort of allelochemical, provided the receiver is not of the same species as the "emitter." In this instance, "emitter" does not have exactly the same meaning as with pheromones, in that the signal may not be volatile and is released by such mechanisms as direct contact, as a spray or injected, or like an intraspecific signal in an airborne form. Such a wide swath of chemicals also means that the origin, composition, receptor (or not), and mode of action of each chemical are not tightly linked. Two unifying features of allelochemicals are (i) their use in interspecific or allospecific communication and (ii) synthesis by exocrine glands. Structurally the exocrine glands are distinct from those that produce pheromones in that the lumen is lined with cuticle. A cuticular lining is a necessary prerequisite to avoid **autointoxication** for those species that produce defensive secretions which are non-discriminatory or which do not evoke receptor-mediated responses. Some allelochemicals are sequestered compounds such as cardiac glycosides, terpenoids, or alkaloids derived from food rather than synthesized *de novo*

(Blum, 1996). Sequestered compounds offer the advantage of already being in "active" form, a common feature of many allelochemicals, yet others must be activated from a precursor (e.g., some venoms) or require an enzymatic reaction like the hot sprays generated by bombardier beetles.

7.5.2   Types of allelochemicals

Allelochemicals represent a diverse range of chemical signals. The signals are categorized based on the impact on the receiver and, to a lesser extent, the effect on the emitter (Nordlund & Lewis, 1976; LeBlanc & Logan, 2010) (Figure 7.4). This section offers a brief examination of each category of allelochemicals.

7.5.2.1   Allomones

Chemical substances produced and released by an individual of one species that has a deleterious effect on an individual of another species are termed **allomones**. The emitter may not benefit from the chemical release, but typically do since allomones are most commonly used as defensive compounds (Gullan & Cranston, 2014). These interspecific signals are represented by a range of compounds, such as the proteinaceous venoms used by Hymenoptera (parasitic and social) and predatory true bugs (salivary venoms), quinolic sprays, alkylpyrazine odors, and aldehyde repellants (Blum, 1996) (Figure 7.5). Sequestered compounds from food also represent allomone signals, and they too are diverse chemically (Gardner & Stermitz, 1988; Frick & Wink, 1995). Perhaps the most entertaining of the compounds are those released by ground beetles in the genus *Chaenius*. When attacked by predatory ants, the beetles emit a volatile "sedative" from glands positioned near the anus. The ant is temporarily immobilized, allowing the would-be prey an opportunity to escape. In some instances, the slumbering attacker is consumed by another predator, a scenario that has led to the phrase "fatal flatulence" for this allomone.

7.5.2.2   Kairomones

Chemicals that are emitted by one species which evoke a positive or beneficial response in the recipient are referred to as **kairomones**. The emitter does not benefit from release of the signal,

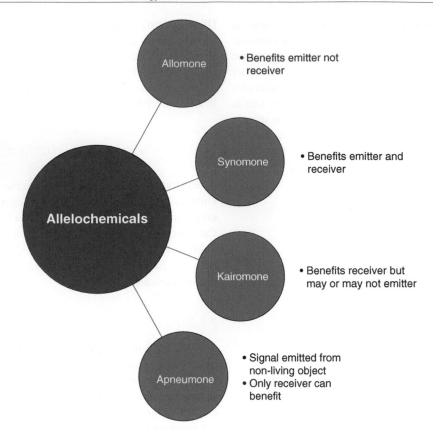

Figure 7.4   Types of allelochemicals used by insects in interspecific communication. *Source*: Based on LeBlanc & Logan (2010).

Figure 7.5   Adult female of the parasitic wasp *Nasonia vitripennis* (Pteromalidae) injecting venom (allomone) into a fly host. *Source*: M.E. Clark/Wikimedia Commons/ Public domain.

often in volatile form, but may be harmed if the receiver uses the chemical cue to track the releaser so that the originating species is consumed as food or used as a host for parasitic species (Grasswitz & Jones, 2002). For some insects, kairomone signals detected by the receiver are pheromones produced by another species (Hölldobler, 1989; Gullan & Cranston, 2014). Oviposition pheromones on fly

eggs or sex pheromones emitted during mate attraction on a carcass may indeed serve as cues to attract predatory silphids, ants, and yellowjacket wasps to carrion. Adults of the yellowjacket *Vespula germanica* are capable of using sex phero- mones of the fruit fly *Ceratitis capitata* (Family Tephritidae) as kairomones (Grasswitz & Jones, 2002), so an extension to carrion breeding flies seems plausible. Phoretic mites, especially obligatory phoretic species, appear to rely on volatile cuticular compounds of necrophagous flies and beetles to select carrier species for trans- port (Perotti *et al.*, 2010; Niogret *et al.*, 2018). In other instances, the chemical cues are derived from plants, frequently synthesized as secondary metabolites, that attract the attention of potential herbivorous insects and other animals (Blum, 1996). Perhaps, the most common example associated with carrion ecology is bacteria- mediated signaling. Quorum sensing and bacterial products such as indole and dimethyl trisulfide attract necrophagous flies and potentially other species to decomposing animal remains and may also influence feeding and oviposition decisions.

These bacterial signals seem to be clear examples of kairomones. However, in most scenarios described, both organisms are speculated to receive benefit (i.e., food for the fly and transport for the bacteria) from the relationship, and if so, the bacterial signals are truly synomones. Since the signals originate from bacteria, the example is also referred to as interkingdom signaling (Ma et al., 2012; Tomberlin et al., 2012).

7.5.2.3 Synomones

When the chemical substance released generates a response in the receiver that is also beneficial to the emitter, the signal is called a **synomone**. The definition does not imply that the chemical substance itself benefits both the originator and recipient. Rather, the response of the receiver leads to a positive outcome for individuals of both species (Nordlund & Lewis, 1976). Numerous examples can be found associated with plant herbivory. For example, bark beetle attack of different pine tree species can result in the injured tree tissue releasing terpene compounds that serve as attractants to parasitic wasps that subsequently locate and parasitize the beetles (Wood, 1982). Thus, the wasp locates a host aided by plant-derived chemicals, and the tree experiences reduced herbivory since the wasps are parasitoids that ultimately kill the beetle hosts. A similar relationship exists with onions attacked by the onion maggot fly, *Delia* (*Hylemya*) *antiqua*: the volatile compound that evokes tear release in humans when slicing onions serves as a synomone for a braconid parasitoid, *Aphaerata pallipes*, that utilizes fly larvae as hosts (Harris & Miller, 1988). The so-called "fly factor" references scenarios in which adult flies inoculate carrion with bacteria via regurgitation and fecal elimination, and then the bacteria generate byproducts that serve as signaling molecules and that facilitate growth and development of the carrier and potentially other species (Holl & Gries, 2018; Uriel et al., 2018). The only difference between the bacterial signaling described here as the fly factor from those functioning as kairomones is the source of microorganisms: the fly transporting and inoculating carrion versus bacteria comprising the microbiome/necrobiome of the deceased. The fly factor can also be considered another example of interkingdom signaling that occurs in carrion communities.

7.5.2.4 Apneumones

Chemical signals emitted from a non-living object that trigger a response in the recipient are known as **apneumones**. The emitter may not be of biological origin, but usually is, and typically the first thought that comes to mind is dead and decaying animals. In the case of animal remains, the interspecific signals are usually in the form of putrid gases and volatile organic compounds (VOCs), yielding the distinctive odors of death (Vass et al., 2002; Statheropoulos et al., 2005). The initial attraction of flies in the families Calliphoridae and Sarcophagidae to a dead body is believed to be triggered by odors emanating from the decomposing tissues of a corpse, prompting such behaviors as adult feeding, mate finding, and oviposition (Archer & Elgar, 2003; Aak et al., 2010). A more detailed examination of chemical attraction of necrophagous insects to carrion can be found Section 7.6. Interestingly, apneumones can be used in intraspecific communication as well. As discussed in Section 7.3, this form of communication relies on pheromones, and in instances in which death signals are recognized from conspecifics, the term **necromone** is frequently used (Yao et al., 2009). In several species of social Hymenoptera that form well-organized colonies or nests, some workers have the task or ability, depending on your point of view, of detecting dead workers and removing their bodies from the colony (Hölldobler & Wilson, 1990). These workers are often identified as the "undertaker caste" and in the case of some ant species, the dead are simply piled onto the colony's trash heap. The undertaker role is critical to the success of the colony by ensuring that disease-causing organisms do not spread throughout the colony or nest (Yao et al., 2009).

7.5.3 Mode of action

As should be expected with the wealth of chemicals that function as allelochemicals, a broad range of mechanisms and pathways are used, often unique to each type of chemical signal. In many cases, the chemical signals operate similarly to pheromones in that the responses are receptor-mediated and stimulate signal transduction pathways. However, some of the chemicals do not require receptor binding to affect the recipient, and still others are non-discriminatory in their action and induce a wide range of cellular responses (Schmidt, 1982; Blum, 1996). The point is that allelochemicals are not as easy to categorize as pheromones, which can be grouped as primers or releasers. For many allelochemicals, no detailed studies have been conducted to decipher which tissues are targeted or the pathways that are modulated.

7.6   Chemical attraction to carrion

In an attempt to develop a paradigm for forensic entomology, Tomberlin *et al.* (2011) proposed a roadmap to unify basic and applied research in this discipline. As part of this paradigm, they offered terminology for each phase of insect interaction with carrion (Figure 7.6). Relevant to this section of the textbook is the distinction between the detection phase and the acceptance phase (Tomberlin *et al.*, 2011). The detection phase occurs after the initial exposure period during which the insects have not or cannot detect the body, while the onset of the acceptance phase begins when insects first make contact with the corpse. During detection, carrion insects must first be activated or become aware of the body (activation phase), presumed to be the result of volatile compounds released from the animal remains. Assuming that necrophagous species are hardwired like other insects, resource-finding behavior or foraging should be triggered by a cascade of neurological events dependent on threshold binding (Mittelstaedt, 1962). This

behavior is part of the searching phase described by Tomberlin *et al.* (2011) as the secondary component to the two-armed detection phase. Both activation and searching are expected to rely on the chemicals released from a cadaver. The remainder of this section is dedicated to what is known for necrophagous insects about such signals.

7.6.1   Initial colonizers of carrion

Adult flies in the families Calliphoridae and Sarcophagidae are usually the first insects to arrive and colonize a dead body (Smith, 1986; Anderson, 2020). In most instances, blow flies arrive before any other insects. Given the topic of this chapter, which of the chemical signal(s) discussed earlier are involved in attraction of calliphorids to a dead body? The knee-jerk response would of course be apneumones. After all it is apparent that a non-living object is emitting odors that draw the attention of necrophagous insects. While intuitively this seems obvious, a case can also be made for kairomones and synomones in the form of bacterial signals. The reality is that the chemical ecology of carrion succession has not been fully worked out (LeBlanc & Logan, 2010). It is true that numerous insects seek out animal remains, usually at specific times during the decomposition process. However, there is an incomplete understanding of the factors that

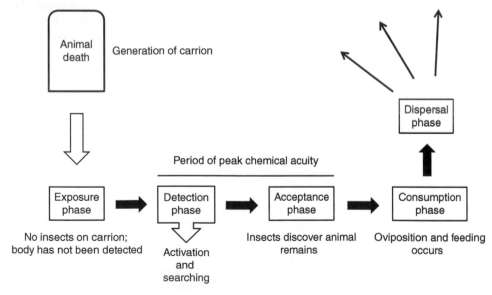

Figure 7.6 Schematic representation of necrophagous insects' association with carrion. *Source*: Modified from Tomberlin *et al.* (2011).

attract carrion-breeding calliphorids, sarcoph-agids, muscids, and other flies and beetles to a corpse. The limited data available suggest that a combination of olfactory and visual cues is needed by some species to locate and land on a carcass (Spivak *et al.*, 1991; Wall & Fisher, 2001), although a keen sense of visual and olfactory acuity is restricted to (needed by) adult females, at least for some species (Aak & Knudsen, 2011). This latter observation is not surprising when remembering from Chapter 6 how competitive the carrion environment can be for an adult female and her progeny. Failure to accumulate the necessary nutriment for egg provisioning or larval development will lower fitness for either the mother or progeny, or possibly both. It is thus imperative that necrophagous flies be highly attuned to the chemical cues associated with carrion to effectively compete for the temporally and spatially limited resources of a dead body (Archer & Elgar, 2003).

7.6.2 Apneumones "signal" the way to carrion

Once an animal dies, decomposition begins immediately. The process is sequential and reasonably predictable, within the confines of environmental conditions, and taking into account the characteristics (i.e., animal type, size, and age) of the animal itself as well as usage by carrion-feeding invertebrate and vertebrate animals (Anderson, 2020). Details concerning the process of body decomposition will be explored in Chapter 10. For now, we are only concerned with the chemicals or odors released from a corpse that "communicate" with insects. Early in decomposition, inorganic gases and sulfur-containing volatiles originate from the decay processes occurring within the alimentary canal, largely facilitated by an array of microorganisms (LeBlanc & Logan, 2010). As a body progresses in decomposition, a variety of gases, liquids, and VOCs are released from soft tissues (Table 7.1) (Vass *et al.*, 2002; Statheropoulos *et al.*, 2005). During putrefaction in soft tissues, a high degree of protein degeneration occurs that is marked by production of sulfur-containing organic compounds (Gill-King, 1997). Conceivably, any of these compounds can function alone or in combination to attract necrophagous insects to a body, thereby serving as apneumones.

Table 7.1 Volatile organic compounds emanating from decaying adult human bodies.

Chemical	Sulfur-containing?
Dimethyl sulfide	Yes
Toluene	No
Hexane	No
1,2,4-Trimethylbenzene	No
2-Propanone	No
3-Pentanone	No
2-Methylpentane	No
1,3,5-Trimethylbenzene	No
Hexanal	No
Ethanol	No
Methyl ethyl disulfide	Yes
Carbon disulfide	Yes
Dimethyl trisulfide	Yes
Propylester acetic acid	No
p-Xylene	No
2-Heptanone	No
Carbonoxide sulfide	Yes

This list represents some of the most abundant compounds emitted during decomposition of more than 100 total detected.
Source: Data from Statheropoulos *et al.* (2005) and Vass *et al.* (2002).

Though hundreds of chemicals are released from a body during the decomposition process, precise olfactory stimulants have not been identified for carrion-breeding insects (LeBlanc & Logan, 2010). Adult females of *Calliphora vicina* are attracted to traps baited with single compounds like dimethyl trisulfide and other sulfur-containing compounds (Aak *et al.*, 2010). Synthetic blends of dimethyl trisulfide, mercaptoethanol, and *o*-cresol are much more attractive to females, but not as effective as authentic odors emitting from various carrion types (fish, mice, fresh liver). Urech *et al.* (2004) found that formulations of 2-mercaptoethanol, indole, butanoic/pentanoic acid, and sodium sulfide were equally attractive as liver to *Chrysomya rufifacies* and the synthetic blend drew more *Lucilia cuprina* than fresh liver. Attraction to single compounds or synthetic blends containing sulfur derivatives (sulfides/sulfurous) has also been reported for the calliphorids *Calliphora vomitoria*, *C. uralensis*, *Protophormia terraenovae*, *Lucilia sericata*, *L. caesar*, *Cynomya* sp., *Cochliomyia hominivorax*, and *Co. macellaria* (Table 7.2) (Mackley & Brown, 1984; Ashworth & Wall, 1994; Nilssen *et al.*, 1996; Stensmyr *et al.*, 2002; Urech *et al.*, 2004; Aak *et al.*, 2010; Chaudhury *et al.*, 2017).

Table 7.2   Chemical attractants to necrophagous flies.

Substance or blend	Species	Laboratory or field tested
Dimethyl disulfide or trisulfide	*Calliphora vicina*	L/F
	Cochliomyia macellaria	L
	Lucilia caesar	L
Butylated hydroxyl toluene or 3-hydroxy-2-butanone or nonanal	*Calliphora vicina*	L
Blend: dimethyl trisulfide, mercaptoethanol, *o*-cresol	*Calliphora vicina*	L/F
	Calliphora uralensis	F
	Calliphora vomitoria	F
	Cochliomyia hominivorax	L/F
	Cynomya spp.	F
	Lucilia sericata	L/F
	Protophormia terraenovae	F
Blend: 2-mercaptoethanol, indole, butanoic/pentanoic acid, sodium sulfide	*Calliphora* spp.	F
	Chrysomya spp.	F
	Lucilia cuprina	F
Blend: dimethyl disulfide, dimethyl trisulfide, phenol, p-cresol, indole	*Cochliomyia hominivorax*	L
	Cochliomyia macellaria	L

Source: Data from Aak *et al.* (2010), Ashworth & Wall (1994), Chaudhury *et al.* (2017), Johansen *et al.* (2014); Mackley & Brown (1984), Nilssen *et al.* (1996), Stensmyr *et al.* (2002) and Urech *et al.* (2004).

Olfactory stimulation by oligosulfides was confirmed in antennal stimulation assays using *C. vicina* and *L. caesar* (Stensmyr *et al.*, 2002). The assays demonstrated that action potentials are generated in sensory neurons located in olfactory receptors of adult flies exposed to the same sulfide-containing compounds used in behavioral tests that elicited attraction in both flies. Antennal stimulation by such compounds also suggests involvement of **allothetic pathways** in the processing of chemical signals emanating from carrion. Allothetic systems are associated with the detection and integration of external stimuli generally via sensory sensilla (Visser, 1986). Transcriptosome analysis with *Calliphora stygia* suggests that male and female blow flies share similar odor coding or recognition capacity (Leitch *et al.*, 2015). However, expression levels of antennal odor receptors differ between the sexes, an indication that sensitivity, and hence responsiveness, may not be equal between males and females. Such differences would be predicted to manifest itself in different species with one sex detecting and accepting carrion first, and in turn, using pheromonal signaling to attract mates and/or other conspecifics.

There is an interesting omission from our list of potential chemical attractants: ammonia-containing compounds. Compounds like ammonia, putrescine and cadaverine are produced during the breakdown of proteins during putrefaction of soft tissues, yielding distinctly strong odors believed to be important in attracting a range of necrophagous insects to carrion (Anderson, 2020). While many species likely use these chemicals for activation or searching (Wardle, 1921), experimental evidence actually demonstrates a lack of response (cadaverine) or repulsive activity (putrescine) toward each compound (Frederickx *et al.*, 2012). As will be seen in Section 7.8, ammonia-rich compounds may be most influential on fly attraction through modulation of quorum sensing behavior of inhabitants of the necrobiome (Ma *et al.*, 2012).

The feeding and reproductive status of adult flies directly correlates with olfactory acuity. For example, anautogenous species appear to demonstrate a stronger odor attraction to carrion than autogenous flies. Non-protein-fed, pre-vitellogenic adult females of anautogenous species are extremely responsive to carrion odors in comparison with satiated flies (Ashworth & Wall, 1994; Kasper *et al.*, 2015). Flies that are yet to acquire nutriment are presumed to be highly protein motivated: they have an absolute need to find protein-rich carrion to provision eggs. Once protein **satiety** is reached, meaning the minimum reproductive threshold is attained (see Chapter 6), vitellogenesis is initiated and adult females display diminished interest in odors from carrion (Wall & Warnes, 1994). However, responsiveness to animal remains elevates once the females are **gravid** as there is now a presumed motivation to locate an oviposition site (Ashworth & Wall, 1994; Archer & Elgar, 2003; Aak & Knudsen, 2012). Autogenous or capital breeders may be expected to show lower responsiveness to carrion odors since the initial clutch is not protein driven. However, gravid females presumably have the same "motivation" for an oviposition site as anautogenous flies and, like other species, high chemical acuity is a competitive advantage in resource location. The reality is that information on olfaction in autogenous species is very limited. The bulk of what is known comes from *Cochliomyia hominivorax*, an obligate parasite that depends on chemical signals derived from wounds, blood, and bacteria for host attraction and oviposition (Hammack, 1991; Chaudhury *et al.*, 2010).

7.7 Chemical attraction to carrion by subsequent fauna

Waves of insect colonization are associated with carrion as decomposition progresses (Smith, 1986). The chemical profile associated with body decay changes over time (Vass *et al.*, 2002), so insects in each wave of colonization/succession are presumably attuned to specific chemical cues indicative of resource suitability. Such chemical acuity likely reflects to a degree tissue location on a corpse since nutrient availability in specific soft tissues is expected to change over time as decomposition and feeding proceeds. Gravid blow fly females are able to distinguish between different chemical profiles and arrive when the cadaver is suited for oviposition and feeding by progeny (Archer & Elgar, 2003; Johansen *et al.*, 2014). As with early or initial colonization, precise chemical signals have not been identified for insects arriving during later phases of decomposition. Once female blow flies have landed on the corpse, moisture content of soft tissues probably contributes to changes in oviposition or larviposition decisions (location on corpse, size of clutch) (Archer & Elgar, 2003). Undoubtedly tactile feedback also contributes to timing and placement of eggs in species that specialize on later stages of decay.

Pheromonal signaling appears to be used by some calliphorids to recruit conspecifics to carrion. Laboratory studies with *Lucilia cuprina* show that gravid females are more likely to oviposit in body folds or openings where large numbers of eggs have been deposited (Barton Browne *et al.*, 1969). The resulting egg clustering is thought to reflect deposition of pheromones on the egg chorion and is consistent with the notion that cooperative feeding by larvae is needed or highly advantageous in the form of larval aggregations, an idea first introduced in Chapter 6. If the same chemical signal is detected by other species of flies that in turn deposit progeny to a growing maggot mass, the cue would be a synomone. Alternatively, if egg pheromones are intercepted by competing fly species or predatory (or parasitic) beetles, ants or wasps, the chemical would be functioning as a kairomone to the recipient. To date, no egg-clustering pheromone has been isolated from any species of calliphorid, but there is compelling research that argues for aggregated oviposition through bacterial signaling (Brundage *et al.*, 2017; Uriel *et al.*, 2018).

7.8 Chemical attraction to bacteria on carrion

Our discussion of the chemicals that attract necrophilous insects to a decomposing body has oversimplified the complexity of the system. The reality is that multiple signals originating from several sources undoubtedly are involved in activation, detection, and ultimately acceptance of carrion as a resource (Tomberlin *et al.*, 2011). Among the sources of chemicals yet to be discussed are microorganisms, specifically bacteria. Several species of flies, not all necessarily considered necrophagous but many saprophagous, preferentially oviposit in microbially rich environments as opposed to those with low levels of bacteria or essentially sterile environments (DeVaney *et al.*, 1973; Hough-Goldstein & Bassler, 1988). Bacteria appear to employ two mechanisms to "communicate" with insects. The communication being implied is not fully understood as to whether it is actively driven by the microorganisms or the end result of other metabolic processes that simply yield byproducts that are attractive to specific necrophiles. One method of communication involves kairomonal signaling in which the bacteria secrete com-pounds during decomposition that lead to activation usually in gravid blow flies. The bac-terially derived products are predominantly pro-teins, VOCs (e.g., indole) and metabolites synthesized as the microorganisms process soft tissues or fluids of the dead animal or tissues (Emmens & Murray, 1983; Hammack, 1991; Robacker & Lauzon, 2002; Lee *et al.*, 2015), or are derived from the bacteria themselves and associated with quorum sensing, such as during swarming[4] (Ma *et al.*, 2012; Tomberlin *et al.*, 2012). These same signals may serve as synomones if the bacteria benefit as well, which they likely do because of transport by the flies, hence mechanical inoculation, of new resources (i.e., other carrion). A second mechanism of sig-naling involves the bacteria enhancing existing chemical attractants originating from some source other than the prokaryotes. Non-sulfur-containing compounds like ethyl acetate, tetra-methylpyrazine, and *n*-haptanal produced by the bacteria *Klebsiella* spp. or *Bacillus subtilis* aug-ment the attractant properties of other chemical signals to a range of non-necrophagous and necrophagous species (Ikeshoji *et al.*, 1980; Emmens & Murray, 1983). Decomposition of soft tissues by bacteria also liberate VOCs, which in turn serve as attractants to necrophagous flies and stimulate higher oviposition output (Liu *et al.*, 2016; Holl & Gries, 2018). The Gram-negative bacterium *Proteus mirabilis* synthesizes enzymes (i.e., urease) that act on substrates of the surrounding environment to produce ammonia and putrescine; compounds known to modulate quorum sensing behavior and subse-quently leads to activation of fly foraging (Ma *et al.*, 2012). Interestingly, in the absence of bacteria, neither ammonia nor putrescine alone are capable of yielding a similar behavioral response in blow flies (Easton & Feir, 1991; Frederickx *et al.*, 2012).

Bacterial signaling appears to be associated with several features of fly oviposition on carrion. Once the acceptance phase has been achieved, many species of flies regurgitate or defecate on the resource (Figure 7.7). The process essentially inoculates carrion with exogenous bacteria. Oviposition can occur at any time before, during and after. Conspecific and allospecific oviposition are thought to be influenced by microbially derived volatile compounds on the egg surface

Figure 7.7 Regurgitation of ingested food serves as a mechanism to inoculate a corpse and other animal remains with bacteria. *Source*: Alvesgasper/Wikimedia Commons/CC BY-SA 3.0.

(Brundage *et al.*, 2017), which may include stimulation of egg clustering. Some species may even utilize a form of mimicry whereby the micro-bial fauna of the eggs resembles that of competitor species (Brundage *et al.*, 2017). Following egg hatch, predatory larvae are unleashed upon the unsuspecting inhabitants of the carrion community. The role of microorganisms in inter-actions with carrion-breeding insects is in its infancy in terms of research and understanding, so our knowledge in this realm is expected to expand dramatically over the next several years.

This chapter is focused on the chemicals used in chemical communication and attraction but necrophagous insects do use vision for location of carrion. Visual cues are used by several species of calliphorids searching for objects during flight as well as when making decisions about landing on a carcass (Wall & Fisher, 2001; Aak & Knudsen, 2011). In some species, like *C. vicina* and *L. sericata*, visual acuity for food location is well developed in adult females but not in males (Aak & Knudsen, 2011). Sexual dimorphism exists in terms of compound eye structure in many species of blow flies (Rognes, 1991), owing to differences in primary function: males use visual cues to identify females at close range while females tend to rely on vision and olfaction in resource location, which again is particularly critical to the fitness of anautogenous species (Aak & Knudsen, 2011).

Chapter review

Insects rely on chemicals in intraspecific and interspecific communication

- Most organisms have the ability to recognize and respond to chemicals found in the environment. By comparison to other animals, insects as a whole are more dependent on chemical stimuli for interspecific and intraspecific communication.
- A wide range of chemicals is used to convey messages among and between insects and other organisms, ranging from highly volatile, complex blends of compounds to those with limited volatility that can only be perceived through direct contact between the emitter and receiver.
- The chemicals used for communication are broadly referred to as semiochemicals. Semiochemicals can be simply defined as chemicals that modify the behavior and/or physiology of the recipient.

Chemical communication requires efficient chemoreception

- Insects depend on chemicals more than most other animal groups to communicate. One way this is apparent is simply in the wide range of semiochemicals that they produce and the elaborate network of glands and ducts used to synthesize and distribute these chemicals to the outside environment. A second piece of evidence revealing the importance of chemicals to communication is the presence of many varied receptors.
- Chemoreception, the detection of chemical stimuli in the environment, is absolutely essential to any organism dependent on chemical signals for communicating messages.
- Chemoreception is classified based on environment. For example, the ability of terrestrial insects to detect chemical signals is referred to as olfaction or the sense of smell, while aquatic insects rely more on a sense of taste, otherwise referred to as contact or gustatory detection. Olfactory receptors do occur in some aquatic insects, and taste or gustatory reception is prevalent with most terrestrial species.
- Insect chemical receptors are referred to as sensilla (sensillum in the singular). Sensilla generally

have the appearance of fine thin hairs (setae) with one or more pore openings to allow chemical penetration. Alternatively, some sensilla are short, squat, and broad yielding a peg-like morphology and still others have the shape of plates or depressions (pits) along the cuticular surface.
- In the case of olfaction, sensilla are either hair-like or peg-like in appearance and each sensillum has numerous pore openings along the thin walls of the receptor. Olfactory chemoreceptors are referred to as multiporous sensilla. Gustatory receptors more commonly represent the entire range of sensillum morphologies (hair, peg, plates, and pits) but possess only a single opening for chemical penetration. Hence, receptors associated with taste are termed uniporous sensilla.
- Perception of chemical stimuli often begins with trapping of chemical substances just inside the pore openings, followed by movement of the stimuli to a recognition site along the dendritic membrane. Binding does not guarantee further action, because a threshold must be surpassed that often requires binding of multiple chemical molecules. Surpassing the binding threshold leads to the generation of an action potential, which then travels the surface of the neuron cell body and axons to the presynaptic membranes at the axonal terminals.

Semiochemicals modify the behavior of the receiver

- Semiochemicals are used for communication in the external environment (as opposed to internally where hormones are the major chemical messengers), conveying messages for intraspecific and interspecific communication. Chemical signals used to communicate with conspecifics are called pheromones, while those utilized in communication with allospecifics are broadly referred to as allelochemicals.
- Nearly all semiochemicals are produced in exocrine glands. A few exceptions do occur such as those compounds that have been sequestered during feeding. Exocrine glands share the same basic characteristics of being derived from epidermal cells and possessing ducts that allow release of the chemical substances to the outside of the emitter's body. Salivary glands do not adhere to this rule as salivary secretions are not typically released directly into the outside environment.

- Chemical composition of semiochemicals is highly varied, as is the amount of substance produced by insects for communication and the manner of distribution into the environment (i.e., airborne release of volatiles, physical contact, and injection). The physical characteristics of the chemical messages largely reflect how the signal is used or the message is to be conveyed.

Pheromones are used to communicate with members of the same species

- Semiochemicals used in intraspecific communication are pheromones. When transferred from one individual to another, pheromones bind to olfactory receptors to modulate behavior and/or physiology of the receiver.
- Pheromones are highly varied in terms of chemical composition. To a degree this variability allows specificity for communication between members of the same species. However, in many instances multiple species use the same or very similar compounds or blends of chemicals as pheromones, which conceivably may be detected by non-related species. Insect pheromones represent a chemical balance or compromise between volatility and specificity.
- Pheromones are typically classified by function or the effect the chemical signal has on the receiver of the message. Releaser pheromones elicit an immediate behavioral response in the recipient, binding directly to sensory neurons in an olfactory sensillum. Such pheromones convey alarm, aggregation, sex attraction, mate recognition, and oviposition stimulation. By contrast, primer pheromones do not cause an immediate reaction by the receiver. Rather, they operate, as the name implies, to "prime" the recipient for a behavioral or physiological action later in time.
- The mode of action of pheromones generally involves binding of the chemical signal to an olfactory or gustatory receptor, followed by transmission of the message to the CNS. In some species, the pheromones are ingested or swallowed, in which case the chemical signal is either absorbed by the cells of the digestive tract, or passes through the hemolymph to a target tissue or organ lying outside the CNS. Following threshold binding, pheromones are generally expected to activate common signal transduction pathways in target tissues.

Allelochemicals promote communication across taxa

- Allelochemicals are the chemical signals used in interspecific communication. Mostly, every property of these chemicals is more varied in comparison with pheromones, owing to the multitude of signals that fall under the umbrella of allelochemicals. Broadly defined, any chemical that elicits a behavioral response in the receiver, favorable or not, is considered some sort of allelochemical, provided the receiver is not of the same species as the "emitter." Two unifying features of allelochemicals are their use in interspecific or allospecific communication and synthesis by exocrine glands. Structurally the exocrine glands are distinct from those that produce pheromones in that the lumen is lined with cuticle.
- Allelochemicals represent a diverse range of chemical signals. The signals are categorized based on the impact on the receiver and, to a lesser extent, the effect on the emitter. Allomones usually have a negative impact on the receiver, while kairomones benefit the receiver. Synomones benefit both the emitter and receiver. Chemical signals originating from non-living objects, such as dead animals, are termed apneumones.
- The wealth of chemicals that function as allelochemicals indicates that a broad range of mechanisms and pathways are used to alter the behavior or physiology of the receiver. Often the mode of action is unique to each type of chemical signal. In many cases, the chemical signals operate similarly to pheromones in that the responses are receptor-mediated and stimulate signal transduction pathways. However, some of the chemicals do not require receptor binding to affect the recipient, and still others are non-discriminatory in their action and induce a wide range of cellular responses. For many allelochemicals, detailed studies examining the mechanism of action or target tissues are lacking.

Chemical attraction to carrion by initial colonizers

- Chemical signals are used by necrophagous insects to locate a body and then to determine if it is acceptable as a resource for feeding, oviposition, and progeny development. The detection phase occurs after the initial exposure period in which the insects cannot detect the body, while the onset of the acceptance phase begins when insects first make contact with the corpse. The detection phase is believed to be a two-armed process: activation and searching. Carrion insects must first be activated or become aware of the body before they will search for the ephemeral resource.

- Numerous insects seek out animal remains, usually at specific times during the decomposition process. Flies in the families Calliphoridae and Sarcophagidae are usually the first to colonize carrion. However, there is an incomplete understanding of the factors that attract carrion-breeding flies and beetles to a corpse. The limited data available suggest that a combination of olfactory and visual cues is needed by some species to locate and land on a carcass.
- Early in decomposition, inorganic gases and sulfur-containing volatiles originate from the decay processes occurring within the alimentary canal, largely facilitated by an array of microorganisms. As a body progresses through decomposition, a variety of gases, liquids, and VOCs are released from soft tissues. During putrefaction in soft tissues, a high degree of protein degeneration occurs that is marked by production of sulfur-containing organic compounds. Conceivably any of these compounds can function alone or in combination to attract necrophagous insects to a body, thereby serving as apneumones.
- Though hundreds of chemicals are released from a body during the decomposition process, precise olfactory stimulants have not been identified for carrion-breeding insects. Laboratory and field trapping studies suggest that a variety of sulfide/sulfur-containing compounds act as attractants.
- Olfactory stimulation by oligosulfides in antennal stimulation assays demonstrates that the same components used in behavioral assays activate sensory neurons in olfactory receptors.
- The feeding status of adult flies directly correlates with olfactory acuity. Anautogenous species appear to demonstrate a stronger odor attraction to carrion than autogenous flies, a feature expected in non-fed, pre-vitellogenic females. Similarly, gravid females also show a strong motivation via enhanced chemical acuity to locate carrion.

Chemical attraction to carrion by subsequent fauna

- Waves of insect colonization are associated with carrion as decomposition progresses. The chemical profile associated with body decay changes over time, so insects in each wave of colonization/succession are presumably attuned to specific chemical cues indicative of resource suitability. Gravid females can distinguish between different chemical profiles and arrive when the cadaver is suited for oviposition and feeding by progeny. As with early or initial colonization, precise chemical signals have not been identified for insects arriving during later phases of decomposition.
- Pheromonal signaling appears to be used by some calliphorids to recruit conspecifics to carrion. The resulting egg clustering is thought to reflect deposition of pheromones on the egg chorion and is consistent with the notion that cooperative feeding by larvae is needed or highly advantageous in the form of larval aggregations. If the same chemical signal is detected by other species of flies that in turn deposit progeny to a growing maggot mass, the cue would be a synomone. Alternatively, if egg pheromones are intercepted by predatory (or parasitic) beetles, ants or wasps, the chemical would be functioning as a kairomone to the recipient.

Chemical attraction to bacteria on carrion

- Several species of flies preferentially oviposit in microbially rich environments as opposed to those with low levels of bacteria or essentially sterile environments. Bacteria appear to employ two mechanisms to "communicate" with these insects, one that involves kairomonal signaling and another in which bacterially derived secretions enhance the attractant properties of chemical signals produced by another animal or source.
- Decomposition of soft tissues by bacteria liberate VOCs, which in turn serve as attractants to necrophagous flies and stimulate higher oviposition output. The Gram-negative bacterium *Proteus mirabilis* synthesizes enzymes (i.e., urease) that act on substrates of the surrounding environment to produce ammonia and putrescine;

both compounds modulate quorum sensing behavior that leads to activation of fly foraging.

- Bacterial signaling appears to be associated with several features of fly oviposition on carrion, including resource acceptance as an oviposition site, egg clustering of conspecifics and allospecifics, and egg mimicry, whereby the microbial fauna of the eggs resembles that of competitor species.

Test your understanding

Level 1: knowledge/comprehension

1. Define the following terms:

   (a) semiochemical
   (b) olfactory
   (c) chemoreception
   (d) apneumone
   (e) gustation
   (f) pheromone
   (g) quorum sensing.

2. Match the terms (i–vi) with the descriptions (a–f).

   (a) Chemical signals that evoke a response in the receiver that is beneficial to both the originator and recipient
       (i) Allelochemical

   (b) Behavioral response requiring movement or change of position in relation to air currents
       (ii) Activation phase

   (c) Sensory receptor located on cuticle
       (iii) Synomone

   (d) Period in which a carrion insect becomes aware of a dead body
       (iv) Saprophagous

   (e) Insect that feeds on decaying organic matter
       (v) Anemotaxic

   (f) Chemical used in interspecific communication
       (vi) Sensillum

   (g) Communication between bacteria and necrophagous flies
       (vii) Interkingdom signaling

3. Assuming that most carrion-breeding sarcophagids rely on pheromones during their visitation of a dead body, describe how adult females use primer and releaser pheromones.

4. Explain the difference between an apneumone and necromone.

Level 2: application/analysis

1. Explain a scenario associated with carrion in which a pheromone produced by one insect can be used as a kairomone for another.

2. If a VOC such as dimethyl trisulfide is released by a decaying pig located in an open grassy field, what physiological events must occur with an olfactory sensillum on the antennae of a female *Protophormia terraenovae* to elicit the initiation of the searching phase?

3. If designing a trap to catch adult blow flies that are parasitic on cattle, what should be included in a bait to lure in the flies?

4. With several anautogenous species of sarcophagids, the adult male typically has a keener olfactory acuity than females in terms of foraging behavior. Explain the adaptive significance of this behavior and describe what other chemical signals are likely used by these same species to locate carrion.

Level 3: synthesis/evaluation

1. Several chemicals emanating from decomposing animal tissues are presumed to serve as chemical attractants to necrophagous insects. Surprisingly, cause–effect relationships between any of these compounds and activation or searching behavior have not been demonstrated. Design an experiment that examines the chemoattraction of an adult necrophagous fly to a chemical compound associated with animal decay. Your response should include a testable hypothesis/ hypotheses as well as all variables.

Notes

1. Telemarketers are salespersons who attempt to make direct sales (telemarketing or inside sales) of products or services to consumers by making unsolicited contact via the telephone or face-to-face interactions. They are generally not referred to favorably in the United States.

2. A taxis is an innate response to a stimulus in which the body of the insect generally changes position toward the stimulus (positive) or away from the stimulus

(negative). Chemicals (chemo) and wind currents (anemo) are two examples of taxis stimuli.

3. Corpora allata are endocrine glands associated with the frontal lobe that synthesize juvenile hormone.

4. Quorum sensing is the regulation of gene expression in bacteria due to the secretion of autoinduction or pheromone signals, which are density dependent. Swarming is the behavior of flagellated bacterial cells that occurs during biofilm formation and is a phenotype of quorum sensing.

References cited

Aak, A. & Knudsen, G.K. (2011) Sex differences in olfaction-mediated visual acuity in blowflies and its consequences for gender-specific trapping. *Entomologia Experimentalis et Applicata* 139: 25–34.

Aak, A. & Knudsen, G.K. (2012) Egg developmental status and the complexity of synthetic kairomones combine to influence attraction behavior in the blow fly *Calliphora vicina*. *Physiological Entomology* 37: 127–135.

Aak, A., Knudsen, G.K. & Soleng, A. (2010) Wind tunnel behavioural response and field trapping of the blowfly *Calliphora vicina*. *Medical and Veterinary Entomology* 24: 250–257.

Abed, D., Cheviet, P., Farine, J.-P., Bonnard, O., le Quéré, J.-L. & Brossut, R. (1993) Calling behavior of female *Periplaneta americana*: behavioural analysis and identification of the pheromone source. *Journal of Insect Physiology* 39: 709–720.

Anderson, G.S. (2020) Factors that influence insect succession on carrion. In: J.H. Byrd & J.L. Castner (eds) *Forensic Entomology: The Utility of Using Arthropods in Legal Investigations*, 3rd edn, pp. 103–140. CRC Press, Boca Raton, FL.

Archer, M.S. & Elgar, M.A. (2003) Effects of decomposition on carcass attendance in a guild of carrion-breeding flies. *Medical and Veterinary Entomology* 17: 263–271.

Ashworth, J.R. & Wall, R. (1994) Responses of the sheep blowflies *Lucilia sericata* and *L. cuprina* to odour and the development of semiochemical baits. *Medical and Veterinary Entomology* 8: 303–309.

Barton Browne, L., Bartell, R.J. & Shorey, H.H. (1969) Pheromone-mediated behaviour leading to group oviposition in the blowfly *Lucilia cuprina*. *Journal of Insect Physiology* 15: 1003–1014.

Blum, M.S. (1996) Semiochemical parsimony in the Arthropoda. *Annual Review of Entomology* 41: 353–374.

Boulay, J., Devigne, C., Gosset, D. & Charabidze, D. (2013) Evidence of active aggregation behavior in *Lucilia sericata* larvae and possible implication of a conspecific mark. *Animal Behaviour* 85: 1191–1197.

Brodie, B.S., Wong, W.H., VanLaerhoven, S. & Gries, G. (2015) Is aggregated oviposition by the blow flies *Lucilia sericata* and *Phormia regina* (Diptera: Calliphoridae) really pheromone-mediated? *Insect Science* 22:651–660.

Brundage, A.L., Crippen, T.L., Singh, B., Benbow, M.E., Liu, W., Tarone, A.M., Wood, T.K. & Tomberlin, J.K. (2017) Interkingdom cues by bacteria associated with conspecific and heterospecific eggs of *Cochliomyia macellaria* and *Chrysomya rufifacies* (Diptera: Calliphoridae) potentially govern succession on carrion. *Annals of the Entomological Society of America* 110: 73–82.

Chapman, R.F. (1998) *The Insects: Structure and Function*, 4th edn. Cambridge University Press, Cambridge, UK.

Chapman, R.F. (2003) Contact chemoreception in feeding by phytophagous insects. *Annual Review of Entomology* 48: 455–484.

Chaudhury, M.F., Skoda, S.R., Sagel, A. & Welch, J.B. (2010) Volatiles emitted from eight wound-isolated bacteria differentially attract gravid screwworms (Diptera: Calliphoridae) to oviposit. *Journal of Medical Entomology* 47: 349–354.

Chaudhury, M.F., Zhu, J.J. & Skoda, S.R. (2017) Physical and physiological factors influence behavioral responses of *Cochliomyia macellaria* (Diptera: Calliphoridae) to synthetic attractants. *Journal of Economic Entomology* 110: 1929–1934.

DeVaney, J.A., Eddy, G.W., Ellis, E.M. & Harrington, H. (1973) Attractancy of inoculated and incubated bovine blood fractions to screwworm flies (Diptera: Calliphoridae): role of bacteria. *Journal of Medical Entomology* 10: 591–595.

Easton, C. & Feir, D. (1991) Factors affecting the oviposition of *Phaenicia sericata* (Meigen)(Diptera: Calliphoridae). *Journal of the Kansas Entomological Society* 64: 287–294.

Emmens, R. & Murray, M. (1983) Bacterial odors as oviposition stimulants for *Lucilia cuprina* (Wiedemann) (Diptera: Calliphoridae), the Australian sheep blowfly. *Bulletin of Entomological Research* 73: 411–416.

Frederickx, C., Dekeirsschieter, J., Verheggen, F.J. & Haubruge, E. (2012) Responses of *Lucilia sericata* Meigen (Diptera: Calliphoridae) to cadaveric volatile organic compounds. *Journal of Forensic Sciences* 57: 386–390.

Frick, C. & Wink, M. (1995) Uptake and sequestration of oubain and other cardiac glycosides in *Danaus plexippus* (Lepidoptera: Danaidae): evidence for a carrier-mediated process. *Journal of Chemical Ecology* 21: 557–575.

Gardner, D.R. & Stermitz, F.R. (1988) Host plan utilization and iridoid glycoside sequestration by *Euphydryas anicia* (Lepidoptera: Nymphalidae). *Journal of Chemical Ecology* 14: 2147–2168.

Gião, J. & Godoy, W. (2007) Ovipositional behavior in predatory and parasitic blowflies. *Journal of Insect Behavior* 20: 77–86.

Gill-King, H. (1997) Chemical and ultrastructural aspects of decomposition. In: W.D. Haglund & M.H. Sorg (eds) *Forensic Taphonomy: The Postmortem Fate of Human Remains*, pp. 93–104. CRC Press, Boca Raton, FL.

Gomperts, B.D., Kramer, I.M. & Tatham, P.E.R. (2002) *Signal Transduction*. Academic Press, San Diego, CA.

Grasswitz, T.R. & Jones, G.R. (2002) Chemical ecology. In: *Encyclopedia of Life Sciences*. John Wiley & Sons, Ltd., Chichester, UK.

Greenfield, M.D. (2002) *Signalers and Receivers: Mechanisms and Evolution of Arthropod Communication*. Oxford University Press, New York.

Gullan, P.J. & Cranston, P.S. (2014) *The Insects: An Outline of Entomology*, 5th edn. Wiley Blackwell, Chichester, UK.

Hallem, E., Ho, M. & Carlson, J.R. (2004) The molecular basis of odor coding in the *Drosophila* antenna. *Cell* 117: 965–979.

Hammack, L. (1991) Oviposition by screwworm flies (Diptera: Calliphoridae) on contact with host fluids. *Journal of Economic Entomology* 84: 185–190.

Happ, G.M., Schroeder, M.E. & Wang, J.C.H. (1970) Effects of male and female scent on reproductive maturation in young female *Tenebrio molitor*. *Journal of Insect Physiology* 16: 1543–1548.

Harris, M. & Miller, J. (1988) Host-acceptance behaviour in an herbivorous fly, *Delia antiqua*. *Journal of Insect Physiology* 34: 179–190.

Holl, M.V. & Gries, G. (2018) Studying the "fly factor" phenomenon and its underlying mechanisms in house flies, *Musca domestica*. *Insect Science* 25: 137–147.

Hölldobler, B. (1989) Communication between ants and their guests. In: J.L. Gould & C.G. Gould (eds) *Life at the Edge*. W.H. Freeman and Company, New York.

Hölldobler, B. & Wilson, E.O. (1990) *The Ants*. Harvard University Press, Cambridge, MA.

Hough-Goldstein, J.A. & Bassler, M.A. (1988) Effects of bacteria on oviposition by seed corn maggots (Diptera: Anthomyiidae). *Environmental Entomology* 17: 7–12.

Howse, P., Stevens, I. & Jones, O. (1998) *Insect Pheromones and Their Use in Pest Management*. Chapman & Hall, London.

Ikeshoji, T., Ishikawa, Y. & Matsumoto, Y. (1980) Attractants against onion maggots and flies, *Hylemya antiqua*, in onions inoculated with bacteria. *Journal of Pesticide Science* 5: 343–350.

Johansen, H., Solum, M., Knudsen, G.K., Hågvar, E.B., Norli, H.R. & Aak, A. (2014) Blow fly responses to semiochemicals produced by decaying carcasses. *Medical and Veterinary Entomology* 28: 26–34.

Kasper, J., Hartley, S. & Schatkowski, S. (2015) The influence of the physiological stage of *Lucilia caesar* (L.)(Diptera: Calliphoridae) females on the attraction of carrion odor. *Journal of Insect Behavior* 28: 183–201.

Leal, W.S. (2005) Pheromone reception. *Topic in Current Chemistry* 240: 1–36.

LeBlanc, H.N. & Logan, J.G. (2010) Exploiting insect olfaction in forensic entomology. In: J. Amendt, C.P. Campobasso, M.L. Goff & M. Grassberger (eds) *Current Concepts in Forensic Entomology*, pp. 205–222. Springer, London.

Lee, J.-H., Wood, T.K. & J. Lee (2015) Roles of indole as an interspecies and interkingdom signaling molecule. *Trends in Microbiology* 23: 707–718.

Leitch, O., Papanicolaou, A., Lennard, C., Kirkbride, K.P. & Anderson, A. (2015) Chemosensory genes identified in the antennal transcriptome of the blowfly *Calliphora stygia*. *BMC Genomics* 16: 255.

Liu, W., Longnecker, M., Tarone, A.M. & Tomberlin, J.K. (2016) Responses of *Lucilia sericata* (Diptera: Calliphoridae) to compounds from microbial decomposition of larval resources. *Animal Behaviour* 115: 217–225.

Ma, Q., Fonseca, A., Liu, W., Fields, A.T., Pimsler, M.L., Spindola, A.F., Tarone, A.M., Crippen, T.L., Tomberlin, J.K. & Wood, T.K. (2012) *Proteus mirabilis* interkingdom swarming signals attract blow flies. *ISME Journal* 6: 1356–1366. https://doi.org/10.1038/ismej.2011.210.

Mackley, J.W. & Brown, H.E. (1984) Swormlure-4, a new formulation of the Swormlure-2 mixture as an attractant for adult screwworms *Cochliomyia hominivorax* (Diptera: Calliphoridae). *Journal of Economic Entomology* 77: 1264–1268.

Mittelstaedt, H. (1962) Control systems of orientation in insects. *Annual Review of Entomology* 7: 177–198.

Nilssen, A.C., Tommeras, B.A., Schmid, R. & Evensen, S.B. (1996) Dimethyl trisulphide is a strong attractant for some Calliphorids and a Muscid but not for the reindeer oestrids *Hypoderma tarandi* and *Cephenemyia trompe*. *Entomologia Experimentali et Applicata* 79: 211–218.

Niogret, J., Lumaret, J.-P. & Bertrand, M. (2018) Comparison of the cuticular profiles of several dung beetles used as host carriers by the phoretic mite *Macrocheles saceri* (Acari: Mesostigamta). *Natural Volatiles and Essential Oils* 4: 8–13.

Nordlund, D.A. & Lewis, W.J. (1976) Terminology of chemical releasing stimuli in intraspecific and interspecific interactions. *Journal of Chemical Ecology* 2: 211–220.

Perotti, M.A., Braig, H.R. & Goff, M.L. (2010) Phoretic mites and carcasses: Acari transported by organisms associated with animal and human decomposition. In: J. Amendt, C.P. Campobasso, M.L. Goff & M. Grassberger (eds) *Current Concepts in Forensic Entomology*, pp. 69–91. Springer, London.

Randall, D., Burggren, W. & French, K. (2002) *Animal Physiology: Mechanisms and Adaptations*. W.H. Freeman and Company, New York.

Robacker, D.C. & Lauzon, C.R. (2002) Purine metabolizing capability of *Enterobacter agglomerans* affects volatiles production and attractiveness to Mexican fruit fly. *Journal of Chemical Ecology* 28: 1549–1563.

Rognes, K. (1991) *Blowflies (Diptera, Calliphoridae) of Fennoscandia and Denmark*. E.J. Brill, Leiden.

Schmidt, J.O. (1982) Biochemistry of insect venoms. *Annual Review of Entomology* 27: 339–368.

Smith, K.G.V. (1986) *A Manual of Forensic Entomology*. British Museum (Natural History), London.

Spindola, A.F., Zheng, L., Tomberlin, J.K. & Thyssen, P.J. (2017) Attraction and oviposition of *Lucilia eximia* (Diptera: Calliphoridae) to resources colonized by the invasive competitor *Chrysomya albiceps* (Diptera: Calliphoridae). *Journal of Medical Entomology* 54: 321–328.

Spivak, M., Conlon, D. & Bell, W.J. (1991) Wind-guided landing and search behaviour in fleshflies and blowflies exploiting a resource patch (Diptera, Sarcophagidae, Calliphoridae). *Annals of the Entomological Society of America* 84: 447–452.

Statheropoulos, M., Spiliopoulou, C. & Agapiou, A. (2005) A study of the volatile organic compounds evolved from the decaying human body. *Forensic Science International* 153: 147–155.

Stensmyr, M.C., Urru, I., Collu, I., Celander, M., Hansson, B.S. & Angioy, A.-M. (2002) Rotting smell of dead-horse arum florets. *Nature* 420: 625–626.

Tomberlin, J.K., Mohr, R., Benbow, M.E., Tarone, A.M. & Van Laerhoven, S. (2011) A roadmap bridging basic and applied research in forensic entomology. *Annual Review of Entomology* 56: 401–421.

Tomberlin, J.K., Crippen, T.L., Tarone, A.M., Singh, B., Adams, K., Rezenom, Y.H., Benbow, M.E., Flores, M., Longnecker, M., Pechal, J.L., Russell, D.H., Beier, R.C. & Wood, T.K. (2012) Interkingdom responses of flies to bacteria mediated by fly physiology and bacterial quorum sensing. *Animal Behaviour* 84: 1449–1456.

Urech, R., Green, P.E., Rice, M.J., Brown, G.W., Duncalfe, F. & Webb, P. (2004) Composition of chemical attractants affects trap catches of the Australian sheep blowfly, *Lucilia cuprina*, and other blowflies. *Journal of Chemical Ecology* 30: 851–866.

Uriel, Y., Gries, R., Tu, L., Carroll, C., Zhai, H., Moore, M. & Gries, G. (2018) The fly factor phenomenon is mediated by interkingdom signaling between bacterial symbionts and their blow fly hosts. *Insect Science*. https://doi.org/10.1111/1744-7917.12632.

Vander Meer, R., Breed, M.D., Winston, M. & Espelie, K.E. (2019) *Pheromone Communication in Social Insects: Ants, Wasps, Bees and Termites*. CRC Press, Boca Raton, FL.

Vass, A.A., Barshick, S.A., Sega, G., Caton, J., Skeen, J.T., Love, J.C. & Synstelien, J.A. (2002) Decomposition chemistry of human remains: a new methodology for determining the postmortem interval. *Journal of Forensic Science* 47: 542–553.

Vickers, N.J. (2000) Mechanisms of animal navigation in odor plumes. *Biological Bulletin* 198: 203–212.

Visser, J.H. (1986) Host odor perception in phytophagous insects. *Annual Review of Entomology* 31: 121–144.

Vogt, R. (2003) Biochemical diversity of odor detection: OBPs, ODEs, and SNMPs. In: Blomquest, G. & R. Vogt (eds) *Insect Pheromone Biochemistry and Molecular Biology*, pp. 391–445. Academic Press, San Diego.

Wall, R. & Fisher, P. (2001) Visual and olfactory cue interaction in resource-location by the blowfly, *Lucilia sericata*. *Physiological Entomology* 26: 212–218.

Wall, R. & Warnes, M.L. (1994) Responses of the sheep blowfly Lucilia sericata to carrion odour and carbon dioxide. *Entomologia Experimentalis et Applicata* 73: 239–246.

Wardle, R.A. (1921) The protection of meat commodities against blowflies. *Annals of Applied Biology* 8: 1–9.

Wood, D.L. (1982) The role of pheromones, kairomones and allomones in host selection and colonization by bark beetles. *Annual Review of Entomology* 27: 411–446.

Wyatt, T.D. (2003) *Pheromones and Animal Behaviour: Communication by Smell and Taste*. Cambridge University Press, Cambridge, UK.

Yao, M., Rosenfeld, J., Attridge, S., Sidhu, S., Aksenov, V. & Rollo, C.D. (2009) The ancient chemistry of avoiding risks of predation and disease. *Evolutionary Biology* 36: 267–281.

Supplemental reading

Burkepile, D.E., Parker, J.D., Woodson, C.B., Mills, H.J., Kubanek, J., Sobecky, P.A. & Hay, M.E. (2006) Chemically mediated competition between microbes and animals: microbes as consumers in food webs. *Ecology* 87: 2821–2831.

Calcagnile, M., Tredici, S.M., Talà, A., & Alifano, P. (2019) Bacterial semiochemicals and transkingdom interactions with insects and plants. *Insects* 10: 441.

Cardé, R.T. & Miller, J.G. (eds) (2004) *Advances in Insect Chemical Ecology*. Cambridge University Press, Cambridge, UK.

Eisemann, C.H. & Rice, M.J. (1987) The origin of sheep blowfly, *Lucilia cuprina* (Wiedemann) (Diptera: Calliphoridae), attractants in media infested with larvae. *Bulletin of Entomological Research* 77: 287–294.

Morris, M.C. (2005) Tests on a new bait for flies (Diptera: Calliphoridae) causing cutaneous myiasis (flystrike) in sheep. *New Zealand Journal of Agricultural Research* 48: 151–156.

Mullen, G. & Durden, L. (2002) *Medical and Veterinary Entomology*. Academic Press, Amsterdam.

Tabata, J. (2018) *Chemical Ecology of Insects: Applications and Associations with Plants and Microbes*. CRC Press, Boca Raton, FL.

Vass, A.A., Bass, W.M., Wolt, J.D., Foss, J.E. & Ammons, J.T. (1992) Time since death determinations of human cadavers using soil solution. *Journal of Forensic Science* 37: 1236–1253.

Vet, L.E.M. & Dicke, M. (1992) Ecology of infochemical use by natural enemies in a tritrophic context. *Annual Review of Entomology* 37: 141–172.

Wall, R., Green, C.H., French, N. & Morgan, K.L. (1992) Development of an active target for the sheep blowfly Lucilia sericata. *Medical and Veterinary Entomology* 6: 67–74.

Additional resources

Center for Chemical Ecology: http://ento.psu.edu/chemical-ecology

International Society of Chemical Ecology: www.chemecol.org

Journal of Chemical Ecology: http://www.springer.com/life+sciences/ecology/journal/10886

Journal of Insect Behavior: http://www.springer.com/life+sciences/entomology/journal/10905

Max Planck Institute for Chemical Ecology: http://www.ice.mpg.de/ext/

The Pherobase: database for pheromones and semiochemicals: https://www.pherobase.com/

Chapter 8
Biology of the maggot mass

Overview

Necrophagous flies are extremely efficient at utilizing carrion. This efficiency is largely realized
through cooperative or group feeding by fly
larvae. Feeding aggregations form shortly after
egg hatch or larviposition, when hundreds to
thousands of individuals are drawn together, creating a microhabitat known as a larval or maggot
mass. These masses embody life at the edge in
that intense competition occurs between conspecifics and allospecifics for nutriment, with
starvation eminently possible for those not
equipped to effectively compete; temperatures
soar at the core of the mass, sometimes elevating
to more than 30 °C above ambient conditions,
creating an environment that should evoke
thermal stress if not death; and predators and parasites attack from the periphery and above, making the center of the aggregations a highly
sought-after refuge, provided it is not too hot!
Why endure such atrocious living conditions?
The answer is simple: there can be enormous benefits to members of the feeding aggregations. In
fact, some fly species likely cannot survive without

relying on cooperative feeding. This chapter is
dedicated to the maggot mass, and begins with an
examination of what is known about the formation
of larval aggregations among the most common
flies present (Calliphoridae and Sarcophagidae),
followed by a discussion of the trade-offs associated with developing in maggot masses. The
ability of larval masses of flies, and even some
beetles, to generate heat will be examined, particularly how mass heterothermy influences the
growth characteristics of larvae and calculations
of a minimum postmortem interval.

The big picture

- Carrion communities are composed largely of fly
  larvae living in aggregations.
- Formation of maggot masses is more complex
  than originally thought.
- Larval feeding aggregations provide adaptive
  benefits to individuals.
- Developing in maggot masses is not always beneficial to conspecifics or allospecifics.
- Larval aggregation may benefit Coleoptera too!

The Science of Forensic Entomology, Second Edition. David B. Rivers and Gregory A. Dahlem.
© 2023 John Wiley & Sons Ltd. Published 2023 by John Wiley & Sons Ltd.
Companion website: www.wiley.com/go/forensicentomology2

8.1 Carrion communities are composed largely of fly larvae living in aggregations

Death of a vertebrate animal triggers excited activity among an array of arthropods, all attempting to capitalize on the appearance of a nutrient-rich resource. The fact that carrion is unpredictable in terms of timing (patchy), location, or duration (ephemeral) as a resource promotes intense competition to locate, colonize, and assimilate the animal remains. Chapter 7 explored the chemical signals used by necrophagous flies that are attracted to a body. Highly attuned olfaction allows flies in the family Calliphoridae to "win" the initial leg of the competition as blow flies typically are the first colonizers of most types of carcasses (Smith, 1986) (Figure 8.1). Upon arrival, gravid females generally will oviposit in natural body openings (e.g., mouth, nose, ears, and anus) or, when present, in wounds and lesions in the skin. Eggs are deposited in locations that favor neonate larval feeding on liquids and soft tissues high in proteins. Remember from Chapter 6, the importance of protein intake as

juveniles, particularly for autogenous or capital breeders, needed for later reproductive events. As larvae progress in development, they will molt into second and eventually the third and final larval stages. These developmental stages are most common when fly larvae assemble together to form feeding aggregations.

Details about how maggot masses form will be addressed in Section 8.2. For now, it is important to understand that depending on the size of the carcass, several hundred to thousands of individual fly larvae can compose a single maggot mass. When considering that multiple maggot masses can exist on a body at the same time, it should be apparent that the fly larvae in these aggregations account for the largest component (by weight)[1] of the invertebrate carrion community (Hanski, 1987). The masses can also be viewed as dynamic microhabitats in that each mass can vary in size, location (on body), and species composition, with each parameter dependent on ambient conditions, geographic location, season, amount of sunlight, and stage of corpse decomposition (Rivers *et al.*, 2011; Anderson, 2020). Despite observations of maggot mass formation, heat generation, and intense larval competition appearing in the research literature for more than 80 years, our understanding of key facets of maggot mass biology, particularly the physiological ecology of the aggregations as a whole and species–species interactions, is rudimentary. With the growing importance of necrophagous flies as evidence in criminal investigations, especially in contributing to estimation of the postmortem interval, there is a need to explore these fascinating microhabitats in depth to better understand **abiotic** and **biotic** influences on the flies present in feeding aggregations. The remainder of the chapter explores what is known about the central elements of maggot mass biology, including formation, heat generation, cooperative feeding, and deleterious effects of developing in large feeding aggregations.

Figure 8.1 Adult blow flies (Family Calliphoridae) are usually the initial colonizers of most types of animal remains, and consequently are the first to form and occupy a developing larval aggregation on a corpse. *Source*: Photo by D.B. Rivers.

8.2 Formation of maggot masses is more complex than originally thought

Fly maggot masses are impressive in terms of the sheer numbers of individuals that come together as one, working cooperatively to process and assimilate

Figure 8.2 A heterogenous maggot mass or larval feeding aggregation formed during faunal succession on a piglet in a wooded lot in southcentral Pennsylvania. *Source*: Photo by D.B. Rivers.

the soft tissues of a corpse (Figure 8.2). The aggregations can be **homogeneous** with only one species present but are more commonly **heterogeneous,** composed of larvae from the families Calliphoridae, Sarcophagidae, Muscidae, and several micro-dipteran groups (Campobasso *et al.*, 2001). Under most conditions examined in the field and in case studies, various species of blow flies dominate the composition of maggot masses (Goodbrod & Goff, 1990; Joy *et al.*, 2006), suggesting that the benefits attained from larval aggregations are more important to their survival then flies from other families. This speculation has not been examined experimentally and it is entirely possible that other explanations (e.g., higher fecundity, timing of oviposition, rates of feeding, and thermal tolerances) can account for the dominance of blow flies in maggot masses.

Precisely how the feeding aggregations form has not been determined, nor is there a complete understanding as to why larvae assemble into large masses. There are some reasonable explanations to the "why" question, which will be addressed in Section 8.3. Here we will explore the question of how the feeding aggregations form. The simplest answer given so far has been **thigmotaxis,** the type of innate behavior in which an animal responds to touch or physical contact with a solid object. In the case of necrophagous fly larvae, the expectation is that they display positive thigmotaxis, seeking contact with a physical object (i.e., other larvae or soft tissues). When several larvae in close proximity engage in this behavior, the result is clustering in aggregations (Gennard, 2012). This explanation

does not account for why the larvae were in close proximity in the first place. As mentioned earlier, larval aggregations typically do not form until the second or third stage of larval development, well after the initial clustering of eggs that occurs during oviposition. For several species, neonate larvae often migrate from the initial oviposition site (Greenberg & Kunich, 2002). Thus, it is more likely that a positive thigmotaxic response is what helps maintain the assembly of larvae (retention) once formed but probably is not the reason for the initial formation of the aggregation (Devigne *et al.*, 2011; Boulay *et al.*, 2013).

Maggot mass formation is likely under the influence of one of three possible mechanisms: (i) clustered oviposition/larviposition by gravid females, (ii) random formation, and (iii) foraging by larvae involving signal cues used to form or find an aggregation. There is limited information about each mechanism and despite observations that supports one versus another mechanism, it is entirely possible that each contributes to the formation of larval feeding aggregations under specific conditions.

8.2.1 Clustered oviposition and larviposition

As discussed in Chapter 7, pheromonal and/or bacterial signaling may be used by some calliphorid species to recruit conspecific females to carrion. The idea of egg clustering is that female flies coat the egg chorion with chemicals or bacteria that function as assembly pheromones and/ or oviposition stimulants, using the chemical signals to initiate activation and searching in conspecifics (review Chapter 7 for details of these phases), which in turn ultimately seek the same oviposition site for egg deposition. Non-related species of flies appear capable of recognizing the aggregation signals as well (thereby functioning as kairomones or synomones), prompting oviposition or larviposition in the growing cluster or in close proximity. The presumption is that egg clustering inevitably leads to the formation of maggot masses as occurs with some other species of Diptera (Jaenike & James, 1991). In this scenario, neonate larvae reside together from the onset in a location on the carcass that favors species-specific development (Norris, 1965; Anderson, 2020). For this assumption to hold true, larvae need to feed at the oviposition site following egg hatch, and they must benefit from gregarious behavior with conspecifics, a feature consistent with the **Allee effect theory** (Allen, 2010; Charabidze *et al.*, 2011). However, only one of the two assumptions are actually met under natural or laboratory conditions. Larvae of most fly species do not always remain at the site of deposition and generally do not form feeding aggregations until late second or early third stages of larval development (Charabidze *et al.*, 2008; Rivers *et al.*, 2010; Boulay *et al.*, 2013). In fact, if ambient temperatures are not optimal (too hot or cold), it is common for neonate larvae to crawl from the oviposition site almost immediately upon hatch. Young larvae appear to seek orifices and folds of skin for refuge, but they do not aggregate until later stages of development. As will be discussed in Section 8.3, conspecific larvae not only seem to benefit from cooperative feeding, but their very survival may also depend on it.

8.2.2   Random formation

An alternative explanation for the formation of larval feeding assemblages is simply due to chance. Fly larvae representing several species feed and compete voraciously for the limited resources of carrion. To compensate for a rapidly diminishing resource, necrophagous flies are expected to oviposit far more eggs or larvae than can be supported by the carcass (review Chapter 6 for a more thorough discussion). Immediately following egg hatch or larviposition, the young larvae feed exclusively on liquids and soft tissue during all stages of immature development. The end result is large numbers of larvae feeding and developing in close proximity to one another since conspecific oviposition yielded clustering of eggs. The neonate larvae also engage in competition for the same food resources and the mere presence of hundreds to thousands of individuals on a finite "nutrient island" will presumably "force" interactions as the available resources dwindle and the larvae increase in size and volume. The mass becomes more of an end result of competition and overcrowding rather than a self-organized pattern or an orchestrated assemblage to facilitate larval development (Charabidze *et al.*, 2011).

This explanation as the sole mechanism for maggot mass formation is not entirely satisfactory for several reasons. Although oviposition by multiple females frequently occurs in the same location on the carcass, following egg hatch, larvae of many

species rapidly disperse from the site of oviposition seeking avenues to penetrate the interior of the carcass (Greenberg & Kunich, 2002; Charabidze *et al.*, 2008). Carcass surface temperatures frequently promote newly deposited larvae to disperse from the initial deposition site, yet assemble later in development into feeding aggregations. When the corpse is a large mammal, larvae may be separated by several centimeters to meters during the first and portions of the second stage of larval development, or in some instances for the entire larval period. As discussed earlier, feeding aggregations do not form for most species until larvae achieve either the second or third stages and, when they do, the larvae form rather tight masses at specific locations on the body. The key here is that the masses did not result from overcrowding since much of the carcass remains unexploited by necrophagous flies, at least during the initial wave of colonization on a large animal.

8.2.3   Foraging by larvae

Overcrowding or food limitations do not account for the formation of feeding aggregations among the first wave of colonizers on large carcasses, and the occurrence of hundreds to thousands of individuals in these maggot masses would argue against random formation. Mounting evidence points to chemical cues, possibly serving as signals akin to pheromone trails used by social Hymenoptera, leading to the formation and/or retention of larvae in feeding assemblages (Boulay *et al.*, 2013; Fouche *et al.*, 2018). We have already discussed in Chapter 7 the refined chemical acuity of adult females from several species of Calliphoridae to detect and then search for a corpse via foraging behavior. A natural extension of this adaptive trait is for the progeny (larvae) to possess the ability to chemically locate food or a feeding aggregation by either olfaction or gustation. Once again, the underlying assumption is that larvae benefit from "finding" each other.

Larvae from at least one species of sarcophagid, *Sarcophaga (Neobelleria) bullata* Parker, demonstrate chemoattraction to carrion, as well as isolated bovine tissues (Christopherson & Gibo, 1996) (Figure 8.3). The attraction appears to be stronger to food that has been previously fed upon, albeit time dependent, by conspecifics. Such olfactory responses may allow larvae to locate specific tissues on a carcass to feed or to detect maggot masses. Foraging by *S. bullata* is evident in late second and

(a)

(b)

Figure 8.3   (a) Adults of the flesh fly, *Sarcophaga bullata,* and (b) a maggot mass of the same species formed on bovine liver. *Source*: Photos by D.B. Rivers.

early third larval stages, consistent with the age of larval development in which maggot masses are most likely to form for this species (Rivers *et al.*, 2010). Observations with the calliphorids *Lucilia sericata* and *Calliphora vomitoria* suggest that juveniles of both species actively aggregate as second and third stage larvae. Assemblages initially are small and expand over time as more larvae are recruited. Recruitment and retention in the mass appear to be achieved through chemical signals released by larvae (Boulay *et al.*, 2013; Fouche *et al.*, 2018). As the size of the aggregation grows, so does the attractiveness of larvae to the mass, even for allospecifics (Fouche *et al.*, 2018). What prompts "recruitment" behavior is not known, nor is it understood if all larvae are capable of releasing cues for aggregation. One observation that may shed light onto the origins of larval signaling is that recruited larvae will stop to feed where the cues have been released by other larvae (Fouche *et al.*, 2018). From this observation, it would be tempting to argue that gustatory reception is involved with mass detection, but gustation appears to be the result of the signal, not the mechanism for recognition. A more plausible explanation seems to be that the signal originates from endogenous bacteria that are then consumed by conspecifics for transport to other locations, including within the forming maggot mass. Sounds familiar? It should; as this scenario is akin to the "fly factor" discussed in Chapter 7, in which, adult flies inoculate a newly discovered carcass with bacteria via regurgitation and fecal elimination. The bacteria then produce byproducts that serve as signaling molecules (Holl & Gries, 2018). If bacteria are the source of larval

signals, this could account for the timing of larval aggregations, since a delay (albeit short) would be expected from the timing of bacterial inoculation until the release of odors or byproducts that serve as attractants in recruitment. The delay conceivably corresponds to the time from egg hatch until late second or early third instar. Similar larval behaviors have been anecdotally observed for the blow fly *Protophormia terraenovae*, suggesting that larval foraging using chemical cues may be common among calliphorids and sarcophagids, and perhaps other families of Diptera.

8.2.4   Maggot mass heat as an agent of recruitment

Separate from the discussion to come in Section 8.3.2 on the adaptive benefits of hetero-thermy is the potential role of heat in larval recruit-ment. Larval movements by *Chrysomya rufifacies* and *Calliphora vicina* in a feeding aggregation have been shown to be nonrandom. Larvae actively migrate to the hottest spots of maggot masses, typi-cally located at the center (Johnson *et al.*, 2014). Interestingly, larvae of *C. vicina* seek out higher temperatures in a maggot mass than when individ-uals are permitted to select along a temperature gra-dient. The observations suggest that maggot masses afford benefits beyond just those reaped from ele-vated temperature effects on growth. As a consequence, larvae may migrate toward or remain in an established maggot mass, either due to thermal perception or heat-induced augmentation of the larval signals discussed in Section 8.2.3 (Figure 8.4).

Figure 8.4 Larvae from several species of calliphorids and other families of Diptera migrating between separate maggot masses on a juvenile pig carcass. The movement presumably reflects seeking optimal thermal zones for growth and development. *Source*: Photo by D.B. Rivers.

For species with high critical thermal maxima as larvae, the ability to recognize higher temperature zones could be an adaptive trait that permits location of the optimal mass sites for development as well as in foraging for prey species with flies that are predatory as larvae. Maggot mass heat as a larval recruitment cue is simply speculation at this point and has not been demonstrated experimentally.

8.3   Larval feeding aggregations provide adaptive benefits to individuals

Understanding how larval feeding aggregations initiate is less clear than why they form. Individuals and perhaps groups receive adaptive benefits from developing as a cooperative feeding mass. The latter is commonly called a "mass effect" or in the case of necrophagous fly larvae, a **larval mass-effect** (Charabidze *et al.*, 2011; Scanvion *et al.*, 2018). The idea is that a group effect – the result of which could be beneficial or deleterious – is induced or caused by alteration of the environment due to the population of organisms. In the scenario of a maggot mass, the "population" can be homogeneous or heterogeneous. The assumption is that maggot masses are generally favorable microhabitats based on the observations that aggregated

oviposition is not avoided, but instead encouraged by conspecific and allospecific species. Within limits, mass-forming flies achieve more rapid rates of growth and development owing to group feeding and heat generation. The production of heat confers advantages that range from enhanced aspects of food processing and assimilation to protection from low temperatures and possibly predators and/ or parasites (Rivers *et al.*, 2011). Intimate details of the physiology of group feeding by necrophagous flies are still unknown. However, several aspects have been revealed about calliphorids and sarcophagids to suggest that maggot masses are adaptive and essential to the survival of several species of necrophagous flies. What follows is a discussion of the key features of group or cooperative feeding that permit fly larvae to compete effectively for the rich yet limited nutrients of carrion (Hanski, 1987).

8.3.1   Group feeding

Group feeding in larval aggregations seems to rely on the fact that (i) for most species, independent living (i.e., a solitary existence) does not lead to completion of development and (ii) larvae are equipped with features to feed in a gregarious environment. These larval adaptations include modified mouthparts for tissue penetration, digestive enzymes, and physiological modifications associated with food assimilation.

8.3.1.1   Mouth hook manipulation

Carrion is a patchy ephemeral resource that is nutrient rich, but fades fast, simply meaning that the window to use it is very short. It is absolutely essential, then, that larvae of necrophagous flies be adapted for maximum utilization of the resource immediately upon egg hatch or following larval deposition on the food substrate. Newly hatched maggots rise to the challenge by penetrating the corpse with **mouth hooks** and begin a period of voracious feeding during the larval stages. Necrophagous flies have a clear need to work together to create avenues for penetrating the body cavity of a corpse and infiltrating soft tissues. Maggot mass formation offers a multitude of mouth hooks to accomplish this initial task of food manipulation, which arguably can be labeled **extra-oral mastication. Mastication,** the process of chewing, biting and/or tearing of food, is accomplished in most insects by large mandibles possessing an array of dentition and with the aid of accessory jaws in the form of maxillae (review Chapter 4 for a description of insect mouthparts). Fly larvae masticate using only a pair of mouth hooks, modified mandibles that can be retracted into the head region during locomotion and food processing (Figure 8.5).

Mouth hooks are generally considered relatively delicate structures for puncturing the integument of a dead vertebrate animal (Greenberg & Kunich, 2002), particularly if the skin is covered by hair or **pelage** (fur). However, some species possess large-mouth hooks that are heavily sclerotized with a sharp distal end designed for puncturing (Szpila, 2010). Dentition is absent on the mouth hooks, meaning that structures akin to incisors or canine teeth are not present to effectively cut or tear the integument of a corpse, although sharp dental sclerites are present on the ventral regions of these structures (Szpila, 2010). Consequently, blow flies tend to consume softer tissues first (e.g., lung and brain), presumably because these tissues are easier to infiltrate and not because they possess a higher nutrient content than other tissues (Kaneshrajah & Turner, 2004; Clark *et al.*, 2006).

In the absence of a direct path (e.g., natural body opening or lesion) to the internal environment of a corpse, a small individual fly larva is not likely capable of penetrating the tissues quickly enough to meet nutritional needs. The exceptions are juveniles of parasitic species of calliphorids that are only facultatively necrophilic, and which can complete development in a solitary lifestyle. The story ends differently for a lone obligatory necrophagous fly: desiccation and starvation ensue for a solitary neonate larva, with death the final outcome. While

this scenario is not definite for all necrophagous fly species, many require a minimum number of individuals to form a larval aggregation so that normal growth and development can occur (Rivers *et al.*, 2010). Probably the more correct statement is to say that a minimum threshold of individuals is needed for successful larval development. Presumably a critical number of larvae are required to manipulate the food using mouth hooks (as well as other digestive features); too few and the larvae are not capable of obtaining sufficient nutriment. There is evidence to support the notion of a minimum threshold in a larval mass in that a critical number of larvae are needed to generate heat above ambient temperatures (Rivers *et al.*, 2010; Charabidze *et al.*, 2011; Heaton *et al.*, 2014). The idea of a minimum threshold also seems to be in agreement with the argument of egg clustering serving as the mechanism for maggot mass formation. The advantage of cooperative feeding, however, appears to be age dependent in some species, as female calliphorids do not oviposit in response to chemical cues emanating from existing larval masses (Erzinçlioglu, 1996). Whether this trend holds true for flies from other families is not known.

8.3.1.2 Cooperative digestion

Once the larval masses break through the skin of a cadaver, the importance of mouth hooks declines (not entirely) and group feeding through larval aggregations becomes dependent on mass release of digestive enzymes. In experiments that compared larval development rates of the blow fly *Calliphora*

(a)

(b)

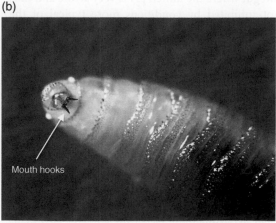

Figure 8.5   Mouth hooks in (a) a "cleared" larva of *Musca domestica*. *Source*: Sikander Kiani/Wikimedia Commons/ Public domain and (b) of the screwworm *Cochilomyia* sp. *Source*: John Kucharsi/Wikimedia Commons/Public domain.

auger when reared on fresh versus frozen (then thawed) sheep liver, the frozen liver was known to have weakened skin from the freeze-thaw cycle and thus was predicted to be more easily accessible to fly larvae if mouth hooks were the key to feeding success. However, no differences in duration of larval development were observed between the food types (Day & Wallman, 2006). Similarly, Clark *et al.* (2006) found that food structure ("liquidized," i.e., presumed homogenized in a food processor, vs. meat chunks) had no significant influence on larval development of *Lucilia sericata*. These observations suggest that features other than mass mouth hook use are more important to cooperative feeding among the flies.

One obvious feature is liberation of nutrients from soft tissues due to the action of digestive enzymes from salivary glands, foregut, and possibly midgut (Greenberg & Kunich, 2002; Anderson, 2020). Larvae of necrophagous calliphorids secrete fluids directly onto the food substrate to initiate extra-oral digestion. The bulk of the digestive enzymes released appear to originate from salivary glands (Price, 1974; Anderson, 1982; Young *et al.*, 1996). Analyses of secretions released onto food reveal that the fluids contain an array of digestive enzymes including trypsin-like and chymotrypsin-like proteases, carbohydrases (i.e., amylase), and a pepsin-like enzyme presumed to be cathepsin D-like proteinase (Pendola & Greenberg, 1975; Bowles *et al.*, 1988; Sandeman *et al.*, 1990; Padilha *et al.*, 2009). Cathepsin D-like proteinase has been speculated to be present in all cyclorrhaphous Diptera that feed on food infested with bacteria. If so, the enzyme is most likely restricted to salivary glands and regions of the foregut where the lumen pH is very acidic and thus not inhibited by conditions more typical of the midgut (Chapman, 1998). Such a spatial distribution of cathepsin D-like proteinase would be in contrast to larvae of the muscid *Musca domestica*, which restricts pepsin-like activity to the acidic environment of the midgut (Padilha *et al.*, 2009). Presumably all larvae in the maggot mass release these digestive enzymes to facilitate tissue digestion and nutrient acquisition.

There is no doubt that large maggot masses break down and consume carrion tissues faster than smaller assemblages (Scanvion *et al.*, 2018; Anderson, 2020). However, group digestion has not been experimentally tested for any species of calliphorid or sarcophagid. Thus, the ideas of mass release of enzymes and cooperative mouth hook penetration of a corpse are intuitive speculation rather than supported by direct evidence. As Greenberg & Kunich (2002) point out, a combination of heavy enzyme output with the "churning" created by constant larval movement should quickly convert corpse tissues into a nutrient soup that bathes the larvae and promotes rapid consumption. More individuals in a feeding aggregation should obviously constitute greater enzyme output and also higher internal temperatures, promoting increased enzymatic activity up to a maximum threshold (35–50 °C; Wigglesworth, 1972) and hence more rapid breakdown and digestion of tissues. Recent work with *L. sericata* seems to support these suppositions as larvae feeding on bovine muscle with the addition of trypsin (from porcine pancreas) developed more quickly and experienced lower mortality than those that did not (Scanvion *et al.*, 2018). In fact, the developmental benefits of trypsin-treated tissue were augmented with higher larval densities, leading the authors to conclude that this likely was a simulation of collective exodigestion that occurs in a large maggot mass (Scanvion *et al.*, 2018). Perhaps even more interesting is that trypsin-digested tissue (Scanvion *et al.*, 2018) is more conducive to larval development than liquidized (Clark *et al.*, 2006) for the same species, a suggestion that larval age and density are more important than food structure in influencing larval growth of *L. sericata* and potentially other fly species.

8.3.1.3 Food assimilation

The dynamics of larval **food assimilation** in naturally formed maggot masses has not been studied. So our understanding and predictions of food assimilation in feeding aggregations is derived from experiments with either isolated individuals or from laboratory studies relying on controlled mass sizes, often with unlimited food. What is known is that efficiency of food assimilation, namely the use of absorbed food nutrients in cellular functions such as metabolism, assembly and as fuel, is a function of the metabolic rate of fly larvae and other insects (Wigglesworth, 1972). Metabolic rate increases with elevated temperatures and since all necrophagous insects are poikilotherms, body temperature is directly influenced by changes in environmental temperature. As discussed later in this section, maggot masses generate internal heat, and as mass size (number of individuals or density) and/or volume increases, internal temperatures of the aggregations elevate, generally several degrees above ambient conditions. With this in mind, the efficiency of food assimilation is expected to increase with mass size up to a maximum threshold level.

When food assimilation efficiency is estimated by the following equation (modified from Karasov & Martínez del Rio, 2007): then internal maggot mass temperatures influence not only the metabolic rate, but also the rate of chemical digestion and transit time of the food bolus in the midgut. Midgut proteases in larvae of *Calliphora vomitoria* reach maximum activity at temperatures above 35 °C and do not decline until fluids exceed 50 °C (Wigglesworth, 1972), temperature ranges commonly associated with feeding aggregations forming through natural faunal succession. Likewise, gut motility, and hence bolus movement, increases with temperature. In some calliphorid species, food moves from mouth to anus in less than 20 minutes at 31 °C (Greenberg & Kunich, 2002). Thus, a combination of high metabolic rate, optimum enzyme activity, and fast transit times should contribute to increased food assimilation efficiency.

$$\text{Assimilation Efficiency} = \frac{\text{digestion rate} \times \text{food retention rate}}{\text{concentration of food} \times \text{midgut volume}}$$

One measure of food assimilation is larval growth rates. Enhanced efficiency in assimilating food derived from the carcass is expected to accelerate rates of larval development, which in fact has been observed in both natural and laboratory conditions. For many species of sarcophagids and calliphorids, as maggot mass size elevates, corresponding temperature increases ensue within the mass, and larval growth rates accelerate up to the maximum size of maggot mass or upper temperature limit (Villet *et al.*, 2010; Rivers *et al.*, 2011). Growth rates for calliphorid and sarcophagid immatures are linear within species-specific physiological limits (Byrd & Butler, 1998; Grassberger & Reiter, 2001, 2002). Provided overcrowding is avoided, rapid development rates do not result in reduced larval body sizes (Rivers *et al.*, 2010; Scanvion *et al.*, 2018), suggesting that necrophagous fly larvae have limited trade-offs between rapid rates of digestion and efficiency of food assimilation (Karasov & Martínez del Rio, 2007). This may be the most significant benefit associated with feeding in large aggregations.

The efficiency of food processing is simply staggering by comparison to other insects. For example, net production (assimilation) efficiency has been calculated for *Lucilia illustris* reared in masses of 10–100 individuals on beef liver at 35 °C to approach 90%. Hanski (1976) criticized his own calculations as being overestimations since the larvae were fed an optimized diet of beef liver, which is likely far more nutrient rich than a natural food source. Work with other blow fly species suggests the efficiency estimate for *L. illustris* is not far off as the values all exceed 75% efficiency (Hanski, 1977; Williams & Richardson, 1984). To put this in perspective, most insects are expected to demonstrate production efficiencies below 50%, with saprophagous species yielding even lower efficiencies of food conversion (Waldbauer, 1968; Heal, 1974). Food assimilation by necrophagous fly larvae appears to rival the extremely high food conversion efficiencies of parasitic insects and other parasitic invertebrates, placing group feeding by these flies as one of the most efficient foraging and conversion strategies among all animal groups (Willmer *et al.*, 2000), and yet another reason that competition is so fierce for animal remains.

8.3.2   *Heterothermy*

Dense aggregations of feeding larvae generate internal heat. Heat production by fly maggots is impressive on at least two levels, the first being the mere fact that a poikilothermic animal can generate heat apparently to meet specific physiological needs, and the second is the degree of heat production. Insects are poikilotherms – their body temperature varies with environmental conditions. They are also classified as **ectotherms,** a broader term used for animals that typically do not generate heat as part of a mechanism to thermoregulate (Randall *et al.*, 2002). All ectotherms are not pokilo-thermic. Thus, heat generation in a larval feeding aggregation is not typical of most insects. Maggot mass heat production also seems to be an example of **heterothermy,** in which a poikilotherm displays **homeothermy** for a short period of time (or vice versa may occur as well) (Willmer *et al.*, 2000). Homeothermic animals are defined as those that maintain a stable internal environment regardless of the conditions outside the body. As will be seen, the maggot mass and larvae that compose it do not fit nicely into this definition, and consequently fall outside the typical definition of heterothermy.

Why? A stable or static temperature is not maintained within feeding assemblages. Temperatures elevate as the masses grow in size and volume, and both of these features change at a rapid rate, as discussed in Section 8.3.1. Perhaps, a more accurate definition of heterothermy associated with maggot masses is "temporary heat production by larvae in a feeding aggregation that elevates above ambient conditions linearly over time but is constrained by physiological limits."

The second facet of heterothermy that is truly impressive is the degree of temperature elevations experienced in feeding aggregations. Heat production can exceed ambient air temperatures by several degrees, and in some instances rise to more than 30 °C above environmental conditions. Some species of calliphorids found near the equator develop in assemblages with temperatures exceeding 50 °C (Richards *et al.*, 2009; Villet *et al.*, 2010). Such temperature conditions are considered a **proteotaxic or thermal stress,** capable of evoking stress responses associated with heat injury or heat shock, and for some species can elicit death. Amazingly, larvae of most calliphorids and sarcophagids appear to thrive in the hot humid microclimate of the maggot mass. Heat production by a maggot masses may be viewed, then, not as just a byproduct of larval metabolism that must be dealt with; rather, it is an adaptive feature of the life-history strategies of carrion-breeding flies.

The generation of elevated temperatures in feeding aggregations raises several fundamental questions about the biology of maggot masses. For example, how is the heat produced and do the flies truly benefit from the elevated temperatures? Some potential benefits of elevated temperatures have already been discussed with regard to the function of digestive enzymes and food assimilation. In Sections 8.3.2.1 and 8.3.2.2, we will explore theories about heat production mechanisms in fly larvae and if the internal temperatures of a maggot mass can confer protection from low temperatures.

8.3.2.1   Heat production

How individual larvae in a feeding aggregation generate the heat that yields such impressive temperature elevations has not been the subject of any systematic experimentation. Nor has the source of heat production been precisely identified, not only in terms of tissues within fly larvae but also whether flies indeed are responsible for elevated temperatures. Heat generation has been attributed to

microbial activity and/or metabolic heat production from the flies. Most investigators attribute production of heat as a byproduct of larval activity including **frenetic movement,** a term defined as the constant locomotion of larvae in the mass, and high metabolism linked to digestion (Campobasso *et al.*, 2001; Greenberg & Kunich, 2002). Larval masses figuratively and literally appear to be "boiling," as the larvae are in constant motion from egg hatch until the onset of pupariation. The digestive processes being referred to encompass muscular movements necessary for extra-oral mastication along with rapid rates of food consumption and muscle contractions associated with food motility along the length of the digestive tube (Williams & Richardson, 1984; Greenberg & Kunich, 2002). All of these events are aerobic processes subject to heat production during catabolism of organic fuels. Maximum temperatures in masses of *L. sericata* are believed to be influenced by heat exchanges (see below) and larval perception of temperature. Rather than temperatures continuing to climb as mass size and weight increase, a maximum threshold temperature is achieved. In which case, the larvae seemingly regulate mass temperature instead of an unregulated mass-effect (Charabidze *et al.*, 2011). Larval movements away from hot spots in aggregations could easily result in a modest yet rapid drop in peak temperatures, thereby functioning as a possible mechanism for self-regulation. Assuming that such regulation is widespread among necrophagous flies, maximum mass temperatures would be expected to align with species-specific, critical thermal maxima of the larval inhabitants.

During natural faunal succession, internal temperatures of maggot masses appear to be influenced by the volume of the larval aggregation, how tightly packed the individuals are in relation to one another in the assemblages, species composition of the aggregation, age of larvae, and also ambient air temperatures (Charabidze *et al.*, 2011). Laboratory studies have also demonstrated that age of the larvae and species result in differences in mass temperature when flies are reared on different tissue types from an array of animals (bovine, pig, sheep, and chicken). However, the temperatures recorded in laboratory-generated maggot masses do not achieve the maximum temperature elevations observed from larval aggregations formed during natural faunal succession (Goodbrod & Goff, 1990; Turner & Howard, 1992; Marchenko, 2001; Heaton *et al.*, 2014). What accounts for such temperature differences has not been determined but it is apparent that multiple factors, including several that are abiotic, contribute to heat generation in the larval aggregations (Joy *et al.*, 2006; Slone & Gruner, 2007; Rivers *et al.*, 2010).

Larval feeding aggregations constitute a microhabitat that offers several challenges from a thermoregulation perspective. Changing conditions of the physical environment of the assemblages stand out as the biggest thermoregulatory obstacle, which in turn accounts for an inability to maintain constant temperatures in masses. The initial habitat of fly larvae is a terrestrial environment that gradually transforms as tissue decomposition progresses and feeding ensues. A liquefied soup of nutrients, wastes, and other organisms envelops the larvae. As the physical structure of the microhabitat changes, so too does the potential for thermoregulation within the mass over time. For example, oviposition and larviposition occur in terrestrial conditions on animal remains, so that any initial heterothermic heat production by the larvae would result as a byproduct of aerobic metabolism and thus be limited mostly by the availability of oxygen (Willmer *et al.*, 2000). Loss of heat by larvae would be expected to occur rapidly due to their large surface area to volume ratio, short diffusional distance, and high integumental conductance due to a lack of insulating barriers (Willmer *et al.*, 2000). The combined attributes likely yield nearly 1:1 direct heat transference to the mass and air surrounding the larvae (Willmer *et al.*, 2000).

The microhabitat becomes increasingly aquatic as tissue decomposition progresses. Heat generated by fly larvae is released by **convection** and the heat

capacity of water greatly limits the potential for temperature increases in the maggot mass (Withers, 1992). Heat is also expected to be lost from the system due to evaporation of liquid from the integument of the maggots exposed to air, a process otherwise known as **evaporative cooling.** Heat convection is further facilitated by the stirring effect created by larval frenetic motion or activity from the center of the assemblages toward the periphery (Anderson & VanLaerhoven, 1996). The direct effect is that the amount of heat collectively produced by all larvae in a given feeding aggregation does not yield maximum temperature gain or elevation in the aggregations (Figure 8.6). By contrast, if the larvae simply fed in stationary, tightly packed positions within the mass, temperature loss from the system would be expected to decrease as no or minimal heat **conduction** between individuals would occur provided body temperatures are **isothermal** in relation to each other. Under such conditions, the rate of evaporative cooling would decline, with a net impact of even higher internal mass temperatures than are typically recorded in succession studies. The closest scenario approximating this "ideal" heat environment occurs during overcrowding, when food resources become diminished and maggots are forced into tightly packed microhabitats.

8.3.2.2 Protection from low temperatures

As already discussed, the internal heat produced in larval aggregations yields a microclimate with temperatures well above ambient. If a feeding aggregation is well established such that internal

Evaporative cooling from integument

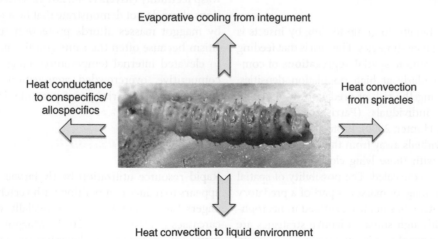

Heat conductance to conspecifics/ allospecifics

Heat convection from spiracles

Heat convection to liquid environment

Figure 8.6   Potential avenues for heat loss from larvae in maggot masses. *Source*: Austinh37/Wikimedia Commons/ CC BY-SA 3.0.

temperatures exceed the surrounding environment, it may be possible that maggot mass heterothermy confers protection from low temperatures or sudden unexpected drops in temperature (i.e., aseasonal temperature changes) (Cragg, 1956; Campobasso et al., 2001). The mechanism(s) behind such low temperature protection have not been studied, but most likely larvae are buffered from sharp depressions in air temperatures by both high internal mass temperatures and the physical barrier of the carcass. For such mechanisms to be effective in protecting against chilling or cold-shock injury, the masses must be large enough or of the appropriate species composition to generate sufficient heat to counter extreme temperature declines and/or prolonged exposures to low temperature.

Maggot heterothermy would not be a suitable method to cope with the harsh conditions of winter. This is an important consideration because although larval masses have been found on carrion during winter months in North America (Deonier, 1940; Cragg, 1956), most species of necrophagous flies residing in temperate regions depend on highly evolved genetic programs that anticipate seasonal changes through adaptive preparatory physiological and morphological mechanisms that protect the flies from the extreme environment of winter (Denlinger, 2002). Heat production from maggot masses alone is thus not a viable strategy for extending the reproductive cycle of necrophagous flies into seasons characterized by unfavorable temperatures for any developmental stage.

8.3.3 Predatory/parasite avoidance strategies

One adaptive benefit to group feeding by insects is predator avoidance strategies. The idea is that feeding aggregations promote **spatial aggregations** of competitors, particularly at high population densities, thereby reducing the risk of predation or attack by parasites to individuals (Parrish & Edelstein-Keshet, 1999; Hunter, 2000). Protection is most evident for individuals away from the periphery of the assemblages, with those lying closest to the center being the most concealed. The possibility of spatial aggregation in maggot masses as part of a predatory avoidance strategy has not been studied in necrophagous flies, although similar adaptive strategies are employed by several species of *Drosophila* to avoid attack by hymenopteran parasitoids (Bernstein, 2000; Rohlfs & Hoffmeister, 2004).

Feeding aggregations may afford predatory avoidance strategies independent of spatial aggregation. For example, elevated maggot mass temperatures may reduce predation on fly larvae by shortening the period of time devoted to immature development (via increased growth rates) on a carcass, thereby decreasing exposure to vertebrate and arthropod predators (Cianci & Sheldon, 1990). Once larval feeding is complete, most species of blow flies and flesh flies wander from the food source to **pupariate** under the protection of soil where they are much less likely to be located by predators or parasites (Greenberg, 1990; Gomes et al., 2006). It is also possible that temperatures within the larval aggregations influence the incidence of parasitism by hymenopteran parasitoids. For example, accelerated larval growth rates due to elevated mass temperatures lead to smaller puparia for some species (Ullyett, 1950; Kamal, 1958; Rivers et al., 2010). Several of these flies can withstand drastic reductions in puparial size, with only a modest reduction in emergence or subsequent fecundity of adults (Kamal, 1958; Williams & Richardson, 1983), yet the nutritional value of small puparia is greatly diminished for some parasitoids, particularly those relying on a gregarious reproductive strategy (Voss et al., 2009). The best-studied example is the gregarious ectoparasitic wasp *Nasonia vitripennis* (Walker) (Hymenoptera: Pteromalidae) that frequently will not use hosts below a minimum threshold size. If oviposition on such flies does occur, wasp clutch sizes are reduced, sex ratios become more male biased, larval development is lengthened in duration, and adult body sizes are stunted, which contributes to reduced wasp fecundity (Rivers et al., 2012). These observations alone do not demonstrate that heat generation by maggot masses affords protection from parasitism because often the same conditions that lead to elevated internal temperatures create intensely competitive, overcrowded aggregations that can produce small puparia by non-proteotaxic stressors.

8.3.4 Suppression of competition

Rapid resource utilization by fly larvae in masses appears to reduce competition with vertebrate scavengers for the same carrion, especially with small carcasses (DeVault et al., 2004). Maggot mass heat may also contribute to deterring or repulsion of vertebrate species from animal remains. Higher ambient temperatures (above 17 °C) tend to result in lower vertebrate scavenger activity on rodent, racoon, and opossum remains (DeVault *et al.*, 2004). So by extension, sufficiently large larval masses that yield internal temperatures well above ambient may in turn repulse vertebrate scavenges. This idea does not preclude the possibility that volatile compounds released during decomposition as the result of fly feeding and/or by the necrobiome convey signals regarding nutrient availability to potential vertebrate scavengers in the same manner as occuring during insect faunal succession. Similarly, rodent carrion colonized by fly larvae inhibits burying beetle interest. Why? Because the beetle larvae are highly susceptible to the lethal effects of bacteria that comprise the necrobiome (Figure 8.7). As an adaptive feature of burying beetle existence on fresh animal remains, both juveniles and adults release antimicrobial compounds onto the surface of the carcass (Rozen *et al.*, 2008; Arce *et al.*, 2013). Now consider the bacterial-fly associations discussed in Chapter 7 and earlier in this chapter; fly larvae and burying beetles seemingly cannot coexist on carrion. Maggot masses magnify the microbe problem for beetle larvae since the high internal temperatures of larval aggregations augments bacterial growth (Cianci & Sheldon, 1990). Flies have opposing issues, for eg., if burying beetles colonize

animal remains first, the oral and anal secretions of the adult beetles are thought to suppress the fly factor from occurring (Hoback *et al.*, 2004). Obviously, the trickle-down effects range from impediment of egg clustering, suppressed liberation of nutrients for fly larvae to consume, to possibly even inhibition of recruitment to larval masses.

8.4 Developing in maggot masses is not always beneficial to conspecifics or allospecifics

Group feeding in large aggregations facilitates larval development for some species that otherwise would not occur and provides several adaptive benefits that prove advantageous during intraspecific and interspecific competition. Living in such large assemblages, however, does pose some unique physiological problems that can negatively impact development and reproduction. The trade-offs include facing overcrowded, highly competitive microhabitats that can generate temperatures consistent with thermal stress. If those were not taxing enough for the young larvae, the odors emanating from the corpse coupled with pheromonal and interkingdom signaling draw the attention of predators and parasites to the larval aggregations, initiating searching behavior for prey or to locate a host for progeny development. Here, we will examine the deleterious consequences associated with feeding and developing in maggot masses.

8.4.1 Attraction of predators and parasitoids

The initial formation of feeding aggregations on animal remains is influenced by chemical signals in the form of apneumones, pheromones, and bacterial signals. Chapter 6 explored how these chemical messages are used by necrophagous flies to locate carrion, promote assembly by conspecifics, and serve as oviposition stimulants. Predatory and parasitic insects can use these same signals to find the carrion and thereby locate fly eggs and larvae indirectly, or they can detect potential prey using direct chemoreception. The fact that several species of predatory beetles in the families Silphidae and Staphylinidae arrive at a carcass either with the first

Figure 8.7 Two adult burying beetles, *Nicrophorus vespilloides* (Family Silphidae) on rodent carrion. *Source:* Calle Eklund/V-wolf/Wikimedia Commons/Public CC BY-SA 3.0.

Table 8.1   Predators of forensically important flies.*

Order	Family	Genus/species
Coleoptera	Cleridae	*Necrobia rufipes*
	Histeridae	*Hister* sp.
		Sapnnus pennsylvanicus
	Silphidae	*Heterosilpha ramose*
		Necrodes surinamensis
		Necrophilia Americana
		Oiceoptoma noveboracense
		Thanatophilus lapponicus
	Staphylinidae	*Creophilus maxillosus*
		Platydracus fossator
		Platydracus maculosus
Diptera	Calliphoridae	*Chrysomya albiceps*
		Chrysomya megacephela
		Chrysomya rufifacies
	Sarcophagidae	*Sarcophaga bullata*[†]
		Sarcophaga haemorrhoidalis[†]
Hymenoptera	Formicidae	Wide range of species
	Vespidae	*Vespula maculifrons*
		Vespula pennsylanica
		Vespula squamosa

* Representative list of predators that feed on eggs and larvae of a variety of forensically important fly species.
[†] Several species of sarcophagids have been reported as facultative predators of allospecifics.
Source: Data obtained from Rivers *et al.* (2011).

wave of colonizers or shortly after suggests that these insects possess a fine-tuned sense of olfaction for recognition of decomposition chemicals comparable to adult flies (Table 8.1).

Predatory insects likely rely on chemical signals released from targeted prey species as well, namely fly eggs and larvae. In these instances, pheromones/bacteria coating eggs to promote clustered oviposition are the most likely source of chemical cues perceived by predaceous insects. The intercepted signals function as kairomones for the predatory species since the receiver benefits at the expense of the emitter (Wertheim *et al.*, 2003). Kairomonal signals may originate from other sources as well, such as oviposition stimulants and chemicals emitted from flies in maggot masses or from the tissues that have been altered as a result of larval feeding in aggregations. A wide range of beetles, ants, wasps, and even other fly species are attracted to a corpse and/or the flies present on carrion, implying that multiple signals are used to locate the resources.

Parasitoids are also attracted to fly-infested carrion and, as with predators, the chemical signals that serve to initiate attraction and searching behavior have not been determined. Parasitoids represent a specialized parasite in the insect world, in which the association between the host and parasitoid always culminates in death of the host. All carrion-specific parasitoids identified to date are parasitic Hymenoptera (Amendt *et al.*, 2000; Disney & Munk, 2004; Turchetto & Vanin, 2004; Rivers, 2016), an incredibly large and diverse group of insects. Fly parasitoids attack either larvae or those stages (pupae and pharate adults) contained within puparia (Figure 8.8). The most intensively studied of these parasitoids is the gregarious ectoparastic wasp *Nasonia vitripennis*, a species that targets fly puparial stages but arrives at the corpse before **wandering** occurs (Whiting, 1967). Wandering is initiated in third-stage larvae that have completed feeding and usually have emptied the crop. Migrating long distances from the corpse has been argued to be an adaptation of some necrophagous flies for avoiding parasitism by *N. vitripennis* and other wasps (Legner, 1977; Greenberg, 1990), either to avoid direct detection through physical contact or to "clean" the integument of food odors that may serve as chemical cues used by parasitic species. Adult female wasps overcome the avoidance strategies of the flies by "riding" the maggots as they wander and then burrow into the soil, the wasps waiting for pupariation and pupation to be completed before ultimately attempting parasitism (Whiting, 1967). For some potential fly hosts, pupariation occurs on the corpse, leaving the fly exposed and seemingly unprotected from parasitoid attack. Despite not burrowing into the soil for pupariation, the incidence of parasitism for these fly puparia is low, indicating that although mass feeding may attract parasitoids, the flies may utilize oviposition deterrents to avoid parasitism (Table 8.2).

Figure 8.8   Adult female of *Muscidifurax raptor* (Family Pteromalidae) laying an egg within a fly puparium. *Source*: Wikimedia Commons/Public domain.

Table 8.2   Parasitoids of forensically important flies.*

Order	Family	Genus/species
Hymenoptera	Braconidae	Aphaereta sp.
		Alysia manducator
	Diapriidae	Spilomicrus sp.
	Encryptidae	Tachinaephagus zealandicus
	Pteromalidae	Muscidifurax raptor
		Muscidifurax zaraptor
		Nasonia vitripennis
		Pachycrepoideus vindemiae
		Spalangia cameroni
		Spalangia nigra
		Spalangia nigroaenea
		Trichomalopsis sarcophagae

* Representative parasitoids that utilize larvae and puparial stages from a variety of forensically important fly species as hosts.
Source: Data obtained from Rivers et al. (2011).

8.4.2   Proteotaxic stress

An obvious source of physiological stress is heat production by large maggot masses. As already discussed, larval aggregations generate heat that causes a linear rise in internal temperatures until reaching some type of physiological ceiling. Elevated internal temperatures undoubtedly evoke thermal stress responses in fly larvae, particularly as rising temperatures approach species-specific upper limits of the **zone of tolerance** or **thermal tolerance range** (Richards & Villet, 2008; Rivers et al., 2010). These terms reflect the range of temperatures that an animal can survive in, indefinitely (Withers, 1992). The upper temperatures in the aggregation do not always reflect the actual internal temperatures of flies in the mass (Prange, 1996). This is an important distinction, as upper lethal temperatures for fly tissues, otherwise known as the **critical thermal maximum,** represent conditions when proteins begin to denature and most features of aerobic metabolism are inhibited (Storey, 2004). Exposure to high temperatures above the zone of tolerance, even for short periods, is able to elicit irreversible damage and possibly cell death.

Fly development at temperatures below the critical thermal maximum can still be detrimental. High internal temperatures of larval aggregations that have been reported under natural and laboratory situations constitute proteotoxic or thermal stress, capable of inducing general stress responses as well as the heat-shock protein (Hsp) response in several insect species. Depending on the severity

of temperature elevation and duration of exposure, Hsp synthesis can occur at the expense of normal protein synthesis, potentially compromising the development of the fly (Feder, 1996). Hsp expression has been observed in at least two species of flies (*Sarcophaga bullata* and *Protophormia terraenovae*) reared in large laboratory-generated maggot masses (Rivers et al., 2010), and certainly must occur in maggot masses where temperatures soar to well above 35 °C. Long before achieving conditions that exceed the critical thermal maximum, larvae in feeding assemblages may experience sufficient exposure to high temperatures that stimulate thermal stress and injury. Non-lethal high-temperature stress can be manifested in necrophagous flies as an inhibition of feeding and/or growth, delay in the onset or completion of pupariation, suppressed puparial sizes, distortions in puparial shape, increased length in pupal and/or pharate adult development, and disruption of extrication behavior, which can include shifting the peak day and time of day for adult eclosion (Chen et al., 1990; Joplin et al., 1990; Yocum et al., 1994).

Does the production of heat by maggot masses mean that overheating is inevitable for larvae in the aggregations? The answer is not a simple yes or no, largely because many aspects of fly thermoregulation are not known. Some speculation exists that the fly larvae have the ability to self-regulate body temperatures by moving in and out of the feeding aggregations (Charabidze et al., 2011; Heaton et al., 2014), the so-called frenetic activity described earlier perhaps reflecting such movements. Maggot movements conceivably would permit larvae the ability to avoid overheating during periods of elevated temperatures by behaviorally driven **spatial partitioning** and/or by releasing heat through evaporation (i.e., evaporative cooling) (Villet et al., 2010). Evaporative cooling is usually restricted to insects that have access to essentially an unlimited water pool (either internally or in their environment) and that can generate heat to facilitate the water loss (Prange, 1996). Fly larvae developing in maggot masses under semi-liquid to liquid conditions seem to fit these criteria. Such thermoregulatory abilities would also appear to depend on the fly larvae first experiencing high, potentially stressful, temperatures before cooling can occur (Prange & Modi, 1990). Despite the merits of the argument, there is no experimental evidence to support the contention that maggots can regulate body temperature to avert heat stress.

8.4.3   Overcrowding

Carrion is a patchy ephemeral resource that represents an isolated nutrient island to thousands of individuals. Competition is intense for the resource, and consequently a corpse is utilized quickly under natural conditions. To compensate for a rapidly diminishing resource, necrophagous flies are predicted to deposit far more eggs or larvae than can be supported by a dead body (Kneidel, 1984), a situation that attempts to maximize maternal fitness. The net effect is that larval aggregations which form during natural faunal succession are typically composed of thousands of individuals from a variety of calliphorid, sarcophagid, muscid, and microdipteran species. In a short period of time, overcrowding occurs. Obviously, overcrowding means decreased food availability per individual, which in turn is expected to increase the length of larval development since it takes longer to acquire the critical weight/nutrients associated with the next molt (Ullyett, 1950; Williams & Richardson, 1984). Overcrowding is likely countered by elevations in internal mass temperatures, since up to an upper threshold temperature limit, increased temperatures accelerate larval growth rates. This helps to explain why some calliphorids appear to have faster rates of development under "crowded" conditions (Saunders & Bee, 1995; Ireland & Turner, 2006).

Once the maggot mass reaches a critically high number of individuals, nutrient availability declines, and larval waste products accumulate. The result is diminished growth rates. Slower growth increases predation on fly larvae since a longer period of time is devoted to immature development on a carcass, and hence the time exposed to predators increases. Reduced nutrient availability contributes to suboptimal body weights, resulting in reduced puparial sizes as well. For several necrophagous fly species, extreme reductions in puparial size yield low mortality during puparial stages and only modestly alter adult fecundity. By contrast, some species like *S. bullata* and *P. terraenovae* display dramatic developmental alterations from intraspecific overcrowding: high larval mortality, reduced puparial weights, and a high incidence of pupal/pharate adult mortality.

Overcrowding has been extensively studied in the vinegar fly, *Drosophila melanogaster* (Diptera: Drosophilidae) and the impact of larval crowding is very similar to that with necrophagous flies: larval development is extended; mortality from egg to adult increases; sizes of larvae, puparia, and subsequent adults are smaller; and adult fecundity is reduced (Scheiring *et al.*, 1984; Zwaan *et al.*, 1991). Depletion of nutrients, as well as build-up of wastes (urea, uric acid), are thought to contribute to the overcrowding effects; perhaps more intriguing, larval crowding triggers the heat-shock response (Buck *et al.*, 1993), which actually leads to prolonged adult longevity and increased thermal hardiness in the resulting adults (Sørensen & Loeschcke, 2001). This appears to be in sharp contrast to the deleterious effects reported for *P. terraenovae* and *S. bullata* reared in overcrowded maggot masses (Rivers *et al.*, 2010), although alterations in longevity and fecundity of adults were not tested.

8.5   Larval aggregation may benefit Coleoptera too!

The focus thus far has been on larval masses produced by flies, mainly because other inhabitants of carrion communities do not create a similar microhabitat. A possible exception to that rule is with some species of Coleoptera. Under laboratory conditions, anyone who has ever maintained a colony of dermestid beetles has undoubtedly witnessed mass clustering when a new carcass or piece of meat is added (Figure 8.9). Larvae of *Dermestes maculatus* (Family Dermestidae) form loose aggregations on fresh and dry carrion, which can produce internal heat, dependent on the number of individuals in the mass (Gunn, 2018). The temperature elevations (approaching 10 °C above ambient) are predicted to

Figure 8.9   Feeding aggregation of larvae and adults of *Dermestes maculatus* (Family Dermestidae) on a dried piece of bovine liver. *Source*: Photo by D.B. Rivers.

accelerate the rate of development for the dermestid larvae but has not been experimentally tested. Interestingly, the experiments were conducted using fresh remains, which dermestid beetles are seldom found on under natural conditions. The fact that the larvae readily consumed fresh tissue and completed development asserts that dermestids may not prefer dried remains over fresh, as currently thought (Smith, 1986). This would be with observations of larvae of *D. maculatus* feeding on live turkeys and showing no preference for dry, fed over meat (from calf and chicken) (Samish *et al.*, 1992). The long gestation period of the egg stage (3–4 days at 23 °C) puts them at a competitive disadvantage with necrophagous flies. For example, several species of flies can progress from egg hatch to the formation of a maggot mass over that same period of time. The absence of large feeding aggregations of dermestid beetles on dried remains in natural environments seemingly supports the possibility of preference for fresh tissue. Alternatively, the lack of tissue water necessary for evaporative cooling mechanisms, if needed by beetle larvae, is a counter argument to the absence of larval masses on later stages of decay. There are some beetle species in which individuals display heterothermy independent of flight or other forms of locomotion (Morgan & Bartholomew, 1982). It is thus conceivable that dermestid larvae generate heat individually during food consumption, independent of cooperative gregarious behavior. Although most aspects discussed are speculative, the ideas do represent intriguing areas for further research.

Chapter review

Carrion communities are composed largely of fly larvae living in aggregations

- Gravid female flies oviposit in natural body openings or, when present, in wounds and lesions in the skin. Eggs are deposited in locations that favor neonate larval feeding on liquids and soft tissues high in proteins. As larvae progress in development, they molt into second and eventually the third larval stages, periods of development most commonly when fly larvae assemble to form feeding aggregations.
- Several hundred to thousands of individual fly larvae can compose a single maggot mass.

Multiple maggot masses can exist on a body at the same time, collectively representing the largest component (by weight) to the invertebrate carrion community.

- Larval aggregations can be viewed as dynamic microhabitats in that each mass can vary in size, location (on body), and species composition, with each parameter dependent on ambient conditions, geographic location, season, amount of sunlight, and stage of corpse decomposition.

Formation of maggot masses is more complex than originally thought

- Fly maggot masses can be homogeneous with only one species present but are commonly heterogeneous, composed of larvae from the families Calliphoridae, Sarcophagidae, Muscidae, and several microdipteran groups. Under most conditions examined in the field and in case studies, various species of blow flies dominate the composition of maggot masses, suggesting that the benefits attained from larval aggregations are more important to their survival then flies from other families.
- Precisely how the feeding aggregations form has not been determined. The simplest answer given so far has been thigmotaxis, the type of innate behavior in which an animal responds to touch or physical contact with a solid object. In the case of necrophagous fly larvae, the expectation is that they display positive thigmotaxis, seeking contact with a physical object (i.e., other larvae). It is more likely, however, that a positive thigmotaxic response is what helps maintain the assembly of larvae once formed but probably is not the reason for the initial formation of the aggregation.
- Maggot mass formation is likely under the influence of one of three possible mechanisms: clustered oviposition/larviposition by gravid females, random formation, or foraging by larvae responding to recruitment cues. There is limited information about all three mechanisms and despite observations that supports one versus another mechanism, it is entirely possible that all three contribute to the formation of larval feeding aggregations.
- Internal heat released from larval aggregations could be recognition cues for larval recruitment and/or retention. Some species are known to seek optimal temperatures zones within a mass, so heat perception is definitely a trait of fly larvae.

Larval feeding aggregations provide adaptive benefits to individuals

- Individuals and perhaps groups receive adaptive benefits from developing as a cooperative feeding mass. Within limits, mass-forming flies achieve more rapid rates of growth and development owing to group feeding and heat generation. The production of heat confers advantages that range from enhanced aspects of food processing and assimilation to protection from low temperatures and possibly predators and/or parasites.
- Group feeding in larval aggregations seems to rely on the fact that (i) for most species, independent living (i.e., a solitary existence) does not lead to completion of development and (ii) larvae are equipped with features to feed in a gregarious environment. These larval adaptations include modified mouthparts for tissue penetration, digestive enzymes, and physiological modifications associated with food assimilation.
- Newly hatched maggots penetrate animal remains with mouth hooks and begin a period of voracious feeding during the larval stages. Necrophagous flies have a clear need to work together to create avenues for penetrating the body cavity of a corpse and infiltrating soft tissues. Maggot mass formation offers a multitude of mouth hooks to accomplish this initial task of food manipulation, which arguably can be labeled extra-oral mastication.
- Once the larval masses break through the skin of a cadaver, the importance of mouth hooks declines (not entirely) and group feeding through larval aggregations becomes dependent on mass release of digestive enzymes.
- A combination of high metabolic rate, optimum enzyme activity, and fast transit times in larvae developing in feeding aggregations should contribute to increased food assimilation efficiency. Growth rates for calliphorid and sarcophagid immatures are linear within species-specific physiological limits. Provided overcrowding is avoided, rapid development rates do not result in reduced larval body sizes, suggesting that necrophagous fly larvae have limited trade-offs between rapid rates of digestion and efficiency of food assimilation.
- Dense aggregations of feeding larvae generate internal heat. If a feeding aggregation is well established such that internal temperatures exceed the surrounding environment, it may be possible that maggot mass heterothermy confers protection from low temperatures or sudden unexpected drops in temperature.
- One adaptive benefit to group feeding by insects is predator avoidance strategies. The idea is that feeding aggregations promote spatial aggregations of competitors, particularly at high population densities, thereby reducing the risk of predation or attack by parasites to individuals. Feeding aggregations may afford predator avoidance strategies independent of spatial aggregation. For example, elevated maggot mass temperatures may reduce predation on fly larvae by shortening the period of time devoted to immature development (via increased growth rates) on a carcass, thereby decreasing exposure to vertebrate and arthropod predators. Heat-stressed flies may also be less suitable for parasitism by parasitic wasps that use larval or puparial stages of fly development as hosts.
- Rapid utilization of carrion coupled with release of heat to the environment by larval masses decreases competition with vertebrate scavengers. It is possible that the large feeding aggregations produce sufficient hear to repulse some scavengers.
- Fly-bacterial associations appear to be incompatible with the survival strategies of burying beetles. Consequently, the two cannot coexist on carrion. Thus, if one species precedes the other in colonization, the other will not attempt resource utilization.

Developing in maggot masses is not always beneficial to conspecifics or allospecifics

- Living in large larval assemblages poses some unique physiological problems that can negatively impact development and reproduction. The trade-offs include facing overcrowded, highly competitive microhabitats that can generate temperatures consistent with thermal stress. Odors emanating from the corpse coupled with pheromonal/bacterial signaling draw the attention of predators and parasites to the larval aggregations, initiating searching behavior for prey or to locate a host for progeny development.
- Predatory and parasitic insects can use the same chemical signals as necrophagous flies to find the carrion and thereby locate fly eggs and larvae indirectly, or they can detect potential prey using direct chemoreception. The fact that several

species of predatory beetles in the families Silphidae and Staphylinidae arrive at a carcass either with the first wave of colonizers or shortly after suggests that these insects possess a finely tuned sense of olfaction for recognition of decomposition chemicals comparable to adult flies.

- Larval aggregations generate heat that causes a linear rise in internal temperatures until reaching some type of physiological ceiling. Elevated internal temperatures undoubtedly evoke thermal stress responses in fly larvae, particularly as rising temperatures approach species-specific upper limits of the zone of tolerance. Exposure to high temperatures above the zone of tolerance, even for short periods, is able to elicit irreversible damage and possibly cell death.
- Fly development at temperatures below the critical thermal maximum can still be detrimental. The high internal temperatures of larval aggregations that have been reported under natural and laboratory situations constitute proteotoxic or thermal stress, capable of inducing general stress responses as well as the Hsp response in several insect species.
- Larval aggregations that form during natural faunal succession are typically composed of thousands of individuals from a variety of calliphorid, sarcophagid, muscid, and microdipteran species. In a short period of time, overcrowding occurs. Obviously, overcrowding means decreased food availability per individual, which in turn is expected to increase the length of larval development since it takes longer to acquire the critical weight/nutrients associated with the next molt. Slower growth increases predation on fly larvae since a longer period of time is devoted to immature development on a carcass, and hence the time exposed to predators increases. Reduced nutrient availability contributes to suboptimal body weight, resulting in reduced puparial sizes.

Larval aggregation may benefit Coleoptera too!

- Larvae of *Dermestes maculatus* form loose aggregations on fresh and dry carrion, which can produce internal heat dependent on the number of individuals in the mass. The temperature elevations (approaching 10 °C above ambient) are predicted to accelerate the rate of development for the dermestid larvae but has not been experimentally tested.

- The fact that the larvae readily consume fresh tissue and complete development argues that dermestids may in fact do not prefer dried remains over fresh as currently thought.
- There are some beetle species in which individuals display heterothermy independent of flight or other forms of locomotion. It is thus conceivable that dermestid larvae generate heat individually during food consumption, independent of cooperative gregarious behavior.

Test your understanding

Level 1: knowledge/comprehension

1. Define the following terms:

   (a) proteotaxic
   (b) thigmotaxis
   (c) heterothermic
   (d) zone of tolerance
   (e) spatial aggregation
   (f) food assimilation
   (g) maggot mass
   (h) larval mass-effect.

2. Match the terms (i–vi) with the descriptions (a–f).

(a) Constant non-directional movement in a feeding aggregation	(i) Mastication
(b) Animal that experiences varying internal temperatures	(ii) Homeothermy
(c) Constant internal body temperatures	(iii) Parasitoid
(d) Insect parasite that always kills its host	(iv) Critical thermal maximum
(e) Physical manipulation or breakdown of food	(v) Poikilotherm
(f) Upper temperature limit of tissues	(vi) Frenetic movement

3. Describe the theories accounting for how maggot masses form on carrion. Explain which idea is the most plausible for the formation of large feeding aggregations.

4. Explain how fly maggots are believed to regulate body temperature in a larval aggregation.

Level 2: application/analysis

1. An analysis of different necrophagous fly species reveals that during formation of homogeneous maggot masses, internal mass temperatures can be species-specific. Explain how species-specific maggot mass temperatures may afford a competitive advantage in heterogeneous larval aggregations.
2. High temperatures in larval aggregations are capable of inducing a stress response in the form of heat-shock protein production in fly larvae. Detail what other stresses are associated with group feeding that conceivably can evoke stress responses.
3. Explain why calliphorid larvae and burying beetles seemingly cannot coexist on small rodent carrion.

Level 3: synthesis/evaluation

1. Provide a detailed explanation for how spatial aggregations can lead to coexistence of mixed fly species in a large maggot mass.
2. Design an experiment which tests whether dermestid beetle larvae display cooperative digestion in a larval mass. Identify the independent, dependent, and control variables.
3. Speculate on why maggot mass temperatures recorded under laboratory conditions are typically much cooler than those observed in field studies.

Note

1. Though maggots may number in the tens of thousands on a corpse, the animal group in highest abundance on animal remains is mites, ranging from phoretic to parasitic to predaceous. Growing research on the necrobiomes argues that bacteria are the dominant life form on the dead.

References cited

Allen, P.E. (2010) Group size effects on survivorship and adult development in the gregarious larvae of *Euselasia chrysippe* (Lepidoptera: Riodinidae). *Insectes sociaux* 57: 199–204.

Amendt, J., Krettek, R., Niess, C., Zehner, R. & Bratzke, H. (2000) Forensic entomology in Germany. *Forensic Science International* 113: 309–314.

Anderson, G.S. (2020) Factors that influence insect succession on carrion. In: J.H. Byrd & J.L. Castner (eds) *Forensic Entomology: The Utility of Using Arthropods in Legal Investigations*, 2nd edn, pp. 201–250. CRC Press, Boca Raton, FL.

Anderson, G.S. & VanLaerhoven, S.L. (1996) Initial studies on insect succession on carrion in southwestern British Columbia. *Journal of Forensic Science* 41: 617–625.

Anderson, O.D. (1982) Enzyme activities in the larval secretion of *Calliphora erythrocephala*. *Comparative Biochemistry and Physiology B* 72: 569–575.

Arce, A.N., Smiseth, P.T. & Rozen, D.E. (2013) Antimicrobial secretions and social immunity in larval burying beetles, *Nicrophorus vespilloides*. *Animal Behaviour* 86: 741–745.

Bernstein, C. (2000) Host-parasitoid models: the story of a successful failure. In: M.E. Hochber & R. Ives (eds) *Parasitoid Population Biology*, pp. 41–57. Princeton University Press, Princeton, NJ.

Boulay, J., Devigne, C., Gosset, D. & Charabidze, D. (2013) Evidence of active aggregation behavior in *Lucilia sericata* larvae and possible implication of a conspecific mark. *Animal Behaviour* 85: 1191–1197.

Bowles, V.M., Carnegie, P.R. & Sandeman, R.M. (1988) Characterization of proteolytic and collagenolytic enzymes from the larvae of *Lucilia cuprina*, the sheep blowfly. *Australian Journal of Biological Science* 41: 269–278.

Buck, S., Nicholson, M., Dudas, S., Wells, R., Force, A., Baker, G.T. & Arking, P. (1993) Larval regulation of adult longevity in a genetically-selected long-lived strain of *Drosophila*. *Heredity* 71: 23–32.

Byrd, J.H. & Butler, J.F. (1998) Effects of temperature on *Sarcophaga haemorrhoidalis* (Diptera: Sarcophagidae) development. *Journal of Medical Entomology* 35: 694–698.

Campobasso, C.P., Di Vella, G. & Introna, F. (2001) Factors affecting decomposition and Diptera colonization. *Forensic Science International* 120: 18–27.

Chapman, R.F. (1998) *The Insects: Structure and Function*, 4th edn. Cambridge University Press, Cambridge, UK.

Charabidze, D., Bourel, B. & Gosset, D. (2011) Larval-mass effect: Characterisation of heat emission by necrophagous blowflies (Diptera: Calliphoridae) larval aggregates. *Forensic Science International* 211: 61–66.

Charabidze, D., Bourel, B., LeBlanc, H., Hedouin, V. & Gosset, D. (2008) Effect of body length and temperature on crawling speed of *Protophormia terraenovae* larvae (Robineau-Desvoidy) (Diptera: Calliphoridae). *Journal of Insect Physiology* 54: 529–533.

Chen, C.-P., Lee, R.E. Jr & Denlinger, D.L. (1990) A comparison of the responses of tropical and temperate flies (Diptera: Sarcophagidae) to cold and heat stress. *Journal of Comparative Physiology B* 160: 543–547.

Christopherson, C. & Gibo, D.L. (1996) Foraging by food deprived larvae of *Neobellieria bullata* (Diptera: Sarcophagidae). *Journal of Forensic Sciences* 42: 71–73.

Cianci, T.J. & Sheldon, J.K. (1990) Endothermic generation by blowfly larvae *Phormia regina* developing in pig carcasses. *Bulletin of the Society of Vector Ecology* 15: 33–40.

Clark, K., Evans, L. & Wall, R. (2006) Growth rates of the blowfly, *Lucilia sericata*, on different body tissues. *Forensic Science International* 156: 145–149.

Cragg, J.B. (1956) The olfactory behaviour of *Lucilia* species (Diptera) under natural conditions. *Annals of Applied Biology* 44: 467–471.

Day, D.M. & Wallman, J.F. (2006) A comparison of frozen/thawed and fresh food substrates in development of *Calliphora augur* (Diptera Calliphoridae) larvae. *International Journal of Legal Medicine* 120: 391–394.

Denlinger, D.L. (2002) Regulation of diapause. *Annual Review of Entomology* 47: 93–122.

Deonier, C.C. (1940) Carcass temperatures and their relation to winter blowfly populations and activity in the southwest. *Journal of Economic Entomology* 33: 166–170.

DeVault, T.L., Brisbin Jr., I.L. & Rhodes Jr., O.E. (2004) Factors influencing the acquisition of rodent carrion by vertebrate scavengers and decomposers. *Canadian Journal of Zoology* 82: 502–509.

Devigne, C., Broly, P., & Deneubourg, J.L. (2011) Individual preferences and social interactions determine aggregation of woodlice, *PloS ONE* 6: e17389. http://dx.doi.org/10.1371/journal.pone.0017389.

Disney, R.H.L. & Munk, T. (2004) Potential use of Braconidae (Hymenoptera) in forensic cases. *Medical and Veterinary Entomology* 18: 442–444.

Erzinçlioglu, Z. (1996) *Blowflies*. Richmond Publishing Co. Ltd., Slough, UK.

Feder, M.E. (1996) Ecological and evolutionary physiology of stress proteins and the stress response. In: I.A. Johnston & A.F. Bennett (eds) *Animals and Temperature: Phenotypic and Evolutionary Adaptation*, pp. 79–102. Cambridge University Press, Cambridge, UK.

Fouche, Q., Hedouin, V. & Charabidize, D. (2018) Communication in necrophagous Diptera larvae: interspecific effect of cues left behind by maggots and implications in their aggregation. *Scientific Reports* 8: 2844. https://doi.org/10.1038/s41598-018-21316x.

Gennard, D.E. (2012) *Forensic Entomology: An Introduction*, 2nd edn. Wiley-Blackwell, Chichester, UK.

Gomes, L., Godoy, W.A.C. & Von Zuben, C.J. (2006) A review of post-feeding larvae dispersal in blowflies: implications for forensic entomology. *Naturwissenschaften* 93: 207–215.

Goodbrod, J.R. & Goff, M.L. (1990) Effects of larval populations density on rates of development and interactions between two species of *Chrysomya* (Diptera: Calliphoridae) in laboratory culture. *Journal of Medical Entomology* 27: 338–343.

Grassberger, M. & Reiter, C. (2001) Effect of temperature on *Lucilia sericata* (Diptera: Calliphoridae) development with special reference to the isomegalen- and isomorphen-diagram. *Forensic Science International* 120: 32–36.

Grassberger, M. & Reiter, C. (2002) Effect of temperature on development of the forensically important holarctic

blow fly *Protophormia terraenovae* (Robineau-Desvoidy) (Diptera: Calliphoridae). *Forensic Science International* 128: 177–182.

Greenberg, B. (1990) Behaviour of postfeeding larvae of some Calliphoridae and a muscid (Diptera). *Annals of the Entomological Society of America* 83: 1210–1214.

Greenberg, B. & Kunich, J.C. (2002) *Entomology and the Law*. Cambridge University Press, Cambridge, UK.

Gunn, A. (2018) The exploitation of fresh remains by *Dermestes maculatus* De Geer (Coleoptera, Dermestidae) and their ability to cause a localised and prolonged increase in temperature above ambient. *Journal of Forensic and Legal Medicine* 59: 20–29.

Hanski, I. (1976) Assimilation by *Lucilia illustris* (Diptera) larvae in constant and changing temperatures. *Oikos* 27: 288–299.

Hanski, I. (1977) An interpolation model of assimilation by larvae of the blowfly, *Lucilia illustris* (Calliphoridae) in changing temperatures. *Oikos* 28: 187–195.

Hanski, I. (1987) Carrion fly community dynamics: patchiness, seasonality and coexistence. *Ecological Entomology* 12: 257–266.

Heal, O.W. (1974) Comparative productivity in ecosystems: secondary productivity. In: *Proceedings of the First International Congress of Ecology. Structure, Functioning and Management of Ecosystems*, p. 37. Centre for Agricultural Publishing and Documentation, Wageningen, The Netherlands.

Heaton, V., Moffatt, C. & Simmons, T. (2014) Quantifying the temperature of maggot masses and its relationship to decomposition. *Journal of Forensic Sciences* 59: 676–82.

Hoback, W.W., Bishop, A.A., Kroemer, J., Scalzitti, J. & Shaffer, J.J. (2004) Differences among antimicrobial properties of carrion beetle secretions reflect phylogeny and ecology. *Journal of Chemical Ecology* 30: 719–729.

Holl, M.V. & Gries, G. (2018) Studying the "fly factor" phenomenon and its underlying mechanisms in house flies, *Musca domestica*. *Insect Science* 25: 137–147.

Hunter, A.F. (2000) Gregariousness and repellant defences in the survival of phytophagous insects. *Oikos* 91: 213–224.

Ireland, S. & Turner, B. (2006) The effects of larval crowding and food type on the size and development of the blowfly, *Calliphora vomitoria*. *Forensic Science International* 159: 175–181.

Jaenike, J. & James, A.C. (1991) Aggregation and the coexistence of mycophagous *Drosophila*. *Journal of Animal Ecology* 60: 913–928.

Johnson, A.P., Wighton, S.J., & Wallman, J.F. (2014) Tracking movement and temperature selection of larvae of two forensically important blow fly species within a "maggot mass". *Journal of Forensic Sciences* 59: 1586–1591.

Joplin, K.H., Yocum, G.D. & Denlinger, D.L. (1990) Cold shock elicits expression of heat shock proteins in the flesh fly *Sarcophaga crassipalpis*. *Journal of Insect Physiology* 36: 825–834.

Joy, J.E., Liette, N.L. & Harrah, H.L. (2006) Carrion fly (Diptera: Calliphoridae) larval colonization of sunlit and shaded pig carcasses. *Forensic Science International* 164: 183–192.

Kamal, A.S. (1958) Comparative study of thirteen species of sarcosaprophagous Calliphorida and Sarcophagidae (Diptera). I. Bionomics. *Annals of the Entomological Society of America* 51: 261–270.

Kaneshrajah, G. & Turner, B. (2004) *Calliphora vicina* larvae grow at different rates on different body tissues. *International Journal of Legal Medicine* 118: 242–244.

Karasov, W.H. & Martínez del Rio, C. (2007) *Physiologcal Ecology*. Princeton University Press, Princeton, NJ.

Kneidel, K.A. (1984) Competition and disturbance in communities of carrion-breeding Diptera. *Journal of Animal Ecology* 53: 849–865.

Legner, E.F. (1977) Temperature, humidity and depth of habitat influencing host destruction and fecundity of muscoid fly parasites. *Entomophaga* 22: 199–206.

Marchenko, M.I. (2001) Medicolegal relevance of cadaver entomo-fauna for the determination of the time since death. *Forensic Science International* 120: 89–109.

Morgan, K.R. & Bartholomew, G.A. (1982) Homeothermic response to reduced ambient temperature in a scarab beetle. *Science* 216: 1409–1410.

Norris, K.R. (1965) The bionomics of blowflies. *Annual Review of Entomology* 10: 47–68.

Padilha, M.H.P., Pimentel, A.C., Ribeiro, A.F. & Terra, W.R. (2009) Sequence and function of lysosomal and digestive cathepsin D-like proteinases of *Musca domestica* midgut. *Insect Biochemistry and Molecular Biology* 39: 782–791.

Parrish, J.K. & Edelstein-Keshet, L. (1999) Complexity, pattern, and evolutionary trade-offs in animal aggregation. *Science* 284: 99–101.

Pendola, S. & Greenberg, B. (1975) Substrate-specific analysis of proteolytic enzymes in the larval midgut of Calliphora vicina. *Annals of the Entomological Society of America* 68: 341–345.

Prange, H.D. (1996) Evaporative cooling in insects. *Journal of Insect Physiology* 42: 493–499.

Prange, H.D. & Modi, J. (1990) Comparative evaporative cooling in grasshoppers and beetles. *Physiologist* 33: A88.

Price, G.M. (1974) Protein metabolism by the salivary glands and other organs of the larva of the blowfly, *Calliphora erythrocephala*. *Journal of Insect Physiology* 20: 329–347.

Randall, D., Burggren, W. & French, K. (2002) *Animal Physiology: Mechanisms and Adaptations*. W.H. Freeman and Company, New York.

Richards, C.S. & Villet, M.H. (2008) Factors affecting accuracy and precision of thermal summation models of insect development used to estimate post-mortem intervals. *International Journal of Legal Medicine* 122: 401–408.

Richards, C.S., Price, B.W. & Villet, M.H. (2009) Thermal ecophysiology of seven carrion-feeding blowflies (Diptera: Calliphoridae) in southern Africa. *Entomologia Experimentalis et Applicata* 131: 11–19.

Rivers, D.B. (2016) Parasitic Hymenoptera as forensic indicator species. In: B.S.K. Setty & J. Rao (eds) *Forensic Analysis- From Death to Justice*, pp. 67–83. InTech, Rijeka, Croatia.

Rivers, D.B., Ciarlo, T., Spelman, M. & Brogan, R. (2010) Changes in development and heat shock protein expression in two species of flies (*Sarcophaga bullata* [Diptera: Sarcophagidae] and *Protophormia terraenovae* [Diptera: Calliphoridae]) reared in different sized maggot masses. *Journal of Medical Entomology* 47: 677–689.

Rivers, D.B., Kaikis, A., Bulanowski, D., Wigand, T. & Brogan, R. (2012) Oviposition restraint and developmental alterations in the ectoparasitic wasp, *Nasonia vitripennis*, when utilizing puparia resulting from different size maggot masses of *Lucilia illustris*, *Protophormia terraenovae*, and *Sarcophaga bullata*. *Journal of Medical Entomology* 49: 1124–1136.

Rivers, D.B., Thompson, C. & Brogan, R. (2011) Physiological trade-offs of forming maggot masses by necrophagous flies on vertebrate carrion. *Bulletin of Entomological Research* 101: 599–611.

Rohlfs, M. & Hoffmeister, T.S. (2004) Spatial aggregation across ephemeral resource patches in insect communities: an adaptive response to natural enemies? *Oecologia* 140: 654–661.

Rozen, D.E., Engelmoer, D.J.P. & Smiseth, P.T. (2008) Antimicrobial strategies in burying beetles breeding on carrion. *Proceedings of the National Academy of Sciences* 105: 17890–17895.

Samish, M., Argaman, O. & Perelman, D. (1992) Research note: The hide beetle, *Dermestes maculatus* DeGeer (Dermestidae), feeds on live turkeys. *Poultry Science* 71: 388–390.

Sandeman, R.M., Feehan, J.P., Chandler, R.A. & Bowles, V.M. (1990) Tryptic and chymotryptic proteases released by larvae of the blowfly, Lucilia cuprina. *International Journal of Parasitology* 20: 1019–1023.

Saunders, D. & Bee, A. (1995) Effects of larval crowding on size and fecundity of the blowfly *Calliphora vicina* (Diptera: Calliphoridae). *European Journal of Entomology* 92: 615–622.

Scanvion, Q., Hedouin, V. & Charabidzé, D. (2018) Collective exodigestion favours blow fly colonization and development on fresh carcasses. *Animal Behaviour* 141: 221–232.

Scheiring, J.F., Davis, D.G., Ranasinghe, A. & Teare, C.A. (1984) Effects of larval crowding on life history parameters in *Drosophila melanogaster* Meigen (Diptera: Drosophilidae). *Experimental Gerontology* 77: 329–332.

Slone, D.H. & Gruner, S.V. (2007) Thermoregulation in larval aggregations of carrion-feeding blow flies (Diptera; Calliphoridae). *Journal of Medical Entomology* 44: 516–523.

Smith, K.G.V. (1986) *A Manual of Forensic Entomology*. British Museum (Natural History), London.

Sørensen, J.G. & Loeschcke, V. (2001) Larval crowding in *Drosophila melanogaster* induces Hsp70 expression, and leads to increased adult longevity and adult thermal stress resistance. *Journal of Insect Physiology* 47: 1301–1307.

Storey, K.B. (2004) Biochemical adaptation. In: K.B. Storey (ed) *Functional Metabolism: Regulation and Adaptation*, pp. 383–414. John Wiley & Sons, Inc., Hoboken, NJ.

Szpila, K. (2010) Key for the identification of third instars of European blowflies (Diptera: Calliphoridae) of forensic importance. In: J. Amendt, C.P. Campobasso, M.L. Goff & M. Grassberger (eds) *Current Concepts in Forensic Entomology*, pp. 43–56. Springer, London.

Turchetto, M. & Vanin, S. (2004) Forensic evaluations on a crime scene with monospecific necrophagous fly population infected by two parasitoid species. *Aggrawal's International Journal of Forensic Medicine and Toxicology* 5: 12–18.

Turner, B. & Howard, T. (1992) Metabolic heat generation in dipteran larval aggregations: a consideration for forensic entomology. *Medical and Veterinary Entomology* 6: 179–181.

Ullyett, G.C. (1950) Competition for food and allied phenomena in sheep blowfly populations. *Philosophical Transactions of the Royal Society of London, Series B* 234: 77–175.

Villet, M.H., Richards, C.S. & Midgley, J.M. (2010) Contemporary precision, bias and accuracy of minimum post-mortem intervals estimated using development of carrion-feeding insects. In: J. Amendt, C.P. Campobasso, M.L. Goff & M. Grassberger (eds) *Current Concepts in Forensic Entomology*, pp. 109–137. Springer, London.

Voss, S.C., Spafford, H. & Dadour, I.R. (2009) Hymenopteran parasitoids of forensic importance: host associations, seasonality and prevalence of parasitoids of carrion flies in Western Australia. *Journal of Medical Entomology* 46: 1210–1219.

Waldbauer, G.P. (1968) The consumption and utilization of food by insects. *Advances in Insect Physiology* 5: 229–288.

Wertheim, B., Vet, L.E.M. & Dicke, M. (2003) Increased risk of parasitism as ecological costs using aggregation pheromones: laboratory and field study of *Drosophila–Leptopilina* interaction. *Oikos* 100: 269–282.

Whiting, A. (1967) The biology of the parasitic wasp *Mormoniella vitripennis*. *Quarterly Review of Biology* 42: 333–406.

Wigglesworth, V.B. (1972) *The Principles of Insect Physiology*, 7th edn. Chapman & Hall, London.

Williams, H. & Richardson, A.M.M. (1983) Life history responses to larval food shortages in four species of necrophagous flies (Diptera: Calliphoridae). *Australian Journal of Ecology* 8: 257–263.

Williams, H. & Richardson, A.M.M. (1984) Growth energetics in relation to temperature for larvae of four species of necrophagous flies (Diptera: Calliphoridae). *Australian Journal of Ecology* 9: 141–152.

Willmer, P., Stone, G. & Johnston, I. (2000) *Environmental Physiology of Animals*. Blackwell Publishing Ltd., Oxford.

Withers, P.C. (1992) *Comparative Animal Physiology*. Saunders College Publishing, New York.

Yocum, G.D., Zdarek, J., Joplin, K.H., Lee, R.E. Jr, Smith, D.C., Manter, K.D. & Denlinger, D.L. (1994) Alteration of the eclosion rhythm and eclosion behaviour in the flesh fly, *Sarcophaga crassipalpis*, by low and high temperature stress. *Journal of Insect Physiology* 40: 13–21.

Young, A.R., Meeusen, E.N.T. & Bowles, V.M. (1996) Characterization of ES products involved in wound initiation by *Lucilia cuprina* larvae. *International Journal of Parasitology* 26: 245–252.

Zwaan, B.J., Bijlsma, R. & Hoekstra, R.F. (1991) On the developmental theory of ageing. I. Starvation resistance and longevity in *Drosophila melanogaster* in relation to pre-adult breeding conditions. *Heredity* 66: 29–39.

Supplemental reading

Ahmad, A. & Omar, B. (2018) Effect of carcass model on maggot distribution and thermal generation of two forensically important blowfly species, *Chrysomya megacephala* (Fabricius) and *Chrysomya rufifacies* (Macquart). *Egyptian Journal of Forensic Sciences* 8: 64.

Charabidze, D. & Hedouin, V. (2019) Temperature: the weak point of forensic entomology. *International Journal of Legal Medicine* 133: 633–639.

Feder, M.E., Blair, N. & Figueras, H. (1997) Natural thermal stress and heat shock protein expression in *Drosophila* larvae and pupae. *Functional Ecology* 11: 90–100.

Hanski, I. & Kuusela, S. (1977) An experiment on competition and diversity in the carrion fly community. *Annales Entomologica Fennici* 43: 108–115.

Heaton, V., Moffatt, C. & Simmons, T. (2018) The movement of fly (Diptera) larvae within a feeding aggregation. *The Canadian Entomologist* 150: 326–333.

Ives, A.R. (1991) Aggregation and coexistence in a carrion fly community. *Ecological Monographs* 61: 75–94.

Korsloot, A., van Gestel, C.A.M. & van Straalen, N.M. (2004) *Environmental Stress and Cellular Response in Arthropods*. CRC Press, Boca Raton, FL.

Kotzé, Z., Villet, M.H. & Weldon, C.W. (2016) Heat accumulation and development rate of massed maggots of the sheep blowfly, *Lucilia cuprina* (Diptera: Calliphoridae). *Journal of Insect Physiology* 95: 98–104.

Kouki, J. & Hanski, I. (1995) Population aggregation facilitates coexistence of many competing carrion fly species. *Oikos* 72: 223–227.

Krebs, J.R. & Davies, N.B. (1996) *Introduction to Behavioral Ecology*. Blackwell Publishing Ltd., Oxford.

Levot, G.W., Brown, K.R. & Shipp, E. (1979) Larval growth of some calliphorid and sarcophagid Diptera. *Bulletin of Entomological Research* 69: 469–475.

Reznik, S.Y., Chernoguz, D.G. & Zinovjeva, K.B. (1992) Host searching, oviposition preferences and optimal synchronization in *Alysia manducator* (Hymenoptera:

Braconidae), a parasitoid of the blowfly, *Calliphora vicina. Oikos* 65: 81–88.

Schmidt-Nielsen, K. (1997) *Animal Physiology: Adaptation and Environment.* Cambridge University Press, Cambridge, UK.

Tarone, A.M. & Benoit, J.B. (2020) Insect development as it related to forensic entomology. In: J.H. Byrd & J.L. Castner (eds) *Forensic Entomology: The Utility of Using Arthropods in Legal Investigations*, 3rd edn, pp. 225–252. CRC Press, Boca Raton, FL.

Terra, W.R. & Ferreira, C. (1994) Insect digestive enzymes: properties, compartmentalization and function. *Comparative Biochemistry and Physiology B* 109: 1–62.

Additional resources

Blow flies: http://ipm.ncsu.edu/AG369/notes/blow_flies.html

Dirty Jobs maggot mass video: http://dsc.discovery.com/videos/dirty-jobs-maggot-mass.html

Forensic entomology resources website: http://forensicentomology.com/

Fly life cycle: http://australianmuseum.net.au/Decomposition-fly-life-cycles

Insect thermoregulation: https://insectantifreeze.weebly.com/

Chapter 9
Temperature tolerances of necrophagous insects

Overview

Insects residing in temperate and cold regions must cope with harsh winter conditions to survive. Daily temperatures can fluctuate widely, dropping below freezing for extended periods, and coupled with severely desiccating conditions the threat of low-temperature injury or mortality is high over the long duration of winter. At the opposite end of the spectrum, insects living in regions where the climate is characterized by periods of extreme heat are challenged with surviving in conditions that inhibit normal cellular processes and can approach temperatures that literally melt membrane lipids. Some species of necrophagous insects may encounter both types of environmental extreme during their life cycle, while others endure less "extreme" upper and lower temperatures but do experience dramatic shifts in daily and/or seasonal temperatures. Stressful high-temperature exposure is a self-induced feature for fly larvae developing in large feeding aggregations, in that the larvae themselves appear to generate the heat that elevates temperatures in the microhabitat otherwise known as a maggot mass. Regardless of whether life is threatened by heat or cold, natural or self-derived, necrophagous insects possess an array of adaptive traits that provide the means to endure temperature insults associated with seasonal change, aseasonal fluctuations, or the unique internal environment of a maggot mass. This chapter will examine the general principles of insect adaptations to aseasonal and seasonal temperature changes, as well as how these are relevant to insect colonization and development on a corpse in different biogeographical and seasonal conditions.

The big picture

- Necrophagous insects face seasonal, aseasonal, and self-induced (heterothermy) temperature extremes.
- Temperature challenges do not mean death: necrophagous insects are equipped with adaptations to survive a changing environment.
- Life-history features that promote survival during proteotaxic stress.
- Deleterious effects of high temperatures on necrophagous flies.
- Life-history strategies and adaptations that promote survival at low temperatures.
- Deleterious effects of low-temperature exposure.

The Science of Forensic Entomology, Second Edition. David B. Rivers and Gregory A. Dahlem.
© 2023 John Wiley & Sons Ltd. Published 2023 by John Wiley & Sons Ltd.
Companion website: www.wiley.com/go/forensicentomology2

9.1 Necrophagous insects face seasonal, aseasonal, and self-induced (heterothermy) temperature extremes

All insects are poikilothermic ectotherms. What this means is that, as ecothermic animals, insects do not regulate internal body temperature through production of internal heat. Unlike many ectotherms, internal body fluids are not maintained within any specific temperature range. Rather, as poikilotherms (synonymous with eurytherm), the internal environment varies with changing ambient temperatures (Figure 9.1). The net effect is that living in climates prone to dramatic shifts in daily or even hourly conditions, or residing in regions that undergo seasonal change, insects must possess

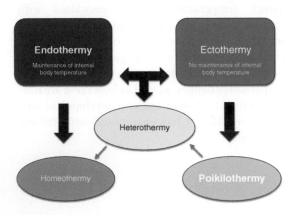

Figure 9.1 Relationship between ability/capacity to maintain body temperature in animals for long or short periods of time.

the ability to maintain metabolic activity over a wide range of temperatures. In general, insects possess enzymes that operate under a broad range of conditions, including varying temperature and pH (Randall *et al.*, 2002; Storey, 2004). The range of temperatures over which insects can maintain metabolic processes or survive indefinitely is referred to as the **zone of tolerance** or thermal tolerance range (Figure 9.2) (Withers, 1992). Outside the temperature zone, namely at temperatures above the upper thermal limit (**critical thermal maximum**) or below the lower thermal limit (**critical thermal minimum**), are conditions that will initially evoke inhibition of cellular reactions, thereby retarding most aspects of growth and development. If exposure is sufficiently long or movement out of the thermal tolerance range occurs unexpectedly or rapidly, injury or death may be the end result.

In the context of a necrophagous lifestyle, life is good provided temperatures remain within the zone of tolerance. For example, corpse decomposition is temperature dependent (Mann *et al.*, 1990), becoming a more favorable habitat for carrion insect colonization with increasing temperatures, up to an upper threshold limit (Campobasso *et al.*, 2001). Efficiency of searching behavior for carrion, mate location, egg provisioning, and oviposition elevate with temperature. Insect growth on carrion is essentially linear with increasing ambient temperatures, although continuous elevations in carrion temperature may be restricted to within the large feeding aggregations created by fly larvae. Postfeeding developmental events such pupariation, pupation, and adult emergence are all linked to environmental temperatures, with the length of each decreasing as temperature increases up to a critical ceiling. What happens if the ceiling is breached? Or conversely, what happens if temperature drops, either gradually

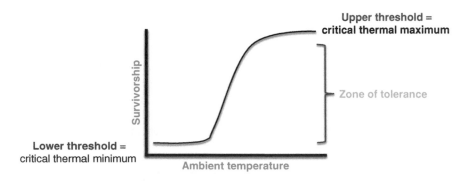

Figure 9.2 Typical survival curve for necrophagous insects in relation to environmental temperatures. Thresholds or limits are species-specific and tend to coincide with biogeographic distributions.

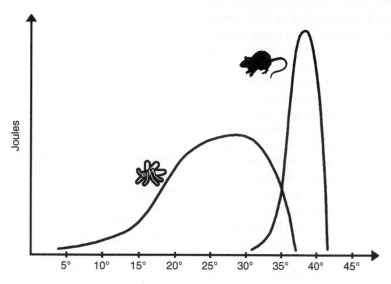

Figure 9.3 Sustained energy output (Joules) of a poikilotherm (an insect) and a homeotherm (a mouse) as a function of core body temperature. The homeotherm has a much higher output but can only function over a very narrow range of body temperatures. *Source*: Modified from Petter Bockman at https://commons.wikimedia.org/wiki/File:Homeothermy-poikilothermy.png.

or suddenly, below a critical lower limit? Once ambient temperatures lie outside the thermal range for a given species, growth and development are compromised. For necrophagous larvae of flies and beetles, the temperature threshold below which development does not occur is referred to as the **base temperature** or **developmental limit** (Villet *et al.*, 2010). The base temperature is generally not considered the same as the lower lethal limit. However, if exposure to these unfavorable conditions lasts long enough, irreversible injury leading to death may result. In fact, lengthy exposure may not be needed to induce lethal conditions in an insect that experiences a sharp decline (**cold shock**) or rapid elevation (**heat shock**) in temperature. In these cases, exposure to temperature extremes lasting as short as 20 minutes or less are sufficient to permanently and irreversibly injure the insect.

Conditions that foster temperature elevations or depressions beyond thermal limits include seasonal changes, unpredictable aseasonal fluctuations in environmental conditions and, in some instances, self-induced heating. The latter example is mostly associated with heterothermic heat production within maggot masses, which was a major focus of Chapter 8. Heterothermy associated with maggot masses can be defined as temporary heat production by larvae in a feeding aggregation that elevates temperature above ambient conditions linearly over time but is constrained by physiological limits. Heterothermy is also associated with some Coleoptera species; prior to flight in adults and as a feature of larval aggregations formed by at least one family (Dermestidae) (Verdú *et al.*, 2006; Gunn, 2018). The remainder of this chapter will focus on how insects in general, and necrophagous Diptera specifically, deal with temperature extremes, regardless of the causation. We will also discuss what happens to necrophagous insects that are not properly equipped for sudden temperature swings or changing seasonal temperatures. As you will see by the end of the chapter, temperature is among the most important factors shaping the lives of any insects, rivaled in significance only by nutrient availability (Figure 9.3).

9.2   Temperature challenges do not equal death: necrophagous insects are equipped with adaptations to survive a changing environment

Changing environmental conditions, particularly temperature, do not necessarily lead to the demise of an insect. Certainly temperature swings, gradual or sudden, can elicit stress, injury or even death. However, in many instances, if not most, insects possess an array of adaptations to overcome or avoid extreme heat or cold. The adaptations may be as subtle as behavioral modifications whereby gravid calliphorid females oviposit in sheltered locations such as body cavities and openings or under clothing on a human corpse to buffer against rapid changes in temperatures, or where young fly larvae migrate to concealed environments on a corpse or form maggot masses within a body cavity so that the internal heat of the aggregation affords protection from low temperatures or the effects of the wind (Campobasso *et al.*, 2001). **Spatial partitioning**, simply defined as a physical separation of two or more species using the same resource as means to coexist, appears to be the key to survival for some necrophagous fly species living in mixed species maggot masses that display differential heat zones (Richards *et al.*, 2009). Many insects rely on physiological and biochemical mechanisms to acquire thermotolerance or cold hardiness. Acquisition of cold hardiness is frequently a component of a highly evolved genetic program associated with seasonality in which insects anticipate seasonal change and prepare for the arrival of low temperatures through a series of physiological, morphological, and/or behavioral modifications. The adaptations are phenotypic expressions of the genetic program(s) that allows insects in temperate, arctic, or near-arctic regions to avoid or survive unfavorable conditions (Denlinger & Lee, 2010). A highly efficient stress response system that often initiates synthesis of heat-shock proteins (Hsps) in response to an array of environmental insults is typical of all insects and appears to be especially critical to insects in which specific stages of development are exposed to high temperatures for an extended period of time. Expression of heat proteins is not restricted to seasonal or aseasonal temperature changes as the heat-shock response appears to afford protection to the self-induced heterothermy of feeding aggregations (Rivers *et al.*, 2010), a proteotaxic stressor that becomes increasingly more critical with size of the maggot mass or carcass (Charabidze *et al.*, 2011; Rivers *et al.*, 2011; Scanvion *et al.*, 2018).

Our approach for examining extreme temperatures in the environment will be to explore the adaptations used by necrophagous insects to overcome the deleterious effects of high and low temperatures, as well as those associated with heterothermic heat production in larval feeding aggregations.

9.3   Life-history features that promote survival during proteotaxic stress

Proteotaxic or high-temperature stress is far more common to necrophagous insects than low-temperature exposure. Why? Recall our earlier discussion of the relationship between insect activity and temperature: the corpse becomes a more favorable habitat with increases in temperature. By extension, this obviously means that carrion-inhabiting insects peak in abundance during warm months, when temperatures of a corpse can reach their zenith due to a combination of factors: solar radiation, tissue decomposition, microbial processes, and insect (fly) activity. Insects generally avoid low temperatures associated with seasonal change either by initiating **diapause,** a physiological state of dormancy, or by leaving the area before the cold of winter arrives. Even milder low temperatures can lead to cold avoidance as several species of calliphorids, sarcophagids, and dermestids refuse to oviposit or larviposit on cold tissue (Campobasso *et al.*, 2001; Martín-Vega *et al.*, 2017). Thus, necrophiles are far more likely to be active during periods when temperatures exceed the zone of tolerance than when climatic conditions drop temperatures below the critical thermal minimum. However, it is also important to recognize that several species of necrophagous flies (i.e., *Cynomya cadaverina*, *Protophormia terraenovae*, and *Calliphora vicina*) and to a lesser extent beetles, are active at low temperatures that other insects avoid.

Heterothermic heat production in maggot masses is also a critical feature of development for flies belonging to the families Calliphoridae and Sarcophagidae. Some of these species appear to tolerate higher temperatures than others in the feeding aggregation, leading to speculation that induced high-temperature elevations promote a competitive advantage in heterogeneous maggot masses (Richards *et al.*, 2009; Villet *et al.*, 2010). For the more temperature-sensitive species, the changing internal temperatures of the mass may be viewed as an aseasonal change in the microhabitat that constitutes a proteotaxic stress that can lead to thermal injury or death.

Heat stress is a serious threat to necrophagous insects, particularly flies developing in feeding aggregations. Thousands of individual insects may feed on a single large mammal following death, all potentially dealt the challenge of feeding under proteotaxic conditions, yet the vast majority successfully complete development, propagate, and contribute to continuation of the species. For these insects, survival depends on the ability to overcome or avoid thermal stress. Avoidance is not entirely possible if residing on or in the corpse for an extended period. So developmental success appears to be more a matter of combating high-temperature stress. How do necrophagous insects accomplish this difficult task? There appears to be little variation in the maximum temperatures (40–50 °C) that most insects can tolerate (Heinrich, 1993), so the solutions to heat-inducible stress are shared among most groups. Many rely on several highly evolved adaptations that promote acquisition of thermotolerance, protection from heat stress, or mechanisms to quickly dissipate excessive heat. The mechanisms range from a highly conserved heat-shock response to evaporate cooling and behavioral mechanisms such as spatial partitioning and locomotion. Sections 9.3.1–9.3.5 will examine some of the more common adaptations employed by necrophagous insects to deal with extreme heat in the environment.

9.3.1 Heat-shock response

Insects possess a highly efficient general stress response that often functions to produce a series of stress proteins that are believed to confer protection and repair under a wide range of environmental and artificial insults (Lindquist & Craig, 1988). The proteins are typically referred to as **Hsps,** not because they are only produced in response to heat but simply because they were first characterized following high-temperature stress. Some Hsps are produced under non-stress conditions and thus are constitutively expressed. Such Hsps are referred to as cognates (heat-shock cognate or Hsc) and have several roles in normal cellular functioning. However, when an insect is exposed to temperature elevations that exceed an upper threshold, tissue-specific synthesis of Hsps begins at the expense of normal protein production (Feder & Hoffman, 1999). For many necrophagous flies, expression of Hscs continues until achieving temperatures at or near 35 °C; above this temperature Hsps are expressed (Korsloot et al., 2004). Further increases in temperature can lead to upregulation of other Hsps while earlier stress protein production may depress or stop

(Feder & Hoffman, 1999). The absolute temperatures that evoke Hsp expression are species-specific, and thus synthesis of stress proteins is dependent on a number of factors for a given insect, likely including phylogeny (Villet et al., 2010).

Most Hsps generally afford protection to proteins and other cellular constituents during heat stress. Hsps may also be linked to acquisition of thermotolerance, a topic discussed in Section 9.3.2. Stress proteins are grouped into families based on mass, and thus functions during heat stress are associated with particular Hsp families (Table 9.1).

The major families of Hsps are:

- Hsp or sp[1] 90 (kDa[2]) family;
- Hsp 70 (kDa) family;
- Hsp 60 (kDa) family;
- Small Hsp family.

During non-stress conditions, proteins in the Hsp 60 and Hsp 70 families have roles as molecular chaperones, functioning to help proteins fold correctly or maintain native configurations. These Hsps, along with proteins from other families, also aid in translocation of components throughout a cell, apoptotic mechanisms, and an array of pathways associated with cell growth and development (Vayssier & Polla, 1998). Though the Hsp 90 family is highly conserved in eukaryotes, there is no evidence that these proteins contribute to thermal stress responses in necrophagous insects, at least not in sarcophagids (Tammariello et al., 1999). In contrast, temperatures surpassing 35 °C evoke synthesis of the Hsp 70 family, with Hsp70 being the dominant protein produced in response to most stressors. The heat-inducible Hsp70 protein functions as a molecular chaperone by binding to denatured or native proteins and guiding them to lysosomes for removal (Korsloot et al., 2004) or to aid in protein folding following high-temperature exposure (Neven, 2000). Proteins in the Hsp 60 family are also expressed in response to sudden temperature elevations (heat shock). Hsps from both families (60 and 70) are expressed in response to thermal stress in larvae, pupae, and adults of calliphorids and sarcophagids (Yocum & Denlinger, 1992; Sharma et al., 2006; Rivers et al., 2010).

Small Hsps are the least conserved of the stress proteins in eukaryotes. Four small Hsps (22, 23, 26, and 27) appear to be synthesized in flies, and during heat stress in *Drosophila melanogaster* every cell

Table 9.1   Stress protein families expressed in necrophagous insects.

Family of stress protein	Type of stress protein	Function (known or presumed)
Hsp 90	Hsp90	Upregulated in non-diapausing larvae and during quiescent phase of pupal diapause, and in larvae residing in large maggot masses
	Hsp83	Molecular chaperone during stress and non-stress
Hsp 70	Hsp70(72)	Induced due to range of stresses including heat, cold, anoxia, and desiccation; upregulated during pupal but not larval diapause as well as recovery phase of rapid cold hardening; upregulated in larvae in response to maggot mass size and temperature
	Hsc70	Constitutively produced during stress and non-stress, protein-binding
Hsp 60	Hsp68(60)	Upregulated during pupal but not larval diapause; upregulated in larvae in response to maggot mass size and temperature; upregulation in larvae due to environmental insults
Small Hsps	Hsp27	Upregulated in pupa diapause and in larvae in response to maggot mass size and temperature
	Hsp23	Not altered by larval diapause
	Unidentified	Upregulated during pupal diapause

Pupal diapause is in reference to *Sarcophaga crassipalpis* (Sarcophagidae) while larval diapause is in association with calliphorids that have been examined. Stress in maggot masses has been documented with calliphorid and sarcophagid species.

type examined expresses large quantities of the four small Hsps (Arrigo & Landry, 1994). Hsp27 is expressed in all stages of larval development in *Sarcophaga bullata* and *Protophormia terraenovae* when larvae are reared in large maggot masses, and hence high internal heat (Rivers *et al.*, 2010). Expression of the small Hsps is likely associated with maintenance of muscle function following high-temperature exposure (Parsall & Lindquist, 1994). They also appear to be involved in microfilament stabilization and protection against inhibition of protein synthesis during heat stress (Korsloot *et al.*, 2004).

9.3.2   Acquisition of thermal tolerance

The expression of Hsps following a mild temperature insult (i.e., a brief exposure to non-damaging temperatures) confers thermotolerance to necrophagous flies. Synthesis of proteins in the Hsp 70 family is most commonly associated with acquisition of thermotolerance. In the case of the flesh fly *Sarcophaga crassipalpis*, a brief exposure to 40 °C for 2 hours affords protection when pupae or pharate adults are then subsequently exposed to the normally lethal conditions of 45 °C for 90 minutes (Yocum & Denlinger, 1992). Similar protection occurs with the closely related

S. bullata and the blow fly *Calliphora vicina* (El-Wadawi & Bowler, 1995). The thermal tolerance is short-lived and cannot be lengthened by increasing the duration of preconditioning. In *D. melanogaster*, larval development under sublethal high temperatures evokes thermotolerance in the subsequent adults. Overcrowded conditions that generate heat production and that trigger the heat-shock response can also confer thermal tolerance to the imago stage (Sørensen & Loeschcke, 2001).

The capacity for acquisition of thermal tolerance differs between the sexes in the adult stages of some flies. In the best-studied case, *S. crassipalpis*, exposure to 45 °C during the pharate adult stages or as adults is lethal after a relatively short treatment (90 minutes) but can be overcome with a brief pretreatment at a milder temperature (Yocum & Denlinger, 1992). Thermotolerant adult females remain receptive to mating but experience a reduction in egg provisioning (Rinehart *et al.*, 2000). Adult males, however, are more severely affected in that sperm transfer is completely abolished, apparently due to heat damage to the testes. This differential thermal response may be pervasive throughout necrophagous Diptera in that adult females from seven species of African calliphorids survived high-temperature exposure at much higher rates than males from the same species (Richards *et al.*, 2009).

Whether the differences in high-temperature survival are linked to differential Hsp expression or acquisition of thermotolerance has not been tested.

Investigation of thermal tolerance acquisition in forensically important species of Coleoptera is sparse. "Young" but not "old" larvae of *Tribolium castaneum* display enhanced thermal tolerance following a short exposure (1 hour) to 40 °C, which corresponds to upregulation of Hsp70 (Mahroof *et al.*, 2005). Interestingly, no other stage of development for the beetle shows an increase in expression of this Hsp following high-temperature exposure. In fact, Hsp70 expression in eggs is downregulated in response to heat-treatment at 40 °C. Correspondingly, eggs, old larvae, pupae and adults all remain heat-susceptible. Similar studies with necrophagous beetles have not been performed.

9.3.3 Behavioral mechanisms

Exposure to high temperatures causes some insects to engage in behaviors that promote favorable temperatures by returning to conditions lying within the zone of tolerance or by moving away from threatening temperatures. Some social insects can provide ventilation to the colony so that excess heat dissipates prior to induction of thermal stress and/or use collective wing beating as fans to cool the internal environment of the hive (Heinrich, 1993) (Figure 9.4). Necrophagous insects do not utilize such mechanisms, which means that behavioral responses to heat are generally in the form of locomotion to avoid or move away from high temperatures. For this to be true, two prerequisites must be met:

Figure 9.4 Adult honey bees *(Apis mellifera ligustica)* using their wings to fan the hive. *Source*: Ken Thomas/ Wikimedia Commons/Public domain.

1. Insects possess thermal receptors that recognize specific temperatures associated with critical thermal minima and maxima or which can detect changes in temperature.
2. Movement away from the carcass is not potentially more harmful than exposure to high temperatures on the carcass.

There is evidence to suggest that some carrion-inhabiting insects do in fact possess thermal receptors. Most of the data is associated with adult necrophagous flies, and the details will be discussed in Section 9.3.4. As to the second criterion, feeding larvae will not survive for long if they crawl from the corpse, so such developmental stages would appear to be prohibited from utilizing locomotor responses as a means to cope up with heat. An exception may be larval movement through maggot masses to avoid high internal temperatures (Rivers *et al.*, 2011). This proposition has not been demonstrated experimentally, and thus it remains a speculation. However, some calliphorid larvae will crawl toward "hot" spots in maggot masses presumably to reach optimal temperatures that promote feeding, growth, and development (Heaton *et al.*, 2014). The larval movements appear to confirm the capacity for thermal perception in these species. By extension, it also seems logical to assume that larvae can move away from regions that are too hot (Charabidze *et al.*, 2011). Adult flies appear to have the capacity to walk across the surface of carrion, stopping to feed, mate, or oviposit in regions on the corpse that meet appropriate conditions such as temperature and moisture content (Ashworth & Wall, 1994). They obviously can fly away at any point, suggesting that thermal detection is associated with tarsal **pulvilli,** the soft cushion-like pads or footpads located between terminal claws of some Diptera.

Spatial partitioning can be considered a behavioral mechanism for heat avoidance. The mechanism appears to be utilized by fly maggots with lower upper thermal limits in comparison with competing species that commonly coexist in maggot masses. Some species with high critical thermal maxima may be able to dominate a feeding aggregation by driving internal mass temperatures above threshold limits of competing species (Richards *et al.*, 2009). This competitive adaptation appears to be exploited by *Chrysomya marginalis* and *Chrysomya rufifacies* (Williams & Richardson, 1984; Richards *et al.*, 2009). Survival of more heat-sensitive species depends on cooperative feeding in the assemblages (see Chapter 8), and thus these flies feed at the periphery of the maggot mass as a compromise between avoiding overheating but depending on group feeding.

9.3.4 Evidence for thermoreceptors in flies

The capacity to respond to high (or low) temperature insults seemingly requires the ability to detect absolute temperatures or temperature changes in the environment or local habitats, as on carrion. Thermal- and humidity-sensitive neurons have been identified on the antennae of an array of insects representing several different insect orders. The receptors are typically arranged within a single sensillum[3] that is consistently peg-shape in morphology, lacking pores and possessing multiple (usually three) sensory neurons (Chapman, 1998). Dendrites from two of the neurons extend through the peg sensillum to the cuticle surface while the third dendritic extension does not enter the body of the peg. The two longer dendrites are believed to be associated with hygroreception, with one detecting increases in relative humidity while the other responds to drops in air water content (Chapman, 1998). Since relative humidity is directly influenced by changes in temperature, some necrophagous insects may be able to assess changing temperatures via hygroreception. The third dendrite is presumed to be thermosensitive. The thermal receptor functions by perceiving declines in temperature, and consequently is referred to as a cold receptor. In some insects, the cold receptors can detect temperature changes of less than 1 °C (Altner & Loftus, 1985).

Thermal reception is also associated with some olfactory receptors (chemoreceptors) located on the maxillary palps, such as occurs with the American cockroach *Periplaneta americana* or the antennae of some species of cockroaches and grasshoppers (Tominaga & Yokohari, 1982; Altner & Loftus, 1985). Whether such thermal or humidity receptors are present on any stage of necrophagous flies has not been shown. However, several anecdotal observations argue that one or both types of sensory detectors do occur on adults and larvae. As discussed earlier, adult calliphorids likely rely on receptors on footpads to detect temperatures and moisture content of carcass surfaces to assess whether to feed or oviposit (Cragg, 1956; Ashworth

& Wall, 1994). Spatial partitioning displayed by heat-sensitive species like *Chrysomya albiceps* suggests that larvae can detect absolute temperatures or heat zones as part of a heat avoidance strategy (Richards *et al.*, 2009). Similarly, larvae of *Lucilia sericata* avert heat stress by entering larval diapause when temperatures exceed 35 °C (Greenberg, 1991). In cases of necrophagous fly larvae, temperature detection may rely on both thermal receptors and hygroreceptors since there is linkage between maggot mass heterothermy and increasing fluid release from tissue decomposition. Presumably these conditions lead to elevations in relative humidity in the microclimate associated with feeding aggregations.

9.3.5 Evaporative cooling, conductance, and convection

Necrophagous insects forced to contend with high-temperature stress must rely on physiological mechanisms to dissipate excess heat or face injury and death. Excessive heat is particularly challenging for maggots for, as discussed earlier, simply crawling far away from the source of heat, the maggot mass, is not an option. Some protection is gained through activation of the heat-shock response, but this is an energetically and developmentally expensive system since it relies on production of stress proteins and the inhibition of normal protein synthesis. What this means for the long term is that development will be delayed or impaired if exposure to the heat stressor lasts for a long period. A less demanding process for dealing with high temperatures is evaporative cooling. Evaporative cooling is the physical process in which evaporation of a liquid into the surrounding environment (air) cools an object that is in contact with it. Heat is transferred from the object, in this case an insect, to the liquid, and the **latent heat of vaporization,** the heat necessary to evaporate the liquid, is typically derived from either the air or insect's body (Chapman, 1998). The greater the difference between the temperature of the insect and the environment, the faster the rate of evaporative cooling.

Evaporative cooling is restricted to insects that are large enough to rely on internal water pools, or insects that can derive the liquid from the environment, which is size independent (Prange, 1996). Although it has been suggested that saliva and digestive fluids from necrophagous fly larvae could be the source of liquid for evaporative cooling (Villet *et al.*, 2010), considering their small size it is much more likely that the water pool is the semi-liquid to liquid conditions of a maggot mass resulting from tissue decomposition. Coupled with the high conductance (i.e., transference of heat due to lack of insulating barriers of a fly larva's body) excess internal heat should be easily transferred to the surrounding fluids. Internal mass temperatures have been reported to exceed ambient air temperatures by 10–30 °C on a regular basis, which should be sufficient latent heat to promote evaporation from the integumental surfaces of fly larvae. However, the humidity of the microhabitat is ordinarily high (Rivers *et al.*, 2011), which will lower the rate of evaporation (Prange, 1996). As with so many of the physiological topics discussed in this book, evaporative cooling in necrophagous flies has not been tested experimentally and thus must still be considered conjecture.

Heat loss via convection will occur if the internal temperature of the body is above ambient temperature and air is moving over the integumental surface. This form of heat dissipation is probably most important when larvae are younger, feeding aggregations form in non-concealed locations, and before significant tissue decomposition occurs to change the microhabitat from terrestrial to semi-aquatic (see Chapter 8 for details). Young fly larvae are expected to lose heat by convection at a faster rate simple because they have a higher surface area to volume ratio than older larger larvae (Willmer *et al.*, 2000).

The heat of a corpse appears to be more likely to impact feeding stages of flies than other life stages or with non-fly insects that frequent carrion (Villet *et al.*, 2010). That said adult necrophagous flies and carrion beetles can dissipate excess heat via convection during flight (Chapman, 1998). Bubbling behavior of adult flies is another avenue of heat dissipation. During post-feeding, several species of flies engage in oral extrusion of digestive fluids from the foregut that appear as droplets or "bubbles" dangling from the proboscis (Figure 9.5). When the droplet is extruded, evaporative cooling occurs provided the ambient air is cooler than the bubble temperature (Gomes *et al.*, 2018). If the droplet is reabsorbed, the cooler temperature of the ingested liquid facilitates lowering of the body temperature. This provides the additional benefit of food concentration (Hendrichs *et al.*, 1992).

Figure 9.5   An adult anthomyiid displaying bubbling behavior. *Source*: Alvesgaspar/Wikimedia Commons/CC BY-SA 3.0.

9.4   Deleterious effects of high temperatures on necrophagous flies

Insects that lack the appropriate adaptations to deal with high temperatures, regardless of the source of heat, will suffer some type of heat injury. If exposure is for a long enough period or the temperature sufficiently exceeds the upper thermal limit or critical thermal maximum for a given species, the conditions will likely be lethal. The temperature elevations associated with large maggot masses on a corpse achieve both conditions. In some instances, internal temperatures of feeding aggregations can reach 45–50 °C or more (Anderson & VanLaerhoven, 1996; Richards & Goff, 1997), even when ambient temperatures are below 20 °C (Deonier, 1940; Waterhouse, 1947). These temperatures are more than 10–15 °C higher than the upper thermal limits for many calliphorid and sarcophagid species (Table 9.2), so the chance of heat injury exists despite mechanisms to cope with less severe proteotaxic stress.

The consequences of high-temperature exposure are generally classified into two broad categories: lethal and sublethal effects. Lethal effects need no explanation other than understanding the onset of mortality in different developmental stages and the mechanisms responsible for death. Sublethal effects of high temperature are also dependent on the stage of development at the time of exposure as well as the absolute temperature encountered and the

Table 9.2   Upper thermal limits for necrophagous flies.

Species	Stage	Critical thermal maximum (°C)
Calliphora croceipalpis	Larval	42.9
Calliphora hilli	Larval	35–40
Calliphora stygia	Larval	35–40
Calliphora vicina	Larval	28–39
Calliphora vomitoria	Larval	>39
Chrysomya albiceps	Larval	48.8
Chrysomya marginalis	Larval	≥50.1
Chrysomya megacephala	Larval	49.0
Chrysomya putoria	Larval	48.5
Chrysomya ruffiacies	Larval	>45
Cochliomyia macellaria	Larval	40–45
Lucilia cuprina	Larval	40–45
Lucilia illustris	Larval	>39.5
Lucilia coeruleiviridis	Larval	>38.9
Lucilia sericata	Larval	35–47.5
Phormia regina	Larval	38–45
Protophormia terraenovae	Larval	>35
Sarcophaga bullata	Larval	>35
	Pupal/ pharate adult	42–45
Sarcophaga crassipalpis	Larval	42–45
	Pupal/ pharate adult	45
	Adult	>35
Sarcophaga haemorrhoidalis	Larval	>35
Sarcophaga tibialis	Larval	>35

Upper temperature limits for larvae were determined experimentally, or extrapolated from maggot mass temperatures associated with faunal succession or laboratory rearing. In the case of faunal succession, internal temperatures in larval aggregations may have been from heterogonous masses.
Source: Adapted from Rivers et al. (2011).

rate at which temperatures elevated. Much of what is known with regard to necrophagous insects stems from laboratory experiments using two closely related Nearctic sarcophagids, *Sarcophaga crassipalpis* and *S. bullata*.

9.4.1   Lethal effects

Insects that experience temperatures above the critical thermal maximum will eventually stop movement, a condition called **heat stupor,** and if not removed from the area or temperatures do not

decline, death will result. The lethality of high temperatures is dependent on (i) species, indicative of a phylogenetic component to thermal tolerance, (ii) age of the insect, (iii) absolute temperature above the upper thermal limit, (iv) length of exposure, and (v) rate of warming, i.e., how rapidly ambient temperatures elevate in relation to body temperature (Neven, 2000; Villet *et al.*, 2010). Developmentally, the egg stage is the most sensitive to heat in terms of critical thermal maxima (lower than other stages) and length of high-temperature exposure that is tolerable before death ensues. For flesh flies, the pharate adult stages and adults are nearly identical in their sensitivity to high heat in that neither can tolerate long durations in the presence of temperatures exceeding 45 °C (Chen et al., 1990; Yocum & Denlinger, 1992). In these experiments, flies were exposed to rapidly rising temperatures that constituted heat shock, and thus the duration of tolerance to 45 °C is higher when elevations occur more gradually (Rinehart et al., 2000). Regardless of the rate of temperature increase, flies remaining at temperatures above 35 °C for a sufficiently long time will eventually die. By contrast, third instar larvae, wandering larvae (crop empty), or pupae (**phanerocephalic stage**) can withstand the same temperature extremes for much longer periods than other development stages whether heating is gradual or constitutes heat shock (Chen et al., 1990).

How flies die following high-temperature stress is not clear. As temperatures approach upper thermal limits for a given species, proteins will begin to denature provided protection in the form of Hsps or other factors is absent. If temperatures continue to climb, respiratory metabolism is totally inhibited (Meyer, 1978). Stress protein synthesis is oxygen-dependent, and thus Hsp production will cease. Similarly, prolonged exposure to high temperatures will deplete metabolic reserves resulting in the same effect on Hsp synthesis. One of the most important influences of temperature in general is on enzyme activity. High-temperature insults induce conformational changes in enzyme structure, interfere with binding to substrates, and alter membrane fluidity and thereby change enzyme activity (Neven, 2000). Any of these conditions will greatly alter cellular physiology and compromise cell homeostasis. Changes in membrane fluidity are likely to evoke a loss of membrane integrity and, once this occurs, unregulated flux in and out of the cell will ensue. The resulting osmotic imbalance will trigger a cascade of events that may include nuclear and/or plasma membrane blebbing or lysis, cytoskeleton rearrangement and collapse, autolysis, and activation of signal transduction pathways generating death signals (Gomperts *et al.*, 2002; Korsloot *et al.*, 2004). The precise pathways involved in heat-induced death have not been deciphered but quite likely involve multiple death pathways including **apoptosis,** non-apoptotic programmed cell death and **oncosis**[4].

9.4.2 Sublethal effects

Exposure to high temperatures above the upper thermal limit is not always lethal. If the proteotaxic stressor is of short duration or below the critical thermal maximum, the insect may recover yet suffer some form of heat injury. The effects of sublethal exposure to heat can be far-reaching, ranging from disruption or shifts in timing of key developmental events, malformations in later stages, to depressions in longevity or reproductive output. When larvae experience heat stress, the rate of development generally accelerates up to a physiological threshold, after which larval development slows, possibly delaying the initiation of pupariation (Campobasso *et al.*, 2001; Marchenko, 2001) or triggering a precocious larval diapause (Greenberg, 1991). In some instances, formation of puparia is abnormal with larvae failing to form the characteristic barrel shape or smooth cuticle (Rivers *et al.*, 2010). Such abnormalities are likely due to temporal and spatial disruption of patterned muscular contractions regulated by central motor programs (Zdarek *et al.*, 1987; Rivers *et al.*, 2004). Even if pupariation is not affected, pupal development can be retarded (Byrd & Butler, 1998; Rivers *et al.*, 2010), which in turn can delay the onset of pharate adult development and subsequent adult emergence (Rivers *et al.*, 2010). Heat exposure during larval development or in puparial stages of S. *bullata* and S. *crassipalpis* can also shift the timing of adult emergence, resulting in asynchronous eclosion of males and females (Yocum *et al.*, 1994; Rivers *et al.*, 2010). Severe heat injury in the larval or puparial stages may not be lethal immediately, allowing progression of pharate adult development but resulting in failure to initiate or complete **extrication** (Denlinger & Yocum, 1998), the complex series of behaviors associated with emergence from a puparium and soil.

Reproductive success may diminish as a result of high heat during any stage of development but is most pronounced when pharate adults or adult flies are subjected to high temperatures. Wing deformations certainly shorten the longevity of adult flies and may also disrupt the courting behavior of either sex. Either scenario will constitute a threat to fecundity. Heat shock of pharate adults of S. *crassipalpis* does not depress copulation in comparison with non-heat-stressed flies. However, adult females display a marked decrease in egg provisioning and fertilization. The former is attributed to direct heat injury to oocytes and ovary formation while the reduction in fertilization is due to suppression of spermatogenesis in males (Rinehart *et al.*, 2000). In some instances, males are rendered sterile.

9.5 Life-history strategies and adaptations that promote survival at low temperatures

During the peak of carrion insect activity, the threat of low temperatures is not common or expected. We know this because insect activity increases on a corpse with elevations in temperature. Warmer climatic conditions also favor tissue decomposition, and hence emission of chemical signals that activate searching behavior in necrophiles and opportunists (predators and parasites) (see Chapter 7 for a discussion of chemical signaling). Generally, low temperatures are associated with seasonality, and most insects residing in regions in which seasonal change is characterized by cold possess the ability to respond to **environmental tokens** that forecast climatic shifts. An environmental token is a feature of the environment like humidity, photoperiod, temperature, oxygen content, or barometric pressure that signifies changing climatic conditions. For example, a photoperiod of 15 hours of daylight and 9 hours of darkness typifies summer months in temperate regions of North America at 42° N latitude, while impending seasonal change can be perceived by a shortened daylength (down to 13.5 hours by mid-August) as the fall equinox gradually arrives before the onset of winter. The significance of this foreshadowing of events is that the insects can anticipate the arrival of unfavorable environmental conditions and then prepare by migrating to a more suitable region, initiating a physiological state of hibernation or, in the case of temperature, gradually acclimating to a changing environment so that by the time freezing temperatures are experienced the insect is protected from injury. Such **acclimation** or **acclimatization** in temperate regions is unique to low temperatures in that conditions of winter often drop temperatures below the lower thermal limit or critical thermal minimum of most necrophagous insects; in contrast, typical summer temperatures do not exceed the upper thermal limits for insects active during warmer months. Thus, low temperature is typically seasonal and can be anticipated, whereas high temperature extremes are more often aseasonal or self-induced and thus arrive suddenly and unexpectedly. This is not to say that aseasonal drops in temperature do not occur: sudden declines in temperature can evoke chilling injury or cold shock (Figure 9.6).

As with heat, cold induces stress and is a serious threat to necrophagous insects. For carrion insects to survive the harsh conditions of winter or aseasonal drops in temperatures, survival is dependent on the ability to overcome or avoid cold stress. Avoidance by migration is not really possible for necrophagous insects residing in temperate or extreme low temperature zones such as arctic regions, in that the life cycles of adult flies are typically too short to support long periods of locomotion. Thus, survival depends mainly on preparatory physiological and biochemical programs that are implemented prior to the arrival of low temperatures. In other words, necrophagous insects that face seasonal cold temperatures rely on specific strategies to combat cold and, more specifically, to deal with

the threat of freezing of body fluids. The strategies use an array of highly evolved adaptations that either promote the acquisition of **cold hardiness** (i.e., acclimation to low temperatures as a means to avoid chilling injury or freezing) or which induce freezing of extracellular body fluids. These strategies and associated adaptations are discussed here.

9.5.1 Strategies for seasonal low temperatures

How insects cope with subzero temperatures basically depends on whether they can tolerate formation of ice in body fluids. Some cannot survive ice and thus rely on mechanisms to avoid freezing while others are tolerant of ice nucleation and actually promote freezing of extracellular fluids. The strategies used may change with life stages (i.e., a larva may be freeze tolerant while the adult must avoid freezing) and there does not appear to be any phylogenetic relationship between the low-temperature strategies employed by a particular group of insects (Willmer *et al.*, 2000).

9.5.1.1 Freeze avoidance

Formation of ice in body fluids and the subsequent expansion of ice crystals will damage cells beyond repair in several species of insects. Consequently, such species are labeled "freeze intolerant" and must avoid ice nucleation to survive the long duration of winter, or any period during which temperatures fall below freezing. The temperature at which

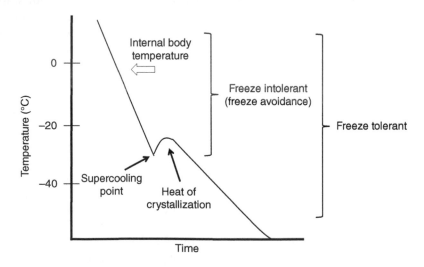

Figure 9.6   Insect body temperature as ambient conditions become cold. Bars on the right represent common responses of necrophagous insects to low temperatures. *Source*: Adapted from Lee (2010).

spontaneous formation of ice occurs in biological fluids is termed the **supercooling point** (SCP), also known as nucleation temperature (in reference to ice nucleation) or temperature of crystallization (reviewed by Wilson *et al.*, 2003). In species that employ **freeze avoidance,** one mechanism to enhance cold hardiness is to adjust the composition of body fluids so that the SCP is lowered, meaning that the temperature in which ice forms in extracellular fluids is much colder. Two processes generally achieve this: lowering the overall water content of extracellular fluids and removal of **ice nucleating agents** (ICNs) from fluids prone to freeze. Water content can be modified through induced dehydration whereby the internal water pool is lowered or by increasing the solute composition of body fluids.[5] The second process involves removal of ICNs. An ICN can be any object, living or not, that can serve as a nucleus for water condensation and subsequent ice crystallization. Some large macromolecules, undigested or partially digested food, and microorganisms are common examples of ICNs. One method for removing these potential nucleating agents is purging of the gut prior to low temperature arrival, an approach employed by calliphorids and sarcophagids that utilize larval or pupal diapause, respectively, to cope with winter. The net effect of these processes is that the body fluids can undergo **supercooling,** meaning the insect remains unfrozen despite exposure to very low temperatures (Table 9.3).

Antifreeze or **cryoprotectant** compounds are also significant to the freeze avoidance strategy. In theory, any solute added to a body fluid should depress the SCP due to the colligative properties of the solute in question (Storey, 2004). However, some solutes are more effective than others in

Table 9.3   Upper and lower thermal limits for forensically important beetles.

Species	Stage	Critical thermal minimum (°C)	Critical thermal maximum (°C)
Dermestes frischii	Egg	15–20	>35
	Larval	15–20	>35
	Pupal	15–20	>35
Dermestes maculatus	Egg	<15	>35
	Larval	<15	>35
	Pupal	15–20	>35
Dermestes undulatus	Egg	15–20	30–35
	Larval	15–20	30–35
	Pupal	15–20	30–35
Sciodrepoides watsoni	First instar larvae	<15	>28
	Second instar larvae	<15	>28
	Third instar larvae	<15	>28
	Pupae	<15	>28
Tribolium castaneum	Young larvae (6-days)	<23	>40
	Old larvae (22-days)	<23	>40
	Pupal	<23	>40
Creophilus maxillosus	Egg-adult emergence	<9.6	>32.5
Oxelytrum discicolle	Egg-adult emergence	<15	>28
Euspilotus azureus	Egg	<9.3	30–35
	First instar larvae	<4.2	30–35
	Second instar larvae	<5.9	30–35
	Pupal	<3.7	30–35
Thanatophilus micans	Egg	<15	>40
	Larval	<15	>28

Temperature limits were determined experimentally in laboratory studies.[5]
Source: Adapted from Mahroof *et al.* (2005), Midgley & Villet (2009), Velásquez & Viloria (2009), Jakubec (2016), Wang *et al.* (2017), Caneparo *et al.* (2017), Martín-Vega *et al.* (2017).

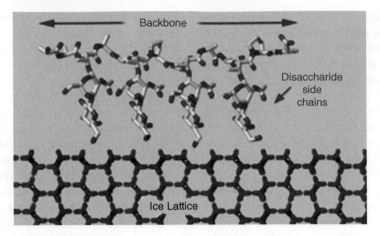

Figure 9.7　Cartoon of an antifreeze protein. The side chains of the protein bind to an existing ice lattice to prevent further growth of the crystal. *Source:* Image created by Olan001 and available at https://commons.wikimedia.org/wiki/File:Mechanism_at_work.pdf.

functioning as cryoprotectants. For example, low-molecular-weight osmolytes such as polyols (polyhydric alcohols) and sugars are commonly used by insects to raise the osmotic concentration of extracellular fluids, thereby reducing the water content so that very little is available for potential freezing (Danks, 2000). Generally, polyols such as glycerol or sorbitol accumulate in tissues and fluids prior to the onset of low temperatures as part of the preparatory programs of seasonality that will be discussed later. Accumulation of polyols occurs in a number of calliphorids and sarcophagids exposed to low temperatures as well as other stresses like anoxia, desiccation, and oxidative insults (Meyer, 1978; Kukal *et al.*, 1991; Yoder *et al.*, 2006).

Antifreeze proteins are often used not to prevent initial ice formation but to prevent expansion or further crystallization of ice crystals that have already formed (Duman, 2001). They generate an effect known as **thermal hysteresis** in which the freezing point of the extracellular fluids is lowered below the melting point (Figure 9.7). The proteins function by physically binding to ice lattices and thus preventing new water molecules from attaching (Willmer *et al.*, 2000). Protection afforded by antifreeze proteins extends to cold stress during sudden unexpected drops in temperature (Denlinger & Lee, 2010).

9.5.1.2　Freeze tolerance

Formation of ice in body fluids is an adaptive strategy employed by some species. For these insects, extracellular fluids are characterized by high subzero crystallization temperatures (close to 0 °C), thereby encouraging ice formation but in a gradual process, which in turn decreases the likelihood of damage due to stretching of membranes. Species employing **freeze tolerance** generally use processes that oppose the strategies of freeze-avoiding species: SCPs are elevated and ICNs acquired in extracellular fluids. The lipid composition of plasma membranes is also altered to permit stretching during ice expansion and phase transitions (Kostál, 2010). Ice crystallization is tolerated only in extracellular fluids and not intracellularly.

The internal environment of a cell is protected from freezing in a similar fashion as during freeze avoidance strategies. Colligative cryoprotectants (i.e., protection is based on concentration) in the form of polyols and sugars accumulate in high concentration within cells, lowering the water content (**osmotic dehydration**) and decreasing the chance of freezing. Compounds used in the intracellular environment share the trait of being **compatible solutes,** meaning that they do not interfere with metabolic processes and are not detrimental to intracellular constituents (Withers, 1992). Noncolligative cryoprotectants (i.e., protection is based on chemical species not concentration) accumulate in much lower concentrations and function to stabilize membranes during freezing. Agents such as trehalose and proline bind to plasma membranes in place of water, protecting membranes as well as proteins during phase transitions (Sinclair *et al.*, 2003).

Intracellular enzymes of freeze-tolerant insects must operate in relatively harsh conditions for this strategy to be successful. As the intracellular environment is modified to protect against freezing, the internal fluids become osmotically highly concentrated due to accumulation of colligative cryoprotectants. The polyols are literally replacing intracellular water, thereby concentrating all cellular constituents. Metabolic pathways must remain operational, albeit in a very limited capacity, for cells to survive the duration of cold-temperature exposure, and consequently enzymes must be able to function over a wide range of conditions, a feature previously mentioned with high-temperature insults.

9.5.1.3   Cryoprotective dehydration

For a relatively small number of insects (mostly soil inhabiting), neither freeze tolerance nor freeze avoidance characterizes the strategy for cold. An example is the collembolan *Onychiurus arcticus*, which resides in moist moss habitats found in arctic regions. Ambient temperatures drop below −20 °C during winter and the moss surrounding the collembolan freezes yet this insect remains unfrozen and displays no signs of increased cold hardiness prior to the onset of harsh weather (Sinclair *et al.*, 2003). Instead, these insects rely on a unique cold temperature strategy known as **cryoprotective dehydration.** This strategy takes advantage of an integument that is "leaky," permitting water loss to the surrounding ice even in relatively modest desiccating conditions (Lee, 2010). Concurrently, glycogen reserves are converted to the non-colligative cryoprotectant trehalose. As body water pools decline due to high water loss rates, dehydration ensues. In fact, for another arctic collemobolan, *Megaphorura arctica*, up to 90% of its body water is lost during cryoprotective dehydration (Lee, 2010). This in turn concentrates the levels of trehalose in extracellular fluids, depressing the melting point of body fluids (Sinclair *et al.*, 2003). Water content responds dynamically to changes in temperature, in this case associated with the microhabitat, and thus at any given temperature in the air surrounding the collembolan the extracellular fluids are in vapor pressure equilibrium with the surrounding ice. The net result is that each individual has a reduced risk of ice formation, despite not relying on traditional freeze avoidance mechanisms. Cryoprotective dehydration is speculated to be widespread in certain alpine and polar invertebrates (Lee, 2010) but probably has no role in insects of forensic importance.

9.5.2   *Hibernation or diapause*

Unfavorable environmental conditions ranging from extreme high or low temperatures, or during drought or rainy seasons can prompt a period of dormancy. Dormant periods can occur at almost any time of the year depending on the insect and biogeographical location, and can include summer dormancy or **aestivation,** winter hibernation, or involve **quiescence** or diapause (Gullan & Cranston, 2020). Quiescence is distinctive in that an insect may lower its metabolism during a period of unfavorable environmental conditions, and then resume normal activity once the environment becomes favorable for growth and development or, in the case of many adult calliphorids, begin a period of feeding.[6] This is in sharp contrast to diapause, which is initiated in response to environmental tokens prior to the advent of unfavorable seasonal weather and is a set genetic program that must run its course until termination is triggered by appropriate stimuli. A detailed discussion of factors that regulate the onset, maintenance, and termination of diapause in necrophagous insects is beyond the scope of this book. In fact, an understanding of the diapause program has only been worked out for a few necrophagous fly species (Denlinger, 2002; Saunders, 2002).

A discussion of strategies for combating cold temperatures goes hand in hand with diapause (Figure 9.8). Diapause is a dynamic physiological state of dormancy characterized by a reduced yet active metabolism, with the primary function of protecting the individual from predictable and unfavorable environmental conditions, most often winter. A primary feature of winter is extended periods of low temperatures, often with several days at or below 0 °C. To survive such conditions, the diapause program of insects includes one or more of the low-temperature strategies discussed in Section 9.5.1. The physiological acclimatization may also include morphological adaptations such as enhanced cuticular hydrocarbon deposition in the integument or puparium, or specific behavioral patterns that may include seeking a protective refuge like burial in soil that accompanies larval or pupal diapause in many necrophagous fly species. Calliphorids that overwinter as adults in reproductive diapause often remain concealed to avoid suboptimal temperatures, but the details of cold hardiness during this stage are poorly understood.

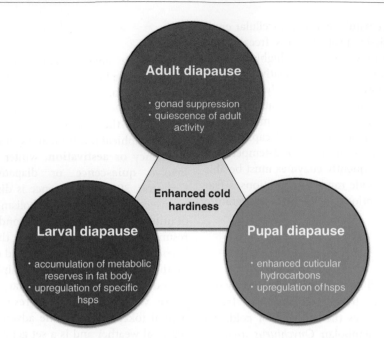

Figure 9.8   Relationship between diapause in various stages of development in necrophagous flies.

The winter survival plan also includes diapause-specific expression of Hsps. For example, during pupal diapause of *S. crassipalpis*, genes in the Hsp 60 (1) and 70 (2) families, four small Hsps, and one small Hsp pseudogene are upregulated while Hsp90 is downregulated and an Hsp70 cognate (Hsc70) is not influenced by induction of diapause (Hayward *et al.*, 2005; Rinehart *et al.*, 2007). Upregulation of these Hsps does not contribute to the induction of diapause but seems to be critical to cold hardening and thus survival during low-temperature exposure (Rinehart *et al.*, 2007). Interestingly, once the pattern of Hsp gene expression has been established, transcripts of all these Hsps do not change in response to temperature throughout the entire duration of pupal diapause (Hayward *et al.*, 2005). The expression of Hsps in *S. crassipalpis* differs from the typical upregulation of all Hsps during other forms of environmental stress such as heat shock, anoxia, and desiccation (Tammariello *et al.*, 1999; Hayward *et al.*, 2004), suggesting that the roles of Hsps during diapause are unique from a general stress response (Feder & Hoffman, 1999). In contrast to the pattern of gene expression in *S. crassipalpis*, Hsp23, Hsp70, and Hsp90 are not altered during larval diapause in *Lucilia sericata* (Tachibana *et al.*, 2005), which again represents a situation that differs from a typical stress response. The ubiquitous nature and highly conserved functions of Hsps in general would argue against

the differences in expression between the two flies as being phylogenetically related. Much more investigation is needed to clarify the role of stress proteins in conferring cold hardiness during diapause.

9.5.3   Aseasonal low-temperature adaptations

Sudden unexpected drops in temperature that are independent of seasonal change are referred to as aseasonal. The occurrence of such conditions is typically of short duration (minutes to hours) and the actual temperatures frequently remain above freezing. Nonetheless, in the absence of protection sudden drops in temperature can evoke direct **chilling injury** or cold shock; if the environmental insult lasts an extended period (days or weeks), indirect chilling injury may result. An examination of the types of low-temperature injury can be found in Section 9.6. Here, we explore the adaptive features of necrophagous flies that respond to aseasonal temperature change. Aseasonal adaptations differ from seasonal low-temperature strategies in at least two ways:

1. Responses to unexpected temperature declines must be much more rapid in mobilizing protection as there is no time for acclimatization in anticipation of unfavorable conditions as occurs with seasonality.

2. Protection is not long-lasting, corresponding to the expectation that aseasonal temperature drops are of short duration.

Necrophagous flies utilize at least four mechanisms to combat sudden unpredicted low temperatures. The mechanisms range from a rapid cold hardening (RCH) response that depends on synthesis of cryoprotective compounds to heterothermy in larval feeding aggregations. The latter example arguably provides the longest length of protection (days) yet it is the only one that is not activated in response to environmental cues. Rather, low-temperature protection is merely a byproduct of internal heat production that has a primary role in larval growth, development, and food assimilation (Chapter 8 provides an in-depth examination).

9.5.3.1 Rapid cold hardening

RCH is very similar to thermotolerance in that brief exposure to a non-damaging low-temperature acclimates the insect when subsequently exposed to extreme cold that ordinarily induces injury or death (Denlinger & Lee, 2010). This adaptive response occurs in insects from at least eight orders represented by 26 families, and in every developmental stage of insects. Perhaps the best-studied examples of RCH are with two species of flesh flies, *S. crassipalpis* and *S. bullata* (Chen *et al.*, 1987; Lee *et al.*, 1987), in which the response can be induced by exposure to temperatures above freezing for as short as 1 hour. Pretreatment stimulates synthesis of glycerol and sorbitol, as well as production of the amino acids alanine and glutamine (Yoder *et al.*, 2006). Each of these compounds can function as colligative cryoprotectants, and thus protection is gained through the same mechanisms associated with freeze-avoidance and freeze-tolerant strategies. Hsps may have a functional role in RCH as well but none appear to be upregulated in response to RCH, although Hsp70 is believed to be associated with the recovery phase (Denlinger *et al.*, 2001).

9.5.3.2 Thermal hysteresis

As discussed during seasonal strategies, antifreeze proteins can be synthesized in response to cold to prevent ice crystals from growing. This response is also critical to insects that experience cold shock where temperatures drop below freezing. Protection is realized through protein binding to small ice crystals that have formed in body fluids, which in turn lowers the temperature that allow ice crystals to grow. This lower temperature is referred to as the **hysteretic freezing point,** and the difference between this condition and the melting point is termed thermal hysteresis. Thus, thermal hysteresis does not prevent spontaneous ice nucleation in body fluids but it does inhibit growth of existing ice crystals. In necrophagous insects not acclimated for low temperatures and ice formation (see discussion of freeze-tolerant strategies, Section 9.5.1.2), this mechanism of protection minimizes the potentially damaging effects of ice crystallization and phase transitions during cold shock.

9.5.3.3 Heat-shock response

Unlike diapause, which displays unique patterns of Hsp synthesis and which is unmodified by temperature change, aseasonal low temperatures evoke a typical stress response in that Hsps from multiple families of stress proteins are upregulated (Feder & Hoffman, 1999). Cold shock elicits a similar Hsp response as high-temperature stress in flesh flies in terms of expression patterns and cellular protection during and following stress. The differences between high and low temperatures responses appear to lie in which specific Hsp genes are immediately activated. For example, cold-shock treatment of pharate adults of *S. crassipalpis* stimulates synthesis of Hsps from the Hsp 90, Hsp 70 and small Hsp families, which is similar to the heat-shock response. However, heat stress evokes differential expression of Hsps based on the developmental stage encountering thermal stress (Joplin & Denlinger, 1990), whereas low-temperature Hsp synthesis does not show such variation (Joplin *et al.*, 1990; Yocum *et al.*, 1998; Rinehart *et al.*, 2000).

9.5.3.4 Heterothermy

Heat production in maggot masses may be a source of low-temperature protection when fly larvae are exposed to non-freezing temperatures. This speculation is derived from observations of blow fly larvae on carrion either during winter or when ambient temperatures were near or below the developmental threshold or base temperature (Deonier, 1940; Cragg, 1956; Huntington *et al.*, 2007). The basic argument for protection is that if maggot masses are already established prior to the arrival of low temperatures, then the internal heat of the assemblage will sustain conditions in the microhabitat to prevent freezing of body fluids and to sustain metabolic activity (Campobasso *et al.*, 2001; Marchenko, 2001). This would account for how larval aggregations on a corpse placed in a morgue can generate temperature elevations well above the standard refrigeration conditions of −15.5 to −12 °C (4–10 °F) (Huntington *et al.*, 2007; Higley & Haskell, 2010) and explain continued larval development of *Calliphora vicina* at temperatures between 1 and 6 °C (Ames & Turner, 2003). Maggot masses developing in concealed locations on animal remains are presumably buffered from fluctuations in temperature and wind during low-temperature episodes, and thus should be capable of generating even higher internal temperatures that can be sustained for longer periods. However, heat production in larval aggregations would not be expected to protect larvae from chilling or freezing injury if the masses were not large enough or of the appropriate species composition to generate sufficient heat to counter extreme temperature declines and/or long exposures to low temperature, nor would maggot heterothermy be a suitable method to cope with the harsh conditions of winter (Rivers *et al.*, 2011).

9.6   Deleterious effects of low-temperature exposure

Temperatures on either side of the freezing point of water are especially critical to insects (Sinclair *et al.*, 2003). When temperatures reach the critical thermal minimum for a particular species, the insect enters **chill-coma,** a state characterized by inhibition of neuromusculature and leading to a total halt in movement and other events dependent on muscle activity (MacMillan & Sinclair, 2011). The condition is reversible but those species not prepared for the arrival of low temperatures and/or extended exposure to cold, whether associated with seasonality or aseasonality, may be harmed or killed. Thus, as with high-temperature stress, the effects of cold may be grouped as either sublethal or lethal, with mortality further divided into immediate versus delayed following exposure. More commonly, the detrimental consequences of suboptimal temperatures are classified by the absolute temperatures experienced (subzero or non-freezing) and the

duration of exposure (very brief, minutes to hours; prolonged, days to weeks). The resulting scheme arranges low-temperature effects into chilling injury, which is the damage that occurs from low-temperature episodes near freezing, or **freezing injury,** which is the affliction that occurs as a result of subzero exposure.

9.6.1   Chilling injury

Necrophagous insects that are not supercooled or acclimated for cold will be injured if challenged with low temperatures, even if the environment does not drop below freezing. Such damage can result when the ambient conditions decline to a low temperature so rapidly that the insect has no time to initiate any type of cold defense. The situation described is termed cold shock and the detrimental effects on cells and tissues are classified as direct chilling or cold-shock injury. It is the rapid rate of cooling that induces harm more so than the actual low temperature experienced, particularly when body fluids remain above 0 °C. Cold shock elicits a broad range of damage, from impairment of the nervous system to reduction of fecundity in both sexes and inducement of deformities in adult structures, which likely originate as a consequence of cell membrane damage and subsequent osmotic and ionic imbalances (Lee, 2010). Neuromuscular functions associated with proboscis reflex extension and **ptilinum** expansion in *S. crassipalpis* are disrupted following cold shock during puparial stages (Yocum *et al.*, 1994; Kelty *et al.*, 1996). Such injury may be attributed to an inability to reestablish electro-chemical gradients of sodium and potassium ions after low-temperature exposure (Kostál *et al.*, 2007). Wing deformation, delayed ovarian development, damage to testes, and egg mortality all occur with cold shock and the severity increases with the length of exposure to cold-shock conditions (Lee, 2010).

Indirect chilling injury results from prolonged exposure to low temperatures above freezing, as opposed to direct chilling injury where temperatures may or may not drop below freezing. The damaging effects of indirect chilling injury also appear to be associated with distortions of plasma membranes, loss of membrane integrity, and chaos in terms of cellular homeostasis, as flux of solutes can no longer be regulated. Similar impairment of nervous system function as described for direct chilling injury will occur, including the inability to coordinate neuromusculature or generate membrane action potentials (Denlinger & Lee, 1998). Depending on the severity of impairment, the insect will likely die but the time before death may be extended if the ambient temperatures remain low.

9.6.2 Freezing injury

Subzero temperatures are more invasive than non-freezing because the damage inflicted is due to ice formation. The consequences of ice are not just from mechanical damage to membranes and organs, which will obviously occur during phase transitions (which includes both formation and ice floats during melting). Freezing injury also leads to several physiological disruptions that will minimally harm cells, but conceivably create injurious conditions that are not reversible. For example, as body fluids freeze, the amount of water available as a solvent decreases, thereby concentrating any solutes in the solutions that are freezing. Excessively high concentrations of particular constituents may induce membrane damage or inhibit organelle functions. Similarly, loss of cell volume beyond a critical threshold as extracellular or intracellular fluids freeze may be irreversible or evoke damage to cellular membranes that lead to leakage or a total loss of integrity. As detailed in Section 9.6.1, such membrane change will lead to a myriad of events that injure the cell and/or lead to cell death.

Mechanical injury is very common when ice forms in body tissues. The keys to mechanical injury are which compartments in the body form ice (intracellular compartments are generally intolerant of any ice), the extent of ice crystallization, and how rapidly the ice expands. If ice nucleation occurs rather slowly, damage may be minimal if crystals are limited to extracellular compartments and antifreeze proteins can be synthesized from fat body to generate protection in the form of thermal hysteresis. However, rapid crystallization will outpace protein synthesis, leading to membrane stretching or tearing. Continued growth of ice lattices can lead to shredding of tissues, displacement of organs, and potentially lesions in the exoskeleton as ice expands (Lee, 2010). Severe mechanical injury will be irreversible and the insect will die.

Chapter review

Necrophagous insects face seasonal, aseasonal, and self-induced (heterothermy) temperature extremes

- All insects are poikilothermic ectotherms. As ectothermic animals, insects do not regulate internal body temperature through production of internal heat, and as poikilotherms (synonymous with eurytherm), the internal environment varies with changing ambient temperatures. The net effect is that living in climates prone to dramatic shifts in daily or evenly hourly conditions, or residing in regions that undergo seasonal change, insects must possess the ability to maintain metabolic activity over a wide range of temperatures.
- The range of temperatures over which insects can maintain metabolic processes or survive indefinitely is referred to as the zone of tolerance or thermal tolerance range. Outside the temperature zone, namely at temperatures above the upper thermal limit (critical thermal maximum) or below the lower thermal limit (critical thermal minimum), are conditions that will initially evoke inhibition of cellular reactions, thereby retarding most aspects of growth and development.
- In the context of a necrophagous lifestyle, life is good provided temperatures remain within the zone of tolerance. Once ambient temperatures lie outside the thermal range for a given species, growth and development are compromised. For necrophagous fly larvae, the temperature threshold below which development does not occur is referred to as the base temperature or developmental limit. The base temperature is generally not considered the same as the lower lethal limit. However, if exposure to these unfavorable conditions lasts for a long enough period of time, irreversible injury leading to death may result.
- Conditions that foster temperature elevations or depressions beyond thermal limits include seasonal changes, unpredictable aseasonal fluctuations in environmental conditions, and, in some instances, self-induced heating.

Temperature challenges do not mean death: necrophagous insects are equipped with adaptations to survive a changing environment

- Changing environmental conditions, particularly temperature, do not necessarily lead to the demise of an insect. Certainly, temperature swings, gradual or sudden, can elicit stress, injury or even death. But in many instances, if not most, insects possess an array of adaptations to overcome or avoid extreme heat or cold.
- The adaptations may be as subtle as behavioral modifications in which gravid calliphorid females oviposit in sheltered locations such as body cavities and openings or under clothing on a human corpse to buffer against rapid changes in temperature, or young fly larvae migrate to concealed environments on a corpse or form maggot masses within a body cavity so that the internal heat of the aggregation affords protection from low temperatures or the effects of the wind.
- The adaptations are phenotypic expressions of the genetic program(s) that allows insects in temperate, arctic, or near-arctic regions to avoid or survive unfavorable conditions. A highly efficient stress response system that often initiates synthesis of Hsps in response to an array of environmental insults is typical of all insects and appears to be especially critical to insects with specific stages of development that are exposed to high temperatures for an extended period of time.

Life-history features that promote survival during proteotaxic stress

- Proteotaxic or high-temperature stress is far more common to necrophagous insects than low-temperature exposure. Why? The corpse becomes a more favorable habitat with increases in temperature. By extension, this obviously means that carrion-inhabiting insects peak in abundance during warm summer months, when the temperature of a corpse can reach a zenith due to a combination of factors: solar radiation, tissue decomposition, microbial processes, and insect activity.
- Heat stress is a serious threat to necrophagous insects, particularly flies developing in feeding aggregations. Tens of thousands of individual insects may feed on a single large mammal following death, all potentially dealt the challenge of feeding under proteotaxic conditions, yet the vast majority successfully complete development, propagate, and contribute to continuation of the species. For these insects, survival is dependent on the ability to overcome or avoid thermal stress.
- Insects possess a highly efficient general stress response which often functions to produce a series of stress proteins that are believed to confer protection and repair under a wide range of environmental and artificial insults. The proteins are typically referred to as Hsps.
- The expression of Hsps following a mild temperature insult confers thermotolerance to necrophagous flies. The thermal tolerance is short-lived and cannot be lengthened by increasing the duration of preconditioning.
- Exposure to high temperatures causes some insects to engage in behaviors that promote favorable temperatures by returning to conditions lying within the zone of tolerance or by moving away from threatening temperatures. Necrophagous insects utilize behavioral responses in the form of locomotion to avoid or move away from high temperatures. Spatial partitioning can be considered a behavioral mechanism for heat avoidance. The mechanism appears to be utilized by fly maggots with lower upper thermal limits in comparison with competing species that commonly coexist in maggot masses.
- The capacity to respond to high (or low) temperature insults seemingly requires the ability to detect absolute temperatures or temperature changes in the environment or local habitats, as on carrion. Thermal- and humidity-sensitive neurons have been identified on the antennae of an array of insects representing several different insect orders, and thermoreception is also associated with olfactory receptors of some species. The receptors are typically arranged within a single sensillum that is peg-shape in morphology, lacking pores and possessing multiple (usually three) sensory neurons.
- Necrophagous insects forced to contend with high-temperature stress must rely on physiological mechanisms to dissipate excess heat or face injury and death. The mechanisms used include evaporative cooling, convection, and conduction.

Deleterious effects of high temperatures on necrophagous flies

- Insects that lack the appropriate adaptations to deal with high temperatures, regardless of the source of heat, will suffer some type of heat injury. If exposure is for a long enough period or the temperature sufficiently exceeds the critical thermal maximum for a given species, the conditions will likely be lethal. The temperature elevations associated with large maggot masses on a corpse achieve both conditions.

- The consequences of high-temperature exposure are generally classified into two broad categories: lethal and sublethal effects. Lethal effects vary in terms of onset in different developmental stages following heat stress. Sublethal effects of high temperature are also dependent on the stage of development at the time of exposure as well as the absolute temperature encountered and the rate at which temperatures elevated.

- Insects that experience temperatures above the critical thermal maximum will eventually stop moving, a condition called heat stupor, and if not removed from the area or temperatures do not decline, death will result. The lethality of high temperatures is dependent on (i) species, indicative of a phylogenetic component to thermal tolerance, (ii) age of the insect, (iii) absolute temperature above the upper thermal limit, (iv) length of exposure, and (v) rate of warming, i.e., how rapidly ambient temperatures elevate in relation to body temperature.

- How flies die following high-temperature stress is not clear. As temperatures approach upper thermal limits for a given species, proteins will begin to denature provided protection in the form of Hsps or other factors are absent. If temperatures continue to climb, respiratory metabolism is totally inhibited. Stress protein synthesis is oxygen-dependent, and thus Hsp production will cease.

- Exposure to high temperatures above the upper thermal limit is not always lethal. If the proteotaxic stressor is of short duration or below the critical thermal maximum, the insect may recover yet suffer some form of heat injury. The effects of sublethal exposure to heat can be far-reaching, ranging from disruption or shifts in timing of key developmental events, malformations in later stages, to depressions in longevity or reproductive output.

Life-history strategies and adaptations that promote survival at low temperatures

- During the peak of carrion insect activity, the threat of low temperatures is not common or expected. Warmer climatic conditions favor carrion insect activity and tissue decomposition,

and hence emission of chemical signals that activate searching behavior in necrophiles and opportunists (predators and parasites). Low temperatures are generally associated with seasonality, and most insects residing in regions in which seasonal change is characterized by cold possess the ability to respond to environmental tokens that forecast climatic shifts. The significance of this foreshadowing of events is that the insects can anticipate the arrival of unfavorable environmental conditions and then prepare by migrating to a more suitable region, initiating a physiological state of hibernation or, in the case of temperature, gradually acclimating to a changing environment so that by the time freezing temperatures are experienced, the insect is protected from injury.

- For carrion insects to survive the harsh conditions of winter or aseasonal drops in temperatures, survival depends on the ability to overcome or avoid cold stress. Avoidance by migration is not really possible for necrophagous insects and thus survival depends mainly on preparatory physiological and biochemical programs that are implemented prior to the arrival of low temperatures.

- The cold strategies use an array of highly evolved adaptations that either promote the acquisition of cold hardiness or which induce freezing of extracellular body fluids. Necrophagous insects commonly rely on freeze avoidance to allow body fluids to supercool without forming ice. Less frequent are species that are freeze-tolerant and actually use adaptations to promote ice nucleation. A third strategy of cryoprotective dehydration is rare, having only been characterized in polar and alpine soil-inhabiting invertebrates.

- Unfavorable environmental conditions, from extreme high or low temperatures or during drought or rainy seasons, can prompt a period of dormancy. Dormant periods can occur at almost any time of year depending on the insect and biogeographical location, and can include summer dormancy or aestivation, winter hibernation, or involve quiescence or diapause. Diapause is a dynamic physiological state of dormancy characterized by a reduced yet active metabolism, with the primary function of protecting the individual from predictable and unfavorable environmental conditions, most often winter. To survive such conditions, the diapause program of insects includes a low-temperature strategy such as freeze avoidance or freeze tolerance.

- Sudden unexpected drops in temperature that are independent of seasonal change are referred to as aseasonal. Aseasonal adaptations differ from seasonal low-temperature strategies in at least two ways: (i) responses to unexpected temperature declines must be much more rapid in mobilizing protection as there is no time for acclimatization in anticipation of unfavorable conditions as occurs with seasonality; and (ii) protection is not long-lasting, corresponding to the expectation that aseasonal temperature drops are of short duration.

- Necrophagous flies utilize at least four mechanisms to combat sudden unpredicted low temperatures. The mechanisms include RCH, the heat-shock response, thermal hysteresis, and heterothermy in larval feeding aggregations.

Deleterious effects of low-temperature exposure

- Temperatures on either side of the freezing point of water are especially critical to insects. When temperatures reach the critical thermal minimum for a particular species, the insect enters chill-coma. The condition is reversible, but those species not prepared for the arrival of low temperatures and/ or extended exposure to cold, whether associated with seasonality or aseasonality, may be harmed or killed.
- The detrimental effects of suboptimal temperatures are classified by the absolute temperatures experienced and the duration of exposure. The resulting scheme arranges low-temperature effects into chilling injury or freezing injury.
- Necrophagous insects that are not supercooled or acclimated for cold will be injured if challenged with low temperatures, even if the environment does not drop below freezing. Such damage can result when the ambient conditions decline to a low temperature so rapidly that the insect has no time to initiate any type of cold defense. This is termed cold shock and the detrimental effects on cells and tissues are classified as direct chilling or cold-shock injury.
- Indirect chilling injury results from prolonged exposure to low temperatures above freezing, as opposed to direct chilling injury where temperatures may or may not drop below freezing. The damaging effects of indirect chilling injury also appear to be associated with distortions of plasma membranes, loss of membrane integrity, and chaos in terms of cellular homeostasis, as flux of solutes can no longer be regulated.
- Subzero temperatures can be more invasive than non-freezing temperatures because the damage inflicted is due to ice formation. The consequences of ice are not just from mechanical damage to membranes and organs, which will obviously occur during phase transitions. Freezing injury also leads to several physiological disruptions that will minimally harm cells, but conceivably create injurious conditions that are not reversible.

Test your understanding

Level 1: knowledge/comprehension

1. Define the following terms:

   (a) chilling injury
   (b) critical thermal minima
   (c) diapause
   (d) heat-shock response
   (e) cryoprotectant
   (f) thermal hysteresis
   (g) evaporative cooling
   (h) antifreeze protein.

2. Match the terms (i–vi) with the descriptions (a–f).

(a) Cessation of movement resulting from high temperatures	(i) Poikilotherm
(b) Physiological response to factors forecasting seasonal change	(ii) Chill-coma
(c) Lowering of body fluids to well below freezing with no ice formation	(iii) Homeothermy
(d) Maintenance of an internal body temperature	(iv) Ice nucleating agent
(e) Cessation of movement resulting from low temperatures	(v) Heat stupor
(f) Serves as physical site for water condensation	(vi) Supercooling
(g) Internal environment varies with ambient conditions	(vii) Acclimatization

3. Why are environmental tokens not used to anticipate conditions that lead to heat or cold shock?

4. Provide examples of when aseasonal temperature changes occur with insects on carrion.

5. How do antifreeze proteins protect insects from freezing injury?

Level 2: application/analysis

1. Explain how colligative cryoprotectants provide low-temperature protection via osmotic dehydration.

2. High temperatures in larval aggregations offer the physiological conundrum of being a thermal stressor and an aseasonal adaptation. Describe the scenarios in which each of these conditions occurs.

3. Describe how regurgitation by an adult fly can be used as a mechanism for cooling body temperature.

4. Explain why a postmortem interval using insect development can be far more challenging to do during winter months than during the summer.

Level 3: synthesis/evaluation

1. Cryoprotective dehydration is an apparently rare strategy used by certain polar and alpine-inhabiting invertebrates, including a few species of insects. Provide a possible explanation for why this mechanism is used over a classic freeze avoidance strategy that can evoke a similar yet distinct osmotic dehydration.

2. Speculate on how the thermal history of fly maggots can influence their utility in criminal investigations, particularly in terms of estimating a postmortem interval.

3. Explain why larval mass heterothermy is not an effective means for coping with seasonal temperature changes.

Notes

1. Families of Hsps are traditionally referred to as "Hsp" but more recently have been identified by "sp" in reference to stress proteins to reflect the broader stress response roles of these proteins.

2. kDa refers to kilodaltons and is a standard measure of protein molecular mass based on amino acid composition.

3. Sensilla are sensory structures in insects used for perceiving changes in the environment. A discussion of sensilla was first presented in Chapter 7.

4. Various forms of cell death have been reviewed by Jaeschke and Lemasters (2003), Manjo and Joris (1995), and Green (2018).

5. There is some disagreement about whether solute adjustments do indeed change the SCPs (Zachariassen & Kristiansen, 2000; Wilson *et al.*, 2003).

6. Several species of adult calliphorids overwinter in reproductive diapause in which the gonads are dormant but the individuals are quiescent.

References cited

Altner, H. & Loftus, R. (1985) Ultrastructure and function of insect thermo- and hygroreceptos. *Annual Review of Entomology* 30: 273–295.

Ames, C. & Turner, B. (2003) Low temperature episodes in development of blowflies: implications for postmortem interval estimation. *Medical and Veterinary Entomology* 17: 178–186.

Anderson, G.S. & VanLaerhoven, S.L. (1996) Initial studies on insect succession on carrion in southwestern British Columbia. *Journal of Forensic Science* 41: 617–625.

Arrigo, A.-P. & Landry, J. (1994) Expression and function of the low-molecular weight heat shock proteins. In: R.I. Morimoto, A. Tissieres & C. Georgopolous (eds) *The Biology of Heat Shock Proteins and Molecular Chaperones*, pp. 335–373. Cold Spring Harbor Laboratory Press, Cold Spring Harbor, NY.

Ashworth, J.R. & Wall, R. (1994) Responses of the sheep blowflies *Lucilia sericata* and *L. cuprina* to odour and the development of semiochemical baits. *Medical and Veterinary Entomology* 8: 303–309.

Byrd, J.H. & Butler, J.F. (1998) Effects of temperature on *Sarcophaga haemorrhoidalis* (Diptera: Sarcophagidae) development. *Journal of Medical Entomology* 35: 694–698.

Campobasso, C.P., Di Vella, G. & Introna, F. (2001) Factors affecting decomposition and Diptera colonization. *Forensic Science International* 120: 18–27.

Caneparo, M.F.C., Fischer, M.L. & Almeida, L.M. (2017) Effect of temperature on the life cycle of *Euspilotus azureus* (Coleoptera: Histeridae), a predator of forensic importance. *Florida Entomologist* 100: 795–801.

Chapman, R.F. (1998) *The Insects: Structure and Function*, 4th edn. Cambridge University Press, Cambridge, UK.

Charabidze, D., Bourel, B. & Gosset, D. (2011) Larval-mass effect: characterisation of heat emission by necrophagous blowflies (Diptera: Calliphoridae) larval aggregates. *Forensic Science International* 211: 61–66.

Chen, C.-P., Lee, R.E. & Denlinger, D.L. (1987) Cold-shock injury and rapid cold hardening in the flesh fly, *Sarcophaga crassipalpis*. *Physiological Zoology* 60: 297–304.

Chen, C.-P., Lee, R.E. Jr & Denlinger, D.L. (1990) A comparison of the responses of tropical and temperate flies (Diptera: Sarcophagidae) to cold and heat stress. *Journal of Comparative Physiology B* 160: 543–547.

Cragg, J.B. (1956) The olfactory behaviour of *Lucilia* species (Diptera) under natural conditions. *Annals of Applied Biology* 44: 467–471.

Danks, H.V. (2000) Insect cold hardiness: a Canadian perspective. *Cryobiology Letters* 21: 297–308.

Denlinger, D.L. (2002) Regulation of diapause. *Annual Review of Entomology* 47: 93–122.

Denlinger, D.L. & Lee, R.E. Jr (1998) Physiology of cold sensitivity. In: G.J. Hallman & D.L. Denlinger (eds) *Temperature Sensitivity in Insects and Application in Integrated Pest Management*, pp. 55–97. Westview Press, Boulder, CO.

Denlinger, D.L. & Lee, R.E. Jr (2010) *Low Temperature Biology of Insects*. Cambridge University Press, Cambridge, UK.

Denlinger, D.L. & Yocum, G.D. (1998) Physiology of heat sensitivity. In: G.J. Hallman & D.L. Denlinger (eds) *Temperature Sensitivity in Insects and Application in Integrated Pest Management*, pp. 7–54. Westview Press, Boulder, CO.

Denlinger, D.L., Rinehart, J.P. & Yocum, G.D. (2001) Stress proteins: a role in insect diapause?. In: D.L. Denlinger, J. Giebultowicz & D.S. Saunders (eds) *Insect Timing: Circadian Rhythmicity to Seasonality*, pp. 155–171. Elsevier Science, Amesterdam.

Deonier, C.C. (1940) Carcass temperatures and their relation to winter blowfly populations and activity in the southwest. *Journal of Economic Entomology* 33: 166–170.

Duman, J.G. (2001) Antifreeze and ice nucleator proteins in terrestrial arthropods. *Annual Review of Physiology* 63: 327–357.

El-Wadawi, R. & Bowler, K. (1995) The development of thermotolerance protects blowfly flight muscle mitochondrial function from heat damage. *Journal of Experimental Biology* 198: 2413–2421.

Feder, M.E. & Hoffman, G.E. (1999) Heat-shock proteins, molecular chaperones, and stress response: evolutionary and ecological physiology. *Annual Review of Physiology* 61: 243–282.

Gomes, G., Köberle, R., Von Zuben, C.J. & Andrade, D.V. (2018) Droplet bubbling evaporatively cools a blowfly. *Scientific Reports* 8: 5464. https://doi.org/10.1038/s41598-018-23670-2.

Gomperts, B.D., Kramer, I.M. & Tatham, P.E.R. (2002) *Signal Transduction*. Academic Press, San Diego, CA.

Green, D.R. (2018) *Cell Death: Apoptosis and Other Means to an End*, 2nd edn. Cold Spring Harbor Press, New York.

Greenberg, B. (1991) Flies as forensic indicators. *Journal of Medical Entomology* 28: 565–577.

Gullan, P.J. & Cranston, P.S. (2020) *The Insects: An Outline of Entomology*, 5th edn. Wiley Blackwell, Chichester, UK.

Gunn, A. (2018) The exploitation of fresh remains by *Dermestes maculatus* De Geer (Coleoptera, Dermestidae) and their ability to cause a localised and prolonged increase in temperature above ambient. *Journal of Forensic and Legal Medicine* 59: 20–29.

Hayward, S.A.L., Rinehart, J.P. & Denlinger, D.L. (2004) Desiccation and rehydration elicit distinct heat shock protein transcript responses in flesh fly pupae. *Journal of Experimental Biology* 207: 963–971.

Hayward, S.A.L., Pavlides, S.C., Tammariello, S.P., Rinehart, J.P. & Denlinger, D.L. (2005) Temporal expression patterns of diapause-associated genes in flesh fly pupae from the onset of diapause through post-diapause quiescence. *Journal of Insect Physiology* 51: 631–640.

Heaton, V., Moffatt, C. & Simmons, T. (2014) Quantifying the temperature of maggot masses and its relationship to decomposition. *Journal of Forensic Sciences* 59: 676–682.

Heinrich, B. (1993) *The Hot-blooded Insects: Strategies and Mechanisms for Thermoregulation*. Harvard University Press, Cambridge, MA.

Hendrichs, J., Cooley, S.S. & Prokopy, R.J. (1992) Post-feeding bubbling behavior in fluid-feeding Diptera: concentration of crop contents by oral evaporation of excess water. *Physiological Entomology* 17: 153–161.

Higley, L.G. & Haskell, N.H. (2010) Insect development and forensic entomology. In: J.H. Byrd & J.L. Castner (eds) *Forensic Entomology: The Utility of Using Arthropods in Legal Investigations*, 2nd edn, pp. 389–406. CRC Press, Boca Raton, FL.

Huntington, T.E., Higley, L.G. & Baxendale, F.P. (2007) Maggot development during morgue storage and its effect on estimating the post-mortem interval. *Journal of Forensic Science* 52: 453–458.

Jaeschke, H. & Lemasters, J.J. (2003) Apoptosis versus oncotic necrosis in hepatic ischemia/reperfusion injury. *Gastroenterology* 125: 1246–1257.

Jakubec, P. (2016) Thermal summation model and instar determination of all developmental stages of necrophagous beetle, *Sciodrepoides watsoni* (Spence) (Coleoptera: Leiodidae: Cholevinae). *PeerJ* 4: e1944. https://doi.org/10.7717/peerj.1944.

Joplin, K.H. & Denlinger, D.L. (1990) Developmental and tissue specific control of the induced 70 kDa related proteins in the flesh fly, *Sarcophaga crassipalpis*. *Journal of Insect Physiology* 36: 239–245.

Joplin, K.H., Yocum, G.D. & Denlinger, D.L. (1990) Cold shock elicits expression of heat shock proteins in the flesh fly *Sarcophaga crassipalpis*. *Journal of Insect Physiology* 36: 825–834.

Kelty, J.D., Killian, K.A. & Lee, R.E. (1996) Cold shock and rapid cold-hardening of pharate adult flesh flies (*Sarcophaga crassipalpis*): effects on behaviour and neuromuscular function following eclosion. *Physiological Entomology* 21: 283–288.

Korsloot, A., van Gestel, C.A.M. & van Straalen, N.M. (2004) *Environmental Stress and Cellular Response in Arthropods*. CRC Press, Boca Raton, FL.

Kostál, V. (2010) Cell structure modifications in insects at low temperatures. In: D.L. Denlinger & R.E. Lee Jr (eds) *Low Temperature Biology of Insects*, pp. 116–140. Cambridge University Press, Cambridge, UK.

Kostál, V., Renault, D., Mehrabianova, A. & Bastl, J. (2007) Insect cold tolerance and repair of chill-injury at fluctuating thermal regimes: role of ion homeostasis. *Comparative Biochemistry and Physiology A* 147: 231–238.

Kukal, O., Denlinger, D.L. & Lee, R.E. (1991) Developmental and metabolic changes induced by anoxia in diapausing and non-diapausing flesh fly pupae. *Journal of Comparative Physiology B* 160: 683–689.

Lee, R.E. Jr (2010) A primer on insect cold-tolerance. In: D.L. Denlinger & R.E. Lee Jr (eds) *Low Temperature Biology of Insects*, pp. 3–34. Cambridge University Press, Cambridge, UK.

Lee, R.E., Chen, C.-P. & Denlinger, D.L. (1987) A rapid cold hardening process in insects. *Science* 238: 1415–1417.

Lindquist, S. & Craig, E.A. (1988) The heat-shock proteins. *Annual Review of Genetics* 22: 631–677.

MacMillan, H.A. & Sinclair, B.J. (2011) Mechanisms underlying insect chill-coma. *Journal of Insect Physiology* 57: 12–20.

Mahroof, R., Zhu, K.Y., & Subramanyam, B. (2005) Changes in expression of heat shock proteins in *Tribolium castaneum* (Coleoptera: Tenebrionidae) in relation to developmental stage, exposure time, and temperature. *Annals of the Entomological Society of America* 98: 100–107.

Manjo, G. & Joris, I. (1995) Apoptosis, oncosis, and necrosis: an overview of cell death. *American Journal of Pathology* 146: 3–15.

Mann, R.W., Bass, W.M. & Meadows, L. (1990) Time since death and decomposition of the human body: variables and observations in case and experimental field studies. *Journal of Forensic Science* 35: 103–111.

Marchenko, M.I. (2001) Medicolegal relevance of cadaver entomo-fauna for the determination of the time since death. *Forensic Science International* 120: 89–109.

Martín-Vega, D., Díaz-Aranda, L.M., Baz, A. & Cifrián, B. (2017) Effect of temperature on the survival and development of three forensically relevant *Dermestes* species (Coleoptera: Dermestidae). *Journal of Medical Entomology* 54: 1140–1150.

Meyer, S.G.E. (1978) Effects of heat, cold, anaerobiosis and inhibitors of metabolite concentrations in larvae of *Calliphora macellaria*. *Insect Biochemistry* 8: 471–477.

Midgley, J.M. & Villet, M.H. (2009) Development of *Thanatophilus micans* (Fabricius 1794) (Coleoptera: Silphidae) at constant temperatures. *International Journal of Legal Medicine* 123: 285–292.

Neven, L.G. (2000) Physiological responses of insects to heat. *Postharvest Biology and Technology* 21: 103–111.

Parsall, D.A. & Lindquist, S. (1994) Heat shock proteins and stress tolerance. In: R.I. Morimoto, A. Tissieres & C. Georgopoulos (eds) *The Biology of Heat Shock Proteins and Molecular Chaperones*, pp. 467–494. Cold Spring Harbor Laboratory Press, Cold Spring Harbor, NY.

Prange, H.D. (1996) Evaporative cooling in insects. *Journal of Insect Physiology* 42: 493–499.

Randall, D., Burggren, W. & French, K. (2002) *Animal Physiology: Mechanisms and Adaptations.* W.H. Freeman and Company, New York.

Richards, C.S., Price, B.W. & Villet, M.H. (2009) Thermal ecophysiology of seven carrion-feeding blowflies in Southern Africa. *Entomologia Experimentalis et Applicata* 131: 11–19.

Richards, E.N. & Goff, M.L. (1997) Arthropod succession on exposed carrion in three contrasting tropical habitats on Hawaii Island. *Journal of Medical Entomology* 34: 328–339.

Rinehart, J.P., Yocum, G.D. & Denlinger, D.L. (2000) Thermotolerance and rapid cold hardening ameliorate the negative effects of brief exposure to high or low temperatures on fecundity in the flesh fly, *Sarcophaga crassipalpis*. *Physiological Entomology* 25: 330–336.

Rinehart, J.P., Li, A., Yocum, G.D., Robich, R.M. & Hayward, S.A. (2007) Upregulation of heat shock proteins is essential for cold survival during insect diapause. *Proceedings of the National Academy of Sciences of the United States of America* 104: 11130–11137.

Rivers, D.B., Zdarek, J. & Denlinger, D.L. (2004) Disruption of pupariation and eclosion behavior in the flesh fly, *Sarcophaga bullata* Parker (Diptera: Sarcophagidae), by venom from the ectoparasitic wasp *Nasonia vitripennis* (Walker) (Hymenoptera: Pteromalidae). *Archives of Insect Biochemistry and Physiology* 57:78–91.

Rivers, D.B., Ciarlo, T., Spelman, M. & Brogan, R. (2010) Changes in development and heat shock response in two species of flies (*Sarcophaga bullata* [Diptera: Sarcophagidae] and *Protophormia terraenovae* [Diptera: Calliphoridae]) reared in different sized maggot masses. *Journal of Medical Entomology* 47: 677–689.

Rivers, D.B., Thompson, C. & Brogan, R. (2011) Physiological trade-offs of forming maggot masses by necrophagous flies on vertebrate carrion. *Bulletin of Entomological Research* 101: 599–611.

Saunders, D.S. (2002) *Insect Clocks*, 3rd edn. Elsevier Science, Amesterdam.

Scanvion, Q., Hedouin, V. & Charabidzé, D. (2018) Collective exodigestion favours blow fly colonization and development on fresh carcasses. *Animal Behaviour* 141: 221–232.

Sharma, S., Reddy, P.V.J., Rohilla, M.S. & Tiwari, P.K. (2006) Expression of HSP60 homologue in sheep blowfly Lucilia cuprina during development and heat stress. *Journal of Thermal Biology* 31: 546–555.

Sinclair, B.J., Vernon, P., Klok, C.J. & Chown, S.L. (2003) Insects at low temperatures: an ecological perspective. *Trends in Ecology and Evolution* 18: 257–262.

Sørensen, J.G. & Loeschcke, V. (2001) Larval crowding in *Drosophila melanogaster* induces Hsp70 expression, and leads to increased adult longevity and adult thermal stress resistance. *Journal of Insect Physiology* 47: 1301–1307.

Storey, K.B. (2004) Biochemical adaptation. In: K.B. Storey (ed) *Functional Metabolism: Regulation and Adaptation*, pp. 383–414. John Wiley & Sons, Inc., Hoboken, NJ.

Tachibana, S.-I., Numata, H. & Goto, S.G. (2005) Gene expression of heat-shock proteins (Hsp23, Hsp70, and Hsp90) during and after larval diapause in the blow fly *Lucilia sericata*. *Journal of Insect Physiology* 51: 641–647.

Tammariello, S.P., Rinehart, J.P. & Denlinger, D.L. (1999) Desiccation elicits heat shock protein transcription in the flesh fly, *Sarcophaga crassipalpis*, but does not enhance tolerance to high or low temperatures. *Journal of Insect Physiology* 45: 933–938.

Tominaga, Y. & Yokohari, F. (1982) External structure of the sensillum capitulum, a hygro- and thermoreceptive sensillum of the cockroach, *Periplaneta americana*. *Cell and Tissue Research* 226: 309–318.

Vayssier, M. & Polla, B.S. (1998) Heat shock proteins chaperoning life and death. *Cell Stress Chaperones* 3: 221–227.

Velásquez, Y. & Viloria, A.L. (2009) Effects of temperature on the development of the Neotropical carrion beetle *Oxelytrum discicolle* (Brullé, 1840) (Coleoptera: Silphidae). *Forensic Science International* 185: 107–109.

Verdú, J.R., Arellano, L., & Numa, C. (2006) Thermoregulation in endothermic dung beetles (Coleoptera: Scarabaeidae): Effect of body size and ecophysiological constraints in flight. *Journal of Insect Physiology* 52: 854–860.

Villet, M.H., Richards, C.S. & Midgley, J.M. (2010) Contemporary precision, bias, and accuracy of minimum post-mortem intervals estimated using development of carrion-feeding insects. In: J. Amendt, C.P. Campobasso, M.L. Goff & M. Grassberger (eds) *Current Concepts in Forensic Entomology*, pp. 109–137. Springer, London.

Wang, Y., Yang, J.B., Wang, J.F., Li, L.L., Wang, M., Yang, L.J., Tao, L.Y., Chu, J. & Hou, Y.D. (2017) Development of the forensically important beetle *Creophilus maxillosus* (Coleoptera: Staphylinidae) at constant temperatures. *Journal of Medical Entomology* 54: 281–289.

Waterhouse, D.F. (1947) The relative importance of live sheep and of carrion as breeding grounds for the Australian sheep blowfly *Lucilia cuprina*. CSIRO Bulletin 217. CSIRO, Australia.

Williams, H. & Richardson, A.M.M. (1984) Growth energetics in relation to temperature for larvae of four species of necrophagous flies (Diptera: Calliphoridae). *Australian Journal of Ecology* 9: 141–152.

Willmer, P., Stone, G. & Johnston, I. (2000) *Environmental Physiology of Animals*. Blackwell Publishing Ltd., Oxford.

Wilson, P.W., Heneghan, A.F. & Haymet, A.D.J. (2003) Ice nucleation in nature: supercooling point (SCP) measurements and the role of heterogeneous nucleation. *Cryobiology* 46: 88–98.

Withers, P.C. (1992) *Comparative Animal Physiology*. Saunders College Publishing, New York.

Yocum, G.D. & Denlinger, D.L. (1992) Prolonged thermotolerance in the flesh fly, *Sarcophaga crassipalpis*, does not require continuous expression or persistence of the 72 kDa heat-shock protein. *Journal of Insect Physiology* 38: 603–609.

Yocum, G.D., Zdarek, J., Joplin, K.H., Lee, R.E. Jr, Smith, D.C., Manter, K.D. & Denlinger, D.L. (1994) Alteration of the eclosion rhythm and eclosion behaviour in the flesh fly, *Sarcophaga crassipalpis*, by low and high temperature stress. *Journal of Insect Physiology* 40: 13–21.

Yocum, G.D., Joplin, K.H. & Denlinger, D.L. (1998) Upregulation of a 23 kDa small heat shock protein transcript during pupal diapause in the flesh fly, *Sarcophaga crassipalpis*. *Insect Biochemistry and Molecular Biology* 28: 677–682.

Yoder, J.A., Benoit, J.B., Denlinger, D.L. & Rivers, D.B. (2006) Stress-induced accumulation of glycerol in the flesh fly, *Sarcophaga bullata* Parker (Diptera: Sarcophagidae): evidence indicating anti-desiccant and cyroprotectant functions for this polyol during aseasonal stress. *Journal of Insect Physiology* 52: 202–214.

Zachariassen, K.E. & Kristiansen, E. (2000) Ice nucleation and antinucleation in nature. *Cryobiology* 41: 257–279.

Zdarek, J., Fraenkel, G. & Friedman, S. (1987) Pupariation in flies: a tool for monitoring effects of drugs, venoms, and other neurotoxic compounds. *Archives of Insect Biochemistry and Physiology* 4: 29–46.

Supplemental reading

Bowler, K. & Terblanche, J.S. (2008) Insect thermal tolerance: what is the role of ontogeny, ageing and senescence? *Biological Reviews* 83: 339–355.

Calderwood, S.K. (2010) *Cell Stress Proteins*. Springer, New York.

Danks, H.V. (2004) Seasonal adaptations of arctic insects. *Integrative and Comparative Biology* 44: 85–94.

Dillon, M.E., Wang, G., Garrity, P.A. & Huey, R.B. (2009) Thermal preference in *Drosophila*. *Journal of Thermal Biology* 34: 109–119.

Donovan, S.E., Hall, M.J.R., Turner, B.D. & Moncrieff, C.B. (2006) Larval growth rates of the blowfly, *Calliphora vicina*, over a range of temperatures. *Medical and Veterinary Entomology* 20: 106–114.

Kabakov, A.E. & Gabai, V.L. (1997) *Heat Shock Proteins and Cryoprotection: ATP-deprived Mammalian Cells*. Springer, New York.

Kashmeery, A.M.S. & Bowler, K. (1977) A study of the recovery from heat injury in the blowfly (*Calliphora erythrocephala*) using split-dose experiments. *Journal of Thermal Biology* 2: 183–184.

Kim, H.G., Margolies, D. & Park, Y. (2015) The roles of thermal transient receptor potential channels in thermotactic behavior and in thermal acclimation in the red flour beetle, *Tribolium castaneum. Journal of Insect Physiology* 76: 47–55.

Lee, R.E. Jr. & Denlinger, D.L. (1991) *Insects at Low Temperatures.* Springer, New York.

Neven, L.G. (2000) Physiological responses of insects to heat. *Postharvest Biology and Technology* 21: 103–111.

Rivers, D.B., Lee, R.E. Jr & Denlinger, D.L. (2000) Cold hardiness of the fly pupal parasitoid *Nasonia vitripennis* is enhanced by its host *Sarcophaga bullata. Journal of Insect Physiology* 46: 99–106.

Terblanche, J.S., Deere, J.A., Clusella-Trullas, S. & Chown, S.L. (2007) Critical thermal limits depend on methodological contexts. *Proceedings of the Royal Society of London Series B* 274: 2935–2043.

Waldbauer, G. (1998) *Insects Through the Seasons.* Harvard University Press, Cambridge, MA.

Additional resources

Insect Molecular Physiology Laboratory: http://oardc. osu.edu/denlingerlab/t01_pageview2/About_Us.html

Insect Stress Biology Laboratory: http://www.teetslab.com/

Journal of Insect Physiology: http://www.journals.elsevier. com/journal-of-insect-physiology/

Journal of Thermal Biology: http://www.journals.elsevier. com/journal-of-thermal-biology/

Laboratory of Insect Diapause and Environmental Physiology: http://www.entu.cas.cz/kostal/

Physiological Entomology: http://www.royensoc.co.uk/ publications/Physiological_Entomology.html

Society for Cryobiology: www.societyforcryobiology.org

Thermal Biology Group: http://www.sebiology.org/ animal/thermobiology.html

Chapter 10

Postmortem decomposition of human remains and vertebrate carrion

Overview

Immediately following death, a corpse begins the processes of decomposition that will eventually culminate in the complete disintegration of the body, essentially returning all nutrients to the environment. Between the stopping of the heart and formation of skeletal remains, a milieu of physio-chemical events facilitated by the deceased's own enzymes and those of an array of microorganisms promote catabolism of all soft tissues. Outwardly, the physical appearance of the body changes along a relatively predictable continuum, and which can be used to estimate stages of decay either of the whole or portions of the remains. All phases of decomposition are influenced by a multitude of factors, with temperature, moisture, microbes, and insect activity being among the principal influences on the processes of decay. The combined forces of natural decay influenced by human intervention and coupled with the characteristics that define the deceased result in decomposition events that are both unique to the individual yet comprised of a series of predictable processes. This chapter examines the postmortem changes that occur with animal remains decomposing in a terrestrial environment, with emphasis placed on physical alterations that aid in estimating the time since death.

The big picture

- Decomposition of human and other vertebrate remains is a complex process.
- Numerous factors affect the rate of body decomposition.
- When the heart stops internal changes occur almost immediately but are not outwardly detectable.
- Body decomposition is characterized by a series of recognizable changes in physical appearance.

10.1 Decomposition of human and other vertebrate remains is a complex process

The moment that a human or any other animal dies, immediate changes begin to occur that start the processes of decomposition. **Taphonomy** is the study of the processes of decomposition. It is focused on "how" organisms decay following death and in turn become fossilized. As originally defined, the field is broadly focused on the conversion of organismal remains, parts, and

---

The Science of Forensic Entomology, Second Edition. David B. Rivers and Gregory A. Dahlem.
© 2023 John Wiley & Sons Ltd. Published 2023 by John Wiley & Sons Ltd.
Companian website: www.wiley.com/go/forensicentomology2

products from the biosphere to the lithosphere (Efremov, 1940). More recently, **forensic taphonomy** has emerged as a subdiscipline that focuses on processes of human (or other animals) decomposition within a legal context. Forensic taphonomy utilizes taphonomic methods to estimate time of death, reconstruct the circumstances before and after decomposition, and to discriminate the "products" of human behavior from those derived from abiotic or biotic factors as they contribute to decay or deterioration of the remains (Haglund & Sorg, 2001).

Animal decay is a continuous process that can be characterized by distinct physical and chemical changes that are unique for each organism, yet the events are mostly sequential and relatively predictable. What this means is that no two organisms decompose in exactly the same manner, but the processes involved are the same, at least when the comparison is made within similar environmental conditions and contexts (Vass, 2001; Forbes & Carter, 2018). The progress of decomposition is typically associated with stages which represent physical milestones achieved over time. The stages are a convenient means for summarizing physio-chemical changes in animal remains. However, they are determined subjectively rather than empirically and thus can vary greatly based on the experience and training of the death investigator (Suckling et al., 2015). Similarly, the use of stage assignment does not necessarily represent discrete or sequential ecological stages during animal decomposition (Schoenly & Reid, 1987). Progression of body decomposition is very much an individualistic experience, influenced by the circumstances of death, environment where the remains are located, organisms that interact directly or indirectly with the corpse, and the physical and chemical composition of the deceased. Deterioration of the corpse reflects a continuum of events, in which the rate of decay differs from one region of the body to another. As a consequence, decomposition of the body as a whole utilizes predictable processes but discrete stages of physical change do not necessarily occur, outside of the initial stage (frequently termed "fresh" stage) that is associated with the moment of death, otherwise known as the **agonal period**.

Features of the environment such as temperature, moisture levels, whether the body is found in a terrestrial versus aquatic environment, geographic location, time of year (seasonality), and if vertebrate scavengers and insects have access to the corpse can all change the rate of body decomposition as well as

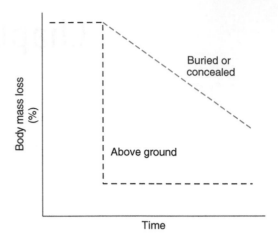

Figure 10.1 Body mass changes of corpse over time based on location. Mass changes with concealed bodies vary dependent on the means of concealment (i.e., wrapped, container) or location. Figure is based on Carter et al. (2007) and Payne (1965).

other aspects of tissue decay (Goff, 2010) (Figure 10.1). A discussion of physical decomposition stages for bodies lying in terrestrial habitats can be found in Section 10.4, while Chapter 11 presents aspects of decomposition under a wide range of conditions and how these contexts influence insect utilization (i.e., foraging behavior, colonization, rates of development, etc.) of animal remains.

Chemical decomposition is more narrowly classified into two categories: **autolysis** and **putrefaction**. Autolysis describes the self-digestion or destruction of a cell due to the action of enzymes found within that cell. Self-digestion rarely occurs in living cells or antemortem (prior to death), and thus autolysis is associated with postmortem or after-death alterations in cells and tissues. Putrefaction is broadly defined as the chemical degeneration of soft tissues, predominantly catalyzed by microbial action and yielding an array of byproducts, including strongly odoriferous gases, liquids, and small organic molecules (Vass et al., 2002). In some instances, putrefaction is classified specifically with breakdown of proteins in soft tissues, but regardless of which organic molecules are referenced, putrefaction generally begins somewhat later than autolysis simply because of the reliance on microorganisms (Figure 10.2).

The types of chemical pathways associated with postmortem biochemistry point to yet another feature of animal decay: the breakdown of cells and tissues is dependent on both abiotic and biotic decomposition. Abiotic decomposition relies on the

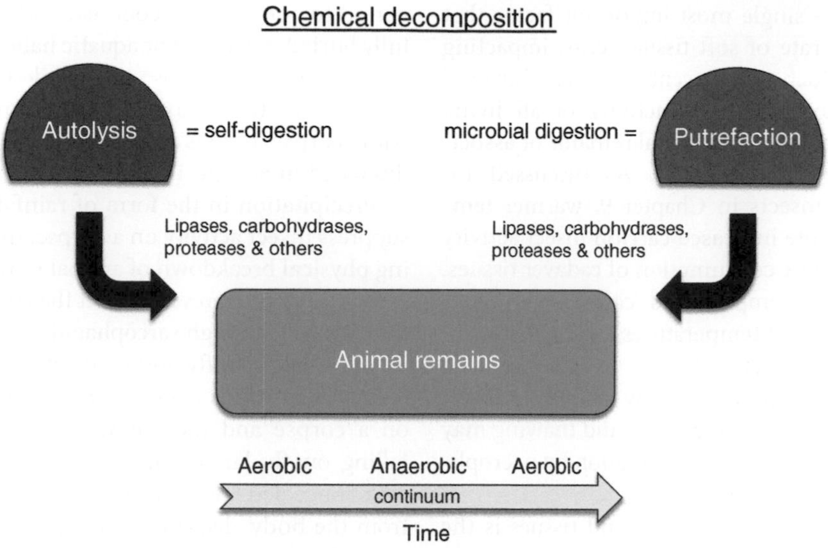

Chemical decomposition

Figure 10.2   Relationship between autolysis and putrefaction in the chemical decomposition of vertebrate carrion.

chemical action of autolysis and putrefaction as well as physical processes associated with the environment and microhabitat of the corpse. The latter can expand over time to become what is known as a **cadaver decomposition island** (CDI), which essentially encompasses the animal remains and the soil (and inhabitants) that has become saturated with expelled body fluids (Carter *et al.*, 2007). By contrast, biotic decomposition or biodegradation is the catabolic degradation of organic material facilitated by living organisms. Several organisms play significant roles as necrophagous and saprophagous nutrient recyclers and include bacteria, fungi, necrophagous insects, and invertebrate and vertebrate scavengers (Barton *et al.*, 2020). Such organisms are pivotal to human decomposition as their presence has a direct effect on the rate of chemical and physical decay.

matter of days. However, temperature directly affects insect activity, including attraction to a corpse and all facets of growth and development when feeding on carrion. Body and ambient temperatures also regulate microbial activity on and in the corpse, which in turn directly influences the pace of most abiotic and biotic decomposition processes. Changing environmental conditions can modify the impact of a number of factors influencing decomposition, including temperature. When temperature is the dominant feature modulating the decay processes in a terrestrial environment, the rate of decomposition of soft tissues can be predicted using the following formula (developed by Vass, 2001):

$$y = \frac{1285}{x}$$

10.2 Numerous factors affect the rate of body decomposition

Even just a cursory examination of the physical and chemical processes involved in animal degradation should reveal that decomposition is quite complex and can be influenced by a wide range of factors at each step. Temperature is the most important factor affecting chemical and physical decomposition of soft tissues. At first, the principal importance of temperature may seem surprising since insect activity can quickly remove all soft tissues on a corpse in a

where, y is equal to the number of days for a body to become skeletonized or mummified, and x is the average temperature in degrees Celsius during the decomposition process. This is a very broad and oversimplified estimate of the rate of soft tissue decay as there are many factors in the corpse's environment that profoundly shape the decomposition process.

10.2.1 Abiotic factors

Physical or non-living features of the climate and immediate environment in which the corpse is found directly impact the chemical and physical processes of body degradation. Temperature is regarded as the single most important factor that influences the rate of soft tissue decay, impacting the rate of all biochemical events within and around the corpse, as well as the activity of all living organisms colonizing the animal remains or associated with the eventual CDI. As discussed for necrophagous insects in Chapter 9, warmer temperatures promote increased carrion insect activity and thus accelerate consumption of cadaver tissues. Conversely, low temperatures can greatly delay decomposition or, if temperatures are cold enough, decay ceases completely (Mann *et al.*, 1990). In some instances, cold may actually facilitate biodegradation since periods of freezing and thawing may aid tissue penetration and assimilation by necrophagous fly larvae (Micozzi, 1986).

Moisture content of the cells and tissues is the second most critical feature of decomposition. Obviously water is a critical solvent and medium for enzyme function and the pathways associated with autolysis and putrefaction. Environmental moisture in the form of relative humidity, precipitation, and soil conditions modulate several features of corpse decay. For example, bodies disposed of in tropical environments can be skeletonized in as short as 2 weeks (Ubelaker, 1997), whereas those placed in arid hot conditions experience rapid bloating followed by **mummification** (Galloway *et al.*, 1989) (Figure 10.3). Mummification is the process of rapid drying of soft tissues under high heat and low moisture or highly desiccating conditions, like dissolved salt in water vapor (i.e., "salty" air) that results in dried, shrunken, preserved soft tissues (Goff, 2010). Subzero temperatures with very low relative humidity can also mummify tissues but the process requires a much longer period to complete. In contrast, bodies partially or fully buried in wet soil or aquatic habitats under the appropriate conditions can lead to **adipocere** formation, a thick waxy layer that can envelop the entire corpse. Details of adipocere formation will be discussed in Section 10.3.

Precipitation in the form of rainfall can greatly suppress insect activity on a corpse, thereby retarding physical breakdown of animal remains. In most cases, heavy rain prevents adult flies from searching for carrion, although sarcophagid species appear to be more likely to fly and oviposit in the rain than calliphorids (Catts & Goff, 1992). Pooling of water on a corpse and the physical force of raindrops falling on fly larvae may force the immatures to seek concealed locations or even temporarily crawl from the body, delaying decomposition of soft tissues. Though infrequent, sudden unexpected snowfall or frozen precipitation can injure or kill necrophagous insects feeding on animal remains.

The position of a body in relation to sunlight influences the rate of heat loss immediately after death as well as insect colonization and the temperatures of maggot masses that form on the body (Joy *et al.*, 2006). Solar radiant heating alters water loss from tissues, and consequently carrion mass decreases more rapidly in direct sun than in partial or full shade. Similarly, some vertebrate scavengers prefer animal remains that are in more concealed shaded areas, while others such as vultures or crows do not discriminate. Blow fly oviposition follows a similar pattern in that some species prefer carcasses in bright sunny locations while others seek carrion in partial or full shade as oviposition sites (Campobasso *et al.*, 2001).

10.2.2   Biotic factors

Living organisms are essential to carrion degradation and recycling of nutrients locked up in the cells and tissues of an animal's body. Organisms that frequent carrion are usually classified as decomposers (Table 10.1). More specifically, necrophagous insects and scavengers are chiefly responsible for macro-decomposition: penetration of the intact body and removal of soft tissues, which facilitates rapid disintegration of biomass. Microorganisms are responsible for micro-decomposition, specifically the events of chemical decomposition in the form of putrefaction. Volatile organic molecules and gases liberated from the corpse due to microbial and fungal activity serve as signals to additional

Figure 10.3   Mummified remains of a Peruvian male. *Source*: Photo by Ashley Van Haeften. Image available at https://commons.wikimedia.org/wiki/File:Peruvian_mummified_male_A31655,_face_detail.jpg.

Table 10.1   Biotic components of a typical carrion community.*

Classification	Organisms	Role in carrion community
Accidentals	Non-predictable insect species	No real role on carrion but for whatever reason are found on corpse
Adventive species	Some non-necrophagous insects and arthropods	Species that use corpse as habitat or dwelling
Decomposers	Microorganisms (bacteria, fungi, protozoa)	Responsible for micro-decomposition and putrefaction
	Necrophagous insects	Responsible for macro-decomposition of carrion
	Saprophagous insects	Responsible for macro-decomposition of cadaver decomposition island
Omnivores	Ants, wasps, beetles, some non-insect arthropods	Feed on a wide range of organisms on and around body
Predators/parasites	Ants, wasps, beetles, some non-insect arthropods	Consume early stages of necrophagous insects, especially fly eggs and larvae
Scavengers	Mostly vertebrate species (birds, rodents, fox, raccoons)	Feed on corpse tissues and carrion insects

* Classification of organisms is based largely on Goff (2009, 2010).

organisms to make use of the animal remains as degradation proceeds. Fluids released into the soil create a unique microcosm occupied by soil-inhabiting bacteria, fungi, nematodes, insects, and an array of other invertebrate species (Forbes & Carter, 2018). A similarly distinct habitat exists at the soil–carcass interface, serving as home to a distinct fauna of microorganisms and invertebrates. Most of these soil-dwelling organisms are saprophagous rather than being truly necrophagous, and their impact on body decomposition is important but much less significant since direct consumption of tissues does not occur.

Necrophagous insects clearly play the most significant roles in macro-decomposition of vertebrate carrion in terrestrial habitats. The impact of necrophagous flies and beetles on the physical decay of large vertebrate remains under a wide range of conditions will be examined in Chapter 11.

10.3 When the heart stops internal changes occur almost immediately but are not outwardly detectable

Long before physical signs of decomposition are evident externally, the corpse has begun to deteriorate internally. Cessation of the heart generally marks the onset of death. This is an oversimplification but serves as a starting point to characterize death physiology and immediate internal changes that occur. Within minutes of the last heartbeat, the oxygen content of blood begins to decline, resulting in an elevation in carbon dioxide levels and a concurrent drop in fluid pH. Oxygen trapped in body cavities and the lungs also becomes finite (Carter *et al.*, 2007), and as long as the corpse remains a closed cylinder (i.e., the skin is intact) the internal cadaver environment gradually shifts from one that favors aerobic metabolism to one that favors anaerobic metabolism. Cessation of circulation leads to pooling of blood and other fluids in low-lying locations due to the forces of gravity. Cellular waste products build up, exceeding critical threshold levels that ultimately disrupt cellular homeostasis and driving the intracellular environment toward chaos.

The initial postmortem alterations are generally biochemical events occurring within cells, a process referred to as **necrosis**[1], simply defined as the physiological changes associated with dead cells and also throughout extracellular fluids. It is important to understand that although death signifies the total loss of cellular regulation, enzymatic activity does not shut off, and as long as suitable substrates are present and the intracellular environment is conducive (i.e., pH and oxygen levels are not inhibitory), several biochemical pathways remain functional for a short period of time. Eventually, the chemical reactions will no longer proceed along "normal" pathways and the processes become a biochemical frenzy. Intracellular enzymes (lipases, proteases, carbohydrases, and others) are released by lysosomes into the cytoplasm to begin self-digestion of the cells or autolysis. Autolysis is not an active process since enzyme release is facilitated by cell death, and consequently the rate of enzyme destruction is completely dependent on ambient conditions. The action of lysosomal enzymes will ultimately force the disintegration of the cytoskeleton and thus inward collapse of plasma membranes (loss of cell volume), or rupturing (lysis) of cellular membranes. Either form of cellular demise results in the release of intracellular constituents into the surrounding fluids, perpetuating a chain reaction of enzymatic destruction of adjacent cells and tissues (Figure 10.4).

The pace of autolysis obviously depends on temperature as well as abundance of enzymes. For example, some tissues such as the liver or regions of the gastrointestinal tract possess high levels of digestive enzymes or, in the case of brain tissues, have high water content. Either scenario facilitates rapid self-digestion. Tissues with low enzymatic activity and/or constituents located intracellularly or luminally that inhibit or degrade protein structure (e.g., kidney, bladder) deteriorate more slowly.

In the same way that cell death facilitates autolysis by unregulated enzyme release, autolysis in turn promotes putrefaction through pore formation in plasma membranes or total rupture of cells. The result is efflux of intracellular fluids and components (molecules, organelles, and blebs both micro and macro) into the extracellular environment. Intracellular fluids are nutrient-rich and consequently fuel surges in microbial growth and activity under increasingly anaerobic conditions. Enhanced bacterial abundance and diversity promotes microbe-mediated putrefaction of soft tissues, leading to end products in the form of liquefied tissues, gases, and an array of volatile organic compounds (VOCs) (Vass *et al.*, 2002). Putrefaction also relies on the activity of other microorganisms including fungi and protozoa, as well as bacteria that are inoculated onto or into the corpse via the necrophagous insects and scavengers that visit the host (Thompson *et al.*, 2012; Holl & Gries, 2018).

Some of the more prominent internal changes that occur after death include algor mortis and macromolecular decomposition.

10.3.1   Algor mortis

Algor mortis is the condition whereby the body temperature of the deceased gradually cools to reflect ambient temperatures (Figure 10.5). In essence, the corpse becomes a poikilotherm. Measurement of body temperature rectally can be used to obtain a rough estimate of the time of death if performed during the early stages (within 18 hours) (Goff, 2010). Body temperature cools in a relatively predictable manner, barring widely fluctuating environmental conditions, in which a 2 °C decline is expected in the first hour after death, and cooling continues at a rate of 1 °C per hour until nearing ambient (Saferstein, 2017). The Glastier equation can be used to estimate the time elapsed since death based on measurements of rectal temperature:

$$PMI = \frac{98.6\,°F - \text{rectal temperature in}\,°F}{1.5}$$

The equation is based on the assumption that temperature declines can be modeled as a linear process even though cooling follows an exponential decay curve (Guharaj, 2009) and that core body temperature is represented by a static set point. Rectal temperature actually represents a range (34.4–37.8 °C, 94–100 °F) rather than a single value as suggested by the equation (Sund-Lavender *et al.*, 2002).

The rate of body cooling is influenced by numerous factors, including ambient temperatures, wind, fat composition of the deceased, and whether the body is clothed, wrapped, encased, or buried. As decomposition progresses, body temperature will rise due to microbial activity, solar radiation, and formation of maggot masses. For buried bodies, maggot masses generally do not form and solar radiation has only minimal or no effect on body temperature.

10.3.2   Macromolecule decomposition

Degradation of the major molecules (proteins, lipids, and carbohydrates) of tissues is elicited by the same chemical processes, autolysis and putrefaction, that we have already discussed. Since both types of chemical decomposition rely on enzymatic action, the natural conclusion to protein, fat, or carbohydrate breakdown is that degradation proceeds in a similar fashion to that of chemical digestion along the length of the alimentary canal. In reality, the end products of macromolecule decomposition can be quite different and yield compounds that are unique to death physiology. As should be expected,

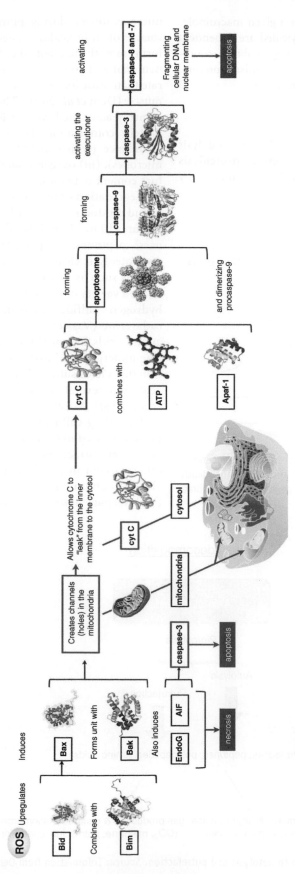

Figure 10.4 Complexity of cellular pathways associated with apoptotic and necrotic death. Oncosis rather necrosis is the more typical term for non-programmed cell death. Necrosis is the general physiological changes that occur following any type of cell death. *Source:* Bsseifer/Wikimedia Commons/CC BY-SA 4.0.

the rate of decomposition for a given macromolecule and the final products yielded are dependent on numerous factors, including ambient conditions, oxygen availability, microorganisms present, and specific tissues undergoing decay.

10.3.2.1 Protein decomposition

The decay of proteins is initiated during autolysis by cellular enzymes, a process termed **proteolysis** (Evans, 1963), and is further facilitated by

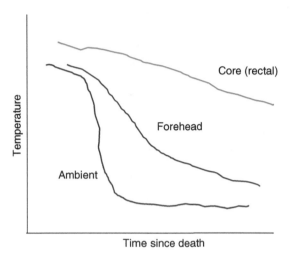

Figure 10.5 Relationship between body temperature and ambient temperature over time following death. Eventually postmortem body temperature will equilibrate with ambient temperatures.

microorganisms during putrefaction of soft tissues. Protein breakdown does not occur at a uniform rate throughout the body. For example, neuronal and epithelial tissues degrade at a faster rate than those located in epidermis, reticulin, and muscle (Dent *et al.*, 2004). The differences are presumably associated with at least four factors: (i) protein composition, (ii) enzyme content and abundance (which includes those derived from microbes), (iii) oxygen availability, and (iv) moisture content of tissues.

Degradation of proteins following death leads to end products that differ from protein digestion (Figure 10.6). Initial breakdown yields proteoses[2], peptones, polypeptides, and amino acids. As proteolytic cleavage continues, phenolic compounds such as skatole and indole are produced, as are a variety of gases including carbon dioxide, hydrogen sulfide, methane, putrescine, and cadaverine (Vass *et al.*, 2002). Sulfur-containing amino acids like methionine, cysteine, and cystine undergo desulfhydrylation during putrefaction that yields sulfides, ammonia, thiols, pyruvic acid, and several gases (Vass *et al.*, 2002; Dent *et al.*, 2004). Many of these compounds contain foul-smelling sulfhydryl groups (−SH) and have been discussed in Chapter 7 as possible chemical cues that attract waves of necrophagous and other carrion insects to animal remains. Under alkaline conditions, such as that occurring with fluid seepage into soil, ammonia can undergo

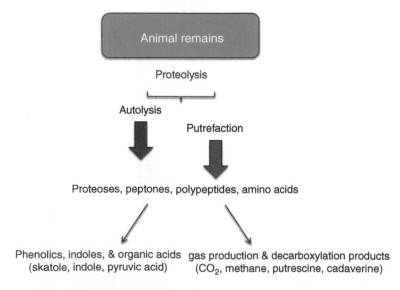

Figure 10.6 Protein decay due to autolysis and putrefaction. *Source*: Information from Dent *et al.* (2004).

volatilization, converting NH_4^+ to NH_3 (Dent *et al.*, 2004), yielding a familiar odor of death (Table 10.2).

Table 10.2 Early postmortem changes in blood proteins and metabolites.

Blood protein	Concentration postmortem
Fibrinolysins	Increase
Lactic dehydrogenase	Increase
Acid phosphatase	Increase
Alkaline phosphatase	Increase
Hemoglobin	No change
Lipoproteins	No change
Immunoglobins	No change
Blood metabolite	
Glucose	Increase
Ammonia	Increase
Uric acid	Increase
Lactic acid	Increase
Bilirubin	Increase
Hypoxanthine	Increase
Xanthine	Increase

Early postmortem changes refer to the first three days following death.
Source: Donaldson & Lamont (2014).

10.3.2.2 Fat decomposition

Lipid decomposition after death involves a series of hydrolysis and oxidation reactions mediated by endogenous lipases as well as enzymes derived from bacteria and other microorganisms (Figure 10.7). Neutral fats generally undergo hydrolysis by lipases released via autolysis to yield fatty acids (stearic, oleic, and palmitic) (Dent *et al.*, 2004), which can then be hydrogenated or oxidized. The initial byproducts are a mixture of saturated and unsaturated fatty acids. However, continued hydrolysis and/or hydrogenation gradually convert the unsaturated fatty acids to saturated (Evans, 1963). As long as the moisture content of the tissues is not depleted (tissue type dependent) so that lipases remain functional, neutral fats present in adipose tissue can be completely degraded. The final result is a mass of fatty acids that, under the right conditions, may yield adipocere (Dent *et al.*, 2004). If the lipid masses react with liberated potassium or sodium ions present in body fluids, the fatty acids can be converted to salts provided the fluid pH is neutral to alkaline.

A unique fatty acid death product is adipocere, created when fatty acids in adipose tissue form a grayish-white wax-like substance, which can become a thick, almost armor-like mass surrounding all or a portion of the corpse (Notter *et al.*, 2009). The fatty acids are generated by intrinsic lipases

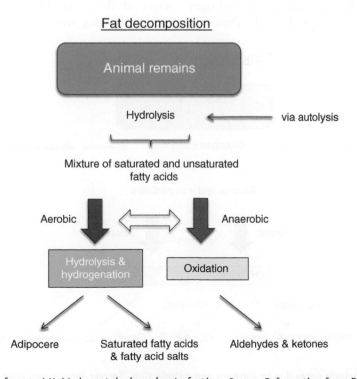

Figure 10.7 Decay of neutral lipids by autolysis and putrefaction. *Source*: Information from Dent *et al.* (2004).

released during autolysis. If the environmental conditions are favorable (e.g., soil moisture content and pH), the resulting unsaturated fatty acids are hydrogenated to saturated fatty acids by microbial enzymes. Formation of stearic and palmitic fatty acids dominate the end products and, in turn, can undergo β-oxidation (Evershed, 1992). Binding of fatty acids to sodium or potassium ions may lead to hardening of the adipocere (Notter et al., 2009), forming a brittle insoluble barrier that can effectively delay or prevent insect colonization.

10.3.2.3 Carbohydrate decomposition

Decomposition of carbohydrates begins in a similar way to carbohydrate catabolism in that glycogen in liver is broken down into glucose molecules (Figure 10.8). Liberation of glucose occurs via the activity of endogenous carbohydrases and through the action of microbial enzymes. The generated glucose has at least two fates: (i) to serve as organic fuel for microbial proliferation and (ii) wherein 6-carbon molecules are completely oxidized to yield carbon dioxide and water (Forbes & Carter, 2018). As oxygen availability in the cadaver becomes limited (hypoxic to anoxic), aerobic metabolism is inhibited, thus preventing further oxidation of glucose. Under increasingly anaerobic conditions, the glycogen monomeric units are converted to butyric acid, lactic acid, acetic acid, alcohols, and gases such as hydrogen sulfide and methane (Dent et al., 2004). These end products further inhibit oxygen-dependent pathways. When oxygen levels again rise in the corpse as soft tissues decompose, skin blisters, or are consumed and thereby create avenues for air movement into the body cavity, aerobic pathways are utilized by microorganisms to oxidize glucose into carbon dioxide and water, the latter contributing to decomposition liquids that accumulate with tissue decay (Cabriol et al., 1998; Dent et al., 2004).

10.4 Body decomposition is characterized by a series of recognizable changes in physical appearance

Physical deterioration of human remains or any type of vertebrate carrion represents a continuum of events subject to wide variation associated largely with environmental influences (Mann et al., 1990; Barton et al., 2020). For convenience, decomposition is classified by a series of discrete stages which summarize the physiochemical changes that occur during decay, as well as to provide a point of reference for periods of insect activity (Schoenly & Reid, 1987; Carter et al., 2007). A range of sequential stages of decay has been proposed by several authors, ranging from six (fresh, bloated, active

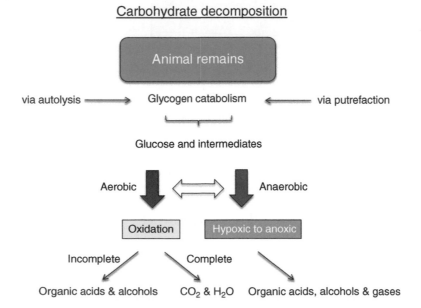

Figure 10.8 Carbohydrate decay due to autolysis and putrefaction. *Source*: Information from Dent *et al.* (2004).

putrefaction, advanced putrefaction, dry putrefaction, and remains) originally proposed by Payne (1965) when using pigs as models for decomposition, to as few as four (Reed, 1958; Johnson, 1975), with the top prize going to Galloway *et al.* (1989) who described five phases with 21 stages. Each classification scheme has merits (reviewed by Micozzi, 1991) but for consistency with the majority of research in forensic entomology, here we will adopt the five stages described by both Goff (2010) and Kreitlow (2010), in which soft tissue decomposition proceeds sequentially through the fresh, bloated, decay, postdecay, and skeletal or remains stages (Table 10.3). As we begin this discussion, it is important to again emphasize that categorizing

Table 10.3 Comparison of decay stages used to characterize human decomposition.

Stage of decomposition	Description
Fresh Initial	Begins at moment of death. Characterized by internal changes associated with autolysis followed by microbial enzymatic degradation.
Bloated Putrefaction Early decomposition	Anaerobic conditions promote microbial conversion of macromolecules into gaseous byproducts resulting in internal pressure increases, ballooning of tissues, and fluid purging from cadaveric openings.
Active putrefaction Active decay Black putrefaction Decay Advanced decomposition	Release of cadaveric fluids to create cadaver decomposition island, gases escape generating very strong odor and body collapse, rapid mass loss from tissues, skin slippage, and sloughing occurs with.
Advanced putrefaction Advanced decay Butyric fermentation Post decay	Tissues continue to lose water due to evaporation and scavenger consumption, body dries out, microbial activity yields byproducts such as butyric acid producing "cheesy" smell, most soft tissue removed by end of stage.
Dried putrefaction Dry	Transition to "dried putrefaction" and to "remains" is difficult to assess, mass changes are very slow, tissue moisture is almost non-existence, soft tissues
Remains Skeletal Dry remains Skeletonization	are dried or removed, remaining tissues are bone, hair and cartilage, decay continues until bone degradation yields elemental forms.

Many authors rely on either a five- or six-stage classification system for animal decomposition, and despite similar descriptions, utilized different names to describe the similar physical stages of decay. Consequently, stages three and/or four above overlap or envelop stage five in the table. Stages are based on Megyesi *et al.* (2005); Carter *et al.* (2007), Goff (2009, 2010).

patterns of decomposition into stages is for the convenience of investigators, since decay of a corpse is a continuous event that is highly variable.

10.4.1 Fresh stage

The fresh stage begins at the moment of death and continues until the onset of bloating is detectable externally (Figure 10.9). Death physiology is initiated after the heart stops, with oxygen immediately becoming a finite resource, and in turn the internal environment gradually shifts from aerobic catabolism to anaerobic pathways. A cascade of processes is triggered that leads to pallor mortis, livor mortis, rigor mortis, algor mortis, and autolysis, the timing of which are dependent on numerous factors, largely under the influence of temperature and moisture. External signs of death are often not evident until rigor mortis or discoloration of skin occurs with livor mortis.

The initial moments after death represent the exposure phase for carrion insects in which the corpse becomes available for colonization but has yet to be detected (Tomberlin *et al.*, 2011). This period does not last long, however, as several species of calliphorids can detect (detection phase) and land on carrion (acceptance or discovery phase) within minutes after death (Smith, 1986; Tomberlin *et al.*, 2011). Adult feeding and oviposition by the flies follow shortly after detection. The first wave of colonization by necrophagous insects also includes adult sarcophagids, generally arriving in the first minutes to hours

Figure 10.9 An adult pig, *Sus scrofa*, in the fresh or initial stage of decomposition. *Source*: Photo by Hbreton19 and available at https://commons.wikimedia.org/wiki/File:Example_of_a_pig_carcass_in_the_fresh_stage_of_decomposition.jpg.

of the fresh stage (Goff, 2010). Depending on the season, adult beetles from the family Silphidae will arrive at any time during the fresh stage.

The fresh stage is subjectively determined to be complete when bloating is evident (Table 10.3). Anaerobic bacteria within the digestive tract metabolize organic molecules under hypoxic to anoxic conditions, yielding several byproducts including gases such as methane, ammonia, and hydrogen sulfide (Clark *et al.*, 1997). Generally the initial signs of bloating are evident by distension of the abdomen, marking the end of the fresh stage and beginning of bloat. What follows is a brief description of several physiochemical changes that occur during the fresh stage, most of which are manifested as physical signs of decomposition.

10.4.1.1 Pallor mortis

Pallor mortis is postmortem paleness, sometime referred to as "paleness of death." It is the immediate (within 15–20 minutes) change in tone associated with light colored skin resulting from the cessation of capillary circulation. Because it occurs so rapidly after death, pallor mortis is generally not useful in estimation of time of death, with the exception of a corpse that still has "normal" color when discovered.

10.4.1.2 Livor mortis

When the heart stops, blood circulation cannot continue. Non-moving fluids like blood will begin to settle in lower portions of the body due to gravity. This process is referred to as **livor mortis**, but terms such as "lividity," "postmortem hypostasis," and "suggilations" are also used interchangeably (Goff, 2010). As the blood leaves capillary beds in upper regions, the tissues become pale in color, whereas in regions in which the blood accumulates the tissues become discolored, turning red to purple (Figure 10.10). Discoloration does not occur where skin makes contact with the ground or some other solid object as the capillaries are compressed from the weight of the body. Further settling occurs via gravitational forces as red blood cells precipitate out of plasma. Generally livor mortis becomes fixed around 12 hours after death dependent on temperature. Prior to fixation, if the body is moved, a second pattern of livor mortis can occur. As putrefaction advances, pooled blood will appear green due to formation of sulhemoglobin (Vass *et al.*, 2002). Eventually livor mortis fades during advanced stages of decomposition.

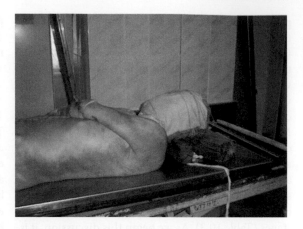

Figure 10.10 Livor mortis is evident when circulation stops and blood settles due to gravitation forces. The appearance of livor mortis is characteristically red to purple where the blood pools. *Source*: goga312/ Wikimedia Commons/CC BY-SA 3.0.

10.4.1.3 Rigor mortis

One of the first visible signs of death is stiffening of the limbs and other extremities due to chemical changes in muscles. The stiffening is called **rigor mortis** and has been reported to be associated with gelling of cytoplasm as intracellular pH drops (Vass, 2001). More often, rigor in striated muscle (skeletal and cardiac) is believed to result from Ca^{2+} efflux from terminal cisternae of sarcoplasmic reticulum located in muscle cells into regions of lower concentration (i.e., sarcomeres[3]). The contractile fibers respond by initiating a muscle contraction. As contractile fibers slide past each other, the diffusional movement of calcium ions into sarcomeres activates troponin, a protein complex that promotes the formation of crossbridges between myosin and actin proteins in myofibrils[4], essentially locking muscles in a state of contraction (Martini *et al.*, 2011) (Figure 10.11). In a living cell, ATP is required for Ca^{2+}-ATPase pumps to transfer calcium ions back to the lumen (terminal cisternae) of the sarcoplasmic reticulum, thereby allowing muscle relaxation to commence. After death, ATP is not available to drive the pumps, which in turn prevents muscle cells from disengaging the couple between actin and myosin, and thus muscles remain in a state of contraction until enzymes associated with autolysis and/or putrefaction degrade myofibril proteins (Gill-King, 1997).

Rigor mortis typically begins in skeletal muscles 2–4 hours after death (timing is temperature and pH dependent), and reaches maximum stiffness in approximately 12 hours. The condition typically lasts

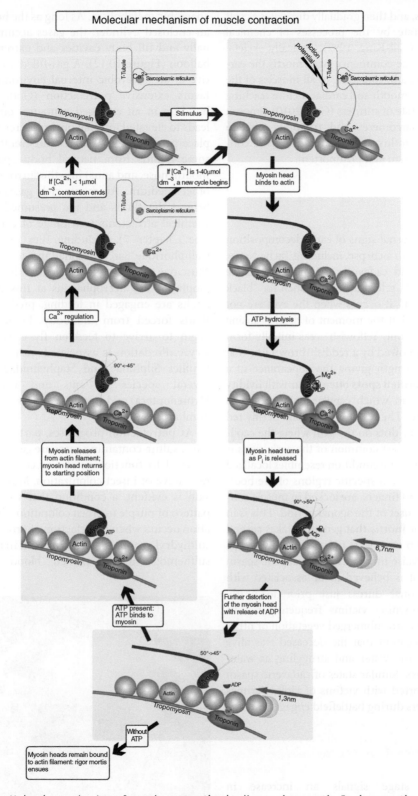

Figure 10.11 Molecular mechanism of muscle contraction leading to rigor mortis. In rigor mortis, ATP-dependent muscle relaxation cannot occur. *Source*: Hank van Helvete; Modified by GravityGilly/CC BY-SA 3.0.

for 72–80 hours, and then gradually dissipates as soft tissues deteriorate by the processes of chemical decomposition (Gill-King, 1997). Typically, skeletal muscles of the face commence rigor mortis the earliest following death, spreading to all muscles of the body, including smooth and cardiac, before reaching the maximum state of stiffness (Goff, 2010). Smooth muscles lack sarcomeres and troponin, but do depend on Ca^{2+} influx for initiation of muscle contraction antemortem and postmortem.

10.4.1.4 Other physical signs of early decomposition

Several other external signs of early decomposition can be evident with a corpse, including Tache Noire, Tardieu spots, and cadaveric spasm. **Tache noire** (Tache noire de la sclerotique) is French for "black spot of the sclera." It occurs when the eyes are not completely closed at the moment of death, causing the sclera to dry out. Yellowish areas initially form on the sclera, followed by a reddish-brown line that transverses the length giving the appearance of a hemorrhage. **Tardieu spots** often occur with lividity or livor mortis, in which capillaries burst due to pooling of blood. The result is formation of tiny red to purple spots or dots on the skin where the capillaries burst. The least common of the early signs is **cadaveric spasm**. The condition resembles localized rigor mortis in which specific regions of the body, usually hands and fingers, are locked in muscle contraction at the onset of the agonal period. This is in contrast to rigor mortis that generally takes several hours to develop or at least to be outwardly recognizable. The precise mechanism of cadaveric spasm is unknown but is believed to be associated with extreme emotional duress just prior to death. Autopsy of drowning victims frequently reveals hands clutched onto submerged vegetation or other objects, an indication that the deceased was alive when entering the water and struggling as water filled the airways. Similar states of cadaveric spasm have been reported with victims of violent crimes and with soldiers during battlefield engagements.

10.4.2 Bloated stage

The bloated stage signals an increase in decomposition activity. Anaerobic bacteria in the gastrointestinal tract intensify the catabolism of organic substrates since the inter-cadaveric environment is becoming increasingly favorable for anaerobic metabolism. As long as the body remains an enclosed cylinder, the gases accumulate internally and fill body cavities and extremities like a balloon (Figure 10.12). A gas-filled body is indicative of an anaerobic internal environment, which favors extensive putrefaction (Goff, 2010). Gas build-up in an enclosed container obviously also leads to elevated internal pressures, resulting in displacement of internal fluids and eventually forcing some liquids from natural body openings (e.g., mouth, nose, and anus) into the surrounding soil or other medium. Ammonia-rich gases also escape behind the fluids and are presumed to serve as chemical attractants to a range of carrion insects (see Chapter 7), including flies in the families Calliphoridae, Sarcophagidae and, to a lesser extent, Muscidae and Piophilidae. Neither the muscids nor piophilids are necrophagous at this point as the adults are engaged in feeding, predominantly on fluids forced from the body. Predatory insects begin to arrive to feed on fly eggs and young larvae. Predation is associated with beetles in the families Silphidae and Staphylinidae, as well as several species of ants and wasps (Order Hymenoptera) and non-insect arthropods (Byrd & Tomberlin, 2020).

As putrefaction progresses, particularly of proteins, sulfur-containing gases are generated that are believed to function as strong cues to signal the next wave of insect colonization. **Marbling** of the skin is evident, a condition that yields a mosaic pattern of purple to green coloration. This discoloration occurs when putrefaction of proteins liberates sulfhydryl groups that in turn result in formation of sulfhemoglobin in pooled blood located in

Figure 10.12 An adult pig, *Sus scrofa*, in the bloat stage of decomposition. *Source*: Hbreton19/Wikimedia Commons/CC BY-SA 3.0.

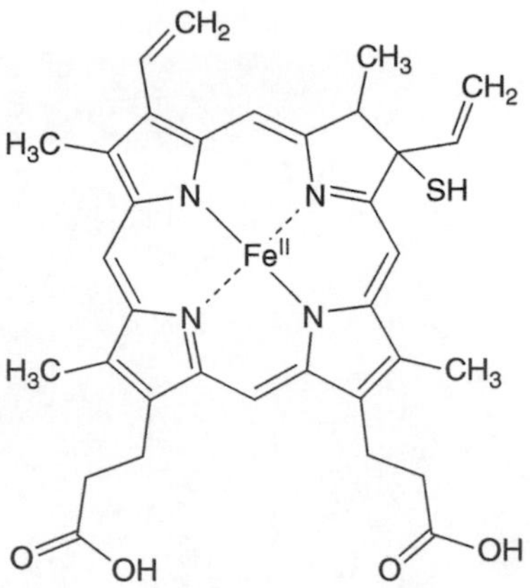

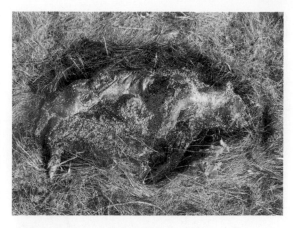

Figure 10.13 Structure of sulfhemoglobin, the compound responsible for the greenish coloration of capillaries evident as marbling. *Source*: Yikrazuul/ Wikimedia Commons/Public domain.

Figure 10.14 An adult pig, *Sus scrofa*, in the decay or advanced decay stage of decomposition. *Source*: Hbreton19/ Wikimedia Commons/CC BY-SA 3.0.

subcutaneous capillaries (Gill-King, 1997; Goff, 2010) (Figure 10.13). Skin blistering and hair slippage often occur toward the end of the stage. The end of the stage is also marked by highly elevated gas pressure that purges large volumes of decomposition fluids through cadaveric openings into the soil around the remains, driving the soil pH toward alkaline and forming a CDI that links soil-inhabiting invertebrates to the corpse (Carter *et al.*, 2007; Forbes & Carter, 2018). Soft tissues weakened by internal gas pressure, loss of fluids and/or feeding by necrophagous insects split and crack, providing an avenue for gases to escape. The result is deflation of the body and the subjective end of the bloated stage.

10.4.3 Decay stage

Some investigators divide the decay stage into two distinct stages: active and advanced (Table 10.3). Regardless of the classification scheme followed, the stage begins with a post-bloat cadaver. The preceding release of gases (ammonia and sulfur-containing) and VOCs produces a strong smell of death. Soft tissues exposed to air have darkened (brown to black) due to necrosis and have lost a significant amount of moisture but are not dry yet. Thus, autolysis and microbial decomposition still occur.

Entomologically, the decay stage represents the peak of insect colonization and assimilation of corpse tissues (Figure 10.14). Consequently, this stage is referred to as the consumption phase with regard to carrion insect activity (Tomberlin *et al.*, 2011). Calliphorid and sarcophagid larvae (possibly some microdipteran species as well) have formed large feeding aggregations initially near cadaveric orifices, but as the decay stage continues the maggot masses spread over most of the body. The feeding aggregations are responsible for rapid consumption of soft tissues, contributing to loss of most of the flesh mass by the end of the stage. Significant predatory activity is also associated with this stage, as more staphylinids and silphids are attracted to the remains, as are beetles in the family Histeridae. The latter portions of decay draw the interest of several species of microdipterans that now experience far less competition from calliphorids and sarcophagid larvae, which typically complete feeding and wander from the corpse by the end of the decay stage. This also marks the onset of the dispersal phase with regard to blow flies and flesh flies since larvae migrate from the remains once feeding is complete (Tomberlin *et al.*, 2011).

10.4.4 Postdecay stage

The postdecay stage begins with the remains having most of the soft tissue removed, exposing bone and cartilage (Figure 10.15A). Any remaining flesh dries quickly, significantly decreasing the nutritional value for most species of necrophagous insects. Maggot masses totally dissipate, and as the resulting postfeeding larvae wander from the corpse to pupariate in the soil (Greenberg, 1991), an array of vertebrate and invertebrate predators as

(a)

(b)

Figure 10.15   An adult pig, *Sus scrofa*, in the (a) postdecay and (b) skeletal (remains) stages. *Source*: Hbreton19/
Wikimedia Commons/CC BY-SA 3.0 and Ashley Van Haeften/Wikimedia Commons/CC BY 2.0.

well as parasitic wasps (predominantly from the families Braconidae and Pteromalidae) take advantage of the vulnerability of exposed prey/hosts. Diminished water content of soft tissues effectively concentrates the remaining macromolecules, particularly fats and protein, which in turn increases the nutritional value of the remains for some species of necrophagous insects. This is most evident with beetles in the families Cleridae and Dermestidae, and piophilid and possibly syrphid flies (Goff, 2010; Heo *et al.*, 2020; Byrd & Tomberlin, 2020). In both cases, although adults arrive during earlier stages of decomposition, oviposition does not occur until the onset of the postdecay stage. Larval development is completed before the end of the stage. The postdecay stage subjectively comes to an end with the dried flesh removed completely or nearly so through the feeding activity of beetles and flies, leaving behind animal remains in the form of bone, cartilage, and hair (Goff, 2010).

10.4.5   Skeletal or remains stage

At the onset of this stage, the cadaver has been reduced to bone and hair (Figure 10.15B). If the remains are exposed to the elements, the bone will become dry and bleached. What is left of the corpse has virtually no nutritional value to most carrion-feeding insects, and thus by this stage of decomposition insect activity is nonexistent (Goff, 2009). However, traces of past insect activity may still be evident. For example, fly puparia are often present all around the remains and in the soil up to several meters from the body (Greenberg, 1991). The empty puparia have some utility in estimating the postmortem interval or can be used for toxicological screening in instances where drugs or toxins are suspected to be involved in the death (Goff *et al.*, 1997; Introna *et al.*, 2001). Early in this stage, puparia parasitized by parasitic wasps such as *Nasonia vitripennis* (Hymenoptera: Pteromalidae) offer use in estimating periods of insect activity and also seasonality depending on whether any wasps remain in the fly hosts, particularly if diapausing larvae are present. Soil-dwelling invertebrates such as mites may inhabit the skeletal remains and can possibly be used to make broad estimates of the postmortem interval (Goff, 2010). Unlike other phases of decomposition, this stage has no definitive end and lasts until the bones and hair completely disintegrate. This may require months to years to reach completion.

Chapter review

Decomposition of human and other vertebrate remains is a complex process

- The moment that a human or any other animal dies, immediate changes begin to occur that start the processes of decomposition.
- Taphonomy is the study of decomposing or decaying organisms over time, including the processes leading to fossilized remains. Forensic taphonomy is the subdiscipline that focuses on

processes of human decomposition within a legal context. Forensic taphonomy utilizes taphonomic methods to estimate time of death, reconstruct the circumstances before and after decomposition, and to discriminate the "products" of human behavior from those derived from abiotic or biotic factors as they contribute to decay or deterioration of the remains

- Animal decay is a continuous process that can be characterized by distinct physical and chemical changes that are unique for each organism, yet the events are sequential and relatively predictable. What this means is that no two organisms decompose in exactly the same manner, but the processes involved are the same.

- Features of the environment such as temperature, moisture levels, whether the body is found in a terrestrial versus aquatic environment, geographic location, time of year (seasonality), and if vertebrate scavengers and insects have access to the corpse can all change the rate of body decomposition as well as other aspects of tissue decay.

- Chemical decomposition is more narrowly classified into two categories: autolysis and putrefaction. Autolysis describes the self-digestion or destruction of a cell due to the action of enzymes found within that cell. Putrefaction is broadly defined as the chemical degeneration of soft tissues, predominantly catalyzed by microbial action and yielding an array of byproducts including strongly odoriferous gases, liquids, and small organic molecules.

- The breakdown of cells and tissues is dependent on both abiotic and biotic decomposition. Abiotic decomposition relies on the chemical action of autolysis and putrefaction as well as physical processes associated with the environment and microhabitat of the corpse. The latter can expand over time to become what is known as a CDI, which essentially encompasses the animal remains and the soil (and inhabitants) that has become saturated with expelled body fluids. By contrast, biotic decomposition or biodegradation is the catabolic degradation of organic material facilitated by living organisms.

Numerous factors affect the rate of body decomposition

- The physical and chemical processes involved in animal degradation reveal that decomposition is quite complex and can be influenced by a wide range of factors at each step. Temperature is the most important factor affecting chemical and physical decomposition of soft tissues.

- Physical or non-living features of the climate and immediate environment in which the corpse is found directly impact the chemical and physical processes of body degradation. Temperature is regarded as the single most important factor that influences the rate of soft tissue decay, impacting the rate of all biochemical events within and around the corpse, as well as the activity of all living organisms colonizing the animal remains or associated with the eventual CDI.

- Moisture content of the cells and tissues is the second most critical feature of decomposition. Environmental moisture in the form of relative humidity, precipitation, and soil conditions modulate several features of corpse decay.

- The position of a body in relation to sunlight influences the rate of heat loss immediately after death as well as rate of water loss, insect colonization, and the temperatures of maggot masses that form on the body.

- Living organisms are essential to carrion degradation and the recycling of nutrients locked up in the cells and tissues of an animal's body. Organisms that frequent carrion are usually classified as decomposers. More specifically, necrophagous insects and scavengers are chiefly responsible for macro-decomposition: penetration of the intact body and removal of soft tissues, which facilitates rapid disintegration of biomass. Microorganisms are responsible for micro-decomposition, specifically the events of chemical decomposition in the form of putrefaction.

When the heart stops internal changes occur almost immediately but are not outwardly detectable

- Cessation of the heart generally marks the onset of death. Within minutes of the last heartbeat, the oxygen content of blood begins to decline, resulting in an elevation in carbon dioxide levels and a concurrent drop in fluid pH; cessation of circulation leads to pooling of blood and other fluids in low-lying locations due to the forces of gravity; and cellular wastes build up, ultimately disrupting cellular homeostasis and driving the intracellular environment toward chaos.

- Although death signifies the total loss of cellular regulation, enzymatic activity does not shut off, and several biochemical pathways remain functional for a short period of time. Eventually, the chemical reactions will no longer proceed along "normal" pathways and the processes become a biochemical frenzy. Intracellular enzymes (lipases, proteases, carbohydrases, and others) are released by lysosomes into the cytoplasm to begin self-digestion of the cells or autolysis. The action of lysosomal enzymes will ultimately force the disintegration of the cytoskeleton and thus inward collapse of plasma membranes (loss of cell volume), or rupturing (lysis) of cellular membranes. Prior to cellular demise, carbohydrates, proteins and lipids are degraded to yield products distinctive from typical chemical digestion in living cells and tissues.
- When the heart stops, blood circulation cannot continue. Non-moving fluids like blood will begin to settle in lower portions of the body due to gravity. This process is referred to as livor mortis.
- One of the first visible signs of death is stiffening of the limbs and other extremities due to chemical changes in muscles, a process called rigor mortis. Rigor mortis is believed to be the result of the unregulated influx of calcium ions into sarcomeres of striated muscle cells, stimulating typical events of muscle contraction: the protein complex troponin forms linkages between actin and myosin proteins on myofibrils. The end result is that the muscles remain locked in a state of contraction because the dead muscle cells are no longer capable of "relaxation." Rigor mortis lasts until enzymes associated with autolysis degrade the intracellular proteins.
- Algor mortis is the condition in which the body temperature of the deceased gradually cools to reflect ambient temperatures. In essence, the corpse becomes a poikilotherm.

Body decomposition is characterized by a series of recognizable changes in physical appearance

- Physical deterioration of human remains or any type of vertebrate carrion represents a continuum of events subject to wide variation associated largely with environmental influences. For convenience, decomposition is classified by a series of discrete stages that summarize the physiochemical changes that occur during decay, as well as providing a point of reference for periods of insect activity. A range of sequential stages of decay has been proposed by several authors, ranging from six to as few as four. The majority of research in forensic entomology has adopted a five-stage scheme, in which soft tissue decomposition proceeds sequentially through the fresh, bloated, decay, postdecay, and skeletal or remains stages.
- The fresh stage begins the moment of death and continues until the onset of bloating is detectable externally. A cascade of processes is triggered that lead to livor mortis, rigor mortis, algor mortis, and autolysis, the timing of which are dependent on numerous factors, largely under the influence of temperature and moisture. External signs of death are usually not evident until rigor mortis or discoloration of skin occurs with livor mortis, although pallor mortis may be evident within a few minutes in individuals with light skin tone. The initial moments after death represent the exposure phase for carrion insects in which the corpse becomes available for colonization but has yet to be detected. This period does not last long, however, as several species of calliphorids can detect (detection phase) and land on carrion (acceptance or discovery phase) within minutes after death.
- The bloating stage signals an increase in decomposition activity. Anaerobic bacteria in the gastrointestinal tract intensify the catabolism of organic substrates since the intercadaveric environment is becoming increasingly favorable for anaerobic metabolism. As long as the body remains an enclosed cylinder, the gases accumulate internally and fill body cavities and extremities like a balloon. A gas-filled body is indicative of an anaerobic internal environment, which favors extensive putrefaction. Ammonia-rich sulfur containing gases also escape behind the fluids and are presumed to serves as chemical attractants to a range of carrion insects, including flies in the families Calliphoridae, Sarcophagidae and, to a lesser extent, Muscidae and Piophilidae. Predatory insects (mostly Coleoptera) begin to arrive to feed on fly eggs and young larvae.
- The decay stage begins with a post-bloat cadaver. The preceding release of gases and VOCs produces a strong smell of death. Soft tissues exposed to air have darkened due to necrosis and have lost a significant amount of moisture but are not dry yet. Thus, autolysis and microbial decomposition still occur. The decay stage represents the peak of

insect colonization and assimilation of corpse tissues and is referred to as the consumption phase with regard to carrion insect activity. Calliphorid and sarcophagid larvae (possibly some microdipteran species as well) have formed large feeding aggregations initially near cadaveric orifices, but as the decay stage continues the maggot masses spread over most of the body. The feeding aggregations are responsible for rapid consumption of soft tissues, contributing to loss of most of the flesh mass by the end of the stage. Significant predatory activity is also associated with this stage. Prior to the end of the stage, blow flies and flesh flies enter the dispersal phase as post-feeding larvae migrate from the remains to initiate pupariation.

- The postdecay stage begins with the remains having most of the soft tissue removed, exposing bone and cartilage. Any remaining flesh dries quickly, significantly decreasing the nutritional value for most species of necrophagous insects. Diminished water content of soft tissues effectively concentrates the remaining macromolecules, particularly fats and proteins, which in turn increases the nutritional value of the remains for some species of necrophagous insects, particularly beetles in the families Cleridae and Dermestidae and piophilid flies. The stage subjectively comes to an end with the dried flesh removed completely or nearly so through the feeding activity of beetles and flies, leaving behind animal remains in the form of bone, cartilage, and hair.
- At the onset of this stage, the cadaver has been reduced to bone and hair. What is left of the corpse has virtually no nutritional value to most carrion-feeding insects, and thus by this stage of decomposition insect activity is nonexistent. However, traces of past insect activity may still be evident in the form of fly puparia. Soil-dwelling invertebrates such as mites may inhabit the skeletal remains, and can possibly be used to make broad estimates of the postmortem interval. Unlike other phases of decomposition, this stage has no definitive end and lasts until the bones and hair completely disintegrate.

Test your understanding

Level 1: knowledge/comprehension

1. Define the following terms:

   (a) taphonomy
   (b) forensic taphonomy
   (c) adipocere

   (d) autolysis
   (e) putrefaction
   (f) bloat stage
   (g) rigor mortis
   (h) cadaveric spams.

2. Match the terms (i–vi) with the descriptions (a–f).

   (a) Begins with onset of death
   (b) Peak period of insect activity on corpse
   (c) Decay of proteins enzymatically following death
   (d) Events that occurred prior to death
   (e) Pooling of blood in lower body regions due to gravity
   (f) Self-digestion of cells by intracellular enzymes
   (g) Moment of death

   (i) Autolysis
   (ii) Antemortem
   (iii) Livor mortis
   (iv) Agonal period
   (v) Proteolysis
   (vi) Consumption phase
   (vii) Fresh stage

3. Describe the chemical decomposition processes that occur following death that are associated with decay of animal remains.

4. Explain how it is possible for the conditions of decomposition for a corpse to move from aerobic to anaerobic to again aerobic.

5. Discuss how rigor mortis, bloating, and cadaveric spasm can be distinguished.

Level 2: application/analysis

1. Explain why the events of putrefaction do not begin at the same time as the onset of autolysis.

2. Describe how the stages of physical decay of a human body can be used as predictors of periods of insect activity.

3. Adult flies from the family Piophilidae are often observed walking or flying in the vicinity of a fresh corpse, yet larvae are still feeding on the body during postdecay. What accounts for the seemingly long association between this fly species and the cadaver?

4. Several features of postmortem decomposition are tied to cessation of aerobic metabolism. Describe which of those features result in obvious physical manifestations of decay.

Level 3: synthesis/evaluation

1. Detail the processes that allow muscle stiffening to occur after death. Are all muscles in the body expected to undergo rigor mortis at the same rate and by the same mechanism(s)? Explain your answer.
2. Speculate on how placement of human remains in a sealed plastic container would impact the rate of decomposition in terms of autolysis, putrefaction, and end products of protein degradation.

Notes

1. Necrosis is variously defined throughout the literature, ranging from a form of cell death or mechanism of death, to the physiology of cells postmortem (Manjo & Joris, 1996).
2. A proteose results from the partial hydrolysis of a protein.
3. A sarcomere is the contractile unit of a myofibril. During a contraction, long fibrous proteins in the sarcomere slide past each other to shorten the cell or return it to a relaxed state.
4. Myofibrils are the basic unit or muscle fibers of muscle cells, myocytes.

References cited

Barton, P.S., Archer, M.S., Quaggiotto, M.-M. & Wallman, J.F. (2020) Invertebrate succession in natural and terrestrial environments. In: J.H. Byrd & J.K. Tomberlin (eds) *Forensic Entomology: The Utility of Arthropods in Legal Investigations*, 3rd edn, pp. 141–154. CRC Press, Boca Raton, FL.

Byrd, J.H. & Tomberlin, J.K. (2020) Insects of forensic importance. In: J.H. Byrd & J.K. Tomberlin (eds) *Forensic Entomology: The Utility of Arthropods in Legal Investigations*, 3rd edn, pp. 15–62. CRC Press, Boca Raton, FL.

Cabriol, N., Pommier, M.T., Gueux, M. & Payne, G. (1998) A comparison of lipid composition in two types of human putrefaction liquid. *Forensic Science International* 94: 47–54.

Campobasso, C.P., Di Vella, G. & Introna, F. (2001) Factors affecting decomposition and Diptera colonization. *Forensic Science International* 120: 18–27.

Carter, D.O., Yellowless, D. & Tibbett, M. (2007) Cadaver decomposition in terrestrial ecosystems. *Naturwissenschaften* 94: 12–24.

Catts, E.P. & Goff, M.L. (1992) Forensic entomology in criminal investigations. *Annual Review of Entomology* 37: 253–272.

Clark, M.A., Worrell, M.B. & Pless, J.E. (1997) Postmortem changes in soft tissues. In: W.D. Haglund & M.H. Sorg (eds) *Forensic Taphonomy: The Postmortem Fate of Human Remains*, pp. 151–164. CRC Press, Boca Raton, FL.

Dent, B.B., Forbes, S.L. & Stuart, B.H. (2004) A review of human decomposition processes in soil. *Environmental Geology* 45: 576–585.

Donaldson, A.E. & Lamont, I.L. (2014) Estimation of post-mortem interval using biochemical markers. *Australian Journal of Forensic Sciences* 46: 8–26.

Efremov, E.A. (1940) Taphonomy: a new branch of paleontology. *Pan-American Geology* 74: 81–93.

Evans, W.E.D. (1963) *The Chemistry of Death*. Thomas Publishers, Springfield, IL.

Evershed, R.P. (1992) Chemical composition of a bog body adipocere. *Archaeometry* 34: 253–265.

Forbes, S.L. & Carter, D.O. (2018) Process and mechanisms of death and decomposition of vertebrate carrion. In: M.E. Benbow, J.K. Tomberlin & A.M. Tarone (eds) *Carrion Ecology, Evolution and Their Applications*, pp. 13–30. CRC Press, Boca Raton, FL.

Galloway, A., Birkby, W.H., Jones, A.M., Henry, T.E. & Parks, B.O. (1989) Decay rates of human remains in an arid environment. *Journal of Forensic Sciences* 34: 607–616.

Gill-King, H. (1997) Chemical and ultrastructural aspects of decomposition. In: W.D. Haglund & M.H. Sorg (eds) *Forensic Taphonomy: The Postmortem Fate of Human Remains*, pp. 93–104. CRC Press, Boca Raton, FL.

Goff, M.L. (2009) Early post-mortem changes and stages of decomposition in exposed cadavers. *Experimental and Applied Acarology* 49: 21–36.

Goff, M.L. (2010) Early postmortem changes and stages of decomposition. In: J. Amendt, C.P. Campobasso, M.L. Goff & M. Grassberger (eds) *Current Concepts in Forensic Entomology*, pp. 1–24. Springer, London.

Goff, M.L., Miller, M.L., Paulson, J.D., Lord, W.D., Richards, E. & Omori, A.I. (1997) Effects of 3,4-methylenedioxymethamphetamine in decomposing tissues on the development of *Parasarcophaga ruficornis* (Diptera: Sarcophagidae) and detection of the drug in postmortem blood, liver tissue, larvae and puparia. *Journal of Forensic Sciences* 42: 276–280.

Greenberg, B. (1991) Flies as forensic indicators. *Journal of Medical Entomology* 28: 565–577.

Guharaj, P.V. (2009) *Forensic Medicine*, 2nd edn. Universities Press, Hyderabad, India.

Haglund, W.D. & Sorg, M.H. (2001) *Advances in Forensic Taphonomy: Method, Theory, and Archaeological Perspectives*. CRC Press, Boca Raton, FL.

Heo, C.C., Rahimi, R., Mengual, X., Isa, M.S.M., Zainal, S., Khofar, P.N. & Nazni, W.A. (2020) *Eristalinus arvorum* (Fabricius, 1787)(Diptera: Syrphidae) in human skull: A new fly species of forensic importance. *Journal of Forensic Sciences* 65: 276–282.

Holl, M.V. & Gries, G. (2018) Studying the "fly factor" phenomenon and its underlying mechanisms in house flies, *Musca domestica*. *Insect Science* 25: 137–147.

Introna, F. Jr, Campobasso, C.P. & Goff, M.L. (2001) Entomotoxicology. *Forensic Science International* 120: 42–47.

Johnson, M.D. (1975) Seasonal and microseral variations in the pest populations on carrion. *American Midland Naturalist* 93: 79–90.

Joy, J.E., Liette, N.L. & Harrah, H.L. (2006) Carrion fly (Diptera: Calliphoridae) larval colonization of sunlit and shaded pig carcasses. *Forensic Science International* 164: 183–192.

Kreitlow, K.L.T. (2010) Insect succession in natural environment. In: J.H. Byrd & J.L. Castner (eds) *Forensic Entomology: The Utility of Arthropods in Legal Investigations*, 2nd edn, pp. 251–270. CRC Press, Boca Raton, FL.

Manjo, G. & Joris, I. (1996) *Cells, Tissues and Disease*. Blackwell Publishing Ltd., Malden, MA

Mann, R.W., Bass, W.M. & Meadows, L. (1990) Time since death and decomposition of the human body: variables and observations in case and experimental field studies. *Journal of Forensic Sciences* 35: 103–111.

Martini, F.H., Nath, J.L. & Bartholomew, E.F. (2011) *Fundamentals of Anatomy and Physiology*, 9th edn. Benjamin Cummings, San Francisco, CA.

Megyesi, M.S., Nawrocki, S.P. & Haskell, N.H. (2005) Using accumulated degree-days to estimate the postmortem interval from decomposed human remains. *Journal of Forensic Sciences* 50: 618–626.

Micozzi, M.S. (1986) Experimental study of postmortem change under field conditions: effects of freezing, thawing and mechanical injury. *Journal of Forensic Sciences* 31: 953–961.

Micozzi, M.S. (1991) *Postmortem Changes in Human and Animal Remains. A Systematic Approach*. Thomas Publishers, Springfield, IL.

Notter, S.J., Stuart, B.H., Rowe, R. & Langlois, N. (2009) The initial changes of fat deposits during decomposition of human and pig remains. *Journal of Forensic Sciences* 54: 195–201.

Payne, J.A. (1965) A summer carrion study of the baby pig *Sus scrofa* Linnaeus. *Ecology* 46: 592–602.

Reed, H.B. (1958) A study of dog carcass communities in Tennessee, with special reference to the insects. *American Midland Naturalist* 59: 213–245.

Saferstein, R. (2017) *Criminalistics: An Introduction to Forensic Science*, 12th edn. Prentice Hall, Boston.

Schoenly, K. & Reid, W. (1987) Dynamics of heterotrophic succession in carrion arthropod assemblages: discrete series or a continuum of change. *Oecologia* 73: 192–202.

Smith, K.G.V. (1986) *A Manual of Forensic Entomology*. British Museum (Natural History), London.

Suckling, J.K., Spradley, M.K. & Godde, K. (2015) A longitudinal study of human outdoor decomposition in central Texas. *Journal of Forensic Sciences* 61: 19–25.

Sund-Lavender, M., Forsberg, C. & Wahren, L.K. (2002) Normal oral, rectal, tympanic and axillary body temperatures in adult men and women: a systematic literature review. *Scandinavian Journal of Caring Science* 16: 122–128.

Thompson, C., Brogan, R. & Rivers, D.B. (2012) Microbial interactions with necrophagous flies. *Annals of the Entomological Society of America* 106: 799–809.

Tomberlin, J.K., Mohr, R., Benbow, M.E., Tarone, A.M. & VanLaerhoven, S. (2011) A roadmap bridging basic and applied research in forensic entomology. *Annual Review of Entomology* 56: 401–422.

Ubelaker, D.H. (1997) Taphonomic applications in forensic anthropology. In: W.D. Haglund & M.H. Sorg (eds) *Forensic Taphonomy: The Postmortem Fate of Human Remains*, pp. 77–90. CRC Press, Boca Raton, FL.

Vass, A.A. (2001) Beyond the grave: understanding human decomposition. *Microbiology Today* 28: 190–193.

Vass, A.A., Barshick, S.-A., Sega, G., Caton, J., Skeen, J.T., Love, J.C. & Synstelien, J.A. (2002) Decomposition chemistry of human remains: a new methodology for determining the postmortem interval. *Journal of Forensic Sciences* 47: 542–553.

Supplemental reading

Benbow, M.E., Tomberlin, J.K. & Tarone, A.M. (2018) *Carrion Ecology, Evolution, and Their Applications*. CRC Press, Baco Raton, FL.

Black, S. (2018) *All That Remains: A Life in Death*. Black Swan, London.

Early, M. & Goff, M.L. (1986) Arthropod succession patterns in exposed carrion on the island of Oahu, Hawaii. *Journal of Medical Entomology* 23: 520–531.

Forbes, S.L., Stuart, B.H., Dadour, I.R. & Dent, B.B. (2004) A preliminary investigation of the stages of adipocere formation. *Journal of Forensic Sciences* 49: JFS2002230-9.

Olea, P.P., Mateo-Tomás, P. & Sánchez-Zapata, J.A. (2019) *Carrion Ecology and Management*. Springer, Switzerland.

Parks, C.L. (2011) A study of the human decomposition sequence in central Texas. *Journal of Forensic Sciences* 56: 19–22.

Rodriguez, W.C. & Bass, W.M. (1983) Insect activity and its relationship to decay rates of human cadavers in East Tennessee. *Journal of Forensic Sciences* 28: 423–432.

Schoenly, K., Goff, M.L., Wells, J.D. & Lord, W.D. (1996) Quantifying statistical uncertainty in succession-based entomological estimates of the postmortem interval in death scene investigations: a simulation study. *American Entomologist* 42: 106–112.

Schotsmans, E.M., Márquez-Grant, N. & Forbes, S.L. (2017) *Taphonomy of Human Remains: Forensic Analysis of the Dead and the Depositional Environment*. Wiley-Blackwell, Sussex.

Sharanwoski, B.J., Walker, E.G. & Anderson, G.S. (2008) Insect succession and decomposition patterns on shaded and sunlit carrion in Saskatchewan in three different seasons. *Forensic Science International* 179: 219–240.

Tabor, K.L., Fell, R.D. & Brewster, C.C. (2005) Insect fauna visiting carrion in Southwest Virginia. *Forensic Science International* 150: 73–80.

Additional resources

Body farm: study of human decomposition video: http://www.youtube.com/watch?v=V_SiqND9bNA

Centre for Forensic Science at University of Western Australia: www.forensicscience.uwa.edu.au

Decomposition of baby pigs video: http://www.youtube.com/watch?v=R1CD6gNmhr0

Forensic Anthropology Center at University of Tennessee: http://fac.utk.edu/

Forensic Anthropology Center at Texas State University: http://www.txstate.edu/anthropology/facts/

Institute for Forensic Anthropology and Applied Sciences: http://forensics.usf.edu/

Stages of pig decomposition: http://australianmuseum.net.au/Stages-of-Decomposition

Chapter 11

Insect succession on carrion under natural and artificial conditions

Synanthropic flies, particularly calliphorids, are initiators of carrion decomposition and, as such, are the primary and most accurate forensic indicators of time of death.

Bernard Greenberg, Emeritus Professor of Biology,
University of Illinois at Chicago[1]

Overview

Necrophagous insects can detect a corpse within minutes of death and then attempt to oviposit/larviposit in natural body openings and/or wounds. The model for understanding necrophagous fly colonization and development on animal remains is based on carrion decomposition in terrestrial environments during ambient conditions (e.g., seasons characterized by warm temperatures) that favor insect activity. Animals die, or bodies are placed, in many locations (including man-made) and habitats that do not emulate the situation of terrestrial decay. In addition, all terrestrial habitats are not the same as they differ by biogeographical location, soil composition, biotic fauna, and climatic conditions. Correspondingly, the insect fauna associated with these natural and artificial conditions of postmortem decomposition differ in many respects (taxa, behavior, and developmental

duration) from those commonly found during sequential colonization on terrestrial carrion decaying in the summertime. This chapter will explore the factors influencing insect succession, namely barriers to oviposition, restriction of necrophagous insect development (e.g., speeding up, slowing down, or under extremes), and shifts in timing and/or species associated with faunal colonization.

The big picture

- What's normal about terrestrial decomposition? Typical patterns of insect succession on bodies above ground.
- Succession patterns under forensic conditions are not typical.
- Several factors serve as barriers to oviposition by necrophagous insects.

The Science of Forensic Entomology, Second Edition. David B. Rivers and Gregory A. Dahlem.
© 2023 John Wiley & Sons Ltd. Published 2023 by John Wiley & Sons Ltd.
Companian website: www.wiley.com/go/forensicentomology2

- The physical conditions of carrion decay can function as a hurdle to insect development.
- Insect faunal colonization of animal remains is influenced by conditions of physical decomposition.

11.1 What's normal about terrestrial decomposition? Typical patterns of insect succession on bodies above ground

When a vertebrate animal dies in a terrestrial habitat, the body will immediately draw the attention of specific species of necrophagous calliphorids that constitute the first wave of colonization. As the physical and chemical decomposition of the corpse progresses, successive waves of arthropod colonization, dominated by insects, will proceed in a sequential and predictable manner (Catts & Goff, 1992). These predictions are based on extensive fieldwork examining insect colonization of animal remains and case study observations, with particular focus on human cadavers in instances of suspicious death and homicide. Much of what is understood in terms of insect succession on carrion is derived from studies using pig carrion as surrogates to human decomposition, placed in terrestrial habitats, above ground, and during ambient conditions characterized by warm temperatures (summertime) in full sunlight (Matuszewski *et al.*, 2011; Anderson, 2020). Thus the ideal, or at least typical, pattern of insect succession on vertebrate carrion that forms the foundation underlying minimum postmortem interval calculations in forensic entomology is based on physical and chemical decay in terrestrial environments when seasonality favors carrion insect activity. Terrestrial decomposition studies are also responsible for providing the bulk of what is known about the formation of carrion communities; their structure and composition; faunal patterns of colonization in terms of species, timing, and oviposition preferences; and trophic interactions occurring within and across phyla (Reed, 1958; Payne, 1965; Johnson, 1975; Hanski, 1987; Greenberg, 1991). Carrion decomposition in terrestrial habitats has thus served as an **ecological model**, helping define the relationships between insect colonizers and animal remains over time under a range of conditions. This ecological model is a major underpinning for medicocriminal entomology.

Before examining patterns of insect colonization in terrestrial locales, remember that neither insect colonization of carrion nor physical decomposition of animal remains is locked into discrete phases or stages. Rather, as we discussed in Chapter 10, these processes occur along a continuum that is subject to considerable variation based on ambient conditions and a whole array of other factors, many of which will be discussed in this chapter as we examine natural and artificial influences on insect succession. What this means in the short term is that though certain insects (e.g., calliphorids and sarcophagids) are predicted to arrive on a corpse or oviposit during preferred periods of decomposition (Smith, 1986), insect faunal succession is continuous, so that any given species may utilize the cadaver during multiple stages of decomposition, and successive waves of colonizers can overlap during their association with carrion (Figure 11.1). In this regard, some investigators believe that Mégnin's original descriptions of waves (death squads) of insect succession[2] has led to an overemphasis in trying to associate insect activity on carrion with specific stages of physical decomposition (Schoenly & Reid, 1987; Greenberg, 1991). Doing so has the unfortunate effect of suggesting that the relationship between necrophagous insects (flies in particular) and the corpse is somewhat rigid and that a given colonizer only has a narrow time period to use the corpse based on the discrete stages of decay, and that once the window of opportunity is closed the species cannot possibly be found on the deceased. This view simply is not correct. For example, a gravid female blow fly that prefers a fresh corpse will often make "compromises" by using later stages of decomposition for oviposition, because from the standpoint of maternal fitness, it is better to use a less suitable resource than to have never oviposited at all.

We already know from Chapter 7 that specific chemical signals must be detected by necrophagous insects for them to initiate foraging behavior. A window of time exists in which a dead animal has yet to be detected, a stage referred to as the exposure phase (Tomberlin *et al.*, 2011), which is expected to be of short duration in warm terrestrial habitats that facilitate rapid tissue decay and a high degree of chemical volatility (Vass *et al.*, 1992).

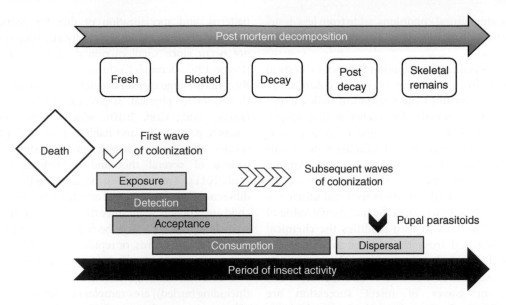

Figure 11.1   Relationship between insect succession and postmortem decomposition for carrion located in a terrestrial environment with warm ambient temperatures. *Source*: Adapted from Lefebvre & Gaudry (2009) and Tomberlin *et al.* (2011).

In contrast, low temperature environments favor longer exposure periods since physical and chemical decay of carrion is slow or inhibited. Once chemical signals derived from either the remains or necrophiles (see Chapter 7) are perceived through sensory recognition (olfaction), detection is said to have occurred (Figure 11.1), and the necrophagous insects, initially dominated by gravid adult calliphorids, enter the searching phase (Tomberlin *et al.*, 2011). Adult blow flies have been suggested to be the insect equivalent of vultures, demonstrating a remarkable ability to locate a patchy ephemeral resource over large landscapes, traveling as much as 20 km a day in search of food (Greenberg, 1991). For insects that are anautogenous or rely on income breeding (see details in Chapter 6), searching represents true foraging behavior since the corpse serves as a required food source for females to provision eggs. Physical contact with the remains leads to probing, tasting, and other forms of resource quality or suitability assessment by the first colonizers (Ashworth & Wall, 1994), eventually culminating in acceptance (or rejection) of the carrion as food and/or as an oviposition site. Egg hatch or larviposition initiates the longest association with the animal remains and carrion community, a phase called consumption (Tomberlin *et al.*, 2011). This phase is characterized by assimilation of carrion tissues by necrophagous fly larvae, resulting in

rapid loss of carrion body mass as soft tissues are consumed, which in turn promotes high rates of water loss. It is during the consumption phase that large feeding aggregations or maggot masses form, a microhabitat characterized by frenzied larval activity, high rates of food intake and assimilation, and generation of internal heat (Rivers *et al.*, 2011; Charabidze *et al.*, 2012).

Completion of feeding by calliphorids and sarcophagids occurs during the third stage of larval development and is followed by wandering or dispersal from the carcass into a protective environment (i.e., buried in soil, under the corpse or carrion) for pupariation (Greenberg, 1991). For many species of flies (in the subfamily Calliphorinae and sarcophagids), dispersal constitutes postfeeding larvae crawling several meters from the body, outside the cadaver decomposition island (Carter *et al.*, 2007), to pupariate in dry soil, buried 1–2 inches in the upper layers of the topsoil (Cammack *et al.*, 2010). Observations of postfeeding larvae from the subfamily Chrysomyinae (e.g., *Phormia regina*, *Protophormia terraenovae*, and *Chrysomya rufifacies*) indicate that these species remain close to the corpse or pupariate on the remains (Norris, 1965; Greenberg, 1990). This behavior should in theory promote higher incidences of parasitism by parasitoids locating exposed puparia. There are, however, no data available comparing parasitism in exposed versus buried hosts under natural conditions, aside from less desirable muscid hosts (Voss *et al.*, 2009), to test this prediction.

The preceding description of insect succession merely follows one path of insect colonization from detection until the association with carrion has ended (dispersal). The reality is that several waves of insect succession occur on a corpse, depending on the ambient conditions of decomposition and a multitude of other factors. Among these other influences are the actual insect colonizers: as the initial colonizers feed and utilize the corpse, their activity alters the nutritional value of the remains. This in turn modifies the chemical signals released from the cadaver as well as the insects that perceive the cues and then are motivated to search for the resource (Goff, 2010). Thus subsequent waves of insect succession are dependent on the preceding colonizers, which are dependent on the initial decay environment. Arguably a case can be made to suggest that faunal succession on each dead animal is unique (even if only slightly) from any other one. This does not mean that patterns of colonization do not occur such as suggested in Figure 11.1. Rather, the differences require caution in over-generalizing the predictability of patterns of decomposition and insect succession.

11.2 Succession patterns under forensic conditions are not typical

After death, carrion represents a rapidly changing microhabitat available for insect colonization. As decomposition progresses through mostly subjectively defined stages, the remains become attractive to different species of necrophagous insects and associated predators, parasites, and adventive species (Johnson, 1975; VanLaerhoven & Anderson, 1999). Insect succession on large vertebrate carcasses located in above-ground terrestrial environments occurs in a relatively predictable sequence (Anderson, 2020). Consequently, such predictability of insect succession is used as a method to estimate periods of insect activity following death in medicocriminal investigations. Relatively modest adjustment to this model, such as switching a large carcass with a small mammal, results in altered community structure: one species of calliphorid or sarcophagid tends to dominate the small carcass along seasonally predictable

patterns, and specialization of flies for particular stages of decomposition (especially late stages) does not occur since small carrion decays very rapidly (Payne, 1965; Denno & Cothran, 1976; Kneidel, 1984). By contrast, large carrion decomposes slowly enough that discrete physical stages can be recognized (Payne, 1965; Goff, 2010), which in turn favors resource partitioning and habitation by multiple fly species, often forming heterogeneous feeding aggregations of several thousand individuals (Rivers *et al.*, 2011). "Large" is a relative term, so that faunal differences in colonization of an adult human versus a child are probably slight since the human condition as a whole represents a large body mass in comparison with say rodents, birds, or reptiles (Figure 11.2). Such factors as carcass size, habitat, seasonality, climate, and environment [e.g., terrestrial, aquatic, and soil (including buried)] are examples of natural influences affecting the rate of carrion decay as well as patterns of insect succession. Each of these factors can vary considerably as to the degree of impact on insect colonization and subsequent development. For instance, carcass size may have only modest influence when decomposition occurs in a terrestrial environment, yet more dramatic shifts in fauna occur with seasonal change in the same location, or an entirely different **guild** of insects is associated with colonization in an aquatic habit versus terrestrial, or when a body is buried. Even in a defined environment such as an aquatic ecosystem, numerous additional factors shape the pattern of colonization including the composition of the water (fresh, marine, brackish), depth, temperature, season, and biogeographical location.

Adding to the complexity of insect succession under natural conditions is the physical structure of the animal. Obviously size matters but so too does the shape (types and location of body openings) and natural coverings of the body. For example, an animal covered by pelage (fur) or plumage (feathers) has a protective barrier that retards or entirely inhibits adult feeding and/or oviposition/larviposition by necrophagous flies. In humans, such natural barriers are replaced by clothing. In cases of homicide, the victim's body is typically not placed in the model environment discussed in Section 11.1. Rather, the perpetrator of the violent crime often attempts to hide the corpse in a concealed location or unique habitat, or modifies the corpse (e.g., burning, dismemberment, and chemical treatment) in an attempt to alter typical patterns of physical decay or insect colonization (Vanin *et al.*, 2013; Scholl & Moffatt, 2017; Malainey & Anderson, 2020). Similarly, decomposition following an accident or

(a)

(b)

(c)

Figure 11.2   Examples of small, medium, and large vertebrate carrion. (a) Dead squirrel. *Source*: Lucarelli/Wikimedia Commons/CC BY-SA 3.0. (b) Dead rabbit. *Source*: H005/Wikimedia Commons/public domain. (c) Dead deer. *Source*: D.B. Rivers.

suicide may occur in a concealed location such as in a vehicle or indoors, and/or involve intoxication of tissues with substances that impact necrophagous insect activity (Goff & Lord, 2010). In these cases, decomposition, and hence insect succession, occurs under non-ideal, possibly artificial conditions.

Although the conditions, whether natural or artificial, that animal remains experience during decomposition can vary considerably, all generally influence succession by necrophagous flies in a manner that represents deviations from the terrestrial ecological model. Patterns of insect succession are altered through the different factors or conditions that act as barriers to detection or progeny deposition, constrain necrophagous insect development (e.g., speeding up, slowing down, or under extremes), and induce shifts in faunal colonization (in terms of timing and/or species) on a given carcass (Figure 11.3).

11.3   Several factors serve as barriers to oviposition by necrophagous insects

Body location after death can have a profound impact on the ability of necrophagous insects to detect, locate, and use animal remains. What this means in a practical sense is that if the corpse is located in a concealed area such as indoors, in a vehicle, or stuffed into a closed trash can, the chemical cues emanating from the body will likely take longer to escape concealment to be subsequently detected in the environment than when carrion decomposes exposed in a terrestrial environment. It follows then that if the concealed location serves as a barrier to odor volatility, it certainly seems to reason that the structure or location will also delay or completely retard insect colonization. These artificial habitats share similar influences on necrophagous insect activity by functioning as (i) physical barriers to activation/searching and (ii) physical deterrents to oviposition/larviposition (i.e., barriers to acceptance, including adult feeding).

Like so many of the other considerations used in the interpretation of corpse decomposition, a concealed location influences insect activity but the degree of influence lies along a continuum. Thus, any given scenario of body decomposition does not allow absolute predictability in terms of timing or species usage of animal remains. The same principles apply to bodies decomposing in natural habits other than above-ground terrestrial ecosystems: release of decomposition gases and access by terrestrial necrophilic species is greatly hampered or totally denied by aquatic and soil environments, or during seasonal change.

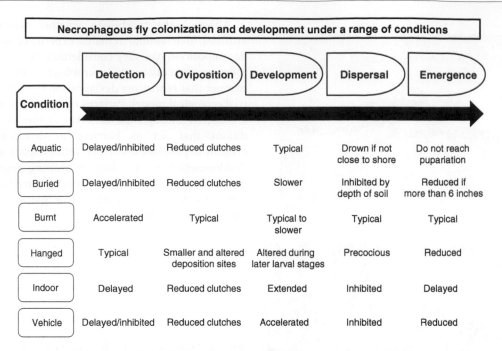

Necrophagous fly colonization and development under a range of conditions

Condition	Detection	Oviposition	Development	Dispersal	Emergence
Aquatic	Delayed/inhibited	Reduced clutches	Typical	Drown if not close to shore	Do not reach pupariation
Buried	Delayed/inhibited	Reduced clutches	Slower	Inhibited by depth of soil	Reduced if more than 6 inches
Burnt	Accelerated	Typical	Typical to slower	Typical	Typical
Hanged	Typical	Smaller and altered deposition sites	Altered during later larval stages	Precocious	Reduced
Indoor	Delayed	Reduced clutches	Extended	Inhibited	Delayed
Vehicle	Delayed/inhibited	Reduced clutches	Accelerated	Inhibited	Reduced

Figure 11.3 Fly colonization and development in comparison with succession in a terrestrial environment (typical). The examples are not exhaustive and are subject to considerable variation based on a number of factors.

11.3.1 Physical barriers to activation/searching

Any location, artificial or natural, that restricts air movement poses a challenge to necrophagous insects in detecting a corpse. The net effect is a lengthening of the exposure phase and a delay in the onset of the activation phase or detection of the body. Foraging behavior cannot begin until activation, so naturally this aspect of fly activity is set back as well. Even under the umbrella of barriers that affect activation, all conditions are not equal in the degree of delay imposed. For example, many conditions appear to retard insect detection of a carcass, while others may abolish activation altogether. Thus, factors or conditions that alter activation can be classified as those that (i) delay odor detection and (ii) abolish odor detection.

11.3.1.1 Factors that delay odor detection

Far more conditions result in retardation of odors, more specifically volatile organic compounds (VOCs), being perceived by necrophagous flies and beetles than those that eliminate detection altogether. Among those that have been reported from field studies or case reports, most are artificial

scenarios such as body decomposition in vehicles, indoors (behind walls, under floorboards, in closets, basements, storage sheds), wrapped (in blankets, carpet or sleeping bags), sealed in trash cans, placed in closed appliances, stuffed in bags (garment or trash), and even concealed in a toolbox (Table 11.1). All share the commonality of reduced air movement, thereby trapping odors or permitting emission only through small openings by the slow process of diffusion. Since the rate of diffusion is dependent on thermal energy (Brownian motion[3]) of a given gas particle (Withers, 1992), the release of VOCs into the environment, and hence detection by insects, is reliant on ambient temperatures. As ambient temperatures elevate, Brownian motion of gas particles increases, and so too does the rate of diffusion. In some instances, such as a body wrapped in plastic coverings, detection can be delayed until the point that elevated gas pressure causes tears in the material or small explosions, sending putrid odor plumes into the environment that facilitate anemotaxic[4] detection by the first wave of insect colonizers.

Similar delays in activation may be associated with some natural habitats. The most obvious are shallow burials that may evoke only a short delay (or none) in carcass recognition by gravid calliphorids. Deeper burials promote extended

Table 11.1 Artificial conditions impeding necrophagous fly detection of a corpse.

Condition	Common examples
Inside a building	In closets, under floorboards, behind a wall, in a chimney, exposed in a basement, in a garage or storage shed, wedge in ventilation duct
Wrapped	Within a blanket, carpet, or plastic sheeting; placed in a sleeping bag or trash bag
Vehicle	In cabin of car or truck, concealed in trunk or storage compartment, located on a boat
Container	Within a sealed trash can, storage container, or toolbox (e.g., in the bed of pickup truck), placed in a closed appliance (refrigerator, freezer, dishwasher, washing machine, or dryer)

Any location that restricts airflow will impede diffusion of volatile gases emitted from a corpse and thereby delay insect detection.

time before fly detection until eventually a depth is reached in which odors no longer escape the soil environment (Rodriguez & Bass, 1985). Gas movement through soil is also dependent on soil type so that loosely packed soil composed of large sand particles permit greater mobility than tightly compact clay particle soils (Forbes & Dadour, 2010). Activation of terrestrial necrophagous insects can also be slowed for a corpse disposed of in an aquatic ecosystem, since if the body of water is deep enough, the remains will sink before resurfacing several days later when gas accumulation induces buoyancy. Odor release from exposed portions of the body leads to detection by some terrestrial species. This scenario is far more common with freshwater decomposition as body recovery from marine ecosystems overtime is extremely rare (Anderson & Hobischak, 2004).

11.3.1.2   Factors that abolish odor detection

Failure to detect a decomposing body in the environment is rare for any animal, not just insects, which relies on carrion for survival. Competition for the food resource is intense and necrophagous insects, particularly those associated with the first waves of colonization, possess highly acute olfaction finely attuned to the odors emitted from animal remains (see Chapter 7 for details of insect olfaction). Thus, scenarios that result in no insect colonization and that indicate failure to detect a corpse imply that the body was located in a habitat that essentially abolished gas release or prevented odor

diffusion. The most likely situations that prevent detection would be locations characterized by anoxic or extreme hypoxic conditions. An anoxic environment is devoid of oxygen but does not necessarily suggest a region that favors anaerobic metabolism; conditions may be unsuitable to both aerobes and anaerobes. Deep burials (>1.8 m or 6 feet) are initially hypoxic since some oxygen is trapped with the body, yet quickly becomes anoxic. Again, soil conditions (type, water content) will influence gas movement, but generally little or no gas escapes from deep within the soil, thereby making detection of a carcass by typical terrestrial colonizers nearly impossible (Figure 11.4). Adult flies from the family Phoridae show a remarkable ability to perceive odors of death from bodies at depths that most other necrophagous insects cannot, but they too have a limit to odor detection based on burial depth.

Bodies found in aquatic ecosystems initially sink to benthic regions, and though several species of aquatic insects may use the corpse, detection by their terrestrial counterparts cannot occur unless the body becomes buoyant and floats. In marine ecosystems, large scavengers typically consume cadavers or the body cavity is pierced before bloating can occur to allow flotation. Under artificial conditions, a body may decompose in such a tightly sealed container that gas cannot escape unless forcibly opened from the outside, or if the materials fail (e.g., crack, break, and tear). Even so, the corpse is likely to be in such an advanced stage of decay that "typical" waves of colonization do not occur. Appliances or containers with sealed gaskets or taped to prevent air exchange, and even some luxury vehicles, can essentially be airtight or prevent nearly all gas exchange with the outside environment, and thereby prevent insect detection and succession.

11.3.2   Physical deterrents to oviposition/larviposition

Conditions that inhibit insect detection of carrion undoubtedly retard or abolish access to the carcass for adult mating, feeding, and oviposition/larviposition. In this regard, an opening or space that permits limited gas exchange with the environment, ultimately leading to activation in necrophagous insects, may be too small for most or any species to gain access to the body. Gasket-sealed appliances/containers, some burials (depth dependent), or any other concealment of the body may fall into this

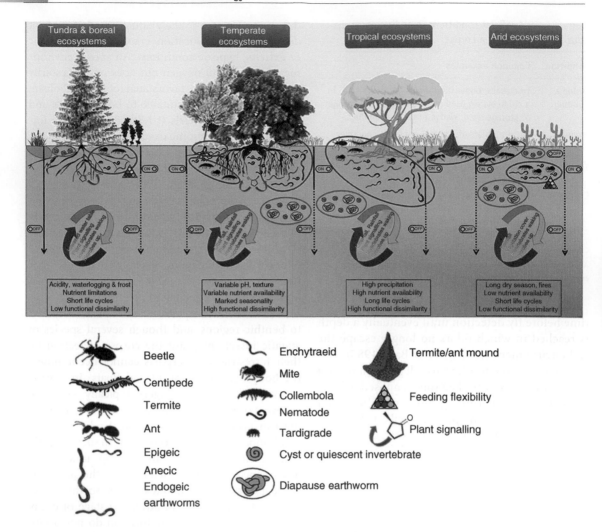

Figure 11.4 Variability in soil composition and biogeographic location influences the fauna and nutrient cycling processes, including insect succession of carrion, associated with specific habitats. *Source*: Briones 2018/Frontiers Media S.A./CC BY 4.0.

category (Gunn & Bird, 2011). Decomposition indoors may also lead to detection by an array of fly species, yet an avenue into the home is unavailable (Pohjoismaki *et al*., 2010; Reibe & Madea, 2010; Anderson, 2011). In such instances, it is common to find numerous adult calliphorids lying dead at points of odor escape, including in closed ductwork or chimneys, a reflection of the insect's desperate attempts to reach the food island. If the body is lying close to a screened door or window, some calliphorid species will oviposit on the screened structure, and following egg hatch neonate larvae display foraging behavior by attempting to migrate toward the corpse. Similar behaviors have been observed with sarcophagid larvae, but only in older stages of development (Christopherson & Gibo, 1996).

At other times, the corpse may be accessible physically, but barriers such as pelage, plumage, clothing, or some type of wrapping disrupt oviposition, thereby restricting egg or larval deposition to natural body openings, exposed skin or, in some cases, on clothing/wrapping materials saturated with body fluids (Erzinçlioglu, 1985; Mann *et al*., 1990; Goff, 1992; Campobasso *et al*., 2001).

A hanged body represents a unique scenario of decomposition in which corpse detection is unaltered but the physical condition of the remains is less conducive to oviposition and subsequent larval development. Hanging leads to much more rapid body mass loss through (i) forced fluid purging due to gravity and (ii) increased evaporative water loss since more corpse surface area is exposed to air than occurs with a cadaver in a **supine** position (lying flat). Flies dependent on moisture and/or tactile stimuli in assessing the resource as an oviposition site deposit reduced clutch sizes or avoid oviposition altogether on a hanged body. A corpse that has been physically or chemically altered is also less attractive to fly oviposition/larviposition, as may occur with a cadaver chemically modified by household cleaning agents, pesticides or petroleum-based products, or one in which the tissue has been charred from intense burning (Anderson, 2005; McIntosh *et al.*, 2017).

11.3.3 Enhancement of detection/oviposition

All the examples described up to this point that deviate from the terrestrial ecological model of insect succession result in a delay in corpse detection and/or oviposition by necrophagous insects. There are times where conditions favor earlier carrion utilization. More specifically, a burnt body may lead to an earlier onset of the activation phase and subsequent acceptance phase, so that oviposition/larviposition occurs sooner than on unburnt cadavers found in the same habitat or location. These statements seemingly contradict our previous discussions that burnt tissue is often rejected as a site for egg deposition (Anderson, 2005). The key to understanding differential fly responses to burned remains is the severity of the tissue damage. Burn damage to human bodies is classified using a subjective 5-level scale referred to as the **Crow–Glassman Scale** (CGS) (Glassman & Crow, 1996). Each level refers to increasing damage to the body that can be assessed by visual inspection (Table 11.2). Early detection of burnt remains and subsequent oviposition by necrophagous flies is associated with bodies displaying burn damage consistent with CGS level 1–2 (Avila & Goff, 1998; Chin *et al.*, 2008), meaning the body has incurred injuries related to smoke death or more severe injury is evident by some tissue charring but the body is otherwise intact (Glassman & Crow, 1996). The attractiveness of such burnt cadavers to calliphorids is attributed to seepage of fluids and gases from the internal environment through cracks and blisters in the skin (Avila & Goff, 1998). By contrast, bodies burned to level CGS 3 or higher are so severely charred that the moisture content of tissues is presumed below a minimum threshold that supports fly oviposition or larval development.

Table 11.2   Crow–Glassman Scale (CGS) for describing extent of burn damage to human bodies.

Level of burn injury	Physical description of body and injuries
CGS 1	Injury due mostly to smoke death; physical appearance similar to non-burn-related death; some blistering of epidermis and singeing of head and facial hair
CGS 2	Body is recognizable but some charring evident; further destruction of body is limited to absence of elements of the feet, hands, genitalia and ears; identification of victim may require aid of forensic odontologist
CGS 3	Major portions of arms and legs missing; the head is present although person's identity is non-recognizable; disarticulation is evident
CGS 4	Extensive burn destruction such that the skull has fragmented and is absent from body; some portions of arms and legs remain articulated to charred body
CGS 5	Destruction is so severe that body has been cremated and little or no tissue remains; the remains are highly fragmented, scattered and incomplete

Source: Adapted from Glassman & Crow (1996).

11.4   The physical conditions of carrion decay can function as a hurdle to insect development

Acceptance of a corpse as an oviposition/larviposition site under any type of artificial or natural conditions increases maternal fitness, regardless of whether progeny deposition has been delayed. However, all animal remains are not equally suited for necrophagous insect development, and depending on the ambient conditions and habitat associated with corpse decomposition, progeny fitness may be compromised, which in turn reduces reproductive success of the mother if her offspring do not reproduce themselves. For instance, any conditions that slow the growth rate of feeding fly larvae potentially increases interspecific and intraspecific competition for carrion resources, elevates the mortality risk[5] associated with predation or parasitism, and potentially increases the chances of injury due to aseasonal or seasonal climatic change. Similarly, microhabitats that accelerate development may evoke stress responses (e.g., heat-shock response) at the expense of normal developmental pathways (Korsloot *et al.*, 2004; Adamo, 2017), thereby disrupting later events such as pupariation, pharate adult development, adult eclosion, or timing of mate finding (Rivers *et al.*, 2010). Still other conditions, often artificial, are hostile or change over time such that the death of all or some of the insect fauna on the remains is the result. Typically, extreme microhabitat temperatures exceeding the critical thermal maxima for a given species are responsible for the induced insect mortality (Rusch *et al.*, 2019, 2020). A wide range of conditions, habitats, and environmental features (amount of sunlight, seasonal change) can function to alter necrophagous insect development on a corpse.

11.4.1   Developmental accelerants

Autolysis, putrefaction, and algor mortis are temperature-dependent processes associated with body decomposition that subsequently influence insect colonization and development on carrion. Insects, as poikilotherms, are also at the mercy of ambient temperatures, most directly those that define the microhabitat of the corpse. Thus, any environment or location that experiences temperature changes will directly influence the growth rate of developing fly larvae on a carcass. Chapter 9 provides specific details of temperature influences on insect development. If conditions favor an elevation in temperature, then corresponding increases in metabolic rate, food consumption, and assimilation, and hence rate of growth, occur in individual larvae and also within larval aggregations. "Artificial" heating of fly larvae can occur inside a vehicle. Heat trapping inside vehicles has essentially the same effect as a greenhouse, as the glass of the windshield and windows allows short wavelengths of visible light to penetrate the interior, with the heat being absorbed by the seats, dashboard, and other objects (Dadour *et al.*, 2011). Longer wavelengths of infrared re-radiated by heated objects cannot pass through glass (Mitchell, 1989), and thus further contribute to heating of the interior air of a vehicle. Even when outside temperatures are cold, temperatures inside the vehicle can elevate to more than 20 °C above ambient, increasing the rate of physical and chemical decomposition of a corpse within the vehicle, as well as accelerating fly larval growth rates up to a thermal maximum. A car trunk or other concealed environment exposed directly to sunlight will have similar interior temperature elevations due to absorption of solar radiation, and the degree of temperature rise is dependent on the composition and color of the encasing material. Elevated temperatures can be induced through direct exposure of cadavers to sunlight in a terrestrial environment: a body will cool more slowly following death in sunlight versus partial or full shade, and internal maggot mass temperatures show a direct correlation with the level of sun exposure (Joy *et al.*, 2006; Majola *et al.*, 2013).

Fly development can be accelerated independent of temperature-mediated influences. In controlled experimental studies in which larvae of the flesh fly *Boettcherisca peregrina* were raised on rabbits injected with various doses of cocaine, sublethal concentrations of the narcotic did not alter the rate of fly development in comparison with control larvae (Goff *et al.*, 1989). However, lethal and twice lethal doses induced more rapid growth in fly larvae until reaching postfeeding stages of larval development; duration of puparial stages was not altered by the presence of cocaine (Goff *et al.*, 1989). Heroin evokes similar increases in development rate of *B. peregrina* and *Lucilia sericata* (Goff *et al.*, 1991; Hedouin *et al.*, 1999), and at least with the sarcophagid also promotes larger body sizes and extended duration of puparial stages (Goff *et al.*, 1991) (Table 11.3). The effect of heroin in the form of morphine on fly development is not simply a stimulation of metabolic rate and subsequent food consumption. Rather, increased larval body sizes beyond an "optimal" weight suggests that the narcotic allows or forces the maggots to ignore a satiation set point to continue feeding. Satiety is the sensory sensation of fullness or suppression of the hunger drive due to acquisition of needed nutrients (Widmaier, 1999). Thus, heroin essentially has the effect of keeping the hunger drive active, conceivably by manipulating neurons that function to either regulate satiety and/or hunger. The significance from a forensic entomology standpoint is that larval development displays aspects independent of temperature, which in turn compromises the use of drug-fed flies in calculation of a postmortem interval. Chapter 18 explores concepts of forensic entomotoxicology in more detail.

11.4.2   Developmental depressants

Any conditions that lower ambient temperatures will create the trickle-down effect of slowing the rate of physical and chemical decomposition of carrion. In turn, this decreases liberation of useable

Table 11.3 Effects of drugs on fly development.

Drug	Method of administration	Insect species	Developmental impact
Cocaine	Injection (postmortem)	*Boettcherisca peregrina* (Diptera: Sarcophagidae)	High doses induced more rapid rate of larval development; duration of puparial stages not affected
Heroin	Injection (postmortem)	*B. peregrina*	Accelerated larval development and increased larval body mass; extended length of puparial stages dependent on dose
	Perfusion (antemortem)	*Lucilia sericata* (Diptera: Calliphoridae)	Accelerated larval development
Methamphetamine	Injection (postmortem)	*Parasarcophaga ruficornis* (Diptera: Sarcophagidae)	High doses induced more rapid rate of larval development; increased puparial mortality; reduced fecundity by second generation
Amitriptyline	Injection (postmortem)	*P. ruficornis* (Diptera: Sarcophagidae)	Extended length of post-feeding wandering stage and increased length of puparial stages

Source: Adapted from Goff *et al.* (1991), Hedouin *et al.* (1999) and Goff & Lord (2010).

nutrients available for consumption by necrophagous insects, which thus negatively impacts growth rate. More directly, low environmental temperatures suppress metabolic rate, food consumption, contractions of the longitudinal and circular muscles regulating gut peristalsis, and assimilation of digested nutrients in individual larvae (Wigglesworth, 1972; Greenberg & Kunich, 2002). Temperatures within the microhabitat of a maggot mass can be suppressed by environmental conditions (Deonier, 1940; Campobasso *et al.*, 2001), evoking slower development for the entire aggregation. The degree of retardation depends on how quickly the temperatures decline and on the absolute low temperatures achieved. Obviously temperature drops are most commonly associated with seasonality which, as discussed in Chapter 9, can be anticipated by necrophagous insects and avoided before the arrival of adverse weather conditions. It is the advent of unexpected (aseasonal) temperature declines that can profoundly alter development of necrophagous flies and beetles. Theoretically, any cold temperatures above the developmental threshold or base temperature[6] of a given species should still promote some development, albeit at a reduced rate, of necrophagous fly larvae (Donovan *et al.*, 2006). Here, again is where the rate at which temperatures decline can be much more significant to disruption of insect development and survivorship than the absolute low temperature reached (Lee, 2010).

Natural temperature declines occur when a body is buried, regardless of whether traditional or felonious[7]; ambient temperatures drop with increasing depth of burial (Forbes & Dadour, 2010). At depths of 1.8 m (6 feet) or more, soil temperature stabilizes to 10–14.4 °C (50–58 °F) during summer months in the United States. Coupled with the inability of maggot masses to form on buried bodies (VanLaerhoven & Anderson, 1999), which in turn means that the benefits of group feeding and high microhabitat temperatures are not realized (see Chapter 8), development of calliphorid larvae may be slowed by days to over a week longer than comparable fly growth rates on bodies decomposing in a terrestrial habitat (Gaudry, 2010; Gunn & Bird, 2011).

Severely burned bodies are less desirable for fly oviposition/larviposition than unburnt ones. If progeny deposition does occur on burnt remains displaying a high degree of charring, larval development may be retarded or halted depending on the nutrient levels of the corpse tissues. Burnt tissues, like fully cooked meat, serves as a poor substrate for fly development, presumably due to a combination of reduced moisture content and a high level of denatured and degraded proteins (Campobasso *et al.*, 2001). The impact of burnt tissues on fly development can occur in at least two scenarios: (i) colonization and subsequent larval development of burnt bodies, and (ii) continued development by flies after the body and carrion insects are exposed to heat and burning (Anderson, 2005). In either situation, fly development on charred remains is expected to be slower than on an unburnt corpse and, though not experimentally tested, is quite likely to be associated with expression of stress responses (e.g., heat-shock response) at the expense of normal developmental pathways. For example, the extreme heat associated with a house fire would at least initially evoke expression of heat-shock proteins (discussed in detail in Chapter 9) in insects feeding on a corpse; extended synthesis of heat-shock proteins inhibits most aspects of normal cellular protein synthesis and depletes metabolic and energy reserves. Insect development would be expected to slow, be delayed, or ultimately lead to the equivalent of heat stupor before the onset of death.

11.4.3 Developmental extremes

Shifts in ambient temperatures near upper or lower thermal limits in terms of the zone of tolerance for growth can retard development. If an individual remains in such conditions for an extended period, a halt in development may ensue, leading to chill-coma or heat stupor. Both conditions are characterized by immobilization of the individual and therefore the inability to move out of potentially damaging conditions, leading to imminent death due to temperatures exceeding critical thermal minima or maxima. Within the context of corpse decomposition, extreme heating may result from fire, the greenhouse effect associated with vehicles, or contained locations placed in direct sunlight. As ambient temperatures elevate, necrophagous insects will respond by attempting to move to cooler conditions and possibly synthesize heat-shock proteins for protection (Rivers et al., 2010). The former is nearly impossible for adult flies that have gained entry to a car via the trunk since they followed an odor plume to find the carcass; no such cues exist to retrace the path to the outside. Fly larvae are thought to reposition themselves repeatedly within a maggot mass to avoid overheating, by crawling to cooler thermal zones as needed (Heaton et al., 2014; Johnson et al., 2014). However, when the ambient temperatures do not provide a respite from the heat, fly larvae become dependent on heat-shock protein production and evaporative cooling. Movement away from the corpse is not really a viable solution for fly larvae for a variety of reasons (threat of desiccation, no avenue for escape, no food outside container/vehicle). Evaporative cooling is only minimally effective within the interior environment of a concealed space like a vehicle or similar container in that air does not circulate and the local humidity climbs as water loss from the corpse proceeds. At high humidity, little or no heat loss occurs via evaporation of water from the insect's body (Willmer et al., 2000).

The lack of air movement also inhibits heat loss by convection. Confined spaces, sealed containers and inside vehicles are dead air spaces subject to rapid temperature elevations. Within a short period of time on a sunny day, the absolute temperature in a vehicle will exceed the critical thermal maximum for a given species of fly. Death will ensue shortly after unless the fly enters heat stupor and the temperatures drop. Internal temperatures will decline as night approaches, but the degree and rate of decrease is dependent on a number of factors including heat conductance of the materials associated with the vehicle or container, season (more rapid in colder climates), amount of sun exposure, and re-radiation of heat from objects located interiorly (i.e., liberation of heat from seats and dashboard of a car) (Dadour et al., 2011). If cooling occurs sufficiently quickly, death may be averted, at least temporarily. Repeated episodes of extreme heating followed by rapid cooling may not induce immediate mortality, but is likely to evoke irreversible damage that is expressed as abnormal puparium formation or inability to eclose. Prolonged exposure to temperatures exceeding the critical thermal maximum (or minimum) will induce immediate death.

A hanged body alters development in fly larvae due to the physical positioning of the corpse (Goff & Lord, 1994). This simply means that as calliphorid and sarcophagid larvae form large maggot masses, thereby accelerating growth rates, the larvae reach a critical size at which they fall from the body due to gravitational forces, unless restrained by clothing or a body cavity of the corpse. Depending on the fly species and age when it unceremoniously drops from the corpse, the fate of the larva may be a quick demise due to starvation, desiccation, or predation. Alternatively, the fly may precociously initiate pupariation if at least an early third instar larva when landing on the ground. The possibility also exists that the immatures resume feeding on pooled body fluids at the ground surface, a situation most likely to occur with clay particle soils or indoors where fluids have collected on non-porous surfaces.

11.5 Insect faunal colonization of animal remains is influenced by conditions of physical decomposition

Deviation from the terrestrial ecological model will result in changes in carrion community structure. Up until this point, a cadaver lying above ground in a terrestrial habitat has served as the model to compare parameters altering detection of the corpse, oviposition/larviposition on animal remains, and developmental constraints for necrophagous insects. The model must be adapted to consider shifts in faunal colonization in that all terrestrial ecosystems are not the same, and consequently do not support the same guild of necrophagous insects. Obviously differences in species composition of carrion communities would be expected when considering global biogeographical differences such as types of biome or, more narrowly, when comparing tropical ecosystems to temperate, alpine or polar ecosystems. Within a given biogeographical area, different habitats (aquatic, terrestrial, or buried) yield entirely different insect fauna utilizing carrion. The same habitat is also able to support species adapted for specific seasons, with little overlap due to quiescence and diapause, both being forms of dormancy used to cope with unfavorable climatic conditions. This section will examine shifts in insect fauna comprising carrion communities due to biogeographical location, habitat type, seasonality, and artificial conditions.

11.5.1 *Biogeographical location*

Insect colonization of carrion can vary considerably based on location. Commonly in forensic entomology literature, necrophagous insect fauna are referenced with regard to global regions, such as the continent or country where the carrion was subjected to insect succession. Providing statements that a blow fly such as *Chrysomya albiceps* is found in parts of Africa or that *Calliphora vicina* is abundant in Central Europe offers only limited utility in understanding the natural distribution or predictability of colonization for a given species.

Information on features such as the climatic conditions and habitats that are preferred or necessary for a species of necrophagous fly reveals much more about the life-history strategies, timing of insect activity on a corpse, and broader distributions. This is necessary since in the examples above neither fly is restricted to a single country or even one continent (Lambiase & Camerini, 2012). Consequently, insect fauna is usually referenced with regard to **biogeoclimatic zones** or **terrestrial ecozones**. A biogeoclimatic zone is defined as a geographic area characterized by a relatively uniform macroclimate with vegetation, soil type (s), moisture content, and zoological life reflective of the climatic conditions. These zones are most often used in reference to vegetation (**climatic climax** or self-perpetuating species) and thus have not been commonly used to account for necrophagous insects, although some successional studies have defined decomposition sites and insect colonizers by biogeoclimatic zones (VanLaerhoven & Anderson, 1999).

Distribution of insect species, whether necrophagous or not, is traditionally aligned with terrestrial ecozones. An ecozone is a broad classification of the Earth's land surfaces based on the distribution patterns of terrestrial organisms. Terrestrial ecozones divide land surfaces based on geographic features such as oceans, high mountain ranges, or large deserts that serve as physical barriers to migration. Consequently, the organisms comprising each ecozone (Figure 11.5) have in theory been isolated so that each displays characteristic adaptive traits necessary for survival in the geographic region comprising a given ecozone. An example is *C. rufifacies*, referenced earlier as being collected from parts of Africa. The fly is actually native to Australia, which places its origin in the Australasia ecozone. As with many common necrophagous flies, this species has expanded its range into several other ecozones (represented by multiple continents and countries) that can support the animals' requirement for warm weather conditions and which experience only modest seasonal change (i.e., weather conditions that typically do not reach subzero temperatures). The calliphorid *P. terraenovae* is at the other end of the spectrum in that it is Holarctic (Palearctic plus Nearctic range) in distribution and adapted to cooler environments than *C. rufifacies*. It, too, has an expanded range so that now the flies can be found in many habitats characterized by seasonally warm conditions.

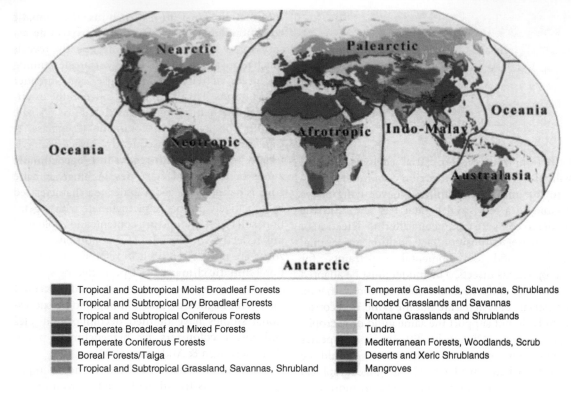

■ Tropical and Subtropical Moist Broadleaf Forests	▨ Temperate Grasslands, Savannas, Shrublands
▨ Tropical and Subtropical Dry Broadleaf Forests	▨ Flooded Grasslands and Savannas
▨ Tropical and Subtropical Coniferous Forests	▨ Montane Grasslands and Shrublands
■ Temperate Broadleaf and Mixed Forests	▨ Tundra
■ Temperate Coniferous Forests	■ Mediterranean Forests, Woodlands, Scrub
▨ Boreal Forests/Taiga	▨ Deserts and Xeric Shrublands
▨ Tropical and Subtropical Grassland, Savannas, Shrubland	■ Mangroves

Figure 11.5   Map of ecozones and biomes. *Source*: Anonymous/Wikimedia Commons/CC BY-SA 2.5.

Within a particular terrestrial ecozone, there is still considerable variation. Take North America for example. The majority of the continent, Canada to parts of Mexico, is classified as Nearctic, yet biogeographic regions comprising this vast area differ in a number of ways: in climate, vegetation, soil types, temperatures, habitats, and thus insect species. The fauna found on the islands of Hawaii are quite different from the rest of the United States, while the species of the Pacific Northwest overlap very little with those of the southeastern regions. Even in a more narrowly defined geographic area in the United States like a region or a state, the biogeoclimate, and hence the zoological profile, can vary substantially across the total area. The state of Maryland in the mid-Atlantic region of the United States is a clear example: the east toward the Atlantic Ocean is dominated by sandy soil with high salt content, which in turn promotes climatic climax vegetation that is salt tolerant; the northern and central regions of the state are composed of clay to silt particle soils with heavily industrialized, urbanized, agricultural, and woody areas; and the extreme western portion of the state is part of the Blue Ridge Mountain range with some areas

designated as alpine. There is clear overlap in calliphorid species inhabiting carrion throughout the state, but geographical isolation is also apparent. The differences must be taken into account in terms of predicting the fauna present on a corpse; it should be obvious that simply referring to any study on "the blow flies of Maryland" is not sufficient to understanding faunal colonization in a particular biogeoclimatic zone within the state (Introna *et al.*, 1991).

The key to understanding the influence of biogeography on insect succession of carrion is that each geographical area supports species adapted to climatic, biotic, and abiotic features of the ecozone. So when a corpse appears in a terrestrial ecozone, it still represents a patchy ephemeral resource subject to intense competition by an array of vertebrate and invertebrate animals, with calliphorids and sarcophagids dominating the first waves of insect colonization. The differences lie in species composition unique to the eight terrestrial ecozones. Lists of necrophagous insects found in each ecozone goes beyond the scope of this book, but such information can be found in a number of excellent resources (see Supplemental reading section for some of those resources).

11.5.2 Habitat type

A given biogeographical area is composed of multiple habitats.[8] As mentioned earlier, aquatic habitats are devoid of true necrophagous insects, so none of the expected terrestrial flies or beetles will or can colonize a body placed in fresh water. This statement makes the assumption that the body of water is sufficiently deep that the fresh corpse sinks to benthic regions. If so, a range of aquatic scavengers will utilize the body as a food source and/or habitat, including larval trichopterans, nymphs of the order Ephemeroptera, and chironomid midges (Merritt & Wallace, 2010). Once internal gas production during putrefaction occurs, the body will become buoyant and thereby the surfaces exposed to air will be subjected to terrestrial insect colonization. Surfaces remaining below the water level are suitable for continued aquatic insect utilization. Similarly, shallow bodies of water such as a creek or stream permit colonization by terrestrial necrophagous insects, while submerged portions of the corpse can be a resource to aquatic invertebrates. Much less is known about insect succession on vertebrate carrion in marine ecosystems since recovery of remains is rare and few insect species reside in such habitats.

Subsurface locations represent a unique habitat that generally excludes most above-surface terrestrial species. As discussed earlier in the chapter, there are some species that can detect odors of decomposition, depending on depth of burial, but most of these insects cannot oviposit on the corpse (Rodriguez & Bass, 1983; VanLaerhoven & Anderson, 1999). Some fly species do appear to specialize on colonization of buried bodies and include some phorids like *Conicera tibialis*, and muscids like *Muscina stabulans* and *Muscina prolapsa* and *Ophyra* spp. (Gunn & Bird, 2011). Often such specialists produce larvae that are adapted to crawl through cracks in the soil and to penetrate tightly sealed containers (i.e., a coffin) to reach the corpse.

Habitat range is defined for some species by filling a **niche** in urban versus rural locations, although overlap frequently occurs, particularly in locations where urban sprawl reaches far into the countryside or, alternatively, where metropolitan areas maintain large green spaces in the form of parks or **riparian buffers**. More specialized within either urban or rural locales are necrophagous flies that demonstrate preferences for animal remains in full sun versus shade. As an example, *Calliphora vomitoria* reportedly prefers large carrion in shade while *C. vicina* and *L. sericata* are attracted to remains in full sun (Joy *et al.*, 2006; Anderson, 2020). As these conditions are not absolute within a given habitat, fly oviposition behaviors can be altered by a range of factors including temperature, altitude, and interspecific and intraspecific competition (Lambíase & Camerini, 2012). The reality is that gravid females desperate for a resource will "violate" habitat preferences if carrion is limited, oviposition is imminent, and a body is detected.

11.5.3 Seasonality

Biogeoclimatic zones experience seasonal change depending on the biogeographic region. Within temperate zones of North American and other regions, predictable climatic change manifests as four seasons, with only one, summer, truly conducive to carrion insect activity. Spring and fall in most areas of the United States display reduced calliphorid activity in terms of species diversity and abundance in comparison with warmer summer months. Winter is nearly devoid of all necrophagous insect activity throughout North America north of the Mexican border (Watson & Carlton, 2003, 2005; Sharanowski *et al.*, 2008), with the exception of some locations in the southeast (e.g., South Carolina and Florida) and southwest (Arizona and California). Some flies (e.g., *Cynomya cadaverine* and *C. vicina*) flourish in the cooler conditions of late winter, spring and fall, but are reportedly not adapted for adult activity during hot ambient temperatures of summer and thus enter quiescence, a state of dormancy characterized by reduced metabolic activity but which can quickly be terminated with the advent of favorable conditions. Sarcophagids are generally absent during these same periods as many species in North America overwinter in a state of pupal diapause. Diapause is also a state of dormancy but is not transient like quiescence. Rather, it is the dynamic phenotype of a genetic program that is initiated and terminated by a series of environmental tokens or signals. Necrophagous species that use diapause to avoid harsh weather conditions of a particular season are essentially "locked" into the genetic program until the appropriate cues signal the changing of seasons (Denlinger, 2002). This ensures that individuals do not prematurely terminate diapause as the result of temporary aseasonal weather patterns, only to be exposed to potentially lethal conditions once typical seasonal temperatures return. Entry into diapause does not guarantee survival during periods of harsh weather, but it is a strategy that helps populations avoid localized extinction.

The key to success for diapause is for an insect species to respond to environmental signals that forecast impending seasonal change, and thus to prepare physiologically and/or morphologically prior to the onset of severe weather. For sarcophagids and calliphorids that rely on pupal, larval, or adult diapause, this is precisely what occurs. The result is that most species that show peak adult activity in summer initiate diapause prior to the end of the season (i.e., before fall). Consequently, neither adults nor developing larvae of these species will be found on carrion during autumn months, and they will remain in diapause typically until late spring. The exception to this rule is the adult diapause of many calliphorids. In such instances, diapause is in reference to reproduction, more specifically to gonads, but the adult is actually in a state of quiescence. Thus, if carrion appears during fall through spring, and temperatures are warm enough to snap the insect back to life (activity), adult flies may enter the activation phase to locate and feed on the corpse. Oviposition will not occur because it cannot since the ovaries or testes are dormant. Thus some species of calliphorids may be observed unexpectedly on a corpse during winter, but their activity is restricted to feeding before returning to a state of quiescence.

Another factor influencing seasonal patterns of succession is the availability of carrion. The occurrence of potential food resources diminishes during unfavorable seasonal conditions such as during cold weather or rainy seasons. Small vertebrate animals in particular are likely to enter a period of hibernation to avoid exposure to harsh weather. It thus makes sense that necrophagous insects synchronize their dormancy with that of vertebrate animals, a feature reminiscent of parasitoid diapause in step with that of their insect hosts. In fact, the seasonal occurrence of many necrophagous sarcophagids may be aligned with the life-history features of small mammals since it appears these flies prefer small carcasses over large (Denno & Cothran, 1976). Likewise, differences in diapause strategies between sarcophagids and calliphorids seem to reflect carcass preferences:

most sarcophagids that enter dormancy in North America rely on pupal or larval diapause meaning they are fixed in hibernation until spring, but several calliphorids that utilize large animal remains depend on adult diapause. Remember that the latter results in dormant gonads but quiescent adults that can become active in response to an aseasonally warm day.

11.5.4   Artificial conditions

Any location that restricts access without abolishing all insect colonization has the potential to influence the species composition of the forming carrion community. This is perhaps most evident with artificial structures like inside a building or home, or confined spaces and vehicles, in which a select group of Diptera reach the remains to successfully colonize it. The conditions obviously overlap with our earlier discussion of factors that restrict or inhibit oviposition/larviposition. Here, the assumption is that some necrophagous species do successfully locate and utilize the corpse, but the species richness is generally less than when the body is fully accessible to a range of invertebrates. For example, the size of the access route, such as holes in window screening, cracks in a car trunk, or openings to ventilation, generally dictate which species gain entry. Studies examining succession under such artificial conditions have consistently revealed that Coleoptera are excluded and a range of smaller Diptera are most likely to colonize the body. Interestingly, size alone cannot be the only factor governing access because in several instances medium- to large-bodied adult sarcophagids have been witnessed to be the dominant or only fly colonizing bodies located indoors. Perhaps enhanced chemical acuity and/or searching behavior are also required for these species.

When species composition is altered on a carcass decomposing indoors, within a vehicle, or some other restricted access location, subsequent decomposition events are affected including the chemical signals emitted. Presumably this leads to modification of the length of decomposition (often extended indoors), duration of association between necrophagous insects and remains, and species that arrive in later stages of decay (Pohjoismaki et al., 2010; Anderson, 2011).

Chapter review

What's normal about terrestrial decomposition? Typical patterns of insect succession on bodies above ground

- When a vertebrate animal dies in a terrestrial habitat, the body will immediately draw the attention of specific species of necrophagous calliphorids that constitute the first wave of colonization. As the physical and chemical decomposition of the corpse progresses, successive waves of arthropod colonization, dominated by insects, will proceed in a sequential and fairly predictable manner. Much of what is understood in terms of carrion insect succession is derived from studies using pig carrion as surrogates to human decomposition, placed in terrestrial habitats, above ground, and during ambient conditions characterized by warm temperatures in full sunlight. Terrestrial decomposition studies are also responsible for providing the bulk of what is known about the formation of carrion communities; their structure and composition; faunal patterns of colonization in terms of species, timing, and oviposition preferences; and trophic interactions occurring within and across phyla.
- The processes of body decomposition occur along a continuum that is subject to considerable variation based on ambient conditions and a whole array of other factors. In terms of insect succession, certain insects (e.g., calliphorids and sarcophagids) are predicted to arrive on a corpse or oviposit during preferred periods of decomposition, so that insect faunal succession is continuous, meaning that any given species may utilize the cadaver during multiple stages of decomposition, and successive waves of colonizers can overlap during their association with carrion.
- A period of time exists in which a dead animal has yet to be detected, a stage referred to as the exposure phase, which is expected to be of short duration in warm terrestrial habitats that facilitate rapid tissue decay and a high degree of chemical volatility. Once chemical signals derived from either the remains or necrophiles are perceived through olfaction, detection occurs, and the necrophagous insects enter the searching phase. Physical contact with the remains leads to

probing, tasting, and other forms of resource quality or suitability assessment by the first colonizers, eventually culminating in acceptance of the carrion as food and/or as an oviposition site. Egg hatch or larviposition initiates the consumption phase. It is during the consumption phase that large feeding aggregations or maggot masses form. Completion of feeding by calliphorids and sarcophagids occurs during the third stage of larval development and is followed by wandering or dispersal from the carcass into a protective environment for pupariation.
- Several waves of insect succession occur on a corpse depending on the ambient conditions of decomposition and a multitude of other factors. Among these other influences are the actual insect colonizers: as the initial colonizers feed and utilize the corpse, their activity alters the nutritional value of the remains. This in turn modifies the chemical signals released from the cadaver as well as the insects that perceive the cues and then are motivated to search for the resource. Subsequent waves of insect succession are dependent on the preceding colonizers, which are dependent on the initial decay environment.

Succession patterns under forensic conditions are not typical

- Insect succession on large vertebrate carcasses located in above-ground terrestrial environments occurs in a relatively predictable sequence. Such predictability of insect succession is used as a method to estimate periods of insect activity following death in medicocriminal investigations. Relatively modest adjustment to this model, such as switching a large carcass with a small mammal, results in altered community structure.
- Factors such as carcass size, habitat, seasonality, climate, and habitat [e.g., terrestrial, aquatic, and soil (buried)] are examples of natural influences affecting the rate of carrion decay as well as patterns of insect succession. Each of these can vary considerably as to the degree of impact on insect colonization and subsequent development.
- Adding to the complexity of insect succession under natural conditions is the physical structure of the animal. Obviously size matters but so too does the shape (types and location of body openings) and natural coverings of the body.

For example, an animal covered by fur or feathers has a protective barrier that retards or entirely inhibits adult feeding and/or oviposition/larviposition by necrophagous flies. In humans, such natural barriers are replaced by clothing.

- Physical decay of a corpse and associated insect succession can be altered by artificial conditions such as when the remains are placed in a concealed location or unique habitat, or the corpse is modified in an attempt to alter typical patterns of physical decay or insect colonization. Similarly, decomposition following an accident or suicide may occur in a concealed location such as a vehicle or indoors and/or involve intoxication of tissues with substances that impact necrophagous insect activity.

Several factors serve as barriers to oviposition by necrophagous insects

- Any location, artificial or natural, that restricts air movement poses a challenge to necrophagous insects to detect a corpse. The net effect is a lengthening of the exposure phase and a delay in the onset of the activation phase or detection of the body. Foraging behavior cannot begin until activation, so naturally this aspect of fly activity is set back as well.
- Artificial scenarios such as body decomposition in vehicles, indoors (behind walls, under floorboards, in closets, basements, and storage sheds), wrapped (in blankets, carpet, or sleeping bags), sealed in trash cans, placed in closed appliances, stuffed in bags (garment or trash), and even concealed in a toolbox all share the commonality of reduced air movement, thereby trapping odors or permitting emission only through small openings by the slow process of diffusion.
- Conditions that inhibit insect detection of carrion undoubtedly retard or abolish access to the carcass for adult mating, feeding, and oviposition/larviposition. In this regard, an opening or space that permits limited gas exchange with the environment, ultimately leading to activation in necrophagous insects, may be too small for most or any species to gain access to the body. Gasket-sealed appliances/containers, some burials (depth dependent), or any other concealment of the body including inside a building may fall into this category.
- A hanged body represents a unique scenario of decomposition in which corpse detection is

unaltered but the physical condition of the remains is less conducive to oviposition and subsequent larval development.

- There are times where conditions favor earlier carrion utilization than expected from the terrestrial ecological model. A burnt body may lead to an earlier onset of the activation phase and subsequent acceptance phase, meaning oviposition/larviposition occurs sooner than on unburnt cadavers found in the same habitat or location.

The physical conditions of carrion decay can function as a hurdle to insect development

- All animal remains are not equally suited for necrophagous insect development, and depending on the ambient conditions and habitat associated with corpse decomposition, progeny fitness may be compromised, which in turn reduces reproductive success of the mother if her offspring do not reproduce themselves. For instance, any conditions that slow or accelerate the growth rate of feeding fly larvae potentially disrupt normal development. In turn, conditions under which corpse decomposition is occurring may change such that the microhabitat reflects an extreme development environment for any type of necrophagous insect. Obviously such conditions reduce the utility of necrophagous insects in estimations of a postmortem interval.
- Insects, as poikilotherms, are at the mercy of ambient temperatures, most directly those that define the microhabitat of the corpse. Thus, any environment or location that experiences temperature changes will directly influence the growth rate of developing fly larvae on a carcass. If conditions favor an elevation in temperature, then corresponding increases in metabolic rate, food consumption and assimilation, and hence rate of growth, occur in individual larvae and also within larval aggregations. Any conditions that lower ambient temperatures will create the trickle-down effect of slowing the rate of physical and chemical decomposition of carrion. In turn, this decreases liberation of useable nutrients available for consumption by necrophagous insects, which thus negatively impacts growth rate. More directly, low environmental temperatures suppress metabolic rate, food consumption, contractions of the longitudinal and circular muscles regulating gut

peristalsis, and assimilation of digested nutrients in individual larvae.

- Severely burned bodies are less desirable for fly oviposition/larviposition than unburnt. If progeny deposition does occur on burnt remains displaying a high degree of charring, larval development may be retarded or halted depending on the nutrient levels of the corpse tissues. Burnt tissues, like fully cooked meat, serves as a poor substrate for fly development, presumably due to a combination of reduced moisture content and high level of denatured and degraded proteins.

- Shifts in ambient temperatures near upper or lower thermal limits in terms of the zone of tolerance for growth can retard development. If an individual remains in such conditions for an extended period of time, a halt in development may ensue, leading to chill-coma or heat stupor. Both conditions are characterized by immobilization of the individual and therefore the inability to move out of harms way, leading to imminent death due to temperatures exceeding critical thermal minima or maxima.

- A hanged body alters development in fly larvae due to the physical positioning of the corpse. What this simply means is that as calliphorid and sarcophagid larvae form large maggot masses, thereby accelerating growth rates, the larvae reach a critical size at which they fall from the body due to gravitational forces, unless restrained by clothing of the corpse.

Insect faunal colonization of animal remains is influenced by conditions of physical decomposition

- A cadaver lying above ground in a terrestrial habitat serves as the model for comparing parameters that alter detection of the corpse, oviposition/larviposition on animal remains, and developmental constraints for necrophagous insects. The model must be adapted to consider shifts in faunal colonization in that all terrestrial ecosystems are not the same, and consequently do not support the same guild of necrophagous insects. Differences in species composition of carrion communities occur when considering global biogeographical differences such as types of biomes or, more narrowly, when comparing tropical ecosystems with temperate, alpine or polar ecosystems. Within a given biogeographical area, different habitats yield entirely different insect fauna utilizing carrion. The same habitat is also able to support species adapted for specific seasons, with little overlap due to quiescence and diapause.

- Distribution of insect species, whether necrophagous or not, is traditionally aligned with terrestrial ecozones. Terrestrial ecozones divide land surfaces based on geographic features such as oceans, high mountain ranges, or large deserts that serve as physical barriers to migration. Consequently, the organisms comprising each ecozone have in theory been isolated by evolution so that each displays characteristic adaptive traits necessary for survival in the geographic region comprising a given ecozone.

- A given biogeographical area is composed of multiple habitats. Habitat range is defined for some species by filling a niche, utilizing or modifying the ecosystem resources in ways specific to a given species or population.

- Biogeoclimatic zones experience seasonal change depending on the biogeographic region. Within temperate zones of North American and other regions, predictable climatic change manifests as four seasons, with only one, summer, truly conducive to carrion insect activity. Spring and fall in most areas of the United States display reduced calliphorid activity in terms of species diversity and abundance in comparison with warmer summer months. Winter is nearly devoid of all necrophagous insect activity throughout North America north of the Mexican border.

- Any location that restricts access without abolishing all insect colonization has the potential to influence the species composition of the forming carrion community. This is perhaps most evident with artificial structures like inside a building or home, confined spaces and vehicles, in which a select group of Diptera reach the remains to successfully colonize it.

Test your understanding

Level 1: knowledge/comprehension

1. Define the following terms:

   (a) climatic climax
   (b) quiescence
   (c) niche
   (d) seasonality
   (e) biogeoclimatic zone
   (f) Crow–Glassman Scale.

2. Match the terms (i–vi) with the descriptions (a–f).

(a) Lowest temperature that insect development can still occur

(i) Anemotaxis

(b) Sensory sensation of fullness

(ii) Terrestrial ecozone

(c) Grouping of insects using the same resource

(iii) Developmental threshold

(d) True hibernation in insects

(iv) Guild

(e) Body orientations in response to wind currents

(v) Satiety

(f) Classification of land surfaces based on distribution of organisms

(vi) Diapause

3. Describe types of natural and artificial conditions that inhibit activation in necrophagous flies.
4. How might the conditions in question 3 be changed to allow detection but prevent oviposition?
5. Describe the impact of a shallow burial on the successional patterns of calliphorids or sarcophagids in comparison with a body exposed above ground in a terrestrial habitat during warm weather conditions.
6. Explain the importance of Brownian motion and diffusion of gas particles to the detection phase of corpse discovery.

Level 2: application/analysis

1. Explain the seemingly paradoxical situation in which burnt remains depress or totally inhibit calliphorid development, yet at times promote early onset of the activation phase.
2. Discuss how ambient temperatures inside an automobile can lead to both chill-coma and heat stupor.
3. Describe the influence of wrapping a body in a carpet or enclosing in a sleeping bag on detection and acceptance of a corpse by adult calliphorids.

Level 3: synthesis/evaluation

1. Explain using entomological evidence how a forensic entomologist would be able to conclude

that a body had been moved from a terrestrial location during the late fresh stage/early bloat to a deep freshwater pond.
2. Discuss how it would be possible for multiple guilds of insects to exist on a corpse decomposing in a supine position above ground in pasture during warm summer conditions.

Notes

1. From Greenberg (1991).
2. Mégnin (1894) identified eight waves (he termed death squads) of insect succession on human cadavers when decomposing above ground in a terrestrial setting.
3. Brownian motion refers to the random drifting of particles distributed in a liquid or gas depending on thermal energy.
4. Anemotaxis is an innate behavior in insects in which they travel toward or away from air currents in response to a stimulus. Foraging behavior in necrophagous flies relies on a positive (toward stimulus) anemotaxic response to decomposition odors (stimulus).
5. Mortality risk for fly larvae from attack by predators or parasites is expected to increase due to lengthened exposure in the environment (Cianci & Sheldon, 1990).
6. Developmental threshold or base temperature represents the lowest temperature at which insect development can occur.
7. Human remains are traditionally buried embalmed and placed in a receptacle (coffin) 1.8 m (6 feet) underground, whereas a felonious burial refers to illegal placement of a body in the ground, usually in a shallow grave, as the felon is trying to hide the corpse without being discovered.
8. A habitat is the natural environment for particular organisms or ecological community.

References cited

Adamo, S.A. (2017) Stress responses sculpt the insect immune system, optimizing defense in an ever-changing world. *Developmental & Comparative Immunology* 66: 24–32.

Anderson, G.S. (2005) Effects of arson on forensic entomology evidence. *Canadian Society of Forensic Science Journal* 38: 49–67.

Anderson, G.S. (2011) Comparison of decomposition rates and faunal colonization of carrion in indoor and outdoor environments. *Journal of Forensic Sciences* 56: 136–142.

Anderson, G.S. (2020) Factors that influence insect succession on carrion. In: J.H. Byrd & J.K. Tomberlin (eds) *Forensic Entomology: The Utility of Using Arthropods in Legal Investigations*, 2nd edn, pp. 103–139. Taylor & Francis, Boca Raton, FL.

Anderson, G.S. & Hobischak, N.R. (2004) Decomposition of carrion in the marine environment in British Columbia, Canada. *International Journal of Legal Medicine* 118: 206–209.

Ashworth, J.R. & Wall, R. (1994) Responses of the sheep blowflies *Lucilia sericata* and *L. cuprina* to odour and the development of semiochemical baits. *Medical and Veterinary Entomology* 8: 303–309.

Avila, F.W. & Goff, M.L. (1998) Arthropod succession patterns onto burnt carrion on two contrasting habitats in the Hawaiian islands. *Journal of Forensic Sciences* 43: 581–586.

Briones M.J.I. (2018) The serendipitous value of soil fauna in ecosystem functioning: The unexplained explained. *Frontiers in Environmental Science* 6: 149.

Cammack, J.A., Adler, P.H., Tomberlin, J.K., Arai, Y. & Bridges, W.C. Jr (2010) Influence of parasitism and soil compaction on pupation of the green bottle fly, *Lucilia sericata*. *Entomologia Experimentalis et Applicata* 136: 134–141.

Campobasso, C.P., Di Vella, G. & Introna, F. (2001) Factors affecting decomposition and Diptera colonization. *Forensic Science International* 120: 18–27.

Carter, D.O., Yellowless, D. & Tibbett, M. (2007) Cadaver decomposition in terrestrial ecosystems. *Naturwissenschaften* 94: 12–24.

Catts, E.P. & Goff, M.L. (1992) Forensic entomology in criminal investigations. *Annual Review of Entomology* 37: 253–272.

Charabidze, D., Bourel, B. & Gosset, D. (2012) Larval-mass effect: characterisation of heat emission by necrophagous blowflies (Diptera: Calliphoridae) larval aggregates. *Forensic Science International* 211: 61–66.

Chin, H.C., Marwi, M.A., Salleh, A.F.M., Jeffery, J., Kurahashi, H. & Omar, B. (2008) Study of insect succession and rate of decomposition on a partially burned pig carcass in an oil palm plantation in Malaysia. *Tropical Biomedicine* 25(3): 202–208.

Christopherson, C. & Gibo, D.L. (1996) Foraging by food deprived larvae of *Neobellieria bullata* (Diptera: Sarcophagidae). *Journal of Forensic Science* 42: 71–73.

Cianci, T.J. & Sheldon, J.K. (1990) Endothermic generation by blowfly larvae *Phormia regina* developing in pig carcasses. *Bulletin of the Society of Vector Ecology* 15: 33–40.

Dadour, I.R., Almanjahie, I., Fowkes, N.D., Keady, G. & Vijayan, K. (2011) Temperature variations in a parked vehicle. *Forensic Science International* 207: 205–211.

Denlinger, D.L. (2002) Regulation of diapause. *Annual Review of Entomology* 47: 93–122.

Denno, R.F. & Cothran, W.R. (1976) Competitive interactions and ecological strategies of sarcophagid and calliphorid flies inhabiting rabbit carrion. *Annals of the Entomological Society of America* 69: 109–113.

Deonier, C.C. (1940) Carcass temperatures and their relation to winter blowfly populations and activity in the southwest. *Journal of Economic Entomology* 33: 166–170.

Donovan, S.E., Hall, M.J.R., Turner, B.D. & Moncrieff, C.B. (2006) Larval growth rates of the blowfly, *Calliphora vicina*, over a range of temperatures. *Medical and Veterinary Entomology* 20: 106–114.

Erzinçlioglu, Y.Z. (1985) The entomological investigation of a concealed corpse. *Medicine, Science and the Law* 25: 228–230.

Forbes, S.L. & Dadour, I. (2010) The soil environment and forensic entomology. In: J.H. Byrd & J.L. Castner (eds) *Forensic Entomology: The Utility of Arthropods in Legal Investigations*, 2nd edn, pp. 407–426. CRC Press, Boca Raton, FL.

Gaudry, E. (2010) The insects colonization of buried remains. In: J. Amendt, C.P. Campobasso, M.L. Goff & M. Grassberger (eds) *Current Concepts in Forensic Entomology*, pp. 273–312. Springer, London.

Glassman, D.M. & Crow, R.M. (1996) Standardization model for describing the extent of burn injury to human remains. *Journal of Forensic Sciences* 41: 152–154.

Goff, M.L. (1992) Problems in estimation of postmortem interval resulting from wrapping of the corpse: a case study from Hawaii. *Journal of Agricultural Entomology* 9: 237–243.

Goff, M.L. (2010) Early postmortem changes and stages of decomposition. In: J. Amendt, C.P. Campobasso, M.L. Goff & M. Grassberger (eds) *Current Concepts in Forensic Entomology*, pp. 1–24. Springer, London.

Goff, M.L. & Lord, W.D. (1994) Entomotoxicology: a new era for forensic investigation. *American Journal of Forensic Medicine and Pathology* 8: 45–50.

Goff, M.L. & Lord, W.D. (2010) Entomotoxicology: insects as toxicological indicators and the impact of drugs and toxins on insect development. In: J.H. Byrd & J.L. Castner (eds) *Forensic Entomology: The Utility of Arthropods in Legal Investigations*, 2nd edn, pp. 427–36. CRC Press, Boca Raton, FL.

Goff, M.L., Omori, A.I. & Goodbrod, J.R. (1989) Effect of cocaine in tissues on the rate of development of *Boettcherisca peregrina* (Diptera: Sarcophagidae). *Journal of Medical Entomology* 26: 91–93.

Goff, M.L., Brown, W.A., Hewadikaram, K.A. & Omori, A.I. (1991) Effects of heroin in decomposing tissues on the development rate of *Boettcherisca peregrina* (Diptera: Sarcophagidae) and implications of this effect on estimations of postmortem intervals using arthropod development patterns. *Journal of Forensic Sciences* 36: 537–542.

Greenberg, B. (1990) Behavior of postfeeding larvae of some Calliphoridae and a muscid (Diptera). *Annals of the Entomological Society of America* 83: 1210–1214.

Greenberg, B. (1991) Flies as forensic indicators. *Journal of Medical Entomology* 28: 565–577.

Greenberg, B. & Kunich, J.C. (2002) *Entomology and the Law*. Cambridge University Press, Cambridge, UK.

Gunn, A. & Bird, J. (2011) The ability of the blowflies *Calliphora vomitoria* (Linnaeus), *Calliphora vicina* (Rob-Desvoidy) and *Lucilia sericata* (Meigen) and the muscid flies *Muscina stabulans* (Fallen) and *Muscina prolapsa* (Harris) (Diptera: Muscidae) to colonise buried remains. *Forensic Science International* 207: 198–204.

Hanski, I. (1987) Carrion fly community dynamics: patchiness, seasonality and coexistence. *Ecological Entomology* 12: 257–266.

Heaton, V., Moffatt, C. & Simmons, T. (2014) Quantifying the temperature of maggot masses and its relationship to decomposition. *Journal of Forensic Sciences* 59(3): 676–682.

Hedouin, V., Bourel, B., Martin-Bouyer, L., Becart, A., Tournel, G., Deveaux, M. & Gossett, D. (1999) Morphine perfused rabbits: a tool for experiments in forensic entomology. *Journal of Forensic Sciences* 44: 347–350.

Introna, F. Jr, Suman, T.W. & Smialek, J.E. (1991) Sarcosaprophagous fly activity in Maryland. *Journal of Forensic Sciences* 36: 238–243.

Johnson, M.D. (1975) Seasonal and microseral variations in the pest populations on carrion. *American Midland Naturalist* 93: 79–90.

Johnson, A.P., Wighton, S.J. & Wallman, J.F. (2014) Tracking movement and temperature selection of larvae of two forensically important blow fly species within a "maggot mass". *Journal of Forensic Sciences* 59(6): 1586–1591.

Joy, J.E., Liette, N.L. & Harrah, H.L. (2006) Carrion fly (Diptera: Calliphoridae) larval colonization of sunlit and shaded pig carcasses. *Forensic Science International* 164: 183–192.

Kneidel, K.A. (1984) Influence of carcass taxon and size on species composition of carrion-breeding Diptera. *American Midland Naturalist* 111: 57–63.

Korsloot, A., van Gestel, C.A.M. & van Straalen, N.M. (2004) *Environmental Stress and Cellular Response in Arthropods*. CRC Press, Boca Raton, FL.

Lambiase, S. & Camerini, G. (2012) Spread and habitat selection of *Chrysomya albiceps* (Wiedemann) (Diptera Calliphoridae) in Northern Italy: forensic implications. *Journal of Forensic Sciences* 57: 799–801.

Lee, R.E. Jr (2010) A primer on insect cold-tolerance. In: D.L. Denlinger & R.E. Lee Jr (eds) *Low Temperature Biology of Insects*, pp. 3–34. Cambridge University Press, Cambridge, UK.

Lefebvre, F. & Gaudry, E. (2009) Forensic entomology: a new hypothesis for the chronological succession pattern of necrophagous insect on human corpses. *Annales de la Societe Entomologique* 45: 377–392.

Majola, T., Kelly, J. & Linde, T.V.D. (2013) A preliminary study on the influence of direct sunlight and shade on carcasses' decomposition and arthropod succession. *Canadian Society of Forensic Science Journal* 46(2): 93–102.

Malainey, S.L. & Anderson, G.S. (2020) Impact of confinement in vehicle trunks on decomposition and entomological colonization of carcasses. *PLoS one* 15(4): e0231207.

Mann, R.W., Bass, W.M. & Meadows, L. (1990) Time since death and decomposition of the human body: variables and observations in case and experimental field studies. *Journal of Forensic Sciences* 35: 103–111.

Matuszewski, S., Bajerlein, D., Konwerski, S. & Szpila, K. (2011) Insect succession and carrion decomposition in selected forests of Central Europe. Part 3: succession of carrion fauna. *Forensic Science International* 207(1–3): 150–163.

McIntosh, C.S., Dadour, I.R. & Voss, S.C. (2017) A comparison of carcass decomposition and associated insect succession onto burnt and unburnt pig carcasses. *International Journal of Legal Medicine* 131(3): 835–845.

Mégnin, P. (1894) *La Fauna de Cadavers. Application de l'Entomologie a la Medicine Legale*. Encyclopdie scientifique des Aides-Memoire. G. Masson and Gauthier-Villars, Paris.

Merritt, R.W. & Wallace, J.R. (2010) The role of aquatic insects in forensic investigations. In: J.H. Byrd & J.L. Castner (eds) *Forensic Entomology: The Utility of Arthropods in Legal Investigations*, 2nd edn, pp. 271–320. CRC Press, Boca Raton, FL.

Mitchell, J.F.B. (1989) The greenhouse effect and climate change. *Review of Geophysics* 27: 115–139.

Norris, K.R. (1965) The bionomics of blowflies. *Annual Review of Entomology* 10: 47–68.

Payne, J.A. (1965) A summer carrion study of the baby pig *Sus scrofa* Linnaeus. *Ecology* 46: 592–602.

Pohjoismaki, J.L.O., Karhunen, P.J., Goebeler, S., Saukko, P. & Saaksjarvi, I.E. (2010) Indoors forensic entomology: colonization of human remains in closed environments by specific species of sarcosaprophagous flies. *Forensic Science International* 199: 38–42.

Reed, H.B. (1958) A study of dog carcass communities in Tennessee, with special reference to the insects. *American Midland Naturalist* 59: 213–245.

Reibe, S. & Madea, B. (2010) How promptly do blowflies colonise fresh carcasses? A study comparing indoor with outdoor locations. *Forensic Science International* 195: 52–57.

Rivers, D.B., Ciarlo, T., Spelman, M. & Brogan, R. (2010) Changes in development and heat shock response in two species of flies (*Sarcophaga bullata* [Diptera: Sarcophagidae] and *Protophormia terraenovae* [Diptera: Calliphoridae]) reared in different sized maggot masses. *Journal of Medical Entomology* 47: 677–689.

Rivers, D.B., Thompson, C. & Brogan, R. (2011) Physiological trade-offs of forming maggot masses by necrophagous flies on vertebrate carrion. *Bulletin of Entomological Research* 101: 599–611.

Rodriguez, W.C. & Bass, W.M. (1983) Insect activity and its relationship to decay rates of human cadavers in East Tennessee. *Journal of Forensic Sciences* 28: 423–432.

Rodriguez, W.C. & Bass, W.M. (1985) Decomposition of buried bodies and methods that may aid in their location. *Journal of Forensic Sciences* 30: 836–852.

Rusch, T.W., Adutwumwaah, A., Beebe, L.E., Tomberlin, J.K. & Tarone, A.M. (2019) The upper thermal tolerance of the secondary screwworm, *Cochliomyia macellaria* Fabricius (Diptera: Calliphoridae). *Journal of Thermal Biology* 85: 102405.

Rusch, T.W., Faris, A.M., Beebe, L.E., Tomberlin, J.K. & Tarone, A.M. (2020) The upper thermal tolerance for a Texas population of the hairy maggot blow fly *Chrysomya rufifacies* Macquart (Diptera: Calliphoridae). *Ecological Entomology* 45(5): 1146–1157.

Schoenly, K. & Reid, W. (1987) Dynamics of heterotrophic succession in carrion arthropod assemblages: discrete series or a continuum of change. *Oecologia* 73: 192–202.

Scholl, K. & Moffatt, C. (2017) Plastic waste sacks alter the rate of decomposition of dismembered bodies within. *International Journal of Legal Medicine* 131(4): 1141–1147.

Sharanowski, B.J., Walker, E.G. & Anderson, G.S. (2008) Insect succession and decomposition patterns in shaded and sunlit carrion in Saskatchewan in three different seasons. *Forensic Science International* 179: 219–240.

Smith, K.G.V. (1986) *A Manual of Forensic Entomology*. British Museum (Natural History), London.

Tomberlin, J.K., Mohr, R., Benbow, M.E., Tarone, A.M. & VanLaerhoven, S. (2011) A roadmap bridging basic and applied research in forensic entomology. *Annual Review of Entomology* 56: 401–422.

Vanin, S., Zanotti, E., Gibelli, D., Taborelli, A., Andreola, S. & Cattaneo, C. (2013) Decomposition and entomological colonization of charred bodies – a pilot study. *Croatian Medical Journal* 54(4): 387–393.

VanLaerhoven, S.L. & Anderson, G.S. (1999) Insect succession on buried carrion in two biogeoclimatic zones of British Columbia. *Journal of Forensic Sciences* 44: 32–43.

Vass, A.A., Bass, W.M., Wolt, J.D., Foss, J.E. & Ammons, J.T. (1992) Time since death determinations of human cadavers using soil solution. *Journal of Forensic Sciences* 37: 1236–1253.

Voss, S.C., Spafford, H. & Dadour, I.R. (2009) Hymenopteran parasitoids of forensic importance: host associations, seasonality and prevalence of parasitoids of carrion flies in Western Australia. *Journal of Medical Entomology* 46: 1210–1219.

Watson, E.J. & Carlton, C.E. (2003) Spring succession of necrophilous insects on wildlife carcasses in Louisiana. *Journal of Medical Entomology* 40: 338–347.

Watson, E.J. & Carlton, C.E. (2005) Insect succession and decomposition of wildlife carcasses during Fall and Winter in Louisiana. *Journal of Medical Entomology* 42: 193–203.

Widmaier, E.P. (1999) *Why Geese Don't Get Obese (and We Do): How Evolution's Strategies for Survival Affect Our Everyday Lives*. W.H. Freeman, New York.

Wigglesworth, V.B. (1972) *The Principles of Insect Physiology*, 7th edn. Chapman & Hall, London.

Willmer, P., Stone, G. & Johnston, I. (2000) *Environmental Physiology of Animals*. Blackwell Publishing Ltd., Oxford.

Withers, P.C. (1992) *Comparative Animal Physiology*. Saunders College Publishing, New York.

Supplemental reading

Al-Qahtni, A.H., Mashaly, A.M., Alajmi, R.A., Alshehri, A.A., Al-Musawi, Z.M. & Al-Khalifa, M.S. (2019) Forensic insects attracted to human cadavers in a vehicular environment in Riyadh, Saudi Arabia. *Saudi Journal of Biological Sciences* 26(7): 1499–1502.

Anderson, G.S. & VanLaerhoven, S.L. (1996) Initial studies on insect succession on carrion in Southwestern British Columbia. *Journal of Forensic Sciences* 41: 617–625.

Arnaldos, I., Romera, E., Garcia, M.D. & Luna, A. (2001) An initial study on the succession of sarcosaprophagous Diptera (Insecta) on carrion in the southeastern Iberian peninsula. *International Journal of Legal Medicine* 114: 156–162.

Baumgartner, D.L. & Greenberg, B. (1985) Distribution and medical ecology of the blow flies (Diptera: Calliphoridae) of Peru. *Annals of the Entomological Society of America* 78: 564–587.

Bourel, B., Martin-Bouyer, L., Hedouin, V., Cailliez, J.C., Derout, D. & Gossett, D. (1999) Necrophilous insect succession on rabbit carrion in sand dune habitats in northern France. *Journal of Medical Entomology* 36: 420–425.

Carvahlo, L.M.L., Thyssen, P.J., Linhares, A.X. & Palhares, F.A.B. (2000) A checklist of arthropods associated with pig carrion in Southeastern Brazil. *Memórias do Instituto Oswaldo Cruz* 95: 135–138.

Hall, D.G. (1948) *The Blowflies of North America*. Thomas Say Foundation, Baltimore, MD.

Hobischak, N.R., VanLaerhoven, S.L. & Anderson, G.S. (2006) Successional patterns of diversity in insect fauna on carrion in sun and shade in the Boreal Forest Region of Canada, near Edmonton, Alberta. *Canadian Entomologist* 138: 376–383.

James, M.T. (1947) *The Flies that Cause Myiasis in Man*. USDA Miscellaneous Publication No. 631. United States Department of Agriculture, Washington, DC.

McAlpine, J.F., Peterson, B.V., Shewell, G.E., Teskey, H.J., Vockeroth, J.R. & Wood, D.M. (eds) (1981) *Manual of Nearctic Diptera*, Vols. 1–3. Research Branch Agriculture Canada Monograph 27.

Matuszewski, S., Bajerlein, D., Konwerski, S. & Szpila, K. (2010) Insect succession and carrion decomposition in selected forests of Central Europe. Part 1: pattern and rate of decomposition. *Forensic Science International* 194(1–3): 85–93.

Tabor, K.L., Fell, R.D. & Brewster, C.C. (2005) Insect fauna visiting carrion in Southwest Virginia. *Forensic Science International* 150: 73–80.

Voss, S.C., Forbes, S.L. & Dadour, I.R. (2008) Decomposition and insect succession on cadavers inside a vehicle environment. *Forensic Science, Medicine, and Pathology* 4(1): 22–32.

Voss, S.C., Spafford, H. & Dadour, I.R. (2009) Annual and seasonal patterns of insect succession on decomposing remains at two locations in Western Australia. *Forensic Science International* 193: 26–36.

Additional resources

Blow flies of eastern Canada and the United States: http://www.biology.ualberta.ca/bsc/ejournal/mwr_11/mwr_11.html

Blow flies of North America: http://www.tolweb.org/treehouses/?treehouse_id=4197

Identification of common calliphorids of Northern Kentucky: http://www.nku.edu/~dahlem/ForensicFly Key/Homepage.htm

Chapter 12

Decomposition in aquatic environments

Overview

Human decomposition in aquatic ecosystems differs substantially from the events of physical and chemical decay that occur in terrestrial habitats. A submerged corpse essentially decays in a three-dimensional environment that can vary in terms of biotic and abiotic composition dependent on position in the water column, location within a particular body of water, and whether in freshwater versus marine habitats. As with decomposition on land, the attributes that define the corpse result in decay in water that are distinct to the individual yet comprised of a series of predictable processes. One major difference between terrestrial and aquatic decomposition is that the degree of predictability is less precise in aquatic ecosystems, in part because the invertebrate fauna is radically different from those found above the water surface, and due to the fact that no aquatic insects have evolved to be exclusively necrophagous on carrion. The latter seemingly diminishes the value of insects in aquatic systems in contributing to forensic investigations. However, as we shall see unfold in this chapter, several species of freshwater insects and marine invertebrates can provide unique insight into human decomposition in aquatic ecosystems, including the length of body submersion, whether the body has been moved from another location, and potentially if the decedent was alive when entering the water. This chapter will also examine the ecological roles of human remains in freshwater ecosystems, discuss theories accounting for the absence of necrophagy in aquatic insects, and compare the postmortem interval and postmortem submersion interval.

The big picture

- Decomposition of human remains in water is different than on land.
- Ecological functions of human remains in aquatic ecosystems.
- Why are there no truly necrophagous aquatic insects on carrion?
- Associations between freshwater insects and aquatic remains.
- Terrestrial insects are still interested in dead bodies in water.
- A postmortem interval cannot be determined with aquatic insects.

The Science of Forensic Entomology, Second Edition. David B. Rivers and Gregory A. Dahlem.
© 2023 John Wiley & Sons Ltd. Published 2023 by John Wiley & Sons Ltd.
Companion website: www.wiley.com/go/forensicentomology2

12.1 Decomposition of human remains in water is different than on land

Most of what is known about human decomposition, especially in a forensic context, is associated with terrestrial habitats. In one sense, this should come as little surprise. After all, the majority of violent crimes occur in terrestrial locations, and in turn, the corpse remains at the scene or is disposed of in a more or less convenient location to the assailant, which is generally on land. Of course, as detailed in Chapter 11, "on land" may entail a wide variety of natural and man-made locations or structures in an attempt to cover up the crime by hiding the remains. As we have discussed throughout this book, no matter where the decedent is placed in terrestrial ecosystems, eventually one or more species of necrophagous insects will detect the corpse via olfactory cues, initiate searching behavior, and potentially accept and utilize the remains for offspring development. Some locations delay or inhibit various steps in this pattern of sequential behaviors (Tomberlin *et al.*, 2011), but the balance sheet favors the success of necrophagous insects, particularly flies in the families Calliphoridae and Sarcophagidae, to find and colonize human and other animal remains decaying anywhere on land.

The situation changes dramatically when a corpse is placed in aquatic ecosystems. In freshwater, a human body breaks the surface tension of water and typically sinks, remaining submerged for period of time dependent on ambient water temperature. Important to the discussion of forensic entomology, there is a window of time in which no portion of the decedent is in contact with air. This feature alone effectively serves as a barrier to prevent detection and subsequent use of human remains by terrestrial insects. The composition of aquatic ecosystems varies tremendously based on the type of water (i.e., marine vs freshwater; stream, pond, lake, and pool), depth, temperature, and season. Detailed studies examining processes of physical and chemical decomposition of humans, pigs or any other animal model are quite limited by comparison to the research conducted with terrestrial decomposition (Stuart & Ueland, 2017). There are several reasons to account for this disparity, including health concerns with deliberate placement of animal remains in natural water supplies, need for specialized equipment and personnel training to

Figure 12.1   The few body farms that exist do not conduct extensive research on human decomposition in aquatic environments. Consequently, most information is derived anecdotally from casework or from research with animal models. *Source*: Tampa Police Department/ Wikimedia Commons/Public domain.

recover human remains from deep water, the overall absence of carrion feeding invertebrates that can provide the same type of developmental data that exclusive necrophagy offers in terrestrial ecosystems (Magni *et al.*, 2013; Fenoglio *et al.*, 2014), and the rarer occasion of human body discovery in aquatic versus terrestrial ecosystems (Figure 12.1).

The lower incidence of human decomposition in aquatic ecosystems is surprising when considering that more than 75% of the planet is covered in water. The potential for discovery of human remains in aquatic ecosystems should be greater than that in terrestrial environments. The reality is just the opposite. Discovery of human remains in aquatic habitats is not rare but also not common (Hobischak & Anderson, 1999). In some locations such as deep bodies of water, recovery of remains occurs infrequently, or only dismembered body parts are discovered (Hyma & Rao, 1991; Haglund and Reay, 1993). In truth, the extent to which human corpses enter aquatic ecosystems in not fully realized. The comparative lack of studies examining human decomposition in a broad array of aquatic ecosystems has resulted in gaps in the knowledge base needed for casework involving corpse recovery from any type of water. This information deficiency extends to understanding aspects of physical and chemical decay of human remains in various aquatic systems as well as the details of the aquatic insects most valuable to deciphering the types and length of associations between a corpse and aquatic carrion communities. What follows in the remainder of the chapter are discussions of what is known about aquatic decomposition of human remains, the types of aquatic insects commonly associated with bodies in freshwater ecosystems, and an examination of the ecological roles of these organisms as they pertain to an understanding of human decomposition and length of submersion in freshwater. Comparisons to marine ecosystems will also be made with the realization that much less information is available for these habitats.

12.1.1 When does aquatic decomposition occur?

Before exploring the process of decomposition in aquatic ecosystems, the first aspect to be addressed is when are human remains found in water environments. We have just established that a human corpse is much more likely to be found in terrestrial environments than aquatic. Are there ever instances in which the roles are reversed? In other words, are there circumstances in which a corpse is more likely to be discovered in aquatic habitats than in terrestrial ecosystems? The answer is "yes," there are specific scenarios in which human remains will be associated with aquatic environments. For example, certain types of natural disasters (e.g., tsunami, hurricane) often result in high causalities, especially along shorelines where sea waves and storm surges evoke the greatest damage in terms of loss of life and destruction of physical structures. Similarly, accident deaths associated with individuals falling from boats or bridges, wrecks involving any type of watercraft or terrestrial motor vehicles involved in crashes near waterways, and aircraft that crash or land in water may yield human remains in marine or freshwater habitats. In some instances, the body of a deceased individual may be disposed of in aquatic ecosystems, either by dumping the body alone or by encasing in some type of object. The latter has been represented by bags made of plastic, cloth or other materials, weighted down or not; and disposal in some type of man-made container including in appliances and vehicles. In reality, the "creativity" of corpse disposal in aquatic environments is equal to that of terrestrial scenarios.

The alternative situations are associated with a person being deliberately alive upon entry into water. This can be a means of suicide or to commit a homicide. As will be discussed in Section 12.1.2, determination of whether a person was alive or not when entering a body of water can be distinguished during autopsy by the presence of key pathological signs. However, recovery of a severely decomposed body places limitations on the ability to determine the cause and manner of death based on the methodology associated with forensic medicine.

12.1.2 Unique aspects of aquatic decomposition

Decomposition in any environment is influenced by ambient temperatures. This is true whether a corpse is positioned in a terrestrial or aquatic ecosystem. Ambient temperatures directly influence the rate of autolysis and putrefaction within the decedent, as well as the metabolic rate of poikilothermic animals such as freshwater insects and marine invertebrates that may feed on or colonize the corpse. The latter is consistent with earlier discussions throughout the book of the necrophagous activity of Diptera and Coleoptera colonizing human remains in terrestrial locations. One important difference does exist between aquatic and terrestrial ecosystems: the temperatures of large or deep bodies of water are very stable by comparison to ambient air. Fluctuating air temperatures, especially during seasonal transition, have only a modest impact on surface temperatures. Deeper layers of water remain unchanged without seasonal upwelling or mixing. Consequently, the rate of decomposition for submerged remains will be relatively constant. In instances in which the water temperatures are cool, decay of soft tissues will occur slowly by comparison to the rate of decomposition that occurs above ground during the same window of time.

The stability of water temperatures would seemingly create an environment that would favor predictability in terms of estimating the time of death for submerged remains. The reality is that aquatic environments differ in substantial ways from terrestrial habitats, leading to less certainty in time estimates (Table 12.1). For example, a body in water is surrounded by the aqueous medium, permitting rapid changes in the surface appearance of the corpse regardless of ambient temperatures. This means that most of the early signs of physical decomposition discussed in Chapter 10 are not evident if the corpse has been submerged. Algor mortis occurs more rapidly due to heat dissipation from all surfaces in contact with the cooler water temperatures. The patterns of livor mortis are not necessarily reflective of the position of the body upon placement in water or if settled to benthic regions

Table 12.1 Unique features of human decomposition in aquatic environments by comparison to terrestrial ecosystems.

Feature	Description
Ambient temperatures	Water temperatures are more stable or constant than air temperatures
Three dimensional medium	Corpse surfaces are in contact with water on all sides, promoting more rapid changes in physical appearance and faster rate of heat loss (algor mortis)
Chemical composition	Chemical makeup of freshwater and marine waters differing from each other and from air
Fluvial transport	Movement of water due to currents and tidal action distribute body and body parts to locations other than point of water entry
Disarticulation	Water movement and increased water mass of floating tissues leads to abrasions, bone breakage, tissues sloughing, and removal of limbs
Saponification	Conversion of triglycerides in soft tissues into adipocere

for a period of time. Why? If the body lacks penetrating wounds or lesions, it will conceivably become buoyant during putrefaction as decomposition gases accumulate, permitting the corpse to not only float but to be moved by currents or tidal action. Dependent on the force of water movement, the body may turn or roll, multiple times, prior to fixation of lividity. The resulting livor mortis patterns generally do not reveal information related to the agonal period or that are relevant to reconstruction efforts.

Movement of a corpse due to currents or tides is referred to as **fluvial transport**. Not only does water movement potentially influence livor mortis, but it also contributes to redistributing the intact body, body parts, and/or bones to other locations from the point of water entry. Thus, an understanding of tidal patterns or flow velocity is necessary to interpret where a body may have entered the ecosystem. Similar considerations are necessary to evaluate how water movements contributed to disarticulation of the corpse. The latter is also an important distinction between aquatic and terrestrial decomposition: disarticulation is much more likely to occur in aquatic ecosystems, independent of invertebrate or vertebrate scavenger activity. In part this can be attributed to the increased tissue biomass associated with water

influx into tissues. The added weight strains tissues and bones, especially as the body becomes buoyant. In turn, tissue slippage occurs, soft tissues slough from the skeleton, and bones disarticulate and/or break.

Outside of temperature influences, the chemical composition of the aquatic system accounts for unique patterns of decay. The high inorganic content of marine environments slows putrefaction (De Donno *et al.*, 2014), whereas tissue flooding or movement of freshwater into tissues can dilute enzymes associated with autolysis and putrefaction, yet induce loss of membrane integrating and sloughing off of tissues much faster than occurs in ambient air, assuming soft tissues are not consumed first by larval feeding activity of necrophagous Diptera. A distinct feature of aquatic decompositions is the formation of adipocere or graves wax. Adipocere results from the saponification of triglycerides in soft tissues into fatty acids and essentially forms soap. This is sometimes referred to as the **Goldilocks effect or phenomenon**, in which the postmortem conditions must be "just right" to favor saponification in soft tissues (O'Brian & Kuehner, 2007). Just right in terms of adipocere formation generally is in reference to aqueous or wet conditions, alkaline pH, and a balanced mix of inorganic ions. Adipocere essentially preserves the underlying soft tissues. If graves wax covers the entire body, scavenging is inhibited and further degradation of soft tissues is slowed or nearly abolished (O'Brian & Kuehner, 2007).

12.1.3 Stages of decomposition in freshwater

Characterization of human decomposition in freshwater ecosystems involves defining distinct stages of physical decay (Figure 12.2). The same was true when examining decomposition of human remains in terrestrial environments. With terrestrial decompositions, stages represent physical milestones achieved over time. The stages are a convenient means for summarizing physiochemical changes in animal remains but are mostly determined subjectively rather than empirically (Suckling *et al.*, 2015). As discussed in Chapter 10, stage assignment does not necessarily represent discrete or sequential ecological stages during animal decomposition (Schoenly & Reid, 1987). Instead, deterioration of the corpse reflects a continuum of events, in which the rate of decay differs from one

Figure 12.2 A piglet placed in a man-made freshwater pond in southcentral Pennsylvania during early spring. The piglet sunk immediately to initiate the submerged fresh stage. *Source*: Photo by D.B. Rivers.

region of the body to another. In aquatic environments, physical stages of decay are frequently associated with where the body is located within the water column – submerged versus floating – and the degree of soft tissue decay that is subjectively determined. Essentially two classification schemes have been devised for characterizing physical decay in aquatic ecosystems. One is based on the natural progression of decomposition associated with a corpse that initially is submerged when entering a body of water and which the corpse is not obstructed by an object below water, meaning that the body can float if it becomes buoyant due to gas production. The second scheme details decomposition for tissues that remain submerged throughout the entire decay process.

The original classification of aquatic decomposition identified six stages based largely on the presence or absence of terrestrial insects on domesticated pigs, *Sus scrofa* (Payne & King, 1972). However, the classification system has been reworked so that five stages of physical decay are now recognized, with each stage characterized by specific associations with freshwater invertebrates (Haefner *et al.*, 2004; Wallace & Merritt, 2020). Arguably, an additional stage may be warranted to separate the initial stage – submerged fresh—into two stages, *submerged fresh* (or early) and *submerged late*. In scenarios (i.e., cold water) in which a corpse remains submerged for an extended period of time (weeks to months), substantial physical changes of the remains may occur that are inconsistent with the stage terminology, i.e. "fresh." However, for purposes of consistency with the

forensic entomology literature, the five-stage scheme is presented here. This classification makes the following assumptions about the corpse:

1. The aquatic ecosystem is freshwater with minimal influx from marine environments;
2. The body is intact at the time it entered the water;
3. The body of water is sufficiently deep to permit full submersion of the corpse.

12.1.3.1   Submerged fresh

The submerged fresh stage is characterized by the corpse entering a body of water, and if sufficiently deep, the body sinks toward the benthic region. This makes the assumption that the decedent has a mass greater than water and that the conditions of the aquatic system favor descent to the bottom. The rate at which the corpse sinks is dependent on numerous abiotic and biotic factors, including the chemical composition of the freshwater system, temperature, currents, and to an extent, degree of immediate scavenging. For example, if vertebrate scavengers removed enough soft tissue to create large lesions in the skin, water can enter the body cavity and in turn accelerate the rate at which the corpse sinks. The submerged fresh stage ends when (if) the body becomes buoyant due to anaerobic gas accumulation and any portion of the body breaks the surface tension of water to make contact with air. In deeper bodies of water, as the abdomen distends the corpse may roll face up during the ascent through the water column so that the ballooned belly breaches the water surface. More often, the body floats face down with limbs dangling downward. This position yields livor mortis in the face, upper chest, neck, arms, hands, feet, and calves.

Essentially the same postmortem changes occur in freshwater as discussed in Chapter 10 for the fresh stage of terrestrial decay. Autolysis begins immediately after death regardless of whether the individual dies on land or in water. Putrefaction follows and is impacted by the degree of **edema**. Edema or movement of water into the tissues is dependent on the chemical composition of water. In the case of freshwater, a net influx into tissues, especially as membrane integrity is lost, is expected to occur. This is a localized effect, occurring first in the outer epidermis. Diffusional movement is temperature dependent, like enzymatic activity, and thus water temperature is the most influential factor on the rate of tissue decay in aquatic environments.

What this also means is that the duration of the submerged fresh stage is dependent on temperature since ultimately anaerobic gas production and accumulation are slowed in cooler water. Thus, in deep water or during seasonal conditions in which water temperatures are colder a corpse will remain submerged much longer than in shallow or warmer waters (Figure 12.3). The size of the remains will also dictate the length of time needed to achieve buoyancy. Total duration of submersion is highly variable, but pig carcasses have been found to need 2–13 days to float in spring and 11–13 in more northern latitudes of North America (Hobischak, 1997; Hobischak & Anderson, 2002). Studies using 20-kilogram pigs in the Columbian Andes from January to April found that bloating occurred after 8 days in a freshwater stream and in only 5 days in a lake (4 meters deep) (Barrios & Wolff, 2011). In man-made ponds, pig remains have been observed to float in as few as three days in warm water (26–29 °C) (Chin *et al.*, 2008).

Algor mortis is achieved more rapidly in freshwater, since water temperature is invariably cooler than ambient air in most scenarios. Frothing or small bubbles around the nasal and mouth openings, termed **froth cone**, may be evident if the decedent drowned or experienced asphyxiation. Wrinkling of the skin due to vasoconstriction and water absorption in the epidermis is common in the early stages of submersion. The presence of cutis anserina or "goose bumps" is an early pathological sign of aquatic decomposition that is associated with individuals who were alive at the time of entering water. Within a matter of hours

(temperature dependent) the skin color generally changes to blue-red to pink, consistent with **cyanosis**. Eventually, the outer epidermis will appear white, wrinkled, and thick. These epidermal changes can obscure many of the early decomposition changes. The onset of rigor mortis is also influenced by water temperature; cooler water slows the onset and systematic progression of unregulated muscle contractions (Figure 12.4). Similarly, the duration of rigor mortis is extended at lower water temperatures. In drowning victims, cadaveric spasm may be evident in the hands or feet.

Submerged remains that descend to the bottom of a body of water draw the interest of a wide range of benthic invertebrates. In freshwater, aquatic insects dominate benthic communities and hence are commonly associated with a human corpse (Keiper & Casamatta, 2001). Immature larvae of various species of caddisflies (Order Trichoptera), midges (Diptera: Chironomidae), and mayflies (Ephemeroptera) will feed and attach to a submerged corpse. Several non-insect invertebrates are also commonly associated with fresh remains and include crustaceans (crayfish, ostracods, copepods, isopods), annelid worms (oligochaetes), snails (Mollusca) and flatworms (Platyhelminthes). Water conditions and composition, as well as seasonality will influence which species will be most likely found on a submerged body in freshwater. More details on specific groups of freshwater insects and the type of associations with human remains will be discussed in Section 12.5.

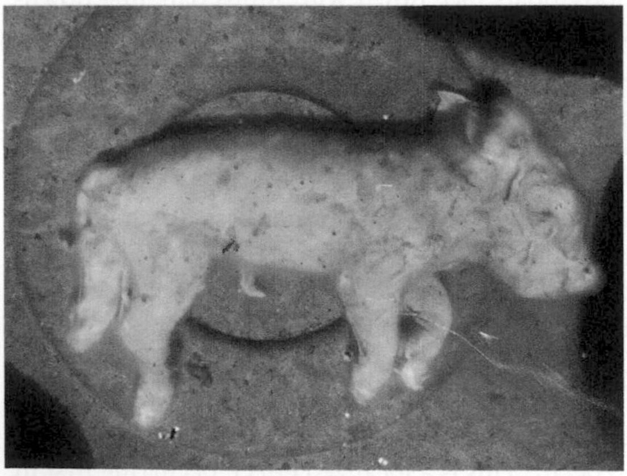

Figure 12.3 Colder spring temperatures delay autolysis and putrefaction. Six days after submersion a thick film formed along ventral region and along legs. *Source:* Photo by D.B. Rivers.

Figure 12.4 Rigor mortis evident in drowning victims whose bodies were discovered after a devastating cyclone struck the island of Sandwip in 1991. *Source:* Rahat/Wikimedia Commons/Public domain.

12.1.3.2   Early floating

The second stage of aquatic physical decay begins when the body floats to the surface and lasts until minor decay of soft tissues becomes evident. Buoyancy or bloating results from gas production via anaerobic bacterial metabolism (Figure 12.5). In a closed container, meaning a corpse lacking penetrating wounds or lesions, gas molecules accumulate. During terrestrial decompositions, the increasing gas pressure eventually forces fluids to be purged from body openings. However, slower rates of tissue breakdown and water pressure outside the body effectively counter the internal gas pressure in aquatic environments, helping to preserve buoyancy for an extended period of time. Most of the skin is discolored in appearance, with the hands and feet displaying washerwoman's skin: pale to white, wrinkled, soft, and easily pealed from underlying layers (Hobischak & Anderson, 1999). Once the body breaks the water surface, the pace of decay as well as the utilization of the corpse by insect colonizers changes dramatically. Portions of the body that remain submerged are still subject to the same abiotic and biotic factors discussed with the initial submerged corpse. Likewise, the same benthic insects and other invertebrates that were attached and/or feeding to the body may continue to do so, provided that their activity is not inhibited by any wave action, increased oxygen content near or at the water surface, or by now being located within the photic zone. The latter generally stimulates increased algal and phytoplankton activity, with mats of algae often forming over the surface of the corpse.

Air exposed surfaces draw the attraction of necrophagous and necrophilous insects. The same waves of colonization discussed in Chapter 11 are not expected, as the chemical composition of the corpse and hence the odor profile are altered by submersion (Figure 12.6). That said intimate details of the chemical signals associated with necrophagous insect attraction to floating remains are unknown. Adults from the families Calliphoridae, Sarcophagidae, Muscidae, and Piophilidae have been observed to feed and oviposit/larviposit on floating human, pig, and rat carcasses (Payne & King, 1972; Tomberlin & Adler, 1998; Barrios and Wolff, 2011). Rove (Staphylinidae) and carrion beetles (Silphidae) typically arrive following the flies, as do an array of flying predatory Hymenoptera (Payne & King, 1972). The feeding activity of necrophagous fly larvae alone accelerates physical decay of soft tissues, leading to the subjectively determined end of the early floating stage and beginning of floating decay (Table 12.2).

12.1.3.3   Floating decay

The third stage of aquatic decomposition is characterized by the corpse initially displaying minor decay of soft tissues and ending when significant decay is evident. Determinations of each state is highly subjective, making it very difficult to assess the duration of floating decay from one study to

Figure 12.5   Early floating stage begins when any portion of the body breaks the surface tension of the water due to buoyancy resulting from accumulation of gas internally. *Source*: Photo by D.B. Rivers.

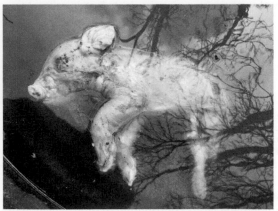

Figure 12.6   Fly activity generally becomes more apparent during the early floating stage when the head breaks the water surface. *Source*: Photo by D.B. Rivers.

Table 12.2 Comparison of decay stages used to characterize aquatic decomposition.

Stage of Decomposition	Description
Submerged fresh	Begins when corpse enters water and submerges below water surface. Characterized by internal changes associated with autolysis followed by microbial enzymatic degradation. Discoloration of skin masks many early physical changes
Early floating	Anaerobic conditions promote microbial conversion of macromolecules into gaseous byproducts resulting in internal pressure increases, ballooning of tissues, and buoyancy of corpse. Body often floats face down, resulting in livor mortis in regions facing benthic regions.
Floating decay	Corpse shows minor to significant decay; flesh sloughs off from underlying layers and some loss in muscle mass. Feeding activity of terrestrial insects causes lesions or holes in soft tissues.
Bloated deterioration (frequently combined with floating decay)	Lesions in tissues result in gas escape, causing remaining tissues to collapse upon self. Corpse remains floating with disarticulate limbs attached.
Advanced floating decay Floating remains	Soft tissues display major deterioration, corpse is unrecognizable; bone exposure, especially ribs, is evident; large larval feeding aggregations are present; some bones break or disconnect from remains. Tissues completely exposed to air may dry and harden.
Sunken remains	Remains sink to benthic regions. Any skin that remains has soup-like consistency with little odor; typically only small bones remain attached to tissues; and terrestrial insect larvae float to surface or drown as remains sink.

Either a five- or six-stage classification system is used for description of carrion decomposition in aquatic environments. Consequently, stages three and/or four above overlap in the table. Stages are based on Payne & King (1972), Haefner *et al.* (2004), and Barrios & Wolff (2011).

another (Figure 12.7). Flesh begins to slough off of underlying layers contributing to disarticulation of soft tissues and bone. Tissues exposed directly to air, dry darken in color (Figure 12.8). Larval feeding by immature calliphorids and sarcophagids

Figure 12.7 The transition from early floating to floating decay is highly variable and subjectively determined. This piglet displays minor tissue decay, especially among submerged limbs. The eye is also bulging due to gas pressure and portions of the epidermis have sloughed off into the water. *Source*: Photo by D.B. Rivers.

increases, resulting in the formation of lesions and holes throughout the soft tissues above water. In some cases, gases along with decomposition fluids may escape from skin lesions (Barrios & Wolff, 2011). Even when the corpse appears to deflate from gas leakage, the remains continue to float. The total duration of the stage is highly variable, being reported to last as short as 8 days to as long as 331 days using pig models (Wallace & Merritt, 2020).

Payne & King (1972) originally described a bloated deterioration stage separating floating decay from advanced floating decay (floating remains), but some investigators (Haefner *et al.*, 2004; Wallace & Merritt, 2020) have cited the difficulty in distinguishing this stage from floating decay based on physical appearance. Even in field studies reporting a bloated deterioration stage, the physical manifestations of decompositions and invertebrate feeding are nearly identical to that of floating decay (Barrios & Wolff, 2011).

12.1.3.4   Advanced floating decay

A corpse displaying severe deterioration characterizes the onset of advanced floating stage. Again, the physical milestones are very subjective. However, the end of the stage is more precise; advanced floating decay ends when the remains sink. This entails all of the remains, which is an important distinction because during the fourth stage of aquatic decomposition, portions of the body will

(a)

(b)

(c)

(d)

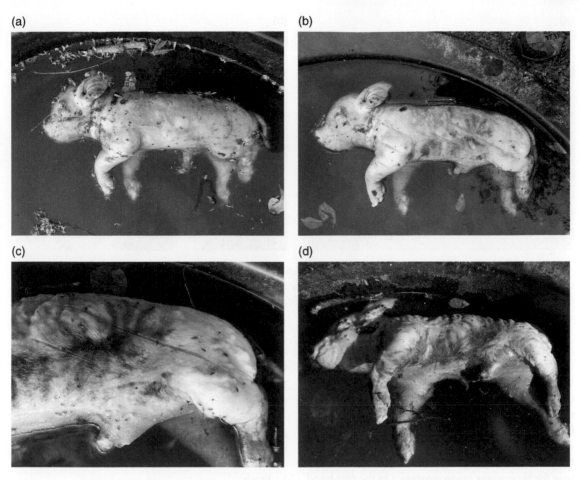

Figure 12.8 Progression of physical decomposition during the floating stage. Adult piophilids began to arrive early (a) and continue to be present on the carcass as the exposed skin dries out (b–d). Bloating of the piglet becomes more evident over time as evidenced by extension of limbs. Algal growth on submerged portions of skin becomes pronounced as the corpse enters the photic zone. *Source*: Photos by D.B. Rivers.

slough off or break and then sink independently of the remainder of the corpse. Though tissue moisture is present, the extent of autolysis or putrefaction that occurs during this stage is not clear. As decay process proceed, the corpse becomes increasingly unrecognizable (Figure 12.9). Contributing to this deterioration is extensive feeding by fly larvae. Large feeding aggregations reach their zenith in terms of size and density during this stage. During warmer months, larval development of the initial colonizers is completed and initiation of wandering or dispersal begins. Of course, the maggots crawl into the water where initially their hydrophobic exoskeleton permits them to float. Some may reach shore dependent on the location of the remains in relation to land and the extent and direction of currents or tidal action. Other larvae may drown relatively quickly or first deplete ATP/oxygen reserves during attempts to crawl in water, and then succumb to drowning.

A final group of larvae are consumed by fish, waterfowl, and other predators associated with the freshwater ecosystem.

As with other stages of decay, the duration of the advanced floating stage is highly variable and dependent on temperature, depth of water, water movement, and season. Estimates with pig carcasses range from approximately 12 days in stream riffles to as long as 171 days in a freshwater pond during summer months in the Midwestern region of the United States. Lower water temperatures and presumably larger massed remains extend the window for this stage of decay.

12.1.3.5 Sunken remains

The final stage is perhaps the easiest to identify. The sunken remains stage begins when the corpse sinks once again to the benthic region. As discussed (a)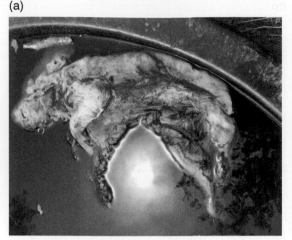

(b)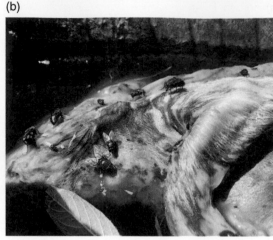

Figure 12.9 During the advanced floating decay, the remains display extensive deterioration and become unrecognizable in appearance. During this spring decomposition, (a) the piglet skin exposed to air became very dry and (b) attracted typical of earlier stages of terrestial decay. *Source*: Photos by D.B. Rivers.

earlier, some tissues and bones may sink at different time intervals depending on the extent of disarticulation and bone breakage. Once submerged, the duration of the stage is highly variable as it begins with some tissue and bone and ends when the bone has degraded to its elemental form. Skeletal remains in terrestrial environments often requires months to years to completely deteriorate; the same holds true for freshwater environments. On the one hand, the temperature of deep bodies of water is generally cooler than ambient air during warmer months, but also does not drop as low during seasonal change associated with winter in temperate environments. Underwater currents can also be abrasive toward tissue and bone, gradually wearing away layers. Crayfish, chironomid larvae, and an array of fish are likely to feed on remaining soft tissues. However, bone will serve more as substrate for grazers than as a food source, with the exception of some boring annelids.

12.1.4 Alternative classification for submerged remains

Several studies have examined physical decomposition of carrion, usually pigs, placed in cages to freshwater and marine ecosystems. For most, the stage classification was essentially the same as the five- or six-stage scheme described in the previous sections. However, at least one has relied on the physical stage descriptors used for terrestrial decompositions. Table 10.3 in Chapter 10 provides an overview the various stages and the description of physical changes that occur from onset to completion of each. Below is the classification scheme used by Dickson *et al.* (2011) to characterize aquatic decomposition of adult pig heads placed in cages submerged in Otago Harbor in southeastern New Zealand:

1. Fresh
2. Early putrefaction
3. Advanced putrefaction
4. Advanced decay
5. Skeletonized remains.

It is worth noting that the study was performed in a marine environment. Thus, if the pig heads were not caged, recovery of any remains would be quite unlikely.

12.1.5 Decomposition in marine environments

The discussion of aquatic decomposition thus far has deliberately separated freshwater from marine ecosystems. As you ponder why the distinction, it is certainly true that the systems differ substantially from each other. Obviously, the chemical composition stands out if for no other reason than the marked difference in salinity. So, too, does depth, temperature, wave action, and biotic composition. But among the primary reasons for considering marine environments as a separate topic is based on the lack of understanding in terms of human decomposition and likelihood of recovery. Information regarding human decay in marine environments is even more sparse than that available for freshwater ecosystems. Why? At least four reasons can account for the lack of baseline information regarding human taphonomy in marine ecosystems (reviewed by Sorg *et al.*, 1997).

1. Human remains in marine waters are relatively rare;
2. Recovery of human remains in marine environments is even more rare;
3. In those instances, in which human remains are recovered from marine ecosystems, scavengers have already dispersed from the body. This makes it challenging to determine which scavengers account for specific physical changes to the corpse and then in turn use this information to relate to time estimates;
4. Very few insects are found in marine environments, none are found in deep marine waters, and none are necrophagous specialists on carrion.

The latter is a limitation mentioned earlier in the chapter for all aquatic insects, but the difference between the number of freshwater insects versus marine can be measured in orders of magnitude. Consequently, direct relationships between marine insect development and human remains do not exist. In fact, the development of any marine invertebrate is not intimately linked to carrion as occurs with terrestrial necrophagous insects. In reality, human remains do not last long when submerged in marine environments. Typically scavenging by invertebrates and vertebrates begins almost immediately upon entering the water. Soft tissues are removed within just a few days, exposing bone and ensuring that the corpse will not become buoyant. Disarticulation and limb removal are common due to the feeding action of several species of vertebrate scavengers, such as sharks and fish with sharp teeth, which are capable of removing large chunks of tissue at a time. Scaling differences of marine invertebrates by comparison to freshwater are significant. In studies examining submerged pig carcasses, echinoderms, crustaceans, and molluscans have been observed covering the entire body (Anderson & Hobischak, 2004). In some instances, just one or two individual sea stars totally envelop the pig remains.

If a body is discovered in shallow water, high saline waters, intertidal zones or washes ashore, colonization by terrestrial insect species will occur. The usual suspects typically arrive, with the exception that in the few studies that have been performed, beetle diversity is far less than occurs with bodies previously exposed to freshwater. No empirical studies have been performed to account for the diminished interest by Coleoptera. Undoubtedly profound changes in osmotic and inorganic ion content occur in soft tissues exposed to or submerged in marine environments. Similarly, no studies have reported on the impact on development from flies feeding on tissues submerged in marine waters.

12.2 Ecological functions of human remains in aquatic ecosystems

Introduction of organic matter from terrestrial systems into aquatic habitats is an expected disruption that occurs with any natural body of water. This is easily illustrated by considering dissolved organic matter (DOM). Dissolved organic matter is the most mobile form of organic matter in soil, and hence in terrestrial environments. It readily makes its way into aquatic ecosystems via numerous mechanisms. The flow or movement of DOM from terrestrial systems to aquatic varies based on soil type, location, season, and other biogeographic influences. Perhaps a more tangible example is to simply consider the vast array of plants that border shorelines of nearly every type of naturally occurring body of water. Solid organic material from trees, shrubs, grasses, flowers or other type of plants land directly in water or often enter aquatic systems by some other means (i.e., in water runoff, wind, animal activity). Once in an aquatic environment, degradation of the plant material begins. Decay of plant matter contributes substantially to the cycling and distribution of energy, carbon, and nutrients within and between ecosystems (Bianchi, 2011; Jansen *et al.*, 2014). This is true for any form or source of organic matter that enters an aquatic ecosystem, regardless of the source.

Animals function as a conduit for nutrient transfer and cycling between terrestrial and aquatic ecosystems. Seabirds, for example, remove fish and invertebrates from marine environments through food consumption. Much of the carbon and nutrients are transferred to nestling young, while waste products accumulate in terrestrial habitats. Similarly, a significant amount of organic matter from terrestrial animal species eventually deposits into

Figure 12.10 Introduction of solid organic matter in the form of pig carrion into an aquatic lake in Hainan, China. *Source*: Anna Frodesiak/Wikimedia Commons/CC0 1.0.

freshwater and marine systems (Figure 12.10). The extent of deposition of solid organic material from terrestrial animals, especially in the form of whole carcasses, is not known. However, the presumption can be made that animal tissue degradation, like with vegetative sources, is a rich resource for nutrient cycling into aquatic environments. That said the dynamics of any type of animal decomposition in aquatic habits is poorly understood by comparison to what is known for plants or microorganisms. Research using salmon and other fish carcasses in decomposition studies have revealed that the fish are important sources of nitrogen and phosphorous cycling in freshwater streams (Kline *et al.*, 1997). The liberated nutrients are consumed by several aquatic insect species, resulting in enhanced growth rates by comparison to when leaf litter serves as the primary source of organic matter (Minakawa *et al.*, 2002). The effects are not limited to isolated individuals as localized populations of caddisflies (Trichoptera) can experience significant increases in overall biomass by feeding on decomposing salmon as opposed to plant-based matter.

The same effects undoubtedly occur with many other aquatic insects that consume animal matter. Decomposition of terrestrial animals, including human corpses, in aquatic ecosystems are expected to evoke similar if not magnified increases in aquatic insect growth rates and biomass. Anecdotal observations from casework supports this prediction for human remains, as large mats of algae commonly cover a floating corpse and support the feeding action of many micro- and macroinvertebrate species. However, as noted earlier in the chapter, empirical research studies using human cadavers in freshwater field research is very limited to non-existent. That said there are several defined ecological roles that human remains serve in aquatic ecosystems that impact a wide array of aquatic organisms. In

freshwater systems, a human corpse can function in the following roles (Wallace & Merritt, 2020):

1. Become an immediate or eventual food source to a variety of aquatic organisms including invertebrates and fish;
2. Provide shelter and/or microhabitat for small invertebrate species;
3. Serve as a substrate for primary producers, especially those in the photic zone;
4. Attract secondary predatory species or primary parasites (on species inhabiting corpse);
5. Serve as substrata for species that graze on exposed bone, primary producers, clothing, or other materials.

The ecological roles presented are not necessarily reflective of discrete successional phases of decomposition, as the attractiveness of human remains is influenced by numerous factors including the type of water, depth of water, season, abiotic characteristics (i.e., temperature, chemical composition, and salinity), and whether the body was clothed. In fact, the latter is important because the extent of skin coverage and the clothing material will influence the ecological roles of the corpse. For example, clothing may cover most exposed skin surfaces, thereby decreasing the potential attachment sites for algae, periphyton, and some aquatic invertebrate species. By contrast, some materials are more conducive as microhabitats. This is true for loose fit clothing with pockets or a hood; each feature reflects more surface area available for shelter. Heavy clothing may also accelerate the pace at which a body initially sinks during the submerged fresh stage or create drag during fluvial transport. The absence of clothing can be equally revealing as it may suggest the possibility of a sexual assault.

12.3   Why are there no truly necrophagous aquatic insects on carrion?

In terrestrial environments, insects are the dominant animal group in all habitats with the exception of the polar ice caps. Insects occupy every conceivable ecological niche, including the specialized roles of saprophagy and necrophagy. In fact, there is rich and diverse guild of terrestrial insects that function in macro-decomposition of organic matter, especially carrion. These insects and the subsequent carrion communities that form are comprised chiefly of Diptera and Coleoptera in terms of abundance and diversity. It should be no surprise, then, that the immatures of such species demonstrate morphological and life history traits that reflect necrophagous specialization. In fact, the evolutionary adaptions permit them to not only survive but also to flourish in the hostile environments of decaying carrion. A review of Chapters 8 and 9 provide details of many of the adaptations utilized by necrophagous fly larvae residing within carrion communities. By contrast, such adaptation for a carrion existence does not exist with aquatic insects, no matter what type of aquatic environment they reside in. The absence of necrophagy in aquatic insects is surprising when considering that up to 70–90% of freshwater benthic communities are comprised of insects (Keiper & Casamatta, 2001; Fenoglio *et al.*, 2014). Putting this into proper context, solid organic matter in the form of animal remains will eventually settle into the benthos of a body water. Freshwater teaming with fish and bird activity seemingly would favor the evolution toward necrophagy by insects living in such environments. Yet that has not occurred, which begs the question "why not?"

There is not a simple answer to the question of why necrophagy is absent in aquatic insects. One hypothesis posed is that among the vast numbers of terrestrial insects, a relatively low proportion have specialized as exclusively necrophagous (Fenoglio *et al.*, 2014). By extension, far fewer aquatic insects should be expected to evolve as specialized necrophagous species. The logic of the argument is founded in the evolution of the Insecta; insects first evolved for a terrestrial existence and then secondarily adapted to freshwater (Rivers, 2016). Among the terrestrial groups that display necrophagy, very few have aquatic representation, and none of those aquatic species are exclusively necrophagous. Several pieces of evidence have been offered as support of this hypothesis (Fenoglio *et al.*, 2014). Firstly, aquatic insects display significant trophic plasticity including the consumption of carrion without specialization. In short, many aquatic insects feed opportunistically on a wide range of food sources. Secondly, carrion is much more difficult to locate in aquatic than terrestrial environments. In this respect, consider the differences between olfaction used in terrestrial environments versus gustation that dominates in aquatic systems (see Chapter 7). Thirdly, there is a greater diversity and richness of animal wastes in terrestrial environments, contributing to specialization of saprophagy and necrophagy on dung and carrion. In contrast, excreta in aquatic systems is typically liquid and consequently dissipates rapidly. Similarly, the idea of carrion being a patchy, ephemeral resource is even more pronounced in aquatic habitats. Physical and chemical breakdown of vertebrate carrion tends to occur much more quickly in water, especially due to currents that cause physical fragmentation and abrasion. Coupled with fluvial transport, portions or all of the animal carcass may be removed from a benthic community soon after it is deposited. The final argument is one of competition. Crustacea display a higher abundance and diversity compared to insects in aquatic environments (Figure 12.11). Consequently, crustacea tend to outcompete insects for vertebrate carrion and inhibit the evolution of necrophagy in aquatic insects. By contrast, few crustacea exist in terrestrial

Figure 12.11 Crustacea like crayfish are the dominant scavengers for vertebrate carrion in freshwater, typically outcompeting most insect species for the same resources. *Source:* Gail Hampshire/Wikimedia Commons/CC BY 2.0.

environments, essentially permitting insects to exploit and dominate terrestrial carrion unmolested by other invertebrate species.

To an extent the underpinnings of the phylogeny hypothesis are that too few species exist in aquatic environments and that there has been insufficient evolutionary time for specialization toward the necrophagous condition. Perhaps some aquatic species are in transition as recent observations suggest that under specific conditions, some species of aquatic insects display preferential necrophagy. For example, the dytiscid *Coptotomus loticus* feeds nearly exclusively on vertebrate carrion in temporary wetlands during the dry-down phase (isolated pools of water that dry up) (McDaniel *et al.*, 2019). The beetle has been observed to display this behavior much more so during years with flooding than when a drought occurs. Other species of dytiscids and hydrophilids readily consume the flesh of floating amphibians and fish (Baena, 2011; Silva-Soares *et al.*, 2019), perhaps a reflection of one of the few locations (e.g., water surface) in aquatic environments that crustaceans do not dominate scavenging of vertebrate carrion (Fenoglio *et al.*, 2014). Interestingly, at least two species of Hemiptera have been observed sucking the contents from newt (Amphibia) carrion (Baena, 2011). This type of feeding behavior by hemipterans on terrestrial carrion is rarely reported.

12.4 Associations between freshwater insects and terrestrial carrion

When examining the associations between freshwater insects and carrion, the term "succession" does not apply as it does in characterizing terrestrial decompositions. This does not mean that aquatic insects do not colonize carrion. They do under specific conditions (Hobischak, 1997; Wallace & Merritt, 2020). However, aquatic insects have not evolved as specialized necrophores. None are exclusively necrophagous on vertebrate carrion, nor is the development of their immatures tied directly to feeding on carrion as it is with necrophagous fly larvae. As we will see in Section 12.6, this does have an enormous impact on the utility of aquatic insects in estimating any portion of a postmortem interval. Instead, aquatic insects demonstrate significant trophic plasticity, feeding on a wide range of organic matter in aquatic systems. Such feeding diversity permits freshwater insects to occupy a wide range of ecological roles in aquatic systems (Figure 12.12). Students are encouraged to consult Merritt & Cummins (1996) for an in-depth discussion of the feeding strategies employed by aquatic insects.

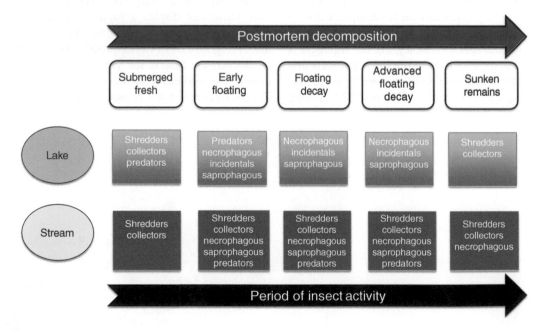

Figure 12.12 Relationship between insect feeding activity and physical decomposition of carrion located in freshwater. Scrapers are expected during floating stages in which algae and plankton are attached to the corpse. Insects classified as necrophagous, saprophagous, and incidentals are in reference to terrestrial species. The information is based on Merritt & Cummins (1996) and Barrios & Wolff (2011).

12.4.1 Ecological roles of freshwater insects on terrestrial carrion

Freshwater insects are often characterized by what they consume as their primary food source. Aquatic insect fauna fall into three major categories: carnivory, herbivory, and detritivory (Merritt *et al.*, 2017). These broad classifications are further divided into functional feeding groups or ecological roles for aquatic species (Merritt & Cummins, 1996) (Figure 12.12). For example, **shredders** encompass species that are detritivores and chewers of particulate organic matter. **Collectors** include detritivores, filter feeders or suspension feeders of fine particulate organic matter. **Scrapers** are herbivores and grazing scrapers of mineral and organic surfaces. Herbivores are often most abundant when at least a portion of the carrion is present within the photic zone. For example, toward the end of the early floating stage and during floating decay, a body often serves as a substrate for algae. In turn, scrapers arrive to feed on the regions located below the water surface. The final category are predators, which include engulfers and carnivores. Predatory species may consume an entire prey species or only parts of animal. By definition, such carnivorous insects can also be considered necrophagous if the animal is deceased when consumed. Several species of dytiscids and hydrophilids fall into this latter category.

12.4.2 Types of freshwater insects found on terrestrial carrion

The utility of aquatic insects to forensic investigations is dependent on an understanding of their life history. Just as discussed for terrestrial species in Chapter 11, the essential information needed includes knowledge of the fauna at the site of body discovery, timing of associations/colonization by aquatic insects with human remains, length of development for each life stage under a variety of conditions, and an understanding of the feeding and reproductive behaviors of the species found on submerged or floating remains (Keiper & Casamatta, 2001). Growth and development of aquatic insects is tied to ambient temperatures, but the details needed for forensic applications exist for only a few species. Detailed life history information for the thirteen orders of insects that have at least some members with aquatic or semi-aquatic life stages is lacking by comparison to terrestrial necrophagous species. Of those insect groups, eight orders have representatives in which one or more developmental stages are likely to be associated with vertebrate carrion discovered in freshwater habitats. The intent here is avoid replication of the excellent works of Merritt & Cummins (1996) and Wallace & Merritt (2020) in providing detailed taxonomic and life history information of freshwater insects typical in North America and associated with forensic investigations. A summary of those insect groups can be found in Table 12.3. The table provides a list of commonly encountered insect orders and families associated with terrestrial vertebrate carrion.

Not surprisingly, the three orders (Coleoptera, Diptera, and Trichoptera) most commonly discovered on human remains in freshwater all display holometabolous development and at least two (Coleoptera and Diptera) are among the most abundant and

Table 12.3 Commonly encountered insect orders and families associated with terrestrial vertebrate carrion.

Order name	Family	Functional feeding groups
Coleoptera		
	Dytiscidae	predators, shredders
	Gyrinidae	collectors
	Hydrophilidae	predators, shredders
Diptera		
	Ceratopogonidae	collectors, predators
	Chironomidae	collectors, scrapers, predators
	Ephydridae	scrapers, predators
	Psychodidae	scrapers, predators
	Simuliidae	collectors
	Syrphidae	scrapers, shredders
Ephemeroptera		
	Ameletidae	shredders
	Baetidae	shredders
	Caenidae	scrapers, shredders
	Heptageniidae	scrapers, shredders
	Leptophlebiidae	scrapers, shredders
	Oligoneuridae	scrapers, shredders
Hemiptera		
	Corixidae	predators
	Leptopodidae	predators
	Naucoridae	predators
	Notonectidae	predators
	Pleidae	predators

(Continued)

Table 12.3   (Continued)

Order name	Family	Functional feeding groups
Megaloptera		
	Corydalidae	predators
Odonata		
	Aeshnidae	predators
	Coenagrionidae	predators
	Libelluidae	predators
	Megapodagrionidae	predators
Plecoptera		
	Taeniopterygidae	shredders
Trichoptera		
	Glossosomatidae	collectors, scrapers, shredders
	Hydropsychidae	collectors, scrapers, shredders
	Lepidostomatida	collectors, scrapers, shredders
	Limnephilidae	collectors, scrapers, shredders
	Polycentropodidae	collectors, scrapers, shredders
	Rhyacophilidae	collectors, scrapers, shredders

Information from Keiper & Casamatta (2001), Baena (2011), Barrios & Wolff (2011), and Wallace & Merritt (2020).

diverse groups of insects in all habitats in which insects can be found. The exception to this trend is Ephemeroptera, which can be very common on bodies located in shallow lotic systems (Haefner *et al.*, 2004). When compared based on functional feeding groups, the majority of aquatic insects associated with vertebrate carrion are collectors or shredders. Both of these feeding strategies overlap with their chief competitors, crustacea. Evidence of resource partitioning between these freshwater insects and crustacea is not readily apparent and likely contributes to the absence of specialized necrophagy in aquatic insects.

12.5   Terrestrial insects are still interested in dead bodies in water

Vertebrate carrion in aquatic environments is not off limits to terrestrial insects. They simply have to wait their turn. In other words, during the float stages of decomposition. Once a corpse becomes buoyant so that any portion of the body breaks the water surface to be exposed to air, the remains attract necrophagous Diptera and Coleoptera and other insects that frequent terrestrial carrion (Figure 12.13). Succession patterns by necrophagous flies are consistent with

Figure 12.13   A wide range of terrestrial necrophagous Diptera are attracted to floating remains in aquatic environments. *Source*: Photo by D.B. Rivers.

those expected for terrestrial decomposition, with the exception that the length of association is shorter with carrion in aquatic environments. The reasons are obvious; detection and acceptance phases cannot occur prior to gas accumulation and the corpse is likely to sink before fly larvae have completed development. This means that in scenarios in which the body fully submerges when entering a body of water, flies, and other terrestrial insects do not have access to animal remains during the fresh stage. Obviously in shallow water, utilization of exposed skin surfaces can occur immediately as occurs in terrestrial decompositions. Depending on the type of carrion, fly diversity may be quite limited. For example, rat carcasses appear to draw the attention of fewer species of flies (only three species, *Cochliomyia macellaria*, *Lucilia sericata*, and *Sarcophaga bullata)* by comparison to piglets (Payne & King, 1972; Tomberlin & Adler, 1998). The latter appears to attract a similar compliment of necrophagous flies in terms of diversity and abundance in man-made ponds and freshwater streams (Payne & King, 1972; Chin *et al.*, 2008). Regardless of the type of carrion, fly larvae are known to consume soft tissues and create lesions within the skin. This feeding activity contributes to gas release from the carrion, leading to the body collapsing upon itself, and eventually facilitates sinking of the remains. When the remains descend under the water surface, the fly larvae present on the corpse either sink, are consumed by fish or other predators, or remain floating for a short period of time. If the larvae are close enough to shore (or moved by wave action), in some instances they can wander on to dry land to initiate pupariation. For most fly species, the successful transition from larva to pupa is dependent on finding a dry substrate to initiate pupariation. Thus, simply making it to shore may not be enough to guarantee survival.

Beetle activity on floating remains is more variable than with necrophagous flies. Some studies have observed large numbers of silphids, staphylinids, and histerid beetles feeding on fly larvae following colonization of floating remains (Payne & King, 1972), while others witnessed few or no beetles (Tomberlin & Adler, 1998; Chin *et al.*, 2008). Part of the differences is believed to be due to the distance of the floating remains in relation to the shoreline. The further from shore, the less likely adult Coleoptera are to use carrion as a resource. In contrast, distance constraints have not been reported for adult Diptera utilizing floating remains. Presumably this reflects differences in olfactory acuity between necrophagous Diptera and

Coleoptera. A similar trend is apparent with predatory Hymenoptera, which can be readily abundant on remains floating close to shore and less frequent on carrion further out on the water.

12.6   A postmortem interval cannot be determined with aquatic insects

In terrestrial environments, estimation of the PMI is based largely on the predictable succession of necrophagous insects, namely flies, that colonize a corpse. Chapter 14 provides a detailed discussion of the theoretical bases for the PMI based on insect development, as well as a description of the "basics" for performing the calculations. This same method does not lend itself well for use with aquatic insects. Why? The absence of exclusive or specialized necrophagy on vertebrate carrion is a major limitation with using aquatic insects. For terrestrial species, necrophagy links larval development with the corpse. Changes in ambient conditions impact both the feeding insects and the corpse. Obviously, such a strong connection between human remains and most aquatic insects does not occur. However, this is not the only problem that aquatic species present. Several biological assumptions must be met in order to use insects in statistical calculations associated with environmental energy known as degree days. A quick examination of Table 14.1 in Chapter 14 reveals that nearly all aquatic insects do not meet two or more of the basic assumptions. One of the problems is really a logistical issue associated with rearing aquatic insects. Much of the developmental data needed to estimate a PMI is not known for most species of aquatic insects that are commonly discovered on a corpse. Ordinarily with terrestrial species, the needed developmental data can be generated from a series of growth experiments raising the insects from egg to adult at range of temperatures. Base temperatures or developmental thresholds can be determined by a similar means. Developmental experiments are not nearly as easy to perform with aquatic species; it is not just a matter of providing food and an incubator. A microhabitat for each species of interest must be created in the laboratory that simulate multiple abiotic and biotic conditions. As might be imagined, such studies have not been performed with the needed frequency required for forensic entomology. The net effect is that it is not appropriate to use these species to estimate a PMI.

The next question that arises is what is the backup plan? That is, if a PMI cannot be calculated, then how can aquatic insects be used to aid forensic investigations. Unfortunately, there are no methods of equivalent precision available for estimation of the PMI associated with aquatic decompositions or water-related deaths (Dickson et al., 2011). That does not mean that aquatic insects offer no utility to death investigations. In fact, the opposite is true. Several features of growth, development, and behavior of aquatic species provide valuable insight into the length of time that a corpse has been associated with an aquatic environment. For instance, with caddisflies, their unique behaviors of case building can be used to estimate the length of time (a range) that the larvae were associated with the corpse or clothing of the victim (Wallace et al., 2008). In turn, such information can contribute to the estimation of the **postmortem submersion interval** or **PMSI**. The postmortem submersion interval is defined as the time in which a body enters a water environment until when the decedent is discovered. At the time of discovery, the body may no longer be completely submerged. However, the key point is that the corpse was wholly submerged initially. The PMSI provides an empirical, albeit less precise than a PMI, method to estimate the length of time that the aquatic insects has been associated with human remains. Most aquatic insects do not immediately attach or colonize a corpse upon entry into a body of water. Thus, a PMSI based on aquatic insect behavior or development generally does not encompass the entire window of time that a corpse is wholly or partially submerged.

Marine environments create a unique challenge for PMSI estimations in that few to no aquatic insects are present on a corpse. This means that insect development or behavior are not used to estimate accumulated degree days, as can occur with a PMI or PMSI in freshwater. An alternative approach is to rely on determination of a total body score (TBS). The TBS provides a numerical scoring system to characterize physical decomposition of a corpse. A PMSI can be estimated from the TBS, but the value lacks the precision of a PMI associated with terrestrial decomposition. Decomposition in water is influenced by salinity, chemical

composition, and temperature, making physical decay in aquatic environments much less predictable by comparison to terrestrial habitats (De Donno et al., 2014). Assessing physical decay is subjective and, in water, is also complicated by the formation of adipocere, which can delay the appearance of taphonomic characters associated with later stages of decay (Heaton et al., 2010).

Chapter review

Decomposition of human remains in water is different than on land

- Decomposition of a corpse is aquatic ecosystems is much different than carrion decay in terrestrial environments. In freshwater, a human body breaks the surface tension of water and typically sinks, remaining submerged for period of time dependent on ambient water temperature. The composition of aquatic ecosystems varies tremendously based on the type of water, depth, temperature, and season.

- Detailed studies examining processes of physical and chemical decomposition of humans, pigs or any other animal model are quite limited by comparison to the research conducted with terrestrial decomposition. There are several reasons to account for this disparity, including health concerns with deliberate placement of animal remains in natural water supplies, need for specialized equipment and personnel training to recover human remains from deep water, the overall absence of carrion feeding invertebrates that can provide the same type of developmental data that exclusive necrophagy offers in terrestrial ecosystems, and the rarer occasion of human body discovery in aquatic versus terrestrial ecosystems.

- Temperatures of large or deep bodies of water are very stable by comparison to ambient air. Fluctuating air temperatures, especially during seasonal transition, have only a modest impact on surface temperatures. Deeper layers of water remain unchanged without seasonal upwelling or mixing. Consequently, the rate of decomposition for submerged remains will be relatively constant.

- The stability of water temperatures would seemingly create an environment that would favor predictability in terms of estimating the time of death for submerged remains. The reality is that aquatic environments differ in substantial ways from terrestrial habitats, leading to less certainty in time estimates.

- Characterization of human decomposition in freshwater ecosystems involves defining distinct stages of physical decay. The same was true when examining decomposition of human remains in terrestrial environments. Essentially two classification schemes have been devised for characterizing physical decay in aquatic ecosystems. One is based on the natural progression of decomposition associated with a corpse that initially is submerged when entering a body of water and which the corpse is not obstructed by an object below water, meaning that the body can float if it becomes buoyant due to gas production. The second scheme details decomposition for tissues that remain submerged throughout the entire decay process.

Ecological functions of human remains in aquatic ecosystems

- In freshwater systems, a human corpse can function as an immediate or eventual food source to a variety of aquatic organisms including invertebrates and fish; to provide shelter and/or microhabitat for small invertebrate species; to serve as a substrate for primary producers, especially those in the photic zone; to attract secondary predatory species or primary parasites (on species inhabiting corpse); and to function as substrata for species that graze on exposed bone, primary producers, clothing, or other materials.

- The ecological roles presented are not necessarily reflective of discrete successional phases of decomposition, as the attractiveness of human remains is influenced by numerous factors including the type of water, depth of water, season, abiotic characteristics, and whether the body was clothed. In fact, the latter is important because the extent of skin coverage and the clothing material will influence the ecological roles of the corpse.

Why are there no truly necrophagous aquatic insects on carrion?

- In terrestrial environments, there is rich and diverse guild of terrestrial insects that function in macro-decomposition of organic matter, especially carrion. These insects and the subsequent carrion communities that form are comprised chiefly of Diptera and Coleoptera in terms of abundance and diversity. By contrast, such adaptation for a carrion existence does not exist with aquatic insects, no matter what type of aquatic environment they reside in. The absence of necrophagy in aquatic insects is surprising when considering that up to 70–90% of freshwater benthic communities are comprised of insects.

- There is not a simple answer to the question of why necrophagy is absent in aquatic insects. One hypothesis posed is that among the vast numbers of terrestrial insects, a relatively low proportion have specialized as exclusively necrophagous. By extension, far fewer aquatic insects should be expected to evolve as specialized necrophagous species. The logic of the argument is founded in the evolution of the Insecta; insects first evolved for a terrestrial existence and then secondarily adapted to freshwater. Among the terrestrial groups that display necrophagy, very few have aquatic representation, and none of those aquatic species are exclusively necrophagous.

- To an extent the underpinnings of the phylogeny hypothesis are that too few species exist in aquatic environments and that there has been insufficient evolutionary time for specialization toward the necrophagous condition. Perhaps some aquatic species are in transition as recent observations suggest that under specific conditions, some species of aquatic insects display preferential necrophagy.

Associations between freshwater insects and aquatic remains

- Detailed life history information for the thirteen orders of insects that have at least some members with aquatic or semi-aquatic life stages is lacking by comparison to terrestrial necrophagous species. Of those insect groups, eight orders have representatives in which one or more developmental stages are likely to be associated with vertebrate carrion discovered in freshwater habitats.

- The three orders most commonly discovered on human remains in freshwater all display holometabolous development and at least two (Coleoptera and Diptera) are among the most abundant and diverse groups of insects in all habitats in which insects can be found. The exception to this trend is Ephemeroptera, which can be very common on bodies located in shallow lotic systems. When compared based on functional feeding groups, the majority of aquatic insects associated with vertebrate carrion are collectors or shredders. Both of these feeding strategies overlap with their chief competitors, crustacea.

Terrestrial insects are still interested in dead bodies in water

- Vertebrate carrion in aquatic environments is not off-limits to terrestrial insects. They simply have to wait for their turn. In other words, during the float stages of decomposition. Once a corpse becomes buoyant so that any portion of the body breaks the water surface to be exposed to air, the remains attract necrophagous Diptera and Coleoptera and other insects that frequent terrestrial carrion.
- Succession patterns by necrophagous flies are consistent with those expected for terrestrial decomposition, with the exception that the length of association is shorter with carrion in aquatic environments.
- In shallow water, utilization of exposed skin surfaces can occur immediately as occurs in terrestrial decompositions. Depending on the type of carrion, fly diversity may be quite limited. For example, rat carcasses appear to draw the attention of fewer species of flies by comparison to piglets.
- Beetle activity on floating remains is more variable than with necrophagous flies. Part of the differences is believed to be due to the distance of the floating remains in relation to the shoreline. The further from shore, the less likely adult Coleoptera are to use carrion as a resource. In contrast, distance constraints have not been reported for adult Diptera utilizing floating remains.

A postmortem interval cannot be determined with aquatic insects

- The absence of exclusive or specialized necrophagy on vertebrate carrion is a major limitation with using aquatic insects. For terrestrial species, necrophagy links larval development with the corpse. Changes in ambient conditions impact both the feeding insects and the corpse. Obviously, such a strong connection between human remains and most aquatic insects does not occur. However, this is not the only problem that aquatic species present. Several biological assumptions must be met in order to use insects in statistical calculations associated with environmental energy known as degree days. The net effect is that it is not appropriate to use these species to estimate a PMI.
- If a PMI cannot be calculated, then how can aquatic insects be used to aid forensic investigations. Unfortunately, there are no methods of equivalent precision available for estimation of the PMI associated with aquatic decompositions or water-related deaths. That does not mean that aquatic insects offer no utility to death investigations. In fact, the opposite is true. Several features of growth, development, and behavior of aquatic species provide valuable insight into the length of time that a corpse has been associated with an aquatic environment such information can contribute to the estimation of the postmortem submersion interval or PMSI. The postmortem submersion interval is defined as the time in which a body enters a water environment until when the decedent is discovered.
- Marine environments create a unique challenge for PMSI estimations in that few to no aquatic insects are present on a corpse. This means that insect development or behavior are not used to estimate accumulated degree days, as can occur with a PMI or PMSI in freshwater. An alternative approach is to rely on determination of a TBS. The TBS provides a numerical scoring system to characterize physical decomposition of a corpse. A PMSI can be estimated from the TBS, but the value lacks the precision of a PMI associated with terrestrial decomposition.

Test your understanding

Level 1: knowledge/comprehension

1. Define the following terms:

  (a) postmortem interval
  (b) postmortem submersion interval
  (c) fluvial transport
  (d) floating decay
  (e) edema
  (f) froth cone
  (g) shredders

2. Match the terms (i–vi) with the descriptions (a–f).

(a) Accumulation of decomposition gases lead to corpse buoyancy	(i) Collectors
(b) Consume fine organic particulate	(ii) Total body score
(c) Movement of bone due to water currents	(iii) Scraper
(d) Numerical system for evaluating physical decay	(iv) Necrophagy
(e) Herbivores	(v) Fluvial transport
(f) Period of time that a corpse is submerged	(vi) PMSI
(g) Consumption of soft tissues of a carrion	(vii) Early float

3. Describe the unique aspects of physical decay of vertebrate carrion that occurs in aquatic environments.

4. Explain the differences between a PMI, PMSI, and TBS.

5. Discuss which type(s) of functional feeding groups would most likely be associated with a corpse in the submerged fresh stage.

Level 2: application/analysis

1. Explain why the events of autolysis and putrefaction are generally delayed in marine environments by comparison to terrestrial habitats.

2. Describe why specialized necrophagy is absent in aquatic insects.

3. Explain why a PMI cannot be estimated based on caddisfly development?

4. Discuss the potential impact of adipocere formation on scrapers and shredders using human remains in freshwater.

Level 3: synthesis/evaluation

1. Discuss what type of information is needed to improve the precision of a PMSI using freshwater insects.

2. Speculate on the types of experiments that could be performed to examine whether a given freshwater predatory insect species is transitioning toward specialized necrophagy.

References cited

Anderson, G.S. & Hobischak, N.R. (2004) Decomposition of carrion in the marine environment in British Columbia, Canada. *International Journal of Legal Medicine* 118(4): 206–209.

Baena, M. (2011) Unusual feeding habits in four Iberian Heteroptera (Hemiptera). *Boletin de la Sociedad Entomológica Aragonesa (S.E.A.)* 48: 399–401.

Barrios, M. & Wolff, M. (2011) Initial study of arthropods succession and pig carrion decomposition in two freshwater ecosystems in the Colombian Andes. *Forensic Science International* 212(1–3): 164–172.

Bianchi, T.S. (2011) The role of terrestrially derived organic carbon in the coastal ocean: a changing paradigm and the priming effect. *Proceedings of the National Academy of Sciences* (USA) 108: 19473–19481.

Chin, H.C., Marwi, M.A., Jeffery, J. & Omar, B. (2008) Insect succession on a decomposing piglet carcass placed in a man-made freshwater pond in Malaysia. *Tropical Biomedicine* 25(1): 23–29.

De Donno, A., Campobasso, C.P., Santoro, V., Leonardi, S., Tafuri, S. & Introna, F. (2014) Bodies in sequestered and non-sequestered aquatic environments: a comparative taphonomic study using decompositional scoring system. *Science and Justice* 54: 439–446.

Dickson, G.C., Poulter, R.T., Maas, E.W., Probert, P.K. & Kieser, J.A. (2011) Marine bacterial succession as a potential indicator of postmortem submersion interval. *Forensic Science International* 209(1–3): 1–10.

Fenoglio, S., Merritt, R.W. & Cummins, K.W. (2014) Why do no specialized necrophagous species exist among aquatic insects?. *Freshwater Science* 33(3): 711–715.

Haefner, J.N., Wallace, J.R. & Merritt, R.W. (2004) Pig decomposition in lotic aquatic systems: the potential use of algal growth in establishing a postmortem submersion interval (PMSI). *Journal of Forensic Science* 49(2): 1–7.

Haglund, W.D. & Reay, D.T. (1993) Problems of recovering partial human remains at different times and locations: concerns for death investigators. *Journal of Forensic Science* 38(1): 69–80.

Heaton, V., Lagden, A., Moffatt, C. & Simmons, T. (2010) Predicting the postmortem submersion interval for human remains recovered from UK waterways. *Journal of Forensic Sciences* 55(2): 302–307.

Hobischak, N.R. (1997) *Freshwater invertebrate succession and decompositional studies on carrion in British Columbia.* Doctoral dissertation, Theses, Dept. of Biological Sciences/Simon Fraser University.

Hobischak, N.R. & Anderson, G.S. (1999) Freshwater-related death investigations in British Columbia in 1995–1996. A review of coroners cases. *Canadian Society of Forensic Science Journal* 32(2–3): 97–106.

Hobischak, N.R. & Anderson, G.S. (2002) Time of submergence using aquatic invertebrate succession and decompositional changes. *Journal of Forensic Science* 47(1): 142–151.

Hyma, B.A. & Rao, V.J. (1991) Evaluation and identification of dismembered human remains. *The American Journal of Forensic Medicine and Pathology* 12(4): 291–299.

Jansen, B., Kalbitz, K. & McDowell, W.H. (2014) Dissolved organic matter: Linking soils and aquatic systems. *Vadose Zone Journal*. https://doi.org/10.2136/vzj2014.05.0051.

Keiper, J.B. & Casamatta, D.A. (2001) Benthic organisms as forensic indicators. *Journal of the North American Benthological Society* 20(2): 311–324.

Kline, T.C., Goering, J.J. & Piorkowski, R.J. (1997) The effect of salmon carcasses on Alaskan freshwaters. In A.M. Milner & M.W. Oswood (eds) *Freshwaters of Alaska*. Springer, New York, NY, pp. 179–204.

Magni, P.A., Borrini, M. & Dadour, I.R. (2013) Human remains found in two wells: a forensic entomology perspective. *Forensic Science, Medicine, and Pathology* 9(3): 413–417.

McDaniel, C.H., McHugh, J.V. & Batzer, D.P. (2019) Colonization of drying temporary wetlands by *Coptotomus loticus* (Coleoptera: Dytiscidae): a unique strategy for an aquatic wetland insect. *Wetlands Ecology and Management* 27: 627–634.

Merritt, R.W. & Cummins, K.W. (eds) (1996) *An Introduction to the Aquatic Insects of North America*. Kendall Hunt, Dubuque, Iowa.

Merritt, R.W., Cummins, K.W. & Berg, M.B. (2017) Trophic relationships of macroinvertebrates. In R.F. Hauer & G. Lamberti (eds) *Methods in Stream Ecology, Volume 1*, pp. 413–433. Academic Press, San Diego.

Minakawa, N., Gara, R.I. & Honea, J.M. (2002) Increased individual growth rate and community biomass of stream insects associated with salmon carcasses. *Journal of the North American Benthological Society* 21(4): 651–659.

O'Brian, T.G. & Kuehner, A.C. (2007) Wxing grave about adipocere: soft tissue change in an aquatic context. *Journal of Forensic Science* 52: 294–306.

Payne, J.A. & King, E.W. (1972) Insect succession and decompostition of pig carcasses in water. *Journal of the Georgia Entomological Society* 7(3): 153–162.

Rivers, D.B. (2016) *Insects: Evolutionary Success, Unrivaled Diversity, and World Domination*. Johns Hopkins Press, Baltimore, MD.

Schoenly, K. & Reid, W. (1987) Dynamics of heterotrophic succession in carrion arthropod assemblages: discrete series or a continuum of change. *Oecologia* 73: 192–202.

Silva-Soares, T., Segadilha, J.L., Braga, R.B. & Clarkson, B. (2019) Necrophagy on *Rhinella granulosa* (Amphibia, Anura, Bufonidae) by the aquatic beetle families Hydrophilidae and Dytiscidae (Insecta, Coleoptera) in Caatinga environment, Northeastern Brazil. *Herpetology Notes* 12: 869–872.

Sorg, M.H., Dearborn, J.H., Monahan, E.I., Ryan, H.F., Sweeney, K.G. & David, E. (1997) Forensic taphonomy in marine contexts. In W.D. Haglund & M.H. Sorg (eds), *Forensic Taphonomy: The Postmortem Fate of Human Remains*, pp. 567–604. CRC Press, Boca Raton.

Stuart, B.H. & Ueland, M. (2017) Decomposition in aquatic environments. In E.M. Schotsmans, N. Márquez-Grant & S.L. Forbes (eds) *Taphonomy of Human Remains: Forensic Analysis of the Dead and the Depositional Environments*, pp. 235–250. Wiley-Blackwell, Sussex.

Suckling, J.K., Spradley, M.K. & Godde, K. (2015) A longitudinal study of human outdoor decomposition in central Texas. *Journal of Forensic Sciences* 61: 19–25.

Tomberlin, J.K. & Adler, P.H. (1998) Seasonal colonization and decomposition of rat carrion in water and on land in an open field in South Carolina. *Journal of Medical Entomology* 35(5): 704–709.

Tomberlin, J.K., Mohr, R., Benbow, M.E., Tarone, A.M. & VanLaerhoven, S. (2011) A roadmap bridging basic and applied research in forensic entomology. *Annual Review of Entomology* 56: 401–422.

Wallace, J.R. & Merritt, R.W. (2020) The role of aquatic organisms in forensic investigations. In J.H. Byrd & J.K. Tomberlin (eds) *Forensic Entomology: The Utility of Arthropods in Legal Investigations*, pp. 155–186. CRC Press, Boca Raton.

Wallace, J.R., Merritt, R.W., Kimbirauskas, R., Benbow, M.E. & McIntosh, M. (2008) Caddisflies assist with homicide case: determining a postmortem submersion interval using aquatic insects. *Journal of Forensic Sciences* 53(1), 219–221.

Supplemental reading

Benbow, M.E., Tomberlin, J.K. & Tarone, A.M. (2018) *Carrion Ecology, Evolution, and Their Applications*. CRC Press, Baco Raton, FL.

Dalal, J., Sharma, S., Verma, K., Dhattarwal, S.K. & Bhardwaj, T. (2017) Data of drowning related deaths with reference to entomological evidence from Haryana. *Data in Brief* 15: 975–980.

Schoenly, K., Goff, M.L., Wells, J.D. & Lord, W.D. (1996) Quantifying statistical uncertainty in succession-based entomological estimates of the postmortem interval in death scene investigations: a simulation study. *American Entomologist* 42: 106–112.

Schotsmans, E.M., Márquez-Grant, N. & Forbes, S.L. (2017) *Taphonomy of Human Remains: Forensic Analysis of the Dead and the Depositional Environment*. Wiley-Blackwell, Sussex.

Sharma, S. & Singh, R. (2014) Forensic aquatic entomology a review. *The Indiana Police Journal* 61: 227–238.

Siri, A., Mariani, R. & Varela, G.L. (2019) Corpses in aquatic environments: two human forensic cases with associated chironomid (Insecta: Diptera: Chironomidae) larvae. *Australian Journal of Forensic Sciences*. https://doi.org/10.1080/00450618.2019.1675761.

Wallace, J.R. (2019) Computer-aided image analysis of crayfish bitemarks—reinterpreting evidence: a case report. *Forensic Science International* 299: 203–207.

Additional resources

Decomposition of baby pigs video: http://www.youtube.com/watch?v=R1CD6gNmhr0

Forensic Anthropology Center at University of Tennessee: http://fac.utk.edu/

Forensic Anthropology Center at Texas State University: http://www.txstate.edu/anthropology/facts/

Forensics in water: http://www.exploreforensics.co.uk/forensics-water.html

Institute for Forensic Anthropology and Applied Sciences: http://forensics.usf.edu/

Stages of pig decomposition: http://australianmuseum.net.au/Stages-of-Decomposition

What happens to a dead body in the ocean? https://www.livescience.com/48480-what-happens-to-dead-body-in-ocean.html

Chapter 13

Microbiomes of carrion and forensic insects

Overview

A decomposing animal is home to a diverse community of prokaryotic and eukaryotic organisms, all united by the dead. Insects are but one member of this diverse and complex community known as a necrobiome. Necrophagous insects are highly touted for their efficiency in consuming corpse tissues and utility in criminal investigations. However, they are poised to be knocked off the perch of being the most useful ecological evidence to crime scene investigations. How can this be? Well, insects have competition, both literally and figuratively, in the form of microscopic cells. A case can be made that microbes are the real stars of decomposition. Bacteria are by far the most abundant members of a carrion community. In fact, a rich and diverse community of microbes, known as a microbiome, is associated with both a living and dead animal. When human death occurs, many of the microorganisms on and in the living host remain, others perish, and a multitude of new arrivals inhabit the corpse. Necrophilous insects are central to the changing microbial community by functioning as transport conduits for several microbial species. The potential of microbes to aid criminal investigations has long

been speculated, but it has been only recently that tools have become available to intimately examine the microbiomes of the dead. Chapter 13 explores the utility of microbiomes in forensic investigations, including discussions of how microbes contribute to the decomposition of carrion, provide clues to the location of a clandestine grave or reveal that a body has been moved from another site, and how bacteria can be used in the estimation of a postmortem interval. The chapter also examines the relationships between bacteria and necrophilous insects, especially in the context of microbiome influences on the attraction to and use of carrion by necrophagous flies.

The big picture

- Microbes of a corpse are important sources of information.
- Microbial communities exist on living and dead hosts.
- Estimation of the postmortem interval using microbes.
- Microbes and insects are intimate members of the necrobiome.

The Science of Forensic Entomology, Second Edition. David B. Rivers and Gregory A. Dahlem.
© 2023 John Wiley & Sons Ltd. Published 2023 by John Wiley & Sons Ltd.
Companian website: www.wiley.com/go/forensicentomology2

13.1   Microbes of a corpse are important sources of information

All higher life forms live in close association with an array of microorganisms. In fact the very existence of plants and animals is dependent on the presence and activity of the microbes that live on and within them. Humans are no different. The Human Microbiome Project launched at the beginning of the twenty-first century has uncovered a wealth of information relating microbial functions to human health and disease. The findings, in short, indicate that humans literally cannot exist without microbes; microorganisms contribute more genes to human survival than human genes do (Human Microbiome Consortium, 2012). Perhaps that should not be a surprise when considering the revelation that microbial cells out number human cells by 10-fold in and on the surface of a human (Human Microbiome Project, 2012). To give this context, it has been estimated that the human body contains somewhere in the vicinity of 40 trillion cells (4×10^{13})[1] (Bianconi et al., 2013). If all human cells and tissues were removed, a complete shell of the individual would still exist in the form of microbes. How do microbes stack up against the most abundant and diverse animal group on the planet? Remember from Chapter 4 that the Class Insecta represents more than 75% of all animal life, with just over 1 million extant species. Simply amazing when viewed alone.

However, those values lose some luster when compared to microbial abundance (Figure 13.1). The total number of microbial cells associated with one person exceeds the total number of all insects on the planet on any given day. Such extraordinary abundance coupled with rich diversity of phyla and taxa allow microbes to occupy a wide array of habitats and fill a broad range of ecological roles, including as decomposers on human corpses and vertebrate carrion. It is here that the intersection between microorganisms and necrophilous insects becomes blatantly obvious, and where **forensic microbiology** – the subdiscipline of microbiology focused on the role of microorganisms in medico-legal and criminal investigations – and forensic entomology become intertwined (Carter et al., 2017). Both disciplines rely on faunal succession patterns and species-specific activity to aid death investigations.

The importance of microbes to death investigations has a long history, dating back to at least the nineteenth century. Early pioneers examined the pathogenicity and contributions of microbes to disease and death. Louis Pasteur and Robert Koch and many others were renowned for their work establishing the linkages between microbes and disease. In fact, any student in a microbiology class has undoubtedly had to learn Koch's Postulates, a set of guidelines used to establish cause–effect relationships between microorganisms and diseases. The postulates were flawed in some respects and have long since been replaced by the Bradford Hill criteria, but nonetheless the original ideas helped lay a foundation for the

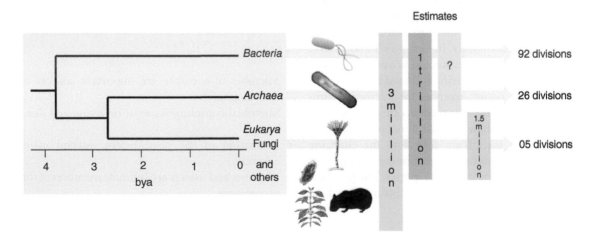

Figure 13.1 Estimates of the number of microbial species present in the three domains of life. Bacteria are the oldest and most biodiverse group, followed by Archaea and Fungi (the most recent groups). *Source*: L. Vitorino/Wikimedia Commons/CC BY-SA 4.0.

emerging field of microbial pathogenicity and later to postmortem microbiology. The former tends to fall more under the guise of medical pathology and not so much with ecological aspects of death investigations. Postmortem microbiology is similar to microbial pathogenicity in that the discipline is focused on diseases or infections induced by microbes, but not with the goal of uncovering etiology of diseases. Instead, postmortem microbiology investigates cause and manner of death as related to microbes. More specifically, microbiological analyses are used to investigate when a suspected but non-confirmed antemortem infection is believed to have contributed to or directly caused death, if signs of microbial infection are evident during autopsy, or when a sudden unexpected death occurred (d'Aleo *et al.*, 2017). Such death investigations are directly linked to the disciplines of forensic medicine and forensic pathology (See Chapter 1).

Forensic microbiology overlaps in scope with postmortem microbiology to the extent that both provide information relevant to death investigations. However, the type of information that forensic microbiology addresses is more consistent with that of forensic entomology. Species-specific microbial activity and patterns of faunal succession by microbes on a decomposing corpse can yield insight into the time of death, reveal whether the body has been moved from another location, or point to the site of a clandestine grave (Pechal *et al.*, 2014; Metcalf *et al.*, 2017). In the case of the latter, microbes associated with a particular soil type or plant can be connected to a specific environment and possibly location (Zhou & Bian, 2018). Microbes have even been used to place a suspect at a crime scene or connect a person to an object such as a weapon (Carter *et al.*, 2017). The potential of microbes in criminal investigations is just now coming to fruition, but the concept of using microorganisms as physical evidence is not new. The famed criminalist Edmond Locard recognized the potential utility of microbes in forensic investigations nearly a century ago. In fact, Locard used microbes as trace evidence, even associating certain microbes with distinct areas of Paris (Metcalf *et al.*, 2017). This example is consistent with **Locard's Exchange Principle**, in which Locard predicted that every contact between two physical objects leaves a trace. Current research in forensic microbiology is certainly confirming this principle, as any contact with a human, dead or

alive, will leave a microbial trace. We will explore this idea more in the remaining sections of the chapter, especially in the context of microbial communities. Much of the insight that can be derived from microorganisms is based on the fact that microbes are ubiquitous and have predictable ecologies (Metcalf, 2019). Sound familiar? It should, the same characteristics are what make necrophagous insects so useful to death investigations.

The emergence of a new array of sequencing tools has elevated the position of microbes as physical evidence. Information is derived from investigating patterns in microbial communities using **next-generation sequencing** of phylogenetically and/or taxonomically specific gene markers coupled with robust statistical and computational methods, including machine learning (Metcalf *et al.*, 2017). The most frequently targeted gene markers include 16S rRNA (bacteria and Archaea), 18S rRNA (microbial eukaryotes) and ITS (fungi) sequences (Belk *et al.*, 2018). Next-generation sequencing or massive parallel sequencing is a broad term that encompasses several modern sequencing techniques. All share the characteristics of permitting sequencing of nucleic acids much more rapidly and cheaply than the previously used Sanger sequencing. The old method of separating nucleic acids by electrophoresis is a slow process. In contrast, non-electrophoretic methods are faster and permit the use of very sensitive detection techniques, such as bioluminescence. All versions of next-generation sequencing are also high-throughput methods. The end result has been a revolution in the study of genomics and molecular biology of any organism. Perhaps nowhere is the pace of research expansion occurring more rapidly than with microbiology and especially with the study of microbial communities residing in and on the human body. Section 13.2 will delve into these microbial communities, with emphasis placed on those that comprise a corpse.

13.2 Microbial communities exist on living and dead hosts

Microorganisms show the potential to be useful physical evidence in a wide range of contexts. In many ways, microbes are more informative than insect evidence. For example, microbes are likely to be present through all seasons, not just during the warm months of summer. Necrophagous insects are predictably absent when temperatures are low, such as occurs during late fall through early spring in the majority of North America. We also know from our discussions of insect succession in Chapters 10 and 11 that in some instances, insects are delayed in detecting or reaching a corpse. If a body is buried, placed in a sealed container, located in certain types of aquatic environments, or discovered in a high-rise building, insect colonization often does not occur at all. Yet none of these scenarios are off limits to microbes. This is not to say that new microbial colonizers may not suffer from the same obstacles that necrophilous insects do in reaching a corpse that is otherwise secluded. However, the difference lies in the fact that the corpse was already "colonized" prior to death.

13.2.1 What is a microbiome?

The human body is home to a community of microorganisms, residing externally on the skin and the internal environment. Microbial communities exist essentially everywhere on and in the body, on every surface of the organs, "tubing" and architecture that makes us functional. Collectively, the community is referred to as the **microbiome** or **microbiota**. The term "microbe" is not just a reference to bacteria. It includes a broad array of microorganisms, including microbes found in all three domains of life (Bacteria, Archaea, and Eukarya) (Metcalf et al., 2017). Thus, the microbial community is composed of a diverse range of microorganisms competing for the resources of the living or dead host (Figure 13.2). Most of what is known about this complex community has only recently been uncovered, largely due to the advent of powerful next-generation sequencing tools. As mentioned in the previous section, these new methods have revolutionized genomics research of all organisms, especially microbes, and not just because of enhanced processing time. The recent advances in next-generation sequencing has permitted obtaining microbial sequence data from organisms that otherwise have been unculturable. This has always been a limitation with microbes discovered in soil

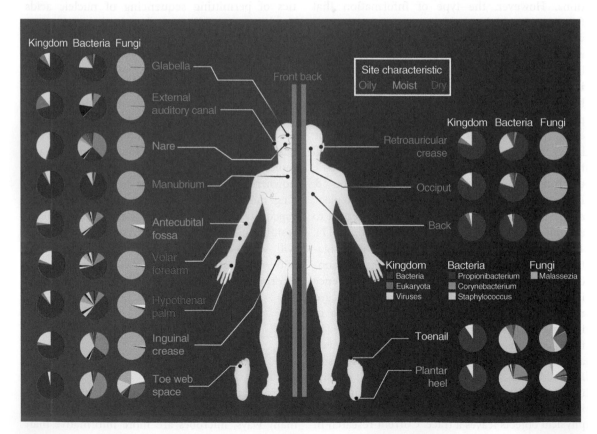

Figure 13.2 Metagenomic analysis of the human skin has shown a wide range of bacteria, fungi, and viruses comprise the microbial community or microbiome. *Source:* National Human Genome Institute/Wikimedia/CC BY 2.0.

and aquatic environments. In fact, some estimates suggest that approximately 99% of existing bacteria cannot be cultured in the laboratory (Thompson *et al.*, 2013). Metagenomic sequencing has overcome this limitation. Investigations of the human microbiome has led to the identification of new taxa and analysis of previously unculturable species. Sequence data has shown that the composition of the microbiome is also not ubiquitous throughout the body and can change over time due to intrinsic (i.e., health) and extrinsic (diet, use of medications, body temperature, and humidity) conditions. Changes in community structure are also evident following death. Sequencing tools have permitted examination of successional patterns in the microbiome, as well as delineating the source (i.e., insects, soil, and vertebrate scavengers) of new microbes entering the community over time. In many respects, next generation sequencing of microbial taxa provides a robust and high-resolution approach to species succession on carrion that forensic entomology does not. The next section will examine features of the microbiome on a human corpse that are useful for criminal investigations.

13.2.2   What is a necrobiome?

At the moment of death or **agonal period**, a deceased person is already colonized by microbes. The microbiome from the living host is ever present, but changes occur rapidly. After death, cells and tissues of the decedent decompose, releasing their contents into the surrounding fluids and tissues. Autolysis initiates these decomposition events, followed in close procession by putrefaction. Putrefaction is largely driven by microbial enzymes. Internally, populations of microbes proliferate, some enter a state of quiescence, and many die off. Outward manifestations of physical and chemical decay can be quite profound and are frequently used as a means to characterize stages of decomposition (See Chapter 10 for more details). As corpse decay progresses, the microbial community changes in composition in both abundance and diversity. In fact, the corpse itself becomes a habitat for a much larger and broader community than just the microbiome. A **necrobiome** forms, a community composed of prokaryotic and eukaryotic organisms united in the attraction and utilization of a decomposing heterotrophic mass, i.e., animal remains (Pechal *et al.*, 2013). Perhaps most recognizable are

the multitudes of necrophilous insects and vertebrate scavengers that arrive, aligned with bacteria, fungi, nematodes, and other microorganisms to form a complex community on and within a decomposing corpse. Within the realm of microbes, the community that is specifically associated with a human corpse or vertebrate carrion is broadly termed the **postmortem microbiome** (Zhou & Bian, 2018). The postmortem microbiome has two major components, the **thanatomicrobiome** or the microbial community located within the corpse including internal organs and cavities enclosed within the body, and the **epinecroticmicrobiome**, the microbial community found on external surfaces (Javan *et al.*, 2016). The external surfaces include superficial epithelial tissues, mucosal membranes of the buccal cavity, and the lining of the digestive tract. The later initially may be surprising when thinking of external surfaces, but the inside of the alimentary canal is continuous with the outside of the body. Neither component of the postmortem microbiome has been extensively studied, although that, too, is changing rapidly. The majority of research studies have been conducted in the last 10 years (actually less), using next-generation sequencing to shed light on phyla and taxa composition and successional changes in community structure. In many instances, only broad generalizations can be made at this time. For our purposes, we will examine the two components separately, focusing on the consistent trends that have been observed with human corpses and animal models.

13.2.2.1   The thanatomicrobiome

Let us begin our discussion of the postmortem microbiome by exploring the internal microbial community or thanatomicrobiome. An examination of the internal microbiome can be thought of as an "inside out" approach (Bucheli & Lynne, 2016). Why inside out? It is based on the investigative approach of examining changes associated with intrinsic microbes after death and determining if and how these microorganisms impact the external microbiome. Far more attention has been given to this component of the postmortem microbiome, largely because microbial samples are routinely collected from a corpse during autopsy, permitting access to a range of tissues for microbial study. Much of the information derived from studies of the internal microbiome have yielded in depth insight into the processes of putrefaction. Traditionally, internal organs are viewed as sterile, with the exception of the gut (Stewart, 2012). So, any microbe discovered on or in an organ postmortem is (i) presumed to be linked to decomposition and (ii) derived from another location from within the body, assuming there is no penetrating trauma associated with the corpse. In terms of specific internal sites, the brain, heart, and spleen are thought to have the most stable or similar microbial community composition over time (Can *et al.*, 2014; Zhou & Bian, 2018), while rapid bacterial growth occurs in blood, bone marrow, liver and prostate (Clement *et al.*, 2016; Thomas *et al.*, 2017). The liver displays the highest microbial diversity during decomposition. There are several possible explanations to account for this diversity (Javan *et al.*, 2016):

1. The liver is the largest solid organ in the body, so increased microbial diversity and abundance are to be expected;
2. Many major metabolic functions including macromolecule storage occur in the liver, meaning an abundance of nutrient-rich compounds are present;
3. The liver receives nutrient-rich blood from two separate blood circuits, increasing the potential association with systemic microbes;
4. Stomach acids, pancreatic and intestinal enzymes, and gallbladder fluids trigger autolysis and other chemical aspects of decomposition following death that immediately accumulate around the liver;
5. The liver is in close proximity to other organs that contain rich microbial flora antemortem and postmortem. So, the potential for microbial transmission following death is quite high.

The presence of microbial communities on other internal organs are believed to occur by similar mechanisms as those associated with the liver. For example, bacterial transport is known to occur via the bloodstream and prior to death (a condition called antemortem bacteremia) to several internal organs. In other cases, microorganisms spread through interstitial or other internal fluids during the process of dying or shortly after. This is viewed as microbial spread during the perimortem window (Javan *et al.*, 2016). Finally, microbial translocation from the epinecroticmicrobiome occurs via blood and other body fluids to internal locations (Morris *et al.*, 2006). Microbial movement occurs relatively rapidly (5–48 hours after death) by this mechanism.

A consistent change in the thanatomicrobiome is a continuous decline in the abundance of anaerobic taxa which were originally associated with the microbiome of the living host. This is especially true for the anaerobes *Lactobacillus* and *Bacteriodetes*. The decline in these genera has been attributed to the release of oxidative gases during autolysis (Zhou & Bian, 2018). In contrast, the thanatomicrobiome experiences a postmortem shift from aerobic to anaerobic bacteria (and presumably other microbes) across all internal body sites (Bucheli & Lynne, 2016). For example, an increase in the abundance of *Clostridiales*, *Pseudomonas*, and *Streptococcus* occur over time following death (Javan *et al.*, 2016). However, the rise in *Pseudomonas* and *Streptococcus* appears to be specific to decomposing females. The most distinct changes are associated with several species of *Clostridium*. At least nine species of *Clostridium* are known to rapidly proliferate on an array of internal body sites, each demonstrating translocation to other areas within the decomposing body. The ability of these bacteria to so effectively utilize a decomposing corpse has led to the concept of the **postmortem clostridium effect** (PCE), essentially a term recognizing the omnipresence of *Clostridium* spp. on a decaying corpse. The defining features of the PCE are the very rapid proliferation rates or short doubling times of these bacteria (in some species < 8 minutes), the ability to digest collagen fibers and thus facilitate translocation to other tissues, and the adaptive capacities to thrive in hypoxic and anoxic conditions that arise once the heart ceases and which lasts until the skin is breached typically sometime post-bloat (Javan *et al.*, 2017).

A unifying feature of many research studies into any aspect of the postmortem microbiome is the search for a biomarker that can serve as a time signal. In other words, are there specific changes in the microbial communities on or in the host that can be used to estimate the postmortem interval? This topic will be discussed in more detail in Section 13.3. For now, insufficient information exists to use changes in the thanatomicrobiome as a microbial clock in death investigations.

13.2.2.2   The epinecroticmicrobiome

By comparison to the thanatomicrobiome, far fewer research studies have focused on the epinecroticmicrobiome (Javan *et al.*, 2016). From the perspective of forensic medicine and autopsy investigations, this bias makes sense. Microbial samples are routinely collected during *in situ* pathological examinations of a corpse. However, from the vantage point of forensic microbiology and carrion ecology, the epinecroticmicrobiome provides more holistic data regarding decomposition and interactions between the microbial communities and the environment. This "outside in" approach to studying the postmortem microbiome is concerned with determining if the microbial community on the skin surface changes over time (Bucheli & Lynne, 2016). If it does, and unquestionably it does, then the next steps are to determine if the changes in community structure occur in a predictable pattern, as occurs with insect succession on a corpse. Microbes on external surfaces are directly influenced by abiotic and biotic factors associated with the environment. Thus, the potential for changes to the microbial community structure and the necrobiome are much greater than with the more concealed, and arguably stable conditions, found inside of the corpse. Consequently, the epinecroticmicrobiome displays more variability in terms of composition over time. This is true in terms of progression of decomposition, on site and geographic differences in which the corpse is discovered or located, and across seasons (Bucheli & Lynne, 2016).

At the time of death, the initial surface microbiome is composed largely of microbes associated with the living host. Changes in the bacterial composition of the community occur from the agonal period through putrefaction or bloat. During the bloat stage and prior to drying of soft tissues, microbes from necrophilous flies are commonly associated with the epinecroticmicrobiome (Bucheli & Lynne, 2016). Bacterial species in the genera *Wohlfahrtimonas* and *Ignatzschineria* represent traces of the association between flies and a cadaver when discovered on a corpse. Despite the fact that hundreds of adult flies and thousands of fly larvae make direct contact with a corpse, frequently for an extended period of time (e.g., several days), only a small percentage of the microbial community is the result of fly deposition on the skin surfaces (Metcalf *et al.*, 2016). It is not clear yet whether fly species that arrive much later in decay (i.e., dry stages) alter the composition of the epinecrotic microbial community. Significant contributions to the changing microbiome can be attributed to the surrounding environment, especially the soil during outdoor decompositions. Several species of *Acinetobacter* are derived from the underlying soil and become quite abundant on corpse tissues devoid

of moisture (Bucheli & Lynne, 2016). Studies examining changes to the epinecroticmicrobiome of submerged remains have observed similar time-dependent increases in Proteobacteria coupled with depressions in Firmicutes (Dickson *et al.*, 2011; Benbow *et al.*, 2015). Proteobacteria are believed to be derived from the environment and thus the increases do not represent proliferation of existing microbes on the skin. In contrast, the decline in Firmicutes species is likely linked to an overall shift from aerobic to anaerobic microbes in the microbiome associated with specific external surfaces (Javan *et al.*, 2017).

Shifts in microbial communities on vertebrate carrion can also be linked to seasonal changes. For example, during seasons characterized by low or no necrophilous fly activity, fly-specific microbes are absent from the epinecroticmicrobiome (Bucheli & Lynne, 2016). Similarly, colonization of pig carcasses by Proteobacteria are season-specific, both in freshwater and marine environments (Dickson *et al.*, 2011; Benbow *et al.*, 2015). Adding to the complexity of the epinecrotic communities is that all of the changes are dependent on the body site sampled. For example, in studies using mouse carcasses, species of Firmicutes increased on the abdomen during the bloat stage, but declined on other areas of the skin, especially if the skin is ruptured, whereas Proteobacteria tended to dominate only on the skin of the head and torso (Metcalf *et al.*, 2013). The take home message is that tremendous variability in microbe composition is found from one site to another on the body. Such variability is exciting from a basic research perspective, but it contributes to the challenge of developing statistical models for predicting successional change on a corpse.

An additional source of "new" microbes to the skin surface is the corpse itself. Thus far, we have discussed the two components of the postmortem microbiome as separate entities, which they are. However, cross colonization does occur. For example, in late stages of bloating, decomposition fluids and gases purge from the body openings, spilling the contents into the environment. These fluids are a source of microorganisms that in turn, can become part of the epinecroticmicrobiome (Bucheli & Lynne, 2016). Likewise, any penetrating lesions or wounds can result in an exchange of microbiomes between the internal and external microbial communities. The same is true for the activity of invertebrate and vertebrate scavengers. Deep penetrating fly larvae may readily spread occupants of the thanatomicrobiome via their

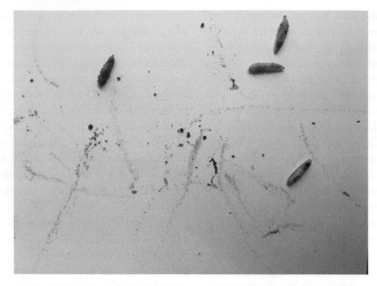

Figure 13.3 Fly larvae can disperse microbes in their secretions and excretions, especially during the wandering phase in which decomposition fluids are spread via the exoskeleton and gut fluids are purged during dispersal from the corpse. *Source*: Photo by D.B. Rivers.

integument and/or through purging gut contents once wandering occurs prior to pupariation (Figure 13.3). The result is distribution of the internal microbiome to epinecrotic communities, the soil environment, or any other surface they make contact with.

13.3 Estimation of the postmortem interval using microbes

Forensic medicine techniques used for determination of the time of death are quite accurate but do have a shelf live. For an individual that has been dead for more than 24 hours, especially if exposed to the environment, it can be challenging to accurately determine when death occurred. The longer the window after death, the more difficult the task becomes. Chapter 14 provides a basic introduction for estimating the time of death, otherwise known as the postmortem interval (PMI), based on insect development and succession patterns of necrophilous insects on a corpse. As with any type of statistical calculation, certain assumptions must be met in order to use insects for the estimation of the PMI. The counter interpretation to that statement is that there are times in which the insects discovered on or with the corpse do

not meet the necessary criteria. This does happen. There are also times in which no insects are associated with the corpse. How is the time of death determined for an advanced decomposition when no insect evidence is available? If this were 10 years ago, the answer would be probably it is not. However, recent advances in postmortem microbiome research suggests that microbes may be the solution. Death investigations using cadavers in terrestrial ecosystems point to the possibility that a microbial clock exists (Beans, 2018). The clock is in reference to the concept that a predictable pattern of microbial succession occurs during mammalian decomposition. The underlying premise is that similar suites or communities of microorganisms colonize and utilize individual hosts in particular environments (Metcalf, 2019). In the case of human death, the prediction is that the pattern of microbial succession on a corpse is repeatable across individuals and in different death scenarios. Thus, like with insect succession patterns, based on the type of microbes discovered on a corpse, in theory, a time estimate could be made of how long the individual has been dead. However, unlike with insects, sufficient baseline data has not been accrued yet to make such predictions. A broad pattern of microbial succession on human cadavers has already been outlined (Figure 13.4). The expectation is that the powerful metagenomic analysis tools being employed for microbiome research will lead to rapid advances

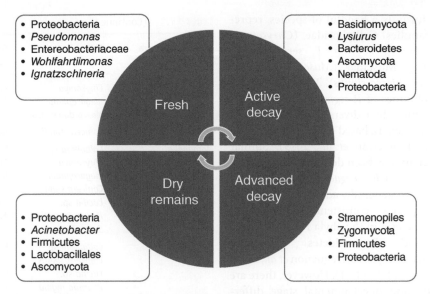

- Proteobacteria
- *Pseudomonas*
- Entereobacteriaceae
- *Wohlfahrtiimonas*
- *Ignatzschineria*

- Basidiomycota
- *Lysiurus*
- Bacteroidetes
- Ascomycota
- Nematoda
- Proteobacteria

Fresh

Active decay

Dry remains

Advanced decay

- Proteobacteria
- *Acinetobacter*
- Firmicutes
- Lactobacillales
- Ascomycota

- Stramenopiles
- Zygomycota
- Firmicutes
- Proteobacteria

Figure 13.4 Characterization of microbial succession on human cadavers placed in a terrestrial ecosystem in the southwestern part of the United States. A broad array of microbes has been detected, including bacteria, fungi, nematodes, and heterokonts or stramenopiles. *Source*: Data derived from Metcalf *et al.* (2016) and Zhou & Bian (2018).

in understanding microbial succession on humans and other vertebrate carrion. The next decade is likely to see the use of microbial succession data used as evidence in a court of law.

13.4 Microbes and insects are intimate members of the necrobiome

Necrophilous insects, especially necrophagous flies, are prone to encountering microbes. This should be obvious for any species of fly that lands or feeds on vertebrate carrion, or otherwise makes contact with a corpse with any part of its body. Many species of flies utilize other septic resources such as feces, spoiled foods, or trash refuge. The point is that any of these food sources promote direct interactions between flies and microbes (Tomberlin *et al.*, 2017). In many instances, the interactions appear to be orchestrated by the microbes themselves. Remember from our discussions of chemical communication in Chapter 7 that there is strong evidence to suggest that bacteria use **quorum sensing** and other cues to lure adult flies to carrion. Bacterial signals are also involved in stimulating oviposition and promoting the formation of larval feeding aggregations. In many cases, flies respond to volatile organic compounds

(VOCs) produced by bacteria or that are by-products of bacterial activity. For example, male calliphorids are attracted to dimethyl disulfide, a compound produced in later stages of decomposition (Paczkowski *et al.*, 2015), which is also a window of time that corresponds to when virgin females are drawn to the same resource (Mohr & Tomberlin, 2015). **Anautogenous** females are presumed to be seeking a protein meal for egg provisioning. Gravid blow fly females show an attraction to phenylacetic acid, a cue that indicates potentially beneficial bacteria are present for larval development (Liu *et al.*, 2016). By contrast, males of the same species do not respond to phenylacetic acid. Each observation is indicative of adaptive benefit to both the microorganisms and insects involved in the associations. Additional examples of bacteria aiding necrophagous flies can be found in Chapters 6 (fly reproduction), 7 (chemical communication), and 8 (maggot masses).

13.4.1 Fly microbiome

Before exploring details of how microbes influence insect development on a corpse, we will examine what is known about which microbes are doing the influencing. To do so, the first step is to unveil current understanding of the microbes that comprise the fly microbiome. Detailed examination of the microbial communities of necrophagous flies has occurred for only a handful of species, representing the families Calliphoridae (*Chrysomyia megacephala*, *Lucilia cuprina*, *L. sericata*, and *Phormia regina*), Muscidae (*Musca domestica*), and Phoridae (*Conicera similis*). The internal and external microbiomes of *P. regina* are represented by impressive microbial diversity: 27 phyla, 171 families, and 533 genera based on mass sequencing of 16S rRNA (Deguenon *et al.*, 2019). Similar microbial diversity has been detected for adults of *M. domestica* and *Ch. megacephala* (Junqueira *et al.*, 2017). More generally, adults and larvae of various fly species consistently display an abundance of bacteria in the phyla Proteobacteria, Firmicutes and Bacteroidetes (Junqueira *et al.*, 2017; Iancu *et al.*, 2018; Deguenon *et al.*, 2019; Pechal *et al.*, 2019) (Table 13.1). However, there are species-specific and developmental stage differences in the microbial species present on or in a given fly. For example, the abundance of Bacteroidetes in larvae of the phorid *C. similis* is

Table 13.1 Microbial species identified in the microbiomes of carrion flies.

Bacterial genera	Fly	Development stage
Acinetobactor	*Chrysomya megacephala*	Adult
	Musca domestica	Adult
	Phormia regina	Adult
Aeromonas	*Chrysomya megacephala*	Adult
	Musca domestica	Adult
Bacillus	*Lucilia cuprina*	Larva
	Sarcophaga sp.	Larva
Clostridium	*Conicera similis*	Adult, larva
Dysgonomonas	*Phormia regina*	Adult
Enterobacter	*Chrysomya megacephala*	Adult
	Musca domestica	Adult
Escherichia	*Chrysomya megacephala*	Adult
	Musca domestica	Adult
	Lucilia cuprina	Larva
	Sarcophaga sp.	Adult, larva
Faecalibacterium	*Phormia regina*	Adult
Ignatzschineria	*Calliphora* sp.	Larva
	Conicera similis	Adult, larva
	Lucilia sp.	Larva
	Phormia regina	Adult
	Sarcophaga sp.	Adult
Hafnia	*Chrysomya megacephala*	Adult
	Musca domestica	Adult

Table 13.1 (Continued)

Bacterial genera	Fly	Development stage
Klebsiella	*Chrysomya megacephala*	Adult
	Musca domestica	Adult
Leucobacter	*Conicera similis*	Adult, larva
Morganella	*Calliphora* sp.	Larva
	Chrysomya megacephala	Adult
	Conicera similis	Adult, larva
	Lucilia sp.	Larva
	Musca domestica	Adult
	Sarcophaga sp.	Adult
Myroides	*Calliphora* sp.	Larva
	Conicera similis	Adult, larva
	Lucilia sp.	Larva
	Phormia regina	Adult
	Sarcophaga sp.	Adult
Paenochrobacterium	*Conicera similis*	Adult, larva
Proteus	*Chrysomya megacephala*	Adult
	Conicera similis	Adult, larva
	Musca domestica	Adult
	Sarcophaga sp.	Adult
Providencia	*Chrysomya megacephala*	Adult
	Musca domestica	Adult
Pseudochrobacterium	*Conicera similis*	Adult, larva
Pseudomonas	*Chrysomya megacephala*	Adult
	Musca domestica	Adult
Psychrobacter	*Musca domestica*	Adult
	Phormia regina	Adult
Serratia	*Chrysomya megacephala*	Adult
	Musca domestica	Adult
Sphingobacterium	*Phormia regina*	Adult
Sporosarcina	*Calliphora* sp.	Larva
	Lucilia sp.	Larva
Staphylococcus	*Lucilia cuprina*	Larva
	Sarcophaga sp.	Adult, larva
Streptococcus	*Lucilia cuprina*	Larva
	Sarcophaga sp.	Adult, larva
Tisseriella	*Conicera similis*	Adult, larva
Vagococcus	*Calliphora* sp.	Larva
	Phormia regina	Adult
Wolbachia	*Lucilia* sp.	Larva
Wohlfahrtimonas	*Calliphora* sp.	Larva
	Lucilia sp.	Larva
	Phormia regina	Adult

The table is not an exhaustive list of all microbes identified in the internal and external microbiomes of necrophagous flies. Information derived from Banjo *et al.* (2006), Iancu *et al.* (2018), Deguenon *et al.* (2019), and Pechal *et al.* (2019).

much higher than in adults, whereas three unique phyla of bacteria, Acidobacteria, Nitrospirae, and Verrucomicrobia, are associated with adults only (Iancu et al., 2018). Larvae and adults of *C. similis* share similar species of bacteria but not in the same relative abundances. The endogenous gut flora of *P. regina* changes over time such that the larval microbiome gradually becomes more reflective of the epinecroticmicrobiome of carrion by the onset of wandering behavior (Weatherbee et al., 2017). A similar trend was observed with calliphorid larvae developing on salmon carcasses (Pechal et al., 2019). The latter example is thought to be the result of acquisition of carrion microbes during larval feeding, or transfer from fly adults, eggs or larvae to salmon carcasses. In turn, the microbes become integrated into the fly larval microbiome through ingestion of carrion tissues (Pechal et al., 2019).

The microbiome of adult flies shows community differences based on location on the body of the fly. For example, legs and wings of adult flies show the highest microbial abundance in comparison to other sites on the exoskeleton (Junqueira et al., 2017). The high microbial density on legs is consistent with mechanical transfer of microbes to flies and other insects that associate with septic environments. Presumably, microbes are transferred to the wings from legs during the extensive grooming behaviors displayed by many species of filth and carrion flies (Figure 13.5). Both body sites are also considered major avenues for fly dispersal of microbes into

the environment. Adult flies are also purported to "regurgitate" gut contents as a mechanism of bacterial transmission (Sasaki et al., 2000; Pava-Ripoll et al., 2012). However, this does not seem likely, as regurgitate represents expulsion of foregut contents via bubbling (Stoffolano & Haselton, 2013). The contents of the bubble or regurgitate are comprised chiefly of salivary gland secretions and food material stored in the crop. Most of the microbes in the crop are believed to be from exogenous sources (El-Bassiony & Stoffolano, 2016). In other words, though microbes may be present in regurgitate, they probably are not derived from the gut microbiome. It seems more plausible that microbes are transferred from the gut either through vomiting—with the resulting material termed vomitus—or through expulsion of liquid feces.

The microbiome of newly emerged flies shows far less diversity than older adult flies. In *M. domestica* and *Ch. megacephala*, the endosymbiotic bacterium *Wolbachia* sp. represents 99% of the microbial fauna in the gut of teneral adults (Junqueira et al., 2017). Very low abundance of *Myroides odoratus*, *Providencia rettgeri*, *P. stuartii*, and *Morganella morganii* comprise the remaining inhabitants of the gut microbiome. The difference in gut microbes between newly emerged and older adults is not the result of feeding status. While it is true that the gut microbial flora increases following feeding as adults, the initial difference is due to holometabolous development. During wandering and prior to the onset of pupariation, the larval gut is purged of the gut contents and a significant portion of the microbiome. The metamorphic transition from larva to adult is characterized by degradation of the larval gut. Any remnants of the larval gut microbiome are presumably lost. Consequently, the pupa and then newly emerged imago display a low complexity gut microbiome (Junqueira et al., 2017).

13.4.2 Microbe-insect interactions

From our earlier discussions of the necrobiome, it is clear that microbes and necrophilous insects make physical contact with one another. These associations lead to exchanges of microorganisms between the exogenous and endogenous microbial communities of carrion and the insects. In

Figure 13.5 An adult sarcophagid using the metathoracic legs to groom the wings. This behavior is considered a means for mechanical transfer of microbes from the legs to the wings. *Source*: Jon Richfield/Wikimedia Commons/CC BY-SA 3.0.

scenarios involving flies, members of the microbiome recruit adult flies to a carcass and flies in turn mechanically transfer microbes on the exoskeleton or ingest the microorganisms during feeding followed by microbial dispersion via secretions and excretions (Tomberlin *et al.*, 2017). The mere presence of fly maggots enhances growth of at least some microorganisms. More specifically, maggot activity alters the localized pH of soft tissues, yielding more alkaline conditions that favor growth of some bacterial species like *Proteus mirabilis*, while inhibiting other species (Barnes *et al.*, 2010). Whether these same conditions inhibit certain species of flies is unknown. However, many fly species release secretions and excretions during larval feeding that display antimicrobial activity (Blueman & Bousfield, 2012). The chemical arsenal of larvae includes several digestive enzymes (e.g., trypsin, chymotrypsin, and cathepsin D), small peptides (lucifensin, sarcotoxin) and an array of non-specific inhibitors such as ammonium carbonate, calcium ions, allantoin, urea, and serine proteases (Horobin *et al.*, 2003; Cuervo *et al.*, 2008; Čeřovský *et al.*, 2010). These compounds presumably suppress or eradicate certain microorganisms that impede larval growth. It is highly conceivable that the action of some of the antimicrobial compounds is species-specific either in terms of the microbes targeted or the fly species that benefit. There is some evidence that some antibacterial compounds (e.g., sarcotoxins) may play a dual role in which they both limit microbial growth and enhance larval development (Matsuyama & Natori, 1988; Matova & Anderson, 2006), most likely through skewing the larval microbiota towards a more favorable population. Several species of necrophagous beetles also demonstrate a need to reduce microbial populations to promote their own growth (Barnes *et al.*, 2010). For example, the elevated tissue pH and/or increased microbial proliferation elicited by fly activity is not conducive to development of necrophagous beetles using the same resource. Thus, beetle colonization is characterized by lowering the pH of the decomposing tissues. Alterations of tissue and fluid pH appear to be a mechanism used by certain beetle species to reduce competition with the more fecund and rapidly developing fly species that utilize the same resources.

Microbes are believed to directly aid the development of necrophagous fly larvae through liberation of nutrients from the corpse or as by-products, or by serving as a direct food source. This view is based largely on the observations that adult flies inoculate carrion with specific microorganisms during feeding and oviposition (Thompson *et al.*, 2013), which are in turn presumed to benefit feeding maggots. Despite ample evidence to support this premise from bacterial interactions with several species of *Drosophila*, surprisingly few studies have examined the direct impact of microbes on carrion fly development. One notable exception is a study examining the role of bacteria in the development of *Cochliomyia macellaria* (Ahmad *et al.*, 2006). Four bacteria were identified as part of the gut microbiota (*Provincia* sp., *Escherichia coli* O157:H7 (an enterohemorrhagic strain), *Enterococcus faecalis*, and *Ochrobactrum* sp.), but each microorganism in isolation or mixed decreased larval development and survival. In contrast, other studies have shown that if fly larvae are rendered axenic due to treatment with antibiotics, growth and maintenance in sterile conditions, or feeding upon sterilized food, they display altered immunological responses, lengthened pupariation time, and decreased mass of pupae (Fitt & O'Brien, 1985; Dillon & Dillon, 2004; Romero *et al.*, 2006).

If microbes are necessary for fly development, what do they do for the maggots? Most likely liberate nutrients from the corpse. This is especially critical during the early stages of development in which the delicate mouth hooks are not capable of penetrating the skin and other tissues of carrion. The reliance on microbial digestion may decrease as larvae progress in development. Formation of a large feeding aggregation permits cooperative feeding that was not possible during earlier stages (See Chapter 8). However, this idea is merely speculation, as no study has examined the role of bacteria in larval nutrition throughout development. Larvae undoubtedly derive nutrition from direct consumption of microbes as well. The larval gut is ideally suited for lysis of microbes. For example, the midgut is highly acidic and possesses lytic enzymes associated with diets rich in bacteria (Terra & Ferreira, 1994). The presence of a pepsin-like digestive enzyme, cathepsin D, in the anterior midgut of some larval muscids and calliphorids is consistent with the idea of bacterial elimination through digestive processes. Microbial cellular contents are thus emptied into the midgut of the larvae and presumed to be used as nutrients to fuel growth and development.

Chapter review

Microbes of a corpse are important sources of information

- The total number of microbial cells associated with one person exceeds the total number of all insects on the planet on any given day. Such extraordinary abundance coupled with rich diversity of phyla and taxa allow microbes to occupy a wide array of habitats and fill a broad range of ecological roles, including as decomposers on human corpses and vertebrate carrion.
- Forensic microbiology provides information relevant to death investigations. Species-specific microbial activity and patterns of faunal succession by microbes on a decomposing corpse can yield insight into the time of death, reveal whether the body has been moved from another location, or point to the site of a clandestine grave. In the case of the latter, microbes associated with a particular soil type or plant can be connected to a specific environment and possibly a location. Microbes have even been used to place a suspect at a crime scene or connect a person to an object such as a weapon.
- The emergence of a new array of sequencing tools has elevated the position of microbes as physical evidence. Information is derived from investigating patterns in microbial communities using next-generation sequencing of phylogenetically and/or taxonomically specific gene markers coupled with robust statistical and computational methods, including machine learning.

Microbial communities exist on living and dead hosts

- The human body is home to a community of microorganisms, residing externally on the skin and the internal environment. Microbial communities exist essentially everywhere on and in the body, on every surface of the organs, "tubing" and architecture that makes us functional. Collectively, the community is referred to as the microbiome or microbiota. The term "microbe" is not just a reference to bacteria. It includes a broad array of microorganisms, including microbes found in all three domains of life (Bacteria, Archaea, and Eukarya). Thus, the microbial community is composed of a diverse range of microorganisms competing for the resources of the living or dead host.

- As corpse decay progresses, the microbial community changes in composition in both abundance and diversity. In fact, the corpse itself becomes a habitat for a much larger and broader community than just the microbiome. A necrobiome forms a community composed of prokaryotic and eukaryotic organisms united in the attraction and utilization of a decomposing heterotrophic mass, i.e., animal remains. Perhaps most recognizable are the multitudes of necrophilous insects and vertebrate scavengers that arrive, aligned with bacteria, fungi, nematodes, and other microorganisms to form a complex community on and within a decomposing corpse. Within the realm of microbes, the community that is specifically associated with a human corpse or vertebrate carrion is broadly termed the postmortem microbiome. The postmortem microbiome has two major components, the thanatomicrobiome or the microbial community located within the corpse including internal organs and cavities enclosed within the body, and the epinecroticmicrobiome, the microbial community found on external surfaces
- In terms of specific internal sites, the brain, heart, and spleen are thought to have the most stable or similar microbial community composition over time, while rapid bacterial growth occurs in blood, bone marrow, liver, and prostate. The liver displays the highest microbial diversity during decomposition.
- A consistent change in the thanatomicrobiome is a continuous decline in the abundance of anaerobic taxa which were originally associated with the microbiome of the living host. This is especially true for the anaerobes *Lactobacillus* and *Bacteriodetes*. The decline in these genera has been attributed to the release of oxidative gases during autolysis. In contrast, the thanatomicrobiome experiences a postmortem shift from aerobic to anaerobic bacteria (and presumably other microbes) across all internal body sites.
- The epinecroticmicrobiome provides more holistic data regarding decomposition and interactions between the microbial communities and the environment. This "outside in" approach to studying the postmortem microbiome is concerned with determining if the microbial community on the skin surface changes over time. If it does, and unquestionably it does, then the net steps are to determine if the changes in

community structure occur in a predictable pattern, as occurs with insect succession on a corpse. Microbes on external surfaces are directly influenced by abiotic and biotic factors associated with the environment. Thus, the potential for changes to the microbial community structure and the necrobiome are much greater than with the more concealed, and arguably stable conditions, found inside of the corpse.

- Shifts in microbial communities on vertebrate carrion can also be linked to seasonal changes, contributions from the environment, and from organisms that visit carrion, such as necrophilous insects and vertebrate scavengers.

- An additional source of "new" microbes to the skin surface is the corpse itself. For example, in late stages of bloating, decomposition fluids and gases purge from the body openings, spilling the contents into the environment. These fluids are a source of microorganisms that in turn, can become part of the epinecroticmicrobiome. Likewise, any penetrating lesions or wounds can result in an exchange of microbiomes between the internal and external microbial communities.

Estimation of the postmortem interval using microbes

- Death investigations using cadavers in terrestrial ecosystems point to the possibility that a microbial clock exists. The clock is in reference to the concept that a predictable pattern of microbial succession occurs during mammalian decomposition. The underlying premise is that similar suites or communities of microorganisms colonize and utilize individual hosts in particular environments. In the case of human death, the prediction is that the pattern of microbial succession on a corpse is repeatable across individuals and in different death scenarios.

- Unlike with insects, sufficient baseline data has not been accrued yet to make such predictions. A broad pattern of microbial succession on human cadavers has already been outlined. The expectation is that the powerful metagenomic analysis tools being employed for microbiome research will lead to rapid advances in understanding microbial succession on humans and other vertebrate carrion. The next decade is likely to see the use of microbial succession data used as evidence in a court of law.

Microbes and insects are intimate members of the necrobiome

- Necrophilous insects, especially necrophagous flies, are prone to encountering microbes. Many species of flies utilize other septic resources such as feces, spoiled foods, or trash refuse. The point is that any of these food sources promote direct interactions between flies and microbes. In many instances, the interactions appear to be orchestrated by the microbes themselves.

- Detailed examination of the microbial communities of necrophagous flies has occurred for only a handful of species, representing the families Calliphoridae (*Chrysomyia megacephala*, *Lucilia cuprina*, *L. sericata*, and *Phormia regina*), Muscidae (*Musca domestica*), and Phoridae (*Conicera similis*). The internal and external microbiomes of *P. regina* are represented by impressive microbial diversity: 27 phyla, 171 families, and 533 genera based on mass sequencing of 16S rRNA. Similar microbial diversity has been detected for adults of *M. domestica* and *Ch. megacephala*. More generally, adults and larvae of various fly species consistently display an abundance of bacteria in the phyla Proteobacteria, Firmicutes, and Bacteroidetes.

- The microbiome of newly emerged flies shows far less diversity than older adult flies. The difference in gut microbes between newly emerged and older adults is not the result of feeding status. While it is true that the gut microbial flora increases following feeding as adults, the initial difference is due to holometabolous development.

- Microbes are believed to directly aid the development of necrophagous fly larvae through liberation of nutrients from the corpse or as byproducts, or by serving as a direct food source. This view is based largely on the observations that adult flies inoculate carrion with specific microorganisms during feeding and oviposition, which are in turn presumed to benefit feeding maggots. Despite ample evidence to support this premise from bacterial interactions with several species of *Drosophila*, surprisingly few studies have examined the direct impact of microbes on carrion fly development.

- If microbes are necessary for fly development, what do they do for the maggots? Most likely liberate nutrients from the corpse. This is especially critical during the early stages of development in which the delicate mouth hooks are not capable of penetrating the skin and other tissues of carrion. The reliance on microbial digestion may decrease as larvae progress in development. Formation of a large feeding aggregation permits cooperative feeding that was not possible during earlier stages. However, this idea is merely speculation, as no study has examined the role of bacteria in larval nutrition throughout development.

Test your understanding

Level 1: knowledge/comprehension

1. Define the following terms:

   (a) microbiome
   (b) necrobiome
   (c) Locard's Exchange Principle
   (d) next-generation sequencing
   (e) forensic microbiology
   (f) postmortem *Clostridium* effect
   (g) quorum sensing.

2. Match the terms (i–vi) with the descriptions (a–f).

(a) Use of microbes in death investigations	(i) Putrefaction
(b) Microbial community of internal organs following death	(ii) Thanato-microbiome
(c) Decay of tissues due to microbial activity	(iii) Microbial pathogenicity
(d) Dominant type of bacteria found in fly microbiome	(iv) Agonal period
(e) Microbiota of corpse skin	(v) Epinecrotic-microbiome
(f) Focused on etiology of diseases caused by microorganisms	(vi) Forensic microbiology
(g) Moment of death	(vii) Proteobacteria

3. Explain how postmortem microbiology and forensic microbiology are similar yet distinct fields of study.

4. Compare and contrast thanatomicrobiome and epinecroticmicrobiome.

5. Indicate which phyla of microorganisms are commonly exchanged between flies and vertebrate carrion.

Level 2: application/analysis

1. Discuss whether the thanatomicrobiome is expected to be altered by necrophagous flies.

2. Under what circumstances are microbes exchanged between the epinecroticmicrobiome and thanatomicrobiome?

3. Explain how a postmortem interval can conceivably be estimated using the epinecroticmicrobiome.

4. Describe how mass parallel sequencing techniques have permitted studies of the microbiome of living and dead hosts that were not previously possible.

Level 3: synthesis/evaluation

1. Several necrophagous fly species are known to produce antimicrobial compounds that are believed to benefit developing fly larvae. Speculate on possible mechanisms of antimicrobial action of these compounds that could permit beneficial microbes to thrive while negatively impacting other microorganisms. How might these compounds display species-specific benefits to one fly species over another?

2. Discuss how the production of antimicrobial compounds by fly larvae could complicate the estimation of a postmortem interval based on microbes.

Note

1. This estimate of human cells is based on an adult weighing approximately 150 pounds and 30 years old.

References cited

Ahmad, A., Broce, A. & Zurek, L. (2006) Evaluation of significance of bacteria in larval development of *Cochliomyia macellaria* (Diptera: Calliphoridae). *Journal of Medical Entomology* 43(6): 1129–1133.

Banjo, A.D., Lawal, O.A. & Adeyemi, A.I. (2006) The microbial fauna associated with the larvae of *Oryctes monocerus. Journal of Applied Sciences Research* 2(11): 837–843.

Barnes, K.M., Gennard, D.E. & Dixon, R.A. (2010) An assessment of the antibacterial activity in larval excretion/secretion of four species of insects recorded in association with corpses, using *Lucilia sericata* Meigen as the marker species. *Bulletin of Entomological Research* 100(6): 635–640.

Beans, C. (2018). News feature: can microbes keep time for forensic investigators? *Proceedings of the National Academy of Sciences* 115(1): 3–6.

Benbow, M.E., Pechal, J.L., Lang, J.M., Erb, R. & Wallace, J.R. (2015) The potential of high-throughput metagenomic sequencing of aquatic bacterial communities to estimate the postmortem submersion interval. *Journal of Forensic Sciences* 60(6): 1500–1510.

Belk, A., Xu, Z.Z., Carter, D.O., Lynne, A., Bucheli, S., Knight, R. & Metcalf, J.L. (2018) Microbiome data accurately predicts the postmortem interval using random forest regression models. *Genes* 9(2): 104. https://doi.org/10.3390/genes9020104.

Bianconi, E. Piovesan, A. Facchin, F., Beraudi, A., Casadei, R., Frabetti, F., Vitale, L., Chiara Pelleri, M., Tassani, S., Piva, F., Perez-Amodio, S., Strippoli P. & Canaider S. (2013) An estimation of the number of cells in the human body. *Annals of Human Biology* 40: 463–471. https://doi.org/10.3109/03014460.2013.807878.

Blueman, D. & Bousfield, C. (2012) The use of larval therapy to reduce the bacterial load in chronic wounds. *Journal of Wound Care* 21(5): 244–253.

Bucheli, S.R. & Lynne, A.M. (2016) The microbiome of human decomposition. *Microbe Wash. DC* 11: 165–171.

Can, I., Javan, G.T., Pozhitkov, A.E. & Noble, P.A. (2014) Distinctive thanatomicrobiome signatures found in the blood and internal organs of humans. *Journal of Microbiological Methods* 106: 1–7.

Carter, D.O., Tomberlin, J.K., Benbow, M.E. & Metcalf, J.L. (2017) *Forensic Microbiology*. Wiley, Sussex, U.K.

Čeřovský, V., Žďárek, J., Fučík, V., Monincová, L., Voburka, Z. & Bém, R. (2010) Lucifensin, the long-sought antimicrobial factor of medicinal maggots of the blowfly *Lucilia sericata*. *Cellular and Molecular Life Sciences* 67(3): 455–466.

Clement, C., Hill, J.M., Dua, P., Culicchia, F. & Lukiw, W.J. (2016) Analysis of RNA from Alzheimer's disease postmortem brain tissues. *Molecular Neurobiology* 53(2): 1322–1328.

Cuervo, P., Mesquita-Rodrigues, C., d'Avila Levy, C.M., Britto, C., Pires, F.A., Gredilha, R., Alves, C.R. & Jesus, J.B.D. (2008) Serine protease activities in *Oxysarcodexia thornax* (Walker)(Diptera: Sarcophagidae) first instar larva. *Memorias do Instituto Oswaldo Cruz* 103(5): 504–506.

d'Aleo, F., Bonanno, R., Bianco, G. & Trunfio, A. (2017) Postmortem microbiology as a routine tool for legal-medicine in Italy: is it time? *Microbiologia Medica* 32(6828): 72–32.

Deguenon, J.M., Travanty, N., Zhu, J., Carr, A., Denning, S., Reiskind, M.H., Watson, D.W., Roe, R.M. & Ponnusamy, L. (2019) Exogenous and endogenous microbiomes of wild-caught *Phormia regina* (Diptera: Calliphoridae) flies from a suburban farm by 16S rRNA gene sequencing. *Scientific Reports* 9(1): 1–13.

Dickson, G.C., Poulter, R.T., Maas, E.W., Probert, P.K. & Kieser, J.A. (2011) Marine bacterial succession as a potential indicator of postmortem submersion interval. *Forensic Science International* 209(1–3): 1–10.

Dillon, R.J. & Dillon, V.M. (2004) The gut bacteria of insects: nonpathogenic interactions. *Annual Reviews in Entomology* 49(1): 71–92.

El-Bassiony, G.M. & Stoffolano Jr, J.G. (2016) Comparison of sucrose intake and production of elimination spots among adult *Musca domestica*, *Musca autumnalis*, *Phormia regina* and *Protophormia terraenovae*. *Asian Pacific Journal of Tropical Biomedicine* 6(8): 640–645.

Fitt, G.P. & O'Brien, R.W. (1985) Bacteria associated with four species of *Dacus* (Diptera: Tephritidae) and their role in the nutrition of the larvae. *Oecologia* 67(3): 447–454.

Horobin, A.J., Shakesheff, K.M., Woodrow, S., Robinson, C. & Pritchard, D.I. (2003) Maggots and wound healing: an investigation of the effects of secretions from *Lucilia sericata* larvae upon interactions between human dermal fibroblasts and extracellular matrix components. *British Journal of Dermatology* 148(5): 923–933.

Human Microbiome Consortium (2012) Structure, function and diversity of the health human microbiome. *Nature* 486: 207–214.

Human Microbiome Project (2012) https://hmpdacc.org/.

Iancu, L., Junkins, E.N. & Purcarea, C. (2018) Characterization and microbial analysis of first recorded observation of *Conicera similis* Haliday (Diptera: Phoridae) in forensic decomposition study in Romania. *Journal of Forensic and Legal Medicine* 58: 50–55.

Javan, G.T., Finley, S.J., Abidin, Z. & Mulle, J.G. (2016) The thanatomicrobiome: a missing piece of the microbial puzzle of death. *Frontiers in Microbiology* 7: 225. https://doi.org/10.3389/fmicb.2016.0225.

Javan, G.T., Finley, S.J., Smith, T., Miller, J. & Wilkinson, J.E. (2017) Cadaver thanatomicrobiome signatures: the ubiquitous nature of *Clostridium* species in human decomposition. *Frontiers in Microbiology* 8: 2096. https://doi.org/10.3389/fmicb.2017.02096.

Junqueira, A.C.M., Ratan, A., Acerbi, E., Drautz-Moses, D.I., Premkrishnan, B.N., Costea, P.I., Linz, B., Purbojati, R.W., Paulo, D.F., Gaultier, N.E. & Subramanian, P. (2017) The microbiomes of blowflies and houseflies as bacterial transmission reservoirs. *Scientific Reports* 7(1): 1–15.

Liu, W., Longnecker, M., Tarone, A.M. & Tomberlin, J.K. (2016) Responses of *Lucilia sericata* (Diptera: Calliphoridae) to compounds from microbial decomposition of larval resources. *Animal Behaviour* 115: 217–225.

Matova, N. & Anderson, K.V. (2006) Rel/NF-κB double mutants reveal that cellular immunity is central to *Drosophila* host defense. *Proceedings of the National Academy of Sciences* 103(44): 16424–16429.

Matsuyama, K. & Natori, S. (1988) Purification of three antibacterial proteins from the culture medium of

NIH-Sape-4, an embryonic cell line of *Sarcophaga peregrina*. *Journal of Biological Chemistry* 263(32): 17112–17116.

Metcalf, J.L. (2019) Estimating the postmortem interval using microbes: knowledge gaps and a path to technology adoption. *Forensic Science International: Genetics* 38: 211–218.

Metcalf, J.L., Parfrey, L.W., Gonzalez, A., Lauber, C.L., Knights, D., Ackermann, G., Humphrey, G.C., Gebert, M.J., Van Treuren, W., Berg-Lyons, D. & Keepers, K. (2013) A microbial clock provides an accurate estimate of the postmortem interval in a mouse model system. *elife* 2: e01104.

Metcalf, J.L., Xu, Z.Z., Bouslimani, A., Dorrestein, P., Carter, D.O. & Knight, R. (2017) Microbiome tools for forensic science. *Trends in Biotechnology* 35(9): 814–823.

Metcalf, J.L., Xu, Z.Z., Weiss, S., Lax, S., Van Treuren, W., Hyde, E.R., Song, S.J., Amir, A., Larsen, P., Sangwan, N. & Haarmann, D. (2016) Microbial community assembly and metabolic function during mammalian corpse decomposition. *Science* 351(6269): 158–162.

Mohr, R.M., & Tomberlin, J.K. (2015) Development and validation of a new technique for estimating a minimum postmortem interval using adult blow fly (Diptera: Calliphoridae) carcass attendance. *International Journal of Legal Medicine* 129(4): 851–859.

Morris, J.A., Harrison, L.M. & Partridge, S.M. (2006) Postmortem bacteriology: a re-evaluation. *Journal of Clinical Pathology* 59(1): 1–9.

Paczkowski, S., Nicke, S., Ziegenhagen, H. & Schütz, S. (2015) Volatile emission of decomposing pig carcasses (*Sus scrofa domesticus* L.) as an indicator for the postmortem interval. *Journal of Forensic Sciences* 60: S130–S137.

Pava-Ripoll, M., Pearson, R.E.G., Miller, A.K. & Ziobro, G.C. (2012) Prevalence and relative risk of *Cronobacter* spp., *Salmonella* spp., and *Listeria monocytogenes* associated with the body surfaces and guts of individual filth flies. *Applied and Environmental Microbiology* 78(22): 7891–7902.

Pechal, J.L., Crippen, T.L., Benbow, M.E., Tarone, A.M., Dowd, S. & Tomberlin, J.K. (2014) The potential use of bacterial community succession in forensics as described by high throughput metagenomic sequencing. *International Journal of Legal Medicine* 128(1): 193–205.

Pechal, J.L., Crippen, T.L., Cammack, J.A., Tomberlin, J.K. & Benbow, M.E. (2019) Microbial communities of salmon resource subsidies and associated necrophagous consumers during decomposition: potential of cross-ecosystem microbial dispersal. *Food Webs* 19: e00114. https://doi.org/10.1016/j. fooweb.2019.e00114.

Pechal, J.L., Crippen, T.L., Tarone, A.M., Lewis, A.J., Tomberlin, J.K. & Benbow, M.E. (2013)

Microbial community functional change during vertebrate carrion decomposition. *PloS one* 8(11): e79035.

Romero, A., Broce, A. & Zurek, L. (2006) Role of bacteria in the oviposition behaviour and larval development of stable flies. *Medical and Veterinary Entomology* 20(1): 115–121.

Sasaki, T., Kobayashi, M. & Agui, N. (2000) Epidemiological potential of excretion and regurgitation by *Musca domestica* (Diptera: Muscidae) in the dissemination of *Escherichia coli* O157: H7 to food. *Journal of Medical Entomology* 37(6): 945–949.

Stewart, E.J. (2012) Growing unculturable bacteria. *Journal of Bacteriology* 194(16): 4151–4160.

Stoffolano Jr, J.G. & Haselton, A.T. (2013) The adult Dipteran crop: a unique and overlooked organ. *Annual Review of Entomology* 58: 205–225.

Terra, W.R. & Ferreira, C. (1994) Insect digestive enzymes: properties, compartmentalization and function. *Comparative Biochemistry and Physiology Part B: Comparative Biochemistry* 109(1): 1–62.

Thomas, T.B., Finley, S.J., Wilkinson, J.E., Wescott, D.J., Gorski, A. & Javan, G.T. (2017) Postmortem microbial communities in burial soil layers of skeletonized humans. *Journal of Forensic and Legal Medicine* 49: 43–49.

Thompson, C.R., Brogan, R.S., Scheifele, L.Z. & Rivers, D.B. (2013) Bacterial interactions with necrophagous flies. *Annals of the Entomological Society of America* 106(6): 799–809.

Tomberlin, J.K., Crippen, T.L., Tarone, A.M., Chaudhury, M.F., Singh, B., Cammack, J.A. & Meisel, R.P. (2017) A review of bacterial interactions with blow flies (Diptera: Calliphoridae) of medical, veterinary, and forensic importance. *Annals of the Entomological Society of America* 110(1): 19–36.

Weatherbee, C.R., Pechal, J.L., Stamper, T. & Benbow, M.E. (2017) Post-colonization interval estimates using multi-species Calliphoridae larval masses and spatially distinct temperature data sets: a case study. *Insects* 8(2): 40.

Zhou, W. & Bian, Y. (2018) Thanatomicrobiome composition profiling as a tool for forensic investigation. *Forensic Sciences Research* 3(2): 105–110.

Supplemental reading

Benbow, M.E., Lewis, A.J., Tomberlin, J.K. & Pechal, J.L. (2013) Seasonal necrophagous insect community assembly during vertebrate carrion decomposition. *Journal of Medical Entomology* 50(2): 440–450.

Benbow, M.E., Tomberlin, J.K. & Tarone, A.M. (2018) *Carrion Ecology, Evolution, and Their Applications*. CRC Press, Baco Raton, FL.

Forbes, S.L., Stuart, B.H., Dadour, I.R. & Dent, B.B. (2004) A preliminary investigation of the stages of adipocere formation. *Journal of Forensic Sciences* 49: JFS2002230–JFS2002239.

Lawrence, A.L., Hii, S.F., Chong, R., Webb, C.E., Traub, R., Brown, G. & Šlapeta, J. (2015) Evaluation of the bacterial microbiome of two flea species using different DNA-isolation techniques provides insights into flea host ecology. *FEMS Microbiology Ecology* 91(12). https://doi.org/10.1093/femsec/fiv134.

Olea, P.P., Mateo-Tomás, P. & Sánchez-Zapata, J.A. (2019) *Carrion Ecology and Management.* Springer, Switzerland.

Pechal, J.L., Schmidt, C.J., Jordan, H.R. & Benbow, M.E. (2017) Frozen: thawing and its effect on the postmortem microbiome in two pediatric cases. *Journal of Forensic Sciences* 62(5): 1399–1405.

Schotsmans, E.M., Márquez-Grant, N. & Forbes, S.L. (2017) *Taphonomy of Human Remains: Forensic Analysis of the Dead and the Depositional Environment.* Wiley-Blackwell, Sussex.

Additional resources

Forensic Anthropology Center at Texas State University: http://www.txstate.edu/anthropology/facts/

Human Microbiome Project: https://www.bcm.edu/departments/molecular-virology-and-microbiology/research/the-human-microbiome-project

Human Postmortem Microbiome Database: https://hpmmdatabase.wixsite.com/hpmmdatabase

Institute for Forensic Anthropology and Applied Sciences: http://forensics.usf.edu/

Necrobiome: Microbial Life After Death: https://video.search.yahoo.com/search/video?fr=yfp-t&p=necrobiome#id=2&vid=9a6c328b347a90f9e4d7407d80567237&action=click

NIH Human Microbiome Project: https://www.hmpdacc.org/

Solving Crimes with the Necrobiome: https://www.biointeractive.org/classroom-resources/solving-crimes-necrobiome

Chapter 14

Postmortem interval

Overview

Insects are useful in several areas of medicocriminal importance, but perhaps the capstone lies with death, most typically homicides and suspicious deaths. It is human death that draws the most attention of criminal and forensic investigators, the lay public, and insects. Necrophagous species can be particularly useful in helping to decipher how long the decedent has been dead. The patterns of colonization and faunal succession during physical decomposition of a corpse, coupled with the unique growth of certain species of necrophagous flies through feeding exclusively (or nearly so) on carrion tissues and having development intimately linked to ambient temperatures, permits the use of these insects to uncover aspects of the time since death of the deceased, otherwise known as the postmortem interval (PMI). Insects, as poikilotherms, are powerful tools for relating development with the temperatures experienced by and on the corpse, which in turn can be linked to time estimates. Using insects to estimate the PMI, or portions of it, is relatively straightforward in terms of

calculation but relies on complex ecological and biological principles, complicated by numerous contributing factors, many of which are not fully understood. This chapter provides an introduction to estimation of the postmortem interval based on insect development through an examination of aspects of necrophagous fly development relevant to determinations of the time since death, and also explore the concepts of thermal units, physiological energy budgets, and the process of calculating a PMI.

The big picture

- The time since death is referred to as the postmortem interval.
- The role of insects in estimating the PMI.
- Modeling growth–temperature relationships.
- Calculating the PMI requires experimental data on insect development and information from the crime scene.
- The evolving PMI: changing approaches and sources of error.
- Total body scores influenced by insect development.

The Science of Forensic Entomology, Second Edition. David B. Rivers and Gregory A. Dahlem.
© 2023 John Wiley & Sons Ltd. Published 2023 by John Wiley & Sons Ltd.
Companion website: www.wiley.com/go/forensicentomology2

14.1 The time since death is referred to as the postmortem interval

Discovery of a human corpse in any context leads to intense investigation predominantly aimed at determining when the individual died and causation of death. Neither process is considered simple, although in some circumstances the onset of death was witnessed or an accurate timeline of when the deceased was last seen alive allows narrowing of the window considerably for estimation of when death occurred. The time since death is not the same as the agonal period, which was discussed in Chapter 10. The agonal period is in reference to the moment of death. In general, the exact moment at which death occurred must be witnessed to know precisely. By contrast, the time of death typically represents a broader period of time, encompassing the perimortem window just before the agonal period until the discovery of the deceased. This time frame is referred to as the **postmortem interval** (PMI). As will be seen later in the chapter, the definition of PMI can be refined based on the information or evidence used to estimate the time of death. Ordinarily, the details surrounding a suspicious death or homicide are not obvious, and thus a large investment of time and resources are needed to decipher the events leading up to and including the discovery of the corpse. You may be asking yourself "why is so much effort put into death investigations?" Especially into the PMI? A carefully developed PMI can be incredibly useful in the identification of the deceased as well as in substantiating or refuting an alibi or eyewitness testimony (Megyesi *et al.*, 2005; Moffat et al., 2016). Other information can also be derived from the PMI but the key is "time;" most aspects of an investigation are dependent on some understanding how long ago or when death occurred.

14.1.1 What is the PMI?

Cause and manner of death investigations are outside the realm of a forensic entomologist and fall under the dominion of those trained in forensic pathology and forensic medicine (see Chapter 1 for more details). As will be discussed throughout this chapter, forensic entomology does offer tools that contribute to defining the time since death or PMI (Tomberlin *et al.*, 2011; Tarone & Sanford, 2017; Wells & LaMotte, 2020). The PMI is commonly viewed as the time from when the deceased was last seen alive until the remains of the decedent are discovered. From a forensic entomological perspective, the PMI has been defined to encompass the time elapsed since death until the time entomological evidence was collected/preserved (Wells & LaMotte, 2020). Other information may also be relevant to the PMI, such as the lack of any necrophagous or necrophilous insects associated with the remains. In reality, the PMI calculation is based on and influenced by numerous abiotic and biotic factors (Catts, 1992; Anderson, 2000), including temperature of the corpse, ambient temperatures, physical decomposition of the body, and a range of biochemical changes that take place within the fluids and tissues of the deceased. The key term in this definition is *estimate*, meaning that an exact value cannot be assigned to when death occurred. Why not? Well there are many reasons for why the PMI is not an exact measurement of death. For one, the moment that death occurs, a series of changes occur in the corpse that is unique to each individual, so that any variable measured will be reflective of that specific death scenario. This should not be surprising considering that when the person was alive, any baseline measurement of fluid osmolytes[1] (like blood glucose), body temperature or heart rate is defined within a homeostatic range that varies based on the physiology of each individual. As well, ecological succession begins almost immediately, especially in natural environments, and consequently the physical and biological conditions of the corpse are modified uniquely based on the ambient conditions and seasonally specific invertebrate, vertebrate, and microbial fauna that colonize, consume, and utilize the deceased (Kreitlow, 2010). The end result is a modified corpse that displays postmortem characteristics that can be quantified as a range of changes, but which does not equate to a single unit of time.

Despite the uniqueness, early after death (the first 24–48 hours) pathological analyses of tissue samples and laboratory testing of biological fluids provide a fairly accurate picture of the onset of death. Other tests are more subjective in the determination of a PMI, such as algor or rigor mortis, lividity (livor mortis), and blood coagulation (Estracanholli *et al.*, 2009), and all are influenced by ambient temperatures. Newer methods of PMI estimation have been developed with the goal of improving precision and accuracy of time of death estimates. Such approaches range from technology-driven protocols like optical fluorescence, specifically designed to measure changes in skin autofluorescence spectra antemortem versus postmortem (Estracanholli *et al.*, 2009), analysis of the necrobiome (Pechal *et al.*, 2014; Abdelhalim *et al.*, 2017), stomach content based time of death (Hubig *et al.*, 2018), to physiological characteristics such as potassium concentrations in fluids of the eye (vitreous fluid) that elevate with increasing time after death, independent of temperature and humidity (Ahi & Garg, 2011). These changes to the body after death do not involve insects, but entomologically derived information can be used during this early period to confirm, add additional precision to, or refute the estimations derived from these physical and biochemical processes. After 48–72 hours, particularly in cases of outdoor exposure of a corpse, techniques associated with forensic medicine become less accurate for assessment of the timing of death, and ecological methodology begins to take center stage (Catts, 1992). As has been discussed throughout this book, necrophagous insects play a unique and important role for PMI determination, especially as the time since death increases past the initial 72-hour window (Goff, 2010).

14.1.2   Why is the PMI important?

Determination of the time since death, and hence the PMI, is extremely valuable in criminal investigations as this information can help identify both the individual responsible for the victim's death and the victim (Gennard, 2012). Such classification and individualization can be done via comparisons testing (Chapter 1) to eliminate suspects (exclusions) and/or connect the deceased with individuals reported missing for the same period of time (inclusions). In a similar fashion, the PMI can be applied in cases of neglect, abuse, and wildlife poaching to link suspects to the crime scene (at least to demonstrate the timing was possible) or remove individuals as suspects. Time of death determinations can be critical to civil matters as well, as in cases of insurance or inheritance (of estates, properties, or residue[2]), where time of death determines beneficiaries. For example, in a scenario in which a married couple who both have children from a previous marriage die together in an automobile accident, the order of death could determine which set of siblings inherit the estate (Jackson, 2016). A PMI based on insect development may at times be in conflict with that from a forensic pathologist, potentially signaling a unique situation had occurred such as antemortem insect colonization (e.g., myiasis) that was not evident during the initial investigation.

14.1.3   The PMI based on entomological evidence

Entomological evidence also represents important factors that contribute to the determination of the time since death. Several species of necrophagous insects can provide a time frame of several hours to months for a PMI estimation depending on how long the deceased has been dead and the length of exposure to biotic and abiotic components of the environment. Specific details of the PMI calculation based on insect development will be presented in Section 14.5. Before diving further into this topic, it is important to recognize two key ideas associated with forensic entomology and the PMI. The first is that there has been considerable debate concerning terminology and application of insect evidence to the estimation of all or a portion of the PMI. The term minimum postmortem interval (PMI_{min}) is based on insect development and is considered to represent the window of time from insect colonization until the developmental stage of the specimen of interest was discovered on the remains (Villet *et al.*, 2010). The maximum postmortem interval (PMI_{max}) is consistent with our earlier definition, as it encompasses the time frame when the decedent was last seen alive until discovery of the corpse. Some authors have proposed other terms accounting for insect association with carrion, including the period of insect activity (PIA), time of colonization (TOC), time of oviposition (TOC), and postcolonization interval (PCI). For some, these terms are preferred over PMI because time periods based on the development of necrophagous insects can be significantly different from the actual time since death, since there is usually a lag between when death occurs and the first wave of colonizers actually detect, locate, and feed on the corpse (Greenberg & Kunich, 2002; Amendt *et al.*, 2007; Tomberlin *et al.*, 2011; Matuszewski & Szafalowicz, 2013; Sanford, 2015). Wells & LaMotte (2020) have countered such arguments by suggesting that the PMI should never be represented as a single value, and thus to estimate a PMI_{min} and PMI_{max} is an estimation of the PMI. Other terms have also been offered with the intent of clarifying the interpretations of insect evidence. This brings us to the second point, which is that this chapter is designed as an introduction to the postmortem interval, especially for the individual with no or limited previous experience with the concepts of accumulated degree days and physiological energy budgets. Consequently, it is not the intent here to address all the differing views of the PMI. Interested students should consult the excellent discussions in Higley & Haskell (2010); Tomberlin *et al.* (2011); Michaud *et al.* (2015); Tarone & Sanford (2017), and Wells & LaMotte (2017, 2020) for advanced aspects of the PMI using entomological evidence.

Two approaches to using carrion insects as indicators of "time" in association with human and other animal death investigations include insect development and the ecological succession of insect carrion communities. Insect development, especially the concept of the oldest insect on a corpse, is the most commonly applied to the estimation of the PMI. In contrast, changes in insect composition or succession on a corpse are used to estimate the **succession internal** (SI), which broadly can estimate the amount of time that human remains were present at a specific location (Perez *et al.*, 2014). Insect-based PMI and SI will be discussed in the following sections.

14.2   The role of insects in estimating the PMI

Generally speaking, forensic medicine techniques are most relevant for the first 72 hours following death, after which time ecological data becomes increasingly more important. Of course this statement must be considered within the context of where the decedent is located (indoors vs. outdoors, aquatic vs. terrestrial, etc.). It should be obvious from the subject matter of this book that insects can help fill the void for PMI estimations by serving as the chief form of ecological (biotic) data. But what makes them so useful? The answers are ones you already know from earlier chapters: the fact that several species of insects are attracted to carrion within a few minutes of death, faunal succession is relatively predictable for specific stages of physical

decomposition, and some species produce larvae whose development is tied to feeding on the corpse (Catts & Haskell, 1990; Cherix *et al.*, 2012). These traits allow necrophagous insects to contribute to all or a portion of the PMI in at least two ways. The first relies on the relative predictability of species that arrive and use carrion in waves during physical decomposition (Schoenly & Reid, 1987). The presence or absence of particular species can convey qualitative information about the length of time the body has been available for insect colonization as well as ambient conditions. For example, the absence of first-wave colonizers, namely calliphorids, signifies that the remains were not assessable for a set period of time, or that environmental conditions did not favor insect activity until later in decomposition. There are other possible explanations but the examples serve to illustrate how the PMI can be influenced by the mere presence or absence of insects. In the second approach, the PMI can be estimated based on insect development. The most common method relies on the age of the oldest fly larva found on the body to make a time estimate of the association between the fly and body. This approach relies on working backward from the developmental stage discovered to oviposition/larviposition. A time can then be assigned (estimated) for how long development would have taken under the environmental conditions associated with the crime scene. If this range of time is then coupled with an estimate of body placement (death) to oviposition, an overall PMI can be predicated (Higley & Haskell, 2010).

The steps described are an oversimplification of the process as there are several contributing pieces of information that are also needed before a time of death estimate can be made. Each will be discussed shortly, but first there are some basic assumptions regarding the use of insects that must be met in order to proceed to the statistical calculations associated with environmental energy known as degree days. (Figure 14.1).

14.2.1   Assumptions for using insects in calculating the PMI

In order to use necrophagous insects to estimate a PMI, "traditionally" there are five basic assumptions that must be fulfilled. Failure to meet any one of the assumptions is likely to lead to error in the form of over- or under-estimation of the time of death. The significance of the error can vary from

Figure 14.1 Two forensically important insects, *Nicrophorus vespillo* and *Lucilia sericata*, utilizing a mole carcass in early stages after death, but each insect differs in meeting the assumptions for estimation of the postmortem interval. *Source:* Walcoford/Wikimedia Commons/CC BY-SA 3.0.

Table 14.1 Common assumptions for using insects in "time" calculations.

Assumption

Adult females did not oviposit/larviposit on a live host

Insects used for PMI estimations actually feed on the body to meet growth and developmental needs

Insects found feeding on body originated from oviposition/larviposition on the same remains

The insects are poikilothermic

A linear relationship exists between temperature and insect growth in terms of immature stages, at least for temperatures lying within the zone of tolerance for a given species and developmental stage

The stage of insect development can be accurately determined

Temperatures at crime scene accurately reflected by nearest weather station

Colonization only occurred during the day

Decedent's tissues did not contain drugs or toxins that would affect insect development

Lifetable data available or can be generated for insect of interest

Insect can be accurately identified to genus species

Source: Information modified from Higley & Haskell (2010), Gennard (2012), and Tarone & Sanford (2017).

rendering the PMI estimate useless to small or introduction of an unknown level of uncertainty (Tarone & Sanford, 2017). The implications of such error are obvious from our earlier discussions of why the PMI is important. The assumptions are as follows:

1. Adult females did not oviposit/larviposit on a live host.
2. Insects used for PMI estimations actually feed on the body to meet growth and developmental needs.
3. The insects are poikilothermic.
4. A linear relationship exists between temperature and insect growth in terms of immature stages, at least for temperatures lying within the zone of tolerance for a given species and developmental stage.
5. The stage of insect development can be accurately determined.

Perhaps an overlooked assumption is that not only must the stage of development be identified, so too must the identity of the genus and species. If the species identity cannot be determined, then it is not possible to estimate the PMI, although there may still be some useful information that can be derived from the insects (Wells & LaMotte, 2020). It may be surprising that the poikilothermic condition even needs to be stated, since all insects fall into this designation. However, as we discuss later in the chapter, some insects can be heterothermic for all or specific stages during development, which can greatly

complicate comparisons to experimental data generated under constant temperature rearing conditions. It is also important to recognize that some species of calliphorids are facultative parasites (see discussion in Chapter 16), and thus if oviposition occurred antemortem, the PMI could be overestimated based on the development data of such species. Finally, insects whose development is linked to or dependent on the corpse are the most useful for calculating the PMI since this generally means a continuous association between the deceased and insect. Necrophagous fly larvae meet this condition, but the adult stages do not, nor do most any other type of insect that frequents decomposing animal remains. Other assumptions have been postulated as important to consider for estimation of the PMI, as well as other parameters identified in Section 14.1.3 (Table 14.1).

From a practical perspective, blow flies and bottle flies (Family Calliphoridae) tend to be more useful in PMI estimations than sarcophagids. This is not because calliphorids are more likely to meet the five assumptions; rather it is a matter of behavior and abundance. Calliphorids usually dominate the first wave of insect succession, being the insects that initiate colonization of a corpse (Smith, 1986). These flies are also easier to identify to genus and species both as adults and larvae; sarcophagids generally require an expert in taxonomy for correct identification (Cherix *et al.*, 2012). Finally, several species of flesh flies are believed to be opportunistic predators which, depending on the degree of predation, may compromise assumption #1. Of course, the same concern can apply to predatory calliphorids like *Chrysomya albiceps* and *C. rufifacies*.

14.2.2 Insect development is linked to ambient temperatures

As discussed in extensive detail in Chapter 9, all insects are poikilothermic ectotherms – internal body temperature cannot be maintained by metabolic heat and thus reflects ambient environmental conditions. This means that elevations or declines in environmental temperatures result in corresponding (proportional) changes in internal temperatures for insects directly exposed to the environment. Insects residing in sheltered locations are somewhat buffered from the ambient temperature changes. However, body temperature even for these species will eventually acclimate, only at a slower rate. The net effect is that insects must possess the ability to maintain metabolic activity over a wide range of temperatures. In general, most species possess enzymes that operate under a broad range of conditions, including varying temperature and pH (Randall *et al.*, 2002; Storey, 2004). Failure to do so will negatively impact growth and development, and can ultimately lead to death.

14.2.2.1   Temperature and death

The range of temperatures over which insects can maintain metabolic processes or survive within indefinitely are referred to as the zone of tolerance or thermal tolerance range (review Chapter 9 for definitions). We will discuss the importance of this range of temperatures for PMI estimations in Section 14.3. Temperatures that fall outside the zone, i.e., at temperatures above the upper thermal limit (critical thermal maximum) or below the lower thermal limit (critical thermal minimum), are conditions that will initially evoke inhibition of cellular reactions, thereby retarding most aspects of growth and development. Temperatures exceeding or which are near these critical thresholds produce insect responses that are the least predictable (Pedigo, 1996), and thus do not provide baseline data with much utility for estimations

of PMI. If necrophagous species are exposed for a sufficiently long period, or movement out of the thermal tolerance range occurs unexpectedly or rapidly, injury or death may be the result.

14.2.2.2   Temperature and growth

Temperature is also the chief abiotic factor regulating development of poikilotherms, which of course includes necrophagous insects. For necrophagous species as opposed to insects in general residing within the environment, the only feature whose definition is somewhat ambiguous is the ambient environment. The temperature of the air, corpse and soil, as well as maggot mass temperatures (and even here the temperature depends on where the fly is located in the mass and which species are involved), can all differ from each other, change over time, and potentially influence the rate of development for a given species during specific stages of development (Villet *et al.*, 2010). How to interpret each of these temperature influences on insect development, and hence the PMI, is a topic receiving a great deal of investigation with no definitive answers as yet. For now, our focus is on the temperatures that comprise the zone of tolerance for a given species in terms of development (as opposed to the zone of tolerance influencing survival). This relationship is curvilinear and can be illustrated by a sigmoidal development curve as shown in Figure 14.2 (Campbell *et al.*, 1974; Higley & Haskell, 2010). Temperatures near the extremes result in reduced growth, until eventually reaching a low or high temperature at which no growth occurs. At low temperatures, this extreme condition results in chill-coma and the threshold is typically referred to as the developmental threshold or limit, also known as the base temperature. Conversely, the high temperature threshold is often called the developmental maximum or upper thermal threshold (Figure 14.2). Induction of heat stupor typically occurs when an insect reaches this upper temperature limit.

Between these thresholds, the rate of development is linear. The linear portion is also the most useful and valid for interpolation and predictions associated with agricultural and forensic entomology applications (Pedigo, 1996; Higley & Haskell, 2010). For the calliphorid *Protophormia terraenovae*, larval development from egg hatch until the onset of post-feeding (mid to late third-stage larva) is linear between the temperatures of 15–35 °C when the rearing temperatures were held

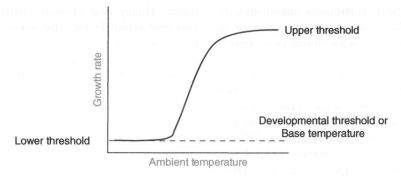

Figure 14.2   Sigmoidal curve depicting the curvilinear relationship between insect development or growth rate and ambient temperatures.

constant (Grassberger & Reiter, 2002a). With subtle differences at the lower end, the range reported for *P. terraenovae* is fairly consistent with many other calliphorid species located in Europe and North America (Greenberg & Tantawi, 1993; Anderson, 2000; Byrd & Allen, 2001; Grassberger & Reiter, 2001, 2002b) whereas the upper end of the range is higher (near or exceeding 45 °C) for species examined from parts of Africa and Australia (Richards *et al.*, 2009; Villet *et al.*, 2010). Far less information is available for species of sarcophagids, owing in large part to issues discussed early in this chapter as to why calliphorids are preferred in PMI estimations.

14.2.2.3   Environmental (thermal) energy is needed by poikilotherms

The relationship between insect growth and temperature can be viewed with an applied goal in mind, namely attempting to relate insect development to environmental conditions so that predictions can be made. What type of predictions? In an agricultural context, usually the primary entomological interest is to understand the dynamics of pest populations, particularly in predicting when surges in pest densities may occur so that control strategies can be developed (Pedigo, 1996). Obviously our interest here is in making predications about the time since death based on insect development data. A key concept essential for achieving this goal is that of environmental energy, also commonly termed thermal energy (Gennard, 2012). The idea is basically that insect growth is dependent on temperature, more specifically that an insect literally uses energy from the environment in the form of heat for its own growth and development. This is obviously in contrast to an endothermic animal like a human, which produces its own heat energy sufficient to meet developmental needs. As we discussed earlier, poikilotherms are not capable of generating sufficient metabolic heat to maintain body temperature or supply developmental demands. Thus the heat is derived from an exogenous source, the environment. Insects require a set amount of heat to develop from one stage to the next. The required heat is referred to as physiological time.

Environmental energy is used as currency referred to as thermal units or degree days (°D). Since the energy is derived from the environment and is temperature dependent, the thermal energy is related to a specific block of time, either hours or days. Within a given day, for example, there is a specific number of thermal units available for insect development. The energy currency accumulates over the unit of time of interest, so that at the end of an hour, day, or series of days, a set number of thermal units or degree days (or degree hours) has accrued. With an understanding of how much energy or thermal units is needed by an insect to reach a specific stage of development under a set of environmental conditions (namely temperatures), an estimate can then be made of the length of environmental exposure needed to accrue the sufficient amount of thermal energy. A discussion of how to calculate degree days and degree hours can be found in Section 14.4.3.

14.3   Modeling growth–temperature relationships

The next step in relating temperature, insect development, and thermal energy is to develop or use an existing growth model that allows accurate predictions of time. Fortunately, insect growth–temperature relationships have been extensively studied, largely in an agricultural pest context, meaning that "new" models or methods are not necessary for application to forensic entomology. Detailed discussions of statistical theory (Arnold, 1960; Pedigo, 1996; Higley & Haskell, 2010) and the many different ways to model animal growth and development (Wagner *et al.*, 1984) go beyond the scope of this book. Our intent is to provide insight into the underlying biological and entomological concepts that allow an understanding of how insects can be used in PMI calculations. Figure 14.2 provides one common way to model insect development using a sigmoidal growth curve that depicts a curvilinear relationship between the rate of growth and the temperature. It is the linear portion of this curve that provides the most utility in estimations of the PMI, provided of course that the assumptions described earlier in Section 14.2.1 are met. As temperatures approach either the developmental threshold or developmental maximum, insect development slows and becomes less predictable because a linear relationship between growth and temperature no longer exists. Growth is frequently modeled as a function of temperature and time (Figure 14.3), resulting a sine-wave curve in which development can be visualized as the area under the curve that occurs above the developmental threshold but below the developmental maximum or upper threshold. Functionally, the sine-wave model is widely used for degree day calculations associated with agricultural pest insects. The area under the curve that lies between the thresholds is assumed to represent accumulated degree days. However, one limitation with this approach is that the area also encompasses development near temperature extremes which, as we discussed earlier, is not

linear. Higley and Haskell (2010) provide an excellent rationale for why linear models make practical sense for use in forensic entomology, and one such approach – physiological energy budgets – is the basis for the remainder of this chapter.

14.3.1   Physiological energy budgets

Physiological energy budgets reflect the total thermal energy needed by an insect to complete development from egg to adult emergence (Gennard, 2012). Actually, the total lifespan does not have to be measured as only a specific life stage, say the second larval instar, may be of interest, so that only the physiological energy budget up to that developmental point is examined. This model relates thermal energy directly to time, specifically examining the relationship between growth and temperature above the developmental threshold. During a 24-hour period (or conversely for 1 hour), the physiological energy budget is represented by a rectangle (Figure 14.4). The vertical sides of the rectangle are defined by the unit of time of interest, in the figure a 24-hour window (1 day), so the sides are at 0 and 24 hours. The horizontal sides are related to temperature thresholds. The bottom side is at the imaginary line of the developmental threshold, since we know that no growth occurs below this temperature. Remember that this lower

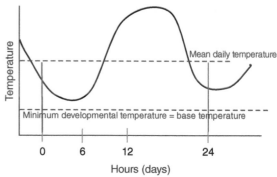

Figure 14.4   Growth model that relates thermal energy to time by showing the relationship between growth and temperature above the developmental threshold. The physiological energy budget is represented by the rectangle formed from the block of time of interest and temperature thresholds. In this case, the upper threshold is the mean daily temperature. *Source*: The graph is based on Gennard (2012) and Higley & Haskell (2010).

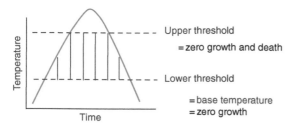

Figure 14.3   Sine-wave model relating development as a function of temperature and time. Insect growth is considered linear between the thresholds. *Source*: Based on Gennard (2012).

threshold is species-specific. Thus, if two species of flies were examined for the same period of time on the same corpse, the modeled rectangle would be expected to be different for each species because of the developmental threshold. The upper horizontal side is derived from a line drawn to represent the mean daily temperature for the 24-hour window. Here is where the relationship takes into account the amount of thermal energy available for a given time period.

The area within the resulting rectangle represents the degree days (more correctly accumulated degree days) or thermal units available for the 24 hours shown. Thus, degree days are the accrued product of time and temperature between the development minimum and maximum. This is the energy available for insect growth. However, as can be seen in Figure 14.5, portions of the rectangle do not fall under the curve, and some areas under the temperature curve are not represented by the rectangle. The method of physiological energy budget calculation depicted here relies on the averaging method of linear estimation, and an important tenet of this technique is that for every point lying outside the rectangle but under the daily temperature curve that overestimates accumulation, there is another point in the rectangle but outside the daily temperatures that underestimates (Gennard, 2012). The net effect is that the outlying points average or cancel each other out. This physiological energy budget is the basis for our use in the coming sections of accumulated degree days (or hours) to account for insect development on a corpse.

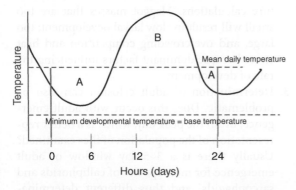

Figure 14.5 The physiological energy budget represented as a rectangle depends on the averaging method of linear estimation. Areas of the rectangle outside the temperature curve (areas "A") that overestimate growth are compensated for by areas under the curve but outside the rectangle that underestimate growth (area "B"). *Source*: The graph is based on Gennard (2012) and Higley & Haskell (2010).

14.4 Calculating the PMI requires experimental data on insect development and information from the crime scene

14.4.1 What is needed to calculate the PMI?

Now that we have a model in place that can be used to relate insect development in the environment to time, we can begin to discuss all the remaining parts needed to calculate a PMI. The information needed includes:

- identification of the fly genus and species and age of development stage;
- experimental development data at relevant temperatures for fly of interest;
- base temperature or developmental threshold for each species of interest;
- temperature data from the crime scene;
- temperature data from a nearby weather station;
- calculation of accumulated degree days representing relevant stages of insect development;
- calculation of accumulated degree days for the crime scene.

Some of the required information has already been discussed. For the remaining pieces, we will now discuss how to obtain or calculate the necessary information.

14.4.2 Base temperature

The base temperature or developmental threshold is the lowest temperature at which insect development can occur (Higley & Haskell, 2010). Temperatures below this threshold result in complete retardation of growth. Consequently, as temperatures approach the base temperature of a given species, growth correspondingly slows until reaching the developmental threshold. For most species, temperature depressions below the base temperature induce chill-coma (discussed in Chapter 9), and if sustained for a sufficient length of time death will result. Base temperatures are experimentally derived and have already been worked out for many of the common blow fly species. However, caution must be exercised in using a base temperature from the literature as many authors have reported differences for the same species. For example, *P. terraenovae* is a Holarctic blow fly that presumably should be more cold tolerant than Nearctic species, yet Greenberg and Tantawi (1993) estimated a base temperature of 12.5 °C for flies collected from the Pacific Northwest of the United States. This is higher than the 9 °C determined for a strain isolated in Vienna, Austria (Grassberger & Reiter, 2002a), and much higher than the developmental threshold (2–6 °C) calculated for *Calliphora vicina* (Vinogradova & Marchenko, 1984; Greenberg, 1991), a species that overlaps in distribution. Further complicating base temperature observations is the fact that Ames and Turner (2003) found no obvious development threshold for *C. vicina* and noticed that some, albeit reduced, development still occurred at temperatures as low as 1 °C. These examples serve to stress the importance of using an appropriate base temperature for the species and location of interest. Failure to use an appropriate base temperature can lead to an overestimation of the amount of thermal energy needed to complete development, and thus the resulting physiological energy budget will yield an inflated estimate of needed accumulated degree days (Oliveira-Costa & de Mello-Patiu, 2004; Van Laerhoven, 2008), meaning that the calculated PMI using such data is overestimated.

14.4.2.1   Calculating the base temperature

What do you do if the base temperature has not been worked out for a particular species, or for a particular geographic location? The base temperature is determined experimentally by rearing the fly from egg hatch to adult eclosion at a series of temperatures lying between the critical minimum and maximum. These developmental values are then graphed as a scatter plot, with the temperature data on the *x*-axis and 1/total days to develop on the *y*-axis (Figure 14.6). A line of best fit or linear regression is then calculated for the data, and the point where the line intersects the *x*-axis is the extrapolated base temperature (Higley & Haskell, 2010). This method of base temperature calculation is called linear approximation estimation (Gennard, 2012).

The method described is relatively straightforward but does have some inherent issues that reflect the lack of a standard operating procedure (SOP) for forensic entomology as a whole, discussed by Tarone and Foran (2008). These issues include the following:

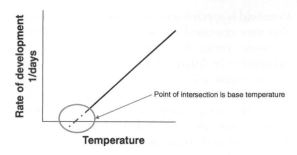

Figure 14.6   Calculation of the base temperature for a given fly species using the linear approximation method. Base temperature is determined by extrapolation as the point that intersects the *x*-axis.

1. Development rate is dependent on available nutrition, which means that the tissue chosen for rearing the flies will influence base temperature. There is no accepted tissue of choice for rearing of necrophagous flies, although many laboratories rely on beef or pork liver. Convincing data are available that has demonstrated slower rates of larval development dependent on type of tissue (Kaneshrajah & Turner, 2004; Day & Wallman, 2006), and thus variability in nutrition undoubtedly will influence base temperature calculations.

2. Developmental rate is also influenced by size, density and/or volume of a maggot mass (Slone & Gruner, 2007; Rivers *et al.*, 2010), which means that base temperature calculations are influenced by the initial size of maggot mass established. Again, there is no SOP for maggot mass size that should be used for base temperature calculations. Maggot masses that are too small will result in slow larval development; too large, and overcrowding competition and heat stress become dominant factors influencing the rate of development.

3. Determination of adult eclosion can also be problematic. Does this occur with adult emergence, or on the peak day, which is a better representation of the population of flies examined? Usually there is a 3–5-day window of adult emergence for most species of calliphorids and sarcophagids, and thus different determinations of eclosion between investigators will result in base temperature differences for the same species.

4. As already discussed in Section 14.3, insect growth demonstrates a curvilinear relationship with ambient temperature. Thus, representation of growth rate versus temperature to estimate the developmental threshold has inherent errors

that likely lead to overestimation of the base temperature for any given fly species.

Establishment of SOPs can easily overcome these issues and ensure more uniform calculations of base temperatures for any species (Amendt *et al.*, 2007).

14.4.3   Accumulated degree days for insect development

Degree days are thermal units that reflect energy currency in the environment which an insect can use for its own growth and development. As we have discussed, the amount of thermal units available on any given day is a direct reflection of the temperatures that occurred within the block of time of interest. For any particular stage of development, the number of thermal units necessary for development is thus dependent on temperature. The span of time needed to complete development then relies on the total number of thermal units that have accrued or accumulated, better known as accumulated degree days (ADD) or accumulated degree hours (ADH). The ADD is calculated using the following formulae:

$$ADD = time(days) \times (temperature - base\ temperature)$$
$$ADH = time(hours) \times (temperature - base\ temperature)$$

The resulting units of measure are °D or °H, but are reported with regard to temperature scale. For example, accumulated degree days calculated based on a Celsius scale would have the units ADD°C (as opposed to ADD°F).

The temperature referred to in the equations is the environmental temperature that influenced a given fly species of interest. Similarly, the base temperature, as we just discussed, is also species-specific and experimentally derived. "Time" in this case is the length of time needed to complete a given stage of development at temperatures comparable to the environmental temperature. Information on the duration of fly development at different temperatures comes from the research literature. Developmental data are available for several of the most common fly species encountered on a corpse (Kamal, 1958; Byrd & Butler, 1997; Anderson, 2000; Byrd & Allen, 2001; Marchenko, 2001; Grassberger & Reiter, 2002a, b; Greenberg & Kunich, 2002) at a range of temperatures. A collection of research papers providing life table data are listed in **Appendix III**. Differences have been reported for the same species collected from multiple geographic regions that may reflect genetic variability or, just as likely, result from the use of different rearing protocols. It is therefore important to only use developmental data for species located geographically near the crime scene. In instances when such data are not available, the parameters of stage development must be worked out in the laboratory under controlled conditions. Here is where some of the same issues described for base temperature calculations also come into play due to a lack of SOP for developmental studies.

14.4.3.1   ADD calculation for insect development

To illustrate how a simple ADD calculation is performed, we will use *Protophormia terraenovae* as an example. Let us say that we are interested in calculating what the ADD would be for the egg stage if the average daily temperature was 25 °C. A search of the literature reveals that a study by Greenberg and Tantawi (1993) contains the development information needed for the calculations. Their paper reported development data at four rearing temperatures (12.5, 23, 29, and 35 °C). Temperature data at 23 °C are most comparable to our conditions, and in which case the egg stage requires 16.8 hours to complete. The final piece of information needed, the base temperature, was also reported in this paper to be 12.5 °C. Substituting the information we have into the equation:

$$ADD = 16.8 / 24 (hours\ in\ a\ day) \times (25 - 12.5)$$
$$= 0.7 \times 12.5 = 8.75°D\ (ADD°C)$$
$$for\ the\ egg\ stage$$

Thus 8.75 ADD°C are needed at 25 °C to complete the egg stage for *P. terraenovae*. Notice in this example that information available for "time" was in hours (16.8) but an ADD defines time in days. Thus, there is a need to convert the hourly data into days, and hence in the example given, 16.8 hours was divided by the number of hours in a day (24). Before moving on, it is important to take note of a couple of considerations associated with the calculation. First, if hourly data is available, an ADH can be calculated and then the resulting value can then be divided by 24 to generate an ADD. If applied to the example above, the information is plugged first into the ADH equation:

$$ADH = 16.8 \times (25 - 12.5)$$
$$= 16.8 \times 12.5$$
$$= 210\ ADH°C$$

This value is then divided by 24 to generate the ADH, which equals 8.75 ADD°C. However, the reverse cannot be done. If all of the time data is in "days" then only an ADD can be determined. The reasoning is that you cannot derive more specific measures (hourly) from a broader data set (days). This is a point that will be repeated later in the chapter.

If interested in some other development stage, say the third stage of larval development, the same procedure would be done, with one exception: the ADD of each developmental stage preceding the one of interest must be calculated and summed to achieve an estimate of accumulated degree days. The logic is simple: the insect first had to complete each proceeding stage to reach the one of interest, in this case the third larval instar. As well, life table data generally reports on the length of time required to complete each individual stage of development. So, in the example above, the egg stage of *P. terraenovae* at 23 °C required 16.8 hours. Greenberg and Tantawi (1993) reported that at the same temperature first instar larvae required 26.4 hours to complete, second stage 27.6, and third stage 44.4 for a total of 98.4 hours. When the time for the egg stage is added in, *P. terraenovae* required 115.2 hours to progress from egg to the third stage of larval development. This in turn gives:

$$ADD = 115.2 / 24 \times (25 - 12.5)$$
$$= 4.8 \times 12.5$$
$$= 60 \, ADD°C$$

Another way of completing the same calculation is to determine ADH or ADD for each developmental stage separately.

$$\textbf{Egg stage} = 8.75 \, ADD°C$$
$$\textbf{First stage larva } ADH = 26.4 \times (25 - 12.5)$$
$$= 26.4 \times 12.5$$
$$= 330 \, ADH°C$$
$$ADD = 330 / 24 = 13.75 \, ADD°C$$
$$\textbf{Second stage larva } ADH = 27.6 \times 12.5$$
$$= 345 \, ADH°C$$
$$ADD = 345 / 24 = 14.38 \, ADD°C$$
$$\textbf{Third stage larvae } ADH = 44.4 \times 12.5$$
$$= 555 \, ADH°C$$
$$ADD = 555 / 24 = 23.13 \, ADD°C$$

Notice in these examples that "time" was for specific stages of development only. Thus, to determine the total ADD°C needed to completed the third stage of larval development, the values for each developmental stage preceding and including the third stage must be summed:

$$\text{Total ADD to completed third stage of larval}$$
$$\text{development} = 8.75 \, (\text{egg}) + 13.75 \, (1^{st} \, \text{larva})$$
$$+ 14.38 \, (2^{nd} \, 1) + 23.13 \, (3^{rd} \, 1)$$
$$= 60.13 \, ADD°C$$

The calculations are simple, but keep in mind that ADD calculations are species-specific in that the base temperature and development data are unique to a given species of fly or other insects. Thus, ADD calculations must be performed for each species of interest on a corpse.

14.4.4   Accumulated degree days for the crime scene

The ADD calculations performed in Section 14.4.3 were specific to a given fly species based on information available in the literature on insect development. What it provided was baseline information that is needed to interpret what is actually discovered at the crime scene. The only way to relate the experimental development data to the insect growth associated with the corpse is to understand how much thermal energy was available at the crime scene. The thermal energy in this context accumulates (as discussed for physiological energy budgets), so that each 24-hour period has a set number of thermal units available for an insect to use based on ambient temperatures. Thus, the thermal energy or ADD available at the crime scene must be estimated for each day between the day of death (or when the corpse was available for insect detection and acceptance) and when the body was discovered. Why? Because this represents how much accumulated thermal energy was available for the oldest fly larva or other stage of interest to use to reach the stage of development found on the body at the time of discovery. Thus in Section 14.4.3 we calculated the ADD needed for specific stages of insect development. Now in this section, we calculate the ADD that was actually available in the environment, specifically at the crime scene. Eventually the two values will be compared, but not yet.

What is needed for the crime scene ADD calculation? Based on the formula for ADD, we need to know time and the relevant temperatures. Let us address each of these factors for the crime scene environment:

1. *Base temperature.* This is still species-specific and based on the insect of interest. The reason this information is still relevant is because the ADD at the crime scene is based on the insect evidence of interest, so temperatures below the developmental threshold are not relevant to estimating available thermal energy for a given fly.
2. *Time.* If calculating the ADD for a 24-hour period, then the unit of time is 1 day, because in fact we are calculating the thermal units available for each day between time of death or until we can account for the stage of insect development and discovery. So the ADD calculation must be done for each day of interest, and time will be equal to 1 on each day. Alternatively, if determining the ADH, the time value is still 1, for 1 hour. In this case ADH is calculated for each hour of interest from the time of discovery and working backwards toward when insect colonization occurred, which of course is what you are trying to estimate.
3. *Temperature.* In this case, temperature is in reference to the crime scene. Obviously, this is problematic because the only temperature we have is from the day of corpse discovery and any day thereafter. Prior to discovery, we have no record of the crime scene temperatures, unless of course the body is located indoors and we can just check the thermostat. The latter presumes that the heating or cooling system was "on" and must be checked to see if a daily temperature regime was pre-programmed.

There are ways to derive or estimate the crime scene temperatures when the data are unknown and that is the topic discussed next.

14.4.4.1 Corrected crime scene temperatures

What if the crime scene temperatures are not known? The answer is to visit a meteorological weather station that is located close to where the body was discovered. In the United States, there are several maintained by the National Oceanic and Atmospheric Administration (NOAA) that provide detailed daily and historic records, many associated with airports (http://www.nws.noaa.gov/climate/). Once at the site, you will need to find a location nearest to the crime scene. Most NOAA weather station sites maintain at least minimum and maximum temperatures for several months to years. Minimum/maximum temperatures can be averaged to provide an approximate daily average temperature for a day of interest. Another valuable

source of historic climate data is Weather Underground (https://www.wunderground.com/) that can provide archived temperature data for cities around the world. However, weather station temperatures cannot be used at face value, unless the corpse is discovered next to the thermometer. The reason is simply that, at this point, it remains unknown how well the weather station data approximate the conditions at the crime scene.

This relationship is determined by measuring temperatures at the crime scene from the time of discovery for 4–5 days (generally such temperatures are measured at approximately 1.2 m or 4 feet from the ground surface to be comparable to the location of temperature recording at NOAA weather stations). The corresponding weather station temperatures for those same days are collected, and the data are then graphed as a scatter plot, with weather station temperatures on the x-axis and crime scene temperatures on the y-axis (Figure 14.7). Simple linear regression is performed, and based on the resulting r-value the strength of the relationship between the two sites can be made. High r-values are indicative of the weather station temperatures being reflective of the crime scene temperatures, while low values suggest that the meteorological data do not approximate the crime scene temperatures well. In the latter scenario, use of such data will lead to over- or under-estimations of the PMI. Of course if no other data are available, then use of the meteorological data may be warranted, but the resulting ADD, and hence PMI, need to have wider time estimate

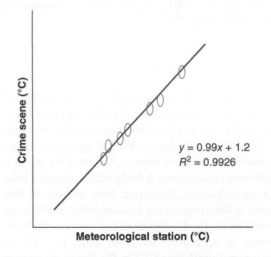

$$y = 0.99x + 1.2$$
$$R^2 = 0.9926$$

Figure 14.7 Determination of relationship between crime scene temperatures and those from a meteorological weather station. The regression equation is used to calculate corrected crime scene temperatures.

brackets to account for the potential error. If there is a strong relationship between the two sites, then the regression equation can be used as a correction factor.

How is the correction factor used? Let us use the data from Figure 14.7, where the regression equation is $y = 0.99x + 1.2$. In the equation, x represents the meteorological data and y the crime scene temperature. Thus, if we know x, in this case weather station temperature data for a given day, we can solve for the unknown value, which is the temperature at the crime scene for any day before discovery. So, if a body were discovered in Baltimore, Maryland on June 15, the unknown crime scene temperatures would be any day before this date. A NOAA weather station is located at the Baltimore-Washington International Airport just to the south of the city, and minimum/maximum temperature data for June 14 was 72 °F/85 °F for a mean daily temperature of 78.5 °F. This temperature can be converted to degrees Celsius using the following equation[3]:

$$T_c = (5/9) \times (T_f - 32)$$

The result is 25.8 °C. This value is x, and substituting into our regression equation yields:

$$y = 0.99(25.8) + 1.2 = 26.7$$

Thus, the corrected crime scene temperature for June 14 is 26.7 °C. This equation is applied to every day of interest for which the crime scene data are not known. The corrected crime scene temperatures also become the values used for calculating the ADD at the crime scene.

14.4.4.2 Calculating the ADD for the crime scene

Now that we have a method for determination of crime scene temperatures for days prior to body discovery, all the necessary information is in place for the ADD calculation. We will again use *P. terraenovae* as the fly of interest, this time for use at a fictitious crime scene. A body was discovered June 15 in Baltimore, Maryland. Our interest at this point is determining the amount of thermal energy available for fly development, and for now the time frame is arbitrarily set from June 15 (day of discovery) until June 10. On the day of discovery, we can use either actual measured temperatures at the crime scene or those from an appropriate weather station. For the sake of illustrating the

correction factor one more time, we will use NOAA minimum/maximum data: 71 °F/88 °F for an average of 79.5 °F or 26.4 °C. This value is then corrected using the determined regression equation for the crime scene and weather station, which in this case will be the same as determined in Section 14.4.4.1 ($y = 0.99x + 1.2$).

Calculation of corrected crime scene temperatures on the day of discovery:

$$y = 0.99(26.4) + 1.2 = 27.3$$

The corrected crime scene temperature for June 15 is 27.3 °C. Returning to the ADD formula, we now have:

$$\begin{aligned} ADD &= time \times (temperature - base\ temperature) \\ &= 1 \times (27.3 - 12.5) \\ &= 14.8\ ADD°C \end{aligned}$$

Thus, 14.8 accumulated degree days were available on June 15. Remember, time is "1" since we are only examining June 15, or one day. Temperature is the corrected crime scene temperature and base temperature is for the fly of interest, in this case *P. terraenovae*. If we had performed ADH calculations instead, we can increase our precision by taking into account the time of day when the body was discovered. However, this increased level of precision can only be done when hourly temperature data are also available from the meteorological site. If only minimum/maximum temperatures can be obtained, the calculation must be an ADD rather than an ADH. As stated in Section 14.4.3.1, you can first calculate an ADH for the crime scene and then convert it to ADD by dividing by 24. However, the reverse conversion (ADD → ADH) is not permissible.

The ADD calculations are repeated for each day from June 15 until June 10 (Table 14.2). The ADD value for June 14 is 14.2 ADD°C; this value is added to the previous day's ADD to yield the Σ or summation ADD. For example:

$$June\ 15\ ADD = 14.8$$
$$June\ 14\ ADD = 14.2$$

The addition of the two generates a ΣADD on June 14 of 29 ADD°C.

Why are the values added? The answer is because the number of thermal units is accumulative over the period of time the insect in question is exposed

Table 14.2 Calculation of accumulated degree days (ADD) at crime scene for *Protophormia terraenovae*.

Date	Weather station temperatures (°C)	Corrected crime scene temperatures (°C)	Base temperature	ADD	ΣADD
June 15*	26.4	27.3	12.5	14.8	—
June 14	25.8	26.7	12.5	14.2	29.0
June 13	25.1	26.0	12.5	13.5	42.5
June 12	26.0	26.9	12.5	14.4	56.9
June 11	26.2	27.1	12.5	14.6	71.5
June 10	25.9	26.8	12.5	14.3	85.8
June 9	25.8	26.7	12.5	14.2	100.0
June 8	25.4	26.3	12.5	13.8	113.8
June 7	25.6	26.5	12.5	14.0	127.8

* The day of corpse discovery. The corrected crime scene temperatures are derived from corresponding meteorological temperature data using the regression equation $y = 0.99x + 1.2$. *Source*: The base temperature was calculated by Greenberg and Tantawi (1993). The table design is based on the ideas of Gennard (2012).

to the environment. This also means that the values relevant to the PMI calculation are those that have been summed, which means the day of discovery is excluded from the estimate (Table 14.2). The process is continued with the ADD calculated for June 13th added to the summation value on June 14th, and so forth for each day of interest. The question arises as to how far back or how many days need to be examined? The answer is that you continue to calculate ΣADD until accounting for the amount of thermal energy needed to reach the stage of insect development discovered on the corpse. This will be illustrated next.

14.4.5 Putting it all together

The checklist of information necessary for PMI estimation is now complete. So it is time to actually estimate the time since death. Returning to the scenario described in the preceding section, a body was discovered on June 15 in Baltimore, Maryland, colonized by multiple species of calliphorids. The oldest larvae identified on the corpse are third-stage feeding larvae of *Protophormia terraenovae* (the reality is that this fly generally is not present during the warmer temperatures of summer in Maryland, but can be found in early June if the preceding spring conditions are mild). The task is to estimate a PMI for the body using the fly evidence. In this case, several necessary pieces of information are available to us: the base temperature based on Greenberg and Tantawi (1993), corrected crime scene data from our efforts in the preceding section (Table 14.2), and developmental data for *P. terraenovae* at temperatures

relevant to the crime scene conditions (Greenberg & Tantawi, 1993). It is important to note that using the base temperature and development data from Greenberg and Tantawi (1993) is making the assumption that populations isolated from the Pacific Northwest of the United States will be very similar to those found in the mid-Atlantic region of the east coast. When calculating the ADD for *P. terraenovae* earlier in the chapter, the developmental data used was for a rearing temperature of 23 °C because this closely approximated the environmental temperature of interest, 25 °C. However, in this crime scene scenario, the corrected crime scene temperatures are closer to the 29 °C conditions used in Greenberg and Tantawi (1993). At that temperature, the egg stage lasts 14.4 hours, first instar larvae 13.2 hours, second stage 18.0 hours, and third stage 49.2 hours, for 94.8 hours total development time. The ADD is calculated as:

$$ADD = 94.8 / 24 \times (29 - 12.5)$$
$$= 3.95 \times 16.5$$
$$= 65.2 \, ADD°C$$

Alternatively, if we had simply used the calculations for 25 °C, then the ADD would have been 60°D. The ADD of 65.2°D is then compared with the ADD calculations available at the crime scene (Table 14.2), specifically checking the ΣADD for when sufficient thermal units were available for *P. terraenovae* to reach the third larval instar. This occurred sometime between June 11 and 12 (Table 14.3). Counting begins on June 14 and then worked backwards until reaching the day with

Table 14.3   Determination of PMI using accumulated degree days (ADD) at crime scene for *Protophormia terraenovae*.

Date	Weather station temperatures (°C)	Corrected crime scene temperatures (°C)	Base temperature	ADD	ΣADD
June 15*	26.4	27.3	12.5	14.8	—
June 14	25.8	26.7	12.5	14.2	29.0
June 13	25.1	26.0	12.5	13.5	42.5
June 12	26.0	26.9	12.5	14.4	56.9
June 11	26.2	27.1	12.5	14.6	**71.5**
June 10	25.9	26.8	12.5	14.3	85.8
June 9	25.8	26.7	12.5	14.2	100.0
June 8	25.4	26.3	12.5	13.8	113.8
June 7	25.6	26.5	12.5	14.0	127.8

*The day of corpse discovery.

June 11 is when sufficient thermal energy accrued to account for the 65.2 ADD °C needed by *P. terraenovae* to reach the 3rd instar larva.

Counting from June 11 toward the day of discovery yields 4 days. With the information provided, this can be considered the PMI.

enough accumulated degree days to account for the stage of larval development, which would be June 11 or 4 days. Keep in mind that the day of discovery does not factor into the PMI estimation. If we had used the ADD from 25 °C, the estimated PMI would have been the same, 4 days.

14.4.6   The PMI should be represented as a range

Estimation of the postmortem interval based on insect development should not be represented as a single value. Instead, it represents a range of time. This may seem confusing since in the last section, we just estimated a PMI as a single unit of time. That example was a simplified scenario to demonstrate the process of PMI estimation. As described by Wells & LaMotte (2020), statistical evaluation of the data generally yields a range as a confidence set. One method that they recommend is called inverse prediction (Wells & LaMotte, 2020). Another way of thinking about how a range of values is more reflective of the PMI is to consider the developmental data used for the carrion insect of interest. In the example using developmental data for *P. terraenovae* in Section 14.4.5, time estimates for the duration for each life stage were given as a single value. In reality,

those were the mean values. In other words, there were individuals that required less time and some that required more time than depicted by the mean. Consider the difference that would result in the ADD calculation for just the egg stage of *P. terraenovae* if we took into account the fastest (10.7 hours) and slowest (18.5 hours) developers at 29 °C[4]:

$$\text{Egg stage : slowest development}$$
$$ADH = 18.5 \times (29 - 12.5)$$
$$= 18.5 \times 16.5$$
$$= 305.25$$
$$ADD = 305.25 / 24 = 12.71 \, ADD°C$$
$$\text{fastest development}$$
$$ADH = 10.7 \times 16.5$$
$$= 176.55$$
$$ADD = 176.55 / 24 = 7.35 \, ADD°C$$

The range of values for the egg stage yields the minimum and maximum thermal units needed for the sample of flies tested. If this were done for each developmental stage leading to the stage discovered at the crime scene, and then compared to the ΣADD for the crime scene, two PMI values would be generated, the minimum and maximum length of time needed to complete development for this species. Together the values represent the range of time representing the PMI based on development of *P. terraenovae*. Other means for calculating a PMI range have been proposed (Schoenly *et al.*, 1996; Ieno *et al.*, 2010).

14.4.7 Estimation of PMI based on insect succession

This approach to estimating the PMI is based on the changes in arthropod assemblages that occur as the corpse undergoes decomposition. The underlying presumption is that the successional changes in insect fauna is relatively predictable for a given geographical location and season. Thus, the presence or absence of expected groups or species can be used to in the estimation of the PMI (Gennard, 2012; Sharma *et al.*, 2015). In order to apply this method, every specimen collected must be accurately identified to at least family (Gennard, 2012), but generally to species, and then is compared to what the predicted or typical pattern of faunal succession expected under the same conditions at the crime scene. If the victim died or was placed immediately after death at the site where discovered, the succession interval may be viewed as the PMI (Perez *et al.*, 2014). Conversely, if the body has been moved from some other location, the successional pattern is not expected to align with that predicted for the crime scene.

14.5 The evolving PMI: changing approaches and sources of error

The examination of the PMI presented in this chapter, including how to calculate accumulated degree days for a given insect species and its application to time of death estimates, was designed to simply present a basic overview. In reality, a practitioner of forensic entomology must take into account several factors that can complicate the basic calculations presented here. Such undertakings require extensive experience gained from case assignments, a thorough understanding of both statistical theory and biological underpinnings, and practice in calculating all facets of the PMI. What we are trying to say is that use of the PMI is more complex than perhaps the overview of this chapter would imply. In this section, some of the complicating factors for calculating the PMI, including sources of error and new directions, will be addressed.

14.5.1 Estimation of larval age

One of the assumptions necessary for using insects in PMI estimations is the ability to correctly identify the age of insects found on the corpse. The feeding stages of necrophagous flies, predominantly calliphorids, are considered the most accurate for estimating larval age, with post-feeding stages (e.g., wandering larvae, prepupae, and puparial stages) being far less precise (Tarone & Foran, 2008). Presently, the most common practices for estimating larval age have depended on measurements of larval length (e.g., isomeglan diagrams), as well as an examination of posterior spiracles (i.e., counting the number of slit openings in each spiracle). While such methods, particularly the former, have been supported experimentally and are considered admissible in court, there are inherent issues. For one, no standard protocols for rearing insects have been defined for use in forensic entomology. Thus, the type of diet, size of maggot masses, rearing conditions (temperature, photoperiod, and humidity) and means to age life stages are left to the discretion of the investigator when attempting to determine base temperature, developmental rates, or length/size to age relationships. As discussed earlier in the chapter, the net effect has been discrepancies between authors in reported values, even for the same species.

The other issue that ties directly to larval size and length is that although development as a whole is considered linear between the developmental threshold and developmental maximum, larval growth is nonlinear in terms of dimensions (length, width, weight) (Wells & LaMotte, 1995). Significant overlap in larval body length of different ages occurs (Anderson, 2000), particularly under the natural conditions of an overcrowded maggot mass on a corpse. This compromises the use of growth curves and isomegalen diagrams generated for several blow fly species with the goal of approximating larval age by length (Grassberger & Reiter, 2001, 2002a, b; Donovan *et al.*, 2006). In addition, as larval development progresses, the duration of each stage increases, meaning larger windows of time must be used to bracket the estimates of development in the environment. The situation leads to less precise estimates of larval age, and hence PMI, with increasing age of the larva (Tarone & Foran, 2011).

One possible solution to the issue of accurately staging larvae and post-feeding flies is coupling traditional methods with use of developmentally regulated gene expression (Ames *et al.*, 2006; Tarone & Foran, 2011). Work conducted with *Lucilia sericata* has shown that at least nine genes have been identified that allow increased precision in aging fly larvae, and at least five permit distinction between feeding and post-feeding stages (Tarone & Foran, 2011). Each of these genes appears to be ecdysone responsive, changing expression levels as development progress, and thus have the potential to be sensitive indicators of age (Tarone & Foran, 2011), provided that environmental tokens[5] other than temperature have an insignificant effect on gene expression. Wide-scale use of such techniques has not yet occurred, but presumably assessment of gene expression will become a standard protocol in the future.

14.5.2 The true ambient temperatures shaping larval development

Temperature is the dominant factor influencing the rate of development of poikilotherms, a topic discussed in great detail here and in Chapter 9. For the purposes of calculating an ADD, the ambient temperatures at the crime scene must be known, either through direct measurement or by estimation using other sources such as meteorological data. The idea no doubt seemed straightforward when presented the first time, but some problems do exist concerning ambient temperatures that need to be accounted for in PMI estimations. These include the fact that (i) temperatures at a crime scene are not static, (ii) experimental development data are typically derived from constant temperature conditions, and (iii) maggot mass temperatures are far higher than ambient air temperatures.

14.5.2.1 Temperatures at a crime scene are not static

Under most environmental conditions in temperate regions, ambient temperatures fluctuate over time. This is generally most evident when comparing daytime to nighttime temperatures (Byrd & Allen, 2001; Clarkson *et al.*, 2004). Rarely, however, do ADD

models take this into consideration and thus static environment temperatures are used to calculate temperatures at the crime scene. The examples provided in this chapter represent the static condition. This issue can be overcome by calculating ADH rather than ADD, so that hourly temperatures are used that accurately reflect the changing environmental conditions. However, ADH calculations are not an option when only minimum/maximum temperature data are available.

14.5.2.2 Experimental development data

In a similar vain, insect development data have generally been derived from constant rearing temperatures. So, the experimentally generated data do not model the conditions fly larvae experience on the cadaver, conceivably leading to over- or under-estimations of fly development used in calculation of the PMI. However, the impact of constant rearing temperatures appears to be species-specific. For example, Greenberg (1991) observed that the developmental period for four species of calliphorids (*Phormia regina*, *L. sericata*, *Chrysomya rufifacies* and *Cochliomyia macellaria*) was longer when reared under fluctuating temperatures than when maintained under constant temperature control. In contrast, development of *Lucilia cuprina*, *P. regina*, *L. sericata*, and *Calliphora vicina* during static laboratory conditions was reported to compare favorably with development on carrion (Dallwitz, 1984; Anderson, 2000; Arnaldos *et al.*, 2005). Once again, the differences observed for the same species by different investigators argues for the need to standardize protocols for rearing and development studies. Such standards have already been proposed for best practices to collect and preserve insect specimens discovered at crime scenes (Amendt *et al.*, 2007; Sanford *et al.*, 2020).

14.5.2.3 Maggot mass temperatures

Chapter 8 details the unique microhabitat created by several species of fly larvae known as the maggot mass. A hallmark of maggot masses is the generation of internal heat that elevates the local temperature to several degrees above ambient temperatures (Turner & Howard, 1992). The amount of heat produced depends on several factors including, but not limited to, the size, density and/or volume of the feeding aggregation, species composition, food source (which may have varying influences depending on where the mass forms on a corpse), microbiota present in the mass, and influence of ambient conditions (Anderson & Van

Laerhoven, 1996; Joy *et al.*, 2006; Slone & Gruner, 2007; Richards *et al.*, 2009; Rivers *et al.*, 2010; Charabidze *et al.*, 2011). In addition, heat production generally changes over time, reaching a zenith early during the third stage of larval development, and then dropping sharply once post-feeding begins (Campobasso *et al.*, 2001). Higley and Haskell (2010) have argued that, if known, maggot mass temperatures should be used as the ambient temperatures for ADD calculations. Unfortunately, rarely are these temperatures known, and they are very difficult to model since so many factors influence the hourly internal mass temperatures (Kotzé *et al.*, 2016). Rivers *et al.* (2012) have proposed that heat-shock expression in fly larvae and puparial stages may allow for evaluation of the thermal history of necrophagous flies on a corpse, but this work is still in its infancy.

14.5.3 Photoperiod influences on larval development

Control of photoperiod during larval rearing is often a neglected component of fly developmental studies. In most studies examining the relationship between temperatures and rate of development, or for purposes of base temperature calculation, the photoperiod has been constant at 24 : 0 light:dark. For at least one species, *P. regina*, constant photoperiod (all light) lengthens development in comparison with cyclic light–dark conditions (Nabity *et al.*, 2007). Failure to account for constant light influences on larval development will lead to the underestimation of the PMI for this species. The counter argument is that constant conditions are needed to avoid phototaxic stage transitions, referred to as emergence gating (Greenberg, 1991; Byrd & Allen, 2001). Emergence gating is when developmental stages transition (e.g., from pharate adult to adult) in synchrony with environmental cycles (e.g., photoperiod). Such synchronous transitions would compromise developmental studies if independent of temperature.

14.5.4 Larval nutrition

Food quantity and quality are known to directly influence the developmental rate for necrophagous flies. This topic has been discussed a few times in this chapter as potentially compromising experimental development data and base temperatures

generated in the laboratory for ADD calculations. No standard diet is recommended for such studies, yet there is an absolute need. Nutritional differences undoubtedly also occur when larvae feed on the corpse, as different tissues are not expected to provide equal nutriment to fly larvae (Clark *et al.*, 2006; Day & Wallman, 2006). Consequently, rates of development might actually be slower for larvae feeding one tissue type (e.g., brain tissue) versus another (e.g., viscera), yet temperature based on degree day calculations do not take this into account.

Quantitative influences on development are expected as maggot mass sizes increase, largely because overcrowding lowers the available nutriment per individual. For *Sarcophaga bullata* and *P. terraenovae*, development slows once a species-specific size threshold for feeding aggregations is exceeded (Rivers *et al.*, 2010). For sarcophagids, the size effect extends into puparial stages, resulting in slower rates of development up until adult emergence (Byrd & Butler, 1998; Rivers *et al.*, 2010). Again, most degree day models do not account for nutritional influences on larval growth, so overestimations of the PMI are indeed possible.

14.6 Total body scores influenced by insect development

Total body score (TBS) determinations are used as a means to quantify physical decomposition of human remains. The method can be coupled with ADD calculations to estimate a postmortem interval (Megyesi *et al.*, 2005; Moffatt *et al.*, 2016). Total body score determinations are performed by forensic anthropologist or others trained in evaluating human pathology. As such, forensic entomologists do not use the technique. However, TBS still has relevance to forensic entomology in that the presence of insects, specifically, fly larvae during specific stages of physical decomposition are used to help evaluate or assign a score to the degree of decay (Keough *et al.*, 2017). To an extent, the approach is similar to PMI estimations based on succession, or succession interval. The difference with TBS is that there is no evaluation of the species present and fly larvae are the only entomological evidence considered. Nonetheless, this represents an indirect method in which entomological evidence can contribute to the estimation of the PMI.

Chapter review

The time since death is referred to as the postmortem interval

- Forensic entomology offers tools that contribute to defining the time since death or postmortem interval (PMI). The PMI is an estimate of when death most likely occurred, and this calculation is based on numerous factors, including temperature of the corpse, ambient temperatures, physical appearance of the body, and a range of biochemical changes that take place within the fluids and tissues of the deceased.

- Determination of the time since death is extremely valuable in criminal investigations as this information can help identify both the individual responsible for the victim's death and the victim. Time of death determinations can also be critical to civil matters as well, as in cases of insurance or inheritance, where time of death determines beneficiaries.

- Several species of necrophagous insects can provide a time frame of several hours to months for a PMI estimation depending on how long the deceased has been dead and the length of exposure to biotic and abiotic components of the environment.

- There has been considerable debate concerning terminology and application of insect evidence to the estimation of all or a portion of the PMI. The term minimum postmortem interval (PMI_{min}) is based on insect development and is considered to represent the window of time from insect colonization until the developmental stage of the specimen of interest was discovered on the remains. The maximum postmortem interval (PMI_{max}) encompasses the time frame when the decedent was last seen alive until discovery of the corpse. Some authors have proposed other terms accounting for insect association with carrion, including the period of insect activity, time of colonization, time of oviposition, and postcolonization interval. For some, these terms are preferred over PMI because time periods based on the development of necrophagous insects can be significantly different from the actual time since death, since there is usually a lag between when death occurs and the first wave of colonizers actually detect, locate, and feed on the corpse.

The role of insects in estimating the PMI

- Generally speaking, forensic medicine techniques are most relevant for the first 72 hours following death, after which time ecological data, chiefly in the form of necrophagous insects, become increasingly more important. What makes them so useful? The answers are ones you already know from earlier chapters: the fact that several species of insects are attracted to carrion within a few minutes of death, faunal succession is relatively predictable for specific stages of physical decomposition, and some species produce larvae whose development is tied to feeding on the corpse.

- Necrophagous insects contribute to a part of the PMI in at least two ways. The first relies on the relative predictability of species that arrive and use carrion in waves during physical decomposition. The presence or absence of particular species can convey qualitative information about the length of time the body has been available for insect colonization as well as ambient conditions. In the second approach, a portion of the PMI can be estimated based on insect, specifically necrophagous fly, larval development. The most common method relies on the age of the oldest larva found on the body to make a time estimate of the association between the fly and body. This approach relies on working backward from the developmental stage discovered to oviposition/larviposition.

- In order to use necrophagous insects to estimate a PMI, there are five basic assumptions that must be fulfilled. Failure to meet any one of the assumptions is likely to lead to significant error in the form of over- or under-estimation of the time of death.

- Perhaps the most important of the assumptions is associated with insect development: a linear relationship exists between temperature and insect growth in terms of immature stages, at least for temperatures lying within the zone of tolerance for a given species and developmental stage.

Modeling growth–temperature relationships

- Insect growth–temperature relationships have been extensively studied, largely in an agricultural pest context, meaning that "new" models or methods are not necessary for application to forensic entomology.

- One common way to model insect development is by using a sigmoidal growth curve that depicts a curvilinear relationship between the rate of growth versus temperature. It is the linear portion of this curve that provides the most utility to estimation of the PMI, provided of course that the five assumptions are met.
- Physiological energy budgets are another growth model that reflect the total thermal energy needed by an insect to complete development from egg to adult emergence. This model relates thermal energy directly to time, specifically examining the relationship between growth and temperature above the developmental threshold.

Calculating the PMI requires experimental data on insect development and information from the crime scene

- The information needed to calculate a PMI includes identification of the fly genus and species and age of development stage, experimental development data at relevant temperatures for fly of interest, base temperature or developmental threshold for each species of interest, temperature data from the crime scene, temperature data from a nearby weather station, calculation of accumulated degree days representing relevant stages of insect development, and calculation of accumulated degree days for the crime scene.
- Insect base temperature and development data come from the research literature and are species-specific.
- Crime scene temperature is needed to calculate the thermal energy known as accumulated degree days available for insect development on the corpse. Generally, this information is not available, so temperature data from some other source, like a meteorological weather station, can be used to calculate corrected crime scene temperature data that approximates the conditions at the crime scene prior to body discovery.
- Once the accumulated degree days have been determined for the crime scene, these values can be compared to the accumulated degree days needed by the insect to reach the oldest stage of development on the corpse, yielding an estimate of the PMI.

The evolving PMI: changing approaches and sources of error

- The examination of the PMI presented in this chapter, including how to calculate accumulated degree days for a given insect species and its application to time of death estimates, is designed to simply present a basic overview. In reality, a practitioner of forensic entomology must take into account several factors that can complicate the basic calculations presented here. Such undertakings require extensive experience gained from case assignments, a thorough understanding of both statistical theory and biological underpinnings, and practice in calculating all facets of the PMI.
- One of the five assumptions necessary for using insects in PMI estimations is the ability to correctly identify the age of insects found on the corpse. The most common practices for estimating larval age have depended on measurements of larval length, as well as an examination of posterior spiracles. While such methods, particularly the former, have been supported experimentally and are considered admissible in court, no standard protocols for rearing insects have been defined for use in forensic entomology. The other issue that ties directly to larval size and length is that although development as a whole is considered linear between the developmental threshold and developmental maximum, larval growth is non-linear in terms of dimensions (length, width, weight).
- For the purposes of calculating an ADD, the ambient temperatures at the crime scene must be known, either through direct measurement or by estimation using other sources such as meteorological data. The idea no doubt seemed straightforward when first presented, but some problems do exist concerning ambient temperatures that need to be accounted for in PMI estimations. These include the fact that temperatures at a crime scene are not static, experimental development data are derived from constant temperature conditions, and maggot mass temperatures are far higher than ambient air temperatures.
- Control of photoperiod during larval rearing is often a neglected component of fly developmental studies. For at least one species, *P. regina*, constant photoperiod (all light) lengthens development in comparison with

cyclic light–dark conditions. Failure to account for constant light influences on larval development will lead to underestimates of the PMI for this species.

- Food quantity and quality are known to directly influence the developmental rate for necrophagous flies. No standard diet is recommended for such studies, yet there is an absolute need. Nutritional differences undoubtedly also occur when larvae feed on the corpse, as different tissues are not expected to provide equal nutriment to fly larvae. Consequently, rates of development are altered based on food quality and quantity, yet temperature based on degree day calculations do not take this into account.

Total body scores influenced by insect development

- Total body score determinations are used as a means to quantify physical decomposition of human remains. The method can be coupled with ADD calculations to estimate a postmortem interval. Total body score determinations are performed by forensic anthropologist or others trained in evaluating human pathology.
- TBS still has relevance to forensic entomology in that the presence of insects, specifically, fly larvae during specific stages of physical decomposition are used to help evaluate or assign a score to the degree of decay. To an extent the approach is similar to PMI estimations based on succession, or succession interval. The difference with TBS is that there is no evaluation of the species present and fly larvae are the only entomological evidence considered.

Test your understanding

Level 1: knowledge/comprehension

1. Define the following terms:

    (a) postmortem interval
    (b) period of insect activity
    (c) accumulated degree day
    (d) succession interval
    (e) base temperature
    (f) physiological energy budget
    (g) Null hypothesis
    (h) thermal unit.

Level 2: application/analysis

1. Describe the five assumptions that must be met for insect use in the calculation of a postmortem interval.
2. Discuss the limitations of estimating a PMI using insect development data that falls outside the linear region of the zone of tolerance.
3. Explain how the averaging method of linear estimation is important to the physiological energy budget model.
4. Construct a graph that depicts the curvilinear relationship between insect growth and temperature. Label the key thresholds associated with insect growth.
5. Convert the following temperatures from degrees Celsius to degrees Fahrenheit:

    (a) 25 °C   (c) 27 °C
    (b) 18 °C   (d) 23 °C.

Level 3: synthesis/evaluation

1. Use the temperature growth data in Table 14.4 to help answer the questions that follow.

Table 14.4 Developmental duration of *Phormia regina* at different temperatures.

Mean (°C)	Duration (hours)				
	Egg	First instar	Second instar	Third instar	Pupae
15.6	32	28	52	154	312
21.2	12	32	28	199	125
25.0	12	18	24	62	124
26.7	16	14	26	56	65
32.2	14	6	16	58	76

Calculate the accumulated degree days in °F (ADD°F) required for *P. regina* to complete the following stages of development when this fly species is reared at 21.1 °C, with the lower threshold temperature for development to take place being 11.4 °C:

    (a) Convert 21.1 °C to °F.
    (b) Egg stage
    (c) First instar larva
    (d) Second instar larva
    (e) Third instar larva
    (f) Pupa.

2. Use the following information and Table 14.5 to calculate the environmental energy available for

Table 14.5   BWI NOAA Station, July 2009.

Day	Minimum temperature (°F)	Maximum temperature (°F)
7/11	69	90
7/10	67	86
7/9	70	89
7/8	71	84
7/7	68	87
7/6	74	88
7/5	72	86
7/4	70	85
7/3	68	85
7/2	68	84
7/1	67	88
6/30	72	90
6/29	69	87
6/28	66	84

larval development of *Calliphora vicina* from June 28 to July 11. Assume that developmental data are available for 26.1 °C, the calculated base temperature is 3 °C, and use the corrected temperature regression of $y = 0.93x + 2.0$. Construct a table to organize the information.

3. The maggot masses of *Lucilia sericata* formed on an animal carcass produce heat that exceeds ambient temperatures by several degrees. Heat production is relatively modest during the first and early second stages of larval development, but then begins to rise in an age-specific fashion until post-feeding, at which time temperatures in the mass begin to decline. Describe how maggot mass temperatures under these conditions can be used in ADD calculations.

4. Using the developmental data for *Musca domestica* in Table 14.6, calculate the range of accumulated degree days in °C (ADD°C) required to complete the second instar of larval development at 23 °C.

Table 14.6   Developmental duration (mean ± SD) of *Musca domestica* at different temperatures.

Mean (°C)	Duration (hours)				
	Egg	First instar	Second instar	Third instar	Pupae
22.0	24.5±0.7	37.8±2.7	43.1±1.9	110.9±4.6	429.0±11.6
25.0	19.1±0.8	27.1±3.4	30.4±2.6	70.4±4.4	301.4±7.8
28.0	16.4±0.6	18.8±2.0	23.2±2.0	55.6±3.8	228.3±9.1

Notes

1. An osmolyte is a solute that contributes to the overall osmotic concentration of a fluid.
2. Residue refers to the personal property of the deceased, such as furniture, antiques, clothing, dishes, appliances, jewelry, etc., but does not include a physical structure like a home or land.
3. To convert a temperature in degrees Celsius to degrees Fahrenheit, use the equation $T_f = [(9/5) \times T_c] + 32$.
4. Developmental data is based on *Protophormia terraenovae* collected in southcentral Pennsylvania (Rivers, unpublished), and not that of Greenberg & Tantawai (1993).
5. An environmental token is an abiotic feature of the environment that alters the behavior, development, or physiology of an insect, often in anticipation of impending climatic change.

References cited

Abdelhalim, K., Khaoula, B., Saddek, B., Abbas, M., & Mokhtar, H. C. (2017) Post mortem interval: Necrobiome analysis using artificial neural networks. *Computational Biology and Bioinformatics* 5(6): 90–96.

Ahi, R.S. & Garg, V. (2011) Role of vitreous potassium level in estimating postmortem interval and the factors affecting it. *Journal of Clinical and Diagnostic Research* 5: 13–15.

Amendt, J., Campobasso, C.P., Gaudry, E., Reiter, C., LeBlanc, H.N. & Hall, M.J.R. (2007) Best practice in forensic entomology: standards and guidelines. *International Journal of Legal Medicine* 121: 90–104.

Ames, C. & Turner, B. (2003) Low temperature episodes in development of blowflies: implications for postmortem interval estimation. *Medical and Veterinary Entomology* 17: 178–186.

Ames, C., Turner, B. & Daniel, B. (2006) Estimating the post-mortem interval (II): the use of differential temporal gene expression to determine the age of blowfly pupae. *International Congress Series* 1288: 861–863.

Anderson, G.S. (2000) Minimum and maximum development rates of some forensically important Calliphoridae (Diptera). *Journal of Forensic Sciences* 45: 824–832.

Anderson, G.S. & Van Laerhoven, S.L. (1996) Initial studies on insect succession on carrion in southwestern British Columbia. *Journal of Forensic Sciences* 41: 617–625.

Arnaldos, M.I., Garcia, M.D., Romera, E., Presa, J.J. & Luna, A. (2005) Estimation of postmortem interval in real cases based on experimentally obtained entomological evidence. *Forensic Science International* 149: 57–65.

Arnold, C.Y. (1960) Maximum-minimum temperatures as a basis for computing heat units. *Proceedings of the American Society of Horticultural Science* 76: 682–692.

Byrd, J.H. & Allen, J.C. (2001) The development of the black blow fly, *Phormia regina* (Meigen). *Forensic Science International* 120: 79–88.

Byrd, J.H. & Butler, J.F. (1997) Effects of temperature on *Chrysomya rufifacies* (Diptera: Calliphoridae) development. *Journal of Medical Entomology* 34: 353–358.

Byrd, J.H. & Butler, J.F. (1998) Effects of temperature on *Sarcophaga haemorrhoidalis* (Diptera: Sarcophagidae) development. *Journal of Medical Entomology* 35: 694–698.

Campbell, A., Frazer, B.D., Gilbert, N., Gutierrez, A.P. & Mackauer, M. (1974) Temperature requirements of some aphids and their parasites. *Journal of Applied Ecology* 11: 431–438.

Campobasso, C.P., Di Vella, G. & Introna, F. (2001) Factors affecting decomposition and Diptera colonization. *Forensic Science International* 120: 18–27.

Catts, E.P. (1992) Problems in estimating the postmortem interval in death investigations. *Journal of Agricultural Entomology* 9: 245–255.

Catts, E., & Haskell, N. (1990) *Entomology and Death: A Procedural Guide.* Joyce's Print Shop, Clemson, South Carolina.

Charabidze, D., Bourel, B. & Gosset, D. (2011) Larval-mass effect: characterization of heat emission by necrophagous blowflies (Diptera: Calliphoridae) larval aggregates. *Forensic Science International* 211: 61–66.

Cherix, D., Wyss, C. & Pape, T. (2012) Occurrences of flesh flies (Diptera: Sarcophagidae) on human cadavers in Switzerland, and their importance as forensic indicators. *Forensic Science International* 220: 158–163.

Clark, K., Evans, L. & Wall, R. (2006) Growth rates of the blowfly, *Lucilia sericata,* on different body tissues. *Forensic Science International* 156: 145–149.

Clarkson, C.A., Hobischak, N.R. & Anderson, G.S. (2004) A comparison of the developmental rate of *Protophormia terraenovae* (Robineau-Desvoidy) raised under constant and fluctuating temperature regimes. *Canadian Society of Forensic Sciences* 37: 95–101.

Dallwitz, R. (1984) The influence of constant and fluctuating temperatures on development and survival rate of pupae of the Australian sheep blowfly, *Lucilia cuprina. Entomologia Experimentalis et Applicata* 36: 89–95.

Day, D.M. & Wallman, J.F. (2006) A comparison of frozen/thawed and fresh food substrates in development of *Calliphora augur* (Diptera Calliphoridae) larvae. *International Journal of Legal Medicine* 120: 391–394.

Donovan, S.E., Hall, M.J.R., Turner, B.D. & Moncrieff, C.B. (2006) Larval growth rates of the blowfly, *Calliphora vicina,* over a range of temperatures. *Medical and Veterinary Entomology* 20: 106–114.

Estracanholli, E.S., Kurachi, C., Vicente, J.R., Campos de Menezes, P.F., Castro e Silva Junior, O. & Bagnato, V.S. (2009) Determination of post-mortem interval using in situ tissue optical fluorescence. *Optics Express* 17: 8185–8192.

Gennard, D.E. (2012) *Forensic Entomology: An Introduction,* 2nd edn., John Wiley & Sons Ltd., Chichester, UK.

Goff, M.L. (2010) Early postmortem changes and stages of decomposition. In: J. Amendt, C.P. Campobasso, M.L. Goff & M. Grassberger (eds) *Current Concepts in Forensic Entomology,* pp. 1–24. Springer, London.

Grassberger, M. & Reiter, C. (2001) Effect of temperature on *Lucilia sericata* (Diptera: Calliphoridae) development with special reference to the isomegalen- and isomorphen-diagram. *Forensic Science International* 120: 32–36.

Grassberger, M. & Reiter, C. (2002a) Effect of temperature on development of *Liopygia* (=*Sarcophaga*) *argyrostoma* (Robineau-Desvoidy) (Diptera: Sarcophagidae) and its forensic implications. *Journal of Forensic Sciences* 47: 1332–1336.

Grassberger, M. & Reiter, C. (2002b) Effect of temperature on development of the forensically important Holarctic blow fly *Protophormia terraenovae* (Robineau-Desvoidy) (Diptera: Calliphoridae). *Forensic Science International* 128: 177–182.

Greenberg, B. (1991) Flies as forensic indicators. *Journal of Medical Entomology* 28: 565–577.

Greenberg, B. & Kunich, J.C. (2002) *Entomology and the Law.* Cambridge University Press, Cambridge, UK.

Greenberg, B. & Tantawi, T.I. (1993) Different developmental strategies in two boreal blow flies (Diptera: Calliphoridae). *Journal of Medical Entomology* 30: 481–483.

Higley, L.G. & Haskell, N.H. (2010) Insect development and forensic entomology. In: J.H. Byrd & J.L. Castner (eds) *Forensic Entomology: The Utility of Using Arthropods in Legal Investigations,* 2nd edn, pp. 389–406. CRC Press, Boca Raton, FL.

Hubig, M., Muggenthaler, H., Schenkl, S., & Mall, G. (2018) Improving stomach content based death time determination by maximum probability estimation. *Forensic Science International* 285: 135–146.

Ieno, E.N., Amendt, J., Fremdt, H., Saveliev, A.A. & Zuur, A.E. (2010) Analysing forensic entomology data using additive mixed effects modeling. In: J. Amendt, C.P. Campobasso, M.L. Goff & M. Grassberger (eds) *Current Concepts in Forensic Entomology,* pp. 139–162. Springer, London.

Jackson, A.R.W. (2016) *Forensic Science,* 4th edn. Pearson, Harlow, UK.

Joy, J.E., Liette, N.L. & Harrah, H.L. (2006) Carrion fly (Diptera: Calliphoridae) larval colonization of sunlit and shaded pig carcasses. *Forensic Science International* 164: 183–192.

Kamal, A.S. (1958) Comparative study of thirteen species of sarcosaprophagous Calliphorida and Sarcophagidae (Diptera). I. Bionomics. *Annals of the Entomological Society of America* 51: 261–270.

Kaneshrajah, G. & Turner, B. (2004) *Calliphora vicina* larvae grow at different rates on different body tissues. *International Journal of Legal Medicine* 118: 242–244.

Keough, N., Myburgh, J. & Steyn, M. (2017) Scoring of decomposition: a proposed amendment to the method when using a pig model for human studies. *Journal of Forensic Sciences* 62(4): 986–993.

Kotzé, Z., Villet, M.H. & Weldon, C.W. (2016) Heat accumulation and development rate of massed maggots of the sheep blowfly, *Lucilia cuprina* (Diptera: Calliphoridae). *Journal of Insect Physiology* 95: 98–104.

Kreitlow, K.L.T. (2010) Insect succession in natural environment. In: J.H. Byrd & J.L. Castner (eds) *Forensic Entomology: The Utility of Arthropods in Legal Investigations*, 2nd edn, pp. 251–270. CRC Press, Boca Raton, FL.

Marchenko, M.I. (2001) Medicolegal relevance of cadaver entomo-fauna for the determination of the time since death. *Forensic Science International* 120: 89–109.

Matuszewski, S. & Szafalowicz, M. (2013) Temperature-dependent appearance of forensically useful beetles on carcasses. *Forensic Science International* 229: 92–99.

Megyesi, M.S., Nawrocki, S.P. & Haskell, N.H. (2005) Using accumulated degree-days to estimate the postmortem interval from decomposed human remains. *Journal of Forensic Sciences* 50: 618–626.

Michaud, J. P., Schoenly, K. G., & Moreau, G. (2015) Rewriting ecological succession history: did carrion ecologists get there first? *The Quarterly Review of Biology* 90(1): 45–66.

Moffatt, C., Simmons, T., & Lynch-Aird, J. (2016) An improved equation for TBS and ADD: establishing a reliable postmortem interval framework for casework and experimental studies. *Journal of Forensic Sciences* 61: S201–S207.

Nabity, P.D., Higley, L.G. & Heng-Moss, T.M. (2007) Light-induced variability in the development of the forensically important blow fly, *Phormia regina* (Meigen) (Diptera: Calliphoridae). *Journal of Medical Entomology* 44: 351–358.

Oliveira-Costa, J. & de Mello-Patiu, C.A. (2004) Application of forensic entomology to estimate of the post-mortem interval (PMI) in homicide investigations by the Rio de Janeiro Police Department in Brazil. *Aggrawal's Internet Journal of Forensic Medicine and Toxicology* 5(1): 40–44.

Pechal, J. L., Crippen, T. L., Benbow, M. E., Tarone, A. M., Dowd, S., & Tomberlin, J. K. (2014) The potential use of bacterial community succession in forensics as described by high throughput metagenomic sequencing. *International Journal of Legal Medicine* 128(1): 193–205.

Pedigo, L. (1996) *Entomology and Pest Management*, 2nd edn. Prentice Hall, Upper Saddle River, NJ.

Perez, A. E., Haskell, N. H., & Wells, J. D. (2014) Evaluating the utility of hexapod species for calculating a confidence interval about a succession based postmortem interval estimate. *Forensic Science International* 241: 91–95

Randall, D., Burggren, W. & French, K. (2002) *Animal Physiology: Mechanisms and Adaptations*. W.H. Freeman and Company, New York.

Richards, C.S., Price, B.W. & Villet, M.H. (2009) Thermal ecophysiology of seven carrion-feeding blowflies (Diptera: Calliphoridae) in southern Africa. *Entomologia Experimentalis et Applicata* 131: 11–19.

Rivers, D.B., Ciarlo, T., Spelman, M. & Brogan, R. (2010) Changes in development and heat shock protein expression in two species of flies (*Sarcophaga bullata* [Diptera: Sarcophagidae] and *Protophormia terraenovae* [Diptera: Calliphoridae]) reared in different sized maggot masses. *Journal of Medical Entomology* 47: 677–689.

Rivers, D.B., Kiakis, A., Bulanowski, D., Wigand, T. & Brogan, R. (2012) Oviposition restraint and developmental alterations in the ectoparasitic wasp *Nasonia vitripennis* (Walker) when utilizing puparia resulting from different size maggot masses of *Lucilia illustris, Protophormia terraenovae* and *Sarcophaga bullata*. *Journal of Medical Entomology* 49: 1124–1136.

Sanford, M. R. (2015) Forensic entomology in the medical examiner's office. *Academic Forensic Pathology* 5(2): 306–317.

Sanford, M. R., Byrd, J.H., Tomberlin, J.K. & Wallace, JR. (2020) Entomological evidence collections methods. In: J.H. Byrd & J.K. Tomberlin (eds) *Forensic Entomology: The Utility of Using Arthropods in Legal Investigations*, 3rd edn, pp. 67–89. CRC Press, Boca Raton, FL.

Schoenly, K. & Reid, W. (1987) Dynamics of heterotrophic succession in carrion arthropod assemblages: discrete series or a continuum of change. *Oecologia* 73: 192–202.

Schoenly, K., Goff, M.L., Wells, J.D. & Lord, W.D. (1996) Quantifying statistical uncertainty in succession-based entomological estimates of the postmortem interval in death scene investigations: a simulation study. *American Entomologist* 42: 106–112.

Sharma, R., Garg, R.K., & Gaur, J.R. (2015) Various methods for the estimation of the post mortem interval from Calliphoridae: A review. *Egyptian Journal of Forensic Sciences* 5: 1–12.

Slone, D.H. & Gruner, S.V. (2007) Thermoregulation in larval aggregations of carrion-feeding blow flies (Diptera; Calliphoridae). *Journal of Medical Entomology* 44: 516–523.

Smith, K.G.V. (1986) *A Manual of Forensic Entomology*. British Museum (Natural History), London.

Storey, K.B. (2004) Biochemical adaptation. In: K.B. Storey (ed.) *Functional Metabolism: Regulation and Adaptation*, pp. 383–414. John Wiley & Sons, Inc., Hoboken, NJ.

Tarone, A.M. & Foran, D.R. (2008) Generalized additive models and *Lucilia sericata* growth: assessing confidence intervals and error rates in forensic entomology. *Journal of Forensic Sciences* 53: 942–948.

Tarone, A.M. & Foran, D.R. (2011) Gene expression during blow fly development: improving the precision of age estimates in forensic entomology. *Journal of Forensic Sciences* 56: S112–S122.

Tarone, A.M. & Sanford, M.R. (2017) Is PMI the hypothesis or the Null hypothesis? *Journal of Medical Entomology* 54: 1109–1115.

Tomberlin, J.K., Mohr, R., Benbow, M.E., Tarone, A.M. & Van Laerhoven, S. (2011) A roadmap for bridging basic and applied research in forensic entomology. *Annual Review of Entomology* 56: 401–421.

Turner, B. & Howard, T. (1992) Metabolic heat generation in dipteran larval aggregations: a consideration for forensic entomology. *Medical and Veterinary Entomology* 6: 179–181.

Van Laerhoven, S.L. (2008) Blind validation of postmortem interval estimates using developmental rates of blow flies. *Forensic Science International* 180: 76–80.

Villet, M.H., Richards, C.S. & Midgley, J.M. (2010) Contemporary precision, bias and accuracy of minimum post-mortem intervals estimated using development of carrion-feeding insects. In: J. Amendt, C.P. Campobasso, M.L. Goff & M. Grassberger (eds) *Current Concepts in Forensic Entomology*, pp. 109–137. Springer, London.

Vinogradova E.B. & Marchenko, M.I. (1984) The use of temperature parameters of fly growth in medico-legal practice. *Sudebno Meditsinkskaya Ékspertiza* 27: 16–19.

Wagner, T.L., Wu, H., Sharpe, P.J.H., Schoolfield, R.M. & Couslon, R.N. (1984) Modeling insect development rates: a literature review and application of a biophysical model. *Annals of the Entomological Society of America* 77: 208–225.

Wells, J.D. & LaMotte, L.R. (1995) Estimating maggot age from weight using inverse prediction. Journal of Forensic Sciences 40: 585–590.

Wells, J.D. & LaMotte, L.R. (2017) The role of a PMI-prediction model in evaluating forensic entomology experimental design, the importance of covariates, and the utility of response variables for estimating time since death. *Insects* 8:47. https://doi.org/10.3390/insects8020047.

Wells, J.D. & LaMotte, L.R. (2020) Estimating the postmortem interval. In: J.H. Byrd & J.L. Castner (eds) *Forensic Entomology: The Utility of Using Arthropods in Legal Investigations*, 3rd edn, pp. 229–240. CRC Press, Boca Raton, FL

Supplemental reading

Adams, Z.J.O. & Hall, M.J.R. (2003) Methods used for the killing and preservation of blowfly larvae, and their effect on post-mortem larval length. *Forensic Science International* 138: 50–61.

Amendt, J., Zehner, R. & Reckel, F. (2008) The nocturnal oviposition behavior of blowflies (Diptera: Calliphoridae) in Central Europe and its forensic implications. *Forensic Science International* 175: 61–64.

Bajerlein, D., Taberski, D. & Matuszewski, S. (2018). Estimation of postmortem interval (PMI) based on empty puparia of *Phormia regina* (Meigen) (Diptera: Calliphoridae) and third larval stage of *Necrodes*

littoralis (L.) (Coleoptera: Silphidae) – advantages of using different PMI indicators. *Journal of Forensic and Legal Medicine* 55: 95–98.

Byrd, J.H. & Butler, J.F. (1996) Effects of temperature on *Cochliomyia macellaria* (Diptera: Calliphoridae) development. *Journal of Medical Entomology* 33: 901–905.

Day, D.M. & Wallman, J.F. (2006) Width as an alternative measurement to length for post-mortem interval estimations using *Calliphora augur* (Diptera; Calliphoridae) larvae. *Forensic Science International* 159: 158–167.

Dabbs, G.R. (2010) Caution! All data are not created equal: the hazards of using National Weather Service data for calculating accumulated degree days. *Forensic Science International* 202(1–3): e49–e52.

Higley, L.G., Pedigo, L.P. & Ostlie, K.R. (1986) DEGDAY: a program for calculating degree days, and assumptions behind the degree day approach. *Environmental Entomology* 15: 999–1016.

Huntington, T.E., Higley, L.G. & Baxendale, F.P. (2007) Maggot development during morgue storage and its effect on estimating the post-mortem interval. *Journal of Forensic Sciences* 52: 453–458.

Introna, F. Jr, Altamura, B.M., Dell'Erba, A. & Dattoli, V. (1989) Time since death definition by experimental reproduction of *Lucilia sericata* cycles in growth cabinet. *Journal of Forensic Sciences* 34: 478–480.

Megyesi, M., Nawrocki, S. & Haskell, N. (2005) Using accumulated degree-days to estimate the postmortem interval from decomposing human remains. *Journal of Forensic Sciences* 50: 618–626.

Michaud, J.-P. & Moreau, G. (2009) Predicting the visitation of carcasses by carrion-related insects under different rates of degree-day accumulation. *Forensic Science International* 185: 78–83.

Michaud, J.-P., Schoenly, K.G. & Moreau, G. (2012) Sampling flies or sampling flaw? Experimental design and inference strength in forensic entomology. *Journal of Medical Entomology* 49: 1–10.

Nuorteva, P. (1977) Sarcosaprophagous insects as forensic indicators. In: C.G. Tedeschi, W.G. Eckert & L.G. Tedeschi (eds) *Forensic Medicine: A Study of Trauma and Environmental Hazards*, Vol. 2, pp. 1072–1095. W.B. Saunders, Philadelphia.

Sharanwoski, B.J., Walker, E.G. & Anderson, G.S. (2008) Insect succession and decomposition patterns on shaded and sunlit carrion in Saskatchewan in three different seasons. *Forensic Science International* 179: 219–240.

Tabor, K.L., Fell, R.D. & Brewster, C.C. (2005) Insect fauna visiting carrion in Southwest Virginia. *Forensic Science International* 150: 73–80.

Tarone, A.M. & Foran, D.R. (2006) Components of developmental plasticity in a Michigan population of *Lucilia sericata* (Diptera: Calliphoridae). *Journal of Medical Entomology* 43: 1023–1033.

Tarone, A.M., Jennings, K.C. & Foran, D.R. (2007) Aging blow fly eggs using gene expression: a feasibility study. *Journal of Forensic Sciences* 52: 1350–1354.

Thevan, K., Ahmad, A. H., Md. Rawi, C. S., & Singh, B. (2010) Growth of *Chrysomya megacephala* (Fabricius) maggots in a morgue cooler. *Journal of Forensic Sciences* 55(6): 1656–1658.

Wang, Y., Yang, L., Zhang, Y., Tao, L. & Wang, J. (2018) Development of *Musca domestica* at constant temperatures and the first case report of its application for estimating the minimum postmortem interval. *Forensic Science International* 285: 172–180.

Wescott, D.J., Steadman, D.W., Miller, N., Sauerwein, K., Clemmons, C.M., Gleiber, D.S., McDaneld, C., Meckel, L. & Bytheway, J.A. (2018) Validation of the total body score/accumulated degree-day model at three human decomposition facilities. *Forensic Anthropology* 1(3): 143–149.

DNA degradation as an indicator of postmortem interval: http://udini.proquest.com/view/dna-degradation-as-an-indicator-of-goid:820481421/

Forensic medicine for medical students. Website that includes videos and information on early PMI determinations:http://www.forensicmed.co.uk/pathology/post-mortem-interval/

Postmortem interval. Brief presentation of various methods for estimating time since death: http://cis201.student.monroecc.edu/~st013/time_of_death_presentation/index.html

What information can a forensic entomologist provide at the death scene: http://www.forensicentomology.com/info.htm

Additional resources

Aboutdegree-days:http://www.ipm.ucdavis.edu/WEATHER/ddconcepts.html

Degree-day calculation: http://ipm.illinois.edu/degreedays/calculation.html

Chapter 15

Insect stains and artifacts: alterations of bloodstain and body fluid evidence

Overview

Blood can tell stories, or at least convey a great deal of information. When released from a human body or a blood-covered object, the fluid serves as invaluable evidence in the investigation of a violent crime, suspicious death, or suicide. Details regarding the identity of individuals, including the victim(s) and possibly those responsible for the criminal act, can be revealed through forensic biology and forensic serology, as can exclusion or inclusion of individuals through blood type analyses. Spatter and stains resulting from a blood-shed event leave characteristic patterns that often unveil insight into the events associated with an act of violence between individuals. Information about the direction blood traveled, location of the body at the time wounding was inflicted, movement of the bleeding individual at the crime scene, and the minimum number of blows necessary to create the bloodstain patterns are just a few examples of what can be interpreted from the size, shape, and angle of impact of bloodstains. However, the utility of blood evidence is compromised when insects attracted to a decomposing body or exuded biological fluids feed, walk, crawl, regurgitate, or defecate in or around bloodstains. The end result of insect activity is

modification of existing bloodstains, deposition of artifacts that are nearly indistinguishable from true bloodstains, and transfer of blood to other locations. Creation of insect-derived stains is not restricted to an association with blood; any type of body fluid and tissue can be sources for insect artifacts. Unique stains are also produced by some juvenile insects, such as those generated by fluid-soaked fly larvae dispersing from a corpse. This chapter will examine the science behind bloodstain pattern analysis and discuss how different types of necrophagous and opportunistic insects confound the use of bloodstain and body fluid evidence in criminal investigations.

The big picture

- Bloodstains are not always what they appear to be at the crime scene.
- Science is the cornerstone of bloodstain pattern analyses.
- Crash course in bloodstain analyses.
- Insect activity can alter blood evidence.
- Insect feeding activity can modify the morphology of bloodstains and yield transfer patterns.

The Science of Forensic Entomology, Second Edition. David B. Rivers and Gregory A. Dahlem.
© 2023 John Wiley & Sons Ltd. Published 2023 by John Wiley & Sons Ltd.
Companion website: www.wiley.com/go/forensicentomology2

- Ingested blood may be expelled as digestive artifacts.
- Parasitic insects can confound blood evidence by leaving spot artifacts.
- How can insect stains be detected?

15.1 Bloodstains are not always what they appear to be at the crime scene

An individual who is violently attacked by another is likely to bleed profusely from any inflicted wounds. Depending on the nature of the wound (e.g., weapon used and where on the body struck), as well as the surface that the expelled blood strikes, distinct bloodstains or spatter[1] will result. For example, bleeding from an open wound on a person in a calm vertical position will yield relatively spherical wet blood drops on a smooth floor surface like linoleum or ceramic tile (Saferstein, 2017) (Figure 15.1). Passive dripping into wet blood produces satellite stains that are readily distinguishable by a trained bloodstain pattern expert from those that occur due to a high-impact force such as a gunshot or bludgeoning. The significance of these examples is that bloodstains resulting from violent crimes leave behind patterns that are interpretable, which in turn

Figure 15.1 Example of human bloodstains. More spherical drops reflect passive falling of blood to the ground, whereas elliptical stains form when droplets hit the surface at an angle. *Source*: Nyki m/Wikimedia Commons/CC BY 3.0.

provide insight into the events of a violent act that allow reconstruction of the crime scene (Bevel & Gardner, 2008). Later in the chapter, we will examine the types of information that can be derived from specific bloodstains.

Blood evidence can provide a wealth of information to a criminal investigator beyond just stain patterns. Forensic serology[2] is the subdiscipline of forensic science that examines biological fluids as they pertain to matters of legal investigation. As blood is the most common bodily fluid that serves as physical evidence, the mere presence of exuded blood often helps to establish that a crime was even committed. Determination of blood type (e.g., type A, B, AB or O) can be used for exclusion or inclusion of individuals in relation to a crime as well as in matters of paternity, and DNA typing of a blood sample is essential for classification, identification, and individualization[3] of the victim and/or perpetrator of a violent crime.

As has been well documented throughout the book, an array of insects is attracted to a decomposing body. Some show interest only in the fluids released from human tissue, independent of whether the person is deceased or not. Still other insects (parasitic species) seek out living hosts to steal a liquid meal, usually in the form of blood. The linkage between these insects with different feeding strategies is that all can potentially confound bloodstain evidence. How? Bloodstain morphology can be altered by insects distorting the shape of wet or dry blood, ingesting blood and then expelling from the digestive tract, or transferring blood to other locations. Any bloodstains resulting from insect activity are termed **insect stains** (SWGSTAIN, 2009). When intermixed with human blood, the insect-derived products create confusion for a crime scene investigator with limited experience processing blood evidence. Insect stains share several biological and physical properties in common with human blood that makes the two nearly impossible to distinguish. In theory any insect that interacts with a corpse or associated exuded body fluids can potentially create stains or artifacts that confuse reconstruction efforts at a crime scene (Rivers & Geiman, 2017). Stains produced as a result of insect activity associated with *any type* of body fluid or tissue are referred to as **insect artifacts**. It should be noted that in some forensic entomology literature, the term "artifact" is used to describe bite marks or alterations of the corpse tissues due to insect activity (Campobasso *et al.*, 2009; Viero *et al.*, 2019), although the use has not been consistently applied to all necrophagous or necrophilous insects. Later in this chapter, a comparison of the features of bloodstains, body fluids, and insect artifacts will be made and placed in the context of specific insect groups or species responsible for generating the artifacts. Emphasis throughout the chapter is placed on insect alterations of bloodstains, since very little information is available regarding other body fluids and tissues.

15.2 Science is the cornerstone of bloodstain pattern analyses

The "science" in forensic analyses is not always obvious. To the casual observer, this is probably true for bloodstain pattern analyses (BPA) and aspects of forensic serology as whole, in part because interpreting types of bloodstains appears straightforward, requiring no scientific knowledge. For example, recognition of different types of bloodstain patterns seems deceptively easy, which no doubt gives the impression that other aspects of bloodstain pattern analysis are as well. The reality is that critical and reliable analyses of blood and bloodstain evidence are quite complex, with the underpinnings of the discipline derived from multiple fields of natural and applied sciences: biology, biochemistry, chemistry, mathematics, and physics (Gaensslen, 2000; James *et al.*, 2014). Effective interpretation of bloodstains relies on the application of the scientific method in conducting carefully controlled experiments, such as when testing bloodstain patterns on materials similar to those found at the crime scene. Remember from our discussion of the scientific method in Chapter 1 that all disciplines engaged in forensic analyses do not have this approach to inquiry/questioning as the core of their training. Consequently, relatively few forensic investigators, let alone the lay public, have the academic foundation to attempt analyses of bloodstains suitable for presentation in court.

Interpretation of bloodstains is a complex process and the intimate details of the theory and practice of BPA go beyond the intended scope of a forensic entomology textbook. The focus of this chapter is to provide a broad and admittedly superficial introduction to this fascinating area of forensic investigation so that we can place the interference of insects in appropriate context. We will begin with an overview of some the basic scientific principles and properties relevant to blood as a biological fluid that shapes BPA.

15.2.1 Biochemistry of blood

Blood is a type of connective tissue composed of a liquid extracellular matrix (plasma) with an array of cells suspended throughout. The liquid fraction is mostly water, which serves as a solvent for a wide range of macromolecules (proteins, carbohydrates, organic acids), salts (electrolytes), and other materials needed by cells and tissues. Once dissolved, the solute content yields relatively high osmotic and ionic concentrations within the plasma (low in comparison with insects), contributing to the **viscosity** and **surface tension** of blood. The solid fraction consists of several cell types (formed elements), including erythrocytes (red blood cells), leukocytes (white blood cells), cellular fragments (platelets or thrombocytes), and other cells in circulation (Figure 15.2). Together, the cellular components of blood and the solutes of plasma contribute to the high viscosity of this connective tissue, so that blood as a fluid is thick, sticky, and resistant to flow. The solute content, specifically macromolecules, also contribute to the surface tension of individual blood droplets. Surface tension is a property of the surface of a liquid whereby it resists an external force. In the case of wet blood, the shape of a drop or pool is maintained by the **cohesion** of molecules dissolved in the blood, resisting the force of atmospheric pressure. The shape of the drop is also influenced by the surface material that the blood landed on when released from the bleeding individual. What is the significance of these blood properties to forensic analysis? As will be seen later in the chapter, surface tension and viscosity of blood provide information on the path or direction that blood traveled when released and the force required to create the blood spatter (Bevel & Gardner, 2008).

The chemical constituents of blood are essential for revealing information about individuals and distinguishing between biological fluids. In the broadest use of blood as evidence, the presence of hemoglobin allows identification of an unknown sample to be classified as a biological fluid, specifically blood (Bevel & Gardner, 2008). Presumptive blood tests are used as a quick preliminary screen of a questioned sample to determine whether a fluid or stain is blood. Catalytic test reagents like Luminol, Hemastix/Heglostix and Sangur rely on

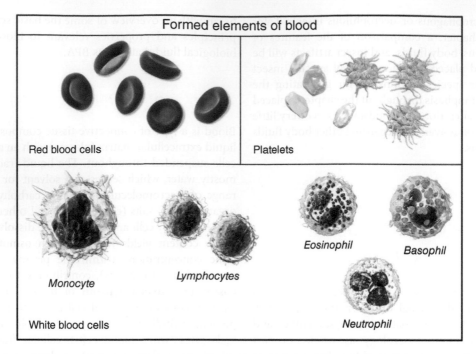

Figure 15.2 Cellular components of human blood contributing to the characteristic high viscosity and resistance to flow of the fluid. *Source*: Bruce Blaus/Wikimedia Commons/CC BY 3.0.

peroxidase activity within hemoglobin molecules to interact with substrates that generate a product (as evidenced by a color change) distinctive for blood (Gaensslen, 2000). A positive test for blood may reveal that a crime was committed, help locate evidentiary objects, or identify the location of the crime scene. In other instances, blood constituents, namely macromolecules in the form of antigenic proteins, are used for classification based on blood types. Person identification does not result from blood typing but exclusion or inclusion of one or more individuals can occur. Linkage of blood to a specific individual (individualization) can be made through the use of DNA analyses of the formed elements of blood. Even traces of DNA are useable if it can be amplified via polymerase chain reaction (PCR) and then compared with DNA fingerprints of a victim, person(s) of interest, or DNA database.

15.2.2 Laws of physics apply to blood droplets

Several principles of physics apply to blood droplets falling passively or in motion due to external forces (e.g., arterial pressure, impact with an object) (Carter, 2001). An understanding of the intimate details of the laws of physics associated with blood in flight is required by an expert in BPA, but not necessarily essential to a forensic entomologist trained in insect artifact analyses. To provide a framework for making comparisons between blood spatter and insect artifacts, a brief description of the relevant physical principles is given below.

15.2.2.1 Fluid mechanics

The high viscosity of blood gives the tissue the consistency of a semi-fluid, characterized as a viscoelastic[4] **non-Newtonian fluid** (Martini *et al.*, 2011). The viscosity of a non-Newtonian fluid is dependent on the shear rate or shear history, which essentially means that a constant coefficient of viscosity cannot be applied to such fluids (Tropea *et al.*, 2007). It also has flow characteristics similar to a thick fluid like ketchup, which is described as a shear thinning fluid. In other words, the viscosity of the fluid decreases with increasing shear stress, simply indicating that the fluid is slow to motion at low deformation rates, but flows easily at high rates (Tropea *et al.*, 2007). Fluid mechanics as a discipline considers matter such as blood in a continuum and generally relies on computational models to describe the motion of an object. What this means to forensic investigation is that bloodstains at a crime scene should be subjected to computational modeling as part of the interpretation of distance, direction, and angle of impact associated with individual blood droplets.

15.2.2.2 Trajectory analysis

Ballistics of blood is the best way to describe this aspect of physics. Blood exiting a body due to application of an external force (e.g., any type of impact weapon) or the piercing of an artery under high pressure will cause droplets of blood to act as projectiles, with the flight path of each individual being the trajectory. Essentially the same factors that influence a bullet's trajectory (wind, gravity, ambient temperature, humidity, and friction) can be applied to blood. Ultimately, trajectory directly influences the angle of impact of blood when striking a surface, which is an important feature of bloodstain pattern analysis used during reconstruction.

15.2.2.3 Gravity

In its simplest interpretation, gravity is the force that causes objects to fall until they come to rest. Obviously, gravitational force has a major impact on the trajectory that blood travels when released from a body, whether due to an external force or during passive bleeding. Gravity is also responsible for the pooling of blood in a decomposing body (otherwise known as livor mortis), independent of whether the fluid has an avenue to escape from the corpse.

15.2.2.4 Terminal velocity

The terminal velocity of a drop of blood simply refers to the object in flight reaching a constant speed. In other words, blood released from a body is no longer accelerating, which is an indication that drag has become equal to the weight of the blood. When a droplet of blood reaches its terminal velocity, the resulting spatter will have a uniform diameter regardless of the height from which it falls (Eckert, 1997), a property that we will see later helps to subjectively distinguish bloodstains from certain types of insect artifacts.

15.2.2.5 Centripetal force

Centripetal force acts on an object moving with uniform speed along a circular path and is directed along the radius toward the center of the path (Tipler & Mosca, 2003). When applied to blood, centripetal force drives the movement of droplets cast from an object covered with wet blood. Cast-off blood will travel in a tangentially straight line from the object when the adhesive forces holding the blood to the object exceed centripetal force. The angle of impact when a droplet makes contact with another surface depends on the location of the object at the time the blood was released from the moving object (Pizzola *et al.*, 1986). Again, the angle of impact and ultimate shape of the bloodstain are used in BPA and in classification of spatter and insect artifacts.

15.3 Crash course in bloodstain analyses

Analysis of bloodstain evidence and the interpretation of bloodstain patterns have evolved into a specialized subfield of forensic science. Relatively few forensic investigators have significant training in this specialty, largely due to the complexity of the analyses (Saferstein, 2017). As mentioned earlier, BPA depends on the application of several disciplines of science to practical problems associated with legal investigations. So, to be proficient in this field, extensive education and training in multiple natural and applied sciences are required. Fortunately, most students interested in forensic entomology have much of the background necessary to at least understand the basics of bloodstain analysis. What follows is a crash course in BPA, since attempting to provide an in-depth background requires a textbook to itself.

Our overview will concentrate on the types of information that can be derived from investigation of bloodstain patterns. A careful examination of the blood evidence has the potential to reveal information about the events before, during, and after bleeding occurred.

15.3.1 Direction of travel

Generally, the direction traveled by a blood droplet is evident from its morphology. The edge characteristics of a bloodstain reveal the path the blood was traveling prior to making contact with an object. When a blood droplet strikes the surface of an object, the droplet collapses from bottom up. Blood in the droplet moves outward creating either a circular or elliptical pattern to the resulting stain (Bevel & Gardner, 2008). One side of the blood spatter will have definite edges (spines or scallops), often terminating in an elongate tail. The tail section (including the stain body) reflects the longitudinal axis of a bloodstain and points toward the direction of travel. The same information can be derived from a stain lacking a definitive tail but that shows jagged or distorted edges (James *et al.*, 2014). Directional interpretation is subject to variation based on the features of the interacting surface; objects that are highly textured (i.e., not smooth) are difficult to interpret, as they can yield stain edges displaying **scalloping, spines,**[5] and satellite stains (Bevel & Gardner, 2008), or the final morphology of the stain is altered following wicking into textiles and fabrics due to capillary action (de Castro *et al.*, 2015; Chang & Michielsen, 2016).

15.3.2   Shape and size of bloodstains

As will be discussed shortly, the shape and size of blood spatter depend on the angle at which wet blood hits a surface. However, equally important are the characteristics of the surface that blood strikes. Texture, absorption qualities, and thickness of the object influence shape and size of a bloodstain (Jackson & Jackson, 2008). For example, blood that strikes a smooth surface like ceramic tile is more likely to form a relatively circular stain than when landing on a rough material like carpet (Figure 15.3). Textured materials can penetrate the surface tension of the blood droplet, disrupting its

integrity, and yielding asymmetrical blood patterns. A drop that lands on an object with high absorptive properties like a cotton fabric will wick the stain into a larger pattern than non-absorptive materials (Figure 15.4). **Wicking** is in reference to the flow of a liquid by capillary action in narrow spaces, without the need for an external force like gravity, to facilitate the movement. In fabrics and textiles in which yarns are "packed" close together, the spaces between the yarns function like small tubes such that a combination of surface tension and adhesive forces between the liquid (e.g., blood) and the yarn act to propel the fluid. Wicking is also apparent with some types of insect artifacts, especially fly regurgitate and defecatory stains, when deposited on absorptive materials.

15.3.3   Angle of impact

A drop of blood approaching an object will strike the surface at a specific angle, referred to in bloodstain analysis as the **angle of impact**. For instance, a drop essentially free falling straight down from a person or object will strike the ground (or other surface) at a right angle or 90° in relation to the underlying surface. The resulting spatter or stain will be circular in shape and typically lack a tail, spikes, or scallops (Saferstein, 2017). As the angle of impact becomes more acute, the blood droplet will become more elliptical in shape, yielding distorted edges in the direction of blood travel (Bevel & Gardner, 2008) (Figure 15.5). Longer tails are evident with decreases in the angle of impact and with increasing velocity at the time of collision. It should be apparent that a

Figure 15.3 Human blood that strike a smooth surface essentially at a 90° angle produce circular stains, often with spikes at the edges and small satellite stains around the circumference. *Source*: Andrew Butko/Wikimedia Commons/CC BY-SA 3.0.

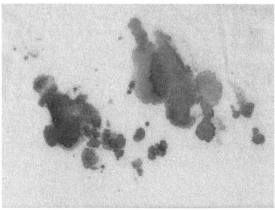

Figure 15.4 Human bloodstains can display extensive wicking on a cotton shirt. *Source*: Ser Amantio di Nicolao/Wikimedia Commons/CC BY-SA 4.0.

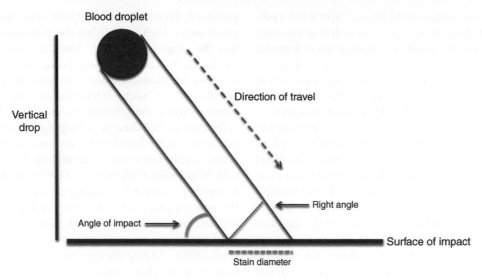

Figure 15.5   Formation of an elliptical bloodstain from a blood droplet striking a surface at an acute angle. *Source*: Redrawn from Bevel & Gardner (2008).

relationship exists between the length and width of a bloodstain and the angle of impact. This relationship is depicted in the following formula:

$$\sin A = \frac{\text{Width of bloodstain}}{\text{Length of bloodstain}}$$

where A is the angle of impact, and width and length measurements are typically measured in millimeters (Bevel & Gardner, 2008).

Measurements of the length of an elliptical bloodstain do not include the tail, spikes, or scallops (Figure 15.6). As an example, the angle of impact for a bloodstain that measures 10 mm in length and 5 mm in width would be calculated as follows:

$$\sin A = 5/10 = 0.50$$

Taking the sine of the width to length ratio as 0.50 yields an angle of impact of 30°. The angle of impact calculated for a series of bloodstains is used to determine the **area of convergence** as well as the **area of origin**.

15.3.4 Origin of impact

The complexity of BPA is evident when attempting reconstruction using calculated angles of impact for bloodstains detected at a crime scene. One aspect of analysis is determination of the area of convergence, an area on a two-dimensional plane that approximates

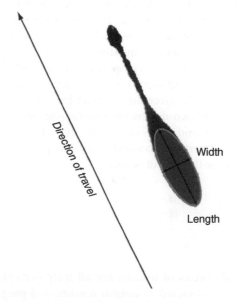

Figure 15.6 An elliptical bloodstain displaying a characteristic tail that points in the direction of movement of blood when striking a smooth surface. *Source*: Kevin Maloney/Wikimedia Commons/Public domain.

the origin of blood (James *et al.*, 2014). It is determined when multiple bloodstains are found at a crime scene; reconstruction depends on determining the relationship between bloodstains showing distinct directionality. The area of convergence can be established by drawing straight lines through the longitudinal axis of multiple bloodstains extending through the tails (Saferstein, 2017). In theory, the point of intersection between several bloodstains represent the source of the blood. There is still a possibility that the crossover is coincidental. However, by increasing the number of bloodstains with similar orientations used in the determination, the likelihood that the point of intersection represents the area of convergence is strengthened, and hence so too are the assumptions regarding the bloodstain source.

Determination of the area of convergence can also be used to estimate the number of blows administered by blunt force to create the blood spatter patterns. How? When an object strikes an individual or some other source of blood multiple times, the release of blood and pattern of stains colliding with a surface are never exactly the same (Wonder, 2007). Thus, the area of convergence can be determined as described above for multiple groups of stains, revealing whether more than one point of intersection is likely. Multiple intersection points suggest more than one blow to the individual.

In attempting to reconstruct where the blood was projected from to create the stain patterns, the area of origin is calculated. This area represents the location in a three-dimensional space that the blood was projected. It gives an estimation of the position of the victim or suspect in space when the events that resulted in blood shed occurred (Saferstein, 2017). Essentially the three-dimensional location is estimated from two-dimensional measurements: angle of impact and area of convergence. Bloodstain pattern experts use computer software packages to determine the area of origin.

15.4 Insect activity can alter blood evidence

Several species of insects, not all truly necrophagous, are attracted to animal remains and purged bodily fluids. As discussed in Chapter 6, carrion serves as a nutrient source for numerous species of necrophagous flies, beetles, and other insects. Applying **Locard's Exchange Principle** to the interaction between necrophilous insects and human remains, evidence of this association will be left behind at the crime scene (James et al., 2014). Those species that dine on a corpse are efficient feeders but sloppy, leaving traces of their feeding activity in numerous locations on and about the site of body decomposition. For instance, an adult female calliphorid will land on or near a corpse, walking along body surfaces or through pools of bodily fluids, sampling the nutritional value of each with

gustatory receptors on footpads and sponging mouthparts. Such activity has the potential to distort the shape of existing bloodstains as well as mechanically transferring small drops of wet blood to other locations, including to sites other than at the crime scene (Striman et al., 2011). Compromising the physical evidence even further is that as a fly **imbibes**,[6] it will regurgitate and defecate some of the liquid diet onto surfaces near the crime scene, sometimes intermixing fly artifacts with bloodstains (Figure 15.7). The fly often moves to another location by walking or flight, in which the new area may serve as a site for deposition of artifacts or transference of wet blood attached to the footpads or tarsi (Durdle et al., 2018; Parker et al., 2020), yielding the appearance of bloodstains in places other than actual crime scenes.

The activity of insects in altering blood evidence confounds reconstruction efforts by crime scene investigators, particularly for blood pattern analysts. Insects introduce several sources of error in reconstruction by (i) altering the shape of existing bloodstains (think about how walking through the blood may alter width–length ratios or modify edges); (ii) transferring wet blood to other locations; and (iii) depositing artifacts that resemble certain types of blood spatter. The latter is especially challenging for forensic investigators because fly regurgitate and feces are virtually indistinguishable from some types of bloodstains based on morphology as well as chemical composition. Regurgitate and defecatory stains are composed principally of the food consumed by the insect. As a consequence, such insect artifacts will test positive for blood via presumptive and confirmatory chemical tests (Fujikawa et al., 2009; Durdle et al., 2015) and, as long as the blood is not fully

Figure 15.7 An adult house fly, *Musca domestica*, with the mouthparts positioned to imbibe food on the surface. *Source*: Muhammad Mahdi Karim/Wikimedia Commons/public domain_GFDL 1.2.

digested, will also yield similar results from DNA typing (hemogenetic individualization) when comparing DNA from the artifact to the bloodstain (Benecke & Barksdale, 2003; Durdle *et al.*, 2013; Kulstein *et al.*, 2015). At present, only one technique has been developed to empirically distinguish insect artifacts and bloodstains (Rivers *et al.*, 2018), but it is not commercially available yet for widespread use at crime scenes. A few subjective measures have been developed (Benecke & Barksdale, 2003; Fujikawa *et al.*, 2011) that rely more on the experience of the investigator rather than quantifiable testing that would not be expected to hold up in court.

What can be easily determined are the behaviors of insects that lead to distorted blood evidence. The next sections will detail the limited information available on how insects modify bloodstains through their feeding activity, release of digestive artifacts in liquid form, transfer of blood via body parts, and the stains associated with insect feeding ante mortem (parasitic) (Table 15.1).

Table 15.1 Types of artifacts produced by necrophilous insects.

Artifact type	Subtype(s)	Mechanism of production
Regurgitate (digestive)	Primary	Released bubble/fluids from foregut
	Secondary	Consumption of fly regurgitate, released from foregut
Defecatory (digestive)	Primary	Passive drop *or* expelled from anus
	Secondary	Consumption of fly feces, released from anus
Transfer patterns	Translocation	Impressions of bloodied legs, abdomen
	Tarsal tracks	Impressions of bloodied footpads (pulvilli)
Meconium	Primary	Expelled pupal metabolic waste after emergence
	Secondary	Consumption of meconium, expelled from mouth
Larval stains	Secretions	Oral secretions from foregut and/or midgut
	Excretions	Excreta released from anus
	Body stains "floor stripes"	Impressions created from fly larvae covered in decomposition fluids dispersing from remains
Peritrophic membrane		Lining of dermestid beetle's midgut released with feces

Source: Data from Rivers & Geiman (2017) and Viero *et al.* (2019).

15.5 Insect feeding activity can modify the morphology of bloodstains and yield transfer patterns

When insects feed on wet blood from a body or on bloodstains, the evidence can be altered by individuals walking through the samples and/or changing the shape due to the action of mouthparts. In the case of opportunistic scavengers like cockroaches, an individual may feed at the periphery of a wet bloodstain, imbibing the liquid with the aid of mandibles and maxillae. As mentioned earlier in the chapter, blood is a viscous fluid that tends to stick to surfaces, including mouthparts and other appendages. While an individual cockroach, ant, fly or other necrophagous insects consumes food, depending on the species it will frequently pause to clean its mandibles and maxillae using legs, palps, and antennae; in the case of viscous blood, this may adhere. If clotting has initiated, the chance of blood adherence to cockroach body parts is increased. The net effect is that an extended period is spent feeding and grooming on wet blood or on blood spatter. This in turn increases the association between the insect and blood evidence, facilitating additional opportunities to modify the liquids or stains. Feeding activity of insects as well as the rate of blood clotting is temperature dependent. Thus, at warmer temperatures, clotting occurs earlier than during cooler conditions and consequently the chance of "sticky" blood clinging to a feeding cockroach or other insect is further enhanced.

A cockroach that walks through blood can in turn mechanically transfer wet blood to other locations as it runs or walks from the crime scene. The blood pattern left behind is commonly streaks of blood as the abdomen drags to maintain balance during walking (Gullan & Cranston, 2014) or patterned impressions of footpads and/or tarsi. This deposition of stains from blood-covered body parts is referred to as **transference** or **transfer patterns** and represents a common type of insect artifact associated with a crime scene (Parker *et al.*, 2020) (Figure 13.5). A distinction can be made between the types of transfer patterns created by insect activity. Stains resulting from dragging a leg or the abdomen through wet blood and then depositing an asymmetrically linear impression are termed

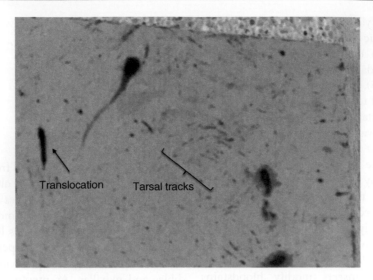

Figure 15.8 A piece of unfinished dry wall covered by many small tarsal tracks produced by adults of *Sarcophaga bullata* following feeding on human feces. *Source*: Photo by D.B. Rivers.

translocation. Such patterns are relatively common when cockroaches or ants feed on blood or tissues but are infrequently produced by adult necrophagous flies (Rivers & Geiman, 2017; Rivers *et al.*, 2020). Bloodied impressions of footpads or tarsi are analogous to "insect footprints" and are termed **tarsal tracks** (Rivers & Geiman, 2017) (Figure 15.8). When adult flies are present at a crime scene in which blood has been shed, tarsal tracks can easily be the most common type of insect artifact contaminating the scene, dependent on the species (Rivers & McGregor, 2018).

Altered bloodstain morphology is not restricted to wet blood. Adult flies will readily "suck" on dried bloodstains or other body fluids, minimally changing the color intensity of the existing stain. Depending on the surface that the blood originally dried upon, feeding action of the flies may yield a cratered appearance (Zuhu *et al.*, 2008), although these can be confused for regurgitate stains that have dried on non-porous surfaces (i.e., plastic) (Durdle *et al.*, 2013).

15.6 Ingested blood may be expelled as digestive artifacts

Necrophagous flies are the biggest culprits in terms of contaminating crime scenes with artifacts. The typical sequences of events include adults discovering a corpse; feeding on blood, tissues, and/or other body fluids; and then either expelling a portion of food via the mouth or as liquid feces. Together the resulting stains are called digestive artifacts, but they can also be distinguished as **regurgitate** and **defecatory stains**. In many instances, one or both types of digestive stains are the dominant type of insect artifact observed at crime scenes or under laboratory conditions (Striman *et al.*, 2011; Kulstein *et al.*, 2015; Rivers & McGregor, 2018; Rivers *et al.*, 2019a).

15.6.1 Regurgitate stains originate from the foregut

The most common type of insect artifact interfering with bloodstain analysis is a regurgitate stain, sometimes referred to as **fly spot** or fly speck. Regurgitate stains originate from the digestive tract of adult flies, typically (but not always) after consuming a liquid meal. Necrophagous flies in the families Calliphoridae, Sarcophagidae, and Muscidae utilize a form of extra-oral digestion[7] whereby a meal is ingested via the sponging mouthparts, transferred to the crop where an array of digestive enzymes (predominantly amylases and proteases) from the salivary glands and possibly foregut and anterior midgut are deposited (Terra & Ferreira, 1994; Rivers *et al.*, 2014), and then regurgitated as a bubble that hangs from the mouthparts before being placed on a surface (Figure 15.9). For a fluid feeder, this method of digestion is incredibly efficient. For example, a major problem encountered

Figure 15.9   Bubbling behavior displayed by *Calliphora vicina*. Note that the bubble is not a clear fluid, which is characteristic of species that use bubbling primarily for evaporative cooling or to decrease mass before flight. *Source*: Alvesgaspar/Wikimedia Commons/CC BY-SA 3.0.

by most animals that feed predominantly on liquids is an initial excess of fluid comprising the food, which means digestive enzymes will be diluted if released into the gut environment, thereby decreasing the likelihood that enzymes will make contact with food molecules. Physical contact between enzymes and substrates relies mostly on passive diffusion, which in simple terms represents a series of chance events dependent on Brownian motion of solute molecules (recall our discussion of Brownian motion's influence on oxygen diffusion in Chapter 11) (Withers, 1992). To overcome the limitations of wet food, many insects depend on morphological adaptations (e.g., cryptonephridial arrangement of Malpighian tubules or a filter chamber) to remove excess water prior to release of enzymes (Chapman, 1998). Some necrophagous flies, by contrast, utilize a physiological mechanism (regurgitation) to achieve the same effect: fly regurgitate will rapidly evaporate water due to the large surface area to volume ratio of the spherical drop (the same principle leading to evaporative cooling discussed in Chapter 9). Consequently, the fluid becomes more concentrated in terms of solutes and enzymes as water is lost from the regurgitate. The adult then returns at a later time to consume the dried blood now composed of digested food products (James *et al.*, 2014). Presumably in order to relocate the expelled food, the adult insect has marked the regurgitate with a pheromone or some other chemical signal, although no studies have been performed to confirm this speculation.

In terms of comparisons with bloodstains at a crime scene, fly regurgitate is virtually indistinguishable morphologically and chemically from several types of blood spatter, particularly medium- and high-impact and expirated stains (Figure 15.9) (Benecke & Barksdale, 2003; James *et al.*, 2014). The size and shape of regurgitate stains varies by species, ranging from circular to asymmetrical in shape, and in size from relatively small (1–2 mm) to approaching 20 mm in diameter; larger stains are more common with larger fly species as well as when deposited on fabrics with high absorbency and wetting characteristics (Rivers & McGregor, 2018; Rivers *et al.*, 2020). The color of regurgitate is also variable (clear to reddish-browns to green) and is influenced by the fly species, type of meal consumed, and surface deposited on (Fujikawa *et al.*, 2011; Striman *et al.*, 2011; Rivers *et al.*, 2019a, b, 2020). Distinctive patterns of staining occur with some species (Striman *et al.*, 2011; Rivers & McGregor, 2018) that in isolated controlled situations may allow recognition of artifacts from particular groups or species but would not stand out if intermixed with blood spatter (Figure 15.10).

15.6.2   Defecatory stains can be produced under high or low velocity

Surprisingly, "liquid frass" is not the name of a grunge rock band out of Seattle.[8] Rather, it is a term referencing defecate from an insect, in this case one that has fed on a fluid diet and hence produces liquid feces. Unlike most other animals, feces of insects are not composed exclusively of undigested food material and other items associated with the alimentary canal. The arrangement of Malpighian tubules, the organs predominantly responsible for osmoregulatory functions like excretion and water balance (Chapman, 1998), are attached to the digestive tract and deposit metabolic wastes in the form of primary urine into the hindgut (a structure that further modifies the urine through absorptive and secretory activities). Thus, the frass of necrophagous flies contains waste materials derived from metabolic processes (urine) and digestive functions (defecate). The consistency of feces (solid, liquid, or semi-solid) is dictated by the composition of the ingested food, water balance characteristics of a given species, and the habitat or environment

(a)   (b)   (c)

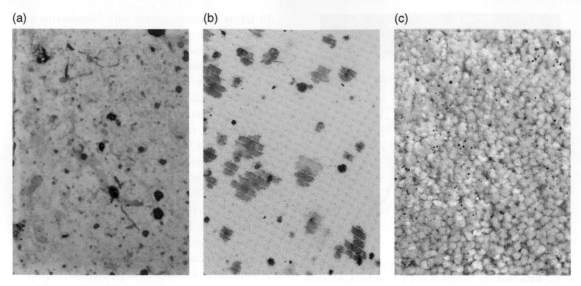

Figure 15.10   Artifacts produced by adult *Calliphora vicina* following feeding on human blood. The majority of artifacts are regurgitate stains but the morphology and color are greatly influenced by the surface material deposited on, as evident by the stains on (a) ceramic tile, (b) dri wick polyester shirt, and (c) plush carpet. *Source*: Photos by D.B. Rivers.

occupied by the insect in question. Ordinarily, most terrestrial insects produce solid frass that contains a bare minimum of water, enough to ensure food passage through the rectum but which promotes maximum water retention. In contrast with the "norm" in a terrestrial environment, necrophagous flies excrete a hypoosmotic urine/feces, indicative of the diet providing copious amount of water to meet nutritive, metabolic, and osmotic needs.

The significance of liquid frass to a criminal investigation is that necrophagous flies attracted to a decomposing body or to bloodstain may deposit this form of insect artifact (referred to as **defecatory stains**) in and around the crime scene. The net effect is similar to fly regurgitate in that the defecatory stains can be confused with some types of blood spatter and share similar characteristics to regurgitate stains in terms of size and color (Benecke & Barksdale, 2003; Rivers & McGregor, 2018). Feces differ from regurgitate in containing high concentrations of nitrogenous wastes in the form of ammonia, uric acid, and allantoin (Rivers & Geimen, 2017). However, both types of stains commonly contain non-digested human blood that permits detection by presumptive and confirmatory tests for blood. The shape of fecal stains is distinct from that of regurgitate in some species, but not in many others (Figure 15.11). For example, dried defecatory stains of some species have a morphological appearance similar to a teardrop, sperm cell, or tadpole in that a distinct tail is evident (Parker *et al.*, 2020). The tail section

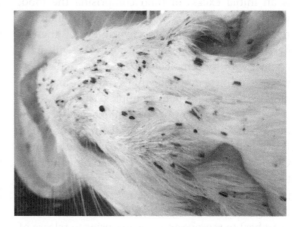

Figure 15.11   Regurgitate and defecatory artifacts produced by adult *Calliphora vicina* following feeding on a guinea pig, *Cavia porcellus*, carcass. The stains are indistinguishable from each other based on morphology. *Source*: Photo by D.B. Rivers.

results from the fly moving or walking as the last of the feces is deposited, which also gives directionality to the stains in terms of fly movement (the tail usually points in the direction traveled by the adult) (Figure 15.12). It is important to note that a tail is not always evident with a fecal stain and for some species, there appear to be geographical differences that result in defecatory stains being distinct from the same species feeding on the same body fluid (Striman *et al.*, 2011; Kulstein *et al.*, 2015; Rivers & McGregor, 2018).

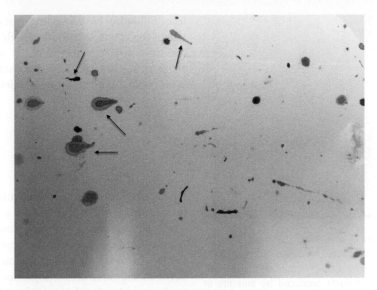

Figure 15.12   Artifacts produced by *Sarcophaga bullata* following consumption of human blood. Several defecatory stains (see arrows) with tails are evident, but they differ in size and color. *Source*: Photo by D.B. Rivers.

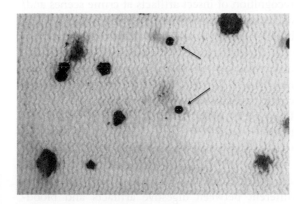

Figure 15.13   Spherical or three-dimensional artifacts (arrows) produced by forcible expulsion of liquid feces. The artifacts shown are on a new knit cotton tee shirt, front face. *Source*: Photo by D.B. Rivers.

Most fly feces are eliminated from the body through passive drop, low velocity mechanisms. The resulting defecatory stains display "round" morphologies with or without tails as already discussed. At other times, small, dark black, nearly spherical artifacts are produced by flies forcibly (high velocity) expelling liquid feces (Figure 15.13). On non-porous surfaces, these stains typically are elliptical in shape with a tail pointing in the direction of fluid movement when striking a surface (Rivers and Geimen, 2017), a feature shared with elliptical human bloodstains. The angle of impact is low in these instances, since feces are typically expelled from the body with the anus aligned parallel or nearly so with the surface that the fly is resting upon. Defecatory stains produced by forceable expulsion did not wet or wick into cotton or polyester fabrics, suggesting that the composition of the defecate differs from other forms of liquid feces (Rivers *et al.*, 2020). Important to a forensic investigation, three-dimensional defecatory stains are readily distinguishable from any type of human bloodstain and thus should be viewed as means to recognize fly activity at a crime scene.

Nearly all attention is devoted to necrophagous flies with respect to defecatory stains. Does that mean other types of insects do not leave behind similar stains? The answer to that question is not clear. Necrophagous flies dominate most scenarios in which human and other animal remains are discovered. Evidence of their interaction with the remains in the form of insect artifacts also dominate in terms of crime scene contamination. The presumption is that any insect that feeds on a corpse or body fluids will likely deposit feces at the scene, provided the (i) association lasts for sufficient duration and (ii) the feces are liquid when eliminated, thus creating a stain. However, there is virtually no information available regarding non-fly species with regard to defecatory stains. Anecdotal observations have shown that larvae and adults of *Dermestes maculatus* both produce liquid feces that resemble defecatory stains of some fly species (Figure 15.14). More research is needed in this area to determine if other species produce similar types of insect artifacts.

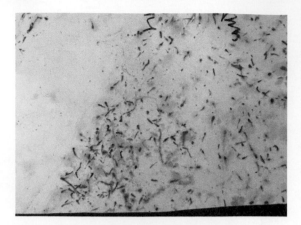

Figure 15.14 A mass of defecatory stains produced by larvae and adults of *Dermestes maculatus* after feeding on dried bovine liver. The artifacts could easily be confused with digestive artifacts produced by muscids or calliphorids. *Source*: Photo by D.B. Rivers.

15.7 Parasitic insects can confound blood evidence by leaving spot artifacts

The occurrence of insect artifacts at a crime scene or location of a suspicious death may be associated with antemortem insect activity. This implies of course that the insects fed on the body, in this case blood, prior to death, which eliminates the necrophagous species discussed thus far. Some parasitic species, namely fleas (Order Siphonaptera) and bed bugs (Order Hemiptera), acquire a blood meal from a human host and then produce liquid feces that leaves spots (also called specks) similar in shape and size to fly regurgitate. Flea and bed bug stains are not expected to occur alongside blood spatter since neither type of insect will feed on a body postmortem. However, these artifacts can confound an investigation because their presence gives the appearance of medium- and high-impact bloodstains, when in fact foul play may not have occurred at all. For instance, severe bed bug infestations have been reported in the homes of some elderly individuals in which furniture, bedding, matrices or other fabrics were found to contain hundreds of defecatory spots, and in the event the individual died of natural causes, the home has the appearance of a crime scene with blood spatter. At present, there is no definitive means to differentiate artifacts derived from bed bugs or fleas from bloodstains of similar size and shape. Adult mosquitoes also deposit liquid

feces following blood feeding, but due to the small size of the stains, distribution high on walls or ceilings, and sparseness in terms of total stains, such artifacts are not considered a challenge to bloodstain pattern analysts or are not detected at all.

15.8 How can insect stains be detected?

Insect artifacts can be problematic in reconstruction efforts at crime scenes if the stains are not recognizable and distinguishable from other types of trace evidence. Based on our discussions of the similarities in morphology and chemical composition between digestive artifacts produced by flies after feeding on blood with actual bloodstains, the potential for confusion seems to be high. This bring us to addressing what methods, if any, are available for recognition of insect artifacts at crime scenes and/ or distinguishing between insect-derived stains and bloodstains, especially when intermixed. A variety of methods have been purported to permit some differentiation of insect artifacts from human bloodstains. The methods can be grouped as visual, contextual and chemical (reviewed by Langer & Illes, 2015). Visual methods attempt to make distinctions based on morphological differences between insect artifacts and human body fluid trace evidence. As we have discussed earlier in the chapter, they are few features that are consistently different between digestive artifacts and bloodstains, especially expirated and impact stains, that can be considered reliable enough to use as the bases for expert testimony in court. Opinions differ in this regard, as some bloodstain pattern experts contend that with the "right" training and experience, insect stains should never be visually confused for any type of human bloodstain (Ristenblatt *et al.*, 2005; Ristenblatt, 2019). The conjecture is based mostly on subjective interpretation rather than empirical analysis, which is an approach that has no place in forensic investigation.

Two qualitative approaches have been developed to classify fly defecatory stains from bloodstains: a method based on stain morphology and use of an alternative light source. The morphological method depends on dried fecal stains assuming a tadpole or teardrop shape so that they can be readily identified in an area intermixed with bloodstains. Measurements of stain tail length (L_{tl}) and body length (L_b) are made and then compared by dividing the tail length by the body length (L_{tl}/L_b). Stains with a ratio greater than one are assumed to be fly artifacts and not bloodstains (Benecke & Barksdale, 2003). Thus, the method essentially relies on the process of elimination of stain suspects to identify defecatory stains. There are a few problems with this approach as the ratios generated to evaluate the technique are from limited sample sizes at crime scenes and from laboratory studies that did not utilize blood as a food source for the flies, and only one fly species was used for validation. Another issue is that for several species, tails are commonly absent from defecatory stains, yielding fecal and regurgitate artifacts that are indistinguishable from each other. The second technique uses an alternate light source equipped with a range of narrow band filters that allow for control of emission wavelength. A wavelength of 465 nm has been demonstrated to allow detection of defecatory stains produced by *Lucilia sericata* (Fujikawa *et al.*, 2011). In reality, a wide range of biological fluids, including all forms of fly artifacts, fluoresce at 465 nm, depending on the tissues or fluids in which interactions occurred (Rivers *et al.*, 2019a). Fluorescence appears to be quenched when artifacts are deposited on several types of textiles and fabrics. Where the technique shows the most promise is being able to distinguish fly digestive artifacts from bloodstains: artifacts will fluoresce on non-porous surfaces, but human blood does not. At this point the method is still in its infancy in that wide-scale examination of other stains and fly species has not occurred nor has the mechanism of illumination been determined. The latter is particularly important in being able to explain in a courtroom why the method is specific for insect artifacts.

Contextual methods rely on visual analysis coupled with where the insect artifacts occur in relation to one another and with respect to human body fluids. Such methods are commonly employed when stains are located in areas away from the corpse, such as on ceilings, at natural (i.e., windows) and artificial (lamps and ceiling lights) light sources, areas that are warm, and especially near food sources (Durdle *et al.*, 2018; Parker *et al.*, 2020). Typical fly behavior of positive phototaxis, foraging, and preference for warm temperatures can account for the placement of insect artifacts in locations away from the primary crime scene. When intermixed with "real" trace evidence, the appearance of unusual morphologies and random stain tails in relation to the majority can suggest the presence of insect artifacts. While these patterns may seem potentially easy to identify, the

methods are not quantifiable nor consistently reliable, posing serious limitations for use in forensic analysis.

The ideal scenario for insect artifact identification is a chemical diagnostic test comparable to confirmatory tests used for classification of human blood, semen, and saliva. Standard methods rely on immunoassay detection that have been developed as lateral flow assays. In other words, an antigen within a given body fluid is detected via antibody binding, providing highly specific identification of the evidentiary object. Lateral flow assays can be used with insect artifacts to determine if a given body fluid has been consumed. However, the assays are not able to distinguish between say human blood and insect artifacts. For example, the absence of antibody detection is not confirmation that the material must be insect derived, as there are other explanations for the same result. An immunological assay currently exists that does permit distinction between fly digestive artifacts and human blood, feces, semen, and saliva (Rivers *et al.*, 2018, 2019a, b). The assay is based on the presence of the digestive enzyme cathepsin D proteinase in the fly gut, and in turn, regurgitate and feces. The enzyme is structurally unique from vertebrate cathepsin D, making it ideally suited for antibody detection of fly artifacts. Digestive artifacts from thirty-one species of necrophagous and saprophagous flies representing ten families have been positively identified using the immunoassay (Rivers *et al.*, 2018). Potentially the method will be developed into a lateral flow assay but presently there is no commercially available chemical assay for detection of insect artifacts.

Chapter review

Bloodstains are not always what they appear to be at the crime scene

- An individual who is violently attacked by another is likely to bleed profusely from any inflicted wounds. Depending on the nature of the wound, as well as the surface that the expelled blood strikes, distinct bloodstains or spatter will result.
- Bloodstains resulting from violent crimes leave behind patterns that are interpretable, which in turn provide insight into the events of a violent act that allow reconstruction of the crime scene.

- Blood evidence can provide a wealth of information to a criminal investigator beyond just stain patterns. As blood is the most common bodily fluid that serves as physical evidence, the mere presence of exuded blood often helps to establish that a crime was even committed. Determination of blood type can be used for exclusion or inclusion of individuals in relation to a crime as well as in matters of paternity, and DNA typing of a blood sample is essential for classification, identification, and individualization of the victim and/or suspect of a violent crime.
- The activity of insects on and around a dead body may lead to the generation of insect artifacts, which can be represented as modified bloodstains, production of stains derived from the digestive tract of the insect in question, and/or transfer of blood via bloodied body parts at or near a crime scene. When intermixed with human blood, the insect-derived products can create confusion for a crime scene investigator with limited experience processing blood evidence. Insect stains share several biological and physical properties in common with blood that make the two nearly impossible to distinguish.
- In theory, any insect that interacts with a corpse or associated exuded body fluids can potentially create stains or artifacts that confuse reconstruction efforts at a crime scene. Stains produced as a result of insect activity associated with *any type* of body fluid or tissue are referred to as insect artifacts.

Science is the cornerstone of bloodstain pattern analyses

- The "science" in forensic analyses is not always obvious. The reality is that critical and reliable analyses of blood and bloodstain evidence are quite complex, with the underpinnings of the discipline derived from multiple fields of natural and applied sciences: biology, biochemistry, chemistry, mathematics, and physics. Effective interpretation of blood spatter relies on the application of the scientific method in conducting carefully controlled experiments, such as when testing bloodstain patterns on materials similar to those found at the crime scene.
- The properties of surface tension and viscosity as well as the actual constituents comprising blood are used in blood pattern analysis as well as in classification and individualization of blood evidence.
- Several laws of physics apply to interpretation of bloodstain evidence, including fluid dynamics, terminal velocity, centripetal force, gravitational forces, and trajectory analyses.

Crash course in bloodstain analyses

- Analysis of bloodstain evidence and the interpretation of bloodstain patterns have evolved into a highly complex specialized subfield of forensic science. The complexity stems in part from the application of several disciplines of science to blood pattern analysis. A careful examination of the blood evidence has the potential to reveal information about events before, during, and after bleeding occurred.
- Generally, the direction traveled by a blood droplet is evident from its morphology. The edge characteristics of a bloodstain reveal the path the blood was traveling in prior to making contact with an object.
- The shape and size of blood spatter is dependent on the angle at which wet blood hits a surface. Equally important are the characteristics of the surface that blood strikes. Texture, absorption qualities, and thickness of the object influence shape and size of a bloodstain.
- A drop of blood approaching an object will strike the surface at a specific angle, known as the angle of impact. A drop essentially free falling straight down from a person or object will strike the ground at a right angle (90°) in relation to the underlying surface. The resulting stain will be circular in shape and typically lack a tail, spikes, or scallops. As the angle of impact becomes more acute, the blood droplet will become more elliptical in shape, yielding distorted edges in the direction of blood travel.
- The complexity of blood pattern analysis is evident when attempting reconstruction using calculated angles of impact for bloodstains detected at a crime scene. One aspect of analysis is determination of the area of convergence, an area on a two-dimensional plane that approximates the origin of blood. The area of origin represents the location in a three-dimensional space that the blood was projected. It gives an estimation of the position of the victim or suspect in space when the events that resulted in blood spatter occurred.

Insect activity can alter blood evidence

- Several species of insects, not all truly necrophagous, are attracted to animal remains and purged bodily fluids. Those species that dine on a corpse are efficient feeders but sloppy, leaving traces of their feeding activity in numerous locations on and about the site of body decomposition. Such activity has the potential to distort the shape of existing bloodstains as well as mechanically transferring small drops of wet blood to other locations, including to sites other than the crime scene.
- The activity of insects in altering blood evidence confounds reconstruction efforts by crime scene investigators, particularly for blood pattern analysts. Insects introduce several sources for error in reconstruction by (i) altering the shape of existing bloodstains, (ii) transferring wet blood to other locations, and (iii) depositing artifacts that resemble several forms of blood spatter.

Insect feeding activity can modify the morphology of bloodstains and yield transfer patterns

- When insects feed on wet blood from a body or on bloodstains, the evidence can be altered by the individuals walking through the samples and/or changing the shape due to the action of mouthparts.
- The deposition of stains from blood-covered body parts is referred to as transference or transfer patterns and represents a common type of insect artifact associated with a crime scene. This type of stain is created when an insect walks through wet blood and then leaves impressions of body parts on a surface or distorts the shape of an existing bloodstain.
- A distinction can be made between the types of transfer patterns created by insect activity. Stains resulting from dragging a leg or the abdomen through wet blood and then depositing an asymmetrically linear impression are termed translocation. Bloodied impressions of footpads or tarsi are analogous to "insect footprints" and are termed tarsal tracks.
- Altered bloodstain morphology is not restricted to wet blood. Adult flies will readily "suck" on dried bloodstains or other body fluids, minimally changing the color intensity of the existing stain.

Ingested blood may be expelled as digestive artifacts

- Necrophagous flies are the biggest culprits in terms of contaminating crime scenes with artifacts. The typical sequences of events include adults discovering a corpse; feeding on blood, tissues, and/or other body fluids; and then either expelling a portion of food via the mouth or as liquid feces. Together the resulting stains are called digestive artifacts, but they can also be distinguished as regurgitate and defecatory stains.
- The most common type of insect artifact interfering with bloodstain analysis is a regurgitate stain, sometimes referred to as fly spot or fly speck. Regurgitate stains originate from the digestive tract of adult flies, typically after consuming a liquid meal.
- In terms of comparisons with bloodstains at a crime scene, fly regurgitate is virtually indistinguishable morphologically and chemically from several types of blood spatter, particularly medium- and high-impact and expirated stains.
- Fly feces differ from regurgitate in containing high concentrations of nitrogenous wastes in the form of ammonia, uric acid, and allantoin. The shape of fecal stains is usually distinct from that of regurgitate as well. Dried frass commonly has a morphological appearance similar to a teardrop, sperm cell, or tadpole in that a distinct tail is evident. The tail section results from the fly moving or walking as the last of the feces is deposited.
- Most fly feces are eliminated from the body through passive drop, low velocity mechanisms. The resulting defecatory stains display "round" morphologies with or without tails as already discussed. At other times, small, dark black, nearly spherical artifacts are produced by flies forcibly (high velocity) expelling liquid feces.

Parasitic insects can confound blood evidence by leaving spot artifacts

The occurrence of insect artifacts at a crime scene or location of a suspicious death may be associated with antemortem insect activity. Some parasitic species, namely fleas and bed bugs, acquire a blood meal from a human host, and then produce liquid feces that leaves spots similar in shape and size to fly regurgitate.

How can insect stains be detected?

- A variety of methods have been purported to permit some differentiation of insect artifacts from human bloodstains. The methods can be grouped as visual, contextual, and chemical. Visual methods attempt to make distinctions based on morphological differences between insect artifacts and human body fluid trace evidence.
- Two qualitative approaches have been developed to classify fly defecatory stains from bloodstains: a method based on stain morphology and use of an alternative light source.
- Contextual methods rely on visual analysis coupled with where the insect artifacts occur in relation to one another and with respect to human body fluids. Such methods are commonly employed when stains are located in areas away from the corpse, such as on ceilings, at natural and artificial light sources, areas that are warm, and especially near food sources. When intermixed with "real" trace evidence, the appearance of unusual morphologies and random stain tails in relation to the majority can suggest the presence of insect artifacts.
- The ideal scenario for insect artifact identification is a chemical diagnostic test comparable to confirmatory tests used for classification of human blood, semen, and saliva. Lateral flow assays can be used with insect artifacts to determine if a given body fluid has been consumed. However, the assays are not able to distinguish between say human blood and insect artifacts.

Test your understanding

Level 1: knowledge/comprehension

1. Define the following terms:
   - (a) forensic serology
   - (b) viscosity
   - (c) scientific method
   - (d) non-Newtonian fluid
   - (e) angle of impact
   - (f) insect artifact.
   - (g) regurgitate stain
   - (h) transfer pattern

2. Match the terms (i–vi) with the descriptions (a–f).

   - (a) Stain resulting from blood-covered object contacting non-bloody surface
   - (b) Consumption of a liquid diet
   - (c) Two-dimensional space where directionality of blood stains intersects
   - (d) Ability of a liquid to resist an external force
   - (e) Jagged edge to blood spatter
   - (f) Dried feces from an insect
   - (g) identification of fly artifact on ceiling light

   - (i) Scallop
   - (ii) Defecatory stain
   - (iii) Transference
   - (iv) Area of convergence
   - (v) Surface tension
   - (vi) Contextual analysis
   - (vii) Imbibe

3. Describe how fluids originating from the alimentary canal of a sarcophagid confound bloodstain evidence.

4. What are the limitations to using morphology of defecatory spots for identification at a crime scene?

5. Can presumptive blood tests like Luminol or Kastle Meyer distinguish between human bloodstains and fly regurgitate stains produced from feeding on blood? What if the fly fed on semen?

Level 2: application/analysis

1. Explain how fly regurgitate can be distinguished from defecatory stains.

2. Explain how defecatory stains can be distinguished from bloodstains.

3. Discuss why or why not regurgitate stains from adult silphids would be expected to occur at a crime scene involving human remains and bloodstains.

4. Describe what features of fly or beetle feces would be useful or unique for developing a confirmatory chemical test to identify defecatory stains.

Level 3: synthesis/evaluation

1. Speculate on how chemical differences associated with blood ante mortem versus postmortem potentially influence insect artifacts.
2. At present, there is no definitive method for identifying fly regurgitate intermixed with blood spatter. Speculate on the types of qualitative and quantitative tests that could be developed for positive identification of insect artifacts produced by regurgitation.

Notes

1. Blood spatter is a type of bloodstain resulting from a blood drop dispersed through the air due to an external force applied to a course of liquid blood.
2. Some experts no longer use the name forensic serology to describe the discipline that examines biological evidence. Forensic biology or forensic biochemistry is used as descriptors by some forensic laboratories (Gaensslen, 2000).
3. Exclusion, inclusion, classification, identification, and individualization are terms used in forensic analyses that were first defined in Chapter 1.
4. A viscoelastic fluid displays the properties of viscosity and elasticity when deformed.
5. Scallops and spines are terms referring to the edges of bloodstains. Both terms suggest that multiple edges have formed along the stain boundary when a blood droplet strikes the surface of an object.
6. Imbibe refers to drinking or absorbing a liquid, such as occurs with necrophagous flies that feed using sponging mouthparts to absorb nutritionally rich liquid foods.
7. Extra-oral digestion is a form of chemical digestion in which the bulk of enzymatic breakdown of foodstuffs occurs outside the mouth or oral cavity.
8. Seattle, Washington has a long history of music innovation and has been home to a number of bands from a range of genres, including Alice in Chains, Candlebox, Foo Fighters, Pearl Jam, and Soundgarden.

References cited

Benecke, M. & Barksdale, L. (2003) Distinction of bloodstain patterns from fly artifacts. *Forensic Science International* 137: 152–159.

Bevel, T. & Gardner, R.M. (2008) *Bloodstain Pattern Analysis, With Introduction to Crime Scene Reconstruction*, 3rd edn. CRC Press, Boca Raton, FL.

Campobasso, C.P., Marchetti, D., Introna, F. & Colonna, M.F. (2009) Postmortem artifacts made by ants and the effect of ant activity on decompositional rates. *The American Journal of Forensic Medicine and Pathology* 30(1): 84–87.

Carter, A.L. (2001) The directional analysis of bloodstain patterns: theory and experimental validation. *Canadian Society of Forensic Science Journal* 34: 173–189.

de Castro, T.C., Taylor, M.C., Kieser, J.A., Carr, D.J. & Duncan, W. (2015) Systematic investigation of drip stains on apparel fabrics: the effects of prior-laundering, fibre content and fabric structure on final stain appearance. *Forensic Science International* 250: 98–109.

Chang, J.Y.M. & Michielsen, S. (2016) Effect of fabric mounting method and backing material on bloodstain patterns of drip stains on textiles. *International Journal of Legal Medicine* 130(3): 649–659.

Chapman, R.F. (1998) *The Insects: Structure and Function*, 4th edn. Cambridge University Press, Cambridge, UK.

Durdle, A., van Oorschot, R.A. & Mitchell, R.J. (2013) The morphology of fecal and regurgitation artifacts deposited by the blow fly *Lucilia cuprina* fed a diet of human blood. *Journal of Forensic Sciences* 58(4): 897–903.

Durdle, A., Mitchell, R.J. & van Oorschot, R.A. (2015) The use of forensic tests to distinguish blowfly artifacts from human blood, semen, and saliva. *Journal of Forensic Sciences* 60(2): 468–470.

Durdle, A., Verdon, T.J., Mitchell, R.J. & van Oorschot, R.A. (2018) Location of artifacts deposited by the blow fly *Lucilia cuprina* after feeding on human blood at simulated indoor crime scenes. *Journal of Forensic Sciences* 63(4): 1261–1268.

Eckert, W. (1997) *Introduction to Forensic Sciences*. CRC Press, Boca Raton, FL.

Fujikawa, A., Barksdale, L. & Carter, D.O. (2009) *Calliphora vicina* (Diptera: Calliphoridae) and their ability to alter the morphology and presumptive chemistry of bloodstain patterns. *Journal of Forensic Identification* 59: 502–512.

Fujikawa, A., Barksdale, L., Higley, L.G. & Carter, D.O. (2011) Changes in the morphology and presumptive chemistry of impact and pooled bloodstain patterns by *Lucilia sericata* (Meigen) (Diptera: Calliphoridae). *Journal of Forensic Sciences* 56: 1315–1318.

Gaensslen, R.E. (2000) Forensic analysis of biological evidence. In: C.H. Wecht (ed.) *Forensic Sciences*, Vol. 1. Matthew Bender and Company, New York.

Gullan, P.J. & Cranston, P.S. (2014) *The Insects: An Outline of Entomology*, 5th edn. Wiley Blackwell, Chichester, UK.

Jackson, A.R.W. & Jackson, J.M. (2008) *Forensic Science*, 2nd edn. Pearson, London.

James, S.H., Nordby, J.J. & Bell, S. (eds) (2014) *Forensic Science: An Introduction to Scientific and Investigative Techniques*, 4th edn. CRC Press, Boca Raton, FL.

Kulstein, G., Amendt, J. & Zehner, R. (2015) Blow fly artifacts from blood and putrefaction fluid on various surfaces: a source for forensic STR typing. *Entomologia Experimentalis et Applicata* 157(3): 255–262.

Langer, S.V. & Illes, M. (2015) Confounding factors of fly artefacts in bloodstain pattern analysis. *Canadian Society of Forensic Science Journal* 48(4): 215–224.

Martini, F.H., Nath, J.L. & Bartholomew, E.F. (2011) *Fundamentals of Anatomy and Physiology*, 9th edn. Benjamin Cummings, San Francisco, CA.

Parker, M.A., Benecke, M., Burd, J.H., Hawkes, R. & Brown, R. (2020) Entomological alteration of bloodstain evidence. In: J.H. Byrd & J.L. Castner (eds) *Forensic Entomology: The Utility of Arthropods in Legal Investigations*, 3rd edn, pp. 399–412. CRC Press, Boca Raton, FL.

Pizzola, P.A., Roth, S. & De Forest, P.R. (1986) Blood droplet dynamics, II. *Journal of Forensic Sciences* 31: 50–64.

Ristenblatt, R.R. (2019) Commentary on: Rivers DB et al. Immunoassay detection of fly artifacts produced by several species of necrophagous flies following feeding on human blood. *Forensic Science International: Synergy* 2019; 1 (1): 1–10. *Forensic Science International: Synergy* 1: 303–304.

Ristenblatt, R.R., Pizzola, P.A., Shaler, R.C. & Sorkin, L.N. (2005) Commentary on: Distinction of bloodstain patterns from fly artifacts by Mark Benecke and Larry Barksdale. Discussion. *Forensic Science International* 149(2–3): 293–294.

Rivers, D. & Geiman, T. (2017) Insect artifacts are more than just altered bloodstains. *Insects* 8(2): 37. https://doi.org/10.3390/insects8020037.

Rivers, D.B. & McGregor, A. (2018) Morphological features of regurgitate and defecatory stains deposited by five species of necrophagous flies are influenced by adult diets and body size. *Journal of Forensic Sciences* 63(1): 154–161.

Rivers, D.B., Acca, G., Fink, M., Brogan, R. & Schoeffield, A. (2014) Spatial characterization of proteolytic enzyme activity in the foregut region of the adult necrophagous fly, *Protophormia terraenovae. Journal of Insect Physiology* 67: 45–55.

Rivers, D.B., Acca, G., Fink, M., Brogan, R., Chen, D. & Schoeffield, A. (2018) Distinction of fly artifacts from human blood using immunodetection. *Journal of Forensic Sciences* 63: 1704–1711.

Rivers, D.B., Cavanagh, G., Greisman, V., Brogan, R. & Schoeffield, A. (2019a) Detection of fly artifacts from four species of necrophagous flies on household materials using immunoassays. *International Journal of Legal Medicine.* https://doi.org/10.1007/s00414-019-02159-1.

Rivers, D.B., Cavanagh, G., Greisman, V., McGregor, A., Brogan, R. & Schoeffield, A. (2019b) Immunoassay detection of fly artifacts produced by several species of necrophagous flies following feeding on human blood. *Forensic Science International: Synergy* 1: 1–10.

Rivers, D.B., Dunphy, B., Hammerschmidt, C. & Carrigan, A. (2020) Characterization of insect stain deposited by *Calliphora vicina* (Diptera: Calliphoridae) on shirt fabrics. *Journal of Medical Entomology*, 57(5): 1399–1406.

Saferstein, R. (2017) *Criminalistics: An Introduction to Forensic Science*, 12th edn. Prentice Hall, Boston.

Striman, B., Fujikawa, A., Barksdale, L. & Carter, D.O. (2011) Alteration of expirated bloodstain patterns by *Calliphora vicina* and *Lucilia sericata* (Diptera: Calliphoridae) through ingestion and deposition of artifacts. *Journal of Forensic Sciences* 56: S123–S127.

SWGSTAIN. (2009) Scientific Working Group on bloodstain pattern analysis: recommended terminology. *Forensic Science Communications* 11(2): 14–17.

Terra, W.R. & Ferreira, C. (1994) Insect digestive enzymes: properties, compartmentalization and function. *Comparative Biochemistry and Physiology – Part B* 109: 1–62.

Tipler, P.A. & Mosca, G. (2003) *Physics for Scientists and Engineers*, 5th edn. Macmillan Publishing, New York.

Tropea, C., Yarin, A. & Foss, J.F. (eds) (2007) *Springer Handbook of Experimental Fluid Mechanics*. Springer, London.

Viero, A., Montisci, M., Pelletti, G. & Vanin, S. (2019) Crime scene and body alterations by arthropods: implications in death investigation. *International Journal of Legal Medicine* 133: 307. https://doi.org/10.1007/s00414-018-1883-8.

Withers, P.C. (1992) *Comparative Animal Physiology*. Saunders College Publishing, New York.

Wonder, A.Y. (2007) *Bloodstain Pattern Evidence: Objective Approaches and Case Applications*. Academic Press, San Diego, CA.

Zuhu, R.M., Supriyani, M. & Omar, B. (2008) Fly artifact documentation of *Chrysomya megacephala* (Fabricius) (Diptera: Calliphoridae) – a forensically important blowfly species in Malaysia. *Tropical Biomedicine* 25(1): 17–22.

Supplemental reading

Anderson, S. (2005) A method for determining the age of a bloodstain. *Forensic Science International* 148: 37–45.

Dethier, V.G. (1976) *The Hungry Fly: A Physiological Study of the Behavior Associated with Feeding*. Harvard University Press, Cambridge, MA.

Durdle, A., Mitchell, R.J. & van Oorschot, R.A.H. (2013) The human DNA content in artifacts deposited by the blowfly *Lucilia cuprina* fed human blood, semen and saliva. *Forensic Science International* 233: 212–219.

Gao, W. Wang, C., Muzyka, K., Kitte, S.A., Li, J., Zhang, W. & Xu, G. (2017) Artemisinin-luminol chemiluminescence for forensic bloodstain detection using a smart phone as a detector. *Analytical Chemistry.* https://doi.org/10.1021/acs.analchem.7b01000.

Gelperin, A. (1971) Regulation of feeding. *Annual Review of Entomology* 16: 365–378.

Karger, B., Rand, S., Fracasso, T. & Pfeiffer, H. (2008) Bloodstain pattern analysis: casework experience. *Forensic Science International* 181: 15–20.

Pelletti, G., Mazzotti, M.C., Fais, P., Martini, D., Ingrà, L., Amadasi, A., Palazzo, C., Falconi, M. & Pelotti, S. (2019)

Scanning electron microscopy in the identification of fly artifacts. *International Journal of Legal Medicine*. https://doi.org/10.1007/s00414-019-02090-5.

Ristenblatt, R.R. III & Shaler, R.C. (1995) A bloodstain pattern interpretation in a homicide case involving an apparent "stomping". *Journal of Forensic Sciences* 40: 139–145.

Rowe, W.F. (2006) Errors in the determination of the point of origin of bloodstains. *Forensic Science International* 191: 47–51.

Van Der Starre, H. (1971) Tarsal taste discrimination in the blowfly, *Calliphora vicina* Robineau-Desvoidy. *Netherlands Journal of Zoology* 22: 227–282.

Wonder, A.Y. (2001) *Blood Dynamics*. Academic Press, San Diego, CA.

Additional resources

Bed bug stains: https://pestseek.com/bed-bug-stains/

Bloodstain pattern analysis: http://murdersunsolved.com/staff/bloodstain-pattern-analysis/

Fly Stains: https://flystains.com/

International Association of Blood Pattern Analysts: http://iabpa.org/

International Association for Identification: www.theiai.org

Interpreting bloodstain patterns: http://www.crimescene-forensics.com/Blood_Stains.html

Scientific Working Group on Bloodstain Pattern Analysis: www.swgstain.org

Chapter 16

Necrophagous and parasitic flies as indicators of neglect and abuse

From the standpoint of medical entomology by far the most important insects are the Diptera or two-winged flies. . . . A great many other species breed in carrion, excrement, or other types of filth, from which they may carry pathogens to our food or drinking water, or directly to the human body. Still others, almost exclusively nonbloodsuckers in the adult stage, may attack the human body as larvae, thus producing the pathogenic condition known as myiasis.

Dr Maurice T. James, Division of Insect Identification
Agricultural Research Administration, U.S. Department of Agriculture[1]

Overview

The utility of flies to forensic investigations usually conjures up images of maggots feeding and crawling on dead bodies. While estimation of a postmortem interval based on fly larval development is one of the most recognized uses of entomological evidence, the reality is that forensically important insects, specifically obligatory and facultatively parasitic flies, can reveal details about a crime before death occurs, including whether a crime was even committed. Flies may infest tissues of a living host, a disease condition referred to as myiasis. Myiasis in humans is relatively rare. So, when it does occur, especially in Western countries, the presence of fly larvae is investigated from a medical and forensic perspective. Forensic cases are typically associated with suspected incidences of neglect and abuse to humans, most commonly with individuals not readily able to care or defend themselves, like children, the elderly, or incapacitated individuals. Necrophagous flies feeding on necrotic tissue; skin encrusted with urine, feces, or blood; exuded bodily fluids; or even food that is later ingested may serve as indicators of neglect or abuse by a caregiver or other individual (s). Fly infestations may reveal that pets or other animals have also suffered from improper or total absence of care. In these instances, species of flies that are typically necrophagous have become opportunistic parasites since they infest a living host. This chapter will examine the biology and conditions that favor necrophagous and parasitic flies invading tissues of humans and other animals prior to death. Particular emphasis will be placed on myiasis involving facultatively parasitic necrophiles, in other words those that prefer necrotic rather than living tissue.

The Science of Forensic Entomology, Second Edition. David B. Rivers and Gregory A. Dahlem.
© 2023 John Wiley & Sons Ltd. Published 2023 by John Wiley & Sons Ltd.
Companian website: www.wiley.com/go/forensicentomology2

The big picture

- Parasitic and necrophagous flies can infest humans, pets, and livestock.
- Myiasis can be classified based on anatomical location or degree of parasitism.
- The conditions that promote facultative myiasis are predictable.
- Chemoattraction of flies to the living does not necessarily differ from the odors of death.
- Necrophagous and parasitic flies display oviposition and development preferences on their vertebrate "hosts."
- Larval myiasis can be fatal.

16.1 Parasitic and necrophagous flies can infest humans, pets, and livestock

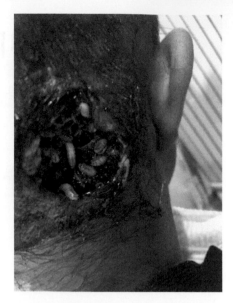

Figure 16.1 Human myiasis of the head and neck (cutaneous myiasis). *Source*: Mohammad2018/Wikimedia Commons/CC BY-SA 4.0.

Necrophagous insects, specifically flies belonging to the families Calliphoridae and Sarcophagidae, search, locate, and utilize animal remains under an array of natural and artificial conditions as a source of nutrition and as a microhabitat for both adults and larvae. Because we have the ability to identify the insect species, have an understanding of carrion resource utilization (e.g., timing of succession, seasonality, tissue preferences for progeny deposition and development), and have access to data on ambient conditions, the study of fly development can help estimate a time interval since death for a discovered corpse (see Chapters 11 and 14 for specific details). It may be surprising to learn, then, that all necrophagous flies do not restrict their activity to dead animals. How can this be? Truly necrophagous species are supposedly adapted for feeding and developing on decomposing or necrotic[2] tissues (Roback, 1951; Norris, 1965) (Figure 16.1). So how or why would a fly adapt for utilization of nutrient-rich carrion feed on a living animal? An evolutionary perspective of the various forms of myiasis will be provided in Section 16.2. However, the short answer for opportunistic necrophagous flies lies in the fact that an animal does not have to be deceased to possess regions with dead or dying tissues. Any type of wound contains dead cells. If the lesion breaches the integument, then the putrid odors emitted from the necrotic tissue are released

into the environment, drawing the attention of many of the same necrophagous flies attracted to a decomposing corpse (Catts & Mullens, 2002). Obviously, the larger the wound, the greater the number of dead cells present, and consequently the more concentrated the chemical signals emanating from the body. This in turn should equate to a higher likelihood of detection by necrophagous flies. Couple such scenarios with flies living synanthropically with humans, or animals in close proximity, and you have a recipe for fly attraction to dead human tissues. Upon discovery of the lesion (s) on the body, oviposition/larviposition occurs, creating a situation that differs from progeny deposition on carrion: the flies are invading dead tissue of a living animal which constitutes a form of parasitism.

Dipteran infestation of living or dead tissue of humans or other vertebrate animals, of body fluids, or of food that is then ingested represents a form of parasitism termed myiasis (also termed fly strike, fly-blown or blow fly strike) (James, 1947; Zumpt, 1965; Francesconi & Lupi, 2012). The impact of myiasis can range from benign to lethal, depending on host conditions, and may be the result of opportunistic associations of adult flies with a vertebrate host, or reflect an obligatory parasitic condition (Stevens *et al.*, 2006). As discussed in the next Section 16.2, myiasis can be classified by location or position of infestation on the body (anatomical classification) or based on the lifestyle or degree of parasitism of the flies involved (ecological classification). This latter aspect is particularly relevant to forensic investigations as infestations by normally necrophagous species *may* be indicative of neglect or abuse (e.g., non-self-inflicted wounds) of persons unable to care for themselves (Benecke & Lessig, 2001; Benecke *et al.*, 2004). Similarly, myiasis of pets, livestock or other domesticated animals often reflects insufficient care or cruel treatment of the animals (Anderson & Huitson, 2004). In cases of neglect, myiasis is not the only condition of entomological importance. The initial association may be between a human or pet and a saprophagous[3] fly that deposits eggs or larvae in, say, feces or urine accumulated in a diaper or on matted hair or fur. When larval feeding occurs in areas pressed against the body, subsequent invasion of host tissues or cavities can follow, and thus myiasis is a secondary result of the initial activity of the flies (James, 1947).

The relevance of insect activity on humans and other animals in terms of forensic entomology is evident in at least three realms: (i) insect infestation of a human, pet, livestock or other domesticated animal implies possible criminal negligence and/or abuse, (ii) fly colonization of injured or incapacitated individuals (i.e., unconscious due to a variety reasons, drug addicts) not associated with neglect, or persons who lack the resources to maintain good hygiene (i.e., homeless) and (iii) development of necrophagous flies ante mortem complicates estimations of a postmortem interval if death ensues following myiasis, regardless of the causation of death. What follows is an examination of the classification of myiasis, conditions that lead to fly infestations of human and other vertebrate animals naturally and during human interference (e.g., neglect and abuse), the chemical signals associated with fly attraction to living hosts, tissue preferences for necrophagous and parasitic flies infesting living humans, and the consequences to human health when necrophagous flies go rogue – flies switch from necrotic to living tissue – and thereby pose a serious challenge to the body's defenses.

16.2 Myiasis can be classified based on anatomical location or degree of parasitism

Competition among necrophagous insects to locate and utilize a patchy ephemeral resource like carrion for feeding and reproduction is intense. As discussed in Chapter 6, the fitness of a female carrion-inhabiting fly is tied to the developmental success of her progeny. Consequently, a gravid female maximizes her fitness by depositing eggs or larvae in locations that favor progeny growth and development, namely a resource that is nutrient-rich and to a degree provides a sheltered environment from predators and parasites. The microhabitat created by decomposition of a corpse fulfills both criteria. So, too, does a living vertebrate under the right conditions. Necrotic tissue is readily abundant in human and other animal populations in both rural and urban areas (Sherman, 2000; Cestari *et al.*, 2007), and is likely on the rise worldwide as human populations are increasingly aging and suffering from medical conditions like diabetes that promote cutaneous lesions or that incapacitate (Batista-da-Silva *et al.*, 2011; Centers for Disease Control and Prevention, 2011). A localized carrion source, meaning a small patch of dead tissue on a vertebrate animal, is the same protein-rich food resource, albeit in lower concentrations, as a corpse, but with less competition as fewer fly species are attracted to wounds or lesions and there is almost no chance of predation or parasitism. Under these conditions, it is easy to visualize how myiasis evolved in necrophagous species. Myiasis, like necrophagy, appears to have evolved as a strategy for efficient acquisition of specific nutriment (protein) for fly larval development and adult reproduction (egg provisioning) (Catts & Mullens, 2002). The earliest forms of myiasis are believed to have evolved from two distinct pathways; one involving saprophagy (of which necrophagy is a subcategory) and another relying on a **sanguinivorous** (blood-sucking) existence (Zumpt, 1965). Beyond these initial insights, the evolutionary pathway (s) leading to myiasis in Diptera is not clear. Why? In part, **convergent evolution** of life history strategies and morphological traits in species that are obligatory or facultatively parasitic has complicated the overall picture (Stevens *et al.*, 2006). Adding to the murkiness is the occurrence of both endoparasitic and ectoparasitic forms of myiasis, and the complexities of the adaptive traits associated with each form of parasitism (Figure 16.2). Similar difficulties have occurred with trying to decipher the evolution of parasitism in parasitic Hymenoptera, which like Diptera, displays endo- and ectoparasitism and as well as **parasitoidism**.

(a)

(b)

Figure 16.2 Morphological differences in fly larvae adapted for (a) facultative myiasis and (b) obligatory myiasis. *Source*: Pavel Krok/Wikimedia Commons/CC BY-SA 3.0. Acarologiste/Wikimedia Commons/CC BY-SA 4.0.

Myiasis is generally defined by anatomical or ecological classification. In the following sections, each form of classification is briefly examined.

16.2.1 Ecological classification of myiasis

Ecological classification identifies myiasis by the type of association or degree of parasitism between the fly and the host (Francesconi & Lupi, 2012). For example, accidental myiasis (also referred to as pseudomyiasis) occurs when fly eggs or larvae infest food of humans or animals and the food is then ingested, or they gain accidental passage into some other bodily opening, including wounds (James, 1947). The key distinction of accidental myiasis from other forms is that the fly involved did not seek out the host. The flies themselves are free-living, typically saprophagous or necrophagous, and express no interest in the animals other than to share their food. When ingested, the flies ordinarily pass through the alimentary canal without evoking any problems to the animal. However, in some cases the larvae infest the alimentary canal once ingested or secondarily after passing out of the anus to reenter through the anal opening (**retroinvasion**). This form of myiasis is relatively uncommon in humans and more likely to be associated with contaminated pet or livestock foods. It can be of legal importance when the food is shown to be fly-infested due to negligence by a primary caregiver or owner of domesticated animals (Benecke & Lessig, 2001; Anderson & Huitson, 2004). Interestingly, or perhaps not surprisingly, accidental myiasis is typically associated only with oral entry (consumption) into the digestive tract. All other forms of fly invasion of the alimentary canal are considered either facultative or obligatory parasitism.

An alternative definition of accidental myiasis is that it represents all instances of larval infestation that do not result in full development during the host-parasite association (Francesconi & Lupi, 2012). In this view, pseudomyiasis is also used in concert with anatomical classification. Thus, in the example of consumption of fly-infested food, the condition would be referred to as accidental intestinal myiasis.

Facultative myiasis is by far the most common of the three types of parasitism (Figure 16.3). In these instances, adult necrophagous or saprophagous flies detect chemical signals originating from wounds or lesions exposed to the external environment, or from exuded bodily fluids such as urine, feces, blood, semen, or interstitial. Consistent with foraging behavior displayed on a corpse, adult females will rely on a variety of stimuli (gustatory, tactile) to assess the resource before making decisions regarding adult feeding and oviposition/larviposition. Unlike with carrion, a living animal is expected to interrupt fly activity, thereby deterring the acceptance phase. In the case of humans, progeny deposition would thus only be expected to occur when an individual is incapacitated or physically unaware of the flies. In reality, however, oviposition can occur under a variety of conditions when an individual is not incapacitated but is presumably unaware of the initial colonization. When discovery of the flies is made, the "normal" or expected response is to remove the flies and certainly before any injury has occurred.

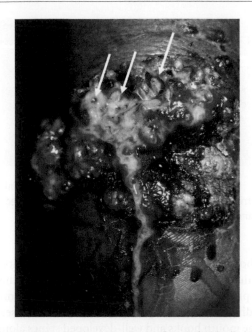

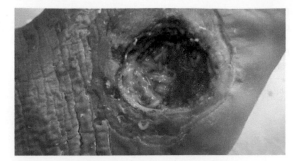

Figure 16.4 Maggot therapy is the controlled use of fly larvae to debride and clean wounds for which other forms of treatment are found to be ineffective. In this case, maggots were used to clean a diabetic ulcer on a foot. *Source*: Alexsey Nosenko/Wikimedia Commons/CC BY 3.0.

Figure 16.3 Facultative myiasis of an individual's heel with several larvae deeply burrowed. *Source*: Courtesy of Gunther Slesak, Saythong Inthalad, Michel Strobel, Matthias Marschal, MJR Hall, and Paul Newton/Wikimedia Commons/CC BY 2.0.

Facultative myiasis represents a paradox from a host–parasitism perspective. For example, the initial invasion of necrotic tissue (s) is not truly parasitism in that the fly does not technically depend on a living host for survival nor does the animal suffer from the interaction. In fact, relatively low numbers of necrophagous larvae feeding on necrotic tissues has been used for years as a means for cleaning wounds by removing dead tissue (debridement) and release of antimicrobial compounds via fly activity and secretions, a treatment known as maggot therapy (Sherman *et al.*, 2000; Sherman, 2009) (Figure 16.4). However, by definition, all forms of myiasis are considered parasitism, and dependent on the length of association and anatomical location, facultative myiasis can lead to severe injury and even death. Death, though, adds to the paradox; true parasites generally do not kill their host.

For most species, facultative myiasis is associated with cutaneous or superficial wounds, short episodes (a few days to a week) of larval feeding on the host, and high pathogenicity levels if larvae feed on live tissues (Stevens *et al.*, 2006). Problems arise when oviposition occurs on tissues that are highly susceptible to the feeding action of fly larvae or the size of the feeding aggregation

exceeds a tolerable limit. Depending on the species and number of larvae present, heat production, larval enzymes, tissue consumption, and build-up of larval waste products induce damage and destruction to the healthy tissue surrounding the wound. Opportunistic secondary invaders in the form of the same or other fly species and microorganisms may also enter the wounds, evoking more severe pathogenic effects on host tissue than that promoted by the initial colonizers (Thompson et al., 2013). Nonetheless, the species responsible for inducing primary myiasis[4] on necrotic tissue serve as facilitators of secondary invaders and other forms of myiasis (Table 16.1). The severity of host damage also elevates when the initial necrophagous species switch from feeding on dead tissues to living as is common with calliphorids like *Lucilia sericata*, *Cochliomyia macellaria* or *Chrysomya rufifacies* (James, 1947; Sukontason et al., 2001). Note that these species are not "committed" to parasitism once they transition to live tissue and retain the ability to switch back to dead tissue as opportunities arise (Catts & Mullens, 2002).

Necrophagy on dead tissues of a living host is believed to represent the selection conditions that favored the evolution of obligatory parasitism among saprophagous Diptera (Catts & Mullens, 2002). It is entirely possible that the need to form larval feeding aggregations to successfully compete for resources and/or high gregariousness[5] in some species led to obligatory parasitism. Why? Both scenarios yield large maggot masses with intense interspecific and intraspecific competition and represent microhabitats that are potentially limiting in that they can produce temperature stress

Table 16.1 Classification of myiasis based on the degree of host parasitism.

Ecological classification	Nature of host association
Obligatory	Fly species is parasitic during larval stages and dependent on a living host
Facultative	
Primary	Fly species is typically free-living on carrion but opportunistically feeds as larvae on necrotic tissue (s) of living vertebrate animal
Secondary	Fly species is typically free-living on carrion but becomes attracted to necrotic tissues infested by primary myiatic species; cannot initiate myiasis themselves
Tertiary	Fly species is typically free-living on carrion but opportunistically feeds on necrotic tissue of severely weakened host
Accidental/ pseudomyiasis	Fly species is free-living and larvae cannot complete development on host

Source: Derived from Kettle (1995), Hall & Wall (1995) and Francesconi & Lupi (2012).

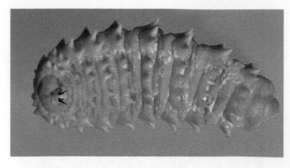

Figure 16.5 Spines on the body of a bot fly (Family Oestridae) larvae. *Source*: Daktaridudu/Wikimedia Commons/CC BY-SA 4.0.

and overcrowding (review Chapter 8 for more details). Adaptations like spatial aggregations or resource partitioning may have driven larvae from the core of the larval aggregation, and hence necrotic tissue, toward living tissues at the periphery. While the precise driving forces that led to obligatory myiasis in calyptrate[6] Diptera cannot be determined experimentally, the process of speculation on evolutionary pathways relies on the same types of scientific inquiry. Further discussions on the evolution of myiasis can be found in Zumpt (1965), Stevens & Wallman (2006) and Stevens *et al.* (2006).

Obligatory myiasis represents "true" parasitism among the various forms of myiasis in that gravid adult females search for a living host for oviposition, and the developing larvae are dependent on healthy host tissue for survival. The dynamics of the host–parasite association are fascinating in that a battle for survival wages between the animals involved, with the host's body defenses embroiled in an effort to rid its body of the parasite, and the fly larvae desperately trying to acquire necessary nutriment while fending off the host immune system (Otranto, 2001). Most obligate species have a comparatively long association with the host, requiring weeks to months to complete larval development. Such a long host–parasite relationship is dependent on unique morphological and physiological adaptations not apparent with accidental or facultative myiasis. In some instances, the parasites rely on

elaborate mechanisms to avoid immunodetection, or the mother and/or offspring inject agents that suppress aspects of the host's defenses that permit feeding, at least temporarily. Larvae also possess large mouth hooks and well-developed spines along the dorsal surface of the body (Figure 16.5). The spines are believed to facilitate migration through the host, often in subdermal locations. By comparison, larvae associated with facultative myiasis do not migrate through host tissues, and correspondingly lack dorsal spines. Digestive enzymes also differ with the various forms of myiasis; in obligatory species the enzymes aid in food acquisition as well as evading host defenses, whereas with facultative species the enzymes function to rapidly degrade and digest host tissues. The differences reflect the coevolutionary adaptions associated with true parasitism (obligate myiasis) versus opportunistic parasitism by otherwise carnivorous species (facultative myiasis). A more detailed examination of the pathogenicity of obligatory and facultative myiasis will occur in Section 16.6.

Obligatory myiasis appears to be restricted to the families Calliphoridae (screwworms like *Chrysomya bezziana* and *Cochliomyia hominivorax*), Sarcophagidae (flesh flies in the genus *Wohlfahrtia*) and Oestridae (bot flies), with the screwworms closely aligned with necrophagous calliphorids in being more generalists whereas bot flies display a high degree of host specificity (Figure 16.6). *Cochliomyia hominivorax* is perhaps the most recognizable species that induces obligatory myiasis. The fly's association with vertebrate hosts seems to represent a bridge between necrophagy to obligatory parasitism. Gravid females oviposit a large (200–500 eggs) clutch into open wounds, larvae feed on dead and

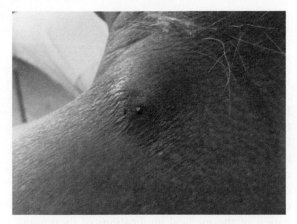

Figure 16.6 Furnuncle formation on skin due to the presence of a bot fly larva. *Source*: Petruss/Wikimedia Commons/CC0 1.0.

Table 16.2 Common forms of myiasis based on anatomical location in or on body of host.

Myiasis type	Description
Auricular	Infestation of ear and auditory canal
Nasopharyngeal	Infestation of mouth, nose or throat
Cutaneous or traumatic	Infestation of wounds or lesions of integument (dermal and subdermal)
Ophthalmomyiasis	Infestation of any aspect of the eye
Oral	Infestation of mouth
Sanguinivorous	Bloodsucking by larvae
Intestinal	Infestation of digestive tract from esophagus to anus
Urogenital	Infestation of urethra, vagina, urinary bladder, or ejaculatory duct
Umbilical*	Infestation of umbilical cord or navel

* Umbilical myiasis is a rare type of myiasis sometimes classified as one occurring in clinical settings.
Source: Based on James (1947), Sherman (2000), and Zumpt (1965).

healthy tissue, eventually the maggots migrate or "screw" into dermal tissues, and larval development is finished within a week. A similar life history is associated with other obligatory species in the Calliphoridae and Sarcophagidae. As will be discussed in Section 16.5, tissue specificity differs between obligate parasites, mostly along the lines of oestrids versus species from other families.

Myiasis in general, but especially with obligatory forms, is typically associated with poor socioeconomic regions of tropical and subtropical countries (Francesconi & Lupi, 2012). In nations where obligatory myiasis is not common, it does frequently appear as a commonly travel-associated skin disease. As a consequence, obligatory myiasis is of major importance to medical and veterinary entomology, but usually only of minor relevance to forensic entomology.

16.2.2 Anatomical classification of myiasis

The classification scheme used thus far to discuss myiasis has relied on the lifestyle of the flies, specifically that of the feeding larvae. Myiasis is frequently placed into categories based on the location on the host's body in which invasion occurred. Four categories are used for broad anatomical classification:

- **sanguinivorous** – includes bloodsucking
- **cutaneous myiasis** – includes furnunclar and migratory myiasis

- **wound myiasis** – which is frequently cutaneous in location
- **cavitary myiasis** – classification is based on the affected organ.

Under this scheme, myiasis is more narrowly classified by inclusion of the specific region, tissue or organ infested. For example, intestinal myiasis refers to fly invasion of the digestive tract, while auricular myiasis is fly infestation of the ears and auditory canal. Table 16.2 provides a list of common forms of myiasis based on anatomical location on the host. Subcategories of anatomical myiasis are quite common for a particular region.

For example, fly infestation associated with the eye can be divided into orbital myiasis, (intraocular invasion), and ophthalmomyiasis interna, which involves maggot development in the anterior or posterior segments of the eye.

16.2.3 Classification of myiasis in clinical settings

Some forms of myiasis are especially uncommon or occur in specific settings. One such form is nosocomial myiasis, which is defined as larval infestations that occur during or after hospital or clinical visitations, which were not present or incubating at the time of admission (Ravasan *et al.*, 2012).

The condition is considered rare, although it is thought to have a higher incidence in tropical and subtropical countries (Hira *et al.*, 2004). In Western nations, nosocomial myiasis is likely under-reported. This form of fly infestation is most frequently associated with patients who are unaware of the invading flies (incapacitated or unconscious) or display some form of immobility or paralysis such that preventing oviposition or removing flies cannot occur. Cases of nosocomial myiasis in North America are not uncommon for patients in facilitated care, nursing homes, or similar healthcare facilities who become infested with flies in diabetic ulcers, wounds, amputated region, pin-site or around a catheter or tube (Szakacs *et al.*, 2007) (Figure 16.7). Other forms of myiasis reported in clinical settings include those associated with leprosy, malignant wounds, and umbilical cord (Francesconi & Lupi, 2012). The latter is not restricted to hospitalization, as childbirth occurs in many different locations based on cultural considerations and circumstances of birthing.

16.3 The conditions that promote facultative myiasis are predictable

Facultative myiasis is by far the most common type of myiasis. It can also be classified as the most important form of myiasis from both a medical and legal perspective. Obviously, larval development in delicate regions like eye orbits or near brain tissue (nasopharyngeal myiasis) can evoke severe pathologies. A more in-depth examination of the medical implications of fly infestations will occur in Section 16.6. But what about the legal importance of facultative myiasis? Many of the reasons are apparent from the earlier discussion of ecological classification of fly infestations. Attraction of flies to dead tissues is a normal or typical response of necrophagous species. The occurrence of the flies does not necessarily represent a breach of law, nor does oviposition on a wound. What crosses the

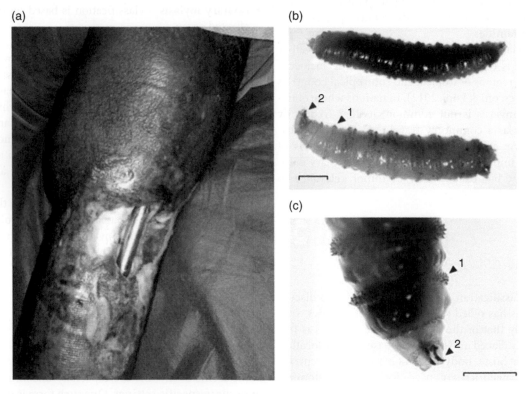

Figure 16.7 Pin-site myiasis in a 77-year-old man 12 years after tibial osteosynthesis, Colombia. (a) Open wound in the man's left leg, showing multiple insect larvae. (b, c) Cochliomyia hominivorax screwworm fly larvae extracted from the wound. Arrow 1 indicates the spinose bands; note the spines arranged in four rows that separate each segment. Arrow 2 indicates its mouth hooks. Scale bars indicate 2 mm (b) and 1 mm (c). *Source*: Center for Disease Control/Wikimedia Commons/public domain.

boundaries from natural myiasis to one of legal concern are the conditions that led to the fly interactions and the duration of the host–parasite association. The latter is reflective of the developmental stage discovered on the individual. For example, newly laid eggs would not be overly surprising during summer months on open wounds or dressings saturated with body fluids. Assuming that routine wound care is performed, the eggs or newly hatched neonate larvae should soon be discovered and disposed of before any damage is inflicted. However, discovery of later stages of larval development cannot be justified for someone under the care of healthcare workers or a parent or guardian. Similarly, the attraction of necrophagous flies to wounds on an individual located indoors and the ease of detection and acceptance behavior by the flies calls into question the environmental conditions associated with where the host (e.g., person infested with flies) resides.

Environmental conditions are in reference to human habitation, and includes the home, dwelling or some other building, and the associated surroundings. The surroundings can vary substantially from a small yard in an urban environment to a rural setting with the possibility of a nearby livestock facility with an ample supply of manure-breeding Diptera. Fly movement from an animal production facility to indoors does not represent negligence, provided that an appropriate manure management plan is in place (See

Chapter 3). However, when the environment of the home and its surroundings promote fly synanthropy, a case can be made for negligence. Classic examples include trash refuge is not disposed of properly or infrequently; improper storage of food, especially leading to spoilage; and human and animal waste accumulation close to human habitation (Figure 16.8). Each of these conditions serve as fly attractants. Once in a home, the chances greatly increase for interactions between gravid flies with an individual not properly cared for in terms of wound care or hygiene (Figure 16.9). Failure to maintain a proper home or dwelling for human habitation constitutes negligence on the part of the homeowner, landlord, or principal occupant.

A negligent environment can also be created through failure to care for individuals who are mentally and/or physically incapable of caring for themselves. This includes children, elderly, and incapacitated individuals. Negligence may be associated with any aspect of personal care, including wound care or hygiene. Obviously an improperly treated wound can lead to many complications derived from facultative myiasis. In a similar manner, diapers or clothing saturated with urine or feces draw the attention of several species of saprophagous and necrophagous flies. In turn, flies may oviposit in or on the soiled diapers or clothing. Under severe conditions, skin lesions may form, creating an avenue for fly infestation of living tissue

Figure 16.8 Accumulation of human or other animal feces near human habitation creates an environment that favors synanthropy by flies. *Source*: Stefan Korner/Wikimedia Commons/CC BY-SA 3.0.

(Figure 16.10). The latter is an example of both negligence and abuse (indirect). Fly infestation at the openings of the urethra, vagina, or anus may occur in the absence of lesions. Direct abuse of an individual is often times detected through bruising and lesions on the body. Of course, any type of open wound can be an opportunity for necrophagous flies in close association.

Facultative myiasis also occurs in conditions that are not associated with neglect or abuse. For example, individuals displaying poor hygiene or that have open wounds but lack the resources for medical treatment are susceptible to fly colonization. These scenarios are represented by persons who are homeless and/or suffer from drug or alcohol addiction. Similarly, natural facultative myiasis can be experienced when a person is injured or attacked. This is even more likely to occur if the person becomes physically or mentally incapacitated as a result of the injuries (Vanin *et al.*, 2017).

16.4   Chemoattraction of flies to the living does not necessarily differ from the odors of death

Death has a distinct and unique odor, whether emanating from a large decomposing body or a small patch of tissue located on a living vertebrate animal. In Chapter 7, we examined the chemical signals used by carrion insects for detection of animal remains and the mechanisms of olfaction employed by adult and juvenile insects. The chemical profile of a corpse changes over time and is influenced by a wide range of abiotic (e.g., environmental

Figure 16.9 Improper food storage and failure to remove trash refuge are major contributors to attracting flies to human habitation and promote an environment for facultative myiasis. *Source*: Userm1970/Wikimedia Commons/CC BY-SA 3.0.

(a)

(b)

Figure 16.10   Flies commonly associated with infestation of soiled diapers include: (a) *Muscina stabulans* (Family Muscidae) and (b) *Fannia canicularus* (Fanniidae). *Source: M. stabulans*/Wikimedia Commons/CC BY-SA 4.0. *F. canicularus*/Wikimedia Commons/CC BY 2.0.

conditions) and biotic (e.g., insect and microbial activity) factors. Presumably, necrotic tissue on a human or pet living indoors is subject to much less variation in terms of environmental conditions, meaning a more stable localized microhabitat, permitting tissue decomposition to occur at closer to a constant rate. Supporting this supposition is the fact that the underlying tissues are alive (whether the closest layers are healthy or not is another factor) and that the animals in question are mammals (i.e., endothermic[7]), which indicates that temperature fluctuations do not occur like those that follow algor mortis[8] or with lesions near the skin surface for animals living outdoors. In short, necrotic tissues on a living vertebrate animal are likely to yield odors associated with autolysis, and less likely to be generated from putrefaction. As we have discussed in Chapter 10, autolysis is the process of self-digestion by enzymes released from dead or dying cells, a process that rarely occurs in living cells and generally begins immediately after the heart stops beating. By contrast, putrefaction is the chemical degradation of soft tissues due to the action of microorganisms and commonly starts sometime after the onset of autolysis (Evans, 1963; Vass et al., 2002). Putrefaction is expected to be limited for localized necrotic tissue since (i) the animal is alive and has a functional immune system and other body defenses for maintaining homeostasis, and (ii) a living person generally practices some aspects of basic hygiene, albeit limited when incapacitated or neglected, such as cleaning the wound or infected area. This does not preclude the possibility of tissue decomposition via putrefaction to some degree, but unregulated microbial activity will be quite limited in living individuals, depending on health.

The implications in terms of fly detection of a wound are that the odors are likely derived predominantly from autolysis of cells within the afflicted tissue (s). These are the same predicted chemical signals associated with a corpse immediately after death (Vass et al., 2002). Consequently, detection of wounds or lesions on the body of humans or other animals would be expected to be by early insect colonizers, which means adult calliphorids. Indeed, several species of blow flies are some of the primary agents inducing facultative myiasis, along with sarcophagids, muscids, and phorids (Catts & Mullens, 2002). As the amount of tissue emitting a chemical signal is quite small in comparison with a corpse, successful "host" detection requires

an especially acute sense of olfaction and close association between the flies and humans and/or animals, simply meaning that wound myiasis is more likely with **synanthropic** species. In the case of obligatory parasitic species, the chemical cues may be unique to a given host and depend on kairomonal[9] signaling by bacteria associated with the wound and microbial modification of existing compounds to attract gravid females (DeVaney et al., 1973; Chaudhury et al., 2010). Undoubtedly bacterial signaling is equally important to facultative myiasis.

Insect succession in and around a wound is expected to occur for animals living outdoors but is not as likely indoors. As the initial colonizers modify the tissues, and assuming the wound/myiasis is not treated, the chemical profile of the necrotic area changes to become more attractive to other fly species. Under these conditions, putrefaction is expected to begin, leading to volatile sulfur-containing compounds to be released from the site. Subsequent colonization of a wound or lesion induced by a facultatively parasitic species is referred to as secondary and tertiary myiasis (Kettle, 1995) (see Table 16.1). Such fly species depend on primary myiasis for successful exploitation of necrotic tissue in much the same way that insect colonizers of later stages of decomposition depend on corpse modification by earlier inhabitants. Similarly, gravid adult screwworms (obligate parasites) are attracted to wounds already infested by other flies (generally of the same species). The chemical cues serving as chemoattractants and as oviposition stimulants are derived from tissues modified by both larval feeding and microbial activity (Eddy et al., 1975; Chaudhury et al., 2002).

16.4.1 Chemoattraction to body fluids

Host detection is not restricted to chemical identification of necrotic tissues. Foraging behavior in several species of flies is activated by signals released from body fluids such as urine, blood, feces, semen, and interstitial fluid. There is no doubt that ammonia-rich compounds in urine serve as powerful attractants (see Chapter 7 for details of chemical attraction to carrion). Strong odors associated with feces are the result of bacteria within the digestive tract producing several sulfur-containing compounds such as skatole, indole, and thiols; these same types of chemicals have been identified as chemical attractants for a range of necrophagous fly species (Mackley & Brown, 1984; Urech *et al.*, 2004; Aak *et al.*, 2010; Chaudhury *et al.*, 2017). Other body fluids are less odiferous than urine and feces, and thus would be presumed to draw less attention from necrophagous and saprophagous flies. The reality, however, is that animal body fluids are not ignored since they are relatively high in a variety of organic molecules, which in turn make them nutrient-rich resources for opportunistic flies. Carbohydrates, organic acids, and amino acids have lower volatility than ammonia- and sulfur-containing compounds, but are readily located by several species of adult calliphorids when present in blood and other body fluids (Hammack, 1991). The precise chemical signals have yet to be determined but possible candidates include long-chain amino acids, particularly those composed of sulfur-containing molecules like methionine and cysteine or that form thiol linkages with other compounds (Brosnan & Brosnan, 2006). Bacterial production of volatile compounds in bodily fluids is likely involved in the detection and location of these food resources by gravid females, since bacterial infection of fluids increases the odors emitted and artificial inoculation of blood and other fluids with bacteria enhance the attractiveness to adult flies (Hammack, 1991; Chaudhury *et al.*, 2002).

16.5 Necrophagous and parasitic flies display oviposition and development preferences on their vertebrate "hosts"

Tissue specificity in terms of oviposition and development sites on or in a host is generally absent with species that induce facultative myiasis. What occurs instead is that gravid females follow odor plumes to locate necrotic tissue regardless of body location, and then factors such as moisture content, tactile feedback, and other stimuli (e.g., olfactory, gustatory and/or visual) influence decisions concerning whether to oviposit, where to deposit eggs, and the size of a clutch. Typically, facultative parasitism among calliphorids,

sarcophagids, phorids, muscids, and other families is restricted to wounds/lesions, constituting various forms of cutaneous myiasis (including dermal and subdermal) and invasion of natural orifices, including urogenital myiasis, oral myiasis, nasopharyngeal myiasis, and intestinal myiasis via the anus (Hall & Wall, 1995; Duro *et al.*, 2007; Goff *et al.*, 2010). Thus, the presence of maggots in a dermal lesion or in feces located in a diaper is not necessarily indicative of a particular species, although adults of *Fannia canicularis, Muscina stabulans, Megaselia scalaris*, and some psychodids (urine) display a stronger affinity for urine and feces than most other necrophagous species (James, 1947; Benecke & Lessig, 2001) (Figure 16.11).

Species responsible for obligate myiasis are more specialists than opportunistic necrophiles and consequently display specificity in terms of host selection and oviposition sites on or in their vertebrate hosts (Table 16.3). There is some overlap in locations selected by obligatory and facultative species for oviposition (i.e., cutaneous and various forms of cavity myiasis), which is not surprising considering the proposed ancestry of obligatory myiasis from necrophagy (Zumpt, 1965). However, several species of oestrids are internal parasites, developing in various regions of the alimentary canal, internal organs, and migrating to subdermal regions from cutaneous lesions (Catts & Mullens, 2002). Because of the location of larval development, obligate parasites must cope with attack from the host immune system (Otranto, 2001), which as we will see in Section 16.6, can have a profound influence on the length of the association between parasite and host. By contrast, fly species engaged in facultative myiasis feed for much shorter durations in comparison with obligatory parasites and consequently are less likely to contend with host immune responses.

16.6 Larval myiasis can be fatal

Fly infestation of tissues can evoke a range of effects on the host, from relatively mild damage if treatment occurs early and the individual (human or other animals) is otherwise relatively healthy, to severe pathological consequences including death. The pathogenicity of the disease (myiasis) is

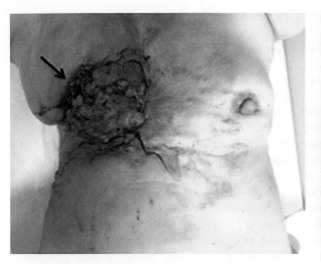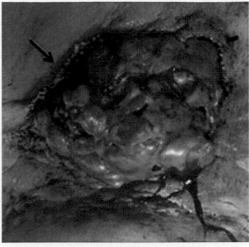

Figure 16.11   A case of human breast myiasis caused by *Lucilia cuprina* (Diptera: Calliphoridae). Ulcerate lesion in the right breast and several larvae in the wound. *Source*: Ozan Akinci, Serhat Sirekbasan, Murat Toksoy and Sefa Ergun/Wikimedia Commons/CC BY 4.0.

Table 16.3   Host specificity and parasitism differences among obligatory and facultative myiasis-inducing flies.

Family	Host tissues	Progeny deposition	Feeding behavior	Penetration of internal organs	Rate of development
Calliphoridae					
FM	Usually body fluids or necrotic tissues	Oviparity	Form feeding aggregations	Larvae do not invade internal organs	Rapid from egg to pupariation
OM	Usually necrotic but obligate parasites	Oviparity	Form feeding aggregations	Larvae do not invade internal organs	Rapid from egg to pupariation
Fanniidae	Usually feces or necrotic tissues	Oviparity	May feed in aggregations but not required	Larvae do not invade internal organs	Rapid from egg to pupariation
Muscidae	Usually body fluids, feces or necrotic tissues	Oviparity	May feed in aggregations but not required	Larvae do not invade internal organs	Rapid from egg to pupariation
Oestridae	Gastrointestinal or subdermal	Oviparity or vivaparous*	Solitary	Larvae commonly invade internal organs	Slow from egg to pupariation
Phoridae	Usually dried necrotic tissues	Oviparity	May feed in aggregations but not required	Larvae do not invade internal organs	Rapid from egg to pupariation
Sarcophagidae					
FM	Usually feces, gut contents or necrotic tissues	Ovoviviparity	Form feeding aggregations	Larvae do not invade internal organs	Rapid from egg to pupariation
OM	Usually necrotic but obligatory parasites	Ovoviviparity	Form feeding aggregations	Larvae do not invade internal organs	Rapid from egg to pupariation

* Viviparous, meaning neonate larvae are deposited by gravid females. FM = facultative myiasis, OM = obligatory myiasis. Calliphoridae and Sarcophagidae have members that induce facultative or obligatory myiasis.
Source: Modified from Sherman *et al.* (2000).

independent of whether myiasis is due to natural parasitism or the result of neglect, although cases resulting from abuse or neglect often go hand in hand with other factors (e.g., malnutrition, unsanitary conditions) affecting the well-being of an individual. Rather, the extent of damage evoked by fly infestation of vertebrate tissues depends on the fly species involved, the host species targeted, the relative health of the individual, and the conditions in which the human or animal resides (Otranto, 2001) (Figure 16.11). The type of association between the host and parasite is also important to the severity of disease resulting from fly invasion. For example, obligatory parasites harm the host but generally the interaction is not fatal. In fact, severely damaging the host is counterproductive to a parasitic lifestyle. By contrast, necrophagous species that opportunistically initiate facultative myiasis have not adopted a parasitic strategy for progeny production. Such species are adapted for rapid utilization of decomposing tissue in an intensely competitive microhabitat, where the focus is on overcoming the competition not on maintenance of the resource (i.e., the corpse).

Parasite–host relationships that have evolved over a long period of time tend to achieve an equilibrium, in which counter initiatives instigated by the parasite offsets host defenses. This implies that vertebrate hosts are not passive to the feeding activity of fly larvae and can mount a powerful immunological defense, at least during certain forms of parasitic invasion. It also suggests that host death is not advantageous to fly parasites. In this section, we will examine the impact of myiasis on the host condition, with reference to the type of fly infestation as well as the host defenses that are used against myiasis and the counter measures employed by fly larvae feeding on vertebrate tissues.

16.6.1 Pathogenicity of myiasis

Different forms of myiasis culminate in contrasting ways in terms of impact on the host species. In the least severe host–parasite association, accidental myiasis, ingested larvae may die while passing through the alimentary canal (Goff et al., 2010), with fly survivorship dependent on the gut environment of the vertebrate host. For instance, the highly acidic stomach of adult humans poses a serious threat to most ingested fly species, whereas the gut of herbivorous hosts is closer to neutrality

(Randall et al., 2002) and consequently fly survivorship is relatively high (Catts & Mullens, 2002). Passage of fly larvae through the digestive system may evoke various gastric problems, including nausea, diarrhea, bloating, and pain, but rarely severe acute or long-term symptoms (Catts & Mullens, 2002).

With facultative myiasis, invasion of necrotic tissue is generally benign with regard to the host condition. However, as the number of individuals in a wound increase, driving feeding activity into healthy tissue, the severity of myiasis escalates. Here lies an important distinction between facultative and obligate parasites associated with myiasis: facultative myiasis is more likely to induce irreversible damage including mortality than obligatory forms (Sandeman, 1996). What accounts for the differences? Obligate myiasis is a condition brought about by true parasites, as opposed to fly species that display characteristics closer to that of parasitoids. A parasitoid is a unique form of parasitism common to the orders Hymenoptera and Diptera in which the host is usually killed as a result of larval feeding (Godfray, 1994). As with facultative myiasis, parasitoidism involves periods of intense feeding by larvae with rapid consumption and assimilation rates, leading to host death during host feeding or occurring not long after the parasitic relationship has ended, when the larvae have completed feeding on the host (Quicke, 1997).

Initial larval infestations may cause discomfort due to pressure from immatures positioning themselves in body cavities or under skin layers, or simply due to the constant movement while feeding on necrotic tissue. Burning sensations are sometimes associated with cutaneous, anal, and genital myiasis possibly due to serous discharge, enzymatic irritation of living tissue and/or bacterial secretions (Cestari et al., 2007). Localized autolysis of healthy tissue occurs as intracellular components are released during larval feeding on necrotic cells. Depending on the location of fly activity, nodule or cyst formation may occur (this is particularly true with obligatory parasites), or there may be induction of cellulitis, sinusitis, pharyngitis, or even meningitis[10] (Sherman et al., 2000). Any type of larval feeding activity near the brain is likely to be irreversibly damaging or fatal. Similarly, diminished hearing may result from damage to the eardrum via auricular myiasis, and short- or long-term vision impairment can occur with larval feeding on the conjunctiva or eyelids (Sherman, 2000; Cestari et al., 2007). The reality is that myiasis has been reported for almost any area of the body, while with any form of facultative myiasis, primary myiasis can lead to severe host pathology due to feeding activity and the action of secondary invaders.

16.6.2   Host responses to myiasis

Vertebrate hosts are not passive to myiasis. Fly larvae trigger both non-specific and specific body defenses. The nature and strength of the host defense response depends on the health of the host, location of fly invasion, and type of antigen encountered. Vertebrate animals recognize various components of the fly integument (exoskeleton), secretions derived from salivary glands and foregut, and waste products released from the digestive tract via the anus as antigenic molecules (Milillo *et al.*, 2010). Induction of inflammation (nonspecific) occurs in response to initial fly colonization, particularly in cases of cutaneous myiasis. An acute inflammatory response is associated with bacterial infection owing to primary myiasis (Sandeman, 1996). Provided that circulation is maintained to the site of necrosis, leukocytes (neutrophils and eosinophils), T cells (lymphocytes), and macrophage-like cells, components of the cellular arm of body defenses (Martini *et al.*, 2011), accumulate below or near the necrotic tissue. These cells function in the non-specific destruction and removal of microorganisms and debris from circulation or sites of infection/injury. The presence of T lymphocytes at myiatic sites is indicative of further cell recruitment, **cytokine** release, and activation of antigen-specific immune responses (Bowles *et al.*, 1992). Minimally these cellular responses help to contain microbial contamination of the wound, at least temporarily, and may also decrease the impact of larval feeding on the host. One avenue that the latter could be realized is through suppressed growth rates if bacterial activity aids digestion and/or food assimilation by fly larvae (Thompson *et al.*, 2013) or if cytokines and other cellular components display cytotoxicity toward the parasites. Cytokines are proteins involved in a variety of cell–cell communication pathways, and are also instrumental in initiation of cell death, including apoptosis, in a number of cell types (Gomperts *et al.*, 2002; Kharroubi *et al.*, 2004). Whether human cytokines display cytotoxicity to fly larvae responsible for any form of myiasis is not clear.

Specific host responses are primarily in the form of antibodies (humoral responses) targeting the antigenic molecules of the fly (Tabouret *et al.*, 2003). Evidence from cutaneous myiasis indicates that high concentrations of host **immunoglobulins** and eosinophils accumulate around the sites of larval feeding, promoting cytotoxicity of secondary microbial invaders and susceptible fly larvae (Otranto, 2001). Interestingly, the cells primarily responsible for immunoglobulin production (B lymphocytes) increase in numbers and expression levels 96–120 hours after initial fly infestation (Bowles *et al.*, 1994). The significance for myiasis is that larval feeding associated with facultative parasitism is generally completed or nearing completion before the initiation of antibody production. Consequently, antibody defenses are effective only toward obligatory myiasis. However, this does not mean that immunoglobulins are not produced in response to facultative myiasis (Thomas & Pruett, 1992). In fact, the timing of antibody production and plasma titer of antibodies specific to necrophagous fly antigens may be useful quantitative tools for assessing whether incidences of facultative myiasis represent cases of neglect or abuse of humans and pets (Figure 16.12).

16.6.3   Larval defenses to host attack

Fly larvae possess an arsenal of weapons to combat the defenses mounted by the host during myiasis. The most direct counterattack is associated with digestive enzymes released via saliva. Larval trypsin and chymotrypsin enzymatically degrade host immunoglobulins, causing a 60–70% reduction in antibody activity under natural and artificial conditions (Kerlin & East, 1992; Sandeman *et al.*, 1995). Total suppression of antibody–antigen responses does not appear to occur via enzymatic degradation. Larvae release other immunoreactive proteins during feeding on host tissue, and unlike the relatively non-specific salivary proteases these proteins specifically bind to T lymphocytes to inhibit early cellular events that follow activation (Elkington *et al.*, 2009).

Excretion of high levels of ammonia into host tissues provides non-specific protection for feeding parasites. Free ammonia in liquid feces is converted to a non-ionized form that is cytotoxic to leukocytes, macrophages, and lymphocytes, depressing localized nonspecific and specific defenses of the host (Guerrini, 1997). The antimicrobial action of larval excretory products (e.g., allantoin, ammonia) suppresses populations of undesirable secondary invaders for the flies, which may in turn dampen antigenic host responses. What this means is that the

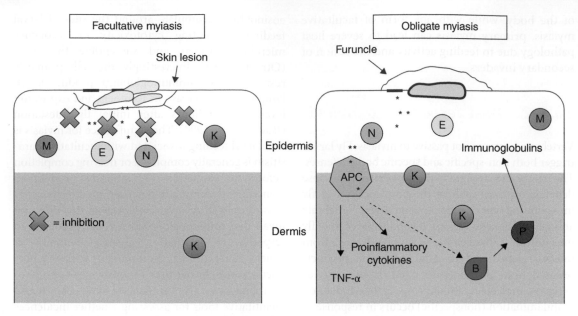

Figure 16.12 Comparison of interactions between fly larvae inducing myiasis and the host. Larval antigens (x) stimulate the activity of macrophages (M), eosinophils (E), neutrophils (N), and natural killer cells (K). Salivary enzymes and excretions can degrade and kill host immunoproteins and defense cells. The host also produces antibodies (immunoglobulins) via the activity of B lymphocytes (B) and plasma cells (P), following activation by antigen-presenting cells (APC). APCs also release regulatory cytokines and tumor necrosis factor (TNF)-α. Information derived from Otranto (2001).

flies may go longer without being detected by host defenses if microbial antigens are suppressed, thereby permitting uninterrupted feeding by the parasites.

Chapter review

Parasitic and necrophagous flies can infest humans, pets, and livestock

- Necrophagous flies do not restrict all their activity to dead animals. How can this be? The short answer lies in the fact that an animal does not have to be deceased to possess regions with dead or dying tissues. Any type of wound contains dead cells. If the lesion breaches the integument, then the putrid odors emitted from the necrotic tissue are released into the environment, drawing the attention of the same necrophagous flies attracted to a decomposing corpse. Upon discovery of the lesion (s) on the body, oviposition/larviposition occurs, creating a situation that differs from progeny deposition on carrion: the flies are invading dead tissue of a living animal which constitutes a form of parasitism.
- Dipteran infestation of living or dead tissue of humans or other animals, of body fluids, or of

food that is then ingested represents a form of parasitism termed myiasis. Myiasis can be classified by location of infestation on the body or based on the lifestyle of the flies involved. This is particularly relevant to forensic investigations as infestations by normally necrophagous species may be indicative of neglect or abuse (e.g., non-self-inflicted wounds) of persons unable to care for themselves, or of pets or other animals that have received improper care.
- The relevance of insect activity on humans and pets in terms of forensic entomology is evident in at least two realms: (i) insect infestation of a human or domesticated animal implies possible criminal negligence and/or abuse, and (ii) development of necrophagous flies ante mortem complicates estimations of a postmortem interval if death ensues following myiasis, regardless of the causation of death.

Myiasis can be classified based on anatomical location or degree of parasitism

- Necrotic tissue is readily abundant in human and other animal populations in both rural and urban areas. A localized carrion source, meaning a small patch of dead tissue on a vertebrate animal, is the same protein-rich food resource, albeit in lower concentrations, as a corpse, but with less competition as fewer fly species are attracted to wounds or lesions and there is almost no chance of predation or parasitism. Under these conditions, it is easy to visualize how myiasis evolved in necrophagous species. Myiasis, like necrophagy, appears to have evolved as a strategy for efficient acquisition of specific nutriment for fly larval development and adult reproduction.

- Myiasis is generally classified by the type of association between the fly (ecological) and the host or location on the body (anatomical).
- Accidental myiasis occurs when fly eggs or larvae infest food of humans or animals and the food is then ingested, or they gain accidental passage into some other bodily opening, including wounds. The key distinction of accidental myiasis from other forms is that the fly involved did not seek out the host.
- Facultative myiasis is by far the most common of the three types of parasitism. In these instances, adult necrophagous or saprophagous flies detect chemical signals originating from wounds or lesions exposed to the external environment, or from exuded bodily fluids such as urine, feces, blood, semen, or interstitial. Consistent with foraging behavior displayed on a corpse, adult females will rely on a variety of stimuli to assess the resource before making decisions regarding adult feeding and oviposition/larviposition. This initial invasion is not truly parasitism in that the fly does not technically depend on a living host for survival nor does the animal suffer from the interaction.
- Necrophagy on dead tissues of a living host is believed to represent the selection conditions that favored the evolution of obligatory parasitism among saprophagous Diptera. Obligatory myiasis represents "true" parasitism among the various forms of myiasis in that gravid adult females search for a living host for oviposition, and the developing larvae are dependent on healthy host tissue for survival.
- Larvae of species engaged in facultative or obligatory myiasis display adaptive morphological and physiological features that facilitate each form of parasitism.

The conditions that promote facultative myiasis are predictable

- Facultative myiasis is by far the most common type of myiasis. It can also be classified as the most important form of myiasis from both a medical and legal perspective.
- The occurrence of flies in human habitation does not necessarily represent a breach of law, nor does oviposition on a wound. However, discovery of later stages of larval development cannot be justified for someone under the care of healthcare workers or a parent or guardian. Similarly, the attraction of necrophagous flies to wounds on an individual located indoors and the ease of detection and acceptance behavior by the flies calls into question the environmental conditions associated with where the host (e.g., person infested with flies) resides.
- When the environment of the home and its surroundings promote fly synanthropy, a case can be made for negligence. Classic examples include trash refuge is not disposed of properly or infrequently; improper storage of food, especially leading to spoilage; and human and animal waste accumulation close to human habitation.
- A negligent environment can also be created through failure to care for individuals who are mentally and/or physically incapable of caring for themselves. This includes children, elderly, and incapacitated individuals. Negligence may be associated with any aspect of personal care, including wound care or hygiene.

Chemoattraction of flies to the living does not necessarily differ from the odors of death

- The chemical profile of a corpse changes over time and is influenced by a wide range of abiotic and biotic factors. Presumably, necrotic tissue on a human or pet living indoors is subject to much less variation in terms of environmental conditions, meaning a more stable localized microhabitat, permitting tissue decomposition to occur at closer to a constant rate. Supporting this prediction is the fact that the underlying tissues are alive and that the animals in question are endothermic, i.e., temperature fluctuations do not occur, unlike following death.
- Fly detection of a wound relies on odors derived predominantly from autolysis of cells within the afflicted tissue (s). These are the same types of chemical signals associated with a fresh corpse, and consequently, would draw the interest of early insect colonizers, i.e., calliphorids.
- In the case of obligatory parasitic species, the chemical cues may be unique to a given host and depend on kairomonal signaling by bacteria

associated with the wound and microbial modification of existing compounds to attract gravid females.

- Subsequent colonization of a wound or lesion induced by a facultatively parasitic species is referred to as secondary and tertiary myiasis. Such fly species depend on primary myiasis for successful exploitation of necrotic tissue in much the same way that insect colonizers of later stages of decomposition depend on corpse modification by earlier inhabitants. Similarly, obligate parasites are attracted to wounds already infested by other flies. The chemical cues serving as chemoattractants and as oviposition stimulants are derived from tissues modified by both larval feeding and microbial activity.

Necrophagous and parasitic flies display oviposition and development preferences on their vertebrate "hosts"

- Tissue specificity in terms of oviposition and development sites on or in a host is generally absent with species that induce facultative myiasis. What occurs instead is that gravid females follow odor plumes to locate necrotic tissue regardless of body location, and then factors such as moisture content, tactile feedback, and other stimuli influence decisions concerning whether to oviposit, where to deposit eggs, and the size of a clutch. Typically, facultative parasitism among calliphorids, sarcophagids, phorids, and muscids is restricted to wounds/lesions and invasion of natural orifices.
- Species responsible for obligate myiasis are more specialists than opportunistic necrophiles and consequently display specificity in terms of host selection and oviposition sites on or in their vertebrate hosts. There is some overlap in locations selected by obligatory and facultative species for oviposition, which is not surprising considering the proposed ancestry of obligatory myiasis from necrophagy. However, several species of oestrids are internal parasites, developing in various regions of the alimentary canal, internal organs, and migrating to subdermal regions from cutaneous lesions.

Larval myiasis can be fatal

- Fly infestation of tissues can evoke a range of effects on the host, from relatively mild damage if treatment occurs early and the individual (human,

pet or livestock) is otherwise relatively healthy, to severe pathological consequences including death. The pathogenicity of the disease is independent of whether myiasis is due to natural parasitism or the result of neglect, although cases resulting from abuse or neglect often go hand in hand with other factors affecting the well-being of an individual. Rather, the extent of damage evoked by fly infestation of vertebrate tissues depends on the fly species involved, the host species targeted, the relative health of the individual, and the conditions in which the human or animal resides.

- The type of association between the host and parasite is also important to the severity of disease resulting from fly invasion. For example, obligatory parasites harm the host but generally the interaction is not fatal. By contrast, necrophagous species that opportunistically initiate facultative myiasis have not adopted a parasitic strategy for progeny production. Such species are adapted for rapid utilization of decomposing tissue in an intensely competitive microhabitat, where the focus is on overcoming the competition not on maintenance of the resource.
- In accidental myiasis, passage of fly larvae through the digestive system may evoke various gastric problems, including nausea, diarrhea, bloating, and pain, but rarely severe acute or long-term symptoms.
- With facultative myiasis, invasion of necrotic tissue is generally benign with regard to the host condition. However, as the number of individuals in a wound increase, driving feeding activity into healthy tissue, the severity of myiasis escalates.
- Myiasis has been reported for almost any area of the body, while with any form of facultative myiasis, primary myiasis can lead to severe host pathology due to feeding activity and the action of secondary invaders.
- Vertebrate hosts are not passive to myiasis. Fly larvae trigger both non-specific and specific body defenses. The nature and strength of the host defense response is dependent on the health of the host, location of fly invasion, and type of antigen encountered.
- Fly larvae possess an arsenal of weapons to combat the defenses mounted by the host during myiasis, ranging from salivary digestive enzymes, immunoreactive proteins, to components in excreta. The defenses of fly larvae effectively degrade immunoglobulins, block activation of cells involved in key aspects of antigenic responses, and indiscriminately kill macrophages, leukocytes, and T lymphocytes.

Test your understanding

Level 1: knowledge/comprehension

1. Define the following terms:

   (a) facultative myiasis
   (b) necrophagy
   (c) sanguinivorous
   (d) gregarious reproduction
   (e) secondary myiasis
   (f) ecological classification
   (g) synanthropic
   (h) non-specific host defenses.

2. Match the terms (i–vi) with the descriptions (a–f).

(a) Larval entry into the anus after passing through digestive tract	(i) Immunoglobulins
(b) Use of fly larvae to debride a wound	(ii) Parasitoid
(c) Accidental entry of maggots into skin lesion	(iii) Nosocomial
(d) Larval infestation occurring in a hospital	(iv) Retroinvasion
(e) Macrophage or leukocyte response to myiasis	(v) Pseudomyiasis
(f) Insect parasite that typically kills its host	(vi) Non-specific defenses
(g) Proteins produced in response to antigens	(vii) Maggot therapy

3. Describe the non-specific defenses used by a human to combat either facultative or obligate myiasis. Which form of the diseases is more effectively regulated by host defenses?

4. Explain how the location of myiasis on the host is related to the severity of the disease.

5. Describe the differences in ecological versus anatomical classification of myiasis.

6. What morphological adaptations are associated with larvae of obligatory myiasis? Facultative myiasis?

7. Would any unique behavioral or anatomical features be expected with larval of pseuodomyiatic species? Explain your answer.

Level 2: application/analysis

1. The presence of necrophagous flies may be an indicator of neglect but it does not mean that myiasis is involved. Provide examples in which non-myiasis cases associated with flies can be linked with neglect or abuse.

2. Explain how "environmental conditions" of a home can be a predictive indicator of facultative myiasis.

3. Compare and contrast natural myiasis versus that associated with negligence.

4. Larval salivary enzymes are capable of suppressing host defenses, but not to the point of yielding complete protection. What are possible explanations as to why digestive enzymes alone do not provide 100% protection from host body defenses?

Level 3: synthesis/evaluation

1. Speculate on how antigenic responses in humans, pets or livestock can possibly be used to determine whether facultative myiasis is the result of neglect by a primary caregiver.

2. Explain whether it is possible for an obligate parasite like *Cochliomyia hominivorax* to induce secondary or tertiary myiasis.

Notes

1. From James (1947).
2. Necrotic tissue refers to cells that are dead and undergoing necrosis, the physiological changes that occur after cell death.
3. An organism, in this case a fly, that feeds on decaying plant or animal matter.
4. Primary myiasis is a form of facultative myiasis induced by fly species that directly infest wounds or lesions (Kettle, 1995).
5. Gregariousness is a term frequently used in reference to clutch sizes greater than one individual (solitary), usually implying that "many" eggs or larvae per batch are deposited in the same location.
6. Calyptrate refers to dipterans that possess calypters, membranous flaps below the hind wings that typically cover the halteres.

7. As discussed in Chapter 9, endothermic animals regulate their internal body temperature, with the set point frequently maintained above ambient temperatures.

8. Algor mortis is the decline in body temperature that occurs after death until reaching equilibrium with ambient conditions.

9. Kairomones are chemical signals that benefit the receiver but generally harm the originator of the message.

10. It is common for bacterial infections to occur in areas of primary myiasis, in which case the suffix "itis" implies a condition involving bacterial invasion of a particular tissue. Hence, sinusitis is a term referring to bacterial infection of the sinuses, in this case facilitated by primary myiasis.

References cited

Aak, A., Knudsen, G.K. & Soleng, A. (2010) Wind tunnel behavioural response and field trapping of the blowfly *Calliphora vicina*. *Medical and Veterinary Entomology* 24: 250–257.

Anderson, G.S. & Huitson, N.R. (2004) Myiasis in pet animals in British Columbia: the potential of forensic entomology for determining duration of possible neglect. *Canadian Veterinarian Journal* 45: 993–998.

Batista-da-Silva, J.A., Moya-Borja, G.E. & Queiroz, M.M.C. (2011) Factors of susceptibility of human myiasis caused by the New World screwworm, *Cochliomyia hominivorax* in Sao Goncalo, Rio de Janeiro, Brazil. *Journal of Insect Science* 11: 1–7.

Benecke, M. & Lessig, R. (2001) Child neglect and forensic entomology. *Forensic Science International* 120: 155–159.

Benecke, M., Josephi, E. & Zweihoff, R. (2004) Neglect of the elderly: forensic entomology cases and considerations. *Forensic Science International* 146 (Suppl): S195–S199.

Bowles, V.M., Grey, S.T. & Brandon, M.R. (1992) Cellular immune responses in the skin of sheep infected with larvae of *Lucilia cuprina*, the sheep blowfly. *Veterinary Parasitology* 44: 151–162.

Bowles, V.M., Meeusen, E.M., Chandler, K., Verhagen, A., Nash, A.D. & Brandon, M.R. (1994) The immune response of sheep infected with larvae of the sheep blowfly *Lucilia cuprina* monitored via efferent lymph. *Veterinary Immunology and Immunopathology* 40: 341–352.

Brosnan, J.T. & Brosnan, M.E. (2006) The sulfur-containing amino acids: an overview. *Journal of Nutrition* 136: 1636S–1640S.

Catts, E.P. & Mullens, G.R. (2002) Myiasis (Muscoidea, Oestroidea). In: G. Mullen & L. Durden (eds) *Medical and Veterinary Entomology*, pp. 318–348. Academic Press, San Diego, CA.

Centers for Disease Control and Prevention (2011) *National Diabetes Fact Sheet: National Estimates and General Information on Diabetes and Prediabetes in the United States, 2011*. U.S. Department of Health and Human Services, Centers for Disease Control and Prevention, Atlanta, GA.

Cestari, T.F., Pessato, S. & Ramos-e-Silva, M. (2007) Tungiasis and myiasis. *Clinics in Dermatology* 25: 158–164.

Chaudhury, M.F., Welch, J.B. & Alvarez, L.A. (2002) Responses of fertile and sterile screwworm (Diptera: Calliphoridae) flies to bovine blood inoculated with bacteria originating from screwworm-infested wounds. *Journal of Medical Entomology* 39: 130–134.

Chaudhury, M.F., Skoda, S.R., Sagel, A. & Welch, J.B. (2010) Volatiles emitted from eight wound-isolated bacteria differentially attract gravid screwworms (Diptera: Calliphoridae) to oviposit. *Journal of Medical Entomology* 47: 349–354.

Chaudhury, M.F., Zhu, J.J. & Skoda, S.R. (2017) Physical and physiological factors influence behavioral responses of *Cochliomyia macellaria* (Diptera: Calliphoridae) to synthetic attractants. *Journal of Economic Entomology* 110: 1929–1934.

DeVaney, J.A., Eddy, G.W., Ellis, E.M. & Harrington, H. (1973) Attractancy of inoculated and incubated bovine blood fractions to screwworm flies (Diptera: Calliphoridae): role of bacteria. *Journal of Medical Entomology* 10: 591–595.

Duro, E.A., Mariluis, J.C. & Mulieri, P.R. (2007) Umbilical myiasis in a human newborn. *Journal of Perinatology* 27: 250–251.

Eddy, G.W., DeVaney, J.A. & Handke, B.D. (1975) Response of the adult screwworm (Diptera: Calliphoridae) to bacteria-inoculated and incubated bovine blood in olfactometer and oviposition tests. *Journal of Medical Entomology* 12: 379–381.

Elkington, R.A., Humphries, M., Commins, M., Maugeri, N., Tierney, T. & Mahony, T.J. (2009) A *Lucilia cuprina* excretory-secretory protein inhibits the early phase of lymphocyte activation and subsequent proliferation. *Parasite Immunology* 31: 750–765.

Evans, W.E.D. (1963) *The Chemistry of Death*. Thomas Publishers, Springfield, IL.

Francesconi, F. & Lupi, O. (2012) Myiasis. *Clinical Microbiology Reviews* 25: 79–105.

Godfray, H.C.J. (1994) *Parasitoids: Behavioral and Evolutionary Ecology*. Princeton University Press, Princeton, NJ.

Goff, M.L., Campobasso, C.P. & Gheraldi, M. (2010) Forensic implications of myiasis. In: J. Amendt, C.P. Campobasso, M.L. Goff & M. Grassberger (eds) *Current Concepts in Forensic Entomology*, pp. 313–326. Springer, London.

Gomperts, B.D., Kramer, I.M. & Tatham, P.E.R. (2002) *Signal Transduction*. Academic Press, San Diego, CA.

Guerrini, V.H. (1997) Excretion of ammonia by *Lucilia cuprina* larvae suppresses immunity in sheep. *Veterinary Immunology and Immunopathology* 56: 311–317.

Hall, M. & Wall, R. (1995) Myiasis of humans and domestic animals. *Advances in Parasitology* 35: 257–334.

Hammack, L. (1991) Oviposition by screwworm flies (Diptera: Calliphoridae) on contact with host fluids. *Journal of Economic Entomology* 84: 185–190.

Hira, P.R., Assad, R.M., Okasha, G., Al-Ali, F.M., Iqbal, J., Mutawali, K.E.H., Disney, R.H. & Hall, M.J.R. (2004) Myiasis in Kuwait: nosocomial infections caused by *Lucilia sericata* and *Megaselia scalaris. American Journal of Tropical Medicine and Hygiene* 70: 386–389.

James, M.T. (1947) *Flies that Cause Myiasis in Man* USDA Miscellaneous Publication 631. United States Department of Agriculture, Washington, DC.

Kerlin, R.L. & East, I.J. (1992) Potent immunosuppression by secretory/excretory products of larvae from the sheep blowfly *Lucilia cuprina. Parasite Immunology* 14: 595–604.

Kettle, D.S. (1995) *Medical and Veterinary Entomology.* Oxford University Press, Oxford.

Kharroubi, I., Ladriere, L., Cardozo, A.K., Dogusan, Z., Cnop, M. & Eizirik, D.L. (2004) Free fatty acids and cytokines induce pancreatic beta-cell apoptosis by different mechanisms: role of nuclear factor-kappaB and endoplasmic reticulum stress. *Endocrinology* 145: 5087–5096.

Mackley, J.W. & Brown, H.E. (1984) Swormlure-4, a new formulation of the Swormlure-2 mixture as an attractant for adult screwworms *Cochliomyia hominivorax* (Diptera: Calliphoridae). *Journal of Economic Entomology* 77: 1264–1268.

Martini, F.H., Nath, J.L. & Bartholomew, E.F. (2011) *Fundamentals of Anatomy and Physiology*, 9th edn. Benjamin Cummings, San Francisco, CA.

Milillo, P., Traversa, D., Elia, G. & Otranto, D. (2010) Analysis of somatic and salivary gland antigens of third stage larvae of *Rhinoestrus* spp. (Diptera: Oestridae). *Experimental Parasitology* 124: 361–364.

Norris, K.R. (1965) The bionomics of blowflies. *Annual Review of Entomology* 10: 47–68.

Otranto, D. (2001) The immunology of myiasis: parasite survival and host defense strategies. *Trends in Parasitology* 17: 176–182.

Quicke, D.L. (1997) *Parasitic Wasps.* Springer, London.

Randall, D., Burggren, W. & French, K. (2002) *Animal Physiology: Mechanisms and Adaptations.* W.H. Freeman and Company, New York.

Ravasan, N.M., Shayeghi, M., Najibi, B. & Oshaghi, M.A. (2012) Infantile nosocomial myiasis in Iran. *Journal of Arthropod-Borne Diseases* 6: 156–163.

Roback, S.S. (1951) A classification of the muscoid calyptrate Diptera. *Annals of the Entomological Society of America* 44: 327–361.

Sandeman, R.M. (1996) Immune responses to mosquitoes and flies. In: S.K. Wikel (ed.) *The Immunology of Host–Ectoparasitic Arthropod Relationship*, pp. 175–203. CAB International, Wallingford, UK.

Sandeman, R.M., Chandler, R.A., Turner, N. & Seaton, D.S. (1995) Antibody degradation in wound exudates from blowfly infections on sheep. *International Journal of Parasitology* 25: 621–628.

Sherman, R. (2000) Wound myiasis in urban and suburban United States. *Archives of Internal Medicine* 160: 2004–2014.

Sherman, R.A. (2009) Maggot therapy takes us back to the future of wound care: new and improved maggot therapy for the 21st Century. *Journal of Diabetes Science and Technology* 3: 336–344.

Sherman, R.A., Hall, M.J.R. & Thomas, S. (2000) Medicinal maggots: an ancient remedy for some contemporary afflictions. *Annual Review of Entomology* 45: 55–81.

Stevens, J.R. & Wallman, J.F. (2006) The evolution of myiasis in humans and other animals in the Old and New Worlds (part I): phylogenetic analyses. *Trends in Parasitology* 22: 129–136.

Stevens, J.R., Wallman, J.F., Otranto, D., Wall, R. & Pape, T. (2006) The evolution of myiasis in humans and other animals in the Old and New Worlds (part II): biological and life-history studies. *Trends in Parasitology* 22: 181–188.

Sukontason, K.L., Sukontason, K., Narongchai, P., Lerttthamnongtham, S., Piangjai, S. & Olson, J.K. (2001) *Chrysomya rufifacies* (Macquart) as a forensically-important fly species in Thailand: a case report. *Journal of Vector Ecology* 26: 162–164.

Szakacs, T.A., Sinclair, B.J., Gill, B.D. & McCarthy, A.E. (2007) Nosocomial myiasis in a Canadian intensive care unit. *Canadian Medical Association Journal* 177: 719–720.

Tabouret, G., Lacroux, C., Andreoletti, O., Bergeaud, J.P., Hailu-Tolosa, Y., Hoste, H., Prevot, F., Grisez, C., Dorchies, P. & Jacquiet, P. (2003) Cellular and humoral local immune responses in sheep experimentally infected with *Oestrus ovis* (Diptera: Oestridae). *Veterinary Research* 34: 231–241.

Thomas, D.B. & Pruett, J.H. (1992) Kinetic development and decline of antiscrewworm (Diptera: Calliphoridae) antibodies in serum of infested sheep. *Journal of Medical Entomology* 29: 870–873.

Thompson, C.R., Brogan, R.S. & Rivers, D.B. (2013) Bacterial interactions with necrophagous flies. *Annals of the Entomological Society of America* 106: 799–809.

Urech, R., Green, P.E., Rice, M.J., Brown, G.W., Duncalfe, F. & Webb, P. (2004) Composition of chemical attractants affects trap catches of the Australian sheep blowfly, *Lucilia cuprina*, and other blowflies. *Journal of Chemical Ecology* 30: 851–866.

Vanin, S., Bonizzoli, M., Migliaccio, M. L., Buoninsegni, L. T., Bugelli, V., Pinchi, V. & Focardi, M. (2017) A case of insect colonization before the death. *Journal of Forensic Sciences* 62(6): 1665–1667.

Vass, A.A., Barshick, S.-A., Sega, G., Caton, J., Skeen, J.T., Love, J.C. & Synstelien, J.A. (2002) Decomposition chemistry of human remains: a new methodology for determining the postmortem interval. *Journal of Forensic Sciences* 47: 542–553.

Zumpt, F. (1965) *Myiasis in Man and Animals on the Old World.* Butterworths, London.

Supplemental reading

Benecke, M. (2010) Cases of neglect involving entomological evidence. In: J.H. Byrd & J.L. Castner (eds) *Forensic Entomology: The Utility of Arthropods to Legal Investigations*, 2nd edn, pp. 627–636. CRC Press, Boca Raton, FL.

Dogra, S.S. & Mahajan, V.K. (2010) Oral myiasis caused by *Musca domestica* larvae in a child. *International Journal of Pediatric Otorhinolaryngology Extra* 5: 105–107.

Goff, M.L., Charbonneau, S. & Sullivan, W. (1991) Presence of fecal material in diapers as a potential source of error in estimations of postmortem interval using arthropod developmental rates. *Journal of Forensic Sciences* 36: 1603–1606.

Gunn, A. (2009) *Essential Forensic Biology*, 2nd edn. John Wiley & Sons Ltd., Chichester, UK.

Huntington, T.E., Voigt, D.W. & Higley, L.G. (2008) Not the usual suspects in human wound myiasis by phorids. *Journal of Medical Entomology* 45: 157–159.

Robbins, K. & Khachemoune, A. (2010) Cutaneous myiasis: a review of the common types of myiasis. *International Journal of Dermatology* 49: 1092–1098.

Steenvoorde, P. & Jukema, G.N. (2004) The antimicrobial activity of maggots: in-vivo results. *Journal of Tissue Viability* 14: 97–101.

Wall, R., Rose, H., Ellse, L. & Morgan, E. (2011) Livestock ectoparasites: integrated management in a changing climate. *Veterinary Parasitology* 180: 82–89.

Williams, R.E. (2009) *Veterinary Entomology: Livestock and Companion Animals*. CRC Press, Boca Raton, FL.

Additional resources

Centers for Disease Control and Prevention, Myiasis: http://www.cdc.gov/parasites/myiasis/

Cochliomyia hominivorax: https://www.cabi.org/isc/datasheet/11753

Department of Medical Entomology, University of Sydney: http://medent.usyd.edu.au/

Journal of Medical Entomology: http://www.entsoc.org/Pubs/Periodicals/JME/

Medical Entomology Centre: http://www.insectresearch.com/home.htm

Myiasis thrash metal band: http://www.myspace.com/myiasis

National Center on Elder Abuse: http://www.ncea.aoa.gov/ncearoot/Main_Site/index.aspx

New World screwworm: https://www.aphis.usda.gov/aphis/ourfocus/animalhealth/animal-disease-information/cattle-disease-information/nws/new-world-screwworm

Screwworms as agents of myiasis: http://www.fao.org/ag/aga/agap/frg/feedback/war/u4220b/u4220b07.htm

United States Air Force Medical Entomology: http://www.afpmb.org/content/united-states-air-force-medical-entomology

What is child neglect and abuse? http://www.childwelfare.gov/pubs/factsheets/whatiscan.cfm

Chapter 17
Wildlife forensics

The United States views the poaching and trafficking of protected wildlife as a threat to good governance, a threat to the rule of law, and a challenge to our stewardship, responsibilities for this good earth.

Attorney General Jeff Sessions, Statement on Behalf
of the United States at the London Illegal
Wildlife Trade Conference, October 11, 2018

Overview

Hunting animals for sport and food has been part of human life from the dawn of time. Hunting animals today involves following federal, state, and local laws governing the number of individuals and time of year that a particular species can be killed or captured. These laws are meant to allow hunting while also regulating the population levels of the wildlife species involved. Some animals cannot be hunted at all. They are protected by laws to preserve endangered or threatened species. When a protected species is killed, or when hunting seasons or bag limits are ignored, criminal and civil consequences can occur. The time, location, and manner of death (or acquisition) of the animal can be key to prosecution of these crimes. Insects can play an important role in

helping us understand what happened and when it happened.

The big picture

- Hunting is governed by laws on what species can be killed, how many can be killed, and when they can be killed.
- Necrophagous insects that feed on and breed in wildlife show similarities to the fauna found on pigs and humans, but there are differences.
- The same methods used to determine minimum PMI for a human case can be used for animals, but the growth rate might be different.
- Insects may be able show evidence of poisons that lead to death of domestic animals and wildlife.
- Insects can provide evidence for crimes involving domestic animals and pets.

The Science of Forensic Entomology, Second Edition. David B. Rivers and Gregory A. Dahlem.
© 2023 John Wiley & Sons Ltd. Published 2023 by John Wiley & Sons Ltd.
Companian website: www.wiley.com/go/forensicentomology2

17.1 Hunting is governed by laws on what species can be killed, how many can be killed, and when they can be killed

Laws governing hunting in the United States vary by state. Generally, someone interested in hunting wildlife must obtain a hunting license for the state where the hunt occurs and agree to comply with all requirements associated with that license. Each state has different types of animals that can be legally hunted. Hunting waterfowl requires a signed federal Migratory Bird Hunting Stamp (Duck Stamp). Depending on the state, a person may be able to kill pest animals on their own property without a license (think rats, moles, pigeons, coyotes), but a person hunting on public land will usually need a hunting license to legally shoot just about any kind of mammal or bird. Large game, like bear or elk, may require a separate permit with much higher restrictions on the number of animals killed and strict date restrictions for when hunting is allowed.

If an individual is caught hunting without a license or breaking the state's hunting laws, there are usually stiff penalties (Figure 17.1). The consequences of illegal hunting vary by area, but they typically involve a fine and revocation of hunting licenses for a period of time, depending on the severity of the offense. Typical fines for illegal hunting can be in the hundreds to thousands of dollars. In the most egregious cases, jail time and forfeiture of firearms may be imposed.

Intentionally killing or harming an endangered species can result in much tougher consequences for those involved. The Endangered Species Act prohibits the import, export, or taking of fish and wildlife that are listed as threatened or endangered species. The term "take" means to harass, harm, pursue, hunt, shoot, wound, kill, trap, capture, or collect, or to attempt to engage in any such conduct (from Definitions in Section 3 of the Endangered Species Act of 1973). The criminal penalties for taking an endangered species in the United States can involve prison time and hefty fines in the tens of thousands of dollars. Exotic pet laws vary by state, but even the more lenient states may require special permits for owning animals like tigers or bears. International trade in species listed by CITES (the Convention on International Trade in Endangered Species) is illegal unless authorized by permit. Exceptions to this rule can occur for institutions like zoos and museums if they can demonstrate that the importation will enhance the propagation or survival of the species.

17.2 Necrophagous insects that feed on and breed in wildlife show similarities to the fauna found on pigs and humans, but there are differences

Most insects are very selective in their food choice for their offspring. When we think of the tight association between herbivorous insects and the plants that they feed on, we can classify them into groups that reflect how selective they are. **Monophagous** herbivores feed on one or a few members of a single genus of plant (e.g., gall making insects or the pitcher plant mining moth, *Exyra semicrocea*). **Oligophagous** herbivores are a bit less selective but only feed on select genera of plants within a single family (e.g., the Colorado potato beetle, *Leptinotarsa decemlineata*, feeds on various genera and species of Solanaceae). **Polyphagous** herbivores can feed on a wide variety of different plant species (e.g., migratory locusts, *Locusta migratoria*, or the spongy moth, *Lymantria dispar*). While these same terms are not generally used, or

Figure 17.1   Local, state, and federal laws govern legal hunting of wildlife. *Source*: MarkBuckawicki/Wikimedia Commons/CC0.

used in the same way, when talking about carnivorous insects, parasitoids, or necrophagous insects, the general idea of degrees of specialization in food selection holds true for all insects. This includes the insects associated with carrion.

Unfortunately, very few investigations into the faunal succession of insects on wildlife carrion have been undertaken and published on. But the few published studies offer some interesting insights. Watson & Carlton (2003, 2005a) looked at three species of wildlife that are often involved in cases of illegal wildlife deaths in the state of Louisiana. They observed the fauna and successional patterns of insects on black bear, white-tailed deer, and American alligator during spring, fall, and winter (Figure 17.2). They also set out pig carcasses at the same time for comparison. They collected 93 arthropod species from 46 families on the seven carcasses they set out in spring, but only 19 insect species from eight families occurred on all four types of animals. The biggest difference that they saw was a reduction in the diversity of necrophilous insects on alligator where 11 of the total 46 families were not collected. A similar pattern of reduced diversity of taxa associated with alligator carcasses occurred in their fall and winter successional studies. A different investigation (Richards *et al.*, 2015) provides a look at the carrion communities associated with wildlife found in a salt marsh habitat in Florida, but at a different level. This research is based on observations of single animals at different times of the year, and it shows similarities between the wildlife carrion communities of Louisiana and Florida. Richards *et al.* (2015) provide information on the carrion community associated with coyote, opossum, raccoon, bobcat, and otter.

What about the 19 insect species found on all animal types by Watson & Carlton (2003) in spring? These would be comparable to the polyphagous herbivores since they feed on a variety of very different taxa of animals. They found several species in families well known for their close association with vertebrate carrion at all of the carcasses. Beetles that feed on fly larvae (e.g., Histeridae, Silphidae, and Staphylinidae) and a variety of flies (e.g., Calliphoridae, Muscidae, Piophilidae, and Sepsidae) were the most significant. Some of these species have been well studied for their use in PMI calculations, while others have not. Published life tables and detailed rearing studies are available for many of the Calliphoridae. Some work has been done on the development of carrion beetles (Silphidae) (Frątczak & Matuszewski, 2016; Watson & Carlton, 2005b) but there are a variety of other taxa that deserve additional attention. These include species like *Hydrotaea leucostoma* (Muscidae) and *Stearibia nigriceps* (Piophilidae).

What about all the species that have only been found at one type of animal? Are they diagnostic of a particular type of animal carrion? It does not appear that there are insects comparable to monophagous herbivores in these carrion communities. A variety of species are documented by Watson & Carlton (2003, 2005a) and Richards *et al.* (2015) as being present on only one type of animal. But many (most?) of these appear to be adventitious species that are using the carcass as a piece of physical environment rather than having any type of attraction to one type of carrion over another. There are some necrophagous species that show a distinct preference to one taxonomic group over others (e.g., mammals instead of alligators) and this should be acknowledged. However, there is not

Figure 17.2 Deer and bear are two common animals involved in wildlife poaching investigations. *Source*: Marshal Hedin/Wikimedia Commons/CC0. Mykola Swarnyk/Wikimedia Commons.

enough current evidence to show the kind of close species-to-species association that we see in symbiotic relationships, like the close associations between host and parasite (e.g., the beaver beetle, *Platypsyllus castoris*, and beavers).

What about the blow flies? Do the species of Calliphoridae found in wildlife differ from the species commonly used in forensic investigations? The species of Calliphoridae found on these various types of animals are what we would expect to find on pig or human corpses in a similar location at a similar time of year. Some of the most important and abundant species found by Watson & Carlton (2003, 2005a) include *Calliphora vicina*, *Chrysomya rufifacies*, *Cochliomyia macellaria*, *Lucilia coeruleiviridis*, *L. sericata*, and *Phormia regina*. Many of these same species also show up as important in Richards *et al.* (2015) experiments in Florida. These observations support the use of blow flies as a primary choice for determining minimum PMI in wildlife investigations during early stages of decomposition.

17.3 The same methods used to determine minimum PMI for a human case can be used for animals, but the growth rate might be a little different

Most of our information on the behavior and growth rates of calliphorid species for use in human casework comes from rearing flies on non-human organs and tissues. This is because necrophagous insects can be successfully reared on a wide variety of different kinds of animals and animal tissue is much easier to obtain than human. Pigs are commonly used as models for successionary studies in the field. Side-by-side comparisons of pig and human show negligible differences (Schoenly *et al.*, 2007) in the composition of their carrion communities over time. Beef or pork liver is often used as a rearing media for laboratory studies. Decaying chicken is often used as a bait when surveying the carrion fly fauna of a particular area. The volatile organic compounds (VOCs) released during early decomposition of chicken influences the types of flies, including a variety of familiar blow fly species, that are attracted to the decaying

tissues (Recinos-Aguilar *et al.*, 2020). In a study of low cost and easily accessible alternative food sources, Weidner *et al.* (2020) found that canned tuna or cat food was comparable to beef liver for successful rearing of blow flies. An investigation on the flies that colonize chicken and fish carrion found that these very different animals attract the same general necrophagous insect communities that we would expect to see on pigs or humans (Bujang *et al.*, 2020). Obviously, the common species of Calliphoridae associated with vertebrate carrion are not too picky about their food.

We know that the insect communities associated with vertebrate carrion (usually pig) vary with the season of the year (e.g., Benbow *et al.*, 2013 and Chapter 11). The wildlife carrion studies mentioned in the previous section also show distinct differences in the Calliphoridae species composition and abundance in spring versus fall and winter. Other studies have shown that the rate of growth of blow fly species can be significantly different when reared on different tissues of the same animal (e.g., muscle versus lung, liver, or heart) (Clark *et al.*, 2006; Boatright & Tomberlin, 2010). When we look at growth rates of various blow fly larvae on muscle tissue of different kinds of animals, we also find some differences and similarities. Boatright & Tomberlin (2010) found growth rates of *Cochliomyia macellaria* to be similar when raised on horse and pig muscle. Clark *et al.* (2006) found that larvae of *Lucilia sericata* grew significantly faster when reared on pig tissue as compared to cow. In a study directly related to wildlife issues, Wilson *et al.* (2014) found that larvae of *Phormia regina* grew significantly faster on a diet of venison as compared to pig (Figure 17.3). Further investigations on the effect of diet (host species and tissue type) would be very useful for minimum PMI estimations used as evidence in wildlife crimes.

One of the most famous cases using minimum PMI estimations based on entomological evidence resulted in the conviction of two suspects for the death of two black bear cubs in Manitoba, Canada (Anderson, 1999). The estimation of the time of death of the two cubs was noted to be "most compelling" by the presiding judge. The only insects collected from the carcasses were blow fly eggs and adults. The eggs were closely observed to determine their hatching time and the larvae were subsequently reared to determine which species were involved. Three species of Calliphoridae were reared from the disemboweled bear cubs (*Lucilia illustris*, *L. sericata*, and *P. regina*). The determination of minimum PMI

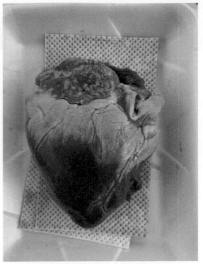

Figure 17.3   Fly larvae may mature at slightly different rates on different animals or on different tissues, like pig heart and cow liver. *Source*: Courtesy of G.A. Dahlem.

successfully linked the suspects to the scene of the crime. It should be noted that the evidence was especially compelling because it involved the timing of eggs to first-instar larvae, where evidence shows that this will not differ much from one type of animal host to another. Wilson *et al.* (2014) points out that if the evidence was third-instar larvae, the minimum PMI would have been much more difficult to estimate due to the larvae feeding on bear tissue rather than pig.

17.4   Insects may be able show evidence of poisons that lead to death of domestic animals and wildlife

Poisoning of wildlife and domestic animals is a major problem in the United States and around the world. Such poisonings can be intentional or accidental and the laws pertaining to killing wildlife or domestic animals with poisons are often different from one state to another (Figure 17.4). If nationally protected species are involved, then federal legislation comes into force. Poisoning can be direct when an animal comes in contact with or consumes a poison or poisoned bait and others can be indirect when it involves animals feeding on the remains of a poisoned animal.

In a study of pesticide poisoning in wildlife and domestic animals in Portugal (Grilo *et al.*, 2021), toxicology results showed that molluscicides and carbamates were the most common pesticides detected, followed by rodenticides. **Molluscicides** are pesticides that are often used in agriculture to control unwanted snails and slugs. The **carbamates** are a group of insecticides derived from carbamic acids. Carbamates, such as carbamyl, methomyl, and carbofuran, are commonly used in agriculture for legal control of damaging insects. **Rodenticides** are pesticides that are commonly used to control rat and mice pest problems. These chemicals have approved uses but are sometimes used in illegal attempts to poison unwanted animals. They are often secondarily associated with the death of pets, birds, domestic, and wild animals. In the United States and Canada, it is not legal to use these sorts of pesticides to kill coyotes, but some people break this law. This can lead to the death of a wide range of non-target species including eagles, foxes, skunks, etc. (Wobeser *et al.*, 2004).

You will find that Chapter 18 of this book is devoted to the topic of forensic entomotoxicology, which deals with the detection and analysis of chemical compounds present in arthropods that fed on a contaminated corpse. This avenue of forensic investigation may provide answers to questions about the cause of death rather than (or in addition to) potential time of death. While insects have not been used in this way for many wildlife death investigations, they do show great potential.

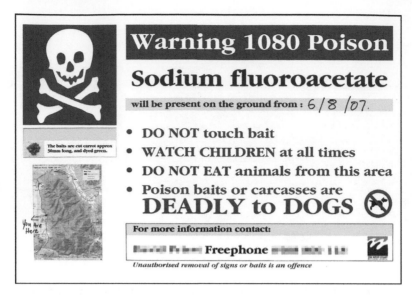

Figure 17.4 Sign warning of the presence of 1080 poison (sodium fluoroacetate) near Lake Kaniere on the West Coast of New Zealand. *Source*: I, Greg O'Beirne/Wikimedia Commons/CC BY-SA 3.0.

One question that comes up from studies of necrophagous insects that feed on poisoned baits or poisoned wildlife is: Will the poison kill the insects or stop them from colonizing such resources? This especially holds true when we are talking about many of these poisons being used as insecticides, chemicals particularly developed to kill insects. A variety of studies have looked the effects of toxic chemicals on survivorship and development of several species of blow flies. It appears that calliphorid larvae are pretty tough and can survive on carrion that has been contaminated with insecticides like malathion, methyl parathion, carbofuran, and even alkaloids like strychnine (Green *et al.*, 2002, Liu *et al.*, 2009; Aubernon *et al.*, 2015; Esquivel *et al.*, 2015).

Another question that arises from poisoned carrion situations is whether or not the necrophagous insects themselves can serve as secondary causes of poisoning. A variety of wildlife may visit carcasses to feed on the insects involved with the carcass rather than the carcass itself (e.g., insectivorous birds). An example of this deals with the transfer of toxic methylmercury from contaminated fish to developing blow flies (Nuorteva & Häsänen, 1972; Sarica *et al.*, 2005). In this case it appears that feeding on fly larvae could cause mercury poisoning to beetles, juvenile salmon, spiders, and birds, but predators of the adult calliphorids are less likely to show serious harm (Sarica *et al.*, 2005). The larvae and pupae contain high levels of mercury but most of this is eliminated by the newly emerged adults when they leave their puparia.

17.5   Insects can provide evidence for crimes involving domestic animals and pets

Domestic animals and pets can be criminally abused and neglected by their owners. It may be possible to use the same sorts of techniques described above to provide evidence against owners in animal cruelty and neglect cases (Brundage & Byrd, 2016). For example, in a case where a farm was sold and the owners abandoned the livestock, leaving them to starve, insect evidence was able to provide information on the date of death of the animals (Defilippo *et al.*, 2015).

In the case of neglect, myiasis can occur when a wound is left untreated and/or the fur is matted or covered in excrement (Figure 17.5). Chapter 16 of this book contains a summary of the topic of necrophagous and parasitic insects as indicators of neglect and abuse. The results of a survey of veterinarians in British Columbia highlights how common myiasis can be with animals brought in for care, with dogs as the species most involved (Anderson & Huitson, 2004). In many cases, the myiasis is more a result of owner

Figure 17.5 Out of all the common house pets, dogs show myiasis most often. *Source*: Courtesy of G.A. Dahlem.

ignorance than deliberate abuse, but insects can be used to provide evidence against those owners who willfully and knowingly abuse their animals.

Chapter review

Necrophagous insects that feed on and breed in wildlife show similarities to the fauna found on pigs and humans, but there are differences

- Most insects are very selective in their choice of food, especially for their offspring.
- The terms monophagous, oligophagous, and polyphagous are commonly used to indicate the degree of selectivity of herbivorous insects. These terms are not generally used for carnivorous, parasitic, or necrophagous insects but the idea of different levels of food selectivity can be applied to the carrion community associated with wildlife.
- Successional studies on wildlife show similarities and differences in the insects associated with different kinds of animals.
- There is currently no good evidence for necrophagous insect specialization with a single species of vertebrate animal.
- The common fly and beetle faunas (e.g., Calliphoridae, Silphidae, etc.) of pig carrion studies are also found feeding on wildlife carcasses. The forensic entomologist working on a wildlife case should consider these as good choices for answering questions relating to the death of the animal under investigation.

The same methods used to determine minimum PMI for a human case can be used for animals, but the growth rate might be a little different

- Most of our information on the behavior and growth rates of blow fly species for use in human casework comes from rearing flies on non-human organs and tissues.
- Side-by-side comparisons of pig and human carrion succession and fly development times show negligible differences, making pigs a preferred model for human forensic studies.
- Blow flies successfully develop on a wide variety of different types of animals and animal tissues.
- Developmental times of maggots fed on various organ tissues (e.g., heart, lung, liver) may be different from times reported on maggots fed on muscle tissue.
- Some differences have been documented on developmental differences of calliphorids fed on one type of animal versus another. One study showed that larvae of the common blow fly *Phormia regina* grew significantly faster on a diet of venison as compared to pig.
- Determination of minimum PMI in a wildlife case should take these differences into account.

Insects may be able to show evidence of poisons that lead to death of domestic animals and wildlife

- Accidental and deliberate poisoning of domestic animals and wildlife are a major cause of concern around the world. Such poisoning can be direct when an animal comes in contact with or consumes poisoned materials. Poisoning can also be indirect when an animal consumes poison by eating an animal that was directly poisoned (e.g., a fox eating a dead rat that died from eating rat poison).
- Principles of forensic entomotoxicology (covered in Chapter 18) may provide answers to questions about the cause of death rather than (or in addition to) potential time of death for wildlife investigations.
- Many of the common poisons involved in wildlife and domestic animal deaths are insecticides.

Insecticides are poisons developed to kill insects, but blow flies are remarkably resistant to many of these highly toxic chemicals and can successfully develop in poisoned animal carcasses.

- It is possible that maggots feeding on a poisoned carcass could sequester enough poison in their bodies to secondarily poison insectivorous animals that feed on them rather than the carrion itself (e.g., a wide range of wild birds).

Test your understanding

Level 1: knowledge/comprehension

1. Explain how the ideas behind the terms used for herbivore feeding selectivity (monophagous, polyphagous and oligophagous) can be applied to necrophagous insects feeding on dead animals.
2. What is the difference between indirect and direct poisoning of an animal like a fox.
3. What are the names of three major types of pesticides and what are they used against?

Level 2: application/analysis

1. How would you set up an experiment to see if a common blow fly, like *Phormia regina*, grows at a different rate on chicken as compared to pork?
2. How is a dead rat that has poison injected in it to make it into a poison bait for another animal different as a food source for flies than a rat that dies from eating poisoned cheese?

Level 3: synthesis/evaluation

1. There is a need for standardized and repeatable methods for forensic entomology research (see Tomberlin *et al.,* 2012 in the supplemental reading section). With that in mind, how would you set up an investigation on the faunal succession of insects on dead gray squirrels to get the most data that could be compared to other wildlife or human casework?

References cited

Anderson, G.S. (1999) Wildlife forensic entomology: determining time of death in two illegally killed black bear cubs. *Journal of Forensic Science* 44: 856–859.

Anderson, G.S. & Huitson, N.R. (2004) Myiasis in pet animals in British Columbia: the potential of forensic entomology for determining duration of possible neglect. *Canadian Veterinary Journal* 45: 993–998.

Aubernon, C., Charabidzé, D., Devigne, C., Delannoy, Y. & Gosset, D. (2015) Experimental study of *Lucilia sericata* (Diptera Calliphoridae) larval development on rat cadavers: effects of climate and chemical contamination. *Forensic Science International* 253: 125–130.

Benbow, M.E., Lewis, A.J., Tomberlin, J.K. & Pechal, J.L. (2013) Seasonal necrophagous insect community assembly during vertebrate carrion decomposition. *Journal of Medical Entomology* 50: 440–450.

Boatright, S.A. and Tomberlin, J.K. (2010) Effects of temperature and tissue type on the development of *Cochliomyia macellaria* (Diptera: Calliphoridae). *Journal of Medical Entomology* 47: 917–923.

Brundage, A. & Byrd, J.H. (2016) Forensic entomology in animal cruelty cases. *Veterinary Pathology* 53: 898–909.

Bujang, R., Alias, M.A.S., Samsudin, S.H., Razi, I.A.N.A., Zasli, N.A.A. & Arun, N. (2020) Insect diversity and decomposition pattern of chicken (*Gallus gallus*) and fish (*Oreochromis niloticus*) carrion in Hutan Sri Gading, UiTM Pahang. *ASM Science Journal* 13 Special Issue 6, 2020 for InCOMR2019: 73–81.

Clark, K., Evans, L. & Wall, R. (2006) Growth rates of the blowfly, *Lucilia sericata*, on different body tissues. *Forensic Science International* 156: 145–149.

Defilippo, F., Rubini, S., Dottori, M. & Bonilauri, P. (2015) The use of forensic entomology in legal veterinary medicine: a case study in the north of Italy. *Journal of Forensic Science & Criminology* 3(5): 501.

Esquivel, E.S., Suárez, A.E.F., Quintero, A.C., Luna, C.E.H., Hernández, R.M., Castro, V.A.R. & Martinez, H.O. (2015) Susceptibility of larvae of *Chrysomya rufifacies* (Diptera: Calliphoridae) to the insecticides methyl parathion and carbofuran, 2011. *Arthropod Management Tests* 2015: 1–2.

Frątczak, K. & Matuszewski, S. (2016) Classification of forensically-relevant larvae according to instar in a closely related species of carrion beetles (Coleoptera: Silphidae: Silphinae). *Forensic Science, Medicine, and Pathology* 12(2): 193–197.

Green, P.W.C., Simmonds, M.S.J. & Blaney, W.M. (2002) Toxicity and behavioural effects of dier-borne alkaloids on larvae of the black blowfly, *Phormia regina*. *Medical and Veterinary Entomology* 16: 157–160.

Grilo, A., Moreira, A., Carrapiço, B, Belas, A. & São Braz, B. (2021) Epidemiological study of pesticide poisoning in domestic animals and wildlife in Portugal: 2014–2020. *Frontiers of Veterinary Science* 7: 616293. https://doi.org/10.3389/fvets.2020.616293.

Liu, X., Shi, Y., Wang, H. & Zhang, R. (2009) Determination of Malthion levels and its effect on the development of *Chrysomya megacephala* (Fabricius) in South China. *Forensic Science International* 192: 14–18.

Nuorteva, P. & Häsänen, E. (1972) Transfer of mercury from fishes to sarcosaprophagous flies. *Annales Zoologici Fennici* 9: 23–27.

Recinos-Aguilar, Y.M., Garcia-Garcia, M.D., Malo, E.A., Cruz-Lopez, L., Cruz-Esteban, S. & Rojas, J.C. (2020) The succession of flies of forensic importance is influenced by volatiles organic compounds emitted during the first hours of decomposition of chicken remains. *Journal of Medical Entomology* 57: 1411–1420.

Richards, S.L., Connelly, C.R., Day, J.F., Hope, T. & Ortiz, R. (2015) Arthropods associated with carrion in a salt marsh habitat in southeastern Florida. *Florida Entomologist* 98: 613–619.

Sarica, J., Amyot, M., Bey, J. & Hare, L. (2005) Fate of mercury accumulated by blowflies feeding on fish carcasses. *Environmental Toxicology and Chemistry* 24: 526–529.

Schoenly, K.G., Haskell, N.H., Hall, R.D. & Gbur, R. (2007) Comparative performance and complementarity of four sampling methods and arthropod preference tests from human and porcine remains at the Forensic Anthropology Center in Knoxville, Tennessee. *Journal of Medical Entomology* 44: 881–894.

Watson, E.J. &nd Carlton, C.E. (2003) Spring succession of necrophilous insects on wildlife carcasses in Louisiana. *Journal of Medical Entomology* 40: 338–347.

Watson, E.J. & Carlton, C.E. (2005a) Insect succession and decomposition of wildlife carcasses during fall and winter in Louisiana. *Journal of Medical Entomology* 42: 193–203.

Watson, E.J. & Carlton, C.E. (2005b) Succession of forensically significant carrion beetle larvae on large carcasses (Coleoptera: Silphidae). *Southeastern Naturalist* 4: 335–346.

Weidner, L.M., Nigoghosian, G., Hanau, C.G. & Jennings, D.E. (2020) Analysis of alternative food sources for rearing entomological evidence. *Journal of Medical Entomology* 57: 1407–1410.

Wilson, J.M., Lafon, N.W., Kreitlow, K.L., Brewster, C.C. & Fell, R.D. (2014) Comparing growth of pork- and venison-reared *Phormia regina* (Diptera: Calliphoridae) for the application of forensic entomology to wildlife poaching. *Journal of Medical Entomology* 51: 1067–1072.

Wobeser, G., Bollinger, T., Leighton, F.A., Blakley, B. & Mineau, P. (2004) Secondary poisoning of eagles following intentional poisoning of coyotes with anticholinesterase pesticides in western Canada. *Journal of Wildlife Diseases* 40: 163–172.

Supplemental reading

Cooper, J.E. & Cooper, M.E. (2013) *Wildlife Forensic Investigation: Principles and Practice.* CRC Press.

Huffman, J.E. & Wallace, J.R. (2011) *Wildlife Forensics: Methods and Applications.* John Wiley & Sons Inc.

Moore, M.K. & Frazier, K. (2019) Humans are animals, too: critical commonalities and differences between human and wildlife forensic genetics. *Journal of Forensic Science* 64: 1603–1621.

Neme, L.A. (2009) *Animal Investigators: How the World's First Wildlife Forensics Lab Is Solving Crimes and Saving Endangered Species.* Scribner

Rolo, E.A., Oliveira, A.R., Dourado, C.G., Farinha, A., Rebelo, M.T. & Dias, D. (2013) Identification of sarcosaprophagous Diptera species through DNA barcoding in wildlife forensics. *Forensic Science International* 228: 160–164.

Tomberlin, J.K., Byrd, J.H., Wallace, J.R. & Benbow, M.E. (2012) Assessment of decomposition studies indicates need for standardized and repeatable research methods in forensic entomology. *Journal of Forensic Research* 3: 147. https://doi.org/10.4172/2157-7145.1000147.

Additional resources

Purchase a hunting license. Links to each state provided by the US Fish and Wildlife Service: https://www.fws.gov/initiative/purchase-hunting-license

Endangered Species Act of 1973: https://www.fws.gov/media/endangered-species-act

CITES: https://cites.org/eng

National Pesticide Information Center – Where to start with pesticide incidents: http://npic.orst.edu/incidents.html#env

EPA – How to report pesticide incidents involving wildlife or the environment: https://www.epa.gov/pesticide-incidents/how-report-pesticide-incidents-involving-wildlife-or-environment

Fact sheet on the "Collection of entomology samples to assist in the investigation of wildlife crime" by M. Hall and A. Whitaker at the Natural History Museum. https://www.tracenetwork.org/wp-content/uploads/2018/05/Information-sheet-on-evidence-collection-for-forensic-entomology-analysis-in-wildlife-crime-2017.pdf

Chapter 18
Forensic Entomotoxicology

Overview

The use of insects in death investigations often extends beyond estimation of a postmortem interval. When a corpse is discovered in an advanced state of decay or is skeletonized, little or no soft tissues are available to perform toxicological analysis. One viable solution to the problem is literally close by: insects that fed on the corpse. In many instances, as necrophagous insects consume human tissues they accumulate chemical compounds and their metabolites found in the food source. The insect in turn can be used to extract and detect a wide range of xenobiotics acquired from human remains, which is the primary basis for the field of entomotoxicology. In essence, insects function as proxy samples for human tissues in qualitative toxicological analyses. Many species of insects are potentially useful in this regard, but necrophagous flies standout as excellent surrogates for human tissues since they are early colonizers and are readily abundant where the remains have been discovered. The reality is that drugs, toxins, and pollutants bioaccumulate throughout food trophic levels. Thus, predatory and parasitic insects also play a role in detection of chemical compounds. Chapter 18 explores the use of insects in detection of xenobiotics from human remains, especially within the context of criminal investigations. The chapter will also examine the limitations of entomotoxicology in quantitative toxicological analysis and discuss the impact that various types of drugs (illicit and prescription) have on insect development. The latter will be examined in the context of estimating a postmortem interval based on insect development.

The big picture

- Insects can serve as surrogates for human tissues.
- Detection of drugs, toxins, and pollutants in carrion-feeding insects.
- Detection of gunshot and explosive residues in or using insects.
- Chemical impact on insect succession and development.

18.1 Insects can serve as surrogates for human tissues

Insects are known to be useful silent witnesses to violent crimes, but generally their utility has been viewed predominantly as a means to estimate a

The Science of Forensic Entomology, Second Edition. David B. Rivers and Gregory A. Dahlem.
© 2023 John Wiley & Sons Ltd. Published 2023 by John Wiley & Sons Ltd.
Companian website: www.wiley.com/go/forensicentomology2

portion or all of the postmortem interval (see Chapter 14 for a discussion of the postmortem interval). That view does not capture the value that many insects offer as toxicological samples. In instances in which a corpse or set of remains is not suitable for typical toxicological analysis associated with autopsy, insects can serve as viable surrogates for human tissues (Bourel *et al.*, 2001; Campobasso *et al.*, 2004). The conditions that favor insect proxies are when body fluids and soft tissues are limited or not available at all such as in advanced stages of decay, or when the remains are skeletonized, severely burnt, or mummified (Pounder, 1991; Introna *et al.*, 2001). Presumably tissues that have undergone saponification or that have been submerged in aquatic environments for an extended period of time are also of limited value to toxicological analyses. Since necrophagous species consume tissues of the deceased, and thus acquire the chemical compounds contained within those tissues, the insects can, in turn, be used to extract and detect any type of drug, toxicant, pollutant or metabolite present in the food source (Chophi *et al.*, 2019). In other words, toxicological screening of the decedent can still occur using necrophagous insects. There are two key facets to consider in the examples described. First, human sources such as organs (e.g., heart, liver, stomach) and fluids (blood, urine, bile, vitreous humor) are the best direct matrices for performing toxicological analyses in death investigations. However, as mentioned earlier, there are many times in which direct sources are not available for testing. In which case, the second point is that some insects can serve as indirect forensic matrices for toxicological analysis when a direct source is not available. As Chapter 18 unfolds, details of which types of insects are useful for toxicological analysis and in what capacity will be discussed.

The term **entomotoxicology** was coined to describe the subfield of forensic entomology that deals with analysis of chemical compounds present in arthropods that fed on a human corpse. To an extent, the term is somewhat problematic in that it really requires a thorough understanding and use of techniques from forensic toxicology, coupled with forensic entomology. Entomotoxicology also implies a broader scope that encompasses any type of chemical detection in insects and related arthropods, from a wide variety of sources and not restricted to just a human corpse. For example, it can be argued that insect toxicology and entomotoxicology overlap in focus. Of course, the term insect toxicology can be confusing when compared to the discipline of **insecticide toxicology** that has a primary focus on *any* organism affected by insecticides, not just insects. Perhaps the highest clarity is derived from the term **forensic entomotoxicology**, which has emerged as a refined version of entomotoxicology in that it is concerned with the use of insects in legal contexts as indirect sources of toxicological evidence in the absence of human tissues or other direct forensic samples (da Silva *et al.*, 2017). The focus is also not restricted to xenobiotics found in human tissues. Rather, the discipline uses insects as evidence in legal investigations requiring toxicological evaluation of a broad array of sources, which may include a corpse, an aquatic environment or any other aspect of the environment. Thus, the field of forensic entomotoxicology overlaps with other disciplines, including insecticide toxicology and environmental forensic toxicology (da Silva *et al.*, 2017). Chemical detection is not the only aim for this field either. Forensic entomotoxicology examines the impact of consumed xenobiotics on insect development, especially in relation to how the toxicants influence determination of a postmortem interval based on development of necrophagous species (Campobasso *et al.*, 2020).

Forensic entomotoxicology is organized around a death investigation with two central themes: What caused death of the individual and when did death occur? (da Silva *et al.*, 2017). Obviously, those themes are investigated using insects as direct or indirect evidence for toxicological analysis. Details of "what" is investigated and "how" will be described in the proceeding sections of the chapter. For now, our focus turns to establishing what makes insects useful to serve as proxies for human tissues. The simple answer to this question is necessity. If no other options are available, well then, of course analysis must be performed on the only soft tissues remaining at the crime scene, the insects. But simply "being present" does not guarantee that insects are useful surrogates for qualitative or quantitative toxicological analysis. What follows, then, is a brief examination of some of the traits of necrophilous insects that make them useful matrices for "human" toxicological analysis.

18.1.1   Necrophagous activity

The most significant feature of any insect to serve as a surrogate for human tissue is necrophagy (Figure 18.1). This does not preclude the possibility that non-necrophagous species can serve as

Figure 18.1 Necrophagous insects, especially fly larvae, are the most useful surrogates for human tissue in toxicological analysis. *Source*: Photo by D.B. Rivers.

Figure 18.2 A large aggregation of *Dermestes maculatus* feeding on rat carrion. In addition to serving as an ample source for toxicological analysis, the larvae and adults also demonstrate concurrent feeding. *Source*: Photo by D.B. Rivers.

surrogates as well. We shall examine the utility of insects that do not exhibit necrophagy as toxicological samples in Section 18.1.3. However, necrophagous insects are generally recommended for forensic entomotoxicology investigations (Gosselin *et al.*, 2011). Necrophagy is an obvious central premise to medico-legal entomology as a whole. In fact, many of the features that characterize necrophagous Diptera, especially members of the family Calliphoridae, and Coleoptera colonizing vertebrate carrion contribute to the utility of these insects as indirect matrices for toxicological analysis. For example, necrophagous flies tend to be early colonizers, and thus are more likely to ingest the parent compound and metabolites, at higher concentrations (i.e., before chemical degradation) than at later stages of postmortem decomposition. Flies, and to lesser extent some Coleoptera, produce large numbers of offspring, all feeding directly on human remains (Gosselin *et al.*, 2011). This constitutes a potentially huge source of insect soft tissue, particularly when large feeding aggregations form, to use for toxicological analysis (Figure 18.2). The fact that even after dispersal, most fly species pupariate in substrate relatively close to the remains permits collection of flies for analyses several days to weeks after a corpse is completely skeletonized. Equally important is that the biology of many necrophagous species is well-characterized which allows for correlation of drug detection with the timing of insect association with the corpse under a variety of conditions.

18.1.2 Detection in multiple developmental stages

Xenobiotics derived from human tissue can be detected in multiple life stages of necrophagous species. This is a result of at least two feeding scenarios. The first is that for some insect species, multiple life stages feed concurrently on a corpse, and then in turn bioaccumulate chemical compounds in the tissues or fluids ingested. Concurrent feeding is relatively common among adults and larvae of some beetle families, such as silphids, staphylinids, and dermestids. Consequently, multiple development stages of the same species are useful sources for toxicological analysis. Non-concurrent feeding (as well as concurrent) is frequently observed with necrophagous Diptera in which adults acquire nutrients from a corpse or carrion for egg provisioning. As mentioned in Chapter 6, anautogenous species have a need for exogenous proteins before the first clutch of eggs can be produced. Obviously in this example, larval feeding is initiated after the adults have fed. Unlike with many necrophagous beetle species, the two life stages generally do not eat together, at least not for long periods of time (Figure 18.3). Conceivably this could mean that two distinct time stamps of drug levels could be obtained if samples of both adults and larvae are collected. Of course, the likelihood of trapping adults that fed on the corpse prior to egg hatch, depending on the crime scene location, is low.

For many species of flies, a xenobiotic ingested during larval feeding bioaccumulates and is

Figure 18.3   Larvae and adults of most necrophagous fly species generally demonstrate non-current feeding on a corpse, conceivably permitting multiple time stamps of xenobiotic concentrations. *Source*: Photo by D.B. Rivers.

Figure 18.4   Meconium released from flies immediately after adult emergence is a viable indirect matrix for toxicological analysis. *Source*: Photo by D.B. Rivers.

development may last for more than two-three weeks, creating an extended window for sampling pupal tissues and meconium. Species that diapause as pupae conceivably remain viable toxicological sources for 6–10 months depending on biogeographic location. In other instances, chemical compounds have been successfully extracted from exuvia and empty puparia. Both forms of exoskeleton degrade very slowly in natural environments. In fact, fly puparia have been unearthed in nearly pristine condition during excavations of ancient Peruvian burial sites, estimated to be *c*. 1250–1900 years old (Huchet & Greenberg, 2010).

18.1.3   Utility of insect secretions and excretions

If some cases, the absence of insects is not an obstacle to performing toxicological analysis. Feces or frass from dermestid beetles have been used successfully as proxies for human *and* insect tissues (Manhoff *et al.*, 1991; Miller *et al.*, 1994). Secretory and excretory products from other necrophagous species may be equally useful in such analyses, but there are few published reports to suggest that such testing has been attempted. Of potential value are artifacts produced by adult flies in the form of regurgitate and defecatory stains. These artifacts or stains are known to be temporary reservoirs of the victim's blood and DNA (Striman *et al.*, 2011; Durdle *et al.*, 2013), and conceivably of xenobiotics as well. Insect artifacts dry relatively quickly, reducing chemical and enzymatic degradation of the stain constitutes. The fact that flies deposit the artifacts in large numbers around the crime scene and in other locations also means that an abundant source of indirect matrices are available for testing (Rivers & Geiman, 2017), sometimes even after the crime scene has been cleaned (Kulstein *et al.*, 2015).

18.1.4   Bioaccumulation across trophic levels

There is some evidence to suggest that predatory and parasitic species bioaccumulate xenobiotics following consumption of necrophagous insects (Nuorteva, 1977; Faria *et al.*, 1999). Such observations are not surprising in that many insect parasitoids are known to accumulate a wide range of

sequestered in subsequent developmental stages. This may result in drug detection within the pupal or adult stages. During pupal development, excretory products including drug metabolites will accumulate within the **meconium** (Figure 18.4). The meconium is not purged until after adult emergence. For sarcophagid species, intrapuparial compounds from their hosts and in turn sequester or incorporate the acquired chemicals into their own tissues. Bioaccumulation of heavy metals, pesticides and other environmental pollutants through trophic levels has been a widespread concern for decades in many regions of the world. In several instances, insects have been used as toxicological indicator species for environmental forensic toxicology (da Silva *et al.*, 2017). So, an extension to forensic entomotoxicology is not surprising. That said no case reports have been published that indicate the use of secondary or tertiary insect species in forensic toxicological analysis.

18.2 Detection of drugs, toxins, and pollutants in carrion-feeding insects

In theory, any chemical compound ingested by an insect should be extractable from its tissues and be detectable by appropriate analytical methods. Certainly, there is ample evidence in the literature to demonstrate that various narcotics, psychotropics, heavy metals, insecticides, and other toxicants can be detected in necrophilous insects following consumption of human remains (Tracqui *et al.*, 2004; da Silva *et al.*, 2017; Stojak, 2017; Chophi *et al.*, 2019). Similarly, numerous animal models ranging from mice, rats, rabbits, and pigs (large and small), as well as isolated tissues (e.g., beef or pork heart, liver, muscle) have been used in laboratory experiments aimed at chemical detection in necrophagous insects. Table 18.1 provides examples of some of the chemical compounds detected in necrophagous insects. In short, a plethora of evidence exists to demonstrate proof of concept. Insects are viable substitutes for human tissues in terms of positive chemical detection. However, despite widespread acceptance that insects *could* be used for xenobiotic detection in the absence of human tissues, forensic entomotoxicology is not readily embraced by all or possibly most forensic toxicologists. Why not? There are multiple viewpoints to this question, including that insects are invaluable to toxicological analysis under very specific conditions (Bourel *et al.*, 2001). What follows is a discussion of perceived limitations to forensic entomotoxicology in terms of qualitative and quantitative analyses using insects as surrogates for human tissues.

Table 18.1 Partial list of drugs, toxins, and pollutants discovered in necrophagous insects.

Chemical class	Insect species	Developmental stage
Alcohol		
Ethanol	*Phormia regina*	L
Barbiturates		
Amobarbital	Not given	L
Phenobarbital	*Chrysomya megacephala*	L
	Chrysomya putoria	L
	Cochliomyia macellaria	L
Sodium phenobarbitone	*Calliphora vicina*	L, P
Benzodiazepines		
Alprazolam	*C. vicina*	L, P
Bromazepam	*Piophila casei*	L, P, A
Clonazepam	*C. vicina*	L, P, A
Diazepam	*C. vicina*	L, P, A
	Chrysomya albiceps	L, P, A
Lorazepam	Calliphoridae	L
Nordiazepam	Calliphoridae	L
Oxazepam	*C. vicina*	L, P, A
Triazolam	*C. vicina*	L, P, A
CNS stimulants		
Amphetamines	*C. vicina*	L
Cocaine	*Ch. albiceps*	L
	Ch. putoria	L
	Cynomyopsis cadaverina	L
	Lucilia sericata	L
	Sarcophaga peregrina	
Methamphetamine	*Calliphora stygia*	L
	Calliphora vomitoria	L
	Sarcophaga ruficornis	L
Insecticides		
DEET	*Blaesoxipha plinthopyga*	L, A
	L. sericata	L, A
Endosulfan	*C. vomitoria*	L
Malathion	*Chrysomya megacephala*	L, P, PP
	Chrysomya rufifacies	L
Parathion	*Dermestes.* sp.	L
	L. sericata	L
	Coleoptera	A
	Diptera	L, P, PU, A
	Hymenoptera	A

(Continued)

Table 18.1 (Continued)

Chemical class	Insect species	Developmental stage
Opioids		
Codeine	*C. albiceps*	L
	L. sericata	L, P, A
Fentanyl	*P. regina*	L
Heroin	*S. peregrina*	L
Methadone	*L. sericata*	L
Morphine	*Calliphora stygia*	L, P, PU, PP, A
	C. vicina	L, PU
	Dermestes freshi	L, P, A
	L. sericata	L, P, PU, A
	P. terraenovae	L, PU
	Thanatophilus sinuatus	L, P, A
Morphine hydrochloride	*Protophormia terraenovae*	L
Over-the-counter drugs		
Acetaminophen (paracetamol)	*C. vicina*	L
Gentamicin	*Chrysomya putoria*	L
	L. sericata	L
Hydrocortisone	*Chrysomya chlorophyga*	L
	Sarcophaga tibialis	L
Nicotine	*C. vomitoria*	L
Psychotropics		
Amitryptyline	*C. vicina*	L, P
	L. sericata	L
	S. ruficornis	L
Clomipramine	*L. sericata*	L
Dothiepin	Calliphoridae	L
Fluoxent	*Dermestes maculatus*	L, P, A, E
Ketamine	*Ch. megacephala*	L
	L. sericata	L
Methylphenidate	*C. vicina*	L
	L. sericata	L
Nortripyline	*Dermestes maculatus*	E, F
	L. sericata	L
	Megaselia scalaris	E, F
	S. ruficornis	L
Trazodone	*C. vicina*	L
Venlafaxine	Calliphoridae	L
Pollutants/heavy metals		
Antimony	*Calliphora dubia*	L, P, PU, A
	L. sericata	L
Barium	*Calliphora dubia*	L, P, PU, A
	L. sericata	L
Cadmium	*Ch. albiceps*	L
	L. sericata	L, PU, A
Copper	*Musca domestica*	L
Iron	*M. domestica*	L
Lead	*Calliphora dubia*	L, P, PU, A
	L. sericata	L
Methyl mercury	*Creophilus maxillosus*	A
	Tenebrio molitor	A
Mercury	Calliphoridae	L, PU, A
Zinc	*M. domestica*	L

Codes for developmental stages: L = larvae, P = pupae, PU = puparia, PP = prepupa, A = adult, E = exuvia, F = frass. Information derived from Campobasso *et al.* (2004), Tracqui *et al.* (2004), Gosselin *et al.* (2011), Bushby *et al.* (2012), Magni *et al.* (2016), Zanetti *et al.* (2016), da Silva *et al.* (2017), Stojak (2017), and Chophi *et al.* (2019).

18.2.1 Qualitative detection of xenobiotics

From a toxicology perspective, detection of a chemical substance alone is generally insufficient to draw many conclusions regarding a forensic investigation, especially since insects are indirect forensic matrices. In many respects, identification of a xenobiotic can be thought of as the starting point not the conclusion of toxicological analysis. Most forms of forensic toxicological assessment require identification and quantification of chemical compounds. This is especially critical for any toxicological analysis associated with determination of the cause and manner of death. Here lies a critical limitation of forensic entomotoxicology: no reliable correlation has been demonstrated between the concentration of drugs and toxins discovered in insects with the tissues which they fed upon (Campobasso *et al.*, 2004; Tracqui *et al.*, 2004). In fact, drug levels in fly larvae have been found to be highly variable in relation to tissue levels and heavily influenced by where the insect was collected from, and presumably fed, on the corpse. From a qualitative analysis perspective, such extreme variability casts doubt on the reliability of xenobiotic detection. In this respect, you may be asking, why would this cast doubt on qualitative detection when this is a quantitative problem. The issue in question is not when a drug or toxicant has been detected in insect tissues, otherwise known as **positive detection**. As has already been established, xenobiotics can be extracted and detected in necrophagous species. There is little concern that positive detection may actually represent a **false positive**. Instead, the issue is when no xenobiotic is detected or identified, meaning **negative detection** (Tracqui *et al.*, 2004). Is this observation genuine or is the negative result a mere reflection of the concentration variability that exists once a chemical is consumed by a necrophagous insect? In other words, concerns have been raised as to whether negative detection in insect proxies represents a **false negative**, meaning that drugs or toxins may indeed be present but could not be reliably detected in insect tissue. At present, there are no methods available to ascertain when false negatives or false positives occur with insect samples. This actually represents a larger issue that some toxicologists believe plagues many studies performed in the name of forensic entomotoxicology.

What is the larger issue? Forensic entomotoxicology does not emulate forensic toxicology. Any form of toxicology, but particularly with forensic toxicology, is characterized by rigorously tested analytic methods and meticulously followed standard operating procedures (SOP). In the United States, laboratories performing analysis for forensic toxicology undergo accreditation by the American Board of Forensic Toxicology (http://www.abft. org/). Any deviation from an accepted SOP will trigger heavy scrutiny during cross-examination in the courtroom. As part of the SOP used in a toxicology laboratory, the analytic methods employed for extraction and detection are based on the metabolites expected to be present in human tissue (da Silva *et al.*, 2017). This predicates a thorough understanding of how a given drug or toxin is metabolized in the human body and the degree of **tropism** – the preferential movement of the compound and metabolites to specific tissues within the body – for a given chemical or metabolite. Thus, metabolic and pharmacokinetic studies typically precede qualitative detection. By comparison, forensic entomotoxicology is perceived as lacking similar approaches to evidence examination (Tracqui *et al.*, 2004; da Silva *et al.*, 2017). To an extent, there has been somewhat of a cart-before-the horse approach with many studies aimed at detection of xenobiotics in insect tissues. For example, analytic methods seem to be selected based on the parent compound expected to be consumed by the insect rather than based on studies focused on how insects metabolize the

drug. In fact, outside of insecticides, metabolic and pharmacokinetic studies have rarely been performed with necrophagous insects. As a consequence, the baseline information needed for developing analytic methods for xenobiotic detection is generally absent. Similarly, SOPs for forensic entomotoxicology have not been developed or at least not to the extent as they have for forensic toxicology (da Silva *et al.*, 2017). Until such deficiencies are addressed, the full potential of using insects as surrogates for human tissues cannot be realized.

18.2.2 Quantitative detection of xenobiotics

Insect larvae are not considered useful for quantitative toxicological analysis. More to the point, no reliable correlation between xenobiotic levels in insects and the tissues in which they have consumed has been established (Campobasso *et al.*, 2004; Tracqui *et al.*, 2004). This is a major limitation for using insects as proxies for human tissues. In fact, some toxicologists have argued that insect use in any toxicological analysis has almost no value for forensic casework (Tracqui *et al.*, 2004). That view is certainly understandable based on the current state of forensic entomotoxicology in terms of baseline information relevant to metabolism and pharmacokinetics of drugs and toxicants in necrophagous insects. However, insects do offer enormous potential to qualitative and quantitative toxicological analysis. To realize that potential, two basic questions need to be addressed:

1. Can a drug or toxicant be qualitatively detected in insect tissues?
2. Are xenobiotic concentrations in insects relevant to postmortem analysis of human tissues?

These questions may initially offer some confusion because the answer to each seemingly has already been addressed. Presently, the answer is "no" to each. However, at the heart of each question posed is what information is needed to change the responses to "yes." For instance, what is the major limitation currently to using insects for quantitative analysis? The simple answer is virtually no information exists on the fate of a toxicant in which the insect is exposed. In other words, what happens to the drugs, toxins, and metabolites present in human tissues when consumed by the insect? Presumably the compounds are initially ingested, but to what extent are they digested (catalyzed), transported and absorbed throughout the insect's body? The rate of ingestion and digestion of soft tissues in fly larvae is accelerated based on the size and/or density of a maggot mass, which presumably extends to all facets of xenobiotic metabolism. For nearly all of the compounds listed in Table 18.1, metabolic studies have not been performed for any forensically relevant insect species. Undoubtedly for some of the compounds, they are not modified or very little, and are either excreted into the environment or sequestered by the insect (da Silva *et al.*, 2017). If the latter occurs, how much of the compound is retained through subsequent developmental stages? It is also highly likely that some of the compounds interact with other chemicals in the insect, thereby changing the pharmacokinetic properties of the ingested xenobiotic. Each of these features – ingestion, transport, digestion/catalysis, absorption, sequestration – must be considered based on the extent of tropism for each compound, tissue specificity, and influence associated from feeding in larval aggregations (Figure 18.5). When the knowledge base for a given drug or toxicant is established, appropriate analytic methods can then be selected for extraction and detection. In turn, the ability to reliably quantify the parent compound and its metabolites in a given insect species should improve dramatically.

The second question is a bit more problematic. It first assumes that quantitative toxicological analysis can be performed for a given xenobiotic in a specific necrophagous species. Assuming that the concentrations in the insect can be reliably determined, the question then becomes what do the toxicant levels mean to the forensic investigation? Is the concentration determined in the insect relatable to the concentration in the corpse at the agonal period or moment of death? (Tracqui *et al.*, 2004). It would seem unlikely. Why? Typically, insects are only used as surrogates when soft tissues are not available. This effectively means that insect samples are collected several days, to weeks to even months after the individual died. Considering the enormous lag time between death and toxicological analysis, numerous changes are expected to occur in drug or toxicant concentrations first in the decomposing corpse and then in the insect. Many of these changes are largely unpredictable and just as many have not been examined. For example, postmortem redistribution of xenobiotics occurs in a corpse and is influenced by time, ambient conditions, and tissue type (Tracqui *et al.*, 2004). Consequently, drug or toxicant concentrations consumed by necrophagous insects are expected to vary based on human tissues consumed, which will differ based upon stage of decomposition, ambient conditions experienced throughout decomposition, and species and age of the insects. Of course, these levels are further influenced by compound stability in human tissues and subsequently in the insect, as well as whether any drug (or toxicant) interactions occur in the corpse and/or insect. Clearly, deciphering the relevancy of xenobiotic concentrations in insect tissue to the deceased is at the very least complicated. The present view by some that insects are unreliable indirect matrices for quantitative

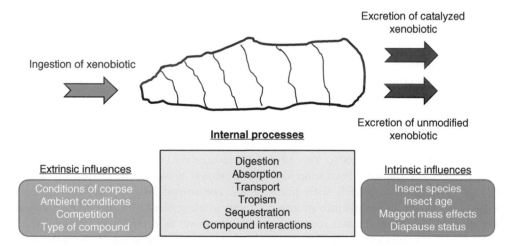

Figure 18.5   Fate of a xenobiotic ingested with human tissue by a necrophagous insect. *Source*: Information based on Norris (1965) and da Silva *et al.* (2017).

toxicological analysis is undeniable. The only way to change this perception is for increased research aimed at understanding the metabolism and pharmacokinetics of major classes of drugs and toxins in an array of necrophagous insects. Even then, the data may reveal that some or all necrophagous insects are quite limited for quantitative toxicological assessments.

18.3 Detection of gunshot and explosive residues in or using insect

Necrophagous fly larvae consuming tissues of a corpse that has been exposed to explosives or bullet fragments can accumulate the chemical signatures that imply foul play. Why this is important is because evidence to implicate a particular bomb-maker (i.e., terrorist organization) often lies in the chemical composition of the explosive device, which may leave a chemical signature in objects found in the explosion area, including the unfortunate victims. Similarly, a gunshot leaves distinctive patterns termed gunshot residues (GSRs) that can be modified during postmortem decomposition and burial and by the activity of insects. These alterations may lead to failure to recognize that wounding, and consequently death, occurred due to a gunshot (s) (LaGoo et al., 2010).

Some of the same insects that confound GSR patterns may also be the solution to the problem of detection. The idea is essentially the same as discussed for qualitative detection of a xenobiotic. In the case of gunshot, residues left behind contain spheroid particles whose morphology can be distinguished from other sources fairly easily to the trained eye. The residues also contain unique elemental combinations dominated by barium (Ba), lead (Pb), and antimony (Sb) (Wolten et al., 1979). The larvae of early colonizing flies generally feed at the skin surface and/or on exposed mucous membranes, areas where GSRs reside (LaGoo et al., 2010). Thus, as the larvae feed, they accumulate the elemental particles associated with the GSRs. The rest of the story is obvious: fly larvae are collected from a corpse suspected of containing GSRs or explosives residues, the residue particles are extracted (by acid or microwave detection), and the extracts examined by techniques (such as inductively coupled plasma mass spectrometry or scanning electron microscopy)

that permit identification of the elements and confirmed by comparisons with known standards (Roeterdink et al., 2004; LaGoo et al., 2010; Abd Rashid and Ahmad, 2012; Labella et al., 2014). One downside to this approach to toxicological detection is that it likely has limited utility with late colonizing insects or those that feed deep within the body cavity (LaGoo et al., 2010). The assumption is that the further from the skin surface or the later in decay, the less residues that are present (Duarte, 2015). The exception would be if the victim did not die immediately after injury, permitting transport of chemical residues throughout the body via circulating blood.

The technique has been tested with four species of calliphorids, *Chrysomya megacephala*, *Lucilia sericata*, *Calliphora dubia*, and *C. vomitoria*. Additional specimens need to be examined to ensure the approaches are suitable for a wide range of necrophagous species. With GSRs, elemental analysis has confirmed that residues are detectable in the pupal stages (LaGoo et al., 2010), long after feeding has been completed, indicating that there is no danger of loss during gut purging prior to pupariation. Somewhat surprisingly, however, empty puparia have not been tested for the presence of explosives or gunshot residues. Puparia are often the only evidence left behind to even indicate that flies had once been present, and once pupariation is complete the puparium becomes physiologically inert (Zdarek & Fraenkel, 1972), essentially locking the chemical profile in place seemingly forever.

18.4 Chemical impact on insect succession and development

Regardless of whether a xenobiotic can be accurately quantified in insect tissues, consumption of any type of "unexpected" compound can potentially alter or disrupt insect development. This statement has several layers that need to be peeled back to understand the true importance to forensic entomotoxicology. First, what is meant by "unexpected" compound? Keep in mind the insect did not intentionally seek out and ingest the xenobiotic. Rather, a necrophagous species is intent on consuming carrion soft tissues and fluids to meet its nutritional needs. Insect feeding is driven by a neuronally-regulated hunger drive and satiety center. For most species, the insect only consumes food that meets the nutritional craving. Once nutriment has been obtained, and assuming the needed nutrients were acquired, the hunger drive is quenched through the activation of the satiety center. A drug, toxin, or pollutant present in human tissues is not associated with hunger activation nor would an exogenous xenobiotic be expected to activate the satiety center. So, intake of a xenobiotic during ingestion of human tissues is indeed unexpected from the perspective of the insect hunger drive. Second, any drug or toxicant present in human tissue was generally consumed by the individual to evoke a specific response, often triggering one or more signal transduction pathways. Many of these same compounds elicit similar, sometimes more pronounced effects when ingested by insect larvae or adults. In the case of juveniles, an ingested and absorbed xenobiotic and/or metabolite has the potential to alter development. As we will discuss later in this section, several illicit drugs are known to accelerate or slow larval development,

prolong intrapuparial stages, and/or evoke a dose-dependent death. Any of these responses leads directly to the third point; any compound that alters insect development has the potential to change the usefulness of that species for estimation of the postmortem interval (Mullany *et al.*, 2014). Thus, there is an absolute need to understand the effects of commonly encountered xenobiotics in human tissues on development of forensically important insects.

Xenobiotics have the potential to impart three potential effects on insect development: accelerate the growth rate of one or more developmental stages, retardation or inhibition of development, and lethality. A fourth outcome is to not alter development at all, which implies the chemical compound is benign towards PMI estimation. Table 18.2 provides some examples of xenobiotic effects on insect development. Not surprisingly, most studies to date have examined the impact on

Table 18.2   Effects of various xenobiotics on fly development.

Drug	Method of administration	Insect species	Developmental impact
Amitriptyline	Injection (postmortem)	*Parasarcophaga ruficornis*	Extended length of post-feeding wandering stage and increased length of puparial stages
Cocaine	Injection (postmortem)	*Sarcophaga peregrina*	High doses induced more rapid rate of larval development; duration of puparial stages not affected
Codeine	Mixed in porcine liver	*Lucilia sericata*	Accelerated larval development in a non-dose-dependent manner; not long-lasting effect as pupal development not altered
Diazepam	Injected (postmortem)	*Chrysomya albiceps* *Chrysomya putoria*	Accelerated larval development; the duration of pupariation and intrapuprial development was extended
Heroin	Injection (postmortem) Perfusion (antemortem)	*Sarcophaga peregrina* *Lucilia sericata*	Accelerated larval development and increased larval body mass; extended length of puparial stages dependent on dose
			Accelerated larval development
Methamphetamine	Injection (postmortem) Mixed with bovine liver	*Parasarcophaga ruficornis* *Calliphora vomitoria*	High doses induced more rapid rate of larval development; increased puparial mortality; reduced fecundity by second generation
			Lengthened development from egg to adult; increased body sizes of larvae and pupae
Nicotine	Mixed with bovine liver	*Calliphora vomitoria*	Larval development not altered but high doses evoked death in intrapuparial stages
Parcetomol	Mixed with porcine liver	*Calliphora vicina*	Accelerated larval development but only on days 3–4 at 20 °C.
Phenobarbital	Mixed in ground beef	*Chrysomya albiceps* *Chrysomya megacephala* *Chrysomya putoria*	Larval development delayed by several hours; *Ch. megacephala* demonstrates more variable responses that are age-dependent

Source: Data from Goff *et al.* (1991), Hedouin *et al.* (1999), Carvalho *et al.* (2001), O'Brien & Turner (2004), Goff & Lord (2010), Magni *et al.* (2014, 2016), and Rezende *et al.* (2014).

necrophagous fly larvae. Though in many studies fly development has been clearly shown to be altered in the presence of certain drugs or toxicants, it is difficult to interpret toxicant influences on insect growth. The problem stems from the fact that drug or toxin influences are generally examined as dose-dependent relationships over time. In death investigations, "time" is the unknown and is represented by the PMI. Obviously necrophagous insects are used to estimate the PMI, but for modeling pharmacokinetic relationships between insect growth and drug influences, the unknown PMI becomes an obstacle (da Silva *et al.*, 2017). Equally difficult to decipher are dose-dependent effects of a given compound on insect development during casework. Fly larvae migrate across the corpse, especially during early larval development or when large maggot masses form and consume soft tissues quickly. Consequently, the maggots may feed on different tissues, which in turn may contain different levels of a drug or metabolite or harbor different xenobiotics (da Silva *et al.*, 2017). In which case, it is not possible to determine whether the resulting alterations in fly development are the result of a single chemical compound nor is it possible to decipher dose-dependent changes in growth. Nearly all of the baseline information on xenobiotic influences on insect development are derived from carefully controlled laboratory studies. However, even in many of those experiments, animal models or isolated tissues were used instead of human tissues. How relatable are those observations to casework scenarios involving human remains? The answer is not known. Presumably at some point during decomposition, xenobiotics no longer exert an influence on insect growth. Is this true? If so, when? The natural inclination is to predict that early colonizers and/or insect waves that arrive during the fresh to early bloat stages are most susceptible to the effects of xenobiotics in human tissues. Again, very little information is available to draw definite conclusions.

18.4.1 Xenobiotics as accelerants

Several types of drugs consumed by fly larvae can accelerate larval development independent of temperature-mediated influences. In controlled experimental studies in which larvae of the flesh fly *Sarcophaga* (=*Boettcherisca*) *peregrina* were raised on rabbits injected with various doses of cocaine, sublethal concentrations of the narcotic did not alter the rate of fly development in comparison with control larvae (Goff *et al.*, 1989). However, lethal and twice lethal doses induced more rapid growth in maggots until reaching post-feeding stages of larval development; duration of puparial stages was not altered by the presence of cocaine (Goff *et al.*, 1989). Heroin evokes similar increases in development rate of *S. peregrina* and *Lucilia sericata* (Goff *et al.*, 1991; Hedouin *et al.*, 1999), and at least with the sarcophagid also promotes larger body sizes and extended duration of puparial stages (Goff *et al.*, 1991). The effect of heroin in the form of morphine on fly development is not simply a stimulation of metabolic rate and subsequent food consumption. Rather, increased larval body sizes beyond an "optimal" weight suggests that the narcotic allows or forces the maggots to ignore a satiation set point to continue feeding. Satiety is the sensory sensation of fullness or suppression of the hunger drive due to acquisition of needed nutrients (Widmaier, 1999). Thus, heroin essentially has the effect of keeping the hunger drive active, conceivably by manipulating neurons that function to either regulate satiety and/or hunger. The significance from a forensic entomology standpoint is that larval development displays aspects independent of temperature, which in turn compromises the use of drug-fed flies in calculation of a postmortem interval. Numerous other drugs and prescription medications including diazepam, methylamphetamine, codeine, ketamine, hydrocortisone and others evoke an accelerated developmental rate in a variety of fly species (Carvalho *et al.*, 2001; Musvasva *et al.*, 2001; Kharbouche *et al.*, 2008; Zou *et al.*, 2013; Mullany *et al.*, 2014). The majority of research studies have relied on calliphorids and to a lesser extent sarcophagids. Differences between families in terms of response to a given compound have not been reported. However, differential responses to the same xenobiotic in the same family has been demonstrated in a limited number of studies (e.g., ketamine and morphine in calliphorid larvae) (Bourel *et al.*, 1999; Zou *et al.*, 2013; Lü *et al.*, 2014). More detailed reviews on xenobiotics as accelerants for fly development include Gagliano-Candela & Aventaggiato (2001), Chophi *et al.* (2019) and Campobasso *et al.* (2020).

18.4.2   Xenobiotics as retardants

A broad array of xenobiotics has been demon-strated to function as developmental depressants in necrophagous insects. For example, phenobarbital, ethanol, methadone, and morphine slow or retard larval development in blow fly larvae (Bourel *et al.*, 1999; Tabor *et al.*, 2005; Gosselin *et al.*, 2011; Rezende *et al.*, 2014). The effect of each compound is dose dependent and varies based on drug/toxin, insect species, and age as to whether a linear rela-tionship between dose and effect is exhibited (Rezende *et al.*, 2014). Methadone is reported to retard larval development of *L. sericata* but does not alter the overall duration of development from egg to adult (Gosselin *et al.*, 2011). This observation suggests that the drug and/or its metabolites pro-duces differential and long-lasting effects on fly development from a single exposure, i.e., consump-tion by larvae. Morphine appears to evoke a differential response dependent on fly species. In the case of *L. sericata,* morphine retards larval development, yet it stimulates accelerated growth rates in *Ch. megacephala* (Bourel *et al.*, 1999).

Several compounds exert their effect during intrapuparial stages. One example is with the psy-chotropic drug amitriptyline, which extends the duration of both the post-feeding wandering stage and length of all intrapuparial stages (Goff & Lord, 2010). Some xenobiotics (e.g., flunitrazepam, phencyclidine) have no effect on larval development but delay the onset and duration of pupal development (Goff *et al.*, 1991; Baia *et al.*, 2016). Presumably pharate adult stages also require longer periods to progress toward adult eclosion. Still other drugs (e.g., diazepam and heroin) display a differential effect in that they accelerate larval development but retard intrapuparial development (Goff *et al.*, 1991; Carvalho *et al.*, 2001). Diazepam's depressant effect becomes evident before pupation, as the drug lengthens the duration of pupariation in *Chrysomya albiceps* and *Ch. putoria* (Carvalho *et al.*, 2001).

18.4.3   Lethal effects of xenobiotics

In some instances, an ingested xenobiotic evokes death. For example, high concentrations of nicotine, sodium pentothal, and methamphetamine results in death either of pupae or pharate adults (Sadler

et al., 1997; Hedouin *et al.*, 1999; Magni *et al.*, 2016). Larval and pupal death have been reported for a variety of insecticides and some heavy metals. The most obvious impact of lethal xenobiotics to forensic entomology is that when no insects survive, estimation of a PMI is seemingly not possible. In those instances, an understanding of when or what developmental stage death occurred could poten-tially still permit estimation of a minimum develop-mental time prior to the onset of death.

Chapter review

Insects can serve as surrogates for human tissues

- In instances in which a corpse or set of remains is not suitable for typical toxicological analysis associated with autopsy, insects can serve as viable surrogates for human tissues. The condi-tions that favor insect proxies are when body fluids and soft tissues are limited or not avail-able at all such as in advanced stages of decay, or when the remains are skeletonized, severely burnt, or mummified. Presumably tissues that have undergone saponification or that have been submerged in aquatic environments for an extended period of time are also of limited value to toxicological analyses. Since necrophagous species consume tissues of the deceased, and thus acquire the chemical compounds contained within those tissues, the insects can, in turn, be used to extract and detect any type of drug, tox-icant, pollutant, or metabolite present in the food source. In other words, toxicological screening of the decedent can still occur using necrophagous insects.

- The term entomotoxicology was coined to describe the subfield of forensic entomology that deals with analysis of chemical compounds present in arthropods that fed on a human corpse. To an extent the term is somewhat prob-lematic in that it really requires a thorough understanding and use of techniques from forensic toxicology, coupled with forensic ento-mology. Entomotoxicology also implies a broader scope that encompasses any type of chemical detection in insects and related arthro-pods, from a wide variety of sources and not restricted to just a human corpse. Perhaps the most clarity is derived from the term forensic entomotoxicology, which has emerged as a refined version of entomotoxicology in that it is concerned with the use of insects in legal contexts as indirect sources of toxicological evidence in the absence of human tissues or other direct forensic samples.

- Forensic entomotoxicology is organized around a death investigation with two central themes: What caused death of the individual and when did death occur? Obviously, those themes are investigated using insects as direct or indirect evidence for toxicological analysis.

Detection of drugs, toxins, and pollutants in carrion-feeding insects

- In theory, any chemical compound ingested by an insect should be extractable from its tissues and be detectable by appropriate analytical methods. Certainly, there is ample evidence in the literature to demonstrate that various narcotics, psychotropics, heavy metals, insecticides, and other toxicants can be detected in necrophilous insects following consumption of human remains. Similarly, numerous animal models ranging from mice, rats, rabbits, and pigs, as well as isolated tissues have been used in laboratory experiments aimed at chemical detection in necrophagous insects. In short, a plethora of evidence exists to demonstrate proof of concept.

- From a toxicology perspective, detection of a chemical substance alone is generally insufficient to draw many conclusions regarding a forensic investigation, especially since insects are indirect forensic matrices. In many respects, identification of a xenobiotic can be thought of as the starting point not the conclusion of toxicological analysis. Most forms of forensic toxicological assessment require identification and quantification of chemical compounds. This is especially critical for any toxicological analysis associated with determination of the cause and manner of death.

- Xenobiotic levels in fly larvae have been found to be highly variable in relation to tissue levels and heavily influenced by where the insect was collected from, and presumably fed, on the corpse. From a qualitative analysis perspective, such extreme variability casts doubt on the reliability of xenobiotic detection. The issue in question is not when a drug or toxicant has been detected in insect tissues, otherwise known as positive detection. Instead, the issue is when no xenobiotic is detected or identified, meaning negative detection. In other words, concerns have been raised as to whether negative detection in insect proxies represents a false negative, meaning that drugs or toxins may indeed be present but could not be reliably detected in insect tissue. At present, there are no methods available to ascertain when false negatives or false positives occur with insect samples.

- Insect larvae are not considered useful for quantitative toxicological analysis. More to the point, no reliable correlation between xenobiotic levels in insects and the tissues in which they have consumed has been established. This is a major limitation for using insects as proxies for human tissues. In fact, some toxicologists have argued that insect use in any toxicological analysis has almost no value for forensic casework.

- Assuming that the concentrations in the insect can be reliably determined, the question then becomes what do the toxicant levels mean to the forensic investigation? Is the concentration determined in the insect relatable to the concentration in the corpse at the agonal period or moment of death? It would seem unlikely. Typically, insects are only used as surrogates when soft tissues are not available. This effectively means that insect samples are collected several days, to weeks to even months after the individual died. Considering the enormous lag time between death and toxicological analysis, numerous changes are expected to occur in drug or toxicant concentrations first in the decomposing corpse and then in the insect. Many of these changes are largely unpredictable and just as many have not been examined.

Detection of gunshot and explosive residues in or using insects

- Necrophagous fly larvae consuming tissues of a corpse that has been exposed to explosives or bullet fragments can accumulate the chemical signatures that imply foul play. Why this is important is because evidence to implicate a particular bomb-maker often lies in the chemical composition of the explosive device, which may leave a chemical signature in objects found in the explosion area, including the unfortunate victims. Similarly, a gunshot leaves distinctive patterns termed gunshot residues that can be modified during postmortem decomposition and burial and by the activity of insects.

- One downside to this approach to toxicological detection is that it likely has limited utility with late colonizing insects or those that feed deep within the body cavity. The assumption is that the further from the skin surface or the later in decay, the less residues that are present. The exception would be if the victim did not die immediately after injury, permitting transport of chemical residues throughout the body via circulating blood.

Chemical impact on insect succession and development

- Regardless of whether a xenobiotic can be accurately quantified in insect tissues, consumption of any type of "unexpected" compound can potentially alter or disrupt insect development. The insect did not intentionally seek out and ingest the xenobiotic. Rather, a necrophagous species is intent on consuming carrion soft tissues and fluids to meet its nutritional needs. A drug, toxin, or pollutant present in human tissues is not associated with hunger activation nor would an exogenous xenobiotic be expected to activate the satiety center. So, intake of a xenobiotic during ingestion of human tissues is indeed unexpected from the perspective of the insect hunger drive. Any drug or toxicant present in human tissue was generally consumed by the individual to evoke a specific response, often triggering one or more signal transduction pathways. Many of these same compounds elicit similar, sometimes more pronounced effects when ingested by insect larvae or adults. Any compound that alters insect development has the potential to change the usefulness of that species for estimation of the postmortem interval
- Xenobiotics have the potential to impart three potential effects on insect development: accelerate the growth rate of one or more developmental stages, retardation or inhibition of development, and lethality. A fourth outcome is to not alter development at all, which implies the chemical compound is benign towards PMI estimation.
- Nearly all of the baseline information on xenobiotic influences on insect development are derived from carefully controlled laboratory studies. However, even in many of those experiments, animal models or isolated tissues were used instead of human tissues.

Test your understanding

Level 1: knowledge/comprehension

1. Define the following terms:

   (a) entomotoxicology
   (b) forensic entomotoxicology
   (c) developmental arrestant
   (d) sequestration
   (e) tropism
   (f) bioaccumulation.

2. Match the terms (i–vi) with the descriptions (a–f).

   (a) An agent that decreases the overall length of insect development with no effect on puparial stages
   (i) Tropism

   (b) Preferential movement of a drug metabolite to fat body tissue
   (ii) nicotine

   (c) Examines bioaccumulation of toxins in aquatic insects
   (iii) codeine

   (d) Evokes death during intrapuparial stages of necrophagous flies
   (iv) gunshot reidue

   (e) Stimulates enhanced body mass in fly larvae
   (v) Forensic entomotoxicology

   (f) Spheroid particles in human tissues
   (vi) heroin

3. Describe the difference between the fields of entomotoxicology and forensic entomotoxicology.

4. What characteristics of insects make them useful to serve as surrogates for human tissues in toxicological analyses?

5. Explain why insects have not been useful for quantitative toxicological analysis of xenobiotics obtained during feeding on human tissues.

6. Discuss the potential problems that arise when necrophagous fly larvae consume a narcotic such as cocaine and in turn is used to estimate a postmortem interval.

Level 2: application/analysis

1. Explain how the inability to use necrophagous insects for quantification impacts the reliability of quantitative detection of xenobiotics.
2. Discuss the limitations with using fly and beetle larvae for detection of gunshot residues. Would the same limitations be expected with residues from other types of munitions?
3. Xenobiotics that are sequestered within necrophagous fly larvae can at times be retained within the puparia. Explain how this can be beneficial to qualitative detection of a chemical compound.
4. Discuss some of the limitations in trying to relate xenobiotic concentrations in fly puparia to those in the corpse.

Level 3: synthesis/evaluation

1. Heroin is known to increase body mass in larvae of *Sarcophaga peregrina*. Extensive testing has not occurred with other insect species. However, when considering the potential impact on estimation of the postmortem interval, speculate on whether this heroin effect would be more disruptive to the use of necrophagous fly or beetle larvae. Provide a biological explanation to justify your answer.
2. Describe some of the experiments that need to be performed to overcome the limitations of quantitative analysis of methadone and its metabolites in silphid larvae.

References cited

Abd Rashid, R. & Ahmad, N.W. (2012) Blowfly, *Chrysomya megacephala* as an alternative specimen in determination of gunshot residue. In: *2012 IEEE Symposium on Business, Engineering and Industrial Applications*, pp. 542–547). IEEE.

Baia, T.C., Campos, A., Wanderley, B.M.S. & Gama, R.A. (2016) The effect of flunitrazepam (Rohypnol®) on the development of *Chrysomya megacephala* (Fabricius, 1794) (Diptera: Calliphoridae) and its implications for forensic entomology. *Journal of Forensic Sciences* 61(4): 1112–1115.

Bourel, B., Hédouin, V., Martin-Bouyer, L., Bécart, A., Tournel, G., Deveaux, M. & Gosset, D. (1999) Effects of morphine in decomposing bodies on the development of *Lucilia sericata* (Diptera: Calliphoridae). *Journal of Forensic Science* 44(2): 354–358.

Bourel, B., Tournel, G., Hedouin, V., Deveaux, M., Goff, M.L. & Gosset, D. (2001) Morphine extraction in necrophagous insects remains for determining ante-mortem opiate intoxication. *Forensic Science International* 120(1–2): 127–131.

Bushby, S.K., Thomas, N., Priemel, P.A., Coulter, C.V., Rades, T. & Kieser, J.A. (2012) Determination of methylphenidate in Calliphorid larvae by liquid–liquid extraction and liquid chromatography mass spectrometry – forensic entomotoxicology using an in vivo rat brain model. *Journal of Pharmaceutical and Biomedical Analysis* 70: 456–461.

Campobasso, C.P., Gherardi, M., Caligara, M., Sironi, L. & Introna, F. (2004) Drug analysis in blowfly larvae and in human tissues: a comparative study. *International Journal of Legal Medicine* 118(4): 210–214.

Campobasso, C.P., Bugelli, V., Carfora, A., Borriello, R. & Villet, M. (2020) Advances in entomotoxicology: weakness and strengths. In: J.H. Byrd & J.K. Tomberlin (eds) *Forensic Entomology: The Utility of Arthropods in Legal Investigations*, 3rd edn, pp. 287–310. CRC Press, Boca Raton, FL.

Carvalho, L.M., Linhares, A.X. & Trigo, J.R. (2001) Determination of drug levels and the effect of diazepam on the growth of necrophagous flies of forensic importance in southeastern Brazil. *Forensic Science International* 120(1–2): 140–144.

Chophi, R., Sharma, S., Sharma S. & Singh, R. (2019) Forensic entomotoxicology: current concepts, trends and challenges. *Journal of Forensic and Legal Medicine* 67: 28–36.

Duarte, M.D.P.N. (2015) Firing distance estimation through the quantification of gunshot residues in blowfly larvae (Calliphoridae family) using inductively coupled plasma-mass spectrometry. Master's Thesis Dissertation, University of Porto.

Durdle, A., van Oorschot, R.A. & Mitchell, R.J. (2013) The morphology of fecal and regurgitation artifacts deposited by the blow fly *Lucilia cuprina* fed a diet of human blood. *Journal of Forensic Sciences* 58(4): 897–903.

Faria, L.D.B., Orsi, L., Trinca, L.A. & Godoy, W.A.C. (1999) Larval predation by *Chrysomya albiceps* on *Cochliomyia macellaria*, *Chrysomya megacephala* and *Chrysomya putoria*. *Entomologia Experimentalis et Applicata* 90(2): 149–155.

Gagliano-Candela, R. & Aventaggiato, L. (2001) The detection of toxic substances in entomological specimens. *International Journal of Legal Medicine* 114(4–5): 197–203.

Goff, M.L. & Lord, W.D. (2010) Entomotoxicology: insects as toxicological indicators and the impact of drugs and toxins on insect development. In: J.H. Byrd & J.L. Castner (eds) *Forensic Entomology: The Utility of Arthropods in Legal Investigations*, 2nd edn, pp. 427–436. CRC Press, Boca Raton, FL.

Goff, M.L., Omori, A.I. & Goodbrod, J.R. (1989) Effect of cocaine in tissues on the rate of development of *Boettcherisca peregrina* (Diptera: Sarcophagidae). *Journal of Medical Entomology* 26: 91–93.

Goff, M.L., Brown, W.A., Hewadikaram, K.A. & Omori, A.I. (1991) Effects of heroin in decomposing tissues on the development rate of *Boettcherisca peregrina* (Diptera: Sarcophagidae) and implications of this effect on estimations of postmortem intervals using arthropod development patterns. *Journal of Forensic Sciences* 36: 537–542.

Gosselin, M., Wille, S.M., Fernandez, M.D.M.R., Di Fazio, V., Samyn, N., De Boeck, G. & Bourel, B. (2011) Entomotoxicology, experimental set-up and interpretation for forensic toxicologists. *Forensic Science International* 208(1–3): 1–9.

Hedouin, V., Bourel, B., Martin-Bouyer, L., Becart, A., Tournel, G., Deveaux, M. & Gossett, D. (1999) Morphine perfused rabbits: a tool for experiments in forensic entomology. *Journal of Forensic Sciences* 44: 347–350.

Huchet, J. B. & Greenberg, B. (2010) Flies, Mochicas and burial practices: a case study from Huaca de la Luna, Peru. *Journal of Archaeological Science* 37(11): 2846–2856.

Introna, F., Campobasso, C.P. & Goff, M.L. (2001). Entomotoxicology. *Forensic Science International* 120(1–2): 42–47.

Kharbouche, H., Augsburger, M., Cherix, D., Sporkert, F., Giroud, C., Wyss, C., Chamod, C. & Mangin, P. (2008) Codeine accumulation and elimination in larvae, pupae, and imago of the blowfly *Lucilia sericata* and effects on its development. *International Journal of Legal Medicine* 122(3): 205–211.

Kulstein, G., Amendt, J. & Zehner, R. (2015) Blow fly artifacts from blood and putrefaction fluid on various surfaces: a source for forensic STR typing. *Entomologia Experimentalis et Applicata* 157(3): 255–262.

Labella, G., Martra, G., Pazzi, M., Testi, R., Vincenti, M., Dadour, I. & Magni, P. (2014) Detection of gunshot residues (GSR) in blowfly larvae using scanning electron microscopy equipped with energy dispersive X-ray microanalysis (SEM-EDX) and inductively coupled plasma mass spectrometry (ICP-MS). In: *11th Meeting of the European Association for Forensic Toxicology*, Vol. 1, No. 1, pp. 74.

LaGoo, L., Schaeffer, L.S., Szymanski, D.W. & Smith, R.W. (2010) Detection of gunshot residue in blowfly larvae and decomposing porcine tissue using inductively coupled plasma mass spectrometry (ICP-MS). *Journal of Forensic Sciences* 55: 624–632.

Lü, Z., Zhai, X., Zhou, H., Li, P., Ma, J., Guan, L. & Mo, Y. (2014) Effects of ketamine on the development of forensically important blowfly *Chrysomya megacephala* (F.) (Diptera: Calliphoridae) and its forensic relevance. *Journal of Forensic Sciences* 59(4): 991–996.

Magni, P.A., Pacini, T., Pazzi, M., Vincenti, M. & Dadour, I.R. (2014) Development of a GC–MS method for methamphetamine detection in *Calliphora vomitoria* L. (Diptera: Calliphoridae). *Forensic Science International* 241: 96–101.

Magni, P.A., Pazzi, M., Vincenti, M., Alladio, E., Brandimarte, M. & Dadour, I.R. (2016) Development and validation of a GC–MS method for nicotine detection in *Calliphora vomitoria* (L.) (Diptera: Calliphoridae). *Forensic Science International* 261: 53–60.

Manhoff, D.T., Hood, I., Caputo, F., Perry, J., Rosen, S. & Mirchandani, H.G. (1991) Cocaine in decomposed human remains. *Journal of Forensic Science* 36(6): 1732–1735.

Miller, M.L., Lord, W.D., Goff, M.L., Donnelly, B., McDonough, E.T. & Alexis, J.C. (1994) Isolation of amitriptyline and nortriptyline from fly puparia (Phoridae) and beetle exuviae (Dermestidae) associated with mummified human remains. *Journal of Forensic Science* 39(5): 1305–1313.

Mullany, C., Keller, P.A., Nugraha, A.S. & Wallman, J.F. (2014) Effects of methamphetamine and its primary human metabolite, p-hydroxymethamphetamine, on the development of the Australian blowfly *Calliphora stygia*. *Forensic Science International* 241: 102–111.

Musvasva, E., Williams, K.A., Muller, W.J. & Villet, M.H. (2001) Preliminary observations on the effects of hydrocortisone and sodium methohexital on development of *Sarcophaga* (Curranea) *tibialis* Macquart (Diptera: Sarcophagidae), and implications for estimating post mortem interval. *Forensic Science International* 120(1–2): 37–41.

Norris, K.R. (1965) The bionomics of blowflies. *Annual Review of Entomology* 10: 47–68.

Nuorteva, P. (1977) Sarcosaprophagous insects as forensic indicators. In: G.G. Tedeshi, W.G. Eckert & L.G. Tedeschi (eds) *Forensic Medicine: A Study in Trauma and Environmental Hazards*, pp. 1072–1095. Saunders, Philadelphia, PA.

O'Brien, C. & Turner, B. (2004) Impact of paracetamol on *Calliphora vicina* larval development. *International Journal of Legal Medicine* 118(4): 188–189.

Pounder, D.J. (1991) Forensic entomo-toxicology. *Journal of the Forensic Science Society* 31(4): 469–472.

Rezende, F., Alonso, M.A., Souza, C.M., Thyssen, P.J. & Linhares, A.X. (2014) Developmental rates of immatures of three *Chrysomya* species (Diptera: Calliphoridae) under the effect of methylphenidate hydrochloride, phenobarbital, and methylphenidate hydrochloride associated with phenobarbital. *Parasitology Research* 113(5): 1897–1907.

Rivers, D. & Geiman, T. (2017) Insect artifacts are more than just altered bloodstains. *Insects* 8(2): 37. https://doi.org/10.3390/insects8020037.

Roeterdink, E.M., Dadour, I.R. & Watling, R.J. (2004) Extraction of gunshot residues from the larvae of the forensically important blowfly *Calliphora dubia* (Macquart) (Diptera: Calliphoridae). *International Journal of Legal Medicine* 118(2): 63–70.

Sadler, D.W., Robertson, L., Brown, G., Fuke, C. & Pounder, D.J. (1997) Barbiturates and analgesics in *Calliphora vicina* larvae. *Journal of Forensic Science* 42(3): 481–485.

da Silva, E.I.T., Wilhelmi, B. & Villet, M.H. (2017) Forensic entomotoxicology revisited – towards a professional standardisation of study designs. *International Journal of Legal Medicine* 131: 1399–1412.

Stojak, J. (2017) Use of entomotoxicology in estimating post-mortem interval and determining cause of death. *Issues of Forensic Science* 295: 56–63.

Striman, B., Fujikawa, A., Barksdale, L. & Carter, D.O. (2011) Alteration of expired bloodstain patterns by *Calliphora vicina* and *Lucilia sericata* (Diptera: Calliphoridae) through ingestion and deposition of artifacts. *Journal of Forensic Sciences* 56: S123–S127.

Tabor, K.L., Fell, R.D., Brewster, C.C., Pelzer, K. & Behonick, G.S. (2005) Effects of antemortem ingestion of ethanol on insect successional patterns and development of *Phormia regina* (Diptera: Calliphoridae). *Journal of Medical Entomology* 42(3): 481–489.

Tracqui, A., Keyser-Tracqui, C., Kintz, P. & Ludes, B. (2004) Entomotoxicology for the forensic toxicologist: much ado about nothing? *International Journal of Legal Medicine* 118: 194–196.

Widmaier, E.P. (1999) *Why Geese Don't Get Obese (and We Do): How Evolution's Strategies for Survival Affect Our Everyday Lives.* W.H. Freeman, New York.

Wolten, G.M., Nesbitt, R.S., Calloway, A.R., Loper, G.L. & Jones, P.F. (1979) Particle analysis for the detection of gunshot residue. I: scanning electron microscopy/energy dispersive X-ray characterization of hand deposits from firing. *Journal of Forensic Sciences* 24: 409–422.

Zanetti, N.I., Ferrero, A.A. & Centeno, N.D. (2016) Determination of fluoxetine in *Dermestes maculatus* (Coleoptera: Dermestidae) by a spectrophotometric method. *Science and Justice* 56: 464–467.

Zdarek, J. & Fraenkel, G. (1972) The mechanism of puparium formation in flies. *Journal of Experimental Zoology* 179: 315–323.

Zou, Y., Huang, M., Huang, R., Wu, X., You, Z., Lin, J., Huang, H., Qiu, X. & Zhang, S. (2013) Effect of ketamine on the development of *Lucilia sericata* (Meigen) (Diptera: Calliphoridae) and preliminary pathological observation of larvae. *Forensic Science International* 226(1–3): 273–281.

Supplemental reading

Bugelli, V., Papi, L., Fornaro, S., Stefanelli, F., Chericoni, S., Giusiani, M., Vanin, S. & Campobasso, C.P. (2017) Entomotoxicology in burnt bodies: a case of maternal filicide-suicide by fire. *International Journal of Legal Medicine* 131(5): 1299–1306.

Dayananda, R. & Kiran, J. (2013) Entomotoxicology. *International Journal of Medical Toxicology and Forensic Medicine* 3: 71–74.

de Lima, L.A.S., Baia, T.C., Gama, R.A., da Silva Gasparotto, L.H. & Lima, K.M. (2014) Near infrared spectroscopy as an emerging tool for forensic entomotoxicology. *NIR News* 25(8): 5–7.

Gião, J.Z., Reigada, C., Moretti, T.C. & Godoy, W.A.C. (2017) Effect of psychoactive drugs on demographic parameters of the blow fly *Chrysomya albiceps* (Wiedemann) (Diptera: Calliphoridae). *Journal of Forensic Research* 8(400). https://doi.org/10.4172/215 7-7145.1000400.

Gosselin, M., Fernandez, M.D.M.R., Wille, S.M., Samyn, N., De Boeck, G. & Bourel, B. (2010) Quantification of methadone and its metabolite 2-ethylidene-1,5-dimethyl-3,3-diphenylpyrrolidine in third instar larvae of *Lucilia sericata* (Diptera: Calliphoridae) using liquid chromatography-tandem mass spectrometry. *Journal of Analytical Toxicology* 34(7): 374–380.

Gunn, J.A., Shelley, C., Lewis, S.W., Toop, T. & Archer, M. (2006) The determination of morphine in the larvae of *Calliphora stygia* using flow injection analysis and HPLC with chemiluminescence detection. *Journal of Analytical Toxicology* 30(8): 519–523.

Oliveira, J.S., Baia, T.C., Gama, R.A. & Lima, K.M. (2014) Development of a novel non-destructive method based on spectral fingerprint for determination of abused drug in insects: an alternative entomotoxicology approach. *Microchemical Journal* 115: 39–46.

Pien, K., Laloup, M., Pipeleers-Marichal, M., Grootaert, P., De Boeck, G., Samyn, N., Boonen, T., Vits, K. & Wood, M. (2004) Toxicological data and growth characteristics of single post-feeding larvae and puparia of *Calliphora vicina* (Diptera: Calliphoridae) obtained from a controlled nordiazepam study. *International Journal of Legal Medicine* 118(4): 190–193.

Samnol, A., Kumar, R., Mahipal Singh, S. & Parihar, K. (2020) Study of insect larva used to detect toxic substance through decomposed bodies. *International Journal of Forensic Science & Pathology* 7(1): 420–422.

Additional resources

American Board of Forensic Toxicology: http://abft.org/

Forensic entomology: https://aboutforensics.co.uk/forensic-entomology/

Identification of Psychoactive drugs: https://www.waters.com/waters/LC-MS-forensic-toxicology?cid=134874113&xcid=ppc-ppc_02585&utm_source=Bing&utm_medium=CPC&utm_campaign=Bing_FOR_Forensic_Toxicology_Non_Brand_US

Overview of forensic toxicology, National Institute of Justice: https://nij.ojp.gov/topics/forensics/forensic-toxicology

Society of Forensic Toxicology: http://www.soft-tox.org/

Toxicology procedural manuals: https://www.dfs.virginia.gov/laboratory-forensic-services/toxicology/procedure-manuals/

Chapter 19

Application of molecular methods to forensic entomology

Co-authored by Evan Wong

Department of Biological Sciences, University of Cincinnati

Overview

It is well known that entomological evidence can aid in investigations of human death, illegal use of natural resources, and even cases of human and animal neglect. The past two decades have been an exciting and important time in forensic entomology with the advent of new molecular methods. These methods can provide detailed information on: (i) the identity of victims or suspects; (ii) the identity of insect species present at the crime scene, and (iii) the location of the crime. Most of the currently used molecular methods involve examination of an organism's DNA, but not all. The aims of this chapter are to discuss several common molecular methods used in forensic entomology, look at some of their strengths (and weaknesses) when compared to "traditional" approaches, and examine some newly developed, molecular-based procedures that may be more widely applied to forensic settings in the future.

The big picture

- Molecular Methods – living things can be defined by their DNA.
- Evidence collection: preserve the DNA integrity.
- Molecular methods of species identification.
- DNA Barcoding Protocol
- Problems encountered in barcoding projects
- Gut content: victim and suspect identifications.
- Molecular methods and population genetics
- Molecular methods: non-DNA based
- Validating molecular methods for use as evidence
- Future directions

19.1 Molecular methods – living things can be defined by their DNA

All living organisms contain genetic information, encoded as a sequence of four bases, or nucleotides, in the double-helix macromolecule known as deoxyribonucleic acid or DNA (Figure 19.1 and Box 19.1). DNA is not simply present in all life forms, but also varies in the information that is present across different species, populations of the same species, and individuals. Recent research has even shown that there are copy-number variations at some loci between identical twins (Bruder *et al.*, 2008). The point is that DNA serves as a unique molecular "fingerprint" of an individual. As such, this provides a good justification for the use of DNA data in forensic

---

The Science of Forensic Entomology, Second Edition. David B. Rivers and Gregory A. Dahlem.
© 2023 John Wiley & Sons Ltd. Published 2023 by John Wiley & Sons Ltd.
Companion website: www.wiley.com/go/forensicentomology2

(a)   (b)

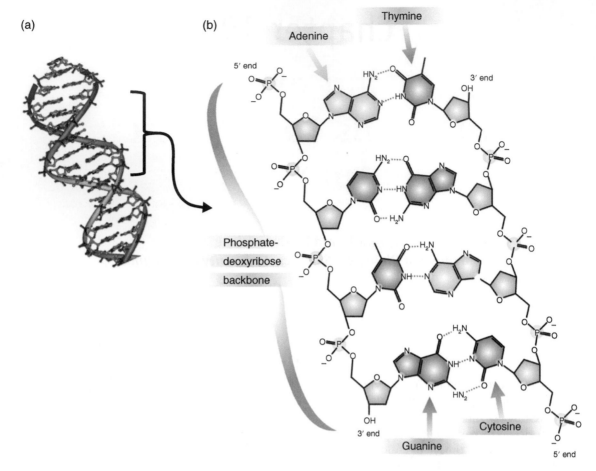

Figure 19.1 Deoxyribonucleic acid (DNA). (a) The three-dimensional structure of DNA, the double helix. *Source*: Jerome Walker,Dennis Myts/Wikimedia Commons/Public domain. (b) The molecular backbone of DNA, with the four nucleotide bases: adenine, thymine, guanine, and cytosine. *Source*: Madeleine Price Ball/Wikimedia Commons/CC BY-SA 3.0.

applications where the identification of the victim or suspect, the identification of species of insects present at the crime scene, and the identification of the location of the crime (among other things) is of the highest importance. Typically, when you hear someone refer to molecular methods used in forensic entomology, they are referring to the utilization of DNA to answer these types of questions.

As discussed in several other chapters, entomological evidence can aid in cases associated with human death (e.g. Goff, 1991), illegal use of natural resources (e.g. Gavin *et al.*, 2010), and instances of human neglect (e.g. de Souza Barbosa *et al.*, 2008). Fly larvae play an important role as physical evidence due to the fact that they are the primary group of decomposers that begin life by directly feeding on the corpse. Therefore, larval age is a useful estimator of the postmortem interval (PMI) (See Chapter 12).

However, development rates of larvae are not static, and are species-specific (e.g. Kamal, 1958). Consequently, the first step in using larvae as evidence is to identify to species of the specimens recovered at the crime scene. An individual insect's DNA sequence remains exactly the same through all life stages, a procedure known as **DNA barcoding** (DNA-based species identification) can be used as a tool to identify specimens from egg to larva to adult (Meiklejohn *et al.*, 2012). DNA barcoding uses short, specific portions of a specimen's DNA sequence to identify it as a member of a particular species. This procedure has two distinct phases. First, it is necessary to have a reference collection of DNA data from specimens whose species identity is known from standard morphological identification. In the case of forensically important Diptera, this confirmed species identification is based on mature

Box 19.1   Human DNA profiles

The most commonly used method for human identification is known as **DNA profiling** (or DNA testing or DNA fingerprinting). DNA profiling techniques involve very different methods from those used for DNA-based species identifications. Individual identifications are based on noncoding regions of the DNA that show short tandem repeats (STR). **Short tandem repeats** (also known as microsatellites) are repeating sequences of DNA, typically 2–6 base pairs in length (e.g. —A—C—G—A—C—G—A—C—G—, represents three repeats of the three base pairs "ACG"). It is estimated that the humans have more than 16,000 STRs across their genome (Rockman & Wray, 2002). However, only a small number of these STRs are used for human identification (typically 20 in the US). This allows for consistent genetic testing across multiple individuals (Butler, 2006). Each person's genome contains two copies (alleles) of each STR locus. At each STR locus used for human identification there are many distinct alleles in the population as a whole, with each allele being characterized by its specific number of copies of the STR repeat motif. In brief, the core STR loci are amplified via PCR from a DNA sample collected from a crime scene and compared to a reference sample in a database to determine if the STR genotypes match at every STR locus tested. An exact match indicates that there is an extremely low probability that the crime-scene DNA sample came from a different individual from the reference sample. Human DNA typing provides a high level of discrimination and probability of identification (1 in 13 trillion for African-Americans, and 1 in 3.3 trillion for Caucasian-Americans for an exact match at 13 STR loci) (Applied Biosystems, 2012). While there are several human DNA databases in use, the largest is the United States' Combined DNA Index System (CODIS), followed by the United Kingdom's National DNA Database (NDNAD), both containing more than 5 million individuals each.

reference (voucher) specimens should be deposited in a recognized natural history museum, where they are available for other scientists to confirm or update the species identifications.

Entomologist traditionally carry out species identifications of immature insects by using larval keys (when available), or adult morphological character keys (e.g. Whitworth, 2006 for Calliphoridae) after the specimens have been reared to adulthood. Problems arise, however, in that larvae of many forensically important species are morphologically similar and may be difficult, if not impossible, to accurately identify (Wells & Sperling, 2001). Rearing larvae to adulthood is not always simple and can take several weeks or more of time when all goes well in the rearing process. If things do not go well, the larvae may not complete development or may die (or be severely damaged) in transit from the collection site to the laboratory (Sperling et al., 1994).

The use of molecular techniques in species identification has the potential to overcome many of the inherent difficulties associated with morphological species identifications. Molecular techniques have even been shown to have the ability to separate **cryptic species** (species that are morphologically indistinguishable) (Lowenstein et al., 2009; Spillings et al., 2009). Several studies over the past decade have successfully used DNA-based methods to identify forensically important flies (e.g. Meiklejohn et al., 2009). Not only are these techniques effective in forensic entomology, but also in criminal investigations involving the illegal use of wildlife, such as whale and dolphin products sold in restaurants (Baker et al., 1996), tracking of the African elephant ivory trade (Wasser et al., 2004), and endangered fish species being served as sushi (Lowenstein et al., 2009).

19.2   Evidence collection: preserve the DNA integrity

An intuitive place to begin is to discuss appropriate insect collection protocols that retain the integrity of the DNA and minimize contamination. In other words, we will discuss means to preserve the evidence for future analyses, a topic that falls under the dominion of **continuity of evidence** discussed in Chapter one. All the techniques discussed in this chapter would be for nothing if the evidence at a crime scene is not collected or stored properly. In general, crime scene investigators are

adult specimens. The second phase is to compare crime-scene material to the reference collection. As described below, the crime-scene material for forensically important Diptera are generally larvae. The aware that a forensic entomologist can help determine the postmortem interval. However, only recently it has become appreciated that DNA-based identification techniques can increase accuracy, speed, and precision in criminal investigations.

As soon as an organism dies, the DNA present in the cells will begin to degrade. This degradation is due partially to an enzyme, known as DNAase, and other chemical reactions that take place (King & Porter, 2004). Thus, fresh, unfrozen samples will give the highest yield of DNA for molecular analyses (Dessauer *et al.*, 1996), but it is a rarity that DNA extraction of specimens will take place directly after collection. If the DNA is to be extracted after a delay, then the specimen itself must be preserved in a way that allows the best possible DNA extraction. The simplest and most used method for preserving insects and their DNA is to preserve live or recently killed specimens, or tissue from the specimens, in ethanol (95–100%), and this is likely to be the method of choice for collection of larvae from a crime scene for DNA-based species identification. However, if they are also to be used for methods requiring RNA isolation (such as estimation of larval age by gene expression levels using RT-PCR or related techniques (see Chapter 12), quick freezing using dry ice (or a dry ice + ethanol bath) at the crime scene (preferred), or as soon as possible after collection, is absolutely necessary to ensure the integrity of the nucleic acids. Preservation agents containing formalin, formaldehyde, or similar fixatives will excessively "harden" insect tissues and complicate DNA extraction, and thus should be avoided.

Ethanol (95–100%) has been recommended as the standard preserving solution due to its ability to quickly penetrate the cellular membranes and inactivate DNAases. Ethanol is the preservation method of choice for collection of larvae at a crime scene, but can be problematic when subsequent morphological identifications are to be carried out (as, for example, during reference database building for calyptrate Diptera). Alcohol preservation of adult flies causes problems involving the manipulation of appendages and genitalia (because of muscle stiffening) and the ability to distinguish color differences, especially of fine setae (see Appendix II). An alternative to immediate placement of adult specimens in alcohol is to keep dead specimens as cool as possible when collected in the field, and then transfer to a freezer (−20° C or lower) until

they can be sorted, pinned, and tissue removed for DNA extraction (See Appendix I).

When storing collected tissue or specimen samples prior to DNA extraction, care must be taken to store in appropriate amount of ethanol solution. It is recommended that a minimum ratio 1 : 3 of insect to ethanol volume be used (e.g. for 10 ml of insects, you should have at least 30 ml of ethanol). Further information on techniques for preserving DNA has been discussed by Hillis *et al.* (1996) and summarized by Arctander (1988).

19.3   Molecular methods of species identification

In 1983, Kary Mullis developed the biochemical technique known as the **polymerase chain reaction** (PCR), which transformed the world of molecular biology, and subsequently made DNA-based forensic entomology feasible. This method, for which Mullis along with Michael Smith were awarded the Nobel Prize in Chemistry, is able to amplify miniscule amounts of DNA (a single or few copies) and generate hundreds of millions of identical copies of the DNA fragment through the thermal cycling technique (Figure 19.2). Prior to the advent of PCR, obtaining DNA sequence data was an expensive, grueling, and time-consuming process. Pre-PCR technology was able to generate sequence data for some fly species, such as *Drosophila yakuba* (Clary & Wolstenholme, 1985) and *Phormia regina* (Goldenthal *et al.*, 1991), but not in an effective manner that could translate directly to forensic entomology.

The first application of molecular methods to the identification of forensically important insects focused on sequencing genes encoding for cyctochrome c oxidase subunits (COI + COII) (Sperling *et al.*, 1994). Mitochondrial DNA (mtDNA) was isolated from three necrophagous fly species, *Protophormia terraenovae*, *Lucilia sericata*, and *L. illustris* and compared to each other. Differences were found at specific locations in the mtDNA where restriction enzymes bind and digest (cut) the mtDNA. **Restriction enzymes** are proteins that bind to the DNA and then digest it at a specific site (location), based upon a match between the DNA sequence and the specific restriction enzyme. For example, the restriction enzyme EcoRI (pronounced "eco R one") recognizes the DNA sequence 5′-**AAGTTC**-3′, and will digest DNA at any site matching this sequence. Theoretically, in a perfect

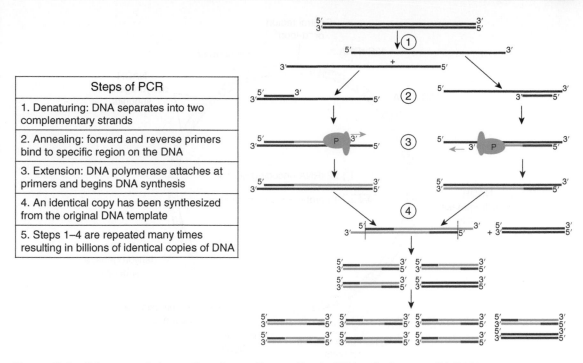

Steps of PCR
1. Denaturing: DNA separates into two complementary strands
2. Annealing: forward and reverse primers bind to specific region on the DNA
3. Extension: DNA polymerase attaches at primers and begins DNA synthesis
4. An identical copy has been synthesized from the original DNA template
5. Steps 1–4 are repeated many times resulting in billions of identical copies of DNA

Figure 19.2 Polymerase chain reaction. *Source*: Magnus Manske/Wikimedia Commons/CC BY-SA 2.5.

world, one could imagine that different species would all possess different restriction enzyme sites because the DNA sequences would be unique (i.e. have varying nucleotide sequences). Thus, when the mtDNA is digested this would produce short DNA fragments of varying lengths. These fragments can be visualized through gel electrophoresis, resulting in a different fragment pattern for each species. Sperling *et al.* (1994) were able to utilize this knowledge to develop an identification technique called PCR restriction fragment length polymorphisms (PCR-RFLPS). The approach uses the unique fragment lengths from the digested PCR products for each species, which has proved to be a fast and an inexpensive method for species identification. This research spearheaded the DNA-based forensics movement, even though the research can be viewed as a validation of a concept and not an actual test.

All eukaryotic organisms (except for a few unicellular groups) contain two independent genomes: the mitochondrial genome and the nuclear genome. Most molecular methods proposed for use in forensic species identification utilize portions of the mitochondrial genome (Figure 19.3), particularly the genes COI + COII (e.g. Malgorn & Coquoz, 1999; Vincent *et al.*, 2000;

Tan *et al.*, 2010; Guo *et al.*, 2012). Mitochondrial DNA has been shown to have particular benefits in its use for species identification. It is easily obtained due to the large quantity present in the cell relative to nuclear DNA and a wide variety of the class Insecta has been sequenced for many of the mitochondrial genes. Another benefit is that a phylogenetic tree that is based on mtDNA sequences tends to match actual species-level distinctions (Wiens & Penkrot, 2002). The reason for this is complicated, but it results from the population genetics of the mitochondrial genome and the fact that mitochondria are haploid and only inherited from the mother (Moore, 1995). Another advantage of the use of mitochondrial genes is that they have relatively high variability (polymorphism) and are flanked by transfer RNA (tRNA) genes that are conserved across species (i.e. many related species share the same DNA sequence for these tRNA genes). These conserved regions allow for universal primers to be created relatively easily that can be amplified in a wide variety of taxa. Primers, also known as oligonucleotides, are short segments of manufactured DNA that bind to specific regions of the genome during the process of a PCR cycle. The ability to connect primers to particular portions of DNA is what

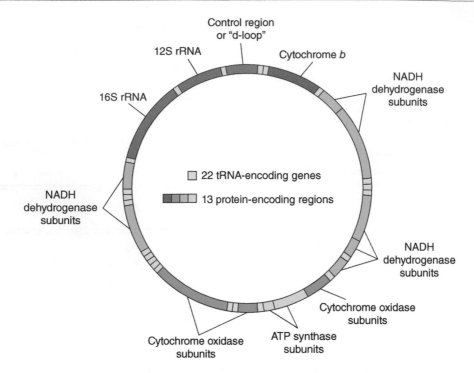

Figure 19.3   Human mitochondrial genome. *Source*: Courtesy of jhc, translated by Knopfkind.

allows amplification, or the production of multiple identical copies of DNA, to take place.

Although commonly used in species identification, the exclusive use of mtDNA is an area of controversy and it has been argued that species should not be separated based on this data alone. For example, it has been found that some forensically important insect species, such as the blow flies *Lucilia coeruleiviridis* and *L. mexicana*, cannot be distinguished based on COI sequences alone. However, *L. coeruleiviridis* and *L. mexicana* are easily separated based on morphological characteristics. To this end, authors have suggested using loci other than COI + II for identification. These other loci include other mtDNA genes (e.g., ND4, ND4L; Wallman *et al.*, 2005), specific nuclear genes (e.g., 28s rRNA; Stevens & Wall, 2001), or a combination of loci (Stevens, 2003; Nelson *et al.*, 2007).

The appropriate type of data analysis and methods used for determination of species identity will be dependent on the type of data used. The PCR-RFLP method, which we have previously described, produces as series of DNA fragments that are dependent on where the restriction enzymes cut the DNA. These DNA fragments are then measured on an agarose or polyacrylamide gel (Figure 19.4) and the fragments of an unknown specimen are referenced to the profile of a known,

vouchered specimen. If there is an exact match, it is hypothesized that they are the same species (Box 19.2). This method, however, does not look at the DNA sequence in its entirety, but only reflects those few base pairs involving the restriction enzyme recognition sites.

The "barcoding" approach to DNA-based species identification uses similar methods to compare an unknown specimen's DNA with unaligned sequences from a reference database. The resulting identification is based on the number of exact matches of nucleotides or by using pair-wise partial alignment (Little & Stevenson, 2007). Similarity methods, such as that derived by alignment using the Basic Local Alignment Search Tool (BLAST), have been an invaluable tool for comparing DNA sequences, but they are known to be inconsistent (Anderson & Brass, 1998; Woodwark *et al.*, 2001) and incorrect in some situations (e.g. Koski & Golding, 2001). Evolutionary relationships are not incorporated into the models that these similarity methods use and fail when DNA sequences from a particular specimen are more similar to sequences of a different species than they are to members of the same species (Little & Stevenson, 2007) due to intraspecific variation.

As an initial comparison, similarity methods (e.g. BLAST) provide useful information, particularly

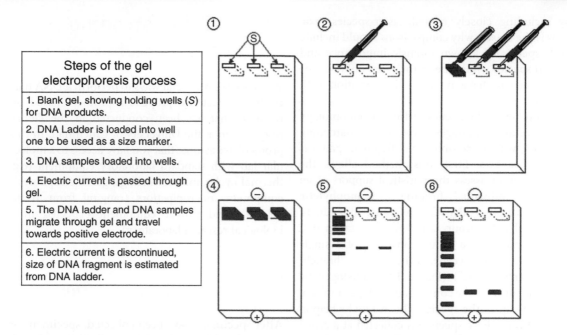

Steps of the gel electrophoresis process
1. Blank gel, showing holding wells (S) for DNA products.
2. DNA Ladder is loaded into well one to be used as a size marker.
3. DNA samples loaded into wells.
4. Electric current is passed through gel.
5. The DNA ladder and DNA samples migrate through gel and travel towards positive electrode.
6. Electric current is discontinued, size of DNA fragment is estimated from DNA ladder.

Figure 19.4   Gel electrophoresis. *Source*: Magnus Manske/Wikimedia Commons/Public domain.

Box 19.2   Barcoding from booze

Mezcal (mescal) is a famous Mexican distilled alcoholic beverage, similar in taste to tequila although stronger and harsher, made from the maguey plant, *Agave parryi*. Interestingly, this drink has a "worm" traditionally added to the drink before it is bottled and sold by the manufacturer. The worm is actually a caterpillar of the moth, *Hypopta agavis*, which lives in the maguey plant. Shokralla *et al.* (2010) attempted to extract DNA of the moth from the mescal, and were able to amplify and sequence the COI gene fragment. In a wider context, Shokralla *et al.* (2010) were able to demonstrate that DNA could potentially be extracted and amplified directly from ethanol of preserved specimens, even if the specimen itself was lost.

when a tentative morphological identification has been obtained. For example, if an investigator were to have a specimen believed to be *Phormia regina* based upon morphological identification, a BLAST search showing results consistent with his findings would add further support. Another instance, where similarity methods would be useful, would be in the identification of larvae. If the DNA obtained from larvae were to be run through a BLAST search, this could narrow the specimen down to the family level,

if not the genus. Knowing which genus the larvae comes from would further expedite the analyses by being able to tell which closely related specimens should be included in the DNA-barcoding library for phylogenetic analysis. Similarity methods appear to work best when sequences used are greater than 200 bases (Anderson & Brass, 1998) and longer sequences (>500 bp) are usually used for phylogenetic investigations on evolutionary relationships.

Phylogenetic methods, which are becoming more common in forensic entomology, utilize the entire DNA sequence of interest (i.e. portions of COI + COII), in a phylogenetic framework to identify unknown specimens to species. In this method, an unknown specimen is identified based upon a statistically supported grouping with vouchered reference specimens of a particular species. While early attempts in species identification utilized only a single individual from a small number of species for the reference database (Chen *et al.*, 2004; Zehner *et al.*, 2004), it is known that species in nature are genetically variable. Therefore, a proper reference database that is used for species identification must include that genetic variability and is described as essential when using DNA barcodes (Ekrem *et al.*, 2007). By including multiple individuals over the geographic range of each target species, the genetic variability can be incorporated. Furthermore, a proper reference database must allow for the recognition of incidental specimens that might be encountered at a crime scene by including closely related, sister-species taxa. This is the reason why comparisons should include both species of known forensic importance and related species whose life histories are unknown or are known to have no normal association with carrion.

Applying phylogenetics in forensic entomology has several advantages over other barcoding methods, which use only DNA fragment patterns on agarose gels. First, this method allows the investigator to assess the statistical support of an unknown individual forming a **monophyletic group** with other identified reference individuals. A monophyletic group is defined as a taxon that forms a **clade**, or a group consisting of a single species and all its decedents (i.e. a single "branch" on the "tree of life") (Figure 19.5). Phylogenetic methods allow for the consideration of geographical variation within the species. This would be helpful in situations where specimens collected at a crime scene are found to be genetically more similar to geographic variants present at a different location, possibly indicating that the victim had been moved. Phylogenetic methods take into account evolutionary processes and are able to accommodate the genetic variability of wild populations in ways other methods cannot (i.e., PCR-RFLP, BLAST, or distance-based).

19.4   DNA barcoding protocol

While there are several ways to perform a DNA barcoding project, we propose a protocol based on research that has been conducted in the DeBry research lab at the University of Cincinnati. In this protocol, we assume that you have access to a molecular lab with standard laboratory equipment (e.g. thermal cycler, microcentrifuge, gel electrophoresis equipment, micropipettors (ranging from 10 to 1000 μl), and similar equipment expected in a biological research laboratory).

19.4.1   Specimen collection

After specimens have been collected, specimens or tissue should be placed into 95–100% ethanol. Microcentrifuge tubes serve well as holding containers for entomological samples. Be sure that each tube holds tissue from only one specimen and that it is appropriately labeled for association with the mounted specimen and/or field collection data. (See **Appendix I** for additional information on collecting and mounting calyptrate Diptera).

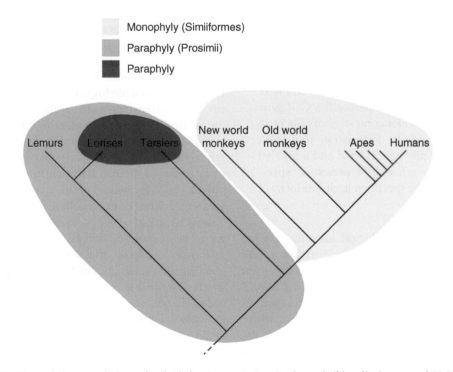

Figure 19.5   Monophyly, paraphyly, and polyphyly. *Source*: Petter Bøckman/Wikimedia Commons/ CC BY-SA 3.0.

19.4.2   Searching databases for sequence data

Much effort has been put forth into making sequence data widely available, free of charge, to the general public. There are two main public reference libraries of species identifiers which can be used to assign unknown specimens to species: (i) The International Nucleotide Sequence Database Collaborative, which is made up of GenBank in the United States, the European Molecular Biology Lab in Germany, and the DNA Data Bank of Japan; (ii Barcode of Life Database (BOLD), which is maintained by the University of Guelph in Ontario.

These databases should be searched systematically looking for individuals that are (i) geographically relevant to the task at hand, and (ii) closely-related to the unidentified specimens. An appropriate starting point for a DNA-barcoding project is to download all the sequences for the particular family of interest. In most situations it is not immediately known what the most closely related individuals will be. Therefore, we recommend that family level sequence data is an appropriate place to start, when obtaining sequence data. Sequence data should be downloaded from the databases in FASTA format and saved in a separate folder along with the accession information indicating where the sequences originated. FASTA format is a text-based format which originated from the FASTA software package (pronounced "fast-A") The format is used for representing either nucleotide sequences or peptide sequences. The nucleotides or amino acids are represented with single-letter codes.

19.4.3   DNA extraction

DNA extraction is accomplished with the use of standard commercial kits, such as "Qiagen DNEasy Blood and Tissue Kit." DNA extraction involves the isolation and purification of the DNA component of the tissue. While kits tend to be more expensive, they produce easily replicated results. Other techniques have been proposed, such as sonication (Hunter *et al.*, 2008), which is a nondestructive DNA extraction method. Nondestructive techniques are those that permit DNA isolation while keeping the entire specimen intact for possible morphological identification. Use of sonication however may not produce overall consistent results, and should be used cautiously. Only a small amount of tissue is necessary for DNA extraction, but the cellular component of the sample must be exposed to chemical reagents. Therefore, an insect leg should be cut into thirds, a larva should be sectioned so that cuticle membrane is ruptured, or a whole insect should be sliced into several pieces so that the extraction chemicals can easily contact the cellular components. The DNA extract should be stored in a mixture of water and TE Buffer (TE Buffer = Tris, a common pH buffer, and EDTA, which stands for ethylenediaminetetraacetic acid) or in molecular grade water at -20 °C or lower. Tissue can be stored under these conditions until ready for the next step of DNA amplification.

19.4.4   DNA amplification

The product of this particular step is an amplified region of DNA, based upon the selected primers for a particular region of the genome using PCR. Research should be conducted to determine which barcoding regions will be used. Cytochrome Oxidase I and II are the preferable barcoding loci used due to the fact that they have highly conserved tRNA genes allowing primers to be easily made, and are appropriately polymorphic for discrimination between most species. While the DNA sequences vary between species, a single primer pair will often work across the family level for a given gene, this particularly holds true for the barcoding COI and COII genes.

The next step is DNA amplification via PCR which require several reagents: (i) Deoxynucleoside triphosphate (DNTPs) (ii) $MgCl_2$ Buffer (iii) DNA Polymerase (iv) Forward and Reverse Oligonucleotide Primers. Primers can be purchased from online sources such as Integrated DNA Technologies (www.idtdna.com). The DNA is amplified via PCR in an automated, three-step process which makes up a cycle. First the DNA is **denatured** (the double helix unwinds into two single strands). This allows for the second step, in which the forward and reverse primers anneal (attach) to the DNA and begin amplification. The final step in a PCR cycle is the extension phase where the enzyme, DNA polymerase, uses the DNTPs to synthesize the DNA fragment from the starting, forward primer to the ending, reverse primer. This three-step process continues for approximately 35 cycles, resulting in billions of identical fragments of DNA, which are in a high enough concentration to be detected by a sequencing platform.

19.4.5   Visualizing DNA product and concentration in gel

Before the PCR products are sent for sequencing it is important to visualize the DNA fragment using gel electrophoresis to verify that the fragment of interest has been amplified and is in an appropriate concentration for the sequencing platform. It is important to verify that there is no contamination, or secondary bands that have been amplified during PCR, which would affect the sequencing results. A good amplification will result in a single, bright band with an approximate size ranging from 500–1000 base pairs for a typical COI gene (Figure 19.6). Refer to Box 19.3 for information regarding gel electrophoresis.

19.4.6   DNA sequencing

DNA sequencing in forensic entomology is still largely based on the Sanger *et al.* (1977) chain-termination method, using PCR-amplified DNA as the template. Those outside of the molecular world often do not know what is involved or where to start. Several of the larger research institutions may have a core facility capable of Sanger DNA sequencing on site; however, it might be less expensive or simpler to send the PCR products to a commercial service for sequencing. You may wish to check with other scientists in your department to see if they have a DNA sequencing facility they recommend (because they might be receiving an institutional discount). PCR products can be

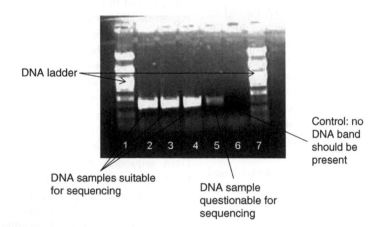

DNA ladder

Control: no DNA band should be present

1   2   3   4   5   6   7

DNA samples suitable for sequencing

DNA sample questionable for sequencing

Figure 19.6   Gel electrophoresis of a PCR product suitable for sequencing. *Source*: Courtesy of Marta Ferreira (MPCF).

Box 19.3   Gel electrophoresis

Gel electrophoresis is a common molecular technique that uses an electric current passed through a semi-solid media, typically agarose, to visualize fragmented DNA of varying nucleotide length (Figure 19.4). This method takes advantage of the fact that DNA is a negatively charged molecule, and as such when an electric current is passed through the agarose gel the fragments of DNA will travel towards the positive electrode. The agarose is a semi-solid media once polymerized and can be viewed as a porous sponge-like material at the molecular level. Larger DNA fragments move more slowly through the gel, while smaller DNA fragments will travel faster through the porous matrix, in a process known as **sieving**. The electric current is applied for a variable amount of time depending on the size of the smallest DNA fragments. At completion, the bands of varying sized fragments will separate from each other forming a unique pattern that is unique to the species in question. In addition to the DNA sample, a DNA ladder is run alongside the sample in the gel. A DNA ladder is a mixture of DNA molecules of known lengths that allow for the estimation of the size of the unknown DNA fragments. For further information regarding gel electrophoresis see the *Supplemental Reading* and *Additional Resources* sections of this chapter.

cleaned and purified by the sequencing facility and are then sequenced.

19.4.7   DNA editing and alignment

The sequence is returned as an **electropherogram** or **chromatogram** (Figure 19.7), in a trace file (which often has an ".abi" extension). One of the most popular DNA editing software programs is FinchTV (available from www.geospiza.com), which is a free, easy to use computer application. For each sequence returned you should view the chromatogram to make sure there is no contamination and that the peaks have good quality scores. The start and end of the sequences will contain more "noise" or "chatter" and result in lower quality scores (Figure 19.7a) (i.e. smaller peaks). Therefore, the start and end of the sequences should be trimmed so that only peaks with good quality scores are included (Figure 19.7b). The edited sequence is then saved as a FASTA file that can be uploaded into a DNA

alignment program. While this process of DNA editing and alignment is mainly automated, the programs require some molecular background knowledge. As such it is suggested that beginners contact and work with someone familiar with the programs and processes.

There are several DNA alignment programs that are readily available from the internet. A **DNA alignment** contains DNA sequences of multiple individuals from corresponding regions in the genome that are lined up so that each individual DNA sequence has the same starting and ending point. This allows direct comparison of DNA variation across species. MacClade (Maddison & Maddison, 2005) and Mesquite (Maddison & Maddison, 2011) are both easy to use and freely available for download and work on PC, Mac, or Unix/Linux systems. Once you have an alignment program installed, begin by uploading the FASTA sequences to the database. The next step is to align all the DNA sequences across all the individuals at corresponding regions in the genome (Figure 19.8). This can be done manually or using alignment software such as

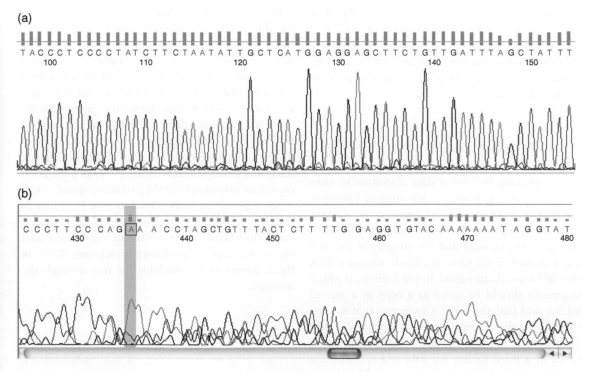

Figure 19.7   DNA chromatograms. (a) Chromatogram showing a high-quality sequence. Each peak represents a sequenced nucleotide. (b) Chromatogram showing contamination. *Source*: Courtesy of Evan Wong at the University of Cincinnati.

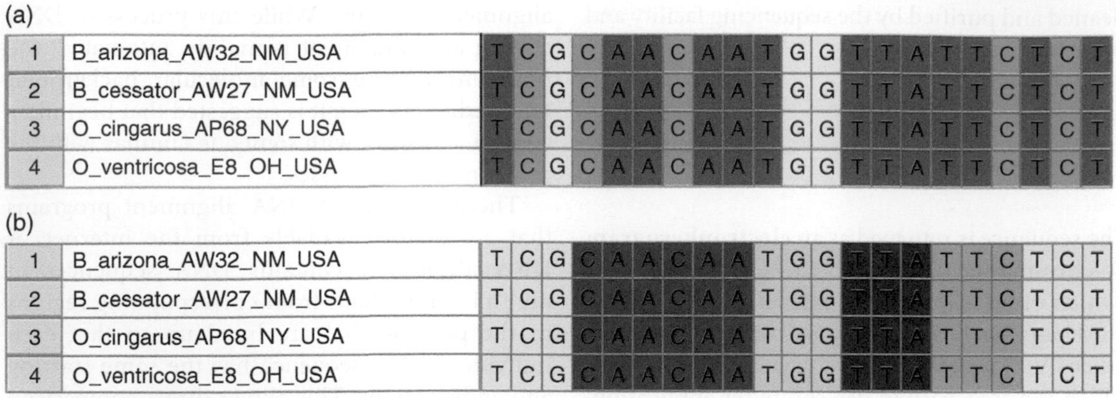

(a)

(b)

Figure 19.8   DNA alignment. (a) Alignment of four specimens by nucleotide. Each color represents a different nucleotide. (b) Alignment of four specimens by codon. Each color represents a different codon. *Source:* Courtesy of Evan Wong at the University of Cincinnati.

Muscle (Edgar, 2004) or ClustalW (Larkin *et al.*, 2007). While using automated alignment software, the alignment output should be rechecked as both the nucleotide sequence and as the translated amino acid sequence (Figure 19.8). The presence of indels (i.e. insertions and deletions) will create gaps in the nucleotide sequences. If indels are present it will become almost a necessity to use alignment software like Muscle or ClustalW, which will automatically calculate and correct the gaps between sequences. Another potential problem with alignment involves sequencing errors of individual nucleotides. Working with someone familiar with these techniques would be essential in cases where alignments are ambiguous.

19.4.8   Data management

It is important to have a data management plan before beginning research. We suggest following the practices proposed by Borer *et al.* (2009). Nucleotide sequence data should go through a three step process to ensure that the integrity of the data are maintained: (i) raw (unedited) sequence files should be saved, un-edited, in one folder; (ii) edited sequences should be saved as a copy in a second folder; and (iii) the FASTA sequence that is converted from the chromatogram format before being uploaded in the DNA alignment program should be saved in a third folder. All data and spreadsheets detailing collection information and life history data should be regularly backed-up on an external hard drive storage unit.

19.4.9   Selecting partitioning scheme and models of DNA evolution

The use of phylogenetic analyses is not simply the comparison between sequences. DNA has varying rates of evolution depending on which section of DNA is used in the analyses. Most of the barcoding examples in this chapter use one or more protein-coding regions of mtDNA. In protein-coding genes, the three codon positions typically are under very different selective constraints and will accumulate new variation at quite different rates. Model-based phylogenetic inference seeks to find subsets within a data set that can be treated with one set of parameter values distinct from the values appropriate for other subsets of the data. Using COI as an example, the sequence data might best be modeled as three distinct subsets, one for each codon position. PartitionFinder (Lanfear *et al.*, 2012) is a program capable of selecting both the partitioning and model scheme simultaneously. This program is similar to the program jModelTest (Darriba *et al.*, 2012), however jModelTest can only estimate the DNA model for a pre-selected partitioning scheme. Both of these programs are available for free through the internet.

19.4.10   Phylogenetic programs: inferring phylogenetic relationships

There are a variety of phylogenetic inference programs that can be used and it is important to note that they each are based on different assumptions and mathematical models. Three freely available programs that appear to work well for this function are: GARLI (Genetic Algorithm for Rapid Likelihood Inference) (Zwickl, 2006); mrBayes (Ronquist & Huelsenbeck, 2003); and MEGA5 (Molecular Evolutionary Genetics Analysis) (Tamura *et al.*, 2011). There is long standing debate over which phylogenetic methods are "better" (i.e. are accurate, consistent, and efficient; see Felsenstein, 1988), and it is not just a biological problem, but also a philosophical and mathematical problem (Yang, 2003). The combined approach of using both a Bayesian Inference and Maximum Likelihood analysis, then comparing the resulting tree topologies (branching patterns of the species) is recommended.

19.4.11 Species assignment and identification

The phylogenetic analysis is the foundation for DNA-based identification projects. After a **DNA barcoding library** is created and maintained (set of vouchered, morphologically identified specimens that have been sequenced for a barcoding locus) an unknown sample is identified to species based upon a statistically supported association (e.g. Bootstrap values for Maximum Likelihood, Posterior Probabilities for Bayesian Inference) with the voucher specimen(s) of a species. Bootstrap values and posterior probabilities mathematically assess the confidence that the species relationships have been correctly inferred. The voucher specimens of a species must form a monophyletic group respective of all other species. The unknown specimen is inferred to be identified from the database if it forms an exclusive monophyletic group with high statistical support. Although not common, occasionally wrong species names have been applied to sequences that are present in publicly available databases (e.g. GenBank). Therefore, caution should be used when selecting sequences to include in the DNA barcode library. To mitigate these problems, it is important to include specimen sequences that: (i) have been published in a peer-reviewed journal; (ii) where the specimen has been deposited into an insect collection, or is retained as a voucher by the researchers and can be referred to in cases of conflict.

19.5 Problems encountered in barcoding projects

Developing and using a DNA barcode library for species identifications will contain a number of potential problems for the aspiring forensic entomologist unfamiliar with the limits at which data can currently be interpreted. Wells and Stevens (2008) discuss problems found in DNA-based species identification which include the amplification of pseudogenes in species from primers created for use in other insect taxa, specimens from published sources being misidentified, and improperly sampled taxa. Pseudogenes are nuclear DNA sequences that are similar to a different gene, but have lost their functionality. The presence of pseudogenes can result in the amplification of both the gene of interest and the pseudogene via PCR if the primers used are not specific enough. If the researcher is not aware of the pseudogene, and mistake it for the genuine mtDNA gene, this could result in an incorrectly identified species. Fortunately, there are methods available for recognizing the presence of pseudogenes and ways to avoid them in barcoding projects. For further information on pseudogenes and techniques to minimize their effects on analyses, see the section on pseudogenes in the *Additional Resources* at the end of this chapter.

If specimens to be included in a DNA barcoding project are improperly sampled, this has the potential to result in species identification errors. Sampling is used to describe the process of selecting the available sequences that will be used for comparative purposes. Sampling can include all available sequences of a particular species and its evolutionary relatives, or a subsample of the available data. Sampling error can be a significant problem when comparisons involve closely related taxa. When investigators begin selecting taxa to include in the DNA database a common misconception is to include only individuals of forensic importance, or species from a single population. Due to the fact that natural populations can vary at a genetic locus, it is possible that species, which appear to be monophyletic, could be paraphyletic after additional sampling. DeBry *et al.* (2010) presented a phylogenetic tree based on partial COI mtDNA sequence data, which inferred that the species *L. coeruleiviridis* and *L. mexicana* were monophyletic species. However, with additional taxon sampling from multiple populations, they were able to show that these two species were actually a paraphyletic group that could not be distinguished using COI mtDNA sequence data. This is significant because if species are not reciprocally monophyletic, it will result in problems for the species identification (Wells & Williams, 2007).

The DNA databases that are available to the general public contain sequence data that has been uploaded by researcher submission. Thus, while these databases have a small level of automated sequence error detection, the majority of the scrutiny is the responsibility of the individual that submits the sequence. Although not common, errors have been identified from these DNA databases (Brunak *et al.*, 1990). To minimize the impact of these errors, it is recommended that the following questions should be explored: (i) who performed the species identification (was it a recognized taxonomic expert in that group); (ii) is a voucher specimen available for further scrutiny; (iii) what is the quality of the published work where the specimen sequence data came from; and (iv) are other independent studies congruent with the investigators findings (Wells & Stevens, 2008). Being able to answer these questions will reduce the potential of including misidentified specimens in a DNA-based, species identification project.

19.6   Gut content: victim and suspect identifications

There are several situations in which knowing what host the insect has fed upon would be beneficial to a criminal investigation. Wells *et al.* (2001) described several situations in which knowing that larvae originated from the corpse would have contributed to the investigation. Knowing the identity of an insect's last meal could reveal a connection between a body and a particular place, even in the absence of a corpse. For instance, suppose that maggots were found without a known food source, such as in the trunk of a car or in an empty garbage can. It is possible that the food source was a carcass of an animal or decayed food items and no further investigation is warranted. However, if the gut content of the maggot showed that the food source was human, further investigation would be needed. If these techniques had been successfully

applied to the phorid flies collected from the trunk of the car in the 2008 Casey Anthony case, it could have provided key evidence for use at the trial (a mother, Casey Anthony, was charged with the murder of her 2 year old daughter, Caylee, in a widely publicized trial where Casey was found to be "not guilty").

In a similar situation, maggots could be found near both a body and an alternate food source (Wells *et al.*, 2001) and mature larvae may disperse from the corpse before burying themselves prior to pupariation. Larval masses of the blow fly *Phormia regina* have been observed to migrate 2–26 m away from a carcass (Lewis & Benbow, 2011). Only the maggots that developed from the corpse would be relevant in establishing a PMI. However, due to the fact that it is not known what food source the maggots have been feeding on, this produces a possible source of error for estimates. Finally, maggots could be found on a victim but the investigator might not be sure if they originated from the body, or crawled to that location (Wells *et al.*, 2001).

Using insect gut contents for victim and suspect identification is not necessarily a complicated process. Evidence collection at the crime scene follows standard protocols with special care taken to preserve molecular and DNA evidence. Larvae collected at the crime scene should be preserved in ethanol, and a subset should be reared to adults if possible. The crop can be dissected and removed. The DNA is extracted from both the crop and its contents using established DNA extraction methods (e.g. Qiagen DNeasy Blood and Tissue Kit). Once the DNA extraction is complete, PCR is conducted to generate sufficient DNA for identification of (i) the insect and (ii) its food source. Primers used by Wells *et al.* (2001) amplified the fly COI and a portion of the human hypervariable region 2 (HV2). The COI fragments are analyzed and compared to published sequences in a barcoding library (see #11 in Box 19.4), thereby allowing a preliminary identification of the species of fly. The HV2 fragments are compared to the reference DNA sample of the victim, and to the standard human sequence. If the HV2 primers do not amplify a DNA product from the crop, this would indicate that its contents were either not human or human but degraded beyond the detection ability of this technique. Wells *et al.* (2001) conclude that maggot crops and their contents can provide a good source for DNA for

Using DNA to discover what a fly has been feeding on is not just a technique for use with larvae. Adult blow flies also feed on carrion and can carry their food's DNA in their crop and on their body. Because adult flies can freely move around in an environment and can be attracted to traps, they can serve as biomonitoring tools to assess the diversity of carrion in a particular area. Calvignac-Spencer et al. (2013) showed that carrion flies in the Calliphoridae and Sarcophagidae can serve as a source of mammal DNA for use in inventories of wild mammal communities. Flies captured in traps in tropical habitats of Côte d'Ivoire and Madagascar contained enough useable mammalian DNA to identify 16 taxa belonging to the orders Artiodactyla (antelope, hippos, etc.), Chiroptera (bats), Eulipotyphla (shrews, moles, etc.), Primates and Rodentia (rodents). They were able to identify 12 of these to the species level. Massey et al. (2022) surveyed vertebrate biodiversity at 39 sites in the southern Amazon using invertebrate-derived DNA (iDNA) metabarcoding based on sequence analysis of the mitochondrial 12S gene using captured carrion flies, sandflies, and mosquitos. They found that carrion fly sampling provided the most representative picture of vertebrate biodiversity, identifying 66 total vertebrate taxa. While it has not been done yet, it may be possible to use a similar approach to capture flies at a potential crime scene to see if they have been feeding on human remains.

use in both the identification of the insect and of the gut contents.

19.7   Molecular methods and population genetics

In recent years, it has become more appreciated that natural populations of species have differing amount of genetic variation (Valle & Azeredo-Espin, 1995; Wasser et al., 2004; Neslon et al., 2007; DeBry et al., 2010). The utility of population genetic methods will be dependent on the loci chosen and **hypervariable loci** are best suited for this

purpose. Hypervariable loci are loci that are likely to be divergent between individuals, or populations, of the same species. While these loci are widely used among biologist to study population-level behavior, phylogeography, and evolutionary events (e.g. Weir, 1996) only recently have hypervariable loci been used by forensic entomologists. It was, and perhaps still is, assumed in forensic entomology that insect species that have wide geographical ranges (Norris, 1965) and/or are highly mobile, such as blow flies (Calliphoridae) and flesh flies (Sarcophagidae), are uniform populations with randomly mixing individuals (Wells & Stevens, 2008).

There are two primary reasons why population genetics would be a beneficial tool in forensic entomology (Wells & Stevens, 2009). In cases where geographic variation exists among populations of the same species, it becomes possible to determine if the body had been moved after death. One can imagine that if a maggot having DNA markers more similar to individuals from a separate geographic location than where a body was discovered. This could provide evidence that the body had been moved. Using population genetic data to determine the geographic origin of a species is called assignment. Assignment is now a routine tool used in many scientific fields, such as conservation and **bioinfestation** studies (Bonizzoni et al., 2001; Manel et al., 2005). Bioinfestation is a term applied to an outbreak of a pest species that causes economic damage and it often involves an invasive foreign pest. These studies allow researchers to discover where a pest originated from.

The second reason for investigating and understanding population level genetic markers is to find sequence differences in mixed populations within one geographical location, or across separate populations in differing geographical locations, which directly relate to different developmental rates. Gallagher et al. (2010) has shown a difference in developmental time related to geographical location for populations of the blow fly L. sericata. Surprisingly, for most insect species of forensic importance, very little research has been devoted to understanding the relationships between temperature and developmental rates in different geographical locations (Wells & Stevens, 2008).

The molecular tools that have been used in population genetics, particularly for forensically important insects, include: microsatellites (Evett & Weir, 1998), amplified fragment length polymorphisms (AFLP), random amplified polymorphic

DNA (RAPDs) (Stevens & Wall, 2001); and single nucleotide polymorphisms (Hahn *et al.*, 2009; Kondakci *et al.*, 2009).

19.8   Molecular methods: non-DNA based

While the majority of this chapter has been devoted to discussing the uses of DNA in forensic investigations, they are not the only molecular tools available. This section will discuss some of the recent advances in the use of non-DNA based molecular methods that can be used in forensic entomology. In particular we will discuss ribonucleic acid (RNA) analysis and gene expression studies and cuticular hydrocarbon analysis.

Ribonucleic acid, together with DNA, comprises the nucleic acids. Although both DNA and RNA are nucleic acids, they perform different functions. RNA is comprised of large biological molecules, which are responsible for the coding, decoding, regulating, and expression of genes. This is significant because conducting an RNA analysis could reveal genes that were active in tissues prior to the time when the organism died and RNAs were processed (Arbeitman *et al.*, 2002; Tarone *et al.*, 2007). RNA analysis shows promise because it has the potential to allow for more precise models of insect development, and thus more accurate PMI estimations. An example of a source of error when using morphological characteristics to determine developmental stage of an organism is seen in a carrion fly third instar larva. Greenberg & Kunich (2002) were able to show that during the third larval instar the size dramatically increases, but once it enters a post feeding stage it will then shrink until the beginning of pupation. This results in a single larva being at an identical size twice during this stage of development. This problem is further exacerbated because of the time spent in the post feeding stage (~1/2 larvae life). This is a significant amount of time that could be a possible source of error in the establishment of a PMI estimate. Furthermore, it has been shown that population and temperature can also affect the body size and the minimal development rate (Tarone *et al.*, 2011).

Estimating the age of insect eggs and larvae is particularly difficult due to the drastic changes in both the size and shape of the organism over time. Entomologists typically use the size and shape characteristics of the individual to determine the age of the specimen and this becomes difficult when insect features rapidly change. Arbeitman *et al.* (2002) showed that three genes: *bicoid, slalom,* and *chitin synthase* were expressed in varying amounts in the fruit fly, *D. melanogaster*, at different stages in development. Using this knowledge, Tarone *et al.* (2007, 2011) showed similar findings in the blow fly *L. sericata*. Zajac *et al.* (2015, 2018) extend the use of mRNA profiling to include developmental aging of *C. vicina* pupae. Utilization of RNA analyses to measure the expression levels of genes present during development has the potential to allow increased precision in differentiation of various larval stages and distinguishing post feeding stages from feeding. This may provide a more accurate PMI.

Cuticular hydrocarbon analyses are a potential tool that may provide a non-DNA based approach to insect identification. Cuticular hydrocarbons are waxy or oily lipids present on the integument of insects that function as signaling compounds or contact pheromones by insects. They are variable between species (interspecific) and are often variable between members of the same species, especially when used for mating recognition (intraspecific) (Byrne *et al.*, 1995; Drijfhout 2010). The use of these chemicals in identifying species has been shown to be useful in Diptera (Urech *et al.*, 2005) and Coleoptera (Page *et al.*, 1997).

One notable study conducted by Byrne *et al.* (1995) used cuticular hydrocarbons profiles of the blow fly *P. regina* and found that geographic differences existed between populations. Results from their study showed that two populations separated by approximately 70 km could be distinguished. This is especially important, given that the mobility of some fly species can be up to an estimated distance of 1500 miles in a given season (Barrett, 1937; Norris, 1965).

A potential use of cuticular hydrocarbon analysis in forensic entomology is as a method to estimate the age of adult flies. Although not generally useful outdoors, obtaining age estimates for adult flies captured near a corpse indoors could extend the timeline over which PMI can be estimated based on forensically important Diptera. Bernhardt *et al.* (2017) showed that the amount of n-pentacosane increased linearly over the first 20 days post-emergence in *L. sericata* and *C. vicina*.

The chemical analysis of hydrocarbons is typically carried out using gas chromatography (GC) because hydrocarbons are non-polar (non-charged molecules) and volatile (evaporate readily, easily

(a)

(b)

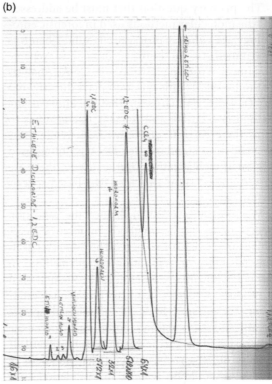

Figure 19.9   Gas chromatography. (a) Gas chromatograph (GC). Courtesy of Mcbort (b) GC chemical profile, peaks indicate chemical bonds. *Source*: Courtesy of Laslovarga.

excited by external energy). Gas chromatography is a standard method that has been used in hydrocarbon analysis and is an inexpensive, fast, and relatively simple method (Figure 19.9a.) The hydrocarbon compounds are detected using a detector (typically a Mass Selective Detector, MSD or Mass Spectrometer). Upon detection a profile is created that shows the presence of unique chemicals as peaks, which can then be identified using chemical analyses (Figure 19.9b). These profiles can then be saved to form a library for comparison to hydrocarbon profiles from unknown specimens (e.g. Drijfhout, 2010).

   A potential area of interest in forensic entomology that has recently been discussed in regards to carrion beetles (Hall *et al.*, 2011) and the blow fly *L. sericata*, (Kruglikova & Chernysh, 2011; Brown *et al.*, 2012) is the association of microorganisms and antimicrobial compounds with these particular species. One could imagine that if particular microorganism flora was determined to be associated with particular insect groups and geographical regions this could expedite species identifications. However, these applications have not been directly tested in a forensic setting.

19.9 Validating molecular methods for use as evidence

The *Daubert* standards were criteria established to evaluate the use of scientific information used in court as evidence or the interpretation of evidence (Solomon & Hackett, 1996). The decision of *Daubert v. Merrell Dow Pharmaceuticals, Inc.* stated that scientific evidence must be (i) testable; (ii) peer-reviewed; (iii) accepted by the scientific community using the technique; and (iv) have a known or potential error rate (Faigman, 2002; Tomberlin *et al.*, 2011). This ruling not only affected the molecular methods that are used in forensic entomology, but the broad landscape of forensic entomology itself. In addition, a recent report released by the National Research Council critically evaluated the forensic sciences and stated that significant improvements were needed in forensic science disciplines to increase accuracy and satisfy the *Daubert* criteria (National Research Council, 2009).

The primary question that must be addressed is if the molecular methods used in forensic entomology are **valid**. A procedure is valid when it is determined to be both accurate and reliable (Technical Working Group on DNA Analysis Methods, 1995). This question plays directly into the *Daubert* standards in that to know if the method is accurate and reliable, one also has to know the error rate of the method. While there are validation standards for DNA profiling used in human identification (Butler, 2005), there are no recognized standards for DNA-based species identifications.

Wells & Stevens (2008) divided the validation aspects of molecular methods in forensic entomology into two distinct parts. The first parts deals with reliability or replication of the method. In particular, how many times does the method need to produce a correct result to be deemed valid? Along these same lines, if DNA-based species identification sometimes produces inaccurate results, should the method as a whole be discarded? Wells & Stevens (2008) suggest that hundreds of specimens should be included in a validation study for DNA-based species identifications as a conservative estimate. However, these questions are new areas in the field of statistical theory and the answers are not known yet.

The second part in the validation of these methods is knowing when a particular reference database includes a sufficient number of individuals and species. If the reference database is lacking in some of the species likely to be encountered, then the method can lead to an incorrect species identification. Since we are not able to state the known error rates associated with species identification, molecular methods can be disputed in the courtroom when used alone. With that said, they can also provide clues and information that cannot be obtained from other types of investigations. It is clear at this point in time that molecular methods should be considered as tools to be used alongside others (e.g. morphology, ecology, etc.) and should not be considered as "stand-alone" solutions (at least not yet).

19.10   Future directions

The future directions of molecular methods in forensic entomology will largely be determined by the progression of technology. While DNA-related technological advances have allowed a lowering of costs and greater ease of use, it is difficult to predict when these methods will become common in actual criminal investigations. There are a few molecular techniques that have recently been developed that are worth mentioning because of the impact that they could have on forensic entomology.

Flow cytometry (FCM) is a method in which a laser is passed through a liquid media and an electronic detection apparatus is able to measure the physical and chemical properties of the fluid. This method has recently been used for genome size measurement for a large number of nuclei in several forensically important fly species (Picard *et al.*, 2012). They were able to conclude that FCM could be a potential new tool in forensic species identification because of the significantly different genome sizes that discriminate closely related species.

Sanger DNA sequencing (described earlier), has largely been supplanted in non-forensic research applications by short-read, massively parallel Next Generation Sequencing (NGS) technologies such as the Illumina platform, or long-read single-molecule technologies such as the platforms developed by Pacific Biosciences and Oxford Nanopore Technologies (Logsdon *et al.*, 2020). These technologies have, to date, found only very limited use in forensic entomology.

Chapter Review

When you hear someone refer to molecular methods used in forensic entomology, they are usually referring to the utilization of DNA

- All life uses DNA as the genetic molecule, but this is a variable molecule. Some parts of the DNA nucleotide sequence vary between individuals, making this an important tool for matching biological material with a single organism – the idea behind DNA fingerprints. Other parts of the molecule remain similar enough to differentiate between members of different geographic populations, species, genera, etc. The more closely related two organisms are by common ancestry and evolution, the more similar their DNA sequence will be.

- The selection of which part of the DNA information we want to use for a forensic application depends on what we want to use it for. If we want to link biological material with a particular individual, we will use extremely variable regions of the DNA (e.g. 13 short tandem repeats in non-coding regions of DNA are used for human DNA fingerprints). The ability to find regions that show differences between geographical populations of the same species may be very important for providing information on the location of a crime. Other parts of the DNA do not vary much within a particular species but shows distinct differences between different species.

- Species-specific portions of the DNA sequence can play an important role in the identification of forensically important insects, in all their varied life stages. This is the idea behind DNA barcoding, which usually involves the sequence of one or two genes of the mitochondrial DNA (mtDNA) genome. For insects, the most commonly used barcode sources are the mtDNA genes for cytochrome c oxidase one and two (COI and COII).

- The use of molecular techniques for species identifications has the potential to overcome many of the problems associated with morphological species identification. In many ways, the use of DNA enables a better and more detailed identification of an organism than any dichotomous key relying on morphological characteristics could ever provide. For example, comparative DNA studies have uncovered previously unknown cryptic species that could not be distinguished on morphological characters alone. The identification of immature stages of insects (which may be very difficult or impossible to do based on current knowledge) becomes as easy as adult identifications when using DNA, since all life stages of an individual contain the same DNA.

- DNA identification skills are transferable across different taxonomic groups. Unlike morphological identification skills, which may require many years of dedicated work to gain an understanding of one particular group of insects, knowledge of DNA identification techniques about one group of insects can be fairly easily applied to a much wider range of organisms. The same basic skills used to identify which species of fly was present as a crime scene can be modified to allow the identification of the species of fish in sushi or allow tracking of illegal elephant ivory.

Preserving DNA

- Biological evidence that undergo DNA analysis must be collected and stored in a way that preserves the integrity of the DNA and minimizes the chance of contamination.

- While fresh, unfrozen samples will give the highest yield of DNA for molecular analyses, it can be difficult or impossible to perform a DNA extraction immediately after collection. A good yield of quality DNA can be obtained from specimens or tissue from specimens that are stored in 95–100% ethanol.

- Exposure to heat and moisture can degrade the DNA molecule after the death of the insect. To minimize this problem when there is a time lag between collection of specimens and the preservation of tissue from the specimens, dead specimens should be kept as cool as possible. Refrigerator temperatures are acceptable for a few days, but ultra-low temperature freezing is suggested if weeks or months may pass before storage in ethanol.

Insects can be identified by their DNA

- The biochemical technique known as PCR (polymerase chain reaction) is the tool that makes DNA-based identification practical. This technique allows the forensic scientist to amplify specific parts of an organism's DNA that show interspecific differences.

- The most commonly used genes for identification of insects are found in the mitochondrial genome rather than in the nuclear genome. These include the genes for cytochrome c oxidase one and two, which are commonly referred to as COI and COII. There are a variety of practical reasons that lead to the preference for using mtDNA for identifications rather than nuclear DNA.

- While COI and COII barcodes work to separate most species, there are forensically important species that they will not work for. In these cases, the mtDNA may not separate closely related species from each other. In those cases, other sequences of DNA may need to be discovered and compared to allow reliable identification.

- Identification based on similarity methods that compare an unknown specimen's DNA with a referenced database (e.g. BLAST search) can provide extra confidence in a morphologically based determination, but should not generally be used as a "stand alone" identification technique. Phylogenetic methods place the specimen's DNA in a broader context by comparing the DNA with a variety of members of the same species and species that show close evolutionary relationship with one another. This is a way to take into account intraspecific variation and allow for the fact that sometimes closely related species have not been sequenced yet.

The steps behind DNA barcode-based identifications

- DNA barcoding requires extraction of DNA from insect tissue and amplification of the specific genetic sequences that will be used as a basis for identification. Only a small portion of the insect's genome is used for this technique, and it usually involves specific regions of the mtDNA.
- A variety of computer software programs are needed to get DNA sequence data into useable form and to allow comparisons of an unknown sequence with identified sequences in the identification process. Virtually all of these programs are freely available in the internet.
- There are several public reference libraries of DNA sequence data that can be freely accessed and downloaded for construction of a focused barcode library for phylogenetic comparisons and identification of specific insect taxa. GenBank is the main DNA library in the United States.
- Phylogenetic analysis is the foundation for DNA-based identification projects. An unknown sample is identified to species based upon a statistically supported association with voucher specimen data from a DNA barcoding library. An unknown specimen is inferred to be identified from the database if it forms a monophyletic group with a previously identified and vouchered species, with high statistical support. A variety of phylogenetic programs are available for such analyses, and they are available for free download from the internet.

Common problems in molecular identifications

- Utilizing a DNA barcoding library can be problematic, particularly if the user is not familiar with how the data should be interpreted.

- The presence of pseudogenes can result in incorrect species identifications if non-specific primers are used, which amplify both the barcoding gene of interest and the pseudogene. The effects of pseudogenes can be reduced if prior research is involved in the selection of the barcoding loci that will be used in the DNA library. There are detailed records available of known pseudogenes in a variety of taxa that should be consulted prior to amplifying barcoding genes.
- Including a limited number of specimens in the DNA barcoding database can result in incorrect species identifications. This is due to natural populations varying in the genetic information present at the barcoding gene. To alleviate the potential problems associated with this, multiple individuals from multiple populations should be included in the barcoding analysis, particularly closely related individuals, which may or may not be of forensic importance.
- Understanding where the sequence data that is included in the barcoding library originates from can reduce the possibility of including misidentified specimens.

Associating insects and victims

- When an adult or larval insect feeds, DNA inside the food material can be used for identification purposes. In a forensic setting, this can serve to link a specific maggot with a specific body, make sure that the maggot was feeding on the body of interest and not something else in the area, and can even serve as a way to identify a potential victim when the body is not present.

Population genetics can provide information about movement of bodies and differential development times

- Natural populations of species of insects have differing amounts of genetic variation. Some species show more intraspecific variation than others, and this variation may be more (or less) pronounced in the DNA than in the organism's phenotypic morphology. Genetic variation is often connected with differences in the geographic distribution of the species.
- Discovery of DNA sequences that vary within a particular species based on geographical location

of the population could provide interesting and useful information on movement of a body from one environment or geographic location to a different place.

- There may be DNA based differences within a single species that result in different developmental rates within a mixed population. It would be important to recognize this before producing PMI estimations.

Molecules other than DNA can have forensic uses

- New molecular techniques are being developed and applied to the forensic setting. Methods using RNA show some promise for establishment of more accurate PMI estimates. Being able to monitor the expression of particular genes may provide a better estimation of the age of insect eggs and larvae.
- Analysis of cuticular hydrocarbons may provide new information that helps to identify a particular species of insect or distinct populations within a single species.

Using molecular based information in the courtroom

- The *Daubert* standards state that scientific evidence must be (i) testable, (ii) peer-reviewed, (iii) accepted by the scientific community using the technique, and (iv) have a known error rate. A procedure is considered to be "valid" when it is determined to be both accurate and reliable. While validation standards have been accepted for human DNA profiling, there are no recognized standards for DNA based species identifications. As such, molecular methods should be considered a tool that is used alongside others (e.g. morphology, ecology, etc.) in forensic entomology, and is not a "stand-alone" solution (at least not yet).

New molecular methods are being developed

- The future directions of molecular methods in forensic entomology will largely be determined by the progression of technology. New technologies will have to show that they can be considered "valid" before they will find wider usage in forensic science.

- Flow cytometry (FCM) is method in which a laser is used passed through a liquid media and an electronic detection apparatus is able to measure the physical and chemical properties of the fluid. This method has been used for genome size measurement for several forensically important fly species. It could be a potential new tool for species identification.
- Next Generation Sequencing (NGS) technologies which can decode large portions of the total genome of an organism could be an even better molecular tool for species or population level determinations than is currently available.

Test your understanding

Level 1: knowledge/comprehension

1. Define the following terms:

   (a) PCR
   (b) restriction enzyme
   (c) phylogenetics
   (d) monophyletic group
   (e) cryptic species
   (f) barcoding library

2. Associate the following terms with one or more of these three applications: (i) species identification; (ii) identification of victim; or (iii) location of a crime

   (a) DNA digestion
   (b) DNA amplification
   (c) DNA alignment
   (d) DNA extraction
   (e) BLAST
   (f) gel electrophoresis

3. Why should tissue be preserved in ethanol if it may be used for DNA analysis?

Level 2: application/analysis

1. Explain how larvae can be identified by comparison with determined DNA sequences from adult specimens.

2. Explain how a single DNA extraction from a maggot can produce a mixture of both fly and human DNA.

Level 3: synthesis/evaluation

1. Explain how a hybrid specimen (offspring of an incorrect mating between two different species) might be identified as one species by morphological

means, a different species by a similarity based approach (like BLAST), and as a hybrid by a phylogenetic approach.

References cited

Anderson, I. & Brass, A. (1998) Searching DNA databases for similarities to DNA sequences: when is a match significant?. *Bioinformatics* 14: 349–356.

Applied Biosystems (2012) *User guide: AmpFISTR SGM Plus PCR Amplification Kit. Life Technologies Corporation.* Publication Number 4309589 Rev. H. Accessed January 17, 2013. http://www3.appliedbiosystems.com/cms/groups/applied_markets_support/documents/generaldocuments/cms:041049.pdf

Arbeitman, M.N., Furlong, E.E.M., Imam, F., Johnson, E., Null, B.H., Baker, B.S., Krasnow, M.A., Scott, M.P., Davis, R.W. & White, K.P. (2002) Gene expression during the life cycle of *Drosophila melanogaster*. *Science* 297: 2270–2275.

Arctander, P. (1988) Comparative studies of avian DNA by restriction fragment length polymorphism analysis: convenient procedures based on blood samples from live birds. *Journal für Ornithologie* 129: 205–216.

Baker, C.S., Palumbi, S.R. & Cipriano, F. (1996) Molecular genetic identification of whale and dolphin products from commercial markets in Korea and Japan. *Molecular Ecology* 5: 671–685.

Barrett, W.L. Jr (1937) Natural dispersion of *Cochliomyia americana. Journal of Economic Entomology* 30: 873–876.

Bernhardt, V., Pogoda, W., Verhoff, M.A., Toennes, S.W. & Amendt, J. (2017) Estimating the age of the adult stages of the blow flies *Lucilia sericata* and *Calliphora vicina* (Diptera: Calliphoridae) by means of the cuticular hydrocarbon *n*-pentacosane. *Science & Justice* 57: 361–365. https://doi.org/10.1016/j.scijus.2017.04.007

Bonizzoni, M., Zheng, L., Guglielmino, C.R., Haymer, D.S., Gasperi, G., Gomulski, L.M. & Malacrida, A.R. (2001) Microsatellite analysis of medfly bioinfestations in California. *Molecular Ecology* 10: 2515–2524.

Borer, E.T., Jones, M.B. & Schildhauer, M. (2009) Some simple guidelines for effective data management. *Bulletin of the Ecological Society of America* 90: 205–214.

Brown, A., Horobin, A., Blount, D.G., Hill, P.J., English, J., Rich, A., Williams, P.M. & Pritchard, D.I. (2012). Blow fly *Lucilia sericata* nuclease digests DNA associated with wound slough/eschar and with *Pseudomonas aeruginosa* biofilm. *Medical and Veterinary Entomology* 26: 432–439.

Bruder, C.E.G., Piotrowski, A., Gijsbers, A.A.C.J., Andersoon, R., Erickson, S., de Ståhl, T.D., Menzel, U., Sandgren, J., von Tell, D., Poplawski, A., Crowley, M., Crastro, C., Partridge, E.C., Tiwari, H., Allison, D.B., Komorowski, J., van Ommen, G.-J.B., Boomsma, D.I., Pederson, N.L., den Dunnen, J.T., Wirdefeldt, K. & Dumanksi, J.P. (2008) Phenotypically concordant and discordant monozygotic twins display different DNA copy-number-variation profiles. *The American Journal of Human Genetics* 82: 763–771.

Brunak, S., Englebrecht, J. & Knudsen, S. (1990) Cleaning up gene databases. *Nature* 343: 123.

Butler, J. M. (2005) *Forensic DNA Typing: Biology, Technology, and Genetics of STR Markers*, 2nd edn, pp. 660. Elsevier Academic Press, New York.

Butler, J.M. (2006) Genetics and genomics of core STR loci used in human identity testing. *Journal of Forensic Sciences* 51: 253–265.

Byrne, A.L., Camann, M.A., Cyr, T.L., Catts, E.P. & Espelie, K.E. (1995) Forensic implications of biochemical differences among geographical populations of the black blow fly, *Phormia regina. Journal of Forensic Science* 40: 372–377

Calvignac-Spencer, S.E., Merkel, K., Kutzner, N., Kühl, H., Boesch, C., Kappeler, P.M., Metzger, S., Schubert, G. & Leendertz, F.H. (2013) Carrion fly-derived DNA as a tool for comprehensive and cost–effective assessment of mammalian biodiversity. *Molecular Ecology* https://doi.org/10.1111/mec.12183.

Chen W.Y., Hung, T.H. & Shiao, S.F. (2004) Molecular identification of forensically important flesh flies (Diptera: Sarcophagidae) in Taiwan. *Journal of Medical Entomology* 41: 47–57.

Clary, D.O. & Wolstenholme, D.R. (1985) The mitochondrial DNA molecular of *Drosophila yakuba*: nucleotide sequence, gene organization, and genetic code. *Journal of Molecular Evolution* 22: 252–271.

Darriba, D., Taboada, G.L., Doallo, R. & Posada, D. (2012) jModelTest 2: more models, new heuristics and parallel computing. *Nature Methods* 9: 772.

DeBry, R.W., Timm, A.E., Dahlem, G.A. & Stamper, T. (2010) mtDNA-based identification of *Lucilia cuprina* (Wiedemann) and *Lucilia sericata* (Meigen) (Diptera: Calliphoridae) in the continental United States. *Forensic Science International* 202: 102–109.

Dessauer, H.S., Cole, C.J. & Hafner, M.S. (1996) Collection and storage of tissues. In: D.M. Hillis, C. Moritz & B.K. Mable (eds) *Molecular Systematics*, 2nd edn, pp. 29–47. Sinauer Assoc., Inc., Suderland, MA.

Valle, J.S. do & de Azeredo-Espin, A.M.L. (1995) Mitochondrial DNA variation in two Brazilian populations of *Cochliomyia macellaria* (Diptera: Calliphoridae). *Revista Brasileira de Genetica* 18: 521–526.

Drijfhout, F.P. (2010) Cuticular hydrocarbons: a new tool in forensic entomology?. In: J. Amendt, M.L. Goff, C.P. Campobasso & M. Grassberger (eds) *Current Concepts in Forensic Entomology*, pp. 179–203. Spinger, New York.

Edgar, R.C. (2004) MUSCLE: multiple sequence alignment with high accuracy and high throughput. *Nucleic Acids Research* 32: 1792–1797l.

Ekrem, T., Willassen, E. & Stur, E. (2007) A comprehensive DNA sequence library is essential for identification with DNA barcodes. *Molecular Phylogenetics and Evolution* 43: 530–542.

Evett, I.W. & Weir, B.S. (1998) *Interpreting DNA Evidence: Statistical Genetics for Forensic Scientists*, pp. 291. Sinuaer, Sunderland, Mass.

Faigman, D.L. (2002) Science and the law: is science different for lawyers?. *Science* 297: 339–340.

Felsenstein, J. (1988) Phylogenies from molecular sequences: inference and reliability. *Annual Review of Genetics* 22: 521–565.

Gallagher, M.B., Sandhu, S. & Kimsey, R. (2010) Variation in developmental time for geographically distinct populations of the common green bottle fly, *Lucilia sericata* (Meigen). *Journal of Forensic Sciences* 55: 438–442.

Gavin, M.C., Solomon, J.N. & Blank, S.G. (2010) Measuring and monitoring illegal use of natural resources. *Conservation Biology* 24: 89–100.

Goff, M.L. (1991) Comparison of insect species associated with decomposing remains recovered inside dwellings and outdoors on the island of Oahu, Hawaii. *Journal of Forensic Science* 36: 748–753.

Goldenthal, M.J., McKenna, K.A. & Joslyn, D.J. (1991) Mitochondrial DNA of the blowfly *Phormia regina*: restriction analysis and gene localization. *Biochemical Genetics* 29: 1–11.

Greenberg, B. and Kunich, J.C. (2002) Entomology and the law: Flies as forensic indictors. Cambridge University Press, Cambridge.

Guo, Y.D., Cai, J.F., Xiong, F., Wang, H.J., Wen, J.F., Li, J.B. & Chen, Y.Q. (2012) The utility of Mitochondrial DNA fragments for genetic identification of forensically important sarcophagid flies (Diptera: Sarcophagidae) in China. *Tropical Biomedicine* 29: 51–60.

Hahn, D.A., Ragland, G.J., Shoemaker, D.D. & Delinger, D.L. (2009) Gene discovery using massively parallel pyrosequencing to develop ESTs for the flesh fly *Sarcophaga crassipalpis*. *BMC Genomics* 10: 234–242.

Hall, C.L., Wadsworth, N.K., Howard, D.R., Jennings, E.M., Farrell, L.D., Magnuson, T.S. & Smith, R.J. (2011) Inhibition of microorganisms on a carrion breeding resource: the antimicrobial peptide activity of burying beetle (Coleoptera: Silphidae) oral and anal secretions. *Environmental Entomology* 40: 669–678.

Hunter, S.J., Goodall, T.I., Walsh, K.A., Owen, R. & Day, J.C. (2008) Nondestructive DNA extraction from blackflies (Diptera: Simuliidae): retaining voucher specimens for DNA barcoding projects. *Molecular Ecology Resources* 8: 56–61.

Kamal, A.S. (1958) Comparative study of thirteen species of sarcosaprophagous Calliphoridae and Sarcophagidae (Diptera). 1. Bionomics. *Annals of the Entomological Society of America* 51: 261–270.

King, J.R. & Porter, S.D. (2004) Recommendations on the use of alcohols for preservation of ant specimens (Hymenoptera, Formicidae). *Insectes Sociaux* 51: 197–202.

Kondakci, G.O., Bulbul, O., Shahzad, M.S., Polat, E., Cakan, H., Altuncul, H., Filoglu, G. (2009) STR and SNP analysis of human DNA from *Lucilia sericata* larvae's gut contents. *Forensic Science International Genetic Supplement Series* 2: 178–179.

Koski, L.B. & Golding, G.B. (2001) The closest BLAST hit is often not the nearest neighbor. *Journal of Molecular Evolution* 52: 540–542.

Kruglikova, A.A. & Chernysh, S.I. (2011) Antimicrobial compounds from the excretions of surgical maggots, *Lucilia sericata* (Meigen) (Diptera, Calliphoridae). *Entomological Review* 91: 813–819.

Lanfear, R., Calcott, B., Ho, S.Y.W. & Guindon, S. (2012) PartitionFinder: combined selection of partitioning schemes and substitution models for phylogenetic analyses. *Molecular Biology and Evolution* 29: 1695–1701.

Larkin, M.A., Blackshields, G., Brown, N.P., Chenna, R., McGettigan, P.A., McWilliam, H., Valentin, F., Wallace, I.M., Wilm, A., Lopez, R., Thompson, J.D., Gibson, T.J. & Higgins, D.G. (2007) Clustal W and Clustal X version 2.0. *Bioinformatics* 23: 2947–2948.

Lewis, A.J. & Benbow, M.E. (2011) When entomological evidence crawls away: *Phormia regina* en masse larval dispersal. *Journal of Medical Entomology* 48: 1112–1119.

Little, D.P. & Stevenson, D.W. (2007) A comparison of algorithms for the identification of specimens using DNA barcodes: examples from gymnosperms. *Cladistics* 23: 1–21.

Logsdon, G.A., Vollger, M.R. & Eichler, E.E. (2020) Long-read human genome sequencing and its applications. *Nature Reviews Genetics* 21:597–614.

Lowenstein, J.H., Amato, G. & Kolokotronis, S.-O. (2009) The real *maccoyii*: identifying tuna sushi with DNA barcodes – contrasting characteristics attributes and genetic distances. *PLoS ONE* 4: e7866.

Maddison, D.R. & Maddison, W.P. (2005) MacClade 4: Analysis of phylogeny and character evolution. Version 4.08a. Available at http://macclade.org.

Maddison, W.P & Maddison, D.R. (2011) Mesquite: a modular system for evolutionary analysis. Version 2.75. Available at http://mesquiteproject.org.

Malgorn, Y. & Coquoz, R. (1999) DNA typing for identification of some species of Calliphoridae. An interest in forensic entomology. *Forensic Science International* 102: 111–119.

Manel, S., Gaggiotti, O.E. & Waples, R.S. (2005) Assignment methods: matching biological questions with appropriate techniques. *TRENDS in Ecology and Evolution* 20: 136–142.

Massey, A.L., de Morias Bronzoni, R.V., da Silva, D.J.F., Allen, J.M., de Lázari, P.R., dos Santos-Filho, M., Canale, G.R., São Bernardo, C.S., Peres, C.A., Levi, T. (2022) Invertebrates for vertebrate biodiversity monitoring: comparisons using three insect taxa as iDNA samplers. *Molecular Ecology Resources* 22: 962–977.

Meiklejohn, K.A., Wallman, J.F. & Dowton, M. (2009) DNA-based identification of forensically important Australian Sarcophagidae (Diptera). *International Journal of Legal Medicine* 125: 27–32.

Meiklejohn, K.A., Wallman, J.F. & Dowton, M. (2012) DNA barcoding identifies all immature life stages of a forensically important flesh fly (Diptera: Sarcophagidae). *Journal of Forensic Science* 58: 184–187.

Moore, W.S. (1995) Inferring phylogenies from mtDNA variation: mitochondrial gene trees versus nuclear-gene trees. *Evolution* 49: 718–726.

National Research Council (U.S.). (2009) *Strengthening Forensic Science in the United States: A Path Forward*, pp 352. National Academy Press, Washington, D.C.

Nelson, L.A., Wallman, J.F. & Dowton, M. (2007) Using COI barcodes to identify forensically and medically important blowflies. *Medical and Veterinary Entomology* 21: 44–52.

Norris, K.R. (1965) The bionomics of blowflies. *Annual Review of Entomology* 10: 47–68.

Page, M., Nelson, L.J., Blomquist, G.J. & Seybold, S.J. (1997) Cuticular hydrocarbons as chemotaxonomic characters of pine engraver beetles (Ips. spp.) in the grandicollis subgeneric group. *Journal of Chemical Ecology* 23: 1053–1099.

Picard, C.J., Johnston, J.S. & Tarone, A.M. (2012) Genome sizes of forensically relevant Diptera. *Journal of Medical Entomology* 49: 192–197.

Rockman, M.V. & Wray, G.A. (2002) Abundant raw material for *Cis*-regulatory evolution in humans. *Molecular Biology and Evolution* 19: 1991–2004.

Ronquist, F. & Huelsenbeck, J.P. (2003) MRBAYES 3: Bayesian phylogenetic inference under mixed models. *Bioinformatics* 19: 1572–1574.

Sanger, F., Nicklen, S. & Coulson, A.R. (1977) DNA sequencing with chain-terminating inhibitors. *Proceedings of the National Academy of Sciences of the United States of America* 74:5463–5467.

Shokralla, S., Singer, G.A.C. & Hajibabei, M. (2010) Direct PCR amplification and sequencing of specimens' DNA from preservative ethanol. *Biotechniques* 48: 233–234.

Solomon, S.M. & Hackett, E.J. (1996) Setting boundaries between science and law: lessons from Daubert v. Merrell Dow Pharmaceuticals, Inc. *Science, Technology & Human Values* 21: 131–156.

de Souza Barbosa, T., Salvitti Sá Rocha, R.A., Guirado, C.C., Rocha, F.J. & Gavião, M.B.D. (2008) Oral infection by Diptera larvae in children: a case report. *International Journal of Dermatology* 47: 696–699.

Sperling, F.A.H., Anderson, G.S. & Hickey, D.A. (1994) A DNA-based approach to the identification of insect species used for postmortem interval estimation. *Journal of Forensic Science* 39: 418–427.

Spillings, B.L., Brooke, B.D., Koekemoer, L.L., Chiphwanya, J., Coetzee, M. & Hunt, R.H. (2009) A new species concealed by *Anopheles funestus* Giles, a major malaria vector in Africa. *American Journal of Tropical Medicine* 81: 510–515.

Stevens, J.R. (2003) The evolution of myiasis in blowflies (Calliphoridae). *International Journal for Parasitology* 33: 1105–1113.

Stevens, J. & Wall, R. (2001) Genetic relationships between blowflies (Calliphoridae) of forensic importance. *Forensic Science International* 120: 116–123.

Tamura, K., Peterson, D., Peterson, N., Stecher, G., Nei, M. & Kumar, S. (2011) MEGA5: molecular evolutionary genetics analysis using maximum likelihood, evolutionary distance, and maximum parsimony methods. *Molecular Biology and Evolution* 28: 2731–2739. https://doi.org/10.1093/molbev/msr121. http://mbe.oxfordjournals.org/content/early/2011/05/04/molbev.msr121.abstract.

Tan, S.H., Rizman-Idid, M., Mohd-Aris, E., Kurahashi, H. & Mohamed, Z. (2010) DNA-based characterization and classification of forensically important flesh flies (Diptera: Sarcophagidae) in Malaysia. *Forensic Science International* 199: 43–49.

Tarone, A.M., Jennings, K.C. & Foran, D.R. (2007) Aging blow fly eggs using gene expression: a feasibility study. *Journal of Forensic Science* 52: 1350–1354.

Tarone, A.M., Picard, C.J., Spiegelman, C. & Foran, D.R. (2011) Population and temperature effects on *Lucilia sericata* (Diptera: Calliphoridae) body size and minimum development time. *Journal of Medical Entomology* 48: 1062–1068.

Technical Working Group on DNA Analysis Methods (1995) Guidelines for a quality assurance program for DNA analysis. *Crime Lab Digest* 22: 21–43.

Tomberlin, J.K., Mohr, R., Benbow, M.E., Tarone, A.M. & VanLaerhoven, S. (2011) A roadmap for bridging basic and applied research in forensic entomology. *Annual Review of Entomology* 56: 401–421.

Urech, R., Brown, G.W., Moore, C.J. & Green, P.E. (2005) Cuticular hydrocarbons of buffalo fly, *Haematobia exigua*, and chemotaxonomic differentiation from horn fly, *H. irritans*. *Journal of Chemical Ecology* 31: 2451–2461.

Vincent, S., Vian, J.M. & Carlotti, M.P. (2000) Partial sequencing of the cytochrome oxidase b subunit gene I: a tool for the identification of the European species of blow flies for postmortem interval estimation. *Journal of Forensic Science* 45: 820–823.

Wallman, F.R., Leys, R. & Hogendoom, K. (2005) Molecular systematics of Australian carrion-breeding blowflies (Diptera: Calliphoridae) based on mitochondrial DNA. *Invertebrate Systematics* 19: 1–15.

Wasser, S.K., Shedlock, A.M., Comstock, K., Ostrander, E.A., Mutayoba, B. & Stephens, M. (2004) Assigning African elephant DNA to geographic region of origin: applications to the ivory trade. *Proceedings of the National Academy of Sciences of the United States of America* 101: 14847–14852.

Weir, B.S. (1996) Intraspecific differentiation. In: D.M. Hillis, C. Moritz & B.K. Mable (eds) *Molecular Systematics*, pp. 385–405. Sinauer, Sunderland, MA.

Wells, J.D. & Sperling, F.A.H. (2001) DNA-based identification of forensically important Chrysomyinae (Diptera: Calliphoridae). *Forensic Science International* 120: 110–115.

Wells, J.D. & Stevens, J.R. (2008) Application of DNA-based methods in forensic entomology. *Annual Review of Entomology* 53: 103–120.

Wells, J.D. & Stevens, J.R. (2009) Molecular methods for forensic entomology. In J.H. Byrd & J.L. Castner (eds) *Forensic Entomology: The Utility of Arthropods in Legal Investigations*, 2nd edn, pp. 437–452. CRC Press, Boca Raton, FL.

Wells, J.D. & Williams, D.W. (2007) Validation of a DNA-based method for identifying Chrysomyinae (Diptera: Calliphoridae) used in death investigation. *International Journal of Legal Medicine* 121: 1–8.

Wells, J.D., Introna, F. Jr., Di Vella, G., Campobasso, C.P., Hayes, J. & Sperling, F.A.H. (2001) Human and insect mitochondrial DNA analysis from maggots. *Journal of Forensic Science* 46: 685–687.

Whitworth, T. (2006) Keys to the genera and species of blow flies (Diptera: Calliphoridae) of America north of Mexico. *Proceedings of the Entomological Society of Washington* 108: 689–725.

Wiens, J.J. & Penkrot, T.L. (2002) Delimiting species based on DNA and morphogical variation and discordant species limits in spiny lizards (*Sceloporus*). *Systematic Biology* 51: 69–91.

Woodwark, K.C., Hubbard, S.J. & Oliver, S.G. (2001) Sequence search algorithms for single pass sequence identification: does one size fit all? *Comparative and Functional Genomics* 2: 4–9.

Yang, Z. (2003) Phylogenetics as applied mathematics. *TRENDS in Ecology and Evolution* 18: 558–559.

Zajac, B.K., Amendt, J., Horres, R., Verhoff, M.A. & Zehner, R. (2015) De novo transcriptome analysis and highly sensitive digital gene expression profiling of *Calliphora vicina* (Diptera: Calliphoridae) pupae using MACE (massive analysis of cDNA ends). *Forensic Science International Genetics* 15: 137–146

Zajac, B.K., Amendt, J., Verhoff, M.A. & Zehner, R. (2018) Dating pupae of the blow fly *Calliphora vicina* Robineau–Desvoidy 1830 (Diptera: Calliphoridae) for post mortem interval – estimation: validation of molecular age markers. *Genes* 9: 153–171.

Zehner, R., Amendt, J. Schutt, S., Sauer, J., Krettek, R. & Povolny, D. (2004) Genetic identification of forensically important flesh flies (Diptera: Sarcophagidae). *International Journal of Legal Medicine* 118: 245–247.

Zwickl, D.J. (2006) *Genetic algorithm approaches for the phylogenetic analysis of large biological sequence datasets under the maximum likelihood criterion*. Ph.D. dissertation. The University of Texas at Austin.

Supplemental reading

Brunstein, J. (2011) "The Quest for the $500 home molecular biology laboratory." One man's quest for a DIY home biology lab for the cheapest price possible. Available at

http://www.mlo-online.com/articles/201112/the-quest-for-the-500-home-molecular-biology-laboratory.php.

Dizon, A., Baker, S., Cipriano, F., Lento, G., Palsbøll, P. & Reeves, R. (2000) Molecular genetic identification of whales, dolphins, and porpoises: proceedings of a workshop on the forensic use of molecular techniques to identify wildlife products in the marketplace. La Jolla, California, U.S. Department of Commerce: xi + 52 pp.

Howard, R.W. & Blomquist, G.J. (2005) Ecological, behavioral, and biochemical aspects of insect hydrocarbons. *Annual Review of Entomology* 50: 371–393.

Lockey, K.H. (1991) Insect hydrocarbon classes: implications for chemotaxonomy. *Insect Biochemistry* 21: 91–97.

Martin, R. (1996) *Gel Electrophoresis: Nucleic Acids*, pp. 175. Bios Scientific, Oxford, UK.

Additional resources

1. Genetic Science Learning Center at the University of Utah (provides a virtual walkthrough of the gel electrophoresis technique): http://learn.genetics.utah.edu/content/labs/gel/

2. Finch TV (a DNA Chromatogram Editing Program): http://www.geospiza.com/Products/finchtv.shtml

3. Examples of DNA editing and alignment programs:
   a. MacClade: Editing and Alignment Program: http://macclade.org
   b. Mesquite: Editing and Alignment Program: http://mesquiteproject.org
   c. "MUSCLE: Multiple sequence alignment": http://www.drive5.com/muscle/
   d. "Clustal: Multiple Sequence Alignment": http://www.clustal.org/

4. Examples of partition and evolutionary model selection programs:
   a. "PartitionFinder" (*Note this is a Python based program, please read installation requirements carefully): http://www.robertlanfear.com/partitionfinder/
   b. "jModelTest" (Does not depend on external programs, runs as a JAVA based program): http://code.google.com/p/jmodeltest2/

5. Examples of phylogenetic programs:
   a. "mrBayes" Phylogenetic Analysis using Bayesian Inference: http://mrbayes.sourceforge.net/
   b. "GARLI" Phylogenetic Inference using Maximum Likelihood: https://code.google.com/p/garli/
   c. "MEGA5" Phylogenetic inference program, with user friendly interface. Can infer phylogeny using Maximum Likelihood, Minimum-Evolution, or Neighbor-Joining Algorithms: http://www.megasoftware.net/
   d. Phylogeny programs posted on a University of Washington website (comprehensive listing of

available programs): http://evolution.genetics.washington.edu/phylip/software.html

6. New Zealand Ministry of Public Health's description of genetic methods used for species identification (description of how to use BLAST): http://www.fos.auckland.ac.nz/~howardross/InspectorFoode/BLAST.html

7. Major DNA Barcode Databases:
   a. GenBank of the United States: http://www.ncbi.nlm.nih.gov/genbank/
   b. European Molecular Biology Lab: http://www.ebi.ac.uk/ena/
   c. DNA Data Bank of Japan: http://www.ddbj.nig.ac.jp/
   d. Barcode of Life Database: http://www.boldsystems.org/

8. Examples of DNA Sequencing Facilities:
   a. Genewiz DNA Seuqencing: http://www.genewiz.com
   b. High Throughput Sequencing (htSEQ): http://www.htseq.org/
   c. SeqWright: http://www.seqwright.com/
   d. GenScript: http://www.genescript.com

9. Examples of websites addressing pseudogenes:
   a. Yale Gerstein Lab: http://www.pseudogene.org
   b. List of pseudogenes: http://www.pseduogene.net

Chapter 20

Archaeoentomology: insects and archaeology

Humanity would probably not survive if all or only certain critically important insects were to disappear from the earth.

Gilbert Waldbauer, Professor Emeritus of Entomology, University of Illinois[1]

Overview

Archaeoentomology, the use of insects to address questions in archaeology, is a non-traditional topic for a forensic entomology book. So the inclusion here may be initially surprising, yet a detailed look reveals that the use of entomology in archaeological exploration relies on many of the same insects and techniques connected to the civil and criminal matters addressed by forensic entomology. In fact much of archaeology as a discipline relies on forensic science, so any linkage to entomological topics should closely align with forensic entomology. This relatively small yet intriguing area draws considerable attention from entomologists and archaeologists alike as it addresses questions about past civilizations and cultures in terms of how and where they lived, ate, worshipped, dealt with disease, and buried their dead. Site excavations in the Old and New World have revealed fossilized insects and remnants of insect activity that provide a window into the types of entomological interactions that occurred with humans and serve as witnesses to the evolution of synanthropy among a wide range of extant insect species. This chapter examines the relationships between insects, archaeology, and forensic entomology, with particular emphasis on ancient peoples and insects in terms of stored products, insect as pests, and funerary practices.

The big picture

- Archaeoentomology is a new "old" discipline.
- Concepts and techniques from forensic entomology can be applied to archaeology.
- Ancient insects and food: connection to stored product entomology.
- Ancient insects as pests: beginnings of synanthropy and urban entomology.
- Ancient insects and burial practices: revelations about past lives and civilizations.
- Forensic archaeoentomology: entomological investigations into extremely "cold" cases.

The Science of Forensic Entomology, Second Edition. David B. Rivers and Gregory A. Dahlem.
© 2023 John Wiley & Sons Ltd. Published 2023 by John Wiley & Sons Ltd.
Companion website: www.wiley.com/go/forensicentomology2

20.1 Archaeoentomology is a new "old" discipline

Insects and **archaeology**. What is the connection? Well, if you started reading this book beginning with this chapter, then your answer might be Indiana Jones!

That is right, Dr. Henry Walton "Indiana" Jones Jr, the fictional archaeologist who stars in five motion pictures and one television series under the direction of Steven Spielberg.[2] Indy's adventures have included searches for the Ark of the Covenant, Holy Grail, Crystal Skull, stolen children, and the Sankara stones. Most important to our interests, in the second film of the franchise, *Indiana Jones and the Temple of Doom* (1984), Dr. Jones becomes covered in insects of all sorts, large and small, as he passes through a hidden tunnel in search of the Thuggee cult. If you ignore the fact that most of the insects depicted in the scene do not occur in India, Asia, or even on the same continent together, then the image projected, that of a rugged athletic academic tromping around in exotic locations, facing unexpected death-defying challenges, and covered head to toe in hideous multilegged beasts, is probably spot on for how most people envision an archaeologist (unfortunately "rugged," "athletic," or even "academic" may not be terms that most use to describe entomologists!) (Figure 20.1). The connection between archaeology and entomology is not about the individuals who are engaged in the academic and intellectual pursuits of discovering our past. Rather, the two are linked through the people of interest – from ancient civilizations to our

Figure 20.1 Harrison Ford as Dr. Henry Walton "Indiana" Jones Jr. on set during filming of the fourth film "Indiana Jones and the Kingdom of the Crystal Skull (2008)." *Source*: John Griffiths/Wikimedia Commons/CC BY-SA 2.0.

most recent past – and the extinct and extant[3] insects that shaped the living conditions of these people.

As we begin our exploration of the intersection between archaeology and entomology, we first need to understand some of the basic tenets of archaeology. Archaeology by definition is the study of human activity over time through the recovery and analysis of individual and cultural materials and environmental data that remain once the people do not. The materials that are studied include artifacts (pottery, tools, clothing, etc.), architecture, recordings (written, symbolic, etc.), biological remains (food, wastes, pets, livestock, **biofacts**, etc.) and cultural landscapes (physical evidence at an excavation site, which overlaps with many of the other materials mentioned) (Bahn & Renfrew, 2008). Study of these materials employs techniques and approaches from scientific and humanities-based disciplines, and thus archaeology is a field that bridges the gap between the arts and the sciences. Much of the exploration and examination of artifacts relies on forensic science investigation, which is why some investigators contend that site excavations and the later study of finds from archaeological contexts should be treated forensically, thereby maximizing the amount of information obtained (Boddington *et al.*, 1981; Hunter *et al.*, 1996). Within the context of this book, the subdiscipline **forensic archaeology** might be the more obvious linkage to discuss. In reality, forensic archaeology is only tangentially connected to the topic of archaeoentomology, in that its focus is on the application of archaeological principles and techniques to matters of legal interest (Dupras *et al.*, 2011). Examples include finding buried items associated with a crime, locating potential gravesites, identifying surface body disposals, and location of mass burial sites.

Archaeoentomology uses insects as evidence and tools to explore the past (Panagiotakopulu, 2001). By "past," we mean the history of people or civilizations, often from ancient or premedieval times, but not restricted to any specific time period in history. Fossilized insects, the remains of unfossilized insects, and evidence of insect activity are used to interpret the environments of past civilizations. This definition serves as a contrast to the closely related fields of **paleoentomology** and **Quaternary entomology**[4] in which fossilized insects serve as the cornerstone for examining questions related to the environment, biology, and activity of insects from long ago (Elias, 1994; Nel *et al.*, 2010). The disciplines also differ in that archaeoentomology is more concerned with using the information to reveal details related to the human condition, whereas paleoentomology is focused on the insects.

What can archaeoentomology reveal about past civilizations? Records from ancient Mesopotamia (~3000 BC) indicate that flies were associated with urban centers, providing some of the earliest indications of synanthropy, the beneficial association for insects living near humans or in the artificial environments humans create. Egyptians understood the concept of maggot therapy as early as 1500–3000 BC, and also developed some of the first insecticides to control urban and stored insect pests (Buckland, 1981). Insect remains on mummies from pharaonic Egypt confirm the presence of malaria and bubonic plague, leading to the speculation that the Black Death or the plague that swept through Europe on three occasions potentially had its origins in ancient Egypt (Panagiotakopulu, 2004a). Details of different human burial practices in the Old and New Worlds have been unearthed from the insect fauna found on mummies and in gravesites (Panagiotakopulu, 2001; Huchet & Greenberg, 2010; Morrow *et al.*, 2016; Vanin & Huchet, 2017). Such studies have also served as some of the few bridging archaeoentomology with forensic entomology, yielding the field of forensic archaeoentomology (Panagiotakopulu & Buckland, 2012) and subfield funerary archaeoentomology (Huchet, 2010; Pradelli *et al.*, 2019; Giordani *et al.*, 2020). The reach of archaeoentomology continues to expand because though the field has its origins as early as 1842 with the Oxford University entomologist F.W. Hope, the number of individuals involved in research in this area is just now beginning to grow.

Our examination of this intriguing area of entomology will continue with a look at how techniques from forensic entomology can be applied to archaeology, and a detailed discussion of how insects influenced foods (stored product entomology), served as pests and vectors of disease in urban centers (urban entomology), and impacted treatment of the dead and influenced burial practices of ancient civilizations (aspects of medicocriminal or medicolegal entomology). Emphasis will be placed on how the remains of ancient insects serve as evidence in revealing information about human history.

20.2 Concepts and techniques from forensic entomology can be applied to archaeology

The use of approaches from forensic entomology in archaeological investigation is considered relatively new, as evidenced by the paucity of research literature related to insects in archaeology, specifically in regard to archaeoentomology (Huchet & Greenberg, 2010). Perhaps the more correct view is that the discipline has existed in some capacity for several decades but has only recently been recognized as an organized field of study. Regardless of the degree of "newness" to the field, archaeoentomology shares several features in common with forensic entomology. Both disciplines have similar goals in terms of applying sound scientific methods in the use of entomological evidence to determine the identity of individuals, reveal details about past environments, and gain insight into the events preceding or occurring after death. What should stand out from comparing these goals is how comparisons testing and application of the scientific method discussed in Chapter 1 are cornerstones to both forensic entomology and archaeoentomology. Concepts and techniques from forensic entomology can be, and have been, applied to address topics in:

- understanding mortuary practices;
- taphonomy of cadavers;
- examining human parasites associated with ancient civilizations;
- deciphering the mode of disease transmission in human civilizations, from ancient to modern times.

Insects serve as the physical evidence for both disciplines, with the fundamental difference being the condition of the specimens: forensic entomology relies predominantly on the collection of living specimens whereas archaeoentomology does not have the luxury of working with live, nor necessarily intact, evidence. The latter discipline deals with fossilized specimens, unfossilized remains of dead insects, and any evidence of insect activity. Just imagine the tedium of unearthing a dried, brittle insect specimen more than 2000 years old, and then attempting to identify genus and species working only with body fragments. The process makes the identification of metallic calliphorid adults from a corpse seem remarkably easy by comparison! Despite the challenge of identification, insects are well-designed for fossilization, much more so than most invertebrate groups. The key is in the composition and construction of the exoskeleton. The insect exoskeleton is predominantly composed of the polysaccharide chitin that in turn can be cross-linked to an array of proteins. Chitin is resistant to decay under most taphonomic conditions, and few organisms possess chitinases necessary to digest the protective outer layer of the body. However, not all specimens are equal in this regard. Some insects, like adult Coleoptera, are heavily sclerotized and thus undergo mineralization or carbonization more readily than soft body specimens like adult flies and larvae (Huchet, 2014; Morrow et al., 2016). Flies still offer utility in archaeoentomological contexts, but typically only as puparia. These differences represent a sharp contrast to forensic entomology, in which Diptera are far and away the most valuable entomological sources of information to a death investigation.

Another important consideration is that reconstruction through hypothesis testing is much more challenging when dealing with ancient test subjects (both entomological and human). An archaeoentomologist cannot bring specimens back to the laboratory to rear to adults for easier identification nor can the conditions of development be worked out in the laboratory for estimating the length of development on various food sources or interactions among or between species. Too many variables are unknown, including the precise environmental conditions that say commodity pests would have been subjected to or the manner in which wheat or barley was raised before harvest to approximate the food quality and other factors when stored and then available for pest insects to use as a resource. Thus, reconstruction through the use of the scientific method in archaeoentomology is in many ways more open-ended (i.e., lacking the ability to narrow hypotheses through repeated experimentation) and speculative than typical of forensic entomology in the twenty-first century.

The term **funerary archaeoentomology** has been coined for the field of research examining insect association with ancient taphonomy and

mortuary practices (Huchet, 2010; Vanin & Huchet, 2017). It is not possible to use insect evidence from archaeological finds to calculate the time of death estimation or postmortem interval as is desired in forensic entomology (see earlier discussion concerning limitations of reconstruction). However, the recovery of insect fauna can potentially provide information on the history of a cadaver. For example, insect evidence may yield insight into the insect fauna associated with ancient burials, such as whether the insects found are independent of human environments or are **exophilic**, represent an assemblage of synanthropes and were thus already in close contact with humans, or were associated only with subterranean burials because the insect themselves lived underground, in other words are **hypogean** species (Panagiotakopulu & Buckland, 2012). Insect remains found on mummies or in excavated graves can also answer questions as to whether the burial was immediately after death or not, and if exposure to the environment, and hence the opportunity for insect colonization, preceded the burial (Panagiotakopulu & Buckland, 2012; Thompson et al., 2018). As discussed later in the chapter, necrophagous species of flies and beetles used by forensic entomologists today have been useful tools for archaeoentomologists examining ancient animal remains (Skidmore, 1995; Funck et al., 2020; Giordani et al., 2020).

In some instances, the utility of specimens collected favors questions of archaeoentomological importance, rather more so than in a criminal investigation context. An example of such differences can be found with a **midden** or trash heap, which serves as an excellent source for finding insects, flies in particular, that relate to contents and thus occupants of a household. The problem for forensic entomology is that the contents relate to the *entire* household from an extended period of time, and thus would be considered contaminated evidence with only limited value to a modern criminal investigation. In contrast, to the archaeoentomologist or archaeologist, the midden is a gold mine of physical evidence useful in reconstructing the past habits of the inhabitants, including gathering information about multiple uses such as a trash heap and garden (Panagiotakopulu, 2004b).

Despite the differences, the similarities between archaeoentomology and forensic entomology are apparent in the approaches and techniques used to investigate insect evidence. The commonality is also evident in how the two disciplines can be subdivided: both examine insects that impact (or have) human food supplies, influence the human condition in urban locations, and examine the association between insects and animal remains, including human. These areas are the focus of the remainder of this chapter as we examine the linkage of ancient insects to stored product entomology, insects in urban centers and the evolution of synanthropy, and what ancient insects on mummies reveal about past civilizations.

20.3 Ancient insects and food: connection to stored product entomology

Figure 20.2 Adult grain weevils, *Sitophilus granaries* demonstrating their destructive potential toward stored grains. *Source:* Entomology, CSIRO/Wikimedia Commons/ CC BY 3.0.

In Chapter 3, we discussed the subdisciplines that comprise forensic entomology: stored product entomology, urban entomology, and medicocriminal entomology. These subdisciplines could easily serve as a means for categorizing the major research areas of archaeoentomology. In the case of stored product entomology, the foundations for the subfield have origins dating back to at least pharaonic Egypt, a period of time (beginning sometime around 3050 BC) in which Egypt was ruled by a series of Pharaohs (Silverman, 2003). What this simply means is that archaeological evidence has revealed that ancient civilizations suffered from insect attack on stored foods and animal products in households and urban centers dating back to at least the early dynastic period (2300 to 3000 BC)[5] in ancient Egypt (Solomon, 1965). This is as true now as it was then: bulk storage of commodities in granaries function as ideal environments to (i) attract large numbers of insects, and (ii) allow the mass aggregation of adult insects at a food source, which is ideal for proliferation. The end result is significant food loss or total destruction.

Among the many insect fossils and remains identified from excavation sites in ancient Greece, Egypt, and the Roman Empire, two extant pest species, the grain weevil *Sitophilus granarius* (Figure 20.2) and the sawtoothed grain beetle *Oryzaephilus surinamensis*, occur in fossil assemblages (Buckland, 1991), often together (Panagiotakopulu, 2001), and thus serve as excellent examples of the types of discoveries from archaeoentomology (Figure 20.3). In the case of *S. granarius*, fossilized and unfossilized remains have

Figure 20.3 An adult sawtootheed grain beetle, *Oryzaephilus surinamensis*, a series pest of stored grains. *Source*: Entomology, CSIRO/Wikimedia Commons/CC BY 3.0.

been uncovered beneath pyramids, in tombs, and in granaries throughout the ancient world (Chaddick & Leek, 1972; Buckland, 1981; Panagiotakopulu & Buckland, 1991) and from parts of Europe dating back to 4000–5000 BC (Buchner & Wolf, 1997). The weevil is cosmopolitan in distribution and one of the most serious pests of stored commodities in the world today. Both of these features are even more impressive when considering that the adults are flightless. Despite the lack of mobility of the insect and the limited forms of locomotion (by comparison with today) available to human civilizations thousands of years ago, *S. granarius* evolved into a synanthropic species even before settled or sustained agriculture developed (Buckland, 1981).

The presence of the weevil is considered an indication of a civilization that relied on mass storage of grains in centralized locations rather than accumulation in the field (Osborne, 1983), as this insect is only known from the former scenarios in today's agricultural practices. Synanthropy of *O. surinamensis* may be considered more recent than that of *S. granarius* simply because adults and juveniles utilize grain already fed upon by primary stored product pests such as the grain weevil. Alternatively, the adaptation of the sawtoothed grain beetle as a secondary pest may reflect resource partitioning akin to the spatial partitioning discussed in Chapter 8 regarding different fly species competing in the same larval aggregation. In which case, *both* species may have been primary commodity pests at one time, developing a close relationship with the human condition over a similar evolutionary time scale. Fossil assemblages of the two species together at an excavation site clearly suggest that synanthropy predates the existence of those ancient civilizations. Again, these two species were selected for discussion to illustrate the types of information that can be extracted from insect remains. Several other stored product species have been identified from a wide range of excavation sites throughout the world and undoubtedly many have had (and still do) a profound impact on the food supply of individuals and civilizations (Buckland, 1991).

Here is an excellent place to examine how hypothesis testing and the scientific method can be applied, or not, to some of the questions that arise in archaeoentomology. The speculation of whether resource partitioning in the form of spatial partitioning accounts for the relationship between the grain weevil and sawtoothed beetle can be tested experimentally under controlled conditions in the laboratory. The fact that both species are extant and easily maintained in colony under laboratory conditions greatly facilitates experimental testing. Repeated revision of the hypotheses through carefully designed experiments can eliminate possible explanations and allow conclusions to be drawn regarding the relationships of the two stored product pests. However, experimentation cannot be used to test hypotheses addressing evolutionary and archaeological questions focused on the origins of synanthropy or biogeographical history of insects identified in archaeological finds. It is important to understand that this does not represent limitations in the use of the scientific method, but instead reflects the limits of working with ancient insects.

The fossilized remains of insects from different regions allow analyses of the distribution and origins of insects. Here is where archaeoentomology and paleoentomology are united in interest. However, the focus of an archaeoentomologist in determining historical biogeography of a particular species (i.e., terrestrial ecozone of origination and global spread) is really aimed at addressing questions surrounding the influence of a given insect on a particular society or civilization(s). In the case of *S. granarius*, its spread across the world in establishing a cosmopolitan distribution led to significant food losses, as much as 25–40% in some regions (e.g., sub-Saharan Africa) (Haile, 2006; Panagiotakopulu & Buckland, 2009). In some instances, starvation was the end result, reflecting one of the major impacts on the human condition. With such severe consequences came attempts at insect control, with excavated insect specimens and even human sarcophagi suggesting the use of natural insecticides. Evidence in the form of preserved writings, charred crops, and fossilized remains indicate that fire ash, nicotine, botanical extracts, and even spells (from the Book of the Dead) were used to repel insect attack from such stored commodities as wheat, barley, lentils, nuts, cereals, fruits, cumin, and dill (Miller, 1987; Panagiotakopulu & Buckland, 2009; Panagiotakopulu *et al.*, 2010). Whether the methods of insect control were effective cannot be ascertained for a particular location, but evidence of the continued use of these natural insecticides certainly indicates that some satisfactory pest management was achieved.

20.4   Ancient insects as pests: beginnings of synanthropy and urban entomology

Our definition of urban entomology from Chapter 3 is the branch of entomology that deals with insects and other arthropods associated with human habitation or the human environment (Hall & Huntington, 2010). A broad interpretation of this definition includes insects that occur in yards and neighborhoods (e.g., urban areas or centers) as well as those of agricultural importance that invade human space. Can we apply this same terminology within an archaeoentomology context? The answer is yes, with some modification. The modification caveat is needed to place insects, particularly from a pest status perspective, into appropriate context. People are uptight in modern highly industrialized nations, displaying a very low tolerance for the presence of any type of insect in their homes. Consequently, several relatively benign species of insects (and other arthropods) are regarded as "pests" out of annoyance rather than due to a real or immediate threat to human health, food, or infestation of building materials (Bain, 2004). A discussion of when an insect is truly a pest with regard to aesthetic and economic injury levels is presented in Chapter 3 and should be reviewed to place this information in appropriate perspective with urban entomology in ancient times. The definition of "urban" in ancient civilizations is much different from the towns, cities, and metropolises of modern times: urban centers sustained much smaller human populations, villages or cities tended to be concentrated along major waterways, and the distinction between rural and urban was blurred as crops and livestock were raised in or just outside of urban centers. Insects were commonplace in buildings and homes, and as long as they did not destroy food supplies or threaten human health, these co-inhabitants likely received very little attention. We have already discussed ancient insects and stored products in Section 16.3, so the focus here is on the relationship between insects and disease.

Insects that vector disease in the modern world are abundant in the fossil record and in other forms discovered throughout sites in the Old and New Worlds. For instance, the bed bug, *Cimex lectularius*, and human flea, *Pulex irritans*, have been collected from mummified heads maintained at the Cairo Museum, Egypt dating back more than 5000 years (Panagiotakopulu, 2004a) (Figure 20.4). Neither insect remains on a host for more than a few minutes after death, so their presence indicates an ectoparasitic association. Similarly, bed bugs as well as lice remains and fossilized specimens were obtained from burial sites, floor materials and even in remnants of roofs during New Kingdom and Thule excavations, again suggesting a parasitic relationship with humans and associated livestock, and allowing speculation about the multipurpose use of bedding and straw for livestock (Panagiotakopulu & Buckland, 1999; Dussault *et al.*, 2014). The worldwide distribution of lice with humans indicates that the relationship between the two has existed for a long period of time (Bain, 2004; Reed *et al.*, 2007), and thus synanthropy among the lice is very old.

Figure 20.4   A bedbug nymph, *Cimex lectularius*, blood feeding on a human host. *Source*: Piotr Naskrecki/ Wikimedia Commons/public domain.

Evolutionary theory predicts that parasites and hosts that have had a long period of time to coevolve should achieve an equilibrium, in which neither species has the upper hand, at least for long stretches of time, in the host–parasite relationship. The result is that despite some injury that might be incurred, both species survive through coexistence. This appears to hold true in the case of the body lice–human association.

Keeping in mind the expected equilibrium of "old" host–parasite relationships, archaeoentomological and biological evidence tell a story about the origins and coevolution of bubonic plague, also known as the Black Death, which challenges contemporary views (Figure 20.5). Bubonic plague swept through parts of Europe on at least three occasions, with the most severe outbreak occurring during the fourteenth century when an estimated 20–30 million people (representing about one-third of the entire European population) succumbed to the disease (McNeill, 1977). The accepted mode of disease transmission has been assumed to rely on bacterial transmission from the rat flea *Xenopsylla cheopis* to humans during blood feeding (Figure 20.6). Why would a rat flea need or use humans as a host? The answer is the key to the mass spread of the plague and also the archaeological intrigue. The causative agent of bubonic plague is the bacterium *Yersinia pestis*, which is transmitted to a new host during blood feeding. The ubiquitous black rat *Rattus rattus* was the host during the European spread of the disease. Poor hygiene practices and easy access to stored food led to a synanthropic relationship between

Figure 20.7   Bubonic plague victims (1720–1721) in a mass grave in Martigues, France. *Source*: S. Tzortzis/Wikimedia Commons/public domain.

Figure 20.5   The Great Plague of London in 1665 (artist unknown). *Source*: Unknown/Wikimedia Commons/public domain.

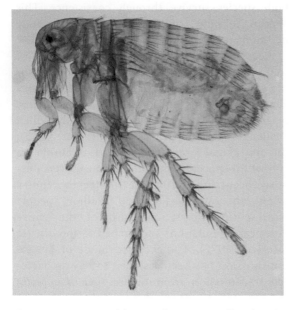

Figure 20.6   An adult rat flea *Xenopsylla cheopis*. *Source*: Katja ZSM/Wikimedia Commons/CC BY-SA 3.0.

the rat and humans (Figure 20.7). The black rat usually dies as a result of the bacterium. Consequently, the rat flea is forced to seek a new host, which during the pandemic[6] plague of Europe was human. What stands out in this scenario is that the primary host *R. rattus* typically dies from bacterial septicemia,[7] a situation not

expected for long-term host–parasite relationships. In fact, the lethality to the black rat argues a new association, and that the bubonic plague did not originate in Europe. Which might be the place of origin? Mummies and fossilized rat fleas from excavations in Amarna from pharaonic Egypt confirm the presence of plague as an endemic disease earlier than the European pandemics (Panagiotakopulu, 2004a). The working hypothesis is that *Y. pestis* coevolved with the Nile rat, *Arvicanthis niloticus*, and that the rat flea *X. cheopis* transmitted the bacterium to a new host, the black rat that stowed away on ships engaged with trade between Egypt and various other ports, including Europe (Panagiotakopulu, 2004a). This intriguing postulate represents another example of a hypothesis that cannot really be tested via experimentation, other than demonstrating that the Nile rat is better equipped than *R. rattus* to deal with the bacterial parasite *Y. pestis*, and thus likely has had a longer host–parasite relationship.

Establishing the presence of insect vectors in ancient civilizations is one thing, but does that necessarily mean that fleas, lice, flies, and mosquitoes always transmitted diseases – were these insects harboring pathogenic parasites 3000–5000 years ago? Good question. How could a question of this type be addressed? Obviously, direct experimentation examining a cause–effect relationship is out of the question since the people and insects are long since deceased. What are available are some forms of comparisons testing, but this is only possible if, say, tissues of human remains and/or insect vectors are preserved well enough for some type of histological analyses or electron microscopy. In the case

Figure 20.8   An adult sand fly, *Phlebotomus papatasi*, engaged in blood feeding on a human host. Sandflies like this species are responsible for the spread of the vector-borne parasitic disease leishmaniasis. *Source*: Frank Collins/Wikimedia Commons/public domain.

Figure 20.9   Egyptian mummy exhibited at the Louvre Museum in Paris. *Source*: Zubro/Wikimedia Commons/CC BY-SA 3.0.

of human mummies from various periods of pharaonic Egypt, isolated tissues have been used to confirm the presence of malarial parasites, *Y. pestis*, and other insect-borne diseases. Remains of known insect vectors have, in turn, been collected from excavation sites of the "infected" mummies. However, the insects were not suitable for confirming that they too were harboring the same parasites as the mummies. In the case of some unrelated insect blood feeders, sand flies (Family Psychodidae) preserved in amber have been shown to harbor leshmanial protozoans from as early as the Cretaceous period (*c.* 145–65 million years ago) (Poinar & Poinar, 2004), and fossilized specimens of some psychodids has led to speculation that these flies have been vectors of disease since the origin of blood feeding in this group (Azar & Nel, 2003). The circumstantial evidence makes it tempting to draw the obvious conclusions for other blood-feeding insects, but we know that such tendencies are not good scientific or archaeological practice (Figure 20.8).

20.5   Ancient insects and burial practices: revelations about past lives and civilizations

A discussion of archaeology often conjures up images of secret passageways in underground temples, leading to a crept containing an ornate sarcophagus

that, once opened, contains a traditionally wrapped Egyptian mummy surrounded by riches beyond measure (Figure 20.9). Well such discoveries rarely happen, but the discovery of a mummy can be the source of great riches in terms of information. From an archaeoentomological perspective, mummification of a corpse can preserve entomological evidence when insects are attached to or located within the deceased, allowing examination of ectoparasites from ancient civilizations, as discussed in Section 20.4. Study of insects associated with the dead, whether mummified or not, from archaeological sites relies on very similar techniques and concepts as medicocriminal entomology. Perhaps the most important unifying feature of the two fields is a reliance on necrophagous flies. The maggot (larval) stage is considered potentially very useful to archaeoentomological sites involving human or other animal remains for the following reasons:

1. Fly larvae give immediate information concerning the living conditions of the individual(s) in question.
2. Larvae usually do not migrate far from the corpse to pupariate and in fact several species of calliphorids pupariate on the dead.
3. Larvae are very responsive to environmental conditions (temperature, humidity, and light), which may be reflected in puparial size and shape (Zdarek *et al.*, 1987; Rivers *et al.*, 2010), or can enter diapause in response to adverse weather, which can be determined by puparial hydrocarbon profile (Yoder *et al.*, 1992).
4. Under optimal conditions, larval development tends to be of short duration, the resulting

puparial sizes are large, and the fecundity (number of larvae) is very high.

5. Well-preserved specimens can be identified at least to family and often to genus and species.

It is important to emphasize "potentially," for as discussed in Section 20.2, the soft body of larval Diptera generally does not preserve well. In contrast, puparia are long lasting and can be evaluated with these features taken into consideration, which may allow estimations of the season when the insects arrived, whether conditions were favorable for fly development, the stage of corpse decomposition when fly colonization occurred, and possibly whether the body had been moved (Skidmore, 1995; Panagiotakopulu, 2004b). These are also important pieces of information gained from flies collected as evidence in a modern criminal investigation and used by a forensic entomologist in comparisons testing and reconstruction.

The types of insects collected from an excavation in which human remains or mummies are found can yield insight into the burial practices of ancient civilizations. We will use an example of burial practices from the Old World, pharaonic Egypt, to contrast the traditions from pre-Columbian Peru (New World). Both the ancient Egyptians and Moche people of Peru (*c.* AD 100–800) practiced mummification, but the methods and reasons were dissimilar (Dunand *et al.*, 2006). Mummification in this context is the deliberate dehydration of a corpse to remove tissue water quickly as a means to preserve the integrity of the body as a whole. This is an oversimplification of the techniques used, and indeed much of the actual practices used by both civilizations remain unknown (Brier, 1996). What is understood about the ancient Egyptians and Moche (and many other civilizations) is that the dead were treated with reverence and with the belief that "souls" lived on in an afterlife. In fact, hieroglyphics associated with the tomb of the pharaoh Tutankhamun (King Tut) show that two flies are used for his after-death journey (Panagiotakopulu, 2004b). As to the physical "shell" of the mortal body left behind after death, the two groups of peoples had quite contrasting views.

To the ancient Egyptians, consumption of the dead by necrophagous insects was not desired. There is evidence that Egyptians understood which insects were commonly associated with carrion and also aware of when these species would attempt to gain access to a body. The process of mummification was performed almost immediately after death. The effect was to eliminate attraction by early colonizers (calliphorids and sarcophagids), and by drying the body and removing the internal organs (placed in canopic jars[8]) few species of insects would show any interest toward the mummified remains at any point (Figure 20.10). At least that was the intent. Several studies have shown that mummified remains may indeed be colonized by flies and beetles just not immediately after death (Huchet, 2010; Huchet, 2014; Morrow *et al.*, 2016). Necrophagous insects are quite resourceful, and insects like dermestids, piophilids and phorids arrived eventually (Curry, 1979), although of the specimens that have been isolated from excavated mummies, whether the insects were contemporary or occurred post excavation is still unresolved (Panagiotakopulu, 2001). Hieroglyphics depict priests spearing beetles, probably reflecting efforts to stop necrophagy, and the Book of the Dead contains a passage that has been interpreted to be a spell cast upon carrion beetles to prevent feeding on the dead (Panagiotakopulu, 2001). Clearly the ancient Egyptians exercised burial practices to protect the bodies from necrophagous insects.

The Moche people of pre-Columbian Peru had a much different view of the relationship between the dead and flies. Bodies were deliberately left exposed to the environment for several days after death to encourage fly colonization. The Moche believed that the feeding fly larvae would engulf the "anima" or spirit of the individual, which would be carried into the resulting adult flies, which in turn would return to live with the people (a recognition of synanthropy among necrophagous flies) (Huchet & Greenberg, 2010). This process of consumption and synanthropy was necessary to complete the human

Figure 20.10 Canopic jars on display in the British Museum. *Source*: Apepch7/Wikimedia Commons/public domain.

cycle. Excavated mummies from Peruvian sites clearly substantiate this cultural practice as fly puparia have been recovered from several unearthed graves, between mats and textiles, within buried skeletons, and even inside ceramic vessels placed as offerings (Donnan & Mackey, 1978). The puparia collected belong to the families Calliphoridae, Sarcophagidae and Muscidae, consistent with the practice of exposing fresh corpses. Among the many calliphorid remains, puparia of *Cochliomyia macellaria* were recovered from pre-Columbian mummies, a fly that is known today to be very abundant in the region, an early colonizer, and synanthropic.

An interesting find from one Peruvian site are sarcophagid puparia (the genus and species were not determined) that show holes consistent with emergence of parasitic wasps (Figure 20.11). The intrigue in this entomological evidence is not so much that parasitic wasps were obviously associated with pre-Columbian civilizations, but instead the interpretation presented as to what this means (Huchet & Greenberg, 2010). The exit holes were believed to be due to the emergence activity of two pteromalid wasps, suspected to be from either the genus *Muscidifurax* or *Spalangia*. What were the bases for focusing in on these two wasp genera? First, species from both genera are known to occur in Peru (Legner, 1988; Geden, 2006). Second, the exit holes were consistent with the body sizes of these wasps, which tend to be much larger than other pteromalids that parasitize carrion-inhabiting flies. The reasoning is logically sound

but may be too limiting in considering other possible scenarios. For example, the rational for focusing in on the two pteromalid species as creating the exit holes was based on the size of the opening. Other wasps are capable of making large exit holes in host puparia, even if their body sizes were much smaller than the holes, including non-pteromalids. *Nasonia vitripennis* (Petromalidae) and *Tachinaephagus zealandicus* (Encyrtidae) are two such species. With *N. vitripennis*, adult female body sizes of this wasp vary considerably with species and size of host, and with the level of intraspecific and interspecific competition, in some cases approaching the size of small *Muscidifurax* spp. and *Spalangia* spp. (Rivers, 1996, 2004). Of most significance to the interpretation from the Peruvian excavation, of the four genera mentioned, only *N. vitripennis* and *T. zealandicus* commonly parasitizes puparia associated with carrion-breeding flies (Lecheta & Luz, 2015); the other two are mostly restricted to muscoid species. The importance of reexamining the archaeoentomological interpretations, as is true in forensic entomology, is to ensure that all aspects of the entomological evidence are considered before drawing conclusions.

It is important to note that several mummified remains of humans and other animals have been found to harbor a wide range of insects. The plethora of examples will not be presented here but specimens include many of forensic interest, including beetles (clerids, dermestids, scarabs) and flies (calliphorids, muscids, phorids, piophilids, sarcophagids)

(a)

(b)

Figure 20.11 Entomological artifacts collected from a grave site in Huaca de la luna, Peru, including (a) an assortment of fly puparia, and (b) puparia displaying emergence holes of hymenopteran parasitoids. *Source*: Courtesy of J.-B. Huchet from the Muséum National d'Histoire Naturelle, Paris, France.

(Hope, 1842; David, 1978; Curry, 1979; Strong, 1981), and even termites (Backwell *et al.*, 2012; Huchet, 2014) and nesting wasps (Thompson *et al.*, 2018).

20.6 Forensic archaeoentomology: entomological investigations into extremely "cold" cases

Examination of entomological evidence from human remains in more recent times overlaps with the scope of forensic entomology. In general, such investigations rely on nearly identical techniques but the scope of archaeoentomology is typically broader, even in instances of forensic importance. As mentioned early in the chapter, much of archaeology is forensic science. However, detailed studies applying archaeoentomology to forensic investigation of human remains have been relatively rare. A recent application of archaeoentomological techniques to a forensic topic involved an examination of the insect fauna associated with the medieval burial of Archbishop Greenfield (Panagiotakopulu & Buckland, 2012). The body was buried in December 1315 in a lead coffin within a stone sarcophagus beneath the floor of York Minster in northern England. Based on examination of the remains of insect fauna within the coffin, it was concluded that the body was buried soon after death as evidenced by a lack of early colonizers. In fact, later-stage fly fauna was not present either, suggesting that the lead coffin was an effective means to protect the corpse from phorids. However, the layers of lead and stone were not sufficient to ensure total exclusion of necrophagous insects: the insect assemblage was dominated by a coffin beetle *Rhizophagus parallelocollis* and a predatory staphylinid *Quedius mesomelinus* (Panagiotakopulu & Buckland, 2012). *Rhizophagus parallelocollis* has been observed in several medieval burial sites, as well as in Roman and post-medieval excavations (Stafford, 1971; Panagiotakopulu & Buckland, 2012). The beetle is not commonly associated with more modern human burials, which has led to the speculation that changes in coffin types and increased burial depths have limited its access to the corpse.

Few other detailed studies have been conducted in forensic archaeoentomology, or the research has been tied directly to forensic entomology. The field will likely remain relatively small but has the potential of addressing topics relevant to both archaeology and forensic entomology, including examination of burial and mortuary practices from pre- and post-medieval times, comparison of rates of decomposition associated with various burials or other means of body disposal based on the insect fauna, and conceivably may allow investigation of the epidemiology of insect-borne diseases through past civilizations (Wylie *et al.*, 1987; Rick *et al.*, 2002; Huchet, 2010).

Chapter review

Archaeoentomology is a new "old" discipline

- The connection between archaeology and entomology is not about the individuals who are engaged in the academic and intellectual pursuits of discovering our past. Rather, the two are linked through the people of interest – from ancient civilizations to our most recent past – and the extinct and extant insects that shaped the living conditions of these people.
- Archaeology by definition is the study of human activity over time through the recovery and analysis of individual and cultural materials and environmental data that remain once the people do not. The materials that are studied include artifacts, architecture, recordings, biological remains, and cultural landscapes.
- Archaeoentomology uses insects as evidence and tools to explore the past. Fossilized insects, the remains of unfossilized insects, and evidence of insect activity are used to interpret the environments of past civilizations.

Concepts and techniques from forensic entomology can be applied to archaeology

- The use of approaches from forensic entomology in archaeological investigation is considered relatively new, as evidenced by the paucity of research literature related to insects in archaeology. Perhaps the more correct view is that the discipline has existed in some capacity for several decades but is only recently being recognized as an organized field of study.
- Insects serve as the physical evidence for both archaeoentomology and forensic entomology,

with the fundamental difference being the condition of the specimens: forensic entomology relies predominantly on the collection of living specimens whereas archaeoentomology does not have the luxury of working with live, nor necessarily intact, evidence. The latter discipline deals with fossilized specimens, unfossilized remains of dead insects, and any evidence of insect activity.

- Reconstruction through hypothesis testing is much more challenging when dealing with ancient test subjects. An archaeoentomologist cannot bring specimens back to the laboratory to rear to adults for easier identification nor can the conditions of development be worked out in the laboratory for estimating length of development on various food sources or interactions among or between species.
- The term "funerary archaeoentomology" has been coined for the field of research examining insect association with ancient taphonomy and mortuary practices. It is not possible to use insect evidence from archaeological finds to calculate the time of death estimation or postmortem interval as is desired in forensic entomology.
- The commonality in the two disciplines is also evident in how both can be subdivided: each examines insects that impact (or have) human food supplies, influence the human condition in urban locations, and examine the association between insects and animal remains.

Ancient insects and food: connection to stored product entomology

- Archaeological evidence has revealed that ancient civilizations suffered from insect attack on stored foods and animal products in households and urban centers dating back to at least the early dynastic period in ancient Egypt. This is as true now as it was then: bulk storage of commodities in granaries function as ideal environments to (i) attract large numbers of insects, and (ii) allow the mass aggregation of adult insects at a food source, which is ideal for proliferation.
- Among the many insect fossils and remains identified from excavation sites from ancient Greece, Egypt, and the Roman Empire, two extant pest species, the grain weevil *Sitophilus granarius* and the sawtoothed grain beetle *Oryzaephilus surinamensis*, occur in fossil assemblages, often together, and thus serve as excellent examples of the types of discoveries from archaeoentomology.

- The fossil and other archaeoentomological evidence suggests that *S. granarius* evolved into a synanthropic species even before settled or sustained agriculture developed. Synanthropy of *O. surinamensis* may be considered more recent than *S. granarius* simply because adults and juveniles utilize grain already fed upon by primary stored product pests such as the grain weevil. Alternatively, the adaptation of the sawtoothed grain beetle as a secondary pest may reflect resource partitioning akin to the spatial partitioning observed in mixed species maggot masses.
- Fossilized remains of insects from different regions allow analyses of the distribution and origins of insects. Here is where archaeoentomology and paleoentomology are united in interest. However, the focus of an archaeoentomologist in determining historical biogeography of a particular species is really aimed at addressing questions surrounding the influence of a given insect on a particular society or civilization(s).

Ancient insects as pests: beginnings of synanthropy and urban entomology

- The definition of urban entomology as it relates to entomology and forensic entomology specifically is the branch of entomology that deals with insects and other arthropods associated with human habitation or the human environment. A broad interpretation of this definition includes insects that occur in yards and neighborhoods (e.g., urban areas or centers) as well as those of agricultural importance that invade human space. This same terminology can be applied to archaeoentomology.
- Insects that vector disease in the modern world are abundant in the fossil record and in other forms discovered throughout sites in the Old and New Worlds. Examples of some of the fossilized and unfossilized remains of several important insect vectors include fleas, lice, and bed bugs.
- Establishing the presence of insect vectors in ancient civilizations is one thing, but that does not necessarily mean that fleas, lice, flies, and mosquitoes always transmitted diseases. Reconstruction and hypothesis testing are not options for ancient insects. So, comparisons

testing through an examination of mummified human tissues and preserved insects must be conducted to identify the pathogenic organisms responsible for evoking diseases like malaria, bubonic plague, and typhus.

Ancient insects and burial practices: revelations about past lives and civilizations

- Study of insects associated with the dead, whether mummified or not, from archaeological sites relies on very similar techniques and concepts as medicocriminal entomology. Perhaps the most important unifying feature of the two fields is a reliance on necrophagous flies. Discovery of fly evidence at a site may allow estimations of the season when the insects arrived, whether conditions were favorable for fly development, the stage of corpse decomposition when fly colonization occurred, and possibly whether the body had been moved.
- The types of insects collected from an excavation in which human remains or mummies are found can yield insight into the burial practices of ancient civilizations.
- Several mummified remains of humans and other animals have been found to harbor a wide range of insects, including many of forensic interest including beetles (clerids, dermestids, scaribs) and flies (calliphorids, muscids, phorids, piophilids, sarcophagids).

Forensic archaeoentomology: entomological investigations into extremely "cold" cases

Few detailed studies have been conducted in forensic archaeoentomology, or the research has been tied directly to forensic entomology. The field will likely remain relatively small but has the potential of addressing topics relevant to both archaeology and forensic entomology, including examination of burial and mortuary practices from pre- and post-medieval times, comparison of rates of decomposition associated with various burials or other means of body disposal based on the insect fauna, and conceivably may allow investigation of the epidemiology of insect-borne diseases through past civilizations.

Test your understanding

Level 1: knowledge/comprehension

1. Define the following terms:
   (a) archaeology
   (b) synanthropy
   (c) taphonomy
   (d) paleoentomology
   (e) cultural landscape
   (f) funerary archaeoentomology.

2. Match the terms (i–vi) with the descriptions (a–f).

   (a) Subterranean insect species   (i) Exophilic
   (b) Process of rapidly removing water from tissues usually during extreme heat   (ii) Midden
   (c) Coexistence in using the same resource by existing in a different location on the resource   (iii) Mummification
   (d) Insects that exist independent of human environments   (iv) Hypogean
   (e) Ancient trash heap
   (f) Infectious disease that spreads across continents   (v) Pandemic plague
                                                               (vi) Spatial partitioning

3. Explain how archaeoentomology, paleoentomology and Quaternary entomology seem very similar, yet remain unique disciplines.
4. Forensic entomology and archaeoentomology overlap with the types of questions asked and techniques/concepts applied to investigations. Explain the similar types of questions and techniques addressed by both disciplines.
5. Explain the limitations of archaeoentomological research that makes it difficult to rely on reconstruction and hypothesis testing in the same way used by a forensic entomologist.

Level 2: application/analysis

1. Archaeoentomology commonly addresses questions of historical biogeography and

synanthropy. Describe what types of entomological evidence would be needed to examine synanthropy in necrophagous calliphorids and Native Americans of the eastern United States.

Notes

1. From Waldbauer (2000).
2. Steven Spielberg is the academy award-winning director who developed the Indiana Jones franchise into four movies: *Raiders of the Lost Ark* (1981), *Indiana Jones and the Temple of Doom* (1984), *Indiana Jones and the Last Crusade* (1989), and *Indiana Jones and the Kingdom of the Crystal Skull* (2008), which starred Harrison Ford as Dr Jones.
3. Extinct and extant are opposing terms that are used in reference to whether an organism is considered no longer living on the planet (extinct) or is a species that does exist today (extant).
4. The field of Quaternary entomology is technically narrower in focus than paleoentomology, concerned with fossilized insects from the Quaternary period, the most recent period of the Cenozoic Era, spanning from approximately 2.5 million years ago to present.
5. This time range also overlaps with the beginnings of the Old Kingdom Period (*c.* 2686–2181 BC) in Egypt.
6. A pandemic plague is one in which an infectious disease has become epidemic and spreads across other regions, countries or even continents.
7. Septicemia occurs when pathogenic organisms, usually bacteria, enter the normally sterile environment of blood, altering homeostasis and potentially leading to sepsis (whole body inflammation) and death.
8. Canopic jars were used during the mummification process for storing the viscera or internal organs for use in the afterlife.

References cited

Azar, D. & Nel, A. (2003) Fossil psychodid flies and their relation to parasitic diseases. *Memórias do Instituto Oswaldo Cruz* 1: 35–37.

Backwell, L.R., Parkinson, A.H., Roberts, E.M., d'Errico, F. & Huchet, J.-B. (2012) Criteria for identifying bone modification by termites in the fossil record. *Palaeogeography, Palaeoclimatology, Palaeoecology* 337–338: 72–87.

Bahn, P. & Renfrew, C. (2008) *Archaeology: Theories, Methods and Practice*, 5th edn. Thames and Hudson Publishers, London.

Bain, A. (2004) Irritating intimates: the Archaeoentomology of lice, fleas, and bedbugs. *Northeast Historical Archaeology* 33: 81–90.

Boddington, A., Garland, A.N. & Janaway, R.C. (1981) *Death, Decay and Reconstruction. Approaches to Archaeology and Forensic Science.* Manchester University Press, Manchester, UK.

Brier, B. (1996) *Egyptian Mummies: Unraveling the Secrets of an Ancient Art.* Harper Publishing, New York.

Buchner, S. & Wolf, G. (1997) Der Kornkafer – *Sitophilus granarius* (Linne) – aus einer bandkeramischen Grube bei Gottingen. *Archaologisches Korrespondenzblatt* 27: 211–220.

Buckland, P.C. (1981) The early dispersal of insect pests of stored products indicated by archaeological records. *Journal of Stored Product Research* 17: 1–12.

Buckland, P.C. (1991) Granaries, stores and insects. The archaeology of synanthropy. In: D. Fournier & F. Sigaut (eds) *La Préparation Alimentaire des Céréals*, pp. 69–81. PACT, Rixensart, Belgium.

Chaddick, P.R. & Leek, F.F. (1972) Further specimens of stored products insects found in ancient Egyptian tombs. *Journal of Stored Product Research* 8: 83–86.

Curry, A. (1979) The insects associated with the Manchester mummies. In: A.R. David (ed.) *The Manchester Mummy Project*, pp. 113–118. Manchester University Press, Manchester, UK.

David, R. (1978) The fauna. In: R. David (ed.) *Mysteries of the Mummies: The Story of the Manchester University Investigations*, pp. 160–167. Book Club Associates, London.

Donnan, C.B. & Mackey, C.J. (1978) *Ancient Burial Patterns of the Moche Valley, Peru.* University of Texas Press, Austin, TX.

Dunand, F., Lichtenberg, R., Lorton, D. & Yoyotte, J. (2006) *Mummies and Death in Egypt.* Cornell University Press, Ithaca, NY.

Dupras, T.L., Schultz, J.J., Wheeler, S.M. & Williams, L.J. (2011) *Forensic Recovery of Human Remains: Archaeological Approaches*, 2nd edn. CRC Press, Boca Raton, FL.

Dussault, F., Bain, A. & LeMoine, G. (2014) Early Thule winter houses: an archaeoentomological analysis. *Arctic Anthropology* 51(1): 101–117.

Elias, S.A. (1994) *Quaternary Insects and Their Environments.* Smithsonian Press, Washington, DC.

Funck, J., Heintzman, P.D., Murray, G.G., Shapiro, B., McKinney, H., Huchet, J.B., Bigelow, N., Druckenmiller, P. & Wooller, M.J. (2020). A detailed life history of a pleistocene steppe bison (Bison priscus) skeleton unearthed in Arctic Alaska. *Quaternary Science Reviews* 249: 106578.

Geden, C.J. (2006) Biological control of pests in livestock production. In: L. Hansen & T. Steenberg (eds) *Implementation of Biocontrol Practice in Temperate Regions: Present and Near Future*, pp. 45–60. Proceedings of the International Workshop at Research Centre Flakkebjerg, Denmark on November 1 to 3, 2005. Available at http://www.bashanfoundation.org/linderman/lindermanbiocontrol.pdf.

Giordani, G., Erauw, C., Eeckhout, P.A., Owens, L.S. & Vanin, S. (2020) Patterns of camelid sacrifice at the site of Pachacamac, Peruvian Central Coast, during the

Late Intermediate Period (ad 1000–1470): perspectives from funerary archaeoentomology. *Journal of Archaeological Science* 114: 105065.

Haile, A. (2006) On farm storage studies on sorghum and chickpea in Eritrea. *African Journal of Biotechnology* 5: 1537–1544.

Hall, R.D. & Huntington, T.E. (2010) Introduction: perceptions and status of forensic entomology. In: J.H. Byrd & J.L. Castner (eds) *Forensic Entomology: The Utility of Arthropods in Legal Investigations*, pp. 1–16. CRC Press, Boca Raton, FL.

Hope, F.W. (1842) Observations on some mummified beetles taken from the inside of a mummified ibis. *Transactions of the Royal Entomological Society of London* 1: 11–13.

Huchet, J.-B. (2010) Archaeoentomological study of the insect remains found within the mummy Namenkhet Amon (San Armenian Monastery, Venice/Italy). *Advances in Egyptology* 1: 59–80.

Huchet, J.-B. (2014) Insect remains and their traces: relevant fossil witnesses in the reconstruction of past funerary practices. *Anthropologie* LII/3 52: 329–346.

Huchet, J.-B. & Greenberg, B. (2010) Flies, Mochicas and burial practices: a case study from Huaca de la Luna, Peru. *Journal of Archaeological Science* 37: 2846–2856.

Hunter, J., Roberts, C. & Martin, A. (1996) *Studies in Crime: An Introduction to Forensic Archaeology*. Batsford Publishing, London.

Lecheta, M.C. & Luz, D.R. (2015) First record of *Tachinaephagus zealandicus* Ashmead, 1904 (Hymenoptera Encyrtidae) parasitizing the blowfly *Sarconesia chlorogaster* (Wiedemann, 1830) (Diptera: Calliphoridae) in Brazil. *Brazilian Journal of Biology* 75: 505–506.

Legner, E.F. (1988) *Muscidifurax raptorellus* (Hymenoptera: Pteromalidae) females exhibit postmating oviposition behavior typical of the male genome. *Annals of the Entomological Society of America* 81: 522–527.

McNeill, W.H. (1977) *Plagues and Peoples*. Penguin Publishers, Harmondsworth, UK.

Miller, R. (1987) Appendix. Ash as an insecticide. In: B.J. Kemp (ed.) *Amarna Reports IV*, pp. 14–16. Egypt Exploration Society, London.

Morrow, J. J., Myhra, A., Piombino-Mascali, D., Lippi, D., Roe, A., Higley, L. & Reinhard, K. J. (2016) Archaeoentomological and archaeoacarological investigations of embalming jar contents from the San Lorenzo Basilica in Florence, Italy. *Journal of Archaeological Science: Reports* 10: 166–171.

Nel, A., Petrulevicius, J.F. & Azar, D. (2010) Palaeoentomology, a young old field of science. *Annales de la Société Entomologique de France* 46: 1–3.

Osborne, P.J. (1983) An insect fauna from a modern cesspit and its comparison with probable cesspit assemblages from archeological sites. *Journal of Archaeological Science* 10: 453–463.

Panagiotakopulu, E. (2001) New records for ancient pests: archaeoentomology in Egypt. *Journal of Archaeological Science* 28: 1235–1246.

Panagiotakopulu, E. (2004a) Pharaonic Egypt and the origins of plague. *Journal of Biogeography* 31: 269–275.

Panagiotakopulu, E. (2004b) Dipterous remains and archaeological interpretation. *Journal of Archaeological Science* 31: 1675–1684.

Panagiotakopulu, E. & Buckland, P.C. (1991) Insect pests of stored products from Late Bronze Age Santorini, Greece. *Journal of Stored Product Research* 27: 179–184.

Panagiotakopulu, E. & Buckland, P.C. (1999) The bed bug, *Cimex lectularius* L. from Pharaonic Egypt. *Antiquity* 73: 908–911.

Panagiotakopulu, E. & Buckland, P. (2009) Environment, insects and the archaeology of Egypt. In: S. Ikram & A. Dodson (eds) *Beyond the Horizon: Studies in Egyptian Art, Archaeology and History in Honour of Barry J. Kemp*, pp. 347–360. The American University in Cairo Press, Cairo, Egypt.

Panagiotakopulu, E. & Buckland, P.C. (2012) Forensic aracheoentomology: an insect fauna from a burial in York Minster. *Forensic Science International* 221: 125–130.

Panagiotakopulu, E., Buckland, P.C. & Kemp, B.J. (2010) Underneath Ranefer's floors: urban environments on the desert edge. *Journal of Archaeological Science* 37: 474–481.

Poinar, G. Jr & Poinar, R. (2004) Evidence of vector-borne disease of Early Cretaceous reptiles. *Vector Borne Zoonotic Diseases* 4: 281–284.

Pradelli, J., Rossetti, C., Tuccia, F., Giordani, G., Licata, M., Birkhoff, J.M., Verzeletti, A. & Vanin, S. (2019) Environmental necrophagous fauna selection in a funerary hypogeal context: the putridarium of the Franciscan monastery of Azzio (northern Italy). *Journal of Archaeological Science: Reports* 24: 683–692

Reed, D.L., Light, J., Allen, J.M. & Kirchman, J.J. (2007) Pair of lice lost or parasites regained: the evolutionary history of anthropoid primate lice. *BMC Biology* 5: 7. https://doi.org/10.1186/1741-7007-5-7.

Rick, F.M., Rocha, G.C., Dittmar, K., Coimbra, C.E.A. Jr, Reinhard, K. Bouchet, F., Ferreira, L.F. & Arauj, A. (2002) Crab louse infestation in pre-Columbian America. *Journal of Parasitology* 88: 1266–1267.

Rivers, D.B. (1996) Changes in the oviposition behavior of the ectoparasitoids *Nasonia vitripennis* and *Muscidifurax zaraptor* (Hymenoptera: Pteromalidae) by different species of fly hosts, prior oviposition experience, and allospecific competition. *Annals of the Entomological Society of America* 89: 466–474.

Rivers, D.B. (2004) Evaluation of host responses as means to assess ectoparasitic pteromalid wasp's potential for controlling manure-breeding flies. *Biological Control* 30: 181–192.

Rivers, D.B., Ciarlo, T., Spelman, M. & Brogan, R. (2010) Changes in development and heat shock protein expression in two species of flies (*Sarcophaga bullata* [Diptera: Sarcophagidae] and *Protophormia terraenovae* [Diptera: Calliphoridae]) reared in different sized maggot masses. *Journal of Medical Entomology* 47: 677–689.

Silverman, D.P. (2003) *Ancient Egypt*. Oxford University Press, New York.

Skidmore, P. (1995) Analysis of fly remains from F23 in the stone-lined pit (F22). In: F. McCormick (ed.) *Excavation at Pluscarden Priory, Moray*, pp. 418–419. Proceedings of the Society of Antiquaries of Scotland, Edinburgh.

Solomon, M.E. (1965) Archaeological records of storage pests: *Sitophilus granarius* (L.) (Coleoptera: Curculionidae) from an Egyptian pyramid tomb. *Journal of Stored Products Research* 1: 105–107.

Stafford, F. (1971) Insects of medieval burial. *Scientific Archaeology* 7: 6–10.

Strong, L. (1981) Dermestids: an embalmer's dilemma. *Antenna* 5: 136–139.

Thompson, J.E., Martin-Vega, D., Buck, L.T., Power, R.K., Stoddart, S. & Malone, C. (2018) Identification of dermestid beetle modification on Neolithic Maltese human bone: Implications for funerary practices at the Xemxija tombs. *Journal of Archaeological Science: Reports* 22: 123–131.

Vanin, S. & Huchet, J.B. (2017). Forensic entomology and funerary archaeoentomology. In: E.M.J. Schotsmans, N. Marquez-Grant, S.L. Forbes (eds.) *Taphonomy of Human Remains: Forensic Analysis of the Dead and the Depositional Environment*, pp. 167–186. Wiley, Oxford, UK.

Waldbauer, G. (2000) *Millions of Monarchs, Bunches of Beetles: How Bugs Find Strength in Numbers*. Harvard University Press, Cambridge, MA.

Wylie, F.R., Walsh, G.L. & Yule, R.A. (1987) Insect damage to aboriginal relics at burial and rock-art sites near Carnarvon in central Queensland. *Australian Journal of Entomology* 26: 335–345.

Yoder, J.A., Denlinger, D.L., Dennis, M.W. & Kolattukudy, P.E. (1992) Enhancement of diapausing flesh fly puparia with additional hydrocarbons and evidence for alkane biosynthesis by a decarbonylation mechanism. *Insect Biochemistry and Molecular Biology* 22: 237–243.

Zdarek, J., Fraenkel, G. & Friedman, S. (1987) Pupariation in flies: a tool for monitoring effects of drugs, venoms, and other neurotoxic compounds. *Archives of Insect Biochemistry and Physiology* 4: 29–46.

Palaeogeography, Palaeoclimatology, Palaeoecology 560: 109989.

Corron, L., Huchet, J.B., Santos, F. & Dutour, O. (2017) Using classifications to identify pathological and taphonomic modifications on ancient bones: do "taphognomonic" criteria exist? *Bulletins et Mémoires de la Société d'Anthropologie de Paris* 29: 1–18.

Forbes, V., Bain, A., Gisladottir, G.A. & Milek, K.B. (2010) Reconstructing aspects of the daily life in late 19th and early 20th-century island: archaeoentomological analysis of the Vatnsfordur farm, NW Iceland. *Archaeologia Islandica* 8: 77–110.

Forbes, V., Britton, K. & Knecht, R. (2015) Preliminary archaeoentomological analyses of permafrost-preserved cultural layers from the pre-contact Yup'ik Eskimo site of Nunalleq, Alaska: implications, potential and methodological considerations. *Environmental Archaeology* 20(2): 158–167.

Huchet, J. B., Pereira, G., Gomy, Y., Philips, T.K., Alatorre-Bracamontes, C.E., Vásquez-Bolaños, M. & Mansilla, J. (2013) Archaeoentomological study of a pre-Columbian funerary bundle (mortuary cave of Candelaria, Coahuila, Mexico). *Annales de la Société entomologique de France (NS)* 49: 277–290.

King, G., Gilbert, M.T., Willerslev, E. & Collins, M.J. (2009) Recovery of DNA from archaeological insect remains: first results, problems, and potential. *Journal of Archaeological Science* 36: 1179–1183.

Kislev, M.E., Hartmann, A. & Galili, E. (2004) Archaeobotanical and archaeoentomological evidence from a well at Atlit-Yam indicates colder, more humid climate on the Israeli coast during the PPNC period. *Journal of Archaeological Science* 31: 1301–1310.

Panagiotakopulu, E. (2003) Insect remains from the collections in the Egyptian Museum of Turin. *Archaeometry* 45: 355–362.

Robbiola, L., Moret, P. & Lejars, T. (2011) A case study of arthropods preserved on archaeological bronzes: micro-archaeological investigation helps reconstructing past environments. *Archaeometry* 53: 1249–1256.

Smith, D. (1996) Thatch, turves and floor deposits: a survey of Coleoptera in materials from abandoned Hebridean blackhouses and the implications for their visibility in the archaeological record. *Journal of Archaeological Science* 31: 1301–1310.

Additional resources

Supplemental reading

Backwell, L., Huchet, J.B., Jashashvili, T., Dirks, P.H. & Berger, L.R. (2020) Termites and necrophagous insects associated with early Pleistocene (Gelasian) *Australopithecus sediba* at Malapa, South Africa.

Archaeological Institute of America: http://www.archaeological.org/societies/

International Paleoentomological Society: http://fossilinsects.net/

Society of American Archaeology: www.saa.org

Society for Historical Archaeology: www.sha.org

Chapter 21

Insects as weapons of war and threats to national security

In [the] world's biggest terror attack in America, terrorists, armed only with box cutters, hijacked planes and brought down the towers of the World Trade Center. Insects are the box cutters of biological warfare – cheap, simple and wickedly effective.

Dr Manas Sarkar, Centre for Medical Entomology and Vector Management
National Centre for Disease Control, India[1]

Overview

The unfortunate reality of the twenty-first century is that the world is not at peace. Civil unrest exists on several continents, which is not unique to any point in mankind's history. What has changed is who is fighting and how wars are waged: small but highly organized terrorist organizations are waging "holy" wars using non-conventional tactics. Traditional weaponry has been supplemented with an arsenal that includes the so-called weapons of mass destruction as well as abiotic and biotic terrorist tools. The latter has led to wide-scale concern that biological and chemical warfare will be waged against western nations that have historically formed political, economic, and military alliances. Though often overlooked in discussions of biological weapons, insects have enormous potential for use in direct assaults on people, as

delivery systems of devastating diseases to humans and other animals, and in targeting agriculture. The entomological terrorism potential stems from the fact that thousands of insects can be raised cheaply and quickly, and can be released on target sites without the need for sophisticated delivery systems. The entomological outlook is not totally bleak, as insects have been recruited to aid nations in their quest to secure their borders and protect the citizens within. In this respect, insects are being used in covert surveillance programs, as tools to locate unexploded munitions like landmines, and in toxicological screening of tissues exposed to bullets or explosives. This chapter examines the roles of insects in issues of national security and terrorism by delving into the use of insects as weapons historically and as impending threats, not just directly to mankind but also with respect to agricultural terrorism. Focus is also

The Science of Forensic Entomology, Second Edition. David B. Rivers and Gregory A. Dahlem.
© 2023 John Wiley & Sons Ltd. Published 2023 by John Wiley & Sons Ltd.
Companian website: www.wiley.com/go/forensicentomology2

placed on how insects can be used as tools to counter threats by terrorist or other organizations, as well as direct application to other legal matters such as discovery of illicit drugs or decomposing bodies.

The big picture

- Terrorism and biological threats to national security are part of today's world.
- Entomological weapons are not new ideas.
- Direct entomological threats to human populations are not all historical.
- Impending entomological threats to agriculture and food safety.
- Insect-borne diseases as new or renewed threats to human health.
- Insects can be used as tools for national security.

21.1   Terrorism and biological threats to national security are part of today's world

September 11, 2001. A date that forever changed the way that citizens of the United States view the world, particularly with respect to the sanctity of the "protective" borders that define the country. In many ways, global peace was cast into chaos with a single day filled with terrorist activity; hijacked planes bringing down the twin towers of the World Trade Center in New York City, a third plane crashing into the symbol of world defense, the Pentagon in Arlington, Virginia, and yet another diverted from its intended target of the Capitol Building in Washington, DC by the brave acts of the captives on board. Nearly 3000 individuals lost their lives that day. The deadly deeds carried out by the terrorist organization Al-Qaeda[2] on the day now known simply as 9/11 represented the first direct enemy attack on the United States' soil since the Japanese bombing of Pearl Harbor, Hawaii on December 7, 1941.[3] These events provided a sobering reality check for citizens of the United States, and perhaps the rest of the world, that no one is totally protected or safe from today's modern warfare, **terrorism**. Terrorism is literally the use of terror, usually through acts of violence, in the name of religion, politics or some other ideological purpose, with no

regard for non-combatants (civilians). Conventional warfare, though barbaric in most ways, has historically followed "unwritten" rules in which non-military civilians were (are) not attacked; the soldiers were left to decide the battles and ultimately the victors and losers of war.[4] In today's global environment, wars are declared not so much by countries but by militant political groups, frequently termed terrorist organizations or cells or non-state groups, who generally hold no ties to any one country and show a lack of respect for life, as soldiers, civilians, men, women, and children are all targets. This unfortunate reality places the entire global community at risk in the twenty-first century (Figure 21.1).

Militant organizations do not rely exclusively on conventional weaponry. Rather, they have heightened the air of fear by developing or threatening to develop biological and chemical weapons that, when utilized, have the potential to impact hundreds of thousands to millions of individuals

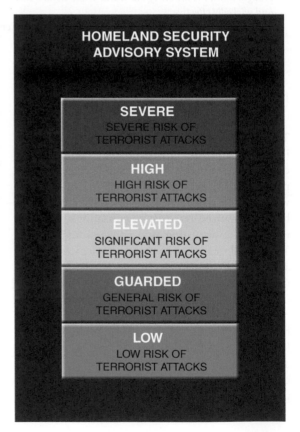

Figure 21.1 United States Department of Homeland Security advisory levels regarding terrorist threats. *Source*: United States Department of Homeland Security/ Wikimedia commons/public domain.

across countries and continents. At the dawn of this century, Iraq,[5] Iran, Syria, China, Libya, North Korea, Russia, Israel, Taiwan, and potentially India, Pakistan, Sudan, and Kazakhstan possess biological weapons (Garrett, 2001). Quite disturbing is the fact that several of these nations are politically unstable or are considered home to radical terrorist organizations, simply meaning that biological weapons in their hands could be used in horrifically unimaginable ways (Frischknecht, 2008). Adding to the impending fear of the new ways of war is the rapid expansion in the arsenal of biological weapons. Prior to 1985, most countries that possessed biological weaponry also possessed the same antidotes, bringing the arms race to a calculated standstill (Garrett, 2001). All that changed beginning in the 1990s with an exponential phase of biological discovery, particularly in the realm of molecular biology and biotechnology. The result is that "old" biological weapons like anthrax are now easier to produce and new bioweapons can be (have been) fashioned for specific targets, with only the creators knowing that the weapons even exist (DaSilva, 2009). The target countries are mostly in the dark as to the nature of the weapons, which leaves most nations poorly equipped to cope with biological terrorist attacks. Today's modern warfare by terrorist organizations rather than by nations does not adhere to the decrees of the Biological and Toxin Weapons Conventions sponsored by the United Nations which outlawed the development and production of biological weapons. The reality is that there are likely more biological weapons in this century than at any other time in history, and the weapons appear to be in the hands of those ready and willing to use them.

To most, biological weapons come in the form of microorganisms that cause disease or are human pathogens. Smallpox, plague (bubonic), anthrax, or Ebola take center stage (Figure 21.2). Each represents disease that conceivably could be contained or an epidemic prevented with proper vaccination programs developed as part of an anti-terrorism or national security effort (Garrett, 2001). However, no western nation has been sufficiently proactive to stave off such attacks (Figure 21.3). The United States has taken some important first steps to prepare for biological warfare. For example, in 1991, the U.S. Congress placed economic embargos on countries believed to be developing biological weapons. This was followed by congressional adoption of the Anti-Terrorism Act of 1996 that permitted federal authorities the right to apprehend and arrest anyone

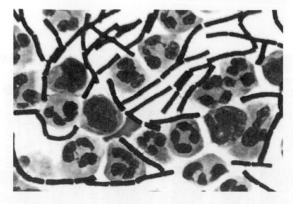

Figure 21.2 Gram stained *Bacillus anthracis* (purple rods), the causative agent of anthrax, in cerebrospinal fluid. *Source*: Photo by CDC/Wikimedia Commons/ public domain. https://commons.wikimedia.org/wiki/ File:Gram_Stain_Anthrax.jpg.

making threats to produce or use biological weapons against the United States citizens. In 1999, over $10 billion was appropriated to combat terrorism in the United States, although the funding was directed specifically to prevent attack via conventional rather than biological weapons (Casagrande, 2000). Since the attacks of 9/11, huge budgetary increases for programs/agencies working on counter-terrorist initiatives have been granted, but only a small proportion of the funding has been earmarked for improving readiness to biological terrorism, particularly those aimed directly at humans or food supplies (Casagrande, 2000; Crutchley *et al.*, 2007). This oversight appears to be due in part to a lack of recognition of where the threats lie and the types of biological weapons most likely to be implemented. What this means specifically is that the primary focus has been on pathogens that elicit human disease. While the kill rates for some can be quite high, developing biological weapons and delivery systems is time-consuming, costly, and dangerous to the terrorist organizations. The more immediate threat seems to be from a source largely ignored to this point. More specifically, the definition of **biological terrorism (bioterrorism)** is almost always limited to microorganisms such as bacteria, viruses, or other pathogens that induce illness or death in humans, animals or plants (Centers for Disease Control and Prevention, 2007). Missing from this definition are macrobiotic organisms, specifically insects.

Does this represent just an oversight or do lawmakers, military leaders, or heads of national security directives in the United States and other western nations truly believe that insects pose no threat as

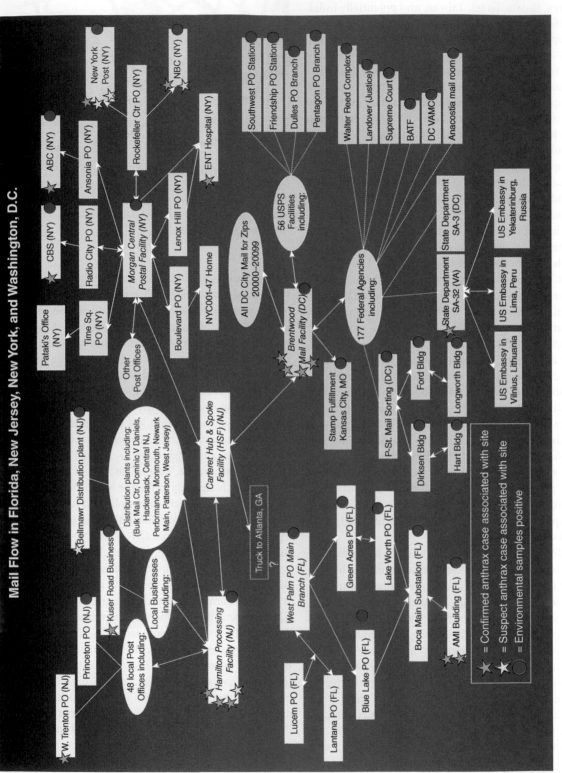

Figure 21.3 Mail flow path of anthrax associated with a case of domestic terrorism in the United States. *Source:* Centers for Disease Control and Prevention/ Wikimedia commons/public domain. https://commons.wikimedia.org/wiki/File:Anthraxmailflow.jpg.

biological weapons? There is likely little comfort that would come from knowing the answers to these questions because they would reveal deficiencies in strategic planning for counter-terrorist defenses or suggest a false sense of security among our leaders and military strategists regarding threats to national security. The last two decades should have taught everyone that extreme caution should be exercised in evaluating whom, where and with what (weapons) in terms of the next potential terrorist attack(s). Insects do represent viable options as biological weapons that could be used by militant groups to inflict large-scale damage, quickly, cheaply, and with much less personal danger to the terrorist cell than when formulating human pathogens into bioweapons (Monthei *et al.*, 2010; Sarkar, 2010). The use of insects as weapons of today's warfare is termed **entomological terrorism** and can be manifested in at least three ways (Lockwood, 2009a):

1. Direct attack on humans using biting, stinging, or other insect species producing toxins or noxious compounds.
2. Agriculturally important species released en masse to target cropping systems or livestock with the intent of crippling food supplies and damaging the economy of a region or country.
3. Mass release of insect vectors harboring pathogens to humans, other animals, or plants, a.k.a., **arboweapons** (Kwak, 2016).

These are not just a list of theoretical examples; entomological weapons have been used for thousands of years in wars to target armies and individuals. There are also accusations that insects have been used to target agricultural systems during various wars of the nineteenth and twentieth centuries (Lockwood, 1987; Kwak, 2016). It is also important to note that examples 2 and 3 above are not mutually exclusive as insect transmission of a plant pathogen can serve as both. The remainder of the chapter will explore how insects are impending threats to national security via the three mechanisms given above: direct attack on humans and other animals, targeting agriculture, and as vectors of disease. We will also examine how insects can be used as tools to aid in national security by thwarting terrorist organizations through surveillance initiatives, roles as biological and chemical sensors, and aiding in detection of explosive or bullet residues via toxicological screening.

21.2 Entomological weapons are not new ideas

The destructive potential of insects as weapons of war has been understood at some level for thousands of years. Perhaps the oldest known descriptions of insects as weapons come from the Christian Bible in which God enlists flies, lice and locusts to inflict pain and suffering on Egypt as a means to prompt Pharaoh (Rameses II) to release the Israelites (Exodus 8–10) (discussed in more detail in Chapter 2). King Menes (*c.* 3407–3346 BC), ruler of the first dynasty of Egypt, practiced **heraldry**[6] using the Oriental wasp, *Vespa orientalis*, as the face of the empire due to its ferocious look and known aggressive behaviors and sting (Berenbaum, 1996). The power of stinging Hymenoptera in ancient Egypt was appreciated beyond just symbolically: entire "bee" nests were placed in porcelain jars to bomb enemies out of entrenchments and fortifications (Lockwood, 2009a). This practice was undoubtedly not unique to Egyptians. The disadvantage of such early ento-weapons was the lack of control over the insects. Once alarm pheromones[7] are released by the agitated bee/wasps, they will sting most any large vertebrate in their path, including the soldiers that designed the bee bombs! More often, stinging and biting insects were used as a means of torture and interrogation. Numerous cultures are known to have tortured captives with insects that could inflict pain and suffering. The interrogation techniques included staking captives to anthills or tying naked individuals to trees in mosquito and biting fly-infested areas (Lockwood, 2009b). Extreme cruelty was exercised in cases where prisoners were force-fed milk and honey to promote anal myiasis, placed in crates filled with blood-sucking bed bugs, or lowered into earthen pits filled with ticks and assassin bugs that inflicted severely painful bites and which possessed venom that promoted digestion of human flesh (Lockwood, 2009b). In some cases, early psychological warfare was apparently adopted, as just the mere threat of the insect was sufficient to extract the desired information (Figure 21.4).

The early beginnings leading to modern biological weapons are evident in multiple military campaigns in which diseased organisms or the insects responsible for transmission were enlisted. Some of the best-known examples include dropping dead or diseased cadavers into water supplies, catapulting plague-infested corpses over the walls

Figure 21.4   The notorious Zindon prison at Bukhara, Uzbekistan, where British officers Stoddart and Conolly were held in the "bug pit" until their public execution outside the Ark in 1842. *Source:* Photo by David Stanley/Wikimedia Commons/CC BY 2.0. https://commons.wikimedia.org/wiki/File:Zindon_Prison_(8145368631).jpg.

of fortified cities to induce a localized plague (fourteenth century), and the British providing smallpox-exposed "gifts" to Native American chiefs during the French and Indian War (1754–1763) as a means to ignite a smallpox epidemic among Indian populations in North America (Kirby, 2005; Lockwood, 2009a). World War I saw the dawn of the use of chemical weapons, and agents such as mustard gas, chlorine, and phosgene were so effective at killing soldiers that it is likely to have delayed the introduction of biological weaponry. There were accusations, however, that the German empire ruled by Kaiser Wilhelm II did engage in the use of biological weapons during this period but no conclusive evidence was ever discovered (Harris & Paxman, 1982). Following the conclusion of World War I, the Geneva Protocol[8] (or treaty) was drafted with the purpose of banning the use of chemical and biological weapons, although it was not until over two decades later that insects were identified and banned as biological agents of war (Lockwood, 1987).

World War II marked a major upturn in biological weaponry research and deployment. The Japanese army developed a major research facility in Manchuria – the infamous home to Unit 731 – that was dedicated to mass production of the causative

agent of bubonic plague, *Yersinia pestis*, and the flea vector *Pulex irritans* L. (Order Siphonaptera: Family Pulicidae), as well as other biological weapons (Brody *et al.*, 2014). Testing was conducted on delivery methods including bombs, the spraying of infested vectors (fleas) into test plots (i.e., Chinese villages), and direct exposure using prisoners of war (Harris & Paxman, 1982) (Figure 21.5). The extent of the program and whether biological weapons were actually used in military campaigns is not fully known, largely because most of the evidence was destroyed prior to the occupation of Manchuria by the Soviet Union. However, based on testimony from a Japanese scientist at the facility, the Manchurian plant was equipped to generate 500 million infested fleas per year (Cookson & Nottingham, 1969). In turn, the Japanese accused the Soviets of possessing an active program of bioweaponry testing in prisoner of war camps (using plague-infested fleas) (Lockwood, 1987). Similarly, Germany, the United States, and Great Britain were involved in the development of entomological weapons in the form of typhus-harboring lice, plague-infested fleas, and mosquitoes vectoring a wide range of pathogens (Harris & Paxman, 1982). Germany and the United States each accused the other of engaging in agricultural sabotage by

Figure 21.5   A photograph released from Jilin Provincial Archives, which, according to Xinhua Press, "shows personnels of 'Manchukuo' attend a 'plague prevention' action which indeed is a bacteriological test directed by Japan's 'Unit 731' in November of 1940 at Nong'an County, northeast China's Jilin Province.". *Source*: Photo by Unknown Source/Wikimedia Commons/public domain. https://commons.wikimedia.org/wiki/File:Unit_731_victim.jpg.

(a)   (b)

Figure 21.6   A purported agent of agroterrorism, (a) larvae and (b) adult of Colorado potato beetle, *Leptinotarsa decemlineata* (Family Chrysomelidae). *Source*: Photo by Tavo Romann/Wikimedia Commons/CC BY 4.0. https://commons.wikimedia.org/wiki/File:Colorado_potato_beetle_larvas_(Leptinotarsa_decemlineata).jpg. Photo by Bj. schoenmakers/Wikimedia Commons/CC0 1.0. https://commons.wikimedia.org/wiki/File:Leptinotarsa_decemlineata_(Colorado_potato_beetle),_Molenhoek,_the_Netherlands.jpg.

releasing the Colorado potato beetle, *Leptinotarsa decemlineata* (Family Chrysomelidae), with the goal of decimating food supplies (Figure 21.6). Though their enemies made accusations against each country, evidence was never found that directly implicated these nations in the actual use of the biological weapons (Table 21.1).

During the period that accounted for the Korean War, the Vietnam War, and the Cold War between the United States and Soviet Union, several countries were actively engaged in the development of biological weapons. Much of the research was dedicated to mosquito-borne diseases such as malaria, yellow fever, and dengue. Why mosquitoes? Mosquitoes are easy to rear in the laboratory, females lay hundreds of eggs at a time, which for some species are easy to store and to transport into another country (e.g., thousands of eggs can be collected on a single paper towel and then hidden away in luggage), and once in the target area they reproduce and become indigenous in a short period of time (Tabachnick *et al.*, 2011) (Figure 21.7). In the case of the latter characteristic, some *Aedes* spp. can develop from egg to adult in less than seven days depending on temperature and other environmental factors, producing three to four generations

Table 21.1 Entomological weapons suspected of being developed for use in war.

Insect	Family name	Military campaign
"Bees"	Unknown	Ancient Egyptians
Murgantia histrionica	Pentatomidae	United States Civil War
Leptinotarsa decemlineata	Chrysomelidae	World War II
Pulex irritans	Pulicidae	World War II, Korean War
Xenopsylla cheopis	Pulicidae	World War II
Body louse, *Pediculus* spp.	Pediculidae	World War II, Korean War
Biting midges in Japan and Korea	Ceratopogonidae	Korean War
Various mosquito species	Culicidae	World War II, Korean War, Cold War, Vietnam War
Lucilia sericata	Calliphoridae	Korean War
Musca domestica	Muscidae	World War II
Muscina stabulans	Muscidae	Korean War
Various black fly species found in Asia	Simuliidae	Korean War
Various species of ticks	Ixodidae	World War II, Korean War

This is an incomplete list of insects considered for use in various military campaigns. Lockwood (2009a) provides comprehensive coverage of insects suspected of being considered or tested for biological weapons development. Much of the work on entomological weapons was conducted by Japan (Unit 731) and the United States (Unit 406).

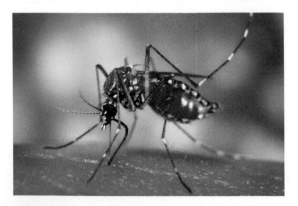

Figure 21.7 An adult female yellow fever mosquito, *Aedes aegypti* (Family Culicidae) taking a bloodmeal from a human host. *Source*: Photo by James Gathany/ Wikimedia Commons/public domain. https://commons. wikimedia.org/wiki/File:Aedes_aegypti_CDC9175.tif.

in a month. No direct evidence exists that any country released biological weapons like infected mosquitoes or other forms but, as with other wars, accusations were made that implicated several nations including the United States, Soviet Union, China, and North Korea (Lockwood, 1987).

The examples given reflect a strong similarity to archaeoentomological research that was discussed in Chapter 20. In fact, most of the investigative techniques that would be employed in such matters would be the same as those used when examining insect activity or remains from an ancient civilization. Attempting to demonstrate that a nation has violated the Geneva Protocol or similar treaties banning biological weapons requires physical evidence, or some means to conduct hypothesis testing. Unfortunately, the evidence available in these cases are historical documents or testimonies from supposed eyewitnesses that serve as circumstantial evidence, but not definitive proof. Some of the circumstantial evidence is in the form of data suggesting an apparently "new" distribution of a pathogen or insect vector in a region, country, or terrestrial ecozone that was previously unknown until the supposed biological attack. Again, as discussed with archaeoentomological evidence, biogeographical histories cannot be subjected to testing via the scientific method to demonstrate, for example, that the insect in question never existed in that region before. So such evidence may lead to a compellingly logical argument to the potential release of a biological weapon, but it cannot serve as undisputable proof.

21.3 Direct entomological threats to human populations are not all historical

Any student of entomology knows several species that, due to their natural activities, can inflict damage to crops and livestock, chew their way through human-constructed structures, serve as vectors for devastating diseases, or cause pain and suffering, even death, through biting and stinging. The number of species that elicit such negative impacts is very small by comparison to the total number of extant species that have been described. Nonetheless, of those that are viewed as destructive or harmful in some way, they are quite efficient at what they do: a hallmark trait of the class Insecta (=Ectognatha)! With the biological revolution that has occurred over the last three to four decades, the potential exists to improve the destructive capabilities of such insects, either through genetic modification or via enhanced delivery systems.

Some of the most frightening possibilities involve insects that directly attack humans (Figure 21.8). Numerous insect species are known

Figure 21.8 The stinging hair caterpillar *Lonomia obliqua* (Family Saturniidae) produces a hemorrhagic venom that can be deadly to humans and other animals. *Source*: Photo by নয়ন জ্যোতি নাথ/Wikimedia Commons/CC BY-SA 4.0. https://commons.wikimedia.org/wiki/File:IMG_lonomia_obliqua123.jpg.

that deliver harmful, even lethal toxins and venoms to victims through their bites, stings, or secretions. The insects represent a wide range of taxa, from stinging (**urticating hair**) caterpillars like the assassin caterpillar *Lonomia oblique*[9] (Family Saturniidae), true bugs containing venomous saliva, beetles that secrete noxious/caustic and sometimes ballistic secretions (e.g., blister and bombardier beetles), to the vast array of social Hymenoptera with salivary and/or sting-associated venoms (Eisner *et al*., 2007) (Table 21.2). The components of these defensive compounds cover the gamut in terms of natural product chemistry: small peptides, polypeptides, midrange proteins, catecholamines, enzymes such as hyaluronidases, phospholipases, hydrolases, lipases and others, and even sequestered toxins (e.g., glycosides, terpenes, and alkaloids) from other sources such as ingested plant material. The utility of such insects as weapons is that generally:

- the active toxic components are non-discriminate (e.g., do not require a receptor-mediated response);
- they evoke an immediate reaction in targeted animals; and
- they yield painful, intense cellular responses that can be long-lasting.

Table 21.2 Venomous or toxic insects that potentially could be biological weapons.

Insect	Family	Mode of attack	Effect of toxin(s)
Coleoptera			
Various blister beetles	Meloidae	Secretions	Skin lesions, intestinal damage, lethality
Paederus spp.	Staphylinidae	Secretions	Festering lesions, pain, blindness, lethality
Hemiptera			
Reduvius personatus	Reduviidae	Bite	Painful, throbbing, burning sensations
Rhynocoris spp.	Reduviidae	Bite	Painful, throbbing, burning sensations
Hymenoptera			
Paraponera clavata	Formicidae	Sting	Intense, throbbing pain
Pogonomyrmex maricopa	Formicidae	Bite, sting	Intense pain, lethal immune reactions
Solenopsis invicta	Formicidae	Bite, sting	Pain, skin pustules
Pepsis spp.	Pompilidae	Sting	Excruciating pain
Various species of wasps	Vespidae	Bite, sting	Pain, repeated aggressive stinging
Lepidoptera			
Lonomia spp.	Saturniidae	Stinging hairs	Pain, hemorrhage
Megalopyge opercularis	Megalopygidae	Stinging hairs	Intense pain, nausea, blisters, abdominal discomfort, difficulty breathing

This is a partial list of toxic or venomous insects that could conceivably be fashioned into biological weapons. The reality is that there are numerous species of insects that could be used, particularly from the order Hymenoptera.

More details on the chemistry and mode of action of these defensive compounds are discussed in Chapter 22.

In fact, depending on the toxins present and the relative sensitivity of an individual, the resulting cellular reactions may be debilitating for hours to days, and some are lethal. The United States' military has recently encountered one potential candidate, rove beetles belonging to the genus *Paederus* (Family Staphylinidae) that has been a source of concern in terms of terrorists forging the insect into a biological weapon. Native to the Middle East and parts of Asia (located in the Palearctic ecozone), the beetles release secretions when agitated that induce a range of calamities including mild skin blisters, painful lesions, temporary blindness if entering eye fluids, and severe intestinal problems if ingested (Monthei *et al.*, 2010). Death can result via oral or injected entry of the toxins. Chance encounter under normal circumstances is relatively low, so a high incidence of intoxication by military personnel of any country would be considered highly suspicious of deliberate release as a biological weapon (Figure 21.9).

Of course the likelihood that any such insects can be fashioned into biological weapons is dependent on the ability to collect thousands of the desired insects (so that in turn they can be raised in large numbers to the desired life stage), the development of an effective delivery system, and prevention of self-infliction (i.e., intoxication of members of one's own organization) (Monthei *et al.*, 2010). Even if each of the above conditions was met, release of an "exotic" insect into a target

country will immediately implicate the warring nation of violating the Geneva Protocol and subsequent treaties of the United Nations. Since terrorist organizations answer to no one, entomological candidates are still an option for development into biological weapons.

21.4 Impending entomological threats to agriculture and food safety

Among the most overlooked threats to national security are those that could directly impact food production. **Agroterrorism**, or the deliberate introduction of animal or plant pathogens or pests that directly attack cropping systems or livestock, with the purpose of instilling fear, causing economic losses, or undermining social stability (Monthei *et al.*, 2010), is a very real threat to nations engaged in agriculture to meet their own needs and/or for export to other countries. A discussion of biological weapons generally focuses on the impact to human life stemming from fear that severely debilitating pathogens will be the primary agents used in such attacks. The reality, however, is that targeting agriculture is likely a more efficient strategy for a non-state (terrorist) group to achieve its end goal of terrorism: the agents of agroterrorism are typically far easier to culture than human pathogens, pose no threat to the individuals responsible for cultivation and preparation of the bioweapons, most nations are underprepared for agro-terrorist attacks, and detection of an induced infection/infestation will likely go undetected for an extended period of time allowing the full effects of the pathogens or pests to be manifested (Horn & Breeze, 1999). It is also very difficult to establish that a deliberate introduction of an agriculture pest or pathogen has occurred, as there are numerous means for accidental entry.

The severity of a well-placed agroterrorism agent in western nations like the United States, United Kingdom, Germany, and France could potentially devastate food production and the economy of entire regions. The United States, as the top producer of agriculture exports per unit acre globally, serves as an excellent example to illustrate the potential impact of agroterrorism (Figure 21.10). Revenue generated from all aspects of agriculture production in the United States accounts for

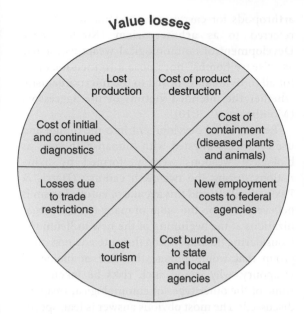

Value losses

- Lost production
- Cost of product destruction
- Cost of containment (diseased plants and animals)
- New employment costs to federal agencies
- Cost burden to state and local agencies
- Lost tourism
- Losses due to trade restrictions
- Cost of initial and continued diagnostics

Figure 21.10 Potential economic losses resulting from agroterrorism. *Source*: Information based on Crutchley *et al.* (2007) and Monthei *et al.* (2010).

approximately 11–13% of gross domestic product or roughly $50 billion annually, with over 15% of the nation's workforce engaged in agricultural-related jobs (Cupp *et al.*, 2004; USDA Economic Research Service, 2005). Obviously, damage to agricultural production as a result of an introduced pathogen or pest will yield huge economic losses to the economy of the United States. Smaller scale "tests" have already occurred at various times in the country's history with accidental introductions of exotic plants and animals (insects specifically) that have caused localized havoc and millions annually in losses and control costs (Pimentel *et al.*, 1989). Current trends toward large corporate farms places the United States at even greater risk to agroterrorism. For example, in terms of hog production, approximately 90% of all pork produced comes from less than 100 producers. Similarly, 50% of finished cattle are raised on fewer than 50 feedlots (Monke, 2007). In terms of cropping systems, over 200 million acres in the United States are planted with only four crops: corn, wheat, soybeans, and hay (U.S. Environmental Protection Agency, 2012). The point is that release of agroterrorist agents as pathogens or direct pests to relatively few types of agricultural systems in concentrated areas could easily and quickly devastate meat or crop production, yielding the intended goal of limiting food availability, weakening the economy, and evoking public fear.

Insects are the ideal agents to carry out acts of agroterrorism. Several species are known to vector plant or livestock pathogens, or directly feed on a variety of crops and livestock. Use in agroterrorism can be as simple as the introduction of an exotic or invasive species[10] into a new region. The effect is the same as accidental releases in which the newly introduced species lacks natural enemies to keep populations in check. Thus for at least a short period of time, the insect wreaks havoc until the natural enemies "catch up" or government agencies develop effective control practices, likely through combined approaches of insecticide application and trapping. The recent introductions of the Brown marmorated stink bug, *Halyomorpha halys* (Family Pentatomidae); emerald ash borer, *Agrilus planipennis* (Buprestidae); and spotted lanternfly, *Lycorma delicatula* (Fulgoridae) into the United States represents the rapid spread and economic toll of a non-anticipated, inoculative pest introduction.

Inoculative releases of insects generally involve the release of small numbers of individuals during a seasonally favorable period that permits establishment of the species in that area. A favorable period may include the appropriate time of year in terms of temperature, moisture, abundant food source, and/or reduced number of potential predators and parasites for one or more developmental stages. In contrast, and more expected of a terrorist use, are **inundative releases** in which masses of individuals are deliberately distributed to essentially overwhelm the area and to produce an immediate response, namely the quick destruction of the targeted agricultural system. During World War II, Germany, Great Britain, and the United States all examined the feasibility of using the Colorado potato beetle, *L. decemlineata*, as a biological weapon via inundative releases (Lockwood, 1987), although there is no evidence that any of the countries actually carried out the plans. Inoculative and inundative releases are more typically associated with **biological control programs** aimed at the use of natural enemies of an agricultural pest to reduce pest densities below economic injury levels.

In the event that an agroterrorist attack occurs utilizing entomological weapons, the national security response will not be through the military. Rather, the same government agencies dedicated to food, crop, and livestock protection using inspection and pest management strategies will be called into action. This represents another reason why a terrorist organization may opt for agroterrorism over other means of attack since such agencies, especially those dedicated to agriculture, are typically underfunded in comparison with other agencies tasked with national security functions, and generally the organizations are not trained for the type of rapid responses needed for such emergencies (Casagrande, 2000). The idea of an agricultural attack is more than just speculation; hundreds of documents related to U.S. agriculture were seized from Al-Qaeda strongholds upon the U.S. invasion of Afghanistan (Crutchley *et al.*, 2007). Once an entomological agent has been identified, particularly in situations of agroterrorism, the most likely recourse is application of insecticides. A well-chosen biological weapon in agricultural applications will be insecticide resistant and costly to control. Some obvious candidates include *L. decemlineata* and several species of beetles in the genus *Diabrotica* (Family Chrysomelidae), which are extremely difficult to control with insecticides and are major crop pests in several regions of the world. In the United States alone, the cost of control for anticipated pests of the major row crops (corn, wheat, and soybeans) can exceed $25 billion annually (Pimentel *et al.*, 2005), highlighting the potential economic losses from entomological weapons in agroterrorism.

21.5   Insect-borne diseases as new or renewed threats to human health

Public fear of biological weapons generally rests with pathogens that cause human disease. The anticipation of severely debilitating symptoms and death from exposure to smallpox, Ebola, plague, anthrax, and other diseases is what makes the terrorist threat so effective at generating a response in the target nation. From an entomological perspective, many of the human pathogens viewed as viable biological weapons are transmitted by insects, usually through blood feeding, and thus the same considerations discussed earlier in the chapter as to what makes insects good candidates for biological weapons development apply here, with one notable exception. The deliberate weaponization of

arthropods for carrying pathogenic microbes is referred to as **arboterrorism** (Kwak, 2016). Development of entomological weapons for use as "disease bombs" does pose a tremendous risk for all individuals exposed to the microorganism, whether the intended victims or the aggressors (Monthei *et al.*, 2010).

Research and development into the design and use of insects infested with human pathogens as biological weapons were performed by many nations during the twentieth century (Harris & Paxman, 1982), and this avenue of entoterrorism is believed to be on the radar of many terrorist organizations at the beginning of the new millennium. Considering the impending threat to any non-state group that would pursue this line of biological weaponry, why would such risks be taken over some of the other types of entomological weapons discussed? The most obvious answer is fear, specifically that associated with the threat of widespread disease throughout major cities in developed countries.[11] Some of the pathogens have extremely high kill rates, so the death toll could easily approach several thousands. Complementing the fear factor is the lack of preparedness of most nations in terms of vaccine development and vaccination programs. So even for diseases with low mortality, the communicability rate is high and thus the risk of an epidemic or pandemic is elevated because few individuals in most countries are immunized, particularly for diseases of the past like smallpox, and the stockpiles of vaccines are insufficient to vaccinate populations at risk (Garrett, 2001). The coronavirus epidemic in China and Japan of 2019–2020 is a clear example of the universal fear propagated by a highly communicable disease that nations were poorly suited to contain or treat (CDC, 2020). Fueling the fear were reports of malign and coercive agents using social media to provide misinformation about the virus in hopes of disorganizing global responses to contain the epidemic. The infrastructure of most any small or large city is not capable of coping with wide-scale disease in terms of sufficient number of trained healthcare workers, facilities to accommodate thousands of disease-stricken citizens, or resources to quickly diagnose the pathogen and mobilize an effective counter-response to contain the spread of disease or vector(s). In short, all municipality resources would quickly be taxed following a bioterrorist attack, leading to enormous financial drains on infected cities, which again is one of the major goals of terrorism.

There is no doubt that the insect vectors available have a lot to do with the development of such biological weapons. In Section 21.4, we discussed the "virtues" of mosquitoes for delivering plant or animal pathogens in acts of agroterrorism; those traits apply with human disease as well. Mosquitoes are particularly attractive choices as well because species exist in many parts of the world that are capable of vectoring the microorganisms that induce yellow fever, dengue, malaria, Rift Valley fever, and a host of other diseases (Monthei *et al.*, 2010) (Table 21.3), and once native species become infected the pathogens can quickly become indigenous. Here again is where ideal insect vectors for forging into biological weapons should be selected for insecticide resistance to dampen the response of targeted nations in controlling the spread of an inundative release of infested mosquitoes. This is a particularly effective approach for mosquitoes and several species of biting flies since control efforts are already hampered by the aquatic

larval stages, which are inherently more difficult to control than terrestrial stages. Bureaucratic battles over funding first line of defense strategies in the form of surveillance programs against infectious disease contributes to the already difficult task of being prepared for arboterrorism (Gu *et al.*, 2008; Vasquez-Prokopec *et al.*, 2010). Research conducted during World War II revealed that other blood-feeding insects such as fleas (*P. irritans*) can be infected with human pathogens, mass reared by the millions quickly, and then dropped via bombs or freely from planes in a mass inundative release (Lockwood, 2009a) (Figure 21.11).

At this point, there is no evidence to indicate that any nation or non-state organization has deliberately used insect vectors as biological agents toward another nation. However, the risk should be viewed as very high to western nations. It is again important to understand that with member nations of the United Nations signing a Biological Weapons Treaty in 1975, little information is available today as to whom (i.e., non-member nations and terrorist organizations) has biological weapons and what types are being developed. Given the vast array of blood-feeding arthropods available for use, and the ability to use biotechnology and molecular biology techniques to potentially enhance the killing or infectivity of the pathogens and/or insect vectors, this form of biological weaponry remains a high risk for all nations targeted by terrorist and non-state organizations.

Table 21.3 Arthropod-borne diseases considered possible biological weapons threats to humans.

Disease	Arthropod vector	Pathogen
Anthrax	Flies	*Bacillus anthracis*
Bubonic plague	Fleas	*Yersinia pestis*
Chikungunya	*Aedes* mosquitoes	*Alphavirus* spp.
Cholera	Flies	*Vibrio cholerae*
Crimean-Congo hemorrhagic fever	Ticks (Ixodidae)	CCHF virus
Dengue fever	*Aedes* mosquitoes	DF virus
Eastern equine encephalitis	Mosquitoes	EEE virus
Epidemic louse-borne typhus	Louse	*Rickettsia prowazekii*
Japanese B encephalitis	*Culex* mosquitoes	*Flavivirus* spp.
Q fever	Ticks (Ixodidae)	*Coxiella burnetii*
Rift Valley fever	Mosquitoes	*Phleobovirus* spp.
Western equine encephalitis	Mosquitoes	WEE virus
Yellow fever	Mosquitoes	Yellow fever virus

This is a partial list of possible arthropod-borne diseases that could be developed into biological weapons.
Source: Information from Monthei *et al.* (2010) and Kwak (2016).

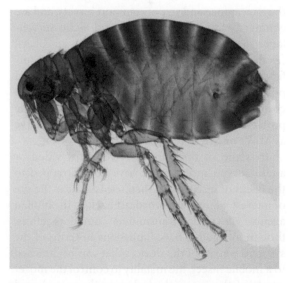

Figure 21.11 Adult female *Pulex irritans. Source*: Photo by Katja ZSM/Wikimedia Commons/CC BY-SA 3.0. https://commons.wikimedia.org/wiki/File:Pulex_irritans_ZSM.

21.6   Insects can be used as tools for national security

Up to this point the discussion of insects' involvement with matters of national security has been dedicated to their use in weapons development and thereby as a global threat. The reality is that insects show no loyalties, and thus have been recruited to protect the nations that are potential targets of all forms of terrorism. The national defense roles range from use in surveillance operations, detection of explosive residues or unexploded munitions, to toxicological applications that allow determination of whether discovered bodies have been subjected to violence involving bullets or explosives. As we will see, the development of such entomological tools also has application to forensic entomology in terms of discovery of decomposing bodies as well as in the determination of foul play with suspicious deaths.

Figure 21.12   Developed by CIA's Office of Research and Development in the 1970s, this micro Unmanned Aerial Vehicle (UAV) was the first flight of an insect-sized aerial vehicle (Insectothopter). *Source*: Photo by Central Intelligence Agency (CIA)/Wikimedia Commons/public domain. https://commons.wikimedia.org/wiki/File:Dragonfly_Insectothopter_-_Flickr_-_The_Central_Intelligence_Agency_(1).jpg.

21.6.1   Surveillance

Cyborg insect spies have been used for years to collect covert information on the comings and goings of criminal and militant organizations, as part of search and rescue missions, in the detection of explosives, and to monitor hazardous environments (Floreano *et al.*, 2010). The spies or micro air or land vehicles (MAV or MLV) are designed to mimic the appearance and size of common insects, and are equipped with high-tech video and audio surveillance devices that provide immediate feedback (Zufferey, 2008). The designs are ingenious but suffer from limitations such as need for an external power source for continuous operation and high cost of production. The obvious solution to the problems has been to recruit living insects to serve as MLVs or MAVs (Figure 21.12).

In some of the "simplest" versions, video or audio surveillance devices have been strapped to the backs of insects, and then released near the site of interest. Signals are broadcast back to the human agents via radio transmitters. Such practices, though intriguing, have limitations in terms of the size and weight of the devices that can be attached to a mobile insect, the inability to control the movements of the entomological spies (hexa-spy), and the distance that the radio signals can be transmitted. To enhance the utility of the insect probes, investigators have developed neurological circuits that can be implanted into the ventral nerve cord and ganglia so that movement of legs and wings can be controlled remotely. This is surprisingly easy in that a relatively small number of neurons (~250) extend from the frontal ganglion (brain) into the hemocoel (Staudacher, 1998), permitting detection and manipulation of central motor programs. Obviously this facilitates control of where the insects move in addition to speed of locomotion. An array of insects (e.g., cockroaches, honey bees, and beetles) has served as surrogates for the biological circuits, which has made this a viable approach for surveillance via land and air (Aktakka *et al.*, 2011).

Living insects do not have the need for an external power source the way that their cyborg counterparts do. However, as a means to extend the flight potential of the living MAV, the heat energy released during aerobic metabolism can be captured using piezoelectric generators attached to the wings (Aktakka *et al.*, 2011). The generators capture the energy and produce electric output in response to insect wing movement. A linkage between the piezoelectric generators and thoracic flight muscles allows energy transfer to power flight. The enhanced energy efficiency of the insect may allow its payload capacity to increase as well, permitting attachment of larger devices for surveillance and detection. At this point, the technology has only been applied to a ground beetle *Cotinis nitida* (L.) (Family Scarabaeidae) but several species of insects are potential candidates for future use.

21.6.2 Biosensors and chemical detection

The utility of insects to national defenses is most refined with their use in chemical detection. The chemicals, whether in volatile or particle form, are critical to the discovery of explosives residues, unexploded munitions, biological weapons, illicit drugs, and even decomposing bodies (Habib, 2007; Rains et al., 2008). Chemical detection relies on the incredibly acute olfactory system possessed by most insects (discussed in more detail in Chapter 7), that is not only among the most sensitive found in animals, but also makes use of hundreds to thousands of sensory receptors distributed across multiple locations on the body (e.g., mouthparts, antennae, legs, and footpads). The fact that insects can be trained to associate odors with a conditioned stimulus, quickly and cheaply, makes them attractive candidates for use in chemical detection systems.

How can insects be fashioned into tools for detecting chemical residues and odorants? Several species have been found suitable for use in odorant detection or sniffer systems, as biosensors, or to trap chemical particles for later toxicological screening. In some instances, the insects are used as free-moving systems, permitted to travel through a region with no constraints, while other forms of detection rely on immobile insects in portable devices, or only sensory cells are used in vitro.[12] Among the best-suited insects for any of these detection systems are various species of social Hymenoptera. Why?

1. The social organization of bees, ants and wasps requires a highly sophisticated system of chemical communication, including the detection of chemical signals in a complex environment.
2. The brain (frontal ganglion) has the capacity for both short- and long-term memory.
3. Many are among the most long-lived of insects, which provides a greater return on the investment of training and fitting tracking devices.
4. They can be trained much faster than other animals, like dogs, and thousands of individuals can be trained simultaneously.
5. Once training is complete, human intervention is not needed other than to interpret the messages.

21.6.2.1 Chemical or odorant detection

Social Hymenoptera, honey bees (*Apis mellifera*) specifically, have been especially useful in the development of sniffer or odorant detection systems. One method of chemical detection relies on **classical (or Pavlovian) conditioning**, in which foraging bees are trained to associate food with the chemical of interest (a form of positive conditioning or reinforcement), say a compound associated with explosives. After an extended training period (the length of which is dependent on the insect species, age, and other physiological parameters), the foragers are fitted with a tracking device and then released into the area to be scanned or searched. Bee detection of the conditioned stimulus (i.e., the chemical of interest) will result in hovering or extended searching in the immediate vicinity of the chemical source, such as an unexploded landmine or decomposing body, allowing optical or radio signal detection at a remote location (Rapasky et al., 2006) (Figure 21.13). In the case of unexploded munitions, one obvious advantage of using an insect like the honey bee is that humans or sniffer animals like dogs do not have to be used in traditional methods of detection which places them in a hazardous situation. A flying insect will also not detonate the unexploded landmine (Habib, 2007).

Other sniffer systems make use of constrained insects in portable devices. The insect is not permitted to roam freely in the search area. Rather the sniffer (i.e., insect) is maintained in a container in which the body movements are restrained by an insect-sized harness, and it must be brought to the area to be scanned for chemical residues. Examples may include searches at an airport or other buildings for explosives or illicit drugs. In the constrained system, the insect used is either naturally responsive to the chemical(s) of interest, or is conditioned as described earlier (King et al., 2004). The insect is then monitored with video or electronic surveillance devices to examine its responses to the volatile chemicals in the area. A specific displayed behavior, muscle movement, or action potential associated with sensory cells can be interpreted as positive detection of the conditioning stimulus or chemical signal of interest. This approach is not particularly useful for the detection of buried munitions, bodies, or other outdoor scenarios (Rains et al., 2008).

Figure 21.13 Canine sniffer systems or sniffer dogs have been used in the past to detect unexploded land mines. More humane approaches using cyborg insects have been developed as viable replacement methods. *Source*: Photo by Norsk Folkehjelp/Wikimedia Commons/CC BY 2.0. https://commons.wikimedia.org/wiki/File:Dog_search_for_mines_in_Bosnia.jpg.

21.6.2.2 Biosensors

Odor detection utilizing sensory cells or gene expression systems is an alternative to the use of the whole insect. The concept originated to eliminate the variable behavioral responses that inherently occur with insects or any other animal. A high degree of variability from one individual to the next is a major source of concern in terms of apparent detection (or not) of hazardous materials, like non-detonated explosives. Consequently, investigation into the feasibility of developing biosensors to replace insect sniffer systems has occurred with a few species of insects. One of the most promising involves the use of the vinegar (fruit) fly, *Drosophila melanogaster* (Family Drosophilidae). This fly offers the advantage of having its odorant detection system characterized in more detail than any other insect: odorant receptor genes have been identified and characterized, and most aspects of the neural circuitry of *D. melanogaster* are understood (Marshall *et al.*, 2010).

With this knowledge in hand, the concept of the biosensor is to express the odor receptor genes in an expression system that permits receptor binding by the chemical signal of interest. Cellular responses can be monitored to determine if receptor binding occurs. Biosensors obviously are analogous to the constrained portable systems discussed earlier but have the added limitation of being an *in vitro* system, and thus it is challenging to maintain functionality when removed from a laboratory setting. The most important limitation is that chemical detection is restricted to the signals innately programmed by the nervous system of the insect. Thus, classical conditioning is not an option. However, this does not preclude the possibility that the chemical sensitivity of the receptor genes can be genetically modified to respond to factitious odors.

21.6.2.3 Trapping chemical particles

This method relies on exactly what the name implies: trapping or collecting chemical particles directly to the body of the insect. Sensory receptors are not involved in the process. In fact, non-specific attachment accounts for most aspects of the process. Conceptually, this method of chemical detection is the least sophisticated of the three types. However, this does not necessarily mean that the techniques of surveillance or chemical detection used by the human investigators are not complex. The trap approach is based on the idea that particles stick or attach to the bodies of insects during natural encounters with objectives, abiotic and biotic, while engaged in locomotion or standing still. One of the most familiar examples is pollen accumulation on the bodies of insects that frequent flowers. In some instances, trapping of particles results due to the morphological architecture of the insects: hairs or setae form baskets or traps on the legs or other locations, or particles become lodged in the articulation between segments. In other cases, branched hairs on the body develop electrostatic charge that trap airborne particles like pollen, pollutants, chemical residues, and possibly even biological warfare agents (Bromenshenk et al., 1985, 2015). The point is that regardless of the method or reason that the particles become attached to the insect, there is a lack of specificity unlike the sniffer systems or biosensors.

What has been described thus far is the role of the insect in collecting particles. How is this useful in detecting explosives or other items of interest? Good question. The tried and true method of old comes from work on pollen analyses associated with foraging honey bees: the insects are individually collected and then the particles can be manually removed for chemical screening in the laboratory (Figure 21.14). This approach is not appealing for a number of reasons in that it depends on catching the "released" insects, is time-consuming, labor intensive, and is not portable so the analyses must occur in a laboratory perhaps a long distance from the search area. Newer, more sophisticated methods are being developed. One of these again involves recruiting foraging honey bees for duty. Prior to leaving the hive, individual bees are equipped with a barcode tag for identification of specific individuals and a tracking device (often GPS) (Habib, 2007). The entrance to the colony is then fitted with chemical detection sensors and barcode readers (Rapasky et al., 2006). Once installation is complete, the bees are allowed to do what they do best: search and collect food. On return to the hive, each forager must pass through the detection devices. If the bees pass through an area in which volatile chemicals or airborne particles associated with explosives, munitions, or biological weapons are present, the chemical sensors installed at the hive entrance will detect them, and the barcode allows matching of the individual with the chemical signals. In turn, the day's travels for the individual bee can be retraced due to the tracking device installed, leading back to the source of the chemical residues.

21.6.3 Toxicological applications

Insects, specifically necrophagous flies, provide another source of utility to national security and criminal investigations: chemical detection as a result of feeding. More to the point, necrophagous fly larvae consuming tissues of a corpse that has been exposed to explosives or bullet fragments can accumulate the chemical signatures that imply foul play. Why this is important is because evidence to implicate a particular bomb-maker (i.e., terrorist organization) often lies in the chemical composition of the explosive device, which may leave a chemical signature in objects found in the explosion area, including the unfortunate victims. Similarly, a gunshot leaves distinctive patterns termed gunshot residues (GSRs) that can be modified during postmortem decomposition and burial and by the activity of insects. These alterations may lead to failure in recognizing that wounding, and consequently death, occurred due to a gunshot(s) (LaGoo et al., 2010).

Some of the same insects that confound GSR patterns may also be the solution to the problem of detection. How? In the case of gunshot, residues left behind contain spheroid particles whose morphology can be distinguished from other sources fairly easily to the trained eye. The residues also contain unique elemental combinations dominated by barium (Ba), lead (Pb), and antimony (Sb) (Wolten et al., 1979). The larvae of early colonizing flies generally feed at the skin surface and/or on exposed mucous membranes, areas where GSRs reside (LaGoo et al., 2010). Thus, as the larvae feed, they accumulate the elemental particles associated with the GSRs. The rest of the story

Figure 21.14 Adult honey bee, *Apis mellifera*, covered with pollen. *Source*: Photo by Andreas Trepte/Wikimedia Commons/CC BY-SA 2.5. https://commons.wikimedia. org/wiki/File:Apis_mellifera_Western_honey_bee.jpg.

is probably obvious: fly larvae are collected from a corpse suspected of containing GSRs or explosives residues, the residue particles are extracted (by acid or microwave detection), and the extracts examined by techniques (such as inductively coupled plasma mass spectrometry) that permit identification of the elements and confirmed by comparisons with known standards (Roeterdink *et al.*, 2004; LaGoo *et al.*, 2010; Motta *et al.*, 2015). One downside to this approach to toxicological detection is that it likely has limited utility with late colonizing insects or those that feed deep within the body cavity (LaGoo *et al.*, 2010). The assumption is that the further from the skin surface or the later in decay, the less residues that are present. The exception would be if the victim did not die immediately after injury, permitting transport of chemical residues throughout the body via circulating blood.

The technique has been tested with a limited number of species – *Lucilia sericata*, *Calliphora dubia*, *Chrysomya albiceps* – so additional specimens need to be examined to ensure the approaches are suitable for a wide range of necrophagous species. With GSRs, elemental analysis has confirmed that residues are detectable in the pupal stages (LaGoo *et al.*, 2010), long after feeding has been completed, indicating that there is no danger of loss during gut purging prior to pupariation. Somewhat surprisingly, however, empty puparia have not been tested for the presence of explosives or gunshot residues. Puparia are often the only evidence left behind to even indicate that flies had once been present, and once pupariation is complete the puparium becomes physiologically inert (Zdarek & Fraenkel, 1972), essentially locking the chemical profile in place seemingly forever.

Chapter review

Terrorism and biological threats to national security are part of today's world

- In today's global environment, wars are declared not so much by countries but by militant political groups, frequently termed terrorist organizations or cells or non-state groups, who generally hold no ties to any one country and show a lack of respect for life, as soldiers, civilians, men, women, and children are all targets.
- Terrorism is literally the use of terror, usually through acts of violence, in the name of religion, politics, or some other ideological purpose, with no regard for non-combatants. Conventional warfare, though barbaric in most ways, has historically followed "unwritten" rules in which non-military civilians were not attacked; the soldiers were left to decide the battles and ultimately the victors and losers of war.
- At the dawn of the twenty-first century, Iraq, Iran, Syria, China, Libya, North Korea, Russia, Israel, Taiwan, and potentially India, Pakistan, Sudan, and Kazakhstan possess biological weapons. Quite disturbing is the fact that several of these nations are politically unstable or are considered home to radical terrorist organizations, simply meaning that biological weapons in their hands could be used in horrifically unimaginable ways.
- Today's modern warfare by terrorist organizations rather than by nations does not adhere to the decrees of the Biological and Toxin Weapons Conventions sponsored by the United Nations which outlawed the development and production of biological weapons. The reality is that there are likely more biological weapons in this century than at any other time in history, and the weapons appear to be in the hands of those ready and willing to use them.
- To most, biological weapons come in the form of microorganisms that cause disease or are human pathogens. Insects also represent viable options as biological weapons that could be used by militant groups to inflict large-scale damage, quickly, cheaply, and with much less personal danger to the terrorist cell than when formulating human pathogens into bioweapons.

Entomological weapons are not new ideas

- The destructive potential of insects as weapons of war has been understood at some level for thousands of years. Insects have been used as means to torture and interrogate, bomb out of fortifications, and through direct attack as vectors of disease and to target agriculture.

- Following the conclusion of World War I, the Geneva Protocol was drafted with the purpose of banning the use of chemical and biological weapons, although it was not until over two decades later that insects were identified and banned as biological agents of war.
- World War II marked a major upturn in biological weaponry research and deployment. The extent of biological weapons development involving insects and whether any ento-weapons were actually deployed is not fully known.
- During the period that accounted for the Korean War, the Vietnam War, and the Cold War between the United States and Soviet Union, several countries were actively engaged in the development of biological weapons. Much of the research was dedicated to mosquito-borne diseases such as malaria, yellow fever, and dengue. No direct evidence exists that any country released biological weapons like infected mosquitoes or other forms, but as with other wars, accusations were made that implicated several nations.

Direct entomological threats to human populations are not all historical

- Numerous insect species are known that deliver harmful or even lethal toxins and venoms to victims through their bites, stings, or secretions. The insects represent a wide range of taxa, from stinging hair caterpillars, true bugs containing venomous saliva, beetles that secrete noxious/caustic and sometimes ballistic secretions (e.g., blister and bombardier beetles), to the vast array of social Hymenoptera with salivary and/or sting-associated venoms.
- The components of these defensive compounds cover the gamut in terms of natural product chemistry: small peptides, polypeptides, mid-range proteins, catecholamines, enzymes such as hyaluronidases, phospholipases, hydrolases, lipases and others, and even sequestered toxins (e.g., glycosides, terpenes, and alkaloids) from other sources such as ingested plant material. From a biological weapons standpoint, they are ideal because each is non-discriminate, produces an immediate reaction in the targeted

individual, and can debilitate the inflicted for hours to days.
- The likelihood that any such insects can be fashioned into biological weapons is dependent on the ability to collect thousands of the desired insects (so that in turn they can be raised in large numbers to the desired life stage), the development of an effective delivery system, and the prevention of self-infliction (i.e., intoxication of members of one's own organization).

Impending entomological threats to agriculture and food safety

- Among the most overlooked threats to national security are those that could directly impact food production. Agroterrorism, or the deliberate introduction of animal or plant pathogens or pests that directly attack cropping systems or livestock, with the purpose of instilling fear, causing economic losses or undermining social stability, is a very real threat to nations engaged in agriculture to meet their own needs and/or for export to other countries.
- Insects are the ideal agents to carry out acts of agroterrorism. Several species are known to vector plant or livestock pathogens, or directly feed on a variety of crops and livestock. Use in agroterrorism can be as simple as the introduction of an exotic or invasive species into a new region.
- Insect releases via terrorist attack are expected to be via inundative releases in which masses of individuals are deliberately distributed to essentially overwhelm the area and to produce an immediate response, namely quick destruction of the targeted agricultural system.
- In the event that an agroterrorist attack occurs utilizing entomological weapons, the national security response will not be via the military. Rather, the same government agencies dedicated to crop protection using pest management strategies will be called into action.
- Once an entomological agent has been identified, particularly in situations of agroterrorism, the most likely recourse is application of insecticides. A well-chosen biological weapon in agriculture applications will be insecticide resistant and costly to control.

Insect-borne diseases as new or renewed threats to human health

- Public fear of biological weapons generally rests with pathogens that cause human disease. The anticipation of severely debilitating symptoms and death from exposure to smallpox, Ebola, plague, anthrax, and other diseases is what makes the terrorist threat so effective in generating a response in the target nation. From an entomological perspective, many of the human pathogens viewed as viable biological weapons are transmitted by insects usually through blood feeding.
- Mosquitoes are particularly attractive choices for use in biological weapons (arboterrorism) development because species exist in many parts of the world that are capable of vectoring the microorganisms that induce yellow fever, dengue, malaria, Rift Valley fever, and a host of other diseases, and once native species become infected the pathogens can quickly become indigenous. Ideally, the insect vectors used should be selected for insecticide resistance to dampen the response of targeted nations trying to control the spread of an inundative release of infested mosquitoes.
- At this point, there is no evidence to indicate that any nation or non-state organization has deliberately used insect vectors as biological agents toward another nation. However, the risk should be viewed as very high to western nations.

Insects can be used as tools for national security

- Though insects have been used to develop biological weapons, they have also been recruited to protect the nations that are potential targets of all forms of terrorism. The national defense roles range from use in surveillance operations, detection of explosive residues or unexploded munitions, to toxicological applications that allow determination of whether discovered bodies have been subjected to violence involving bullets or explosives.
- The utility of insects to national defenses is most refined with their use in chemical detection. The chemicals are critical to the discovery of explosives residues, unexploded munitions, biological weapons, illicit drugs, and even decomposing bodies. Chemical detection relies on the incredibly acute olfactory system possessed by most insects, that is not only among the most sensitive found in animals, but also makes use of hundreds to thousands of sensory receptors distributed across multiple locations on the body. The fact that insects can be trained to associate odors with a conditioned stimulus, quickly and cheaply, makes them attractive candidates for use in chemical detection systems.
- Insect surveillance is used to collect covert information on the comings and goings of criminal and militant organizations, as part of search and rescue missions, in the detection of explosives, and to monitor hazardous environments. In some of the "simplest" versions, video or audio surveillance devices have been strapped to the backs of insects, which are then released near the site of interest. Signals are in turn broadcast back to the human agents via radio transmitters. To enhance the utility of the insect probes, investigators have developed neurological circuits that can be implanted into the ventral nerve cord and ganglia so that movement of legs and wings can be controlled remotely.
- Necrophagous flies provide another source of utility to national security and criminal investigations: chemical detection as a result of feeding. More to the point, necrophagous fly larvae consuming tissues of a corpse that has been exposed to explosives or bullet fragments can accumulate the chemical signatures that imply foul play. Why this is important is because evidence to implicate a particular bomb-maker (i.e., terrorist organization) often lies in the chemical composition of the explosive device, which may leave a chemical signature in objects found in the explosion area, including the unfortunate victims. Similarly, evidence that a gunshot wound was present on a victim may reside within the fly larvae that consumed the tissues of the deceased.

Test your understanding

Level 1: knowledge/comprehension

1. Define the following terms:
   (a) terrorism
   (b) invasive species
   (c) entomological terrorism
   (d) heraldry
   (e) arboterrorism
   (f) inundative release
   (g) biological weapon.

2. Match the terms (i–vi) with the descriptions (a–f).

(a) Bans the use of biological weapons   (i) Agroterrorism

(b) Small-scale release of insects during favorable conditions   (ii) Biosensor

(c) Cultured cells or extracted tissues for odorant detection   (iii) Inoculative

(d) Robotic flying insect   (iv) Sniffer

(e) Release of pathogens or other destructive organisms on cropping systems   (v) Geneva Protocol

(f) Insect used to locate unexploded landmines   (vi) MAV

3. Compare and contrast the terms terrorism, agroterrorism, and entomological terrorism.

4. Describe the features or characteristics of gunshot residues that facilitate the use of necrophagous fly larvae for detection of said residues.

5. Describe the characteristics of an insect that makes it an ideal candidate for development of biological weapons that (i) target humans directly, (ii) can be used in agroterrorism, or (iii) serve to transmit a human pathogen.

6. Describe the challenges that exist for detecting insects used as agents of agroterrorism.

Level 2: application/analysis

1. What are the limitations of using necrophagous insects for toxicological analyses of explosives and gunshot residues?

2. Discuss the characteristics of an insect that make it suitable for use as a vector of microbes used in bioterrorism.

3. Explain the limitations of using insects in a direct attack on human or livestock populations.

Level 3: synthesis/evaluation

1. Explain how the limitations of archaeoentomological research also apply to investigations attempting to determine whether an insect has been used as a biological weapon.

2. Speculate on which types of insects would be best suited for use in the development of biosensors to detect chemical residues associated with explosives and provide explanations that outline your logic.

3. Describe how insects could be trained for detection of illicit drug detection or discovery of decomposing human remains. What are the limitations of such approaches in urban versus rural environments?

Notes

1. From Sarkar (2010).

2. Al-Qaeda is a militant Islamic organization founded by Osama bin Laden in the late 1980s during the former Soviet Union's attempt to occupy Afghanistan. The group has claimed responsibility for the organized attacks on the United States on September 11, 2001.

3. Operation Hawaii was a secret military attack on the United States Naval Station at Pearl Harbor conducted by the Japanese Navy. As a result of the attack, the United States declared war on the Empire of Japan on December 8, 1941 and thus formally joined the allied forces during World War II.

4. This is an admittedly simplified description of conventional warfare, particularly with regard to attacks on unarmed civilians.

5. The Iraq War (2003–2011) waged by the United States, Great Britain and allies against the Iraqi regime of Saddam Hussein effectively eliminated the immediate threat of biological, chemical, and nuclear weapons from Iraq.

6. Heraldry is the practice of using symbols, typically in the form of animals, to depict an army, kingdom, clan, tribe, etc. The development of armor for soldiers in eleventh-century Europe necessitated the need for identifying marks so that soldiers could distinguish friend from foe. In some cases, insects were used as part of the coat of arms.

7. Alarm pheromones are a type of semiochemical used for intraspecific communication to relay a message of danger or attack. With several social Hymenoptera, such pheromones are stored in glands associated with venom production or storage, and are released with exuded venom, triggering a savage group attack.

8. More formally known as the Protocol for the Prohibition of the Use of Asphyxiating, Poisonous or Other Gases, and of Bacteriological Methods of War, the Geneva Protocol was a treaty established after World War I to stop the use of chemical and biological weapons. The treaty did not address the production, storage or transfer of such materials. The United States did not ratify the Geneva Protocol until 50 years after its drafting.

9. The assassin caterpillar or giant silk moth is found in South America. The stinging hairs of the larvae release hemorrhagic toxins that induce uncontrolled bleeding for unfortunate victims that rub up against the urticating hairs.

10. Invasive species refers to non-native or exotic species that when accidentally or deliberately introduced into a new area elicit damage to a cropping system or livestock. Populations are difficult to keep in check due to a lack of established natural enemies from the new association.

11. Terrorism is only effective in evoking fear in cities or countries not accustomed to widespread life-threatening vector-borne diseases, food shortage, or other types of suffering common to poorer countries located in tropical regions or war-torn nations where citizens face strife on a daily basis.

12. *in vitro* refers to conditions that are outside the body, such as cells extracted from a living organism and placed in growth flasks or dishes using media designed to mimic biological fluids.

References cited

Aktakka, E.E., Kim, H. & Najafi, K. (2011) Energy scavenging from insect flight. *Journal of Micromechanics and Microengineering* 21: 95–116.

Berenbaum, M.R. (1996) *Bugs in the System: Insects and Their Impact on Human Affairs*. Helix Press, New York.

Brody, H., Leonard, S.E., Nie, J.B. & Weindling, P. (2014) US responses to Japanese wartime inhuman experimentation after World War II: national security and wartime exigency. *Cambridge Quarterly of Healthcare Ethics* 23(2): 220–230.

Bromenshenk, J.J., Carlson, S.R., Simpson, J.C. & Thomas, J.M. (1985) Pollution monitoring at puget sound with honey bees. *Science* 227: 632–634.

Bromenshenk, J.J., Henderson, C.B., Seccomb, R.A., Welch, P.M., Debnam, S.E. & Firth, D.R. (2015) Bees as biosensors: chemosensory ability, honey bee monitoring systems, and emergent sensor technologies derived from the pollinator syndrome. *Biosensors*, 5(4): 678–711.

Casagrande, R. (2000) Biological terrorism targeted at agriculture: the threat to US national security. *The Nonproliferation Review* 7(3): 92–105.

Centers for Disease Control and Prevention (2007) Bioterrorism overview page. Available at http://www.bt.cdc.gov/bioterrorism/overview.asp. Accessed August 29, 2012.

Centers for Disease Control and Prevention (2020) https://www.cdc.gov/media/releases/2020/s0218-update-diamond-princess.html. Accessed February 22, 2020.

Cookson, J. & Nottingham, J. (1969) *A Survey of Chemical and Biological Warfare*. Monthly Review Press, New York.

Crutchley, T.M., Rodgers, J.B., Whiteside, H.P. Jr., Vanier, M. & Terndrup, T.E. (2007) Agroterrorism: where are we in the ongoing war on terrorism?. *Journal of Food Protection* 70: 791–804.

Cupp, O.S., Walker, D.E. & Hilison, J. (2004) Agroterrorism in the U.S.: key security challenges for the 21st century. *Biosecurity and Bioterrorism* 2: 97–105.

DaSilva, E.J. (2009) Biowarfare and bioterrorism – the dark side of biotechnology. *Biotechnology: Fundamentals in Biotechnology* 13:59–67.

Eisner, T., Eisner, M. & Siegler, M. (2007) *Secret Weapons: Defenses of Insects, Spiders, Scorpions, and Other Many-legged Creatures*. Belknap Press, Cambridge, MA.

Floreano, D., Zufferey, J.-C., Srinivasan, M.V. & Ellington, C. (2010) *Flying Insects and Robots*. Springer, New York.

Frischknecht, F. (2008) The history of biological warfare. In A. Richardt & M.M. Blum (eds) *Decontamination of Warfare Agents: Enzymatic Methods for the Removal of B/C Weapons*. Wiley-VCH Verlag GmbH & Co. KGaA, Weinheim.

Garrett, L. (2001) The nightmare of bioterrorism. *Foreign Affairs* (January/February): 1–7.

Gu, W., Unnasch, T.R., Katholi, C.R., Lampman, R. & Novak, R.J. (2008) Fundamental issues in mosquito surveillance for arboviral transmission. *Transactions of the Royal Society of Tropical Medicine and Hygiene* 102: 817–822.

Habib, M.K. (2007) Controlled biological and biomimetic systems for landmine detection. *Biosensors and Bioelectronics* 23: 1–18.

Harris, R. & Paxman, J. (1982) *A Higher Form of Killing*. Hill and Wang, New York.

Horn, F.P. & Breeze, R.G. (1999) Agriculture and food security. *Annals of the New York Academy of Sciences* 894: 9–17.

King, T.L., Horine, F.M., Daly, K.C. & Smith, B.H. (2004) Explosives detection with hard-wired moths. *IEEE Transactions of Instrumentation and Measurement* 53: 1113–1118.

Kirby, R. (2005) Using the flea as a weapon. *Army Chemical Review* (July–December): 30–35.

Kwak, M.L. (2016) Arboterrorism: doubtful delusion or deadly danger. *Journal of Bioterrorism & Biodefense* 8: 1. doi:10.4172/2157-2526.1000152.

LaGoo, L., Schaeffer, L.S., Szymanski, D.W. & Smith, R.W. (2010) Detection of gunshot residue in blowfly larvae and decomposing porcine tissue using inductively coupled plasma mass spectrometry (ICP-MS). *Journal of Forensic Sciences* 55: 624–632.

Lockwood, J.A. (1987) Entomological warfare: history of the use of insects as weapons of war. *Bulletin of the Entomological Society of America* 33: 76–82.

Lockwood, J.A. (2009a) *Six-legged Soldiers: Using Insects as Weapons of War*. Oxford University Press, New York.

Lockwood, J.A. (2009b) The scary caterpillar. *The New York Times* April 18, 2009.

Marshall, B., Warr, C.G. & de Bruyne, M. (2010) Detection of volatile indicators of illicit substances by the olfactory receptors of *Drosophila melanogaster*. *Chemical Senses* 35: 613–625.

Monke, J. (2007) Agroterrorism: threats and preparedness. Congressional Research Service Report for Congress Order Code RL 32521, March 12, 2007. Available at http://www.fas.org/sgp/crs/terror/RL32521.pdf. Accessed August 20, 2012.

Monthei, D., Mueller, S., Lockwood, J. & Debboun, M. (2010) Entomological terrorism: a tactic in asymmetrical warfare. *The Army Medical Department Journal* (April–June): 11–21.

Motta, L.C., Vanini, G., Chamoun, C.A., Costa, R.A., Vaz, B.G., Costa, H.B. & Romão, W. (2015) Detection of Pb, Ba, and Sb in blowfly larvae of porcine tissue contaminated with gunshot residue by ICP OES. *Journal of Chemistry*. https://doi.org/10.1155/2015/737913.

Pimentel, D., Hunter, M.S., LaGro, J.A., Efroymson, R.A., Landers, J.C., Mervis, F.T., McCarthy, C.A. & Boyd, A.E. (1989) Benefits and risks of genetic engineering in agriculture. *BioScience* 39: 606–617.

Pimentel, D., Zuniga, R. & Morrison, D. (2005) Update on the environmental and economic costs associated with alien species in the United States. *Economic Entomology* 52: 273–288.

Rains, G.C., Tomberlin, J.K. & Kulasiri, D. (2008) Using insect sniffing devices for detection. *Trends in Biotechnology* 26: 288–294.

Rapasky, K.S., Shaw, J.A., Scheppele, R., Melton, C., Carsten, J.L. & Spangler, L.H. (2006) Optical detection of honeybees by use of wing-beat modulation of scattered laser light for locating explosives and land mines. *Applied Optics* 45: 1839–1843.

Roeterdink, E., Dadour, I. & Watling, R. (2004) Extraction of gunshot residues from the larvae of the forensically important blowfly *Calliphora dubia* (Macquart) (Diptera: Calliphoridae). *International Journal of Legal Medicine* 118: 63–70.

Sarkar, M. (2010) Bio-terrorism on six legs: insect vectors are the major threat to global health security. Available at http://www.webmedcentral.com/article_view/1282. Accessed August 22, 2012.

Staudacher, E. (1998) Distribution and morphology of descending brain neurons in the cricket *Gryllus bimaculatus*. *Cell and Tissue Research* 294: 187–202.

Tabachnick, W.J., Harvey, W.R., Becnel, J.J., Clark, G.G., Connelly, C.R., Day, J.F., Linser, P.J. & Linthicum, K.J. (2011) Countering a bioterrorist introduction of pathogen-infected mosquitoes through mosquito control. *Journal of the American Mosquito Control Association* 27: 165–167.

United States Environmental Protection Agency (2012) Land use overview. Available at http://www.epa.gov/agriculture/ag101/landuse.html. Accessed August 29, 2012.

USDA Economic Research Service (2005) Key statistical indicators of the food and fiber sector, April 2005.

Available at http://www.ers.usda.gov/publications/agoutlook/aotables/2005/04apr/. Accessed August 19, 2012. NB: page now removed from website.

Vasquez-Prokopec, G.M., Chaves, L.F., Ritchie, S.A., Davis, J. & Kitron, U. (2010) Unforeseen costs of cutting mosquito surveillance budgets. *PLoS Neglected Tropical Diseases* 4(10): e858. https://doi.org/10.1371/journal.pntd.0000858.

Wolten, G.M., Nesbitt, R.S., Calloway, A.R., Loper, G.L. & Jones, P.F. (1979) Particle analysis for the detection of gunshot residue. I: scanning electron microscopy/energy dispersive X-ray characterization of hand deposits from firing. *Journal of Forensic Sciences* 24: 409–422.

Zdarek, J. & Fraenkel, G. (1972) The mechanism of puparium formation in flies. *Journal of Experimental Zoology* 179: 315–323.

Zufferey, J.-C. (2008) *Bio-inspired Flying Robots*. EPFL Press, Lausanne, Switzerland.

Supplemental reading

Chaudhry, F.N., Malik, M.F., Hussain, M. & Asif, N. (2017) Insects as biological weapons. *Journal of Bioterrorism & Biodefense*. https://doi.org/10.4172/2157-2526.1000156.

Garrett, B.C. (1996) The Colorado potato beetle goes to war. *Chemical Weapons Convention Bulletin* 33: 1–2.

Guillemin, J. (2006) *Biological Weapons: From the Invention of State-sponsored Programs to Contemporary Bioterrorism*. Columbia University Press, New York.

Hay, A. (1999) A magic sword or a big itch: an historical look at the United States biological weapons programme. *Medicine, Conflict, and Survival* 15: 215–234.

Koblentz, G.D. (2011) *Living Weapons: Biological Warfare and International Security*. Cornell University Press, New York.

Moore, A. & Miller, R.H. (2002) Automated identification of optically sensed aphids (Homoptera: Aphiidae) wing-beat forms. *Annals of the Entomological Society of America* 95: 1–8.

Rose, W.H. (1981) An evaluation of the entomological warfare as a potential danger to the United States and European NATO nations. Available at http://www.thesmokinggun.com/archive/mosquito1.html.

Schott, M., Wehrenfennig, C., Gasch, T. & Vilcinskas, A. (2013) Insect antenna-based biosensors for in situ detection of volatiles. In T. Scheper & R. Ulber (eds) *Yellow Biotechnology II*. Springer, Berlin.

Shringarpure, K.A. & Brahme, K. (2018) Biowarfare: where do we stand?. *International Journal of Community Medicine and Public Health*. https://doi.org/10.18203/2394-6040.ijcmph20184022.

Spiers, E.M. (2010) *A History of Chemical and Biological Weapons*. Reaktion Books, Chicago.

Additional resources

Biological/chemical agents and other threats: http://phpartners.org/bioterrorism.html

Bioterrorism: http://www.immed.org/illness/bioterrorism.html

Centers for Disease Control and Preparedness Emergency Responses: http://emergency.cdc.gov/

Counter-terrorism, UK: http://www.homeoffice.gov.uk/counter-terrorism/

Insects as bioweapons: http://abcnews.go.com/Technology/Story?id=99552&page=1#.UED9vFQ6E7A

Institute for BioSecurity: http://www.bioterrorism.slu.edu/bt.htm

National Institute of Justice, Agroterrorism: http://www.nij.gov/journals/257/agroterrorism.html

Nuclear Threat Initiative: http://www.nti.org/threats/

Insect sniffer systems: http://www.technovelgy.com/ct/Science-Fiction-News.asp?NewsNum=1196#pic

United Nations Biological Weapons Convention: http://unog.ch/80256EE600585943/%28httpPages%29/04FBBDD6315AC720C1257180004B1B2F?OpenDocument

United States Department of Homeland Security: www.dhs.gov

Source: Information from Monthei *et al.* (2010) and Kwak (2016).

Chapter 22
Insects and arthropods that cause death

The remarkable dominance of insects and other arthropods on land can be attributed, at least partially, to the extraordinary diversity of their chemical defense mechanisms.

Dr Mario Palma, Center for Study of Social Insects
São Paulo State University, Rio Claro, Brazil[1]

Overview

Insects that bite, sting or are otherwise "defensive" generally evoke fear in humans, who desperately try to avoid them. The beasts are usually not hard to distinguish as most display distinct banding patterns and coloration that are **aposematic**, conveying a message to would-be attackers to back off, because a painful, possibly deadly, response awaits those who choose to get too close. The insects in question could be any of a number of aculeate Hymenoptera that possess a sting apparatus and venom. For these insects, stinging is used for defense, prey capture, and sometimes to aid reproduction. When used for defense, venoms have evolved to elicit an immediate and lasting impression so that an aggressor does not repeat the activity. As a consequence, toxins in the venom may yield a very painful response that may last for hours to days and, in some situations, is lethal to the recipient. Stinging behavior and possession of toxic venoms are not limited to the Hymenoptera. In fact, a wide range of insects representing several taxa produces defensive compounds that can be released through biting, stinging, or secretion from glands or via the exoskeleton. For a small but significant number of these species, the toxins are deadly to humans. Death is not always immediate, but intense debilitating pain frequently characterizes many of these insect-derived products. In such scenarios, insects shift from roles in which they aid forensic investigations of human death to becoming the primary suspects. This chapter examines insects that are capable of inducing death when biting, stinging, or releasing secretions. The process of intoxication and envenomation will be examined for the various methods of insect attack toward humans, as will the chemistry and modes of action of the most common and deadly of toxins.

The Science of Forensic Entomology, Second Edition. David B. Rivers and Gregory A. Dahlem.
© 2023 John Wiley & Sons Ltd. Published 2023 by John Wiley & Sons Ltd.
Companian website: www.wiley.com/go/forensicentomology2

The big picture

- Insects that bite, sting or secrete cause fear, loathing and death.
- Insects that cause death.
- Human envenomation and intoxication by insect-derived toxins.
- Insects that injure humans rely on chemically diverse venoms and toxins.
- Non-insect arthropods that cause death.
- Implications of deadly insects for forensic entomology.

22.1   Insects that bite, sting or secrete cause fear, loathing, and death

Insects cause fear. At times, this entomological trepidation is not deserved as the fearful people display borderline irrational behaviors toward all multilegged creatures, with no willingness to entertain the possibility that not all insects are menacing and some, in fact many, are quite useful if not essential to human existence (Waldbauer, 2000). Fear of insects may not even be linked to any previous event that can account for this loathing of insects or their relatives; it is an innate loathing. For others, the hatred is premeditated. In such instances, the foundation for fear *is* linked to an earlier interaction, undoubtedly caused by a stinging or biting creature that happened to be stepped on, swatted, or in some other way direct contact was made. The mere presence of a yellow-and-black striped insect, for instance, triggers anxiety that often causes afflicted individuals to scream, swat, run, hide, or any other activity to save themselves from the inevitable fate of insect Armageddon (i.e., an attack). Are such responses rational? Probably not in most cases as the insects would rather avoid contact just as much as humans, particularly so as not to "waste" the precious venom or toxins produced for other purposes. However, a small but significant number of insects deserve that fear-laced respect. Why? Because they can be much more aggressive than typical species that are synanthropic with humans, and/or may even produce compounds that are highly toxic. In extreme cases, the interactions can be deadly.

The question that must be asked is why do some insects produce more potent compounds than others? The answer to that question is tied directly to the purpose or function of the toxins and venoms. In other words, what do they do that warrants the need for a potentially deadly toxin(s)? Remember from Chapter 7 that chemical compounds produced by one insect species that are harmful to another are a type of allelochemical known as allomones. Venoms and any type of noxious or toxic chemical released in a volatile, secreted, or injected form through biting, stinging or secretion are all examples of allomones. Generally speaking, allelochemicals are used for chemical defense (including repellency) to capture and/or subdue prey, and to aid in reproduction. The functionality and characteristics of allomones varies to a degree based on the mechanism of distribution or release toward the target species. For example, many species rely on biting, stinging or release of secretions for defensive purposes but the active compounds used by each are chemically diverse, do not operate using the same modes of action, and display different potencies. Correspondingly, the effects on humans are also variable, including the immediate reaction at the time of attack as well as any long-term consequences, including death. Several of these topics, namely pathology, chemical constituents and cellular pathways affected, will be addressed later in the chapter, but for now we are trying to understand why toxin and venoms differ in potency toward humans, potentially being lethal. The defense function is the real key (Figure 22.1).

Biting and stinging (but not secreting defenses) are often aligned with aggressive behavior, and collectively are used to ward off attack by predators or as a preemptive warning to other animals that get too close to either an individual or the colony of

Figure 22.1   Adult yellowjackets on a human hand. *Source*: Photo by Mar S-Hornets and Wasps/Wikimedia Commons/CC BY-SA 4.0. https://commons.wikimedia.org/wiki/File:Vosy_obecn%C3%A9_a_vosy_%C3%BAto%C4%8Dn%C3%A9._Yellow_Jackets._Paravespula.jpg.

social species. Solitary wasps, for example, are highly irritable when foraging while social insects often are most aggressive when the hive, nest or colony is threatened. Of course, the insect cannot distinguish accidental contact from deliberate, so when a bare foot makes contact with a yellowjacket[2] exiting an underground burrow, stinging is inevitable. If truly unlucky, alarm pheromones may be released, drawing the vicious attacks of nearby nest mates (Vetter *et al.*, 1999) (Figure 22.2). Some insects do not even need to use the venom to gain protection; their aposematic coloration or behaviors are sufficient to ward off predators. The example of the yellow-and-black striped insects illustrates common aposematic coloration used by a wide range of aculeate Hymenoptera – the group of ants, bees and wasps belonging to the suborder Aprocrita that all possess a sting apparatus (a modified ovipositor). However, warning alone is not adequate in all cases. Consequently, step two is to back up the warning with aggressive behavior, which in turn may be coupled to the use of a painful allomone. Some insects that are very toxic do not even attempt to warn other animals of their potency, lacking any type of aposematic markings or aggression. Rather, such species synthesize incredibly potent toxins or venoms that elicit such intense pain that the recipient animal is completely or temporarily incapacitated. Almost any contact rapidly debilitates the attacker and lasts for a long period. The idea of extremely painful venoms is thought to yield not just an effective defense from predators, particularly vertebrates, but should also yield

long-lasting protection (Starr, 1985). In other words, the more intense and painful the interaction with potential prey species as a result of stinging, biting or secretion, then the more likely the predatory species will remember that this prey is off limits (Schmidt *et al.*, 1983). Consequently, potency equates to long-term memory!

The unfortunate consequence of high potency can be death. This potential is evident with the full gamut of toxin producers, meaning there are species of biting, stinging, and secreting insects with the capacity to kill a human under the right conditions. As with any topic in the biological sciences, there are always exceptions to the rule. Deadly insects are not different. What we mean is that death may result from toxins or venoms that are not necessarily considered potent. In such instances, constituents of insect fluids contain allergens, usually proteins, that stimulate an immunological response in a person. Not just simple histamine release leading to mild inflammation, but an acute systematic allergic reaction, termed **anaphylaxis,** that reflects hypersensitivity to the venom allergens. This is not a bodily response to toxins, but instead an allergic reaction to insect proteins. Severe responses in an individual may occur, such as airway constriction, hypertension, and gastrointestinal complications (Klotz *et al.*, 2009; Golden, 2015). Any one of these alone or in combination may be fatal to certain individuals. Although several venomous or blood-feeding insects may be responsible for a significant number of human deaths due to anaphylaxis, such species are not the focus of this chapter.

Aggressive and/or deadly insects have two obvious connections to forensic investigations. The first is disruption of crime scene processing. Stinging and biting insects that are attempting to feed near or on the corpse, or that have built nests in close proximity to the corpse/crime scene, often become agitated when anyone or anything gets too close. Some, especially those that fly, will display warning behaviors before attack, affording crime scene investigators an opportunity to potentially avoid insect aggression. However, other species may attack as soon as threatened. Obviously, this creates a difficult situation for anyone attempting to process a crime scene. The second connection is when a deadly outcome results from a human-insect interaction. In other words, an insect(s) injected or released a toxic substance directly responsible for the death of an individual. When a deadly insect encounter is hypothesized to have

Figure 22.2 Adults of *Vespula germanica* at the entrance of underground nest. *Source*: Photo by Gideon Pisanty/Wikimedia Commons/CC BY 3.0. https:// commons.wikimedia.org/wiki/File:Vespula_germanica_ nest_1.jpg.

occurred, investigators should look for signs of envenomation/intoxication or biting and stinging. Small puncture wounds from an ovipositor or stinging hairs or paired bite marks from mandibles serve as trace evidence of the interactions. The remainder of this chapter will focus on truly dangerous insects that synthesize and use deadly toxins and venoms as their means of chemical defenses. Section 22.2 will introduce you to some of these deadly insects, followed by an exploration of what their toxic compounds do to humans and the types of chemicals employed to carry out the deadly deeds.

Figure 22.3   Stinging hair caterpillar, *Lonomia obliqua*. *Source*: Photo by Centro de Informações Toxicológicas de Santa Catarina/Wikimedia Commons/public domain. https://commons.wikimedia.org/wiki/File:Lonomia-obliqua-citsc-1.jpg.

22.2   Insects that cause death

The idea that some insects bite, sting, or secrete noxious compounds is certainly not surprising to anyone, assuming you have not lived a sheltered life devoid of outdoor experience! Insects that synthesize noxious compounds are relatively abundant in most terrestrial ecozones, which means that the average person has likely had contact with one or more species that stings or bites. Secreting insects are not quite as frequent. Regardless, such insects are quite familiar, occur in many regions and, importantly, generally do not cause long-term health concerns for the majority of individuals who have been attacked. Chance encounters with truly dangerous species (again, not considering anaphylaxis) is another matter altogether. Why? For one, there are not as many species of insects that synthesize lethal compounds. Those that do tend to be most abundant in the orders Coleoptera, Hemiptera (Heteroptera), Hymenoptera, and Lepidoptera. Far and away the most common belongs to the aculeate Hymenoptera (Schmidt *et al.*, 1986). However, and this brings us to point number two, the species occur in relatively isolated regions of the world so that human contact is infrequent. This statement needs to be placed in appropriate context: deadly insects are generally not synanthropic. Adding to the isolation is the fact that several of these insects limit toxin production to specific developmental stages. For example, of the lepidopterans that synthesize highly toxic allomones, nearly all do so as juveniles (stinging hair caterpillars) (Figure 22.3). You may be thinking that the monarch butterfly, *Danaus plexippus* (Lepidoptera: Nymphalidae), is a classic example of a butterfly species that is highly

toxic (due to cardiac glycosides) in multiple life stages (larvae, pupae, and adults), which of course is true, but intoxication is realized through consumption and not by the mechanisms of defense described thus far.

Among the Hymenoptera, stinging and biting behavior occurs with adults of bees, ants and wasps, and nearly all of the most potent species rely on aposematic markings and/or behaviors to preempt **envenomation,** the process of injecting venom into a target animal. Thus they attempt to intimidate to avoid using venoms. This latter feature also contributes to fewer encounters with deadly species: venom proteins and toxins can be energetically expensive to make and quantities are limited in terms of how much can be produced on a daily basis. Conservative venom usage is typical of arthropods like scorpions that rarely inject active venom[3] as part of their defense strategy. However, if these insects are provoked continuously or called to action via alarm pheromones, the recipient will realize the full effects of the potent venom.

Secreting species are tougher to conceptualize in that far fewer species rely on such mechanisms for defense. The defensive compounds are released via glands or the exoskeleton in response to attack, most commonly by beetle species such as the blister (Meloidae), rove (Staphylinidae) and bombardier (Carabidae) that inhabit the terrestrial surface. Human interaction is relatively rare, except when adults are attracted to house lights at night or some other accidental encounter, and then if swatted or smashed, toxins may make contact with an individual. Camouflage and aposematism are used by these insects, regardless of the potency of the toxins, an indication that perhaps the allomones in their secretions evolved initially for some other function (i.e., to aid reproduction) than defense. The lack of bright distinctive coloration on some of these beetles may also lead to underestimation of the frequency of human contact since they likely go unnoticed in most instances.

The remainder of this section is dedicated to introducing a few example species that synthesize potentially deadly toxins or venoms. The insects have been grouped by taxa rather than mode of toxin release. However, as should be obvious from our earlier discussions, similar mechanisms of toxin release are commonly used within an order.

22.2.1   Deadly Coleoptera

Toxin production is not widespread among the Coleoptera and appears to be restricted to species that spend the vast majority of their time dwelling on or in the upper layers of soil. For most potent species, toxins are proteins or peptides synthesized in the larval and/or adult stages, although at least one species transfers allomones to the eggs, and are released via secretion or compression (squashing) of the body. Some release non-protein secretions via explosive sprays. Aposematic coloration is used by species in many families, while others tend to have exoskeleton markings that permit blending into the surrounding environment.

22.2.1.1   Blister beetles (Family Meloidae)

Members (males) of this family produce the toxic compound **cantharidin,** an allomone that can cause non-lethal lesions on skin, or if ingested is capable of damaging the lining of the alimentary canal, potentially leading to death in humans and other animals. Cantharidin is also known as a supposed aphrodisiac that stimulates long-lasting erections in males. However, the potential toxicity by ingestion should trump any temptation to test the efficacy. In most instances, human encounters with blister beetles do not result in any type of health problems (Figure 22.4).

22.2.1.2   Bombardier beetles (Family Carabidae)

More than 500 species in this family use a defense mechanism in which hot **quinones**[4] are sprayed from glands in the abdomen. In response to attack

Figure 22.4   An adult blister beetle from the family Meloidae. Some are aposomatic such as the one shown, but others are not, concealing the fact that they are toxic from would be attackers. *Source*: Photo by arian. suresh/Wikimedia Commons/CC BY 2.0. https:// commons.wikimedia.org/wiki/File:Blister_beetle_ (26390828032).jpg.

Figure 22.5   An adult bombardier beetle from the family Carabidae. *Source*: Patrick Coin / Wikimedia Commons / CC BY-SA 2.5. https://commons.wikimedia. org/wiki/File:Brachinus_spPCCA20060328-2821B.jpg.

or a threat, adult beetles release two reactants (hydrogen peroxide and hydroquinone) into a mixing chamber that results in a violent exothermic chemical reaction. The generated mixture is at its boiling point, leading to production of a foul-smelling gas that, when released, produces a loud pop or explosion sound. The latter accounts for the common name of these beetles. When the hot spray or gas makes contact with another insect, death often results. Humans fare much better: the allomones produce a painful dermatitis reaction but no lasting consequences (Figure 22.5).

22.2.1.3   Rove beetles (Family Staphylinidae)

This very large family of beetles is best known for most members being ferocious predators, a feature first discussed in Chapter 5 with regard to predation of necrophagous fly larvae. At least two species in the genus *Paederus* (*P. fuscipes* and *P. riparius*) receive special consideration due to the production of an intra-hemocoelic peptide toxin, pederin (Kellner & Dettner, 1996), that when secreted typically triggers formation of painful blisters on the skin, and which if it penetrates fluids of the eye may induce temporary to permanent blindness. Ingestion or injection into the bloodstream can lead to death. Chance interactions between deadly staphylinids and humans are considered rare, as most species are restricted to regions of the Palearctic ecozone with low human population densities. Despite the potency of the pederin toxin, neither larvae nor adults rely on aposematic coloration or behaviors to deter potential predators (Figure 22.6). This may be due to toxin production being dependent on endosymbiotic bacteria rather than directly by the beetles (Kellner, 2002).

22.2.2   Deadly Hemiptera

True bugs use piercing-sucking mouthparts to feed during nymphal and adult stages, displaying herbivory, carnivory, and parasitism. Parasitic true bugs are blood feeders, like the notorious kissing bugs *Triatoma infestans*, which inject salivary components prior to ingestion of their liquid diet

Figure 22.6   An adult rove beetle, *Paederus littoralis*. *Source*: Photo by Luis Miguel Bugallo Sánchez/Wikimedia Commons/CC BY 3.0. https://commons.wikimedia.org/wiki/File:Paederus_littoralis._Soigrexa,_Bastavales,_Bri%C3%B3n,_Galiza.jpg.

(Amino *et al.*, 2001). However, the **hematophagous** or blood-feeding varieties are not the focus of this chapter. While it is true that parasitic species can serve as vectors of disease and also stimulate anaphylaxis, none are deadly in terms of production of toxins and venoms. This distinction resides with hemipterans that are carnivorous in the family Reduviidae. Feeding stages subdue their prey by piercing with mouthparts and then immediately pumping the captured food with salivary components. For several species, saliva contains a range of powerful digestive enzymes (Table 22.1) and neurotoxins, leading to the term "venomous saliva." A cocktail of carbohydrases, lipases, proteinases, hydrolases, and other enzyme classes is needed for pre-oral digestion of the prey (Cohen, 1995). Several of the digestive enzymes have the capacity, in sufficient concentration, to readily digest cellular membranes and tissue components from a

Table 22.1   Digestive enzymes commonly found in saliva of predatory reduviids.

Enzyme	Substrate
Carbohydrases	
Amylase	Cleaves 1,4-glycosidic bonds in starch and glycogen
Hyaluronidase	Anionic non-sulfated glycosaminoglycans (hyaluronan)
Invertase	Hydrolyzes sucrose
Lipases	
Esterase*	A wide range are present; promote hydrolysis reactions
Lipase	Hydrolyzes a range of fats
Phospholipase A_1	Hydrolyzes phospholipids in fatty acid chains
Phospholipase A_2	Hydrolyzes phospholipids in fatty acid chains
Proteases	
Aminopeptidase	Cleaves amino acids from proteins at the N-terminus
Carboxypeptidase	Cleaves amino acids from proteins at the C-terminus
α-Chymotrypsin	Cleaves peptide bonds formed by aromatic amino acids
Pepsin[†]	Broad specificity toward amino acids
Serine proteases	Broad activity toward amino acids
Trypsin	Cleaves peptide bonds with arginine and lysine

* Technically lipases are a subclass of esterases; here the esterases are specific for lipids.
[†] Only reported for species in the genus *Rhynocoris*.
Source: Data from Cohen (1995) and Zibaee *et al.* (2012).

wide range of animals, including humans. Typically, a single individual like the predator *Reduvius peronatus* or any of several species in the genus *Rhynocoris* is not able to inflict death, although individual bites can be very painful and persist for hours. In the event of a group attack, as employed as a method of torture during pre-medieval times, repeated biting and injecting of saliva can lead to acute toxicity that is lethal. A worse fate, however, is the accumulated effect of concentrated digestive enzymes, which can literally digest soft tissues of an adult human to the point that tissues slough off the bone. Death does not come immediately, meaning the individual endures an unimaginably painful slow decay.

Figure 22.7 An adult harvester ant, *Pogonomyrmex maricopa* (Family Formicidae). *Source*: Photo by Brett Morgan/Wikimedia Commons/public domain. https://commons.wikimedia.org/wiki/File:Maricopa_harvester_ant_(Formicidae-_Pogonomyrmex_maricopa)_(31538269553).jpg.

22.2.3   Deadly Hymenoptera

The Hymenoptera comprises a wealth of deadly insects in comparison with other insect orders. Nearly all stinging and biting species of significance belong to the aculeate group of the suborder Aprocrita. Deadly toxins are administered exclusively through envenomation during the act of stinging. Biting generally has no long-term effect on a large vertebrate. Aposematic coloration and warning behaviors are also common to most of the aculeate Hymenoptera that possess lethal toxins or venoms. The notable exception is several species of ants belonging to the family Formicidae; aposematism is a variable trait in this group. Another formidable trait is that many species are social, relying on an elaborate chemical communication system (discussed in Chapter 7) to convey messages throughout the hive or colony. This efficient form of communication is also utilized as part of individual and colony defense, with the release of alarm pheromones during attack or venom injection mobilizing nest mates into a group attack. Thus, the potency of a given venom or toxin is increased through acute toxicity of multiple stings (Schmidt *et al.*, 1986).

An examination of deadly Hymenoptera requires dividing the order into families because of the species richness within this order. Three families are discussed: Formicidae, Pompilidae, and Vespidae.

22.2.3.1   Ants (Family Formicidae)

Ants represent an incredibly large and diverse family of insects. Most species have the capacity to bite and sting when engaged in prey capture or in warding off a predator. Biting yields only a temporary reprieve as most species inject the relatively weak formic acid during the act, which causes a mild burning sensation (Blum *et al.*, 1958; Touchard *et al.*, 2016). Venom injection via the sting apparatus can be lethal to prey insects, but in most cases evokes a sharp but short-lived pain in large animals like humans. However, a few species of ants synthesize venoms with one or more lethal components (Figure 22.7). For example, *Ectatomma tuberculatum* and *Pogonomyrmex maricopa* produce the most lethal venoms of any known insect (Schmidt *et al.*, 1986). In contrast, the red imported fire ant *Solenopsis invicta* does not technically possess a lethal venom in individual ants, but the ferocity of attack by the colony toward anything in its path, including humans, leads to rapid acute toxicity because hundreds to thousands of ants may sting a single individual within minutes. The fire ant represents a species that does not display aposematic markings and it shows little interest in avoiding confrontation with any organism.

22.2.3.2   Spider wasps (Family Pompilidae)

Wasps in this family produce venom that is not known to be lethal to humans. Inclusion of pompilids in this chapter is because a single sting from one wasp yields one of the most excruciating stings by any insect (Figure 22.8). In fact, only venom from the bullet ant *Paraponera clavata* is considered more painful (Starr, 1985). Adults are aposematic in that brown to black body colors are intermixed with oranges, red and yellow.

Figure 22.8 A pompilid wasp (Family Pompilidae). *Source*: Photo by L. Shyamal/Wikimedia Commons/CC BY 2.5. https://commons.wikimedia.org/wiki/File:Wasp_talakaveri.jpg.

22.2.3.3 Hornets and wasps (Family Vespidae)

Numerous species in the family Vespidae are highly aggressive, demonstrating this most evidently through biting and stinging. Unlike members of the Apidae, these wasps can repeatedly envenomate since the stinger is not barbed (Figure 22.9). The deadliest venoms are produced by several East Asian species (hornets) in the genus *Vespa*. The venoms synthesized are not as lethal as the ant species mentioned earlier. However, members of the genus *Vespa* are among the largest-bodied stinging insects known. Correspondingly, the volume of venom produced is greater than that of any other insect. Thus, if the lethality of the venom is equated based on volume of toxins available, then *V. mandarinia* and *V. tropica* are the most lethal insects on the planet (Schmidt *et al.*, 1986). Even more frightening is the fact that Asian hornets are highly aggressive and frequently attack as a group, delivering a large lethal venom payload to the target animal.

22.2.4 Deadly Lepidoptera

Deadly butterflies and moths. Is that really possibly? Such insects would seem the least likely candidates to possess lethal toxins, at least from the vantage point of ones used during biting and stinging. After all none are considered truly aggressive. Precisely, which is why some insects need powerful chemical defenses. Coupled with the fact that they are not "offensive" and at least the immatures stages are mobile yet slow, potent chemical defenses make sense. We have alluded to caterpillars that produce toxins, but these must be consumed to fully convey the potency message to

Figure 22.9 The stinger of *Vespa crabro* (Family Vespidae). Notice that it lacks barbs that are typical of bees in the Family Apidae. *Source*: Photo by Janek lass/Wikimedia Commons/CC BY 4.0. https://commons.wikimedia.org/wiki/File:Vespa_crabro_needle.jpg.

predators. Several species sequester secondary plant compounds during feeding and retain the toxins throughout larval development. In some instances, toxin sequestration lasts through adulthood and is generally associated with aposematic markings in an attempt to avoid "tasting" by a potential predator. Other species rely on urticating or stinging hairs on caterpillars to thwart attack. Contact with the hairs releases toxins that yield mild to severe reactions in humans. Twelve families of lepidopterans contain stinging hair caterpillars that produce toxins that affect humans on contact, but only two species, both from South America (*Lonomia achelous* and *L. oblique*) and belonging to the family Saturniidae, synthesize toxins that are lethal by means other than anaphylaxis (Diaz, 2005). The venoms of these latter species are hemorrhagic, causing diffuse bleeding that can lead to death. The deadly caterpillars do not display aposematism as the body is adorned in greens and browns that allow them to blend into the environment rather than convey warning to **allospecifics.**

22.3   Human envenomation and intoxication by insect-derived toxins

Our focus in this chapter is on insects that elicit death through the toxins and venoms used in active mechanisms of defense. Although the potencies of some of the toxins employed are among the most lethal known in the insect world and for that matter among arthropods as a whole, they generally do not induce immediate death upon contact. So, what happens between the time of venom injection (envenomation) or toxin exposure (intoxication) until death ensues? The most immediate responses are independent of the toxins involved and occur at the site of insect attack. Skin damage and irritation can result from direct injury due to mouthpart (biting punctures, lacerations) or stinger penetration, or from reactions to non-toxic constituents found in saliva, venom or secretions (Goddard, 1999; Mitpuangchon *et al.*, 2021). Inoculation of the wounds with bacteria, fungi or other microorganisms is not uncommon since the stinger or mouthparts of any insect are certainly not sterile. The result of microbial introduction can range from initiation of a mild histamine release that in turn stimulates a localized inflammatory response, to establishment of a secondary infection that mobilizes a more intense systemic immunological response. As discussed earlier in the chapter, such reactions may initiate anaphylaxis, which is a response that is independent of the lethal venoms and toxins used to envenomate or intoxicate.

The most immediate response directly linked to the toxic allomones of defense is induction of intense pain. A number of pathways are involved in causing the excruciating, throbbing, and lasting pain, but most share in common the destruction of cellular membranes, prompting histamine release and leading to **edema.** Edema is essentially swelling of cells or tissues with interstitial fluids, which can stretch plasma membranes to the point of bursting, facilitating a chain reaction of swelling and bursting in localized regions. Most importantly to the insect, this creates a wave of pain in the recipient. Swelling itself can be painful, applying pressure to nearby neurons, forcing mechanical depolarization, and resulting in generation of sensory signals that lead to recognition of pain and discomfort.

Localized and systemic pathological changes follow the initial damage caused by penetration or contact with skin and are generally concurrent with the onset of pain. Enzymes or other proteins in the toxin cocktail and/or derived from human cells, the latter representing a condition akin to **autocatalysis** (discussed in Chapter 10), may facilitate the damage. The severity of the injury and target tissues affected are dependent on the toxins involved and mode of entry into the body. However, some consistent pathologies are evident: toxins and venoms are cytotoxic at the initial site of entry/contact and in adjacent tissues, serve as metabolic pathway disrupters (generally via signal transduction pathways), and may stimulate lesion formation, localized and/or systemic inflammation and swelling, organ damage, and ultimately death. Examples of toxins that can induce these homeostatic disturbances and the cellular pathways manipulated are the focus of Section 22.4.

22.4   Insects that injure humans rely on chemically diverse venoms and toxins

Understanding the chemical identity of the toxins utilized by insects in defense and the cellular pathways manipulated is essential for developing methods of treatment for individuals who are attacked, or who make contact with noxious secretions. Much like the situation with deadly snake venoms, the timing of appropriate treatment is critical to limiting the severity of symptoms and damage evoked by deadly insect venoms and toxins. Thus, many of the insects discussed in Sections 22.1–22.3 have received a great deal of attention from entomologists, medical personnel, and those engaged in biomedical research and drug development. The work of these individuals has revealed that the active components of venoms tend to be unique for particular insect groups, influenced heavily by the mechanism of envenomation or intoxication. For example, stinging Hymenoptera, other than ants, rely on complex venom cocktails containing proteins and low-molecular-weight molecules and are dominated by linear, polycationic, amphipathic[5] peptides (Palma, 2006; Lai & Liu, 2010). The venoms of fire ants contain saturated and unsaturated **piperidines** (or solenopsins), which are alkaloid compounds that serve as the active ingredients. In contrast, a species that does not need a group attack to be deadly, the ponerine ant *E. tuberculatum*, produces neurotoxic peptides (**ectatomin**) that can be lethal. Biting insects that rely on saliva injection synthesize an array of digestives enzymes and neurotoxins, the latter of which are most often protein in composition. Among the species that secrete or spray allomones, some like rove beetles produce toxic peptides (pederin) that are secreted when attacked, while sprays are frequently released with a high content of quinones.

These examples serve to give a broad overview of the classes of compounds that act as potentially deadly allomones. They do not represent the rich diversity of chemicals used by insects as a whole in chemical defenses, as there is a clear dependence by venomous insects on peptides and proteins as the active components. For most species, the reality is that venoms, saliva and other insect-derived secretions contain a mixture of compounds. Not all are toxic; some merely aid in the distribution (i.e., spreading agents) of other constituents once in contact with prey or host cells, and some function to stabilize venom constituents or prevent autointoxication of the venom-producing insect. Our interests lie with those that can be lethal to humans and thus the remainder of this section is dedicated to examining some of the specific toxins responsible for inducing death.

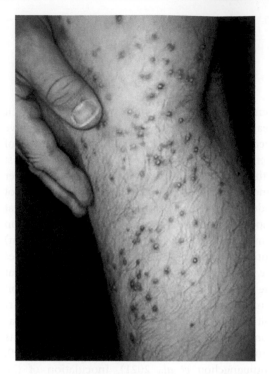

Figure 22.10 Formation of pustules after biting and stinging by the fire ant *Solenopsis invicta* (Family Formicidae). *Source*: Photo by Daniel Wojcik/Wikimedia Commons/public domain. https://commons.wikimedia.org/wiki/File:FireAntBite.jpg.

22.4.1 Ant venoms

Ant venoms are known to display a broad range of activity, including lethality, paralysis, antimicrobial (toward bacteria and fungi), phytotoxic, insecticidal, and hemolytic (Blum *et al.*, 1958; Touchard *et al.*, 2016). Such a wide range of properties clearly indicates that multiple constituents exist in the venoms, and not all contribute to the pathologies associated with human envenomation. Those of most interest to the human condition are the piperidines from fire ant venoms and the extremely toxic peptides derived from other species.

22.4.1.1 Piperidines

Piperidines or solenopsins are a group of alkaloid compounds composed predominantly of 2-methyl-6-alkylpiperdines. Differences in side-chain saturation lead to variation in cocktail mixes and potencies. The relative proportion of the piperidines in venom can vary between nest mates and non-nest mate conspecifics, as well as between different fire ant species (Deslippe & Guo, 2000). The alkaloid content of venom from a single

individual can induce intense pain and **pseudopustule** formation ("pseudo" because the pustules do not contain bacteria) on the skin but is not lethal (Figure 22.10). Mass attack can deliver sufficient piperidine toxicity to kill children and adults. The alkaloids operate non-specifically (i.e., do not require receptor binding), interacting with cellular membranes to induce lysis (Lind, 1982), initially causing histamine release followed by a series of intracellular disturbances associated with unregulated influx and efflux within injured cells.

22.4.1.2 Ectatomin

Ectatomin is a low-molecular-mass (7928 Da) basic protein containing two polypeptide chains linked by a disulfide bridge. After injection, the protein (two molecules are required) embeds into plasma membranes independent of receptor binding to create a non-selective cationic channel that compromises membrane integrity by allowing ion leakage (Pluzhnikov *et al.*, 1999). The damage is only partially reversible *in vitro* and likely permanent following natural envenomation, leading to cell death.

22.4.1.3 Poneratoxin

Poneratoxin is a neurotoxic peptide produced by the ant *Paraponera clavata*. The peptide is small (25 amino acid residues) and amphipathic, the latter a feature that permits diverse and non-specific interactions with plasma membranes of different cell types (Palma, 2006). Voltage-gated sodium channels are disrupted due to toxin binding, blocking synaptic transmission, which in turn induces paralysis and cell death (Figure 22.11).

22.4.2 Wasp venoms

The venom composition of solitary and social wasps is very complex, with a chemical diversity that could easily fill a book by itself if sufficient coverage was permitted to discuss the unique chemistries and modes of action. Our space is more limited so we will examine only some of the components that are known to be among the most toxic to human tissues.

22.4.2.1 Wasp kinins (bradykinin)

These wasp peptides are predominantly found in vespids and are essentially identical to bradykinins in honey bee venom, with the exception that additional amino acids are positioned at the N-terminus[6] (Palma, 2006; Konno *et al.*, 2016). The peptides are principally responsible for long-lasting pain, inducing vasoconstriction, contraction of muscles along the bronchioles, yet relaxation of smooth muscle of the gastrointestinal tract. Kinin fragments bind to at least two receptor types (B_1 and B_2) associated with endothelial or cardiac muscle to stimulate asynchronous muscular contractions, conditions that if sustained for a prolonged period will induce hypertension or hypotension, erythema, and ATP depletion (Levesque *et al.*, 1993).

22.4.2.2 Mandaratoxin

The Asian giant hornet, *Vespa mandarinia*, produces **mandaratoxin,** a large single-chain peptide, in large quantities in its venom (Figure 22.12). Sufficiently high doses of venom induce violent

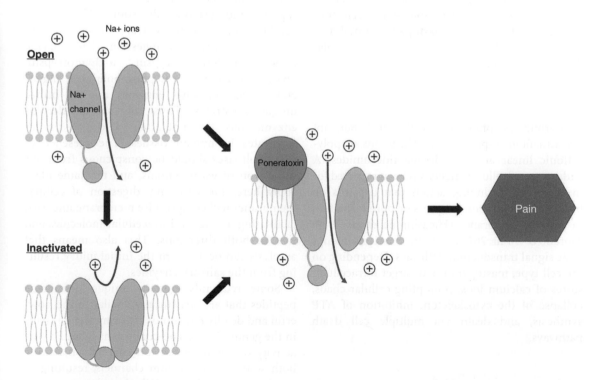

Figure 22.11 Schematic of an open (activated) and closed (inactivated) voltage-gated sodium channel in human skin. Upon binding of poneratoxin, the sodium channel is forced to stay in the activated state. This leads to prolongation of action potentials, which is associated with the pain from bullet ant stings and ultimately contributes to cell death. *Source*: Pchien2/Wikimedia commons/CC BY-SA 4.0.

Figure 22.12   Queen of the Japanese giant hornet, *Vespa mandarinia japonica* (Family Vespidae). *Source*: Photo by Yasunori Koide/Wikimedia Commons/CC BY-SA 4.0. https://commons.wikimedia.org/wiki/File:20200512-P1100051_Vespa_mandarinia_japonica.jpg.

convulsions, constriction of airways, muscle weakness and death, through inhibition of sodium current at postsynaptic neuromuscular junctions (Abe *et al.*, 1982). Adding to the potency of the venom is the presence of a cytotoxin that can stimulate anaphylaxis and of **mastoparan,** a multifaceted peptide that can lead to a wide range of cellular damage in multiple cell types.

22.4.2.3   Mastoparan

This group of peptide toxins is the most abundant constituent in vespid venoms. The toxins are polycationic linear and tetradecapeptide amides. A wide range of cellular effects has been reported for mastoparans, including action that depends on binding to G-protein receptors or pore formation in plasma membranes (Higashijima *et al.*, 1988; Moreno & Giralt, 2015); both mechanisms manipulate signal transduction pathways. Depending on the cell type, mastoparan may target intracellular stores of calcium ions, prompting cellular chaos, collapse of the cytoskeleton, inhibition of ATP synthesis, and death via multiple cell death pathways.

22.4.2.4   Phospholipases

Numerous   phospholipases   (protein-based enzymes) are manipulated via the action of venom components as well as serving as constituents of venom cocktails. In some cell types, mastoparan activates phospholipase C in target cell membranes, which activates several signal transduction pathways that can lead to irreversible cell injury and death via destruction of intracellular calcium homeostasis. Phospholipase A_1 and A_2 are enzymes that digest membrane phospholipids, facilitating loss of membrane integrity, creating chaos within the cell, and ultimately leading to cell death. Enzymatic destruction of cellular membranes can trigger a cascade of events that includes histamine release, inflammation, edema, cytotoxicity, and release of cell-derived phospholipases, which in turn promotes autolysis. The intrinsic phospholipases of venom from *V. mandarinia* nondiscriminately digest human cells, producing mass tissue death dependent on venom concentration.

22.4.3   Salivary venoms

True bugs that produce venomous saliva are predominantly from the family Reduviidae and rely on a salivary composition that is full of digestive enzymes and neurotoxic peptides (Cohen, 1995). Functionally, the saliva is primarily used in prey capture and pre-oral digestion, the latter a condition in which the salivary enzymes liquefy prey tissues prior to ingestion. The enzyme cocktail includes trypsin, α-chymotrypsin, lipases, glucosidases, and phospholipases. In general, insect digestive enzymes do not utilize unique substrates. Thus, any salivary-derived enzyme has the potential to digest similar substrates in human tissues. The presence of phospholipases should be conspicuous from our discussion of wasp venoms, and the same ideas apply here. Likewise, any digestion of cellular membranes will compromise membrane integrity, facilitating leakage of intracellular molecules and flux in both directions. This also means that autolysis can occur from the initial injury resulting from the salivary enzymes.

Some reduviids contain unique neurotoxic peptides that are structurally similar to the powerful and deadly ω-conotoxins from marine snails in the genus *Conus*. The conotoxins modulate the activity of voltage-gated ion channels, including both sodium and calcium channels, resulting in rapid paralysis of prey. Whether the neurotoxic peptides present in saliva operate by the same mechanism is still to be determined (Corzo *et al.*, 2001).

22.4.4   Secretory toxins

Deadly toxins in insect secretions are rare. Most defensive secretions elicit painful yet temporary blisters or lesions of the epidermis but are not lethal. Pederin is a potentially deadly peptide synthesized by rove beetles in the genus *Paederus*. The toxin is a **vesicant**[7] amide possessing two tetrahydropyran rings (Takemura *et al.*, 2002), a structure very similar to the potent cytotoxin psymberin produced by sea sponges in the genus *Psammoncinia*. Pederin production occurs in the hemocoel and requires the endosymbiotic bacteria *Pseudomonas* spp. (Figure 22.13). Active toxin is secreted by larvae during attack or can be transferred from the mother to the eggs to repel predators (Kellner & Dettner, 1996). When making contact with human skin, the toxin induces painful lesions. If pederin is ingested or enters the bloodstream, the peptide blocks mitosis in somatic cells by inhibition of protein and DNA synthesis, culminating in cell death. The same mode of action that makes the toxin lethal also gives it tremendous potential as an anti-tumor/anticancer agent if it can be harnessed to selectively block mitotic events in cancerous cells.

22.4.5   Stinging toxins (non-Hymenoptera)

Several species of urticating caterpillars produce painful but non-lethal venoms. The exceptions to this rule are larvae in the genus *Lonomia* (Family Saturniidae), in which two species synthesize venoms that can evoke death. Venoms of *L. achelous* and *L. oblique* are a complex blend of proteins, peptides, and other components that globally cause diffuse bleeding, renal failure, cerebral damage, and hemorrhage in skin, mucosa, and viscera (Caovilla

& Barros, 2004) (Figure 22.14). Death is an obvious end result depending on the dose of venom injected and body weight of individual envenomated (smaller individuals are more susceptible). Most of the venom components of both species appear to be glycosylated serine proteases[8] (Veiga *et al.*, 2001). Venom from *L. oblique* contains at least four active toxins including phospholipase A_2 and prothrombin activator that contribute to the procoagulant and hemolytic activity (Carrijo-Carvalho & Chudzinski-Tavassi, 2007). In contrast, though the venom of *L. achelous* shares a prothrombin activator in common with *L. oblique*, it lacks phospholipase activity, is somewhat more complex in composition, and stimulates hemorrhage through increased **fibrinolysis** (i.e., digests whole blood clots and fibrin plaques) (Donato *et al.*, 1998).

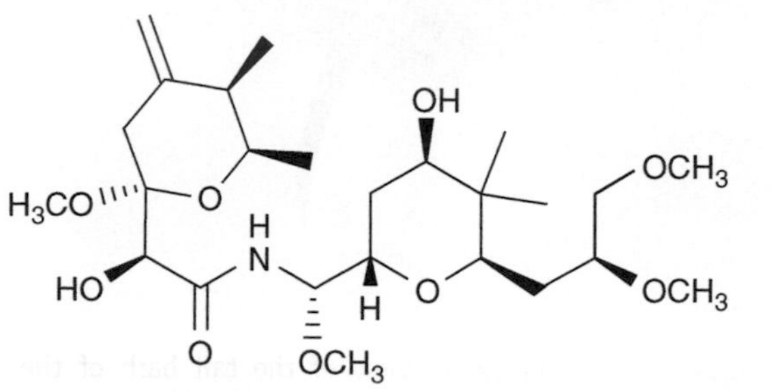

22.5   Non-insect arthropods that cause death

Insects do not own the rights to toxicity. Potent chemical defenses are features that the class Hexapoda shares with many of their arthropod brethren. In fact, several arthropod groups are best known for their ability to inject highly poisonous venoms through bites and stings. Spiders, tarantulas, scorpions, and large menacing centipedes – for lack of a better word – creep people out, and the dangerous varieties, perhaps rightly so, evoke fear. Some of the most deadly animals on the planet are non-insect arthropods, with the ability to deliver a dose of venom in a single bite or sting that may cause severe tissue damage, paralysis, or death in

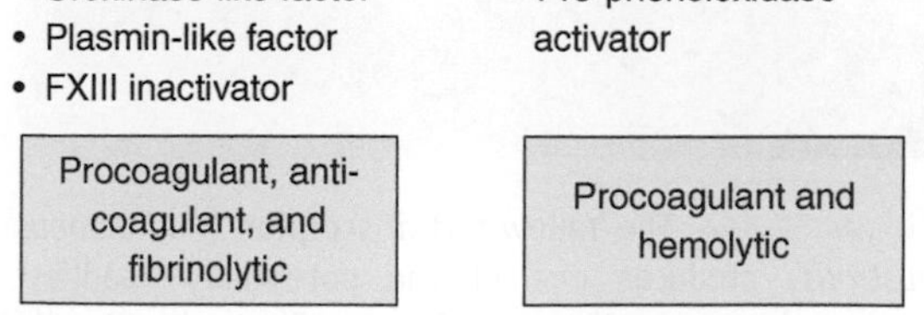

Stinging lepidopteran venoms

Lonomia achelous	*Lonomia oblique*
• Prothrombin activator	• Prothrombin activator
• Factor Xa-like	• Factor X activator
• Factor V activator	• Phospholipase A2
• Factor V inactivator	• Hyaluronidases
• Urokinase-like factor	• Pro-phenoloxidase
• Plasmin-like factor	activator
• FXIII inactivator	
Procoagulant, anti-coagulant, and fibrinolytic	Procoagulant and hemolytic

Figure 22.13 Chemical structure of pederin. *Source:* Edgar181/Wikimedia commons/public domain.

Figure 22.14 Comparison of venoms produced by *Lonomia* species.

just a matter of minutes. The toxicity of such spiders and scorpions, rivals that of the most venomous snake species, truly putting into perspective the rationale for fearing these beasts.

What follows is an examination of some of the most striking (pun intended) non-insect arthropods. The species that will be explored were chosen because they are either among the deadliest of all arthropods or the very mention of their name strikes fear in the heart of humans.

22.5.1 Scorpions

These eight-legged arthropods are from the sub-phylum Chelicerata in the class Arachnida. All species are predatory and depend on aposematic behaviors that display the large **chelae** (pincers) on the **pedipalps** and the curved tail with stinger in aggressive or defensive postures (Figure 22.15). Venom is used for prey capture and can also be employed in chemical defenses. However, many species attempt to conserve their venom for prey-capturing functions by producing a pre-venom that is injected during attack. Often the sting alone, independent of envenomation, is sufficient to ward off a potential predator. For humans, the injection of non-toxic pre-venom triggers a terri-fying set of emotions, since most individuals would have no idea that every scorpion sting is not lethal. At least 25 species of scorpions are known to possess venoms that can kill humans (Polis, 1990). Two of the most notable are pre-sented here.

22.5.1.1 Androctonus australis

Native to desert regions of northern Africa and parts of the Middle East, the yellow fattail scorpion is con-sidered among the most lethal in the world. Venom is reputed to be as toxic as that from the black mamba snake (*Dendroaspis polylepis*). To put this in context, a single black mamba reportedly killed a 7500-pound elephant in Kenya through a single venomous bite (Sheldrick, 2006). This species of scorpion earned its common name from the very large (in relation to body length) and powerful tail, which permits moving objects to be attacked with quick muscular strikes of the stinger. Several human deaths occur each year due to envenomation by *A. australis*.

22.5.1.2 Centruroides sculpturatus

The Arizona bark scorpion is relatively common throughout desert regions of the southwestern United States but is primarily located in the Sonoran Desert. Adults do not burrow and are well suited for withstanding the arid conditions of the desert. However, they tend to avoid the extreme heat of the day, restricting their hunting to nocturnal hours. Consequently, contact with humans is rare. On occasion when they do envenomate a human, intense long-lasting pain occurs that may persist for 2–3 days (Figure 22.16). Localized and temporary numbness, paralysis, and sometimes convulsions are associated with venom injection, but very infre-quently does death result. Reported deaths have most often been associated with small children or adults in poor health. This species is considered the most venomous scorpion species in North America.

Figure 22.15 The Yellow fattail scorpion (*Androctonus australis*) produces one of the potentially deadliest venoms by any arthropod. *Source*: Photo by Quartl/ Wikimedia Commons/CC BY-SA 3.0. https://commons. wikimedia.org/wiki/File:Androctonus_australis_qtl1.jpg.

Figure 22.16 Close up view of the tail barb of the Arizona bark scorpion. *Source*: Photo by JAdams1776/ Wikimedia Commons/public domain. https://commons. wikimedia.org/wiki/File:ScorpionBarb.jpg.

22.5.2   Spiders

The term "arachnophobia" was coined to account for people's fear of spiders. For many, the physical attributes of spiders are what really make them uneasy: eight hairy legs, multiple agglomerate eyes, and the ability to jump (some species). For others, the fact that spiders can inject venom through a bite is sufficient, while still other individuals are freaked out by the slow calculated movement of the legs, perhaps believing that the gaited walk reflects stalking behavior like a cat ready to pounce on its prey. The reality is that movement is more reflective of the lack of extensor muscles beyond the trochanter (leg segment that connects the coxa to the femur) and thus they use hydraulic pressure to extend the legs or other appendages. It is true that nearly all spiders use venom but, as with many other arthropods, venom is used predominantly to subdue prey and in pre-oral digestion. Few species produce a venom cocktail that is harmful to humans, but those that do frequently display warning behaviors (i.e., display fangs) or coloration well in advance of defensive attack. Tarantulas also rely on urticating hairs to defend themselves, throwing the hairs by increasing hemolymph pressure in the abdomen. However, these hairs are not known to be associated with toxins like the urticating caterpillars discussed in Section 22.4.5.

Venom is delivered to prey or during defense through a pair of fangs located at the tip of the **chelicerae** or mouthparts. Since venom is designed for prey capture, the composition of these fluids depends on neurotoxic peptides that can elicit paralysis in other arthropods and an array of digestive enzymes for digestion of the food prior to ingestion. Two of the most harmful species to humans are presented here.

22.5.2.1   Latrodectus mactans

The black widow spider is actually the common name for about 32 species of "widow" spiders in the family Theridiidae. All are venomous, possess some variation of an hourglass shape on the abdominal sterna, and some display aposematic coloration in the form of red and black markings. The southern black widow, *Latrodectus mactans*, is among the best known or infamous species, producing poisonous venom capable of inducing paralysis (Figure 22.17). Venom contains the neu-

Figure 22.17   A female black widow spider, *Latrodectus mactans*, initiating web construction. *Source*: Photo by James Gathany/Wikimedia Commons/public domain. https://commons.wikimedia.org/wiki/File:Black_widow_spider_9854_lores.jpg.

rotoxic protein **latrotoxin,** a high-molecular-mass protein (~130 kDa), of which five different forms are synthesized by *Latrodectus* species. The toxins function presynaptically to release the neurotransmitters acetylcholine, γ-aminobutyric acid (GABA), and norepinephrine, thereby preventing muscle relaxation, a condition termed **tetany.** Prolonged muscle contractions can be painful and lead to muscle cramping in other regions of the body. It is rare for death to result from the bite of any widow species but is possible if appropriate treatment is not received following a single envenomation; if multiple bites occur, acute toxicity can lead to death.

22.5.2.2   Loxosceles reclusa

The brown recluse or violin spider (Family Sicariidae) produces potentially very dangerous venom yet displays no warning coloration or behaviors when startled or facing a potential predatory attack (Figure 22.18.). Rather, the spider relies on nocturnal hunting and avoidance (running away when threatened) as its primary means of defense. When residing indoors, *L. reclusa* hides out in cardboard, piles of clothing, bed sheets on a bed that is not used, shoes, or other areas that generally are not disturbed. Human contact is rare but may occur when an individual slips on a shoe, glove, or clothing that has become the lair for the spider. Adults possess very small fangs, so penetration of most clothing fabrics will not occur. When envenomation does happen, in most instances there are no consequences. However, the venom of the brown recluse contains

Figure 22.18 An adult brown recluse spider, *Loxosceles reclusa*. *Source*: Photo by mattb/Wikimedia Commons/ public domain. https://commons.wikimedia.org/wiki/ File:Loxosceles_reclusa_adult_male_3.jpg.

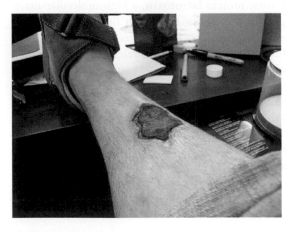

Figure 22.19 Necrotic tissue formed after bite by brown recluse spider. Source: Photo by Jeffrey Rowland/ Wikimedia Commons/CC BY 2.0. https://commons. wikimedia.org/wiki/File:Rowland_recluse_bite.jpg.

a potentially lethal array of enzymes that can cause hemolysis of red blood cells, digestion of cellular membranes, and mass release of histamine and other cellular constituents. In some individuals, the venom is responsible for inducing a condition termed **loxoscelism,** or tissue necrosis. Necrotic lesions may form on the skin (cutaneous necrosis) or be systemic (viscerocutaneous) (Figure 22.19.). The lesions can be very painful and are highly susceptible to secondary infection if proper care and treatment is not received. Generally the conditions are treatable, although death is known to occur in small children (under the age of 7), the elderly, and those with a severely compromised immune system (Wasserman, 2005).

22.5.3   Centipedes

Centipedes are another group of arthropods (Subphylum Myriapoda, Class Chilopoda) that cause humans to get really emotional, largely because of their snake-like appearance (long and cylindrical) and multilegged scurrying behavior. Nearly all species are carnivorous and can be found in a multitude of habitats. However, because they lack a waxy cuticle and some species cannot close their spiracles, centipedes are restricted to moist humid microhabitats. Dark damp basement areas of homes and other buildings mimic such environments, which is where most human encounters occur. When threatened, it is not uncommon for many centipede species to become an aggressor rather than running for cover, and in fact they may run toward the much larger predator or human. Defensive attack relies on a pair of **forcipules,** which are modified legs originating just behind the head to form a pair of pincers or fangs. Venom glands extend through the pincers and open to the outside at the tip of the forcipules. The venom produced is designed for prey capture but is also used in chemical defense, inflicting a painful long-lasting "bite." Since the forcipules are technically not mouthparts, the pincers are not used for biting; rather it is more correct to say pinched or punctured.

The venom of centipedes resembles the composition of many stinging Hymenoptera, particularly that of honey bees. The venom of most species contains an assortment of enzymes including hyaluronidase and phospholipase A (the latter is absent in many *Scolopendra* species), catecholamines like serotonin and histamine, neurotoxic peptides and proteins, and some possess a cardiac toxin (protein). Envenomation can induce pain, swelling, headache, nausea, and fever (Iovcheva *et al.*, 2008; Undheim *et al.*, 2015). Punctures of the skin sometimes lead to secondary bacterial infections and localized necrosis. Death is rare in humans, but when it does occur is often the result of either anaphylactic shock or cardiac irregularities (Bush *et al.*, 2001).

22.6   Implications of deadly insects for forensic entomology

Often overlooked in the discussion of insects in their relationship to forensic entomology is that they can be responsible for death. This should be obvious for any of the insects and other arthropods we have just discussed that deliver toxins or venoms which can be lethal to humans under the right conditions. Arthropod-induced deaths are rare in the United States and Europe, but a small yet significant number occur each year. In some regions of the world, particularly in the tropics, the number of deaths may exceed 1000–2000 annually. Regardless of the frequency of encounters with potential deadly species, it is important to recognize that it is possible for human death to be the result of an interaction with an insect or closely related arthropod. In these instances, no distinction is made between death induced by a toxic compound(s) injected into an individual or the result of anaphylaxis.

How can forensic entomology aid cases in which venomous arthropods are involved? The key is a thorough understanding of the biology of the creatures most likely to be associated with human attacks so that diagnostic bite or sting marks, evidence of arthropod activity, or even distinctive pathologies (e.g., skin lesions or irritations) can be recognized and used to identify the culprit. For example, solitary holes or puncture marks in the skin will narrow the potential suspects to biting Hemiptera or stinging Hymenoptera (and scorpions), whereas paired holes most likely reflect a spider bite, and paired somewhat curved punctures are more suggestive of a rare centipede bite. This is obviously not a comprehensive diagnostic guide but does serve to give you an idea of how bite or sting marks can be used, and when matched up with the life history of species common to a specific region of a particular country may allow the list of candidates to be reduced to just a few species. For species where extensive characterization of allomonal fluids has been performed and/or where significant analyses of immunological responses in envenomated humans has been carried out, antibody testing may reveal the true identity of the beast that caused injury or death. In scenarios in which blood-feeding insects are involved, the victim's DNA signature is present in the captured blood meal (in the insect's gut) and thus can be subjected to molecular analyses to confirm the source of the blood (see Chapter 19 for more details). The latter is an approach still in its infancy but holds promise of revealing several features about the victim, the insect/arthropod, and possibly the timing of their interaction.

Chapter review

Insects that bite, sting or secrete cause fear, loathing and death

- Insects cause fear. At times, this entomological trepidation is not deserved as the fearful people display borderline irrational behaviors toward all multilegged creatures, with unwillingness to entertain the possibility that not all insects are menacing and some, in fact many, are quite useful if not essential to human existence. Fear of insects may not even be linked to any previous event that can account for this loathing of hexapods or their relatives; it is an innate loathing. For others, the hatred is premeditated. In such instances, the foundation for this fear is linked to an earlier interaction, undoubtedly caused by a stinging or biting creature that happened to be stepped on, swatted, or in some other way direct contact was made.

- A small but significant number of insects deserve this fear-laced respect. Why? Because they can be much more aggressive than typical species that are synanthropic with humans, and/or may even produce compounds that are highly toxic. In extreme cases, the interactions can be deadly.

- Venoms and any type of noxious or toxic chemical released in a volatile, secreted, or injected form through biting and stinging are all examples of allomones. Generally speaking, allelochemicals are used for chemical defense, to capture and/or subdue prey, and to aid in reproduction. The functionality and characteristics of allomones varies to a degree based on the mechanism of distribution or release toward the target species. Correspondingly, the effects on humans are also variable including the immediate reaction at the time of attack as well as any long-term consequences, including death.

- Biting and stinging are often aligned with aggressive behavior, and collectively are used to ward off attack by predators or as a preemptive warning to other animals that get too close to either an individual or the colony of social species. Some insects do not even need to use the venom to gain protection; their aposematic coloration or behaviors are sufficient to ward off predators. However, warning alone is not adequate in all cases. Some insects that are very toxic do not even attempt to warn other animals of their potency, lacking any type of aposematic markings or aggression. Rather, such species synthesize incredibly potent toxins or venoms that elicit such intense pain that the recipient animal is completely or temporarily incapacitated. The idea of extremely painful venoms is thought to yield not just an effective defense from predators, particularly vertebrates, but it also should yield long-lasting protection.

- The unfortunate consequence of high potency can be death. This potential is evident with the full gamut of toxin producers, meaning there are species of biting, stinging and secreting insects with the capacity to kill a human under the right conditions.

Insects that cause death

- Insects that bite, sting or secrete noxious and toxic compounds are quite familiar, occur in many regions and, importantly, generally do not cause long-term health concerns for the majority of individuals that have been attacked. A chance encounter with truly dangerous species is another matter altogether. Why? For one, there are not as many species of insects that synthesize lethal compounds. Point number two is that these species occur in relatively isolated regions of the world so that human contact is infrequent. Adding to the isolation is the fact that several of the insects limit toxin production to specific developmental stages.
- Insects that produce lethal compounds released via a bite, sting or secretion are most abundant in the orders Coleoptera, Hemiptera, Hymenoptera, and Lepidoptera. Far and away the most common belong to the aculeate Hymenoptera, namely those that possess an ovipositor modified into a stinger.
- Toxin production is not widespread among the Coleoptera and appears to be restricted to species that spend the vast majority of their time dwelling on or in the upper layers of soil. For most potent species, toxins are proteins or peptides synthesized in the larval and/or adult stages, although at least one species transfers allomones to the eggs, and are released via secretion or compression of the body. Some release non-protein secretions via explosive sprays.
- True bugs use piercing/sucking mouthparts to feed during nymphal and adult stages, displaying herbivory, carnivory, and parasitism. While it is true that parasitic species can serve as vectors of

disease and also stimulate anaphylaxis, none are deadly in terms of production of toxins and venoms. This distinction resides with hemipterans that are carnivorous in the family Reduviidae. Feeding stages subdue their prey by piercing with mouthparts and then immediately pumping the captured food with salivary components. For several species, saliva contains a range of powerful digestive enzymes and neurotoxins, leading to the term venomous saliva.

- The Hymenoptera comprises a wealth of deadly insects in comparison with other insect orders. Nearly all stinging and biting species of significance belong to the aculeate group of the suborder Aprocrita. Deadly toxins are administered exclusively through envenomation during the act of stinging; biting generally has no long-term effect on a large vertebrate. Aposematic coloration and warning behaviors are also common to most of the aculeate Hymenoptera that possess lethal toxins or venoms.
- Several species of caterpillars sequester secondary plant compounds during feeding and retain the toxins throughout larval development. In some instances, toxin sequestration lasts through adulthood and is generally associated with aposematic markings in an attempt to avoid "tasting" by a potential predator. Other species rely on urticating or stinging hairs on caterpillars to thwart attack. Contact with the hairs releases toxins that yield mild to severe reactions in humans.

Human envenomation and intoxication by insect-derived toxins

- What happens between the time of envenomation or intoxication until death ensues? The most immediate responses are independent of the toxins involved and occur at the site of insect attack. Skin damage and irritation can result from direct injury due to mouthpart or stinger penetration, or from reactions to non-toxic constituents found in saliva, venom or secretions. Inoculation of the wounds with bacteria, fungi, or other microorganisms is not uncommon since the stinger or mouthparts of any insect are certainly not sterile. The result of microbial introduction can range from initiation of a mild histamine release that in turn stimulates a localized inflammatory response, to establishment of secondary infection that mobilizes a more intense systemic immunological response. Such reactions may initiate anaphylaxis, a condition more frequently stimulated by venom proteins.

- Localized and systemic pathological changes follow the initial damage caused by penetration or contact with skin and are generally concurrent with the onset of pain. Enzymes or other proteins in the allomonal cocktail and/or derived from human cells, the latter representing a condition akin to autocatalysis, may facilitate the damage. The severity of the injury and target tissues affected are dependent on the toxins involved and mode of entry into the body.

Insects that injure humans rely on chemically diverse venoms and toxins

- Understanding the chemical identity of the toxins utilized by insects in defense and the pathways manipulated to evoke cellular and tissue damage is essential for developing methods of treatment of individuals that are stung, bitten and sprayed, or who make contact with noxious secretions. Much like the situation with deadly snake venoms, the timing of appropriate treatment is critical to limiting the severity of symptoms and damage evoked by deadly insect venoms and toxins. Thus, many of the insects have received a great deal of attention from entomologists, medical personnel, and those engaged in biomedical research and drug development.

- Ant venoms are known to display a broad range of activity, including lethality, paralysis, antimicrobial, phytotoxic, insecticidal, and hemolytic. Such a wide range of properties clearly indicates that multiple constituents exist in the venoms and not all contribute to the pathologies associated with human envenomation. Those of most interest to the human condition are the piperidines from fire ant venoms and the extremely toxic peptides derived from other species.

- The venom composition of solitary and social wasps is very complex, with a chemical diversity that could easily fill a book by itself if sufficient coverage was permitted to discuss the unique chemistries and modes of action. Some of the most important toxins in wasp venoms include kinins, mandaratoxin, mastoparan, and phospholipases.

- True bugs that produce venomous saliva are predominantly from the family Reduviidae and rely on a salivary composition comprising digestive enzymes and neurotoxic peptides. Functionally, the saliva is primarily used in prey capture and pre-oral digestion, the latter a condition in which the salivary enzymes liquefy prey tissues prior to ingestion. The enzyme cocktail includes trypsin, α-chymotrypsin, lipases, glucosidases, and phospholipases. Any salivary-derived enzyme has the potential to digest similar substrates in human tissues. Digestion of cellular membranes will compromise membrane integrity, facilitating leakage of intracellular molecules and flux in both directions. This also means that autocatalysis can occur from the initial injury resulting from the salivary enzymes.

- Deadly toxins in insect secretions are rare. Most defensive secretions elicit painful yet temporary blisters or lesions of the epidermis but are not lethal. Pederin is a potentially deadly peptide synthesized by rove beetles in the genus *Paederus*. Active toxin is secreted by larvae during attack or can be transferred from the mother to the eggs to repel predators. When making contact with human skin, the toxin induces painful lesions. If pederin is ingested or enters the bloodstream, the peptide blocks mitosis in somatic cells by inhibition of protein and DNA synthesis, culminating in cell death.

- Several species of urticating caterpillars produce painful but non-lethal venoms. The exceptions to this rule are larvae in the genus *Lonomia* (Family Saturniidae), in which two species synthesize venoms that can evoke death. Venoms of *L. achelous* and *L. oblique* are a complex blend of proteins, peptides, and other components that globally cause diffuse bleeding, renal failure, cerebral damage, and hemorrhage in skin, mucosa, and viscera.

Non-insect arthropods that cause death

- Insects do not own the rights to toxicity. Potent chemical defenses are features that the class Insecta shares with many of their arthropod brethren. In fact, several arthropod groups are best known for their ability to inject highly poisonous venoms through bites and stings. Spiders, tarantulas, scorpions, and large menacing centipedes – for lack of a better word – creep people out, and the dangerous varieties, perhaps rightly so, evoke fear.

Some of the deadliest animals on the planet are non-insect arthropods, with the ability to deliver a dose of venom in a single bite or sting that may cause severe tissue damage, paralysis, or death in just a matter of minutes.

- Scorpions are eight-legged arthropods from the sub-phylum Chelicerata in the class Arachnida. All species are predatory and depend on aposematic behaviors that display the large chelae on the pedipalps and curved tail with stinger in aggressive or defensive postures. Venom is used for prey capture and can also be employed in chemical defenses. However, many species attempt to conserve their venom for prey-capturing functions by producing a pre-venom that is injected during attack.
- The physical attributes of spiders are what really make them uneasy: eight hairy legs, multiple agglomerate eyes, and the ability to jump. For others, the fact that spiders can inject venom through a bite is sufficient for fear, while still other individuals are freaked out by the slow calculated movement of the legs, perhaps believing that the gaited walk reflects stalking behavior like a cat ready to pounce on its prey. It is true that nearly all spiders use venom but, as with many other arthropods, venom is used predominantly to subdue prey and in pre-oral digestion. Few species produce a venom cocktail that is harmful to humans, but those that do frequently display warning behaviors or coloration well in advance of defensive attack.
- Centipedes are another group of arthropods that cause humans to get really emotional, largely because of their snake-like appearance and multi-legged scurrying behavior. Nearly all species are carnivorous and can be found in a multitude of habitats. Defensive attack relies on a pair of forcipules, which are modified legs originating just behind the head to form a pair of pincers or fangs. Venom glands extend through the pincers and open to the outside at the tip of the forcipules. The venom produced is designed for prey capture but is also used in chemical defense, inflicting a painful long-lasting "bite."

Implications of deadly insects for forensic entomology

- Often overlooked in the discussion of insects in their relationship to forensic entomology is that they can be responsible for death. Arthropod-induced deaths are rare in the United States and Europe, but a small yet significant number occur each year. In some regions of the world, particularly in the tropics, the number of deaths may exceed 1000–2000 annually. Regardless of the frequency of encounters with potential deadly species, it is important to recognize that it is possible for human death to be the result of an interaction with an insect or closely related arthropod.
- How can forensic entomology aid cases in which venomous arthropods are involved? The key is a thorough understanding of the biology of the creatures most likely to be associated with human attacks so that diagnostic bite or sting marks, evidence of arthropod activity, or even distinctive pathologies can be recognized and thus identify the culprit. In scenarios in which blood-feeding insects are involved, the victim's DNA signature is present in the insect's gut, and this can be subjected to molecular analyses to confirm the source of the blood.

Test your understanding

Level 1: knowledge/comprehension

1. Define the following terms:

   (a) allomone
   (b) anaphylaxis
   (c) envenomation
   (d) forcipules
   (e) vesicant
   (f) aculeate Hymenoptera.

2. Match the terms (i–vi) with the descriptions (a–f).

(a) Active ingredient in fire ant venom	(i) Anaphylaxis
(b) Digestion of blood clots by venom	(ii) Glycosides
(c) Potent peptide in many vespid venoms	(iii) Fibrinolysis
(d) Tissue swelling	(iv) Mastoparan
(e) Systemic immunological response to insect proteins	(v) Piperidine
(f) Common plant-sequestered toxins	(vi) Edema

3. Describe some of the common pathological changes in humans following envenomation by vespid wasps.

Level 2: application/analysis

1. Discuss why potent venoms (meaning potentially lethal) are necessary for some arthropod species in terms of their own defense. How does this fit into the context of other species that produce less dangerous toxins yet survive, i.e., their chemical defenses seem to aid in protection from predation without being deadly.
2. Mastoparan is a toxic peptide produced by several species of wasps in the family Vespidae. Interestingly, the toxin is reported to affect a wide range of cell types, inducing different cellular responses. Explain what can account for the different modes of action of the same peptide.
3. Some species that produce potent venoms use aposomatic behavior or coloration to warn would be attack, yet others do not. Explain why it is beneficial for a deadly insect to offer warning before an attack.
4. Provide examples of trace evidence that a death investigator should look for if the deceased is believed to have died to insect envenomation.

Level 3: synthesis/evaluation

1. Explain why anaphylaxis is more likely to occur with honey bee or vespid venoms than with secretions from blister beetles or rove beetles.
2. With several of the dangerous insects discussed, potentially lethal toxins are produced for chemical defense. Would such allomonal secretions be expected to induce anaphylaxis in humans? Explain why or why not.

Notes

1. From Palma (2006).
2. Yellowjacket is the common name given to vespid species in the genera *Vespula* and *Dolichovespula* found throughout North America. They become more aggressive in late summer to early fall, frequently stinging and biting humans, but rarely causing death.
3. Many species rely on a pre-venom to inject when threatened as the initial means to ward off attack.
4. A quinone is a type of organic compound derived from an aromatic precursor like benzene. A variety of quinones are used in the defensive secretions of several arthropod animals.
5. Amphipathic molecules generally have opposing regions that display hydrophobic and hydrophilic properties, sometimes indicating polar versus nonpolar qualities. The polycationic nature of venom peptides refers to the cationic characteristic of the molecule whereby the peptide has more than one positive charge.
6. The N-terminus refers to the amino-terminal end or start of a peptide or protein chain, where the terminal amino acid has a free amine group (NH_2). At the opposite end of the chain is the carboxyl or C-terminus, where the terminal amino acid has a free carboxyl group (COOH).
7. Vesicant is a blister agent, and in the case of pederin refers to formation of painful skin lesions following contact.
8. Serine proteases are enzymes that cleave peptide bonds in proteins where a serine amino acid is recognized in the active site of the enzyme.

References cited

Abe, T., Kawai, N. & Niwa, A. (1982) Purification and properties of a presynaptically acting neurotoxin, mandaratoxin, from hornet (*Vespa mandarinia*). *Biochemistry* 21: 1693–1697.

Amino, R., Tanaka, A.S. & Schenkman, S. (2001) Triapsin, an unusual activatable serine protease from the saliva of the hematophagous vector of Chagas' disease *Triatoma infestans* (Hemiptera: Reduviidae). *Insect Biochemistry and Molecular Biology* 31: 465–472.

Blum, M.S., Walker, R.J., Callahan, P.S. & Novak, A.F. (1958) Chemical, insecticidal, and antibiotic properties of fire ant venom. *Science* 128: 306–307.

Bush, S.P., King, B.O., Norris, R.L. & Stockwell, S.A. (2001) Centipede envenomation. *Wilderness & Environmental Medicine* 12: 93–99.

Caovilla, J.J. & Barros, E.J.G. (2004) Efficacy of two different doses of antilonomic serum in the resolution of hemorrhagic syndrome resulting from envenoming by *Lonomia oblique* caterpillars: a randomized controlled trial. *Toxicon* 43: 811–818.

Carrijo-Carvalho, L.C. & Chudzinski-Tavassi, A.M. (2007) The venom of the *Lonomia* caterpillar: an overview. *Toxicon* 49: 741–757.

Cohen, A.C. (1995) Extra-oral digestion in predaceous terrestrial Arthropoda. *Annual Review of Entomology* 40: 85–103.

Corzo, G., Adachi-Akahane, S., Nagao, T., Kusui, Y. & Nakajima, T. (2001) Novel peptides from assassin bugs (Hemiptera: Reduviidae): isolation, chemical and biological characterization. *FEBS Letters* 499: 256–261.

Deslippe, R.J. & Guo, Y.-J. (2000) Venom alkaloids of fire ants in relation to worker size and age. *Toxicon* 38: 223–232.

Diaz, J.H. (2005) The evolving global epidemiology, syndrome classification, management, and prevention of caterpillar envenoming. *American Journal of Tropical Medicine and Hygiene* 72: 347–357.

Donato, J.L., Moreno, R.A., Hyslop, S., Duarte, A., Antunes, E., Le Bonniec, B.F., Rendu, F. & De

Nucci, G. (1998) *Lonomia oblique* caterpillar trigger human blood coagulation via activation of factor X and prothrombin. *Thrombosis and Haemotasis* 79: 539–542.

Goddard, J. (1999) Skin lesions produced by arthropods. In: W.H. Robinson, F. Rettich & G.W. Rambo (eds) *Proceedings of the Third International Conference on Urban Pests*, pp. 231–234. Graficke Zavody, Russia.

Golden, D. B. (2015) Anaphylaxis to insect stings. *Immunology and Allergy Clinics* 35: 287–302.

Higashijima, T., Uzu, S., Nakajima, T. & Ross, E.M. (1988) Mastoparan, a peptide toxin from wasp venom, mimics receptors by activating GTP-binding regulatory proteins (G proteins). *Journal of Biological Chemistry* 263: 6491–6494.

Iovcheva, M., Zlateva, S., Marinov, P. & Sabeva, Y. (2008) Toxoallergic reactions after a bite from Myriapoda, Genus *Scolopendra* in Varna region during the period 2003–2007. *Journal of IMAB-Annual Proceedings* 1: 79–82.

Kellner, R.L. (2002) Molecular identification of an endosymbiotic bacterium associated with pederin biosynthesis in *Paederus sabaeus* (Coleoptera: Staphylinidae). *Insect Biochemistry and Molecular Biology* 32: 389–395.

Kellner, R.L.L. & Dettner, K. (1996) Differential efficacy of toxic pederin in deterring potential arthropod predators of *Paederus* (Coleoptera: Staphylinidae) offspring. *Oecologia* 107: 293–300.

Klotz, J.H., Klotz, S.A. & Pinnas, J.L. (2009) Animal bites and stings with anaphylactic potential. *Journal of Emergency Medicine* 36: 148–156.

Konno, K., Kazuma, K. & Nihei, K.I. (2016) Peptide toxins in solitary wasp venoms. *Toxins* 8(4): 114.

Lai, R. & Liu, C. (2010). Bioactive peptides and proteins from wasp venoms. In: R.M. Kini, K.J. Clemestson, F.S. Markland, M.A. McLane & T. Morita (eds) *Toxins and Hemostasis*, pp. 83–95. Springer, Dordrecht.

Levesque, L., Drapeau, G., Grose, J.H., Rioux, F. & Marceau, F. (1993) Vascular mode of action of kinin B_1 receptors and development of a cellular model for the investigation of these receptors. *British Journal of Pharmacology* 109: 1254–1262.

Lind, N.K. (1982) Mechanism of action of fire ant (*Solenopsis*) venoms. I. Lytic release of histamine from mast cells. *Toxicon* 20: 831–840.

Mitpuangchon, N., Nualcharoen, K., Boonrotpong, S. & Engsontia, P. (2021) Identification of novel toxin genes from the stinging nettle caterpillar *Parasa lepida* (Cramer, 1799): insights into the evolution of Lepidoptera toxins. *Insects* 12(5): 396.

Moreno, M. & Giralt, E. (2015) Three valuable peptides from bee and wasp venoms for therapeutic and biotechnological use: melittin, apamin and mastoparan. *Toxins* 7: 1126–1150.

Palma, M.S. (2006) Insect venom peptides. In: A. Kastin (ed.) *The Handbook of Biologically Active Peptides*, pp. 409–416. Academic Press, San Diego, CA.

Pluzhnikov, K., Nosyreva, E., Shevchenko, L., Kokoz, Y., Schmalz, D., Hucho, F. & Grishin, E. (1999) Analysis of ectatomin action on cell membranes. *European Journal of Biochemistry* 262: 501–506.

Polis, G.A. (1990) *The Biology of Scorpions*. Stanford University Press, Palo Alto, CA.

Schmidt, J.O., Blum, M.S. & Overal, W.L. (1983) Hemolytic activities of stinging insect venoms. *Archives of Insect Biochemistry and Physiology* 1: 155–160.

Schmidt, J.O., Yamane, S., Matsuura, M. & Starr, C.K. (1986) Hornet venoms: lethalities and lethal capacities. *Toxicon* 24: 950–954.

Sheldrick, D. (2006) *Elephants with broken hearts. Mail Online*, August 16, 2006. Available at http://www.dailymail.co.uk/news/article-400818/Elephants-broken-hearts.html.

Starr, C.K. (1985) A simple pain scale for field comparison of hymenopteran stings. *Journal of Entomological Science* 20: 225–232.

Takemura, T., Nishii, Y., Takahashi, S., Kobayashi, J. & Nakata, T. (2002) Total synthesis of pederin, a potent insect toxin: the efficient synthesis of the right half, (+)-benzoylpedamide. *Tetrahedron* 58: 6359–6365.

Touchard, A., Aili, S.R., Fox, E.G.P., Escoubas, P., Orivel, J., Nicholson, G.M. & Dejean, A. (2016) The biochemical toxin arsenal from ant venoms. *Toxins* 8(1): 30.

Undheim, E.A., Fry, B.G. & King, G.F. (2015) Centipede venom: recent discoveries and current state of knowledge. *Toxins* 7(3): 679–704.

Veiga, A.B.G., Blotchtein, B. & Guimaraes, J.A. (2001) Structures involved in production, secretion and injection of the venom produced by the caterpillar *Lonomia obliqua* (Lepidoptera: Saturniidae). *Toxicon* 39: 1343–1351.

Vetter, R.S., Visscher, P.K. & Camazine, S. (1999) Mass envenomation by honey bees and wasps. *Western Journal of Medicine* 170: 223–227.

Waldbauer, G. (2000) *Millions of Monarchs, Bunches of Beetles: How Bugs Find Strength in Numbers*. Harvard University Press, Cambridge, MA.

Wasserman, G. (2005) Bites of the brown recluse spider. *New England Journal of Medicine* 352: 2029–2030.

Zibaee, A., Hoda, H. & Fazeli-Dinan, M. (2012) Role of proteases in extra-oral digestion of a predatory bug, *Andrallus spinidens*. *Journal of Insect Science* 12: 51. https://doi.org/10.1673/031.012.5101.

Supplemental reading

Chen, L. & Fadamiro, H.Y. (2009) Re-investigation of venom chemistry of *Solenopsis* fire ants. I. Identification of novel alkaloids in *S. richteri*. *Toxicon* 53: 469–478.

dos Santos Pinto, J.R., Fox, E.G., Saidemberg, D.M., Santos, L.D., da Silva Menegasso, A.R., Costa-Manso, E., Machado, E.A., Bueno, O.C. & Palma, M.S. (2012) Proteomic view of the venom from the fire ant

Solenopsis invicta Buren. *Journal of Proteome Research* 11: 4643–4653.

Edwards, J.S. (1961) The action and composition of the saliva of an assassin bug *Platymeris rhadamanthus* Gaerst. (Hemiptera: Reduviidae). *Journal of Experimental Biology* 38: 61–77.

Eisner, T., Eisner, M. & Siegler, M. (2007) *Secret Weapons: Defenses of Insects, Spiders, Scorpions, and Other Many-legged Creatures*. Belknap Press, Cambridge, MA.

Haight, K.L. & Tschinkel, W.R. (2003) Patterns of venom synthesis and use in the fire ant, *Solenopsis invicta*. *Toxicon* 42: 673–682.

Klotz, J.H., Pinnas, J.L., Klotz, S.A. & Schmidt, J.O. (2009) Anaphylactic reactions to arthropod bites and stings. *American Entomologist* 55: 134–139.

Malaque, C.M., Andrade, L., Madalosso, G., Tomy, S., Tavares, F.L. & Seguro, A.C. (2006) Short report: a case of hemolysis resulting from contact with a *Lonomia* caterpillar in Southern Brazil. *American Journal of Tropical Medicine and Hygiene* 74: 807–809.

Nolasco, M., Biondi, I., Pimenta, D.C., & Branco, A. (2018) Extraction and preliminary chemical characterization of the venom of the spider wasp *Pepsis decorata* (Hymenoptera: Pompilidae). *Toxicon* 150: 74–76.

Sahayaraj, K. & Vinothkana, A. (2011) Insecticidal activity of venomous saliva from *Rhynocoris fuscipes* (Reduviidae) against *Spodoptera litura* and *Helicoverpa armigera* by microinjection and oral administration. *Journal of Venomous Animals and Toxins Including Tropical Diseases* 17: 486–490.

Schmidt, J.O. (1982) Biochemistry of insect venoms. *Annual Review of Entomology* 27: 339–368.

Schmidt, J.O. (2016) *The Sting of the Wild*. Johns Hopkins University Press, Baltimore, MD.

Silav-Cardosa, L., Caccin, P., Magnbosco, A., Patron, M., Targino, M., Fully, A., Oliveira, G.A., Pereira, M.H., das Gracas, M., So Carmo, G.T., Souza, A.S., Silva-Neto, M.A.C., Montecucco, C. & Atella, G.C. (2010) Paralytic activity of lysophosphatidylcholine from saliva of the waterbug *Belostoma anurum*. *Journal of Experimental Biology* 213: 3305–3310.

Undheim, E.A. & Jenner, R.A. (2021) Phylogenetic analyses suggest centipede venom arsenals were repeatedly stocked by horizontal gene transfer. *Nature Communications* 12(1): 1–14.

Waldbauer, G. & Nardi, J. (2012) *How To Not Be Eaten: The Insects Fight Back*. University of California Press, Los Angeles, CA.

Walker, A.A., Hernández-Vargas, M.J., Corzo, G., Fry, B.G. & King, G.F. (2018) Giant fish-killing water bug reveals ancient and dynamic venom evolution in Heteroptera. *Cellular and Molecular Life Sciences* 75(17): 3215–3229.

Additional resources

Animal Venom Research International: http://usavri.org/

Australian Venom Research Unit: http://www.avru.org/compendium/biogs/A000088b.htm

Centers for Disease Control and Prevention: http://www.cdc.gov/niosh/topics/insects/

International Society on Toxinology: http://toxinology.org/

Journal of Venom Research: http://www.libpubmedia.co.uk/JVR/JVRHome.htm

National Natural Toxin Research Center: http://www.ntrc.tamuk.edu/venomlist.php

Saudi National Antivenom and Vaccine Production Center: http://antivenom-center.com/

Society for Invertebrate Pathology: http://sipweb.org/

Toxicon: http://www.journals.elsevier.com/toxicon/

Toxins: https://www.mdpi.com/journal/toxins

West Texas Poison Center: http://www.poisoncenter.org/poisonous-critters

Chapter 23
Professional standards and ethics

Overview

Insects can contribute important information for determining the outcome of a variety of civil and criminal cases. However, the mere presence of insects at a crime scene or other location does not automatically mean that the insects will be accepted as physical evidence in a court of law. Rules and regulations exist for assessing the admissibility of scientific evidence. That determination, or the admissibility of entomological evidence, falls under the jurisdiction of the trial judge. Arguably, the same is true with "who" can present or provide analysis of insect evidence in the judicial system. Declaring oneself an expert in forensic entomology or some other area of forensic science does not make it true. Most forensic science disciplines specify standards for certification as an expert. Similarly, standard procedures and codes of conduct have been developed to promote high quality and ethically sound casework. Although the establishment of professional standards does not guarantee that high quality casework is performed, it does provide a framework for assessing the quality of work and comparing the efforts among and between individuals involved in expert forensic analyses. This chapter explores the professional standards and ethical codes commonly put in place in forensic science disciplines, with

special consideration given to recommendations made by the National Academy of Sciences and the National Commission on Forensic Science. Attention will also be given to examining the recommended standards associated with forensic entomology practice in North America and Europe.

The big picture

- General standards and recommendations for forensic science disciplines.
- Guidelines for the practice of forensic entomology.
- Expectations for forensic entomology technicians.
- Ethical considerations for involvement in casework and providing testimony.

23.1 General standards and recommendations for forensic science disciplines

The application of science or any discipline to legal matters in the judicial system requires meeting-specific standards and practices to be allowed do so. This is true for scientific evidence and techniques to be admissible in court, as well as for

The Science of Forensic Entomology, Second Edition. David B. Rivers and Gregory A. Dahlem.
© 2023 John Wiley & Sons Ltd. Published 2023 by John Wiley & Sons Ltd.
Companian website: www.wiley.com/go/forensicentomology2

individuals to be deemed experts within a given discipline. In the United States, federal rules for admissibility of scientific evidence and for establishing the qualifications of an expert have been established through the Frye standard and then superseded by the *Daubert v. Merrell Dow Pharmaceuticals* ruling in 1993. In *Frye v. United States* in 1923, the court ruled that expert testimony must be based on scientific methods that are sufficiently established and tested (Bernstein, 2000). The court specifically ruled:

> Just when a scientific principle or discovery crosses the line between the experimental and demonstrable stages is difficult to define. Somewhere in this twilight zone the evidential force of the principle must be recognized, and while the courts will go a long way in admitting expert testimony deduced from a well-recognized scientific principle or discovery, the thing from which the deduction is made must be sufficiently established to have gained general acceptance in the particular field to which it belongs.

The Frye standard established general rules of admissibility regarding scientific evidence. Essentially, the ruling stated that expert opinion or testimony based on a scientific technique is admissible in court only when the technique or method is considered generally accepted by the relevant scientific community (*Frye v. United States*, 293 F. 1013, D.C. Cir. 1923). It is important to note that as of 1993, the Daubert standard is mandatory in federal courts. However, this ruling is not prescriptive for state courts; seven states still follow the Frye standard for admissibility of scientific evidence. The Daubert standard established the following guidelines for admissibility of scientific evidence and expert testimony in federal courts:

1. Under Rule 702 of the Federal Rules of Evidence, the task of assuring that scientific expert testimony truly proceeds from scientific knowledge rests on the trial judge. This ruling puts the responsibility, burden, or power in the courts to decide what is and is not admissible scientific evidence. To a degree, Rule 702 creates a paradox in that a non-expert (the judge) on the scientific matter at hand decides on whether it should be included in the case. Of course the trial judge is an expert on the rules governing the admissibility of scientific evidence.
2. The trial judge must ensure that the expert's testimony is relevant to the task at hand and that it rests on a reliable foundation (*Daubert v.*

Merrell Dow Pharmaceuticals Inc., 1993). The trial judge must find it more likely than not that the expert's methods are reliable and reliably applied to the facts at hand. Reliability of the technique or method is most often established through peer-reviewed publication, as well as prior use (precedent) in other cases.
3. A conclusion will qualify as scientific knowledge if the proponent can demonstrate that it is the product of sound scientific methodology derived from the scientific method. The latter statement should stand out from our discussion of the scientific method in Chapter 1: Every scientific discipline does not rely on the scientific method.
4. The court defined scientific methodology as the process of formulating hypotheses and then conducting experiments to prove or falsify the hypothesis. This again points to the centrality of the scientific method in determining the admissibility of scientific evidence. A set of illustrative factors are used in the determination of whether these criteria are met:
   a. The theory or technique employed by the expert is generally accepted in the scientific community;
   b. The theory or technique has been subjected to peer review and publication;
   c. The theory or technique can be and has been tested;
   d. The known or potential rate of error is acceptable;
   e. The research was conducted independent of the particular litigation or dependent on an intention to provide the proposed testimony.

The question of acceptable or known error rates is especially problematic for some forensic disciplines. Why? For many, there is no means for determining error rates. Or at least, no agreement has been reached on appropriate statistical measures or models for evaluating certain types of data or for estimating potential error. The comparative sciences (e.g., bloodstain pattern analysis, fingerprints, question documents, tire, and shoe tread) were singled out by the NAS Report (2009) for lacking the ability to estimate error during analyses. Forensic entomology is not immune to this problem.

Under Rule 702 of the Federal Rules of Evidence, if scientific, technical, or other specialized knowledge will assist the trier of fact to understand the evidence or to determine a fact in issue, a witness qualified as an expert by knowledge, skill, experience, training, or education, may testify thereto in the form of an opinion or otherwise, if (i) the testimony is based upon sufficient facts or data, (ii) the testimony is the product of reliable principles and methods, and (iii) the witness has applied the principles and methods reliably to the facts of the case. Does that make perfect sense? Probably not. The language is typical of many law documents and often requires interpretation. In lay person's terms, a witness who is qualified as an expert based on education, skill, knowledge, or training may testify on the matter provided the person is an expert in an area relevant to the issue being considered and that the testimony is based on data or facts derived from appropriate scientific methods or techniques, and that the techniques were used appropriately. Ultimately, the trial judge makes this determination. Hall (2020) provides an excellent review of how the Frye and Daubert standards apply directly to forensic entomology.

How do these statutes impact the training and education in the forensic science subdisciplines? Obviously, there should be a direct connection, especially for preparation of individuals to be practitioners in a given discipline. In other words, professional standards should exist for the education and training to be a forensic practitioner, as well as to govern the practice or application of the discipline by an individual to perform casework. Does this type of linkage actually exist? Well, that has been the subject of controversy in the United States since the publication of the National Academy of Science's 2009 report "Strengthening Forensic Science in the United States: A Path Forward" (NAS, 2009). The report identified a series of issues associated with forensic science, including a lack of scientific standards for most subdisciplines, no or poor regulations governing forensic investigations and analyses, and the absence of centralized oversight regarding policy and practice of forensic science in the United States. The National Research Council (NRC) developed recommendations to address the problems, with the specific goal of improving the use of forensic science analyses in examination of evidence (Table 23.1). Among the most significant recommendations offered by the NRC was the need to remodel training paradigms for preparation of forensic professionals (NAS Report, 2009). This idea was built on the premise that individuals cannot continue to be trained the same way as before and expect differences in how forensic science is practiced.

Table 23.1 Recommendations for standard practices and improvements in the forensic sciences.

Recommendations

#1: To promote the development of forensic science into a mature field of multidisciplinary research and practice, founded on the systematic collection and analysis of relevant data, Congress should establish and appropriate funds for an independent federal entity, the National Institute of Forensic Science (NIFS). NIFS should have a full-time administrator and an advisory board with expertise in research and education, the forensic science disciplines, physical and life sciences, forensic pathology, engineering, information technology, measurements and standards, testing and evaluation, law, national security, and public policy.

#2: The National Institute of Forensic Science (NIFS), after reviewing established standards such as ISO 17025, and in consultation with its advisory board, should establish standard terminology to be used in reporting on and testifying about the results of forensic science investigations. Similarly, it should establish model laboratory reports for different forensic science disciplines and specify the minimum information that should be included. As part of the accreditation and certification processes, laboratories and forensic scientists should be required to utilize model laboratory reports when summarizing the results of their analyses.

#3: Research is needed to address issues of accuracy, reliability, and validity in the forensic science disciplines. The National Institute of Forensic Science (NIFS) should competitively fund peer-reviewed research in the following areas:
   (a) Studies establishing the scientific bases demonstrating the validity of forensic methods.
   (b) The development and establishment of quantifiable measures of the reliability and accuracy of forensic analyses. Studies of the reliability and accuracy of forensic techniques should reflect actual practice on realistic case scenarios, averaged across a representative sample of forensic scientists and laboratories. Studies also should establish the limits of reliability and accuracy that analytic methods can be expected to achieve as the conditions of forensic evidence vary. The research by which measures of reliability and accuracy are determined should be peer reviewed and published in respected scientific journals.
   (c) The development of quantifiable measures of uncertainty in the conclusions of forensic analyses.
   (d) Automated techniques capable of enhancing forensic technologies.

#4: To improve the scientific bases of forensic science examinations and to maximize independence from or autonomy within the law enforcement community, Congress should authorize and appropriate incentive funds to the National Institute of Forensic Science (NIFS) for allocation to state and local jurisdictions for the purpose of removing all public forensic laboratories and facilities from the administrative control of law enforcement agencies or prosecutors' offices.

(Continued)

Table 23.1   (Continued)

#5: The National Institute of Forensic Science (NIFS) should encourage research programs on human observer bias and sources of human error in forensic examinations. Such programs might include studies to determine the effects of contextual bias in forensic practice (e.g., studies to determine whether and to what extent the results of forensic analyses are influenced by knowledge regarding the background of the suspect and the investigator's theory of the case). In addition, research on sources of human error should be closely linked with research conducted to quantify and characterize the amount of error. Based on the results of these studies, and in consultation with its advisory board, NIFS should develop standard operating procedures (that will lay the foundation for model protocols) to minimize, to the greatest extent reasonably possible, potential bias and sources of human error in forensic practice. These standard operating procedures should apply to all forensic analyses that may be used in litigation.

#6: To facilitate the work of the National Institute of Forensic Science (NIFS), Congress should authorize and appropriate funds to NIFS to work with the National Institute of Standards and Technology (NIST), in conjunction with government laboratories, universities, and private laboratories, and in consultation with Scientific Working groups, to develop tools for advancing measurement, validation, reliability, information sharing, and proficiency testing in forensic science and to establish protocols for forensic examinations, methods, and practices. Standards should reflect best practices and serve as accreditation tools for laboratories and as guides for the education, training, and certification of professionals. upon completion of its work, NIST and its partners should report findings and recommendations to NIFS for further dissemination and implementation

#7: Laboratory accreditation and individual certification of forensic science professionals should be mandatory, and all forensic science professionals should have access to a certification process. In determining appropriate standards for accreditation and certification, the National Institute of Forensic Science (NIFS) should take into account established and recognized international standards, such as those published by the International Organization for Standardization (ISO). No person (public or private) should be allowed to practice in a forensic science discipline or testify as a forensic science professional without certification. Certification requirements should include, at a minimum, written examinations, supervised practice, proficiency testing, continuing education, recertification procedures, adherence to a code of ethics, and effective disciplinary procedures. All laboratories and facilities (public or private) should be accredited, and all forensic science professionals should be certified, when eligible, within a time period established by NIFS.

#8: Forensic laboratories should establish routine quality assurance and quality control procedures to ensure the accuracy of forensic analyses and the work of forensic practitioners. Quality control procedures should be designed to identify mistakes, fraud, and bias; confirm the continued validity and reliability of standard operating procedures and protocols; ensure that best practices are being followed; and correct procedures and protocols that are found to need improvement.

#9: The National Institute of Forensic Science (NIFS), in consultation with its advisory board, should establish a national code of ethics for all forensic science disciplines and encourage individual societies to incorporate this national code as part of their professional code of ethics. Additionally, NIFS should explore mechanisms of enforcement for those forensic scientists who commit serious ethical violations. Such a code could be enforced through a certification process for forensic scientists.

#10: To attract students in the physical and life sciences to pursue graduate studies in multidisciplinary fields critical to forensic science practice, Congress should authorize and appropriate funds to the National Institute of Forensic Science (NIFS) to work with appropriate organizations and educational institutions to improve and develop graduate education programs designed to cut across organizational, programmatic, and disciplinary boundaries. To make these programs appealing to potential students, they must include attractive scholarship and fellowship offerings. Emphasis should be placed on developing and improving research methods and methodologies applicable to forensic science practice and on funding research programs to attract research universities and students in fields relevant to forensic science. NIFS should also support law school administrators and judicial education organizations in establishing continuing legal education programs for law students, practitioners, and judges.

#11: To improve medicolegal death investigation:
  (a) Congress should authorize and appropriate incentive funds to the National Institute of Forensic Science (NIFS) for allocation to states and jurisdictions to establish medical examiner systems, with the goal of replacing and eventually eliminating existing coroner systems. Funds are needed to build regional medical examiner offices, secure necessary equipment, improve administration, and ensure the education, training, and staffing of medical examiner offices. Funding could also be used to help current medical examiner systems modernize their facilities to meet current Centers for Disease Control and Prevention-recommended autopsy safety requirements.
  (b) Congress should appropriate resources to the National Institutes of Health (NIH) and NIFS, jointly, to support research, education, and training in forensic pathology. NIH, with NIFS participation, or NIFS in collaboration with content experts, should establish a study section to establish goals, to review and evaluate proposals in these areas, and to allocate funding for collaborative research to be conducted by medical examiner offices and medical universities. In addition, funding, in the form of medical student loan forgiveness and/or fellowship support, should be made available to pathology residents who choose forensic pathology as their specialty.
  (c) NIFS, in collaboration with NIH, the National Association of Medical Examiners, the American Board of Medicolegal Death Investigators, and other appropriate professional organizations, should establish a Scientific Working group (SWG) for forensic pathology and medicolegal death investigation. The SWG should develop and promote standards for best practices, administration, staffing, education, training, and continuing education for competent death scene investigation and postmortem examinations. Best practices should include the utilization of new technologies such as laboratory testing for the molecular basis of diseases and the implementation of specialized imaging techniques.

Table 23.1   (Continued)

(d) All medical examiner offices should be accredited pursuant to NIFS-endorsed standards within a timeframe to be established by NIFS.

(e) All federal funding should be restricted to accredited offices that meet NIFS-endorsed standards or that demonstrate significant and measurable progress in achieving accreditation within prescribed deadlines.

(f) All medicolegal autopsies should be performed or super-vised by a board certified forensic pathologist. This requirement should take effect within a timeframe to be established by NIFS, following consultation with governing state institutions.

#12: Congress should authorize and appropriate funds for the National Institute of Forensic Science (NIFS) to launch a new broad-based effort to achieve nationwide fingerprint data interoperability. To that end, NIFS should convene a task force comprising relevant experts from the National Institute of Standards and Technology and the major law enforcement agencies (including representatives from the local, state, federal, and, perhaps, international levels) and industry, as appropriate, to develop:

(a) standards for representing and communicating image and minutiae data among Automated Fingerprint Identification Systems. Common data standards would facilitate the sharing of fingerprint data among law enforcement agencies at the local, state, federal, and even international levels, which could result in more solved crimes, fewer wrongful identifications, and greater efficiency with respect to fingerprint searches; and

(b) baseline standards – to be used with computer algorithms – to map, record, and recognize features in fingerprint images, and a research agenda for the continued improvement, refinement, and characterization of the accuracy of these algorithms (including quantification of error rates).

#13: Congress should provide funding to the National Institute of Forensic Science (NIFS) to prepare, in conjunction with the Centers for Disease Control and Prevention and the Federal Bureau of Investigation, forensic scientists and crime scene investigators for their potential roles in managing and analyzing evidence from events that affect homeland security, so that maximum evidentiary value is preserved from these unusual circumstances and the safety of these personnel is guarded. This preparation also should include planning and preparedness (to include exercises) for the interoperability of local forensic personnel with federal counterterrorism organizations.

The recommendations are from the National Academy of Sciences report from 2009 "Strengthening Forensic Science in the United States."

Many forensic science disciplines in the United States overhauled training programs to address the educational shortfalls identified in the NAS report. Perhaps the biggest shift in training has been the increased reliance on colleges and universities to provide the initial educational foundation through undergraduate and graduate programs. The majority of these programs are focused narrowly on topics relevant to public forensic science laboratories and crime scene processing, with concentrations commonly offered in molecular biology/forensic DNA and criminal investigation. Several other more specialized programs also exist, including a few geared toward forensic entomology. The Forensic Science Education Program Accreditation Committee (FEPAC) was developed to ensure standardized education in university-based forensic science programs (NAS, 2009). Accredited programs are required to offer standardized curricula and monitored for quality assurance and quality control purposes related to the courses, faculty, and facilities (FEPAC, 2020). Theoretically, graduates of such programs are prepared to begin entry level work in a forensic laboratory upon hire, with only minimal additional training. Similarly, individuals trained in FEPAC accredited programs are also expected to be well versed in the professional standards and codes of conduct associated with the forensic sciences and criminalistics. In reality, it is still too early to tell exactly how much of an impact that FEPAC accreditation has had on improving the education and quality of casework performed by this new generation of forensic practitioners. However, the types of changes made appear to be moving the needle in the right direction to ensure high quality work. Most importantly, these training paradigms are designed to ensure that "science" is indeed the core of the forensic sciences, and that opinions and other avenues for biases are reduced if not removed all together from forensic analyses.

23.2   Guidelines for the practice of forensic entomology

The obvious next question when examining professional policies and standards is how does forensic entomology stack up by comparison to other areas of forensic sciences? In other words, does the discipline have clearly defined guidelines for becoming a forensic entomologist (the education and training piece) and are there standards in place to govern professional practice (the practitioner piece). Actually, the practitioner piece can be divided into several components that entail standards for collection and processing of entomological evidence (Amendt *et al.*, 2007; Sanford *et al.*, 2020), working the case and presentation to the court (Hall, 2020), statistical analyses of entomological evidence and forensic entomology research (Michaud *et al.*, 2012; Tomberlin *et al.*, 2012; Baqué & Amendt, 2013; Wells & LaMotte, 2020), and codes of conduct. While forensic entomology as a discipline was not criticized by the NAS Report (2009), there are nonetheless several themes that resonate. For instance, this discipline does not fall under the umbrella of FEPAC oversight or any other accrediting body, and thus, educational standards are self-regulated by programs and individual instructors. There are no mandatory educational requirements to become a forensic entomologist in North America or Europe. This statement needs some context to be fully appreciated: arguably anyone with some degree of entomological training can conceivably declare themselves a forensic entomologist and solicit casework. Such self-declaration does not mean that the person will be hired by a credible attorney nor recognized as an expert by the court (Hall, 2020). However, there is also no accrediting organization in place that oversees the education and training of forensic entomologists. The point is that unlike some forensic disciplines, mandatory educational standards for forensic entomology do not exist in those nations with the highest concentrations of practitioners.

The key statement is that *mandatory* educational requirements are not a prerequisite to becoming a forensic entomologist. In contrast, there are educational requirements to become a member of professional societies like the European Association of Forensic entomology or the North American Association of Forensic entomology, where individuals generally must have a background in a medical or forensic field and be actively engaged in forensic entomology (http://eafe.org/EAFE_Constitution.html; https://www.nafea.net/about/objectives/). Active engagement in the field is typically demonstrated

through teaching, research or casework, or a combination of the three. The level of membership is usually defined by the highest educational degree obtained, with the PhD serving as the terminal degree in forensic entomology. Board certification is also obtainable for forensic entomology. The American Board of Forensic entomology offers certification for practitioners in medicocriminal entomology and defines specific coursework (systematic entomology, insect taxonomy, immature insects, medical/veterinary entomology, insect physiology, and insect morphology) in their by-laws required for becoming board certified (http://www.forensicentomologist.org/wp-content/uploads/2018/03/2012-By-Laws.pdf). Interestingly, a course in forensic entomology is not a requirement (Table 23.2). Board certification in urban entomology, medical and veterinary entomology, general entomology as well as other areas not related to forensic investigations can be obtained through the Entomological Association of America's Board Certified Entomologist (BCE) programs (https://www.entocert.org/bce-certification). Certification for any of these programs requires demonstration of knowledge, technical skills, and identification of important species relevant to a given area.

Presently only 17 individuals are recognized as active board certified (as either diplomates or associate members) forensic entomologists in North America (https://forensicentomologist.org/members/). However, more individuals than just these seventeen forensic entomologists are engaged in casework in the United States and Canada. This is an indication that the members of the judiciary do not require or expect board certification to establish forensic entomologists as expert witnesses. These current practices are not entirely surprising considering that the field is relatively young with respect to many other forensic vocations in terms of acceptance in the U.S. Judicial System. There are often far fewer criminal cases in which entomological evidence is actually processed or used to establish elements of a case by comparison to other biological evidence. If and when entomological evidence becomes more commonplace in the courts, the requirements may become more stringent for recognition as an expert witness, and, in turn, directly impact the educational standards to become a forensic entomologist.

Table 23.2 Minimum qualifications to become ABFE board certified.

Area	Standard
General	Applicants must be persons of good moral character, high integrity, and good repute, and must possess high ethical and professional standards.
	Only permanent residents of the United States of America and its territories and possessions, or of Canada and its territories, are eligible for Certification
Education	Earned thesis-based Master's degree or Doctoral degree in Entomology, Biology, Ecology, or Zoology. This must include statistics as well as a substantial number of entomology courses including systematic entomology, insect taxonomy, immature insects, medical/veterinary entomology, insect morphology, insect physiology, and insect ecology. The earned degree must be from an accredited institution
Professional experience	Possess at least three years of professional experience that involves the practice of medico-legal forensic entomology casework. This experience must be casework containing associated agency contact information.
	Possess a minimum of one peer reviewed publication on a subject germane to the field of forensic entomology (senior authorship is not required).
	Provide at least three professional presentations to convey research or general education on the subject of forensic entomology at independently-hosted meetings, workshops, or symposia.
	At the time of application, the applicant must submit copies of three case reports prepared by the applicant for review by the Executive Committee of the ABFE. The three medico-legal casework examples acceptable as submission for initial certification for Member status must consist of collected entomological evidence from criminal investigations associated with human death, abuse or neglect. All submitted casework must come from the previous five-year period.
Examinations	Score 80% or higher on the written and practical examinations of ABFE based upon broad principles of medico-legal forensic entomology.

Information is from the By-laws of the American Board of Forensic Entomologists (http://www.forensicentomologist.org/wp-content/uploads/2018/03/2012-By-Laws.pdf).

23.3 Expectations for forensic entomology technicians

Forensic entomologists are rarely called to crime scenes to process insect evidence. There are exceptions of course, such as if a forensic entomologist works for law enforcement, a forensic unit, or a medical examiner to help with death investigations. However, in the majority of instances, entomological evidence is collected by a crime scene technician or police officer. At best, the individual has limited training on how to recognize, collect, and preserve insect evidence. At other times, the person processing the crime scene has no idea which insects to collect or how to process them. The latter can lead to mishandling of the evidentiary sample or contribute to sample degradation. Either scenario potentially compromises critical evidence to a case. This may in turn result in the insect evidence being deemed inadmissible. As means to overcome such deficiencies, the American Board of Forensic Entomology (ABFE) recently established board certification for forensic entomology technicians. The concept of a forensic entomology technician is to recognize individuals who have demonstrated a minimum set of criteria related to the knowledge and technical skills associated with collection and preservation of entomological evidence. The criteria are similar to those for becoming a member of the ABFE. Importantly, technicians are not necessarily employed as forensic entomologists but instead engaged in evidence collection such as crime scene technicians or death investigators. Thus, these are the frontline individuals who must collect insect evidence, preserve it, and then package it for transport to the lab for forensic analysis.

Certification as a forensic entomology technician achieves three outcomes. First, it establishes that an individual possesses the knowledge and technical skills minimally necessary to properly collect, preserve, and ship entomological evidence. The ABFE indicate that the evidence is specifically sent to a diplomate or member of the ABFE. In reality, the entomological evidence may be sent to other individuals as well, who also share the tasks of insect identification, analysis of entomological evidence, and development of case reports. The second outcome is professional development. Individuals achieving certification as forensic entomology technicians strengthen their roles with crime laboratories, medical examiner offices, or similar agencies to investigate scenes of human or veterinary forensic importance where insect evidence is potentially present. By extension, the credentials of the agency are also strengthened as their employee(s) is recognized by the board certifying body for forensic entomology in North America. Board certification of person's collecting entomological evidence also helps promotion of forensic entomology as a discipline. Perhaps the most important aspect is that encouraging certification helps to broadly establish in other forensic science disciplines that the policies, practices and standards of the ABFE and similar organizations should be followed when processing insect evidence. Certification also reinforces the importance of insect evidence to a wide range of civil and criminal investigations.

23.4 Ethical considerations for involvement in casework and providing testimony

All disciplines in forensic science stress the importance of ethical conduct professionally and the need for future practitioners to receive training in the subject of ethics. In fact, accredited forensic science programs generally have ethics training interwoven throughout the curriculum. The situation is somewhat different with forensic entomology in that some instructors, but not the majority, integrate ethical understanding of certain issues or topics into their courses (Butin et al., 2020). Ethics training is generally not a component of graduate entomology programs, which begs the question of how does a future practitioner become well versed in the ethical standards of forensic entomology? In the United States, the former National Commission of Forensic Science (of the National Standards and Technology Council) advocated for a national code of ethics and professional responsibility in all of the forensic sciences. The commission stated that most practitioners in forensic science and forensic medicine are hard-working, ethical professionals but that education and guidance on professional responsibility has been uneven and there was (is) no enforceable

universal code of ethics or professional responsibility. As a first step toward developing a universal code of ethics, the commission reviewed the code of ethics of various forensic science disciplines and found four consistent themes: working within professional competence, providing clear and objective testimony, avoiding conflicts of interest, and avoiding bias and influence, whether real or perceived (NSTC, 2010). Building on these themes, the commission proposed the following national code of ethics for forensic sciences (NSTC, 2010):

1. Accurately represent his/her education, training, experience, and areas of expertise.
2. Pursue professional competency through training, proficiency testing, certification, and presentation and publication of research findings.
3. Commit to continuous learning in the forensic disciplines and stay abreast of new findings, equipment, and techniques.
4. Promote validation and incorporation of new technologies, guarding against the use of non-valid methods in casework and the misapplication of validated methods.
5. Avoid tampering, adulteration, loss, or unnecessary consumption of evidentiary materials.
6. Avoid participation in any case where there are personal, financial, employment-related or other conflicts of interest.
7. Conduct full, fair and unbiased examinations, leading to independent, impartial, and objective opinions and conclusions.
8. Make and retain full, contemporaneous, clear and accurate written records of all examinations and tests conducted and conclusions drawn, in sufficient detail to allow meaningful review and assessment by an independent person competent in the field.
9. Base conclusions on generally accepted procedures supported by sufficient data, standards and controls, not on political pressure or other outside influence.
10. Do not render conclusions that are outside one's expertise.
11. Prepare reports in unambiguous terms, clearly distinguishing data from interpretations and opinions, and disclosing all known associated limitations that prevent invalid inferences or mislead the judge or jury.
12. Do not alter reports or other records or withhold information from reports for strategic or tactical litigation advantage.

13. Present accurate and complete data in reports, oral and written presentations and testimony based on good scientific practices and validated methods.
14. Communicate honestly and fully, once a report is issued, with all parties (investigators, prosecutors, defense attorneys, and other expert witnesses), unless prohibited by law.
15. Document and notify management or quality assurance personnel of adverse events, such as an unintended mistake or a breach of ethical, legal, scientific standards, or questionable conduct.
16. Ensure reporting, through proper management channels, to all impacted scientific and legal parties of any adverse event that affects a previously issued report or testimony.

The goal was to have the code adopted by all societies, boards and other organizations associated with forensic science and forensic medicine disciplines. All members of the organizations would in turn review and sign the code on an annual basis, establishing a yearly reaffirmation or commitment to ethical conduct. Unfortunately, this proposal was never acted upon. Consequently a national code of ethics was not implemented and one currently does not exist today.

What is the status of forensic entomology in terms of professional ethics? A code of conduct or standards for professional responsibility does exist for the major organizations associated with forensic entomology, including EAFE, NAFE, ABFE, and BCE. In fact, the code of ethics for good practice and science for EAFE and BCE (in general) are similar in content to that proposed by the National Standards and Technology Council (http://eafe. org/EAFE_Constitution.html). Adherence to ethical guidelines is mandated for board certified members of ABFE and BCE. The standard for professional societies is less demanding, with terms like "should" prefacing guidelines for professional responsibilities. Nonetheless, organizations associated with forensic entomology are in agreement that professional ethics are needed for its members and practitioners as a whole.

Diplomates or associate members of the ABFE who demonstrate "unethical, immoral, illegal, or unprofessional" violations of the general requirements of the bylaws can be subject to censorship (http://www.forensicentomologist.org/wp-content/uploads/2018/03/2012-By-Laws.pdf). Repeated violations may lead to revocation of board certification.

Enforcement of ethics codes is a challenge for any organization. It is also important to note that violation of a society's or board's guidelines for professional responsibilities does not prohibit an individual from continuing to serve as a forensic entomology practitioner independent of a governing entity. The latter is undoubtedly surprising. However, this is a reflection of current practices in forensic entomology. Professional societies are not the same as governing boards. Thus, in the absence of a mandatory accrediting organization for forensic entomology, enforcement of professional standards and ethical codes can be challenging.

Chapter review

General standards and recommendations for forensic science disciplines

- The application of science or any discipline to legal matters in the judicial system requires meeting specific standards and practices to be allowed do so. This is true for scientific evidence and techniques to be admissible in court, as well as for individuals to be deemed experts within a given discipline. In the United States, federal rules for admissibility of scientific evidence and for establishing the qualifications of an expert have been established through the Frye standard and then superseded by the *Daubert v. Merrell Dow Pharmaceuticals* ruling in 1993.
- A witness who is qualified as an expert based on education, skill, knowledge, or training may testify on the matter provided the person is an expert in an area relevant to the issue being considered and that the testimony is based on data or facts derived from appropriate scientific methods or techniques, and that the techniques were used appropriately. Ultimately, the trial judge makes this determination. Hall (2020) provides an excellent review of how the Frye and Daubert standards apply directly to forensic entomology.
- Professional standards should exist for the education and training to be a forensic practitioner, as well as to govern the practice or application of the discipline by an individual to perform casework. Does this type of linkage actually exist? Well, that has been the subject of controversy in

the United States since the publication of the National Academy of Science's 2009 report "Strengthening Forensic Science in the United States: A Path Forward." The report identified a series of issues associated with forensic science, including a lack of scientific standards for most subdisciplines, no or poor regulations governing forensic investigations and analyses, and the absence of centralized oversight regarding policy and practice of forensic science in the United States.

- The Forensic Science Education Program Accreditation Committee (FEPAC) was developed to ensure standardized education in university-based forensic science programs. Accredited programs are required to offer standardized curricula and monitored for quality assurance and quality control purposes related to the courses, faculty, and facilities. Theoretically, graduates of such programs are prepared to begin entry level work in a forensic laboratory upon hire, with only minimal additional training.

Guidelines for the practice of forensic entomology

- While forensic entomology as a discipline was not criticized by the NAS Report, there are nonetheless several themes that resonate. For instance, this discipline does not fall under the umbrella of FEPAC oversight or any other accrediting body, and thus, educational standards are self-regulated by programs and individual instructors. There are no mandatory educational requirements to become a forensic entomologist in North America or Europe. This is largely due to the fact that there is no accrediting organization in place that oversees the education and training of forensic entomologists.
- There are educational requirements to become a member of professional societies like the European Association of Forensic entomology or the North American Association of Forensic entomology, where individuals generally must have a background in a medical or forensic field and be actively engaged in forensic entomology. Active engagement in the field is typically demonstrated through teaching, research or casework, or a combination of the three. The level of membership is usually defined by the highest educational degree obtained, with the PhD

serving as the terminal degree in forensic entomology. Board certification is also obtainable for forensic entomology. The American Board of Forensic entomology offers certification for practitioners in medicocriminal entomology and defines specific coursework in their bylaws required for becoming board certified.

- If and when entomological evidence becomes more commonplace in the courts, the requirements may become more stringent for recognition as an expert witness, and, in turn, directly impact the educational standards to become a forensic entomologist.

Expectations for forensic entomology technicians

- The concept of a forensic entomology technician is to recognize individuals who have demonstrated a minimum set of criteria related to the knowledge and technical skills associated with collection and preservation of entomological evidence. The criteria are similar to that for becoming a member of the ABFE. Importantly, technicians are not necessarily employed as forensic entomologists but instead engaged in evidence collection such as crime scene technicians or death investigators. Thus, these are the frontline individuals who must collect insect evidence, preserve it, and then package it for transport to the lab for forensic analysis.
- Certification as a forensic entomology technician achieves three outcomes. First, it establishes that an individual possesses the knowledge and technical skills minimally necessary to properly collect, preserve, and ship entomological evidence. The second outcome is professional development. Perhaps the most important aspect is that encouraging certification helps to broadly establish in other forensic science disciplines that the policies, practices, and standards of the ABFE and similar organizations should be followed when processing insect evidence.

Ethical considerations for involvement in casework and providing testimony

- All disciplines in forensic science stress the importance of ethical conduct professionally and the need for future practitioners to receive training in the subject of ethics. In fact, accredited forensic science programs generally have ethics training interwoven throughout the curriculum. The situation is somewhat different with forensic entomology in that some instructors, but not the majority, integrate ethical understanding of certain issues or topics into their courses.

• A code of conduct or standards for professional responsibility does exist for the major organizations associated with forensic entomology, including EAFE, NAFE, ABFE, and BCE. In fact, the code of ethics for good practice and science for EAFE and BCE (in general) are similar in content to that proposed by the National Standards and Technology Council. Adherence to ethical guidelines is mandated for board certified members of ABFE and BCE. The standard for professional societies is less demanding, with terms like "should" prefacing guidelines for professional responsibilities. Nonetheless, organizations associated with forensic entomology are in agreement that professional ethics are needed for its members and practitioners as a whole.

Test your understanding

Level 1: knowledge/comprehension

1. Define the following terms:

   (a) code of ethics
   (b) professional responsibilities
   (c) admissibility of evidence
   (d) FEPAC
   (e) Federal rules of evidence.

2. Explain the criteria that exist for becoming board certified through ABFE.
3. Discuss why certification for forensic entomology technicians is an improvement over the current situation with crime scene investigators.

Level 2: application/analysis

1. Explain how FEPAC accreditation has resulted in improvements in forensic science education at the undergraduate level.
2. Discuss whether forensic entomology as a discipline would benefit from falling under the umbrella of FEPAC or a similar organization.

3. Describe some of the key ethical codes that a forensic entomology practitioner must or should follow when involved in a criminal investigation.

References

Amendt, J., Campobasso, C.P., Gaudry, E., Reiter, C., LeBlanc, H.N. & Hall, M.J. (2007) Best practice in forensic entomology—standards and guidelines. *International Journal of Legal Medicine* 121(2): 90–104.

Baqué, M. & Amendt, J. (2013) Strengthen forensic entomology in court—the need for data exploration and the validation of a generalised additive mixed model. *International Journal of Legal Medicine* 127(1): 213–223.

Bernstein, D.E. (2000) Frye, frye, again: the past, present, and future of the general acceptance test. *Jurimetrics* 41: 385.

Butin, E., Rivers, D. & Wallace, J.R. (2020) Practical considerations for teaching forensic entomology. In: J.H. Byrd & J.K. Tomberlin (eds) *Forensic Entomology: The Utility of Arthropods in Legal Investigations*, pp. 555–574. CRC Press, Boca Raton, FL.

Daubert v. Merrell Dow Pharmaceuticals, Inc. (1993) 509 U.S. 579, 589.

Forensic Science Education Programs Accreditation Commission (2020) Accreditation standards. http://fepac-edu.org/sites/default/files/FEPAC%20 Standards%2002152020.pdf. Accessed February 13, 2021.

Frye v. United States (1923) 293 F. 1013.

Hall, R.D. (2020) The forensic entomologist as expert witness. In: J.H. Byrd & J.K. Tomberlin (eds) *Forensic Entomology: The Utility of Arthropods in Legal Investigations*, pp. 333–348. CRC Press, Boca Raton, FL.

Michaud, J.P., Schoenly, K.G. & Moreau, G. (2012) Sampling flies or sampling flaws? Experimental design and inference strength in forensic entomology. *Journal of Medical Entomology* 49(1): 1–10.

National Research Council Committee on Identifying the Needs of the Forensic Science Community (2009) *Strengthening Forensic Science in the United States: A Pathway Forward*. National Academies Press, Washington, DC.

National Standards and Technology Council (2010) National code of ethics and professional responsibility in the forensic sciences. https://www.justice.gov/archives/ ncfs/page/file/788576/download#:~:text=Forensic%20 Science%20in%20the%20United%20States%3A%20 A%20Path,code%20of%20professional%20 responsibility%20and%20code%20of%20ethics. Accessed February 21, 2021.

Sanford, M.R., Byrd, J.H., Tomberlin, J.K. & Wallace, J.R. (2020) Entomological evidence collection methods: American Board of Forensic Entomology approved methods. In: J.H. Byrd & J.K. Tomberlin (eds) *Forensic*

Entomology: The Utility of Arthropods in Legal Investigations, pp. 63–86. CRC Press, Boca Raton, FL.

Tomberlin, J.K., Byrd, J.H., Wallace, J.R. & Benbow, M.E. (2012) Assessment of decomposition studies indicates need for standardized and repeatable research methods in forensic entomology. *Journal of Forensic Research* 3(5): 147.

Wells, J.D. & LaMotte, L.R. (2020) Estimating the postmortem interval. In: J.H. Byrd & J.K. Tomberlin (eds) *Forensic Entomology: The Utility of Arthropods in Legal Investigations*, pp. 213–224. CRC Press, Boca Raton, FL.

Supplemental reading

Barnett, P.D. (2001) *Ethics in Forensic Science: Professional Standards for the Practice of Criminalistics*. CRC Press, Baco Raton, FL.

Berger, M. (2000) *The Supreme Court's Trilogy on the Admissibility of Expert Evidence. Reference Manual on Scientific Evidence*. Federal Judicial Center, Washington, DC.

Candilis, P.J., Weinstock, R. & Martinez, R. (2007) *Forensic Ethics and the Expert Witness*. Springer Science & Business Media, Switzerland.

Christensen, A.M. & Crowder, C.M. (2009) Evidentiary standards for forensic anthropology. *Journal of Forensic Sciences* 54(6): 1211–1216.

Dixon, L. & Gill, B. (2002) *Changes in the Standards for Admitting Expert Evidence in Federal Civil Cases Since the Daubert Decision*. RAND, Santa Monica, CA.

Giannelli, P. (1994) Daubert: interpreting the federal rules of evidence. *Cardoza Law Review* 15: 1999–2026.

Hofer, I.M., Hart, A.J., Martín-Vega, D. & Hall, M.J. (2017) Optimising crime scene temperature collection for forensic entomology casework. *Forensic Science International* 270: 129–138.

Wecht, C. & Rago, J. (eds) (2005) *Forensic Science and Law*. CRC Press, Boca Raton, FL.

Wells, J.D. & Stevens, J.R. (2008) Application of DNA-based methods in forensic entomology. *Annual Review of Entomology* 53: 103–120.

Wilson-Wilde, L. (2018) The international development of forensic science standards—a review. *Forensic Science International* 288: 1–9.

Additional resources

ASTM Forensic Science Standards: https://www.astm.org/Standards/forensic-science-standards.html

Entomological Society of America's Code of Conduct: https://www.entsoc.org/conduct

NIST Forensic Science Standards Board: https://www.nist.gov/organization-scientific-area-committees-forensic-science/forensic-science-standards-board

Chapter 24
Forensic entomology case studies

Overview

Examples of casework or case studies are excellent tools demonstrating the context in which insects can serve as physical or trace evidence in civil or criminal investigations. This chapter will discuss casework in which forensic entomology was applied to investigations of homicides, toxicology, aquatic decompositions, and medical malpractice associated with myiasis. Most case studies presented revolve around the presence of necrophagous flies, specifically maggots. However, some non-fly examples are also provided to broaden perspective of the types of insects that can serve as physical evidence. A list of published case reports is also included in the chapter to facilitate more in-depth study of particular types of cases involving entomological evidence.

The big picture

- Case studies are important to understanding the context of insect evidence to legal investigations.
- Insect evidence associated with homicides and suspicious deaths.
- Entomological evidence associated with forensic entomotoxicology.
- Postmortem interval determination using insects on a burned corpse.
- Insect evidence associated with an indoor decomposition.
- Entomological evidence associated with aquatic decompositions.

24.1 Case studies are important to understanding the context of insect evidence to legal investigations

There is no substitute for the hands-on experiences associated with real casework. Of course this is not an option for students studying forensic entomology. Student participation in casework is not permitted for several reasons, including possible contamination/alteration of the evidence, loss or misplacement of evidence, and breach of confidentiality between the forensic entomologist serving as an expert and legal counsel/client (Hall, 2020). There is also the issue that once the opposing counsel learns that anyone other than the supposed expert has "handled" the evidence, challenges to

The Science of Forensic Entomology, Second Edition. David B. Rivers and Gregory A. Dahlem.
© 2023 John Wiley & Sons Ltd. Published 2023 by John Wiley & Sons Ltd.
Companian website: www.wiley.com/go/forensicentomology2

chain of custody, as well as expert status can be made. Theoretically it could even lead to the non-expert being called to testify, essentially being viewed as an extension of the expert. Obviously none of these scenarios can be allowed to play out in the judicial system.

While each of these examples are reasonable explanations about the limitations of "who" has access to entomological evidence, it still leads to a sense of frustration for a student. How can you gain practical experience if prevented from even shadowing an expert during real casework? A valid question, and one that does not necessarily have a readily available solution. The most common approach is to provide training through mock crime scenes or decomposition studies using animal surrogates (e.g., pigs, dogs, rabbits, and others) for human corpses. Animal models permit the teaching of techniques and methodologies associated with recognition of physical and trace evidence, collection and preservation of arthropod specimens, and a variety of additional analyses including species identification, determination of specimen age, and estimation of the postmortem interval. To an extent, these type of training exercises allows students to develop contextual understanding of when and how insects utilize a corpse under a range of conditions commonly encountered with violent crimes. However, and this a very important distinction, mock crime scenes are controlled environments by comparison to a crime scene. They lack the contextual intensity of an actual crime scene resulting from a violent crime. Collection of insect evidence is far from the only activity occurring at the crime scene in contrast to a mock scene used for entomological training. It is also a much different visual and emotional experience collecting insects from a human corpse, especially a child, than from an animal model. The same can be said for the "pressure" associated with completion of the forensic analysis. None of these comments are meant to detract from the value of experiential learning obtainable from mock crime scenes. The point is that animal models are not "perfect" substitutes for casework for a variety of reasons (Keough *et al.*, 2017).

An alternative approach to mock crime scenes is the use of case study investigations. Practically speaking, case studies really should be used to augment training with decomposition models. They represent a form of high-impact practices in which students are able to focus on a topic that

bridges knowledge gained from didactic learning and permits application of concepts and techniques to open-ended, real world problems (Kilgo *et al.*, 2015; Rippy, 2020). In this case, the real-life issues are criminal investigations. Students have the ability to apply what they have learned to actual cases. By incorporating different case scenarios involving insects and other arthropods into a course or training, students have opportunities to develop breadth and depth of understanding of casework in forensic entomology. Case studies also allow students to see how a particular case was investigated or solved. Bernard Greenberg used this approach in his excellent text *Entomology and the Law* (2002), outlining his approach to select cases and discussing underlying challenges that may not be immediately obvious from simply reading a published case report. By combining laboratory and field training with case studies, individuals can greatly expand their skill sets before ever dealing with a real case on their own.

What follows are several case reports ranging from homicides, toxicology, aquatic decompositions, and medical malpractice associated with myiasis. Each case presents a specific set of circumstances and/or insects that build upon the factual and conceptual ideas discussed throughout this text[1]. Additional case examples are provided to permit further exploration of casework in forensic entomology.

24.2 Insect evidence associated with homicides and suspicious deaths

24.2.1 Case of insect evidence overturning a wrongful murder conviction

Our start to exploring case reports begins with an infamous case from Ontario, Canada. The *Regina v. Steven Truscott* case is well known in forensic entomology. The case's notoriety is associated with a wrongful conviction that took nearly 50 years to overturn, as well as controversial testimony regarding the entomological evidence. VanLaerhoven & Merritt (2019) provide a detailed account of their work for the defense in the 2006–2007 appeal of the original 1959 conviction of Steven Truscott. A summary of the case, entomological evidence, and key features of the investigation are presented. It is important to note that the prosecution (the Crown) relied on expert analysis from a separate forensic entomologist, and during the original case investigation in 1959, entomological evidence was collected and examined by a biologist working for the Attorney General's Laboratory in Toronto. In all examples discussed in this chapter, students are encouraged to read the cited articles for more complete details of each reported case.

24.2.1.1 Overview of case

On the evening of June 9, 1959 at approximately 7:45 PM, 14-year-old Steven Truscott was seen giving a classmate, 12-year-old Lynne Harper, a ride on his bicycle. He reported that they parted ways and Steven was in the company of witnesses by 8:00 PM throughout the rest of the evening. Later that night, Lynne's father reported her missing. Two days later, Lynne's body was found at 1:15 PM in a nearby wooded lot on the farm of R.S. Lawson, northeast of Clinton, Ontario. She had been sexually assaulted and strangled to death. On June 13, 1959, Steven was charged with Lynne's murder.

Despite being a minor, Steven was ordered to stand trial as an adult. The prosecution's theory of the case was that rather than dropping Lynne off as he claimed to have done, Steven turned off the road and killed Lynne at some point between 7:00 p.m. and 7:45 p.m. that evening. This argument was supported by testimony from various witnesses, who claimed to have seen Steven near the area where Lynne's body was eventually found. Results from autopsy (largely based on stomach contents) also suggested that the time of death occurred in the window of time outlined by the prosecution. Further evidence was in the form of lesions observed on Steven's penis, which, it was argued by the prosecution, could have been sustained by sexually assaulting Lynne (Harland-Logan, 2020).

Steven proclaimed his innocence throughout the trial. He testified that Lynne was fine when he dropped her off at the intersection of the County Road and Highway 8. He further testified that he stopped his bike on a bridge, looked back to where he had dropped off Lynne, and observed her getting into a grey Chevy automobile with a yellow license plate. Several witnesses supported this version of

events, testifying that they had observed Steven and Lynne riding toward the intersection where Steven said he had dropped her off, or that they had seen him standing on the bridge looking in her direction. Witnesses also indicated that Steven seemed "normal" when they saw him on the school grounds at 8 p.m. that evening.

On September 30, 1959, the jury found Steven guilty. The Criminal Code required that a death sentence be imposed for murder, and consequently, the trial judge sentenced 14-year-old Steven to death by hanging. Eventually, his sentence was commuted to life imprisonment. On October 21, 1969, after 10 years behind bars, he was granted parole.

24.2.1.2 Entomological evidence

Photographs (~ 4:00 PM) taken at the crime scene on the day of discovery showed the presence of adult blow flies around the face, with fewer observed on other areas of the body including the groin. The photos also revealed a small maggot mass located around the mouth, nose and eyes. At 5:30 PM, eggs and maggots were collected from the face and abdomen. The provincial pathologist recorded his collection from "around the nose and mouth, innumerable maggots, about 1/16 in. long." At 8:10 PM, the pathologist collected maggots from a lesion on the deceased's left buttock and noted "in the skin lesions and about the vulva, larger maggots, up to 1/4 in long." Autopsy photographs taken between 7:15 and 8:30 PM clearly showed maggots on the victim's buttocks but maggot masses were not evident in this region.

The entomological evidence was transferred to Mr. S.E. Brown, a biologist employed at the Attorney General's Laboratory in Toronto. He examined the maggots between 12:12 and 5:00 PM on June 12, 1959, noting that the sizes of the maggots in the two collections were the same as reported by the provincial pathologist. Mr. Brown reared some of the maggots to adult for identification purposes. He reported that the maggots on the decedent's face as "blue bottles, 1st instar of blow fly, Calliphoridae, genus *Calliphora*." Those collected from the buttocks and vulva were classified as "flesh fly larva, Sarcophagidae, genus *Sarcophaga*." Mr. Brown identified the sarcophagids as "red eyes, three bands on thorax, apical ½ of antennae bristle is bare as opposed to Muscidae, which are plumose to tip." (Figure 24.1)

(a)

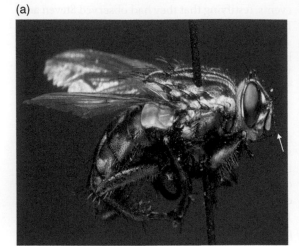

(b)

Figure 24.1   A comparison of the arista on the antennae of (a) an adult sarcophagid with that of (b) an adult muscid. As noted by Mr. Brown in the Truscott case, the apical portion of the arista is bare with adult sarcophagids but plumose the entire length with adult muscids. Note that these are not photos from the case; merely examples to point out the morphological differences cited in the case. *Source*: Photo by 0′.12.1.0.N/Wikimedia Commons/Public domain https://commons.wikimedia.org/wiki/File:Sarcophaga_bullata.jpg. Photo by Muhammad Mahdi Karim/Wikimedia Commons/Public domain https://commons.wikimedia.org/wiki/File:Housefly_musca_domestica_cropped_(2).png.

It should be noted that the entomological evidence was not key to the conviction of Steven Truscott. However, it was important in the 2006–2007 appeal that led to the eventual acquittal of murder charges.

24.2.1.3   Key features

In 2001, the Association in Defence of the Wrongly Convicted (now Innocence Canada) filed an Application for Ministerial Review of Steven's conviction. An investigation was initiated that lead to a 2004 investigative report prepared for the Minister of Justice. The summary finding of that investigation was that there was reasonable basis to conclude that a miscarriage of justice had occurred. This report opened the door for Innocence Canada's legal team to present fresh evidence and new analysis to the Court of Appeal. It was during these proceedings that the entomological evidence become important to the appeal. A key feature of the case was the timeline of when Lynne Harper died. In the original trial, the prosecution's theory was that the girl had been murdered sometime between 7:15 PM and 7:45 PM. The same window of time was proposed in 2006 by a forensic entomologist consulting with the prosecution, based on the original insect evidence examined by Mr. Brown in 1959. These conclusions were derived from photographs and Mr. Brown's notes. The actual

insect specimens, not surprisingly, were no longer available for examination.

The prosecution's forensic entomologist testified that:

a.   If Mr. Brown's original identification of flesh flies were correct, those flies typically would wait 12–24 hours after death before depositing larvae after discovering a corpse. He further explained that if the flies colonized the corpse mid-afternoon on June 10, 1959, the actual estimate of death would be 12–24 hours earlier, meaning June 9, 1959.

b.   However, the forensic entomologist did not believe the flesh fly identifications were correct. He stated that the flies collected from the buttocks/groin area were calliphorids not sarcophagids. The argument was based on the view that the temperatures on the evening of June 9, 1959 were too hot (30 °C) to support the adult activity of *Calliphora* spp. In this theory, the calliphorid larvae would be the oldest on the deceased when collected in 1959.

c.   The forensic entomologist theorized that trauma to the vaginal area caused the calliphorids to preferentially select the groin over the facial area. This theory seemingly created a discrepancy in larval size between the facial area with those in the vagina. The expert witness accounted for the differences by suggesting that there was a delay in colonization between the two locations by about 1 day.

In response to this testimony, the defense team asked a separate forensic entomologist to perform experiments that addressed two questions raised by the prosecution's witness: (i) Could adult *Calliphora* spp. oviposit on a corpse at or near 30 °C in June in Ontario, and (ii) whether or not Lynne Harper could have been killed at 7:30 PM in the evening, with sunset at 9:45 PM and not be colonized by flies prior to dark under those ambient conditions? (VanLaerHoven & Merritt, 2019). The experiment consisted of placing freshly killed pigs in the same woodlot where Lynne Harper's body was discovered. Placement occurred on June 17, 2006 at 7:30 PM under ambient conditions similar to the evening of June 9, 1959. Wounds were created on each pig to simulate those found on Lynne Harper's body.

The defense's forensic entomologist found:

a. Fly oviposition was observed within 30 minutes of corpse placement during the experiments and continued until sunset, at which point the flies left.
b. Large masses were laid on the nose and mouth of each pig, but none were deposited on the buttocks or vaginal area.
c. Adult *Calliphora* spp. were observed and photographed ovipositing in the pig's mouth and nose, and no eggs were laid after sunset.

The forensic entomologist concluded that Lynne Harper's body would likely have been colonized by flies prior to sunset under the ambient conditions of June 9, 1959 if her body had been left between 7:15 PM and 7:45 PM as theorized by the prosecution. Additionally, Mr. Brown's fly identification of blue bottles being present was supported in the re-creation experiments by the observations of oviposition by *Calliphora* spp. at temperatures at or near 30 °C. The defense team also cited ample research literature and personal observations of *Calliphora* spp. and sarcophagids in that part of Ontario in June. An additional set of laboratory experiments were performed by another forensic entomologist to estimate the time, based on thermal units, needed for the calliphorids and sarcophagids to reach the larval lengths originally recorded by the provincial pathologist and Mr. Brown in 1959. This information would shed light on an estimated postmortem interval. It was estimated that for the larvae collected from the face to have reached 2 mm (1/4 in) in length, eggs would have had to have been laid during daylight hours between 11:00 AM on June 10, 1959 and 8:00 AM on June 11, 1959. Similarly, analysis of the flesh fly larvae collected from the buttocks suggested that for larvae to reach 6 mm (1/2 in) in length, larviposition must have occurred during daylight hours between 7:00 AM and 11:00 PM on June 10, 1959. Accumulated degree day calculations were also performed using developmental data from the research literature and species of flies most likely encountered in that part of Ontario. The take home message was that the defense's forensic entomologist established a timeline that differed from that of the prosecutions, and most importantly, coincided with when Steve Truscott was in the company of witnesses.

The Court of Appeal found that the prosecution's entomological expert offered theories which could not be supported by the scientific literature. Ultimately, they disregarded the testimony. The forensic entomology analysis provided by the defense's experts, coupled with several other strong lines of evidence, led to the outright acquittal of Steven Truscott. The Truscott case serves an example of the importance of accurate species identification's for postmortem interval estimates using insect development. It also outlines how information can be derived from reconstruction and experimentation, especially when no specimens are available or that the only evidence of insect activity is derived from photographs. Further details about the Truscott appeal and other non-entomological evidence can be found in Harland-Logan (2020).

24.3 Entomological evidence associated with forensic entomotoxicology

Sometimes the utility of insects to a death investigation is less to do with determination of a postmortem interval and more so on their role as surrogates for human tissue. This can be the case when the corpse is mostly skeletonized, with little to no soft tissue remaining to use in toxicological analysis. Chapter 18 discusses the use of necrophagous insects in forensic entomotoxicology and provides numerous examples of drugs, toxins, and poisons detected in insect tissues after feeding on various animals known to possess xenobiotic compounds. Consumption of such compounds can alter the development of feeding stages of flies and beetles, thereby influencing any subsequent estimation of the postmortem interval. Similarly, detection and/or acceptance of a corpse by foraging insects maybe delayed due to the presence of certain xenobiotics in human tissues. This, in turn, would alter predictions of the timing of insect colonization based on the physical decay of the corpse. In his book "*A Fly for the Prosecution,*" Dr. M. Lee Goff, a pioneer in the field of forensic entomotoxicology, provides several case examples in which drugs or toxins in the corpse altered interpretations of the entomological evidence. In each case, toxicological information contributed to understanding the developmental pattern displayed by fly larvae that had fed on the corpse. Here, we will discuss two of Dr. Goff's cases, one involving illicit drug use (cocaine) by the decedent and the other a suicide resulting from the consumption of the insecticide malathion.

24.3.1 Drug-induced developmental changes

The body of a young woman (early 20's) was discovered in a wooded area outside of Spoken Washington at approximately 15:00 hours on October 12th, 1988. She was found lying face down, shirtless in a clearing surrounded by pines trees near a logging road. The body was in the early bloat stage when discovered, with the face and upper body blackened and marbled. Several stab wounds to the left side chest area were evident. Bloodstain patterns examined near the body indicated that the wounds were created while the decedent was lying on her right side on the ground.

24.3.1.1 Entomological evidence

The body of the decedent was placed in cold storage for 5 days prior to the collection of fly larvae from the corpse. Subsequently, the larvae were reared on bovine kidney (placed on sand in a container) and the resulting adults identified as *Cynomyopsis* (*Cynomya*) *cadaverina* and *Phaenicia* (*Lucilia*) *sericata* (Figure 24.2). Both species of calliphorids are commonly encountered in the state of Washington and many other regions in the United States and Canada. Each fly is typically associated with the initial wave of colonization, with both preferring to lay eggs in natural body openings. Fly larvae were collected from multiple locations on the body, including the head, with some larvae and eggs found in the hair, nose and mouth. The majority of larvae collected were determined to be 6–9 millimeters long. Under the ambient conditions associated with the site of body discovery, both flies would require about 7 days to achieve that range of body sizes. Of the remaining larvae collected, most were smaller,

(a)

(b)

Figure 24.2   The key entomological evidence in this case were eggs and larvae from two calliphorids, (a) *Phaenicia* (*Lucilia*) *sericata* and (b) *Cynomyopsis* (*Cynomya*) *cadaverina*. Note that these are not photos from the case. *Source*: Photo by Amada44/Wikimedia Commons/CC BY-SA 4.0 https://commons.wikimedia.org/wiki/File:Lucilia_sericata_8677.jpg. Photo by Judy Gallagher/Wikimedia Commons/CC BY 2.0 https://commons.wikimedia.org/wiki/File:Blowfly_-_Cynomya_cadaverina,_Meadowood_Farm_SRMA,_Mason_Neck,_Virginia.jpg.

which was presumed to reflect a period of continued egg deposition over several days. A smaller number of larvae were 17–18 millimeters in length, suggesting a developmental window of nearly 3 weeks. All of these large maggots were isolated from either the nose or mouth.

24.3.1.2   Key features

Initial interpretation of the entomological evidence was perplexing. The majority of insect evidence argued that the window of colonization was 7 days or less. However, the larger maggots would seemingly have to have been developing for much longer. The problem with this notion was that the non-entomological evidence did not support the possibility of a longer postmortem interval. The dilemma that arose with this case was trying to account for the discrepancies in maggot size and age. Several possibilities were considered. One was that the larger larvae had crawled from another carrion source and were then discovered on the deceased. This explanation was ruled out when no carrion was discovered in the area. An alternative explanation was suggested by Dr. Goff: some type of drug, specifically cocaine, may be present in the tissues of the decedent. This was the break needed. It turned out that the decedent was known to use cocaine, specifically to snort the drug. High concentrations of cocaine were shown by Dr. Goff to accelerate fly development in another species. The largest maggots were collected from the nose and mouth, where the highest concentrations of the drug would be expected to accumulate if cocaine was snorted just prior to death. Based on the ambient temperatures, flies that consumed cocaine were expected to need only 7 days to achieve 17–18 millimeters. It was later determined that the victim had a long history of cocaine use and had in fact been observed snorting cocaine just before her death.

You may be wondering why the other larvae collected were not larger than normal. After all, habitual drug use would likely result in accumulation of cocaine and metabolites in tissues throughout the body. The key is location of larval feeding. In the case of cocaine, lower drug concentrations and metabolites occur in tissues furthest removed from the site of drug entry. The mode of intoxication also influences which tissues will harbor the high concentrations of the drug. For example, snorting the drug will be associated with high concentrations in the mucosal membranes of the nose, throat and mouth. Thus, the smaller larvae likely did consume the parent drug and/or its metabolites but fed on tissues with lower concentrations. The levels consumed presumably did not exceed a threshold needed to accelerate the rate of development. Hence, larvae of initially the same age developed at different rates based upon where they fed on the deceased.

24.3.2   Insecticide-induced delay in insect colonization

The body of a 58-year-old man was found in the dirt crawl space of his mother's home in Honolulu Hawaii in February 1987. He was in an advanced stage of decomposition and appeared to have died via suicide. Next to the corpse was an 8-ounce bottle of the insecticide Malathion, 6 ounces of which were missing. The decedent had a history of suicidal behavior, having once attempted to shoot himself in the head. The crawl space was considered essentially an outdoor location in which insects and other small animals could have easily gained access to the human remains.

24.3.2.1   Entomological evidence

Fly larvae were collected from multiple locations on the corpse, including the head, neck and torso. Maggot activity was especially prevalent in the eyes, nose, and mouth region. Collected larvae were preserved in ethanol. However, they were not fixed (blanched) in hot water prior to placement in preservative. Consequently, by the time they were observed by a forensic entomologist the larvae were bloated and not useable for the determination of a postmortem interval based upon body length measurements. The larvae were still sufficiently preserved to permit identifications: all larvae were either *Chrysomya megacephala* or *Chrysomya rufifacies*. Specimens of *C. megacephala* were all in the third stage of larval development, while both second and third stage larvae of *C. rufifacies* were collected. This age differentiation is consistent with the oviposition behavior of both flies. *Chrysomya megacephala* is considered one of the earliest colonizers of carrion and adults of *C. rufifacies* arrive during the first wave of colonization but often later than some species. The behavior is associated with the predatory activity of the larvae of *C. rufifacies* (i.e., oviposition once a prey species is present), especially when *C. megacephala* is present. Under the ambient conditions in which the body was discovered, the larval ages were consistent with a postmortem interval of 5 days.

24.3.2.2 Key features

The estimated postmortem interval based on insect development was not in agreement with the 8 days since the decedent had last been seen alive. What could account for the differences in time estimates? The bottle of Malathion was viewed as the potential solution, yet the observations of where the flies were most abundant was surprising. Autopsy results revealed that the man had in fact been intoxicated with the insecticide and was presumed to have consumed it orally. It is this latter aspect that created much of the confusion; fly maggots were collected in large numbers in the mouth. Consequently, it was expected that the larvae would have died consuming the insecticide or at least experienced retarded development. In contrast, there was no evidence of any insecticide influences on these larvae. Toxicology data did not exist at the time of this case related to the effects of oral consumption of Malathion on any species of calliphorid larvae. The only relevant data was associated with contact toxicity via the exoskeleton of flesh fly larvae (*Boettcherisa peregrina*); very high concentrations were required to evoke death. Attention was then turned to the diversity, or more correctly, lack of in terms of insect fauna associated with the corpse. Only two species of blow flies were collected from the deceased, yet the body was exposed to the outdoors for around 8 days. Considering that the body was easily accessible to insects in the crawl space, the lack of insect diversity implied that perhaps the Malathion inhibited colonization and/or induced death in the earliest colonizers. The presumption was that as chemical decomposition of the corpse proceeded, the Malathion in the decedent's tissues was degraded. In turn, colonization was no longer impeded. The forensic entomologist, Dr. Goff, concluded that the window of time was approximately 3 days. Thus, 3 days after the man committed suicide, the Malathion was sufficiently degraded that fly colonization occurred. The body of the deceased was discovered before further insect colonization could occur.

This case represents an example of the application of information from multiple areas of entomology to casework. A broad entomological knowledge base was essential to propose possible scenarios to account for the discrepancy in insect development as well as to apply deductive reasoning to the case. The latter was necessary because several pieces of information that were needed were not available and not reasonably testable.

24.4 Postmortem interval determination using insects on a burned corpse

A burnt corpse offers unique challenges to estimating when insect colonization occurred. For instance, severely burned, charred bodies are rejected by most necrophagous insects as unsuitable for oviposition or progeny development. When the burn injury results in blistering or lesions in the skin without hardening of soft tissues, insect colonization can occur earlier than would be expected with non-burned remains. In a case reported by Pai *et al.* (2007), the burned remains of a young girl who had been beaten and stabbed were found colonized by *Chrysomya megacephala*. Their investigation of the case was aided by a recently conducted field study performed with a burned pig carcass to examine insect colonization and postmortem interval estimations based on insect development. Coincidently, the experiment was performed within 6 kilometers of where the girl's body was discovered.

24.4.1 Overview of case

The body of a young girl was discovered in a sugarcane field in Kaohsiung County, Taiwan on August 29, 2003. The deceased was naked below the waste and the bottom of her shirt had been pulled upward, exposing her breasts. An autopsy revealed that the girl had been stabbed eight times in the neck, chest, abdomen and back. There were numerous lacerations on her face, which were presumed to be the result of an attack with a blunt force object. The body had also been burned with gasoline poured on the body and then ignited. Ambient air temperatures at the scene were reported to be 28.6 °C with a relative humidity of 78.4%. The girl had last been seen alive by her sister on August 27th at approximately 7:00 PM, shortly after she received a phone call.

24.4.1.1 Entomological evidence

Fly larvae were present on the head, face and lower part of the body. A collection of larvae (40 total) was obtained from the deceased's face and lower part of the body. All of the larvae were fixed in KAA (a mixture of kerosene, acetic acid, and ethyl alcohol) before being preserved in 75% ethanol. The preserved specimens were identified as *Chrysomya megacephala* using both identification keys and DNA sequencing. Body length measurements ranged from 4.5 to 12 mm, corresponding to second and third stage larvae. The vast majority (30 of 40) of the larvae collected from the body were late second stage.

24.4.1.2 Key features

The investigators were able to use their recent findings from a pig decomposition study under similar conditions (e.g., burned to level 2 on the Crow-Glassman scale with gasoline under ambient conditions of 30.3 °C and RH of 69.3%) to estimate a PMI. Approximately 50 hours (a range was not provided in the case report) was calculated to be needed for *C. megacephala* to reach 11 millimeters or the third stage of larval development. A suspect was eventually arrested, and, upon confession, admitted to killing the girl at 10 PM on August 27th. This yielded a PMI of 46 hours.

Though the estimated PMI based on insect development closely approximated the known time of death, several issues were apparent with the case. For example, the agonal period was stated by the murderer to be at 10 PM, which would have occurred under nighttime conditions. The investigators reported concern that blow flies generally do not oviposit in darkness. However, based on the estimated PMI, they contended that *C. megacephala* would most definitely have to have laid eggs prior to midnight for third instar larvae to be present at the time the body was discovered. This actually was not a controversial observation at all. While many species of blow flies are believed to not oviposit at night (i.e., in darkness) (Amendt *et al.*, 2008), some, including *C. megacephala* have been observed to lay eggs under nocturnal conditions (Greenberg, 1990; Williams *et al.*, 2017). Even for species that will not fly at night, the presence of tall grass or even the sugarcane stalks could easily have been nighttime resting locations for the flies. If the odors of carrion decomposition are detected, the adult flies will walk to the source and oviposit at night.

Some additional details of the evidence collection were not reported, making it difficult to ascertain whether companion sampling was performed. Rearing a portion of the larval sample to adult would be another means to confirm the identity of the entomological evidence, also solidifying subsequent testimony if needed. It is also not clear whether the investigators searched the soil for dispersed larvae and/or pupae. While it turned out that the oldest insects were collected from the body, that was not evident until after the suspect confessed to the crime. The specific locations of larval collection should also be recorded during processing of the crime scene, as there are tissue differences in developmental rates of different fly species. In this case, it would also be important to note whether the larvae were present in wounds or lesions associated with the blunt force trauma or resulting from the burn injury. Several examples exist in the literature in which some fly species demonstrate preferential oviposition for wounds over natural body openings, as well as earlier colonization for burn wounds over non-burned remains. Without taking unique foraging and oviposition behavior into account, the PMI may be overestimated or underestimated in certain circumstances.

24.5 Insect evidence associated with an indoor decomposition

The insect fauna discovered with indoor decomposition often differs from that typical of body decomposition outdoors in terms of abundance and diversity. The major limiting factors are detection and access to the corpse. Obviously, odors must escape to attract outdoor species. Once detection occurs and foraging behavior is activated, the insects must locate an opening in the dwelling that permits entry. Most modern homes and buildings are designed to prevent heat loss or retain conditioned air. Consequently, insects with less chemical acuity or that are large in size do not commonly colonize human remains in concealed or indoor locations. This often times means that Coleoptera are rare in such cases, but not always, as detailed by Wang *et al.* (2019) for a case in the Guangdon Province of China.

24.5.1   Overview of Case

An adult male was discovered on October 25, 2017 in a temporary shelter under a viaduct in Guangzhou in the Guangdong Province of China. The shelter lacked a door and windows but otherwise was protected from direct exposure to the sun or precipitation. The corpse was found lying on a couch, covered by a thick cotton quilt. Under the covers, the deceased's body was naked and most of the soft tissues were mummified. The exceptions were that the head and a portion of the left arm were uncovered, and the right lower leg was lying on the ground wrapped in a towel. Ambient temperatures at the location averaged 26.6°C for the month prior to discovery of the corpse.

24.5.1.1   Entomological evidence

Several specimens were collected at the death scene, including larvae and empty puparia of *Chrysomya nigripes* (Calliphoridae), empty puparia of *C. megacephala*, adults of *Necrobia rufipes* and *N. ruficollis* (Coleoptera: Cleridae), and larvae and adults of *Dermestes maculatus* (Coleoptera: Dermestidae). Larval age for *C. nigripes* was not mentioned, but the investigators did indicate that the largest and presumably oldest larvae of *D. maculatus* were 11 millimeters in length. The beetle larvae of this length at 26.6°C were estimated to be third or fourth instar.

24.5.1.2   Key features

An autopsy revealed that the decedent had irregular-shaped skull defects on the left tempus and that the cause of death was brain injury. The forensic entomologists involved with this case estimated a PMI using development of *D. maculatus* rather than the fly specimens because they surmised that only empty puparia were present at the scene of body discovery (Figure 24.3). Thus, a period of time could not be accounted for; that is, the time after fly emergence from puparia. It was estimated that third to fourth stage larvae of *D. maculatus* would require 15.2–15.9 days at 26.6°C to reach those stages of development. Previous field research by these investigators at the same location and time of year revealed that *D. maculatus* typically arrive 8 days after death. When this time frame was coupled with beetle development, their estimated PMI was 23.2–23.9 days. It was later determined that the actual PMI was 25 days.

Figure 24.3   An adult *Dermestes maculatus*. In natural situations, the adults are found with human and animal remains in post bloat stages. Note that this photo is not from the case. *Source*: Photo by Paul venter/Wikimedia Commons/CC BY-SA 3.0 https://commons.wikimedia.org/wiki/File:Dermestes_maculatus00.jpg.

The details of the case provided does raise the question of why a PMI estimate was also not made using larvae of *C. nigripes*. Several possibilities come to mind. One is that the larvae were viewed as coming in a later wave of succession than the other fly specimens and beetles or reflected continuous oviposition events occurring after the initial discovery of the corpse by the flies. This would be consistent with the fact the empty puparia of the same species were present. Typically, *C. nigripes* requires 327 hours or approximately 13.5 days to complete development from egg to adult at 26°C (Li *et al.*, 2016). Of course, that does not take into account that most of the body was in a mummified condition, and as such, would not be suitable for larval feeding by most species of calliphorids. This does not preclude the possibility that the larvae were feeding on suitable soft tissues. However, intimate details of entomological collections were not provided in the published case report. Similarly, no discussion is provided of how or if the larvae were preserved properly when collected. It is a possibility that a reliable larval age could not be determined from these larvae and thus a time estimate could not be made.

24.5.2   Evidence of myiasis prior to indoor decomposition

The case report by Sukontason *et al.* (2005) details evidence of human myiasis by two species of calliphorids. The age of the maggots served as key evidence that insects had colonized the decedent prior to the agonal period.

24.5.2.1　Overview of case

The body of a 53-year-old man was discovered indoors on December 8, 2004 in a suburban region of the Muang District within the Chiang Mai Province of northern Thailand. The corpse lacked obvious signs of physical decomposition or mummification. A large lesion was present on the lower right leg of the decedent. Autopsy revealed that the lesion was the result of a squamous cell carcinoma. Ambient air temperatures in the region averaged 20.9 °C (minimum was 14.7 °C, maximum 28.2 °C) from December 4 to 8, 2004 and the mean relative humidity was 72.8%.

24.5.2.2　Entomological evidence

The leg lesion and carcinoma were infested with large numbers of fly larvae. The larvae were determined to be *Chrysomya megacephala* and *C. rufifacies*. All larvae were approximately 1.4 cm in length and estimated to be third stage larvae.

24.5.2.3　Key features

The deceased was identified as a homeless vagabond who likely spent significant time outdoors. Consequently, ambient natural (outdoor) air temperatures were applied to estimating the timing of insect colonization. Based on the ambient outdoor temperatures, both species of flies would require approximately 5 days to reach the third stage of larval development. This PMI estimate was not consistent with the physical decay of the deceased. The man was estimated to have died within 1 day of body discovery. Autopsy findings indicated that death was due to malnutrition associated with the large carcinoma. This cause of death was not surprising considering that the decedent was homeless and most likely did not seek or receive medical attention.

The entomological evidence strongly suggested that ante mortem myiasis of the leg lesion had occurred. In this case, the flies appeared to have colonized the wounds approximately 4 days before the onset of death. There was no evidence that the presence of the flies contributed to the man's death. Instead, the investigators surmised that the lack of

wound care and poor hygiene contributed to the fly infestation. The experience of investigators led them to suspect that myiasis had occurred before death. However, in cases in which a corpse displays obvious physical decomposition, the presence of precocious fly larvae may not be recognized, potentially leading to overestimation of the PMI.

24.5.3　Nosocomial myiasis

Cases of human myiasis have increased over recent years, especially in some tropical regions throughout the world. Of these incidences, a fairly common scenario is a patient who has become infested with fly larvae while located in a hospital or other type of facilitated or health care facility. In addition to the shock and disgust of finding maggots in an open wound, the plaintiff generally seeks monetary restitution for what is viewed from his/her perspective as medical malpractice. Equally frequent is a claim by the defendant that the maggots simply represent natural maggot therapy. An out-of-court settlement is typically reached after the forensic entomologist provides a forensic assessment of the entomological evidence and the opposing lawyer (usually that of the accused) has reviewed, and sometimes consulted with another entomologist, the depositions. Why? Regardless of guilt or innocence, no institution, especially a for-profit one, wants public attention related to myiasis. In a case that one of the authors (DR) of this book worked, the plaintiff was awarded a multimillion-dollar settlement.

In the case described below, Dr. Bernard Greenberg was asked to determine where and when a fly infestation of a leg wound occurred (Greenberg & Kunich, 2002). The patient was convinced that the infestation represented medical malpractice on the part of the hospital or physician that treated the work-site related injury.

24.5.3.1　Overview of case

A construction worker was injured when he fell from his truck on July 3rd. Surgery was performed at a hospital in Chicago to repair a tibial fracture of the right leg. The injury required a bone graft to be performed. The injured man remained in the hospital for nearly 2 weeks, finally being discharged on July 23rd. Six days later (July 29th), the man visited the physician's office to have the sutures removed and to be fitted for a leg cast. On August 2nd, the man returned to the hospital's emergency room complaining of "worms" crawling out from underneath the cast. The emergency room staff described the wound as emitting a strong, unpleasant odor and when the cast was removed, the injured leg displayed a yellow discharge and was infested with maggots. Additional medical procedures were required as a result. Ultimately, a malpractice suit was filed.

24.5.3.2 Entomological evidence

None of the fly larvae were collected or preserved in the emergency room. There were also no photos taken of the maggots. Thus, the only entomological information was based on the medical records from the patient's visit to the emergency room and from the testimony of the patient and medical staff.

24.5.3.3 Key features

The attorneys on both sides of the case were interested in a solitary question: where did the fly infestation originate? In the absence of physical evidence, Dr. Greenberg reasoned that at that time of year in that area of Chicago, the most abundant fly is *Phaenicia (Lucilia) sericata*. He then speculated that for fly larvae to come spilling out from behind the leg cast, they likely had completed feeding and were initiating dispersal behavior. To reach that stage of development, it was estimated that the flies would need slightly more than 3 days if the temperatures were between 29 °C and 35 °C. This temperature estimate was for the ambient conditions under the cast and taking into consideration body temperature. The timing placed fly colonization on July 30th, a period in which the injured construction worker was home convalescing. An examination of his home revealed that there were no screens on the windows, an obvious avenue for flies to enter and oviposit on or near the leg wound. Supporting this timing of colonization was the fact that at no time during his hospitalization or the time prior to suture removal, did he, his physician, or other medical staff report the presence of maggots. This meant that the timing of infestation had to have occurred after July 29th. A medical malpractice suit was filed but based more so on the more serious medical issues associated with the case. The entomological issue became tangential to the case.

This example is fairly typical of many involving nosocomial myiasis in that a treated wound becomes infested with maggots and the key becomes determining where colonization occurred. It is also common that the maggots are discarded. From the vantage point of medical staff, there is no reason to keep the larvae. The presumption is that they expect the patient to be relieved that the insects have been removed and discarded. No one is initially thinking of the "gross, disgusting, worms" as evidence. Unfortunately, without preserved larvae or at least a photo, not too much can be concluded about a case of myiasis without identification of the species. Under most circumstances, a competent attorney will hire a forensic entomologist to counter any argument that is based almost entirely on speculation.

A second common issue with these cases is the argument of maggot therapy. In a case of human myiasis that I (DR) consulted on in Baltimore, the medical staff of the defendant argued that even if fly infestation of the patient occurred at their facility, there would be no harm evoked since fly larvae are used in maggot therapy. In some instances that may well be true. However, there are important distinctions between maggot therapy and natural infestation of a wound. When maggot therapy is used in healthcare, it is a carefully controlled, artificially induced myiasis in which the medical practitioner balances the positive effects of maggot activity on necrotic tissue vs. the live tissue. The physician will monitor the track of the maggots and guide its progress to prevent any loss of healthy tissue. The most important factor is that the maggots used in this therapy have been approved specifically for this treatment [only one species (*Lucilia sericata*) has been approved for use in the United States] and are grown in an FDA approved lab. They are "sterile" and free from contamination. Therefore, there is a clear distinction between the sterile larvae and the actual procedure/process of the maggot therapy to the parasitic condition of human myiasis.

24.6 Entomological evidence associated with aquatic decompositions

Aquatic insects are not considered as useful to death investigations as terrestrial species. This view is based on the fact that exclusive necrophagy does not exist with aquatic insect fauna. Thus, the associations that connect insects with a corpse in terrestrial decompositions, namely insect colonization and subsequent development, are not as intimately tied to human remains in aquatic environments. Colonization of the deceased and development linked to the corpse are two fundamental assumptions of using insects for postmortem interval estimations. In practical terms, aquatic insects are not used to estimate a PMI. Consequently, many times the presence of aquatic species is simply ignored. As Wallace *et al.* (2008) discuss in a case from the Midwest, aquatic insects can yield a great deal of information if you know what to look for.

24.6.1   Overview of case

A sanitary engineer discovered a duck decoy bag partially submerged in a river (freshwater) located in the lower portion of the state of Michigan in the United States on June 13, 2005. Upon recovering the bag, it was discovered that a plastic garbage bag was inside. The bag was pulled to shore and opened. Inside were bones and flesh that were later determined to be human. A strong odor was emanating from the bag indicating that the remains had undergone decomposition.

Further examination of the bag contents revealed that the duck decoy bag was not fully submerged. The plastic bag contained rocks, rib bones, a portion of a chest containing a bone, hair, and other body fragments.

24.6.1.1   Entomological evidence

Insect specimens were collected from inside both bags during autopsy. The specimens were placed in ethanol for preservation and later identified as a single larva of *Musca domestica*, net spinning caddisfly larvae from the family Hydropyschidae, case building caddisfly larvae, *Pycnopsyche guttifer* (Limnelphilidae) and two larvae of *Pycnopysche lepida*.

24.6.1.2   Key features

In analyzing the entomological evidence, the forensic entomologists first needed to consider the typical pattern of human decomposition in freshwater ecosystems. Assuming the corpse was of sufficient mass initially, the body most likely sank or was full submerged when placed in the water. The presence of rocks in the plastic bag were

undoubtedly to add weight to ensure that the bags with the body did not float. In this scenario, terrestrial flies like *M. domestica* would not have access to the corpse until it became buoyant and portions of the bag or body became exposed to air. This would mean that *M. domestica* arrived later in decomposition and would not have been as useful to the investigations as the aquatic species.

The forensic entomologists turned their attention to the aquatic insects to derive a postmortem submersion interval or PMSI. The net spinning caddisfly were immediately ruled out as being useful in estimation of a PMSI in that they have complex taxonomy and life histories. That left the two species of case-building caddisfly larvae. To identify the species of limnephilids, the case construction needed to be examined to determine the type of material used, as well as the size of mineral pieces used in the cases. The size of the stream from where the insects were collected was also necessary context in the determination of the two species collected, *P. guttifer* (stick cases) and *P. lepida* (stone cases). Both species would be expected to occur in that river during the time of year the remains were found. Based upon the fact that larvae of *P. lepida* were not attached to the substrate with a silken strand, it was concluded that the specimens had not yet pupated. Similarly, none of the larvae of *P. guttifer* were attached either, again an indication that pupal development had not been initiated. There were some larval cases that had silken mesh across the anterior end indicating that these individuals had entered a dormant period prior to the onset of pupation. None of the larvae were undergoing a metamorphosis to the pupal stage at the time they were removed from their cases. Based on this information, the forensic entomologists concluded that the body had to have been submerged prior to the dormant period of the caddisflies. Thus, the PMSI was estimated to be between late April and late May of that year. This did not preclude the possibility that the corpse entered the water prior to the window of time encompassing the PMSI. However, the time range did account for the caddisfly activity.

This case example demonstrated not only the utility of aquatic insects in death investigations but how important a thorough understanding of life history information for insect fauna can be in an investigation. There is a clear need for more research associated with aquatic insects associated with decompositions, as well as for more individuals to be trained in this more specialized area of forensic entomology.

Chapter review

Case studies are important to understanding the context of insect evidence to legal investigations

- There is no substitute for the hands-on experiences associated with real casework. Of course this is not an option for students studying forensic entomology. Student participation in casework is not permitted for several reasons, including possible contamination/alteration of the evidence, loss or misplacement of evidence, and breach of confidentiality between the forensic entomologist serving as an expert and legal counsel/client.
- How can you gain practical experience if prevented from even shadowing an expert during real casework? The most common approach is to provide training through mock crime scenes or decomposition studies using animal surrogates for human corpses. Animal models permit the teaching of techniques and methodologies associated with recognition of physical and trace evidence, collection and preservation of arthropod specimens, and a variety of additional analyses including species identification, determination of specimen age, and estimation of the postmortem interval.
- The use of case study investigations is an alternative approach to mock crime scenes. They represent a form of high-impact practices in which students are able to focus on a topic that bridges knowledge gained from didactic learning and permits application of concepts and techniques to open-ended, real world problems. In this case, the real-life issues are criminal investigations. Students have the ability to apply what they have learned to actual cases. By incorporating different case scenarios involving insects and other arthropods into a course or training, students have opportunities to develop breadth and depth of understanding of casework in forensic entomology.

Insect evidence associated with homicides and suspicious deaths

- The *Regina v. Steven Truscott* case is well known in forensic entomology. The case's notoriety is associated with a wrongful conviction that took nearly 50 years to overturn, as well as controversial testimony regarding the entomological evidence. VanLaerhoven & Merritt (2019) provide a detailed account of their work for the defense in the 2006–2007 appeal of the original 1959 conviction of Steven Truscott. A summary of the case, entomological evidence, and key features of the investigation are presented. It is important to note that the prosecution relied on expert analysis from a separate forensic entomologist, and during the original case investigation in 1959, entomological evidence was collected and examined by a biologist working for the Attorney General's Laboratory in Toronto.

Entomological evidence associated with forensic entomotoxicology

Sometimes the utility of insects to a death investigation is less to do with determination of a postmortem interval and more so on their role as surrogates for human tissue. This can be the case when the corpse is mostly skeletonized, with little to no soft tissue remaining to use in toxicological analysis. Consumption of xenobiotic compounds in the corpse's tissues can alter the development of feeding stages of flies and beetles, thereby influencing any subsequent estimation of the postmortem interval. Similarly, detection and/or acceptance of a corpse by foraging insects maybe delayed due to the presence of certain xenobiotics in human tissues. This, in turn, would alter predictions of the timing of insect colonization based on the physical decay of the corpse. In his book "*A Fly for the Prosecution*," Dr. M. Lee Goff, a pioneer in the field of forensic entomotoxicology, provides several case examples in which drugs or toxins in the corpse altered interpretations of the entomological evidence. In each case, toxicological information contributed to understanding the developmental pattern displayed by fly larvae that had fed on the corpse.

Postmortem interval determination using insects on a burned corpse

- A burnt corpse offers unique challenges to estimating when insect colonization occurred. For instance, severely burned, charred bodies are rejected by most necrophagous insects as unsuitable for oviposition or progeny development. When the burn injury results in blistering or lesions in the skin without hardening of soft tissues, insect colonization can occur earlier than would be expected with non-burned remains. In a case reported by Pai *et al.* (2007), the burned remains of a young girl who had been beaten and stabbed were found colonized by *Chrysomya megacephala*. Their investigation of the case was aided by a recently conducted field study performed with a burned pig carcass to examine insect colonization and postmortem interval estimations based on insect development. Coincidently, the experiment was performed within 6 kilometers of where the girl's body was discovered.

Insect evidence associated with an indoor decomposition

- The insect fauna discovered with indoor decomposition often differs from that typical of body decomposition outdoors in terms of abundance and diversity. The major limiting factors are detection and access to the corpse. Obviously, odors must escape to attract outdoor species. Once detection occurs and foraging behavior is activated, the insects must locate an opening in the dwelling that permits entry. Most modern homes and buildings are designed to prevent heat loss or retain conditioned air. Consequently, insects with less chemical acuity or that are large in size do not commonly colonize human remains in concealed or indoor locations. This often times means that Coleoptera are rare in such cases, but not always, as detailed by Wang *et al.* (2019) for a case in the Guangdon Province of China.

Entomological evidence associated with aquatic decompositions

- Aquatic insects are not considered as useful to death investigations as terrestrial species. This view is based on the fact that exclusive necrophagy does not exist with aquatic insect fauna. Thus, the associations that connect insects with a corpse in terrestrial decompositions, namely insect colonization and subsequent development, are not as intimately tied to human remains in aquatic environments. Colonization of the deceased and

development linked to the corpse are two fundamental assumptions of using insects for postmortem interval estimations. In practical terms, aquatic insects are not used to estimate a PMI. Consequently, many times the presence of aquatic species is simply ignored. As Wallace *et al.* (2008) discuss in a case from the Midwest, aquatic insects can yield a great deal of information if you know what to look for.

Test your understanding

Level 1: knowledge/comprehension

1. Define the following terms:

    (a) myiasis
    (b) forensic entomotoxicology
    (c) physical evidence
    (d) nosocomial myiasis
    (e) larviposition
    (f) postmortem interval.

2. Explain the differences between a PMI and PMSI estimation.
3. Discuss the challenges that exist in using insects as PMI indicators for a corpse discovered indoors as opposed to outdoors.
4. Explain why aquatic insects generally cannot be used for estimation of a PMI.

Level 2: application/analysis

1. Explain what procedures are used for preservation of fly larvae collected from a corpse.
2. If fly larvae are collected at a crime scene and placed directly in ethanol, and then mailed to a forensic entomologist 6 days later, can these evidentiary samples be used to estimate a postmortem interval? Explain why or why not.
3. Discuss how myiasis and maggot therapy differ from each other.
4. Explain how the differences identified in question #3 are significant in considering a case of nosocomial myiasis. In other words, is natural maggot therapy a reasonable defense in a medical malpractice suit?
5. Discuss how myiasis complicates the estimation of a PMI.

Level 3: synthesis/evaluation

1. The body of a middle-aged Caucasian male wrapped in a blanket is discovered in a wooded

area of a suburban neighborhood in central Ohio. Under the blanket, the body was clothed in a tee shirt and shorts but no socks or shoes. The corpse was discovered at 11:35 AM on July 5th. The body showed signs of bloating, skin marbling and a distinct odor of decay. Fly maggots were present on the head in the eyes, nose and mouth. Explain how you would proceed with the entomological investigation.

Note

1. The details of all cases are primarily derived from the cited published case reports.

References

Amendt, J., Zehner, R. & Reckel, F. (2008) The nocturnal oviposition behaviour of blowflies (Diptera: Calliphoridae) in Central Europe and its forensic implications. *Forensic Science International* 175(1): 61–64.

Greenberg, B. (1990) Nocturnal oviposition behavior of blow flies (Diptera: Calliphoridae). *Journal of Medical Entomology* 27(5): 807–810.

Greenberg, B. & Kunich, J.C. (2002) *Entomology and the Law*. Cambridge University Press, Cambridge, UK.

Hall, R.D. (2020) The forensic entomologist as expert witness. In: J.H. Byrd & J.K. Tomberlin (eds) *Forensic Entomology: The Utility of Arthropods in Legal Investigations*, pp. 333–348. CRC Press, Boca Raton, FL.

Harland-Logan, S. (2020) Steven Truscott. https://innocencecanada.com/exonerations/steven-truscott/#ftn3. Accessed September 19, 2020.

Keough, N., Myburgh, J. & Steyn, M. (2017) Scoring of decomposition: a proposed amendment to the method when using a pig model for human studies. *Journal of Forensic Sciences* 62(4): 986–993.

Kilgo, C., Ezell Sheets, A. & Pascarella, J. (2015) The link between high-impact practices and student learning: some longitudinal evidence. *Higher Education* 69: 509–525.

Li, L., Wang, Y., Wang, J., Ma, M. & Lai, Y. (2016) Temperature-dependent development and the significance for estimating postmortem interval of *Chrysomya nigripes* Aubertin, a new forensically important species in China. *International Journal of Legal Medicine* 130(5): 1363–1370.

Pai, C.Y., Jien, M.C., Li, L.H., Cheng, Y.Y. & Yang, C.H. (2007) Application of forensic entomology to postmortem interval determination of a burned human corpse: a homicide case report from southern Taiwan. *Journal of the Formosan Medical Association* 106(9): 792–798.

Rippy, M. (2020) Thawing cold cases in the classroom. *Journal of Forensic Science Education* 2(1), https://jfse-ojs-tamu.tdl.org/jfse/index.php/jfse/article/view/16.

Sukontason, K.L., Narongchai, P., Sripakdee, D., Boonchu, N., Chaiwong, T., Ngern-Klun, R., Piangjai, S. & Sukontason, K. (2005) First report of human myiasis caused by *Chrysomya megacephala* and *Chrysomya rufifacies* (Diptera: Calliphoridae) in Thailand, and its implication in forensic entomology. *Journal of Medical Entomology* 42(4): 702–704.

VanLaerhoven, S.L. & Merritt, R.W. (2019) 50 years later, insect evidence overturns Canada's most notorious case—Regina v. Steven Truscott. *Forensic Science International* 301: 326–330.

Wallace, J.R., Merritt, R.W., Kimbirauskas, R., Benbow, M.E. & McIntosh, M. (2008). Caddisflies assist with homicide case: determining a postmortem submersion interval using aquatic insects. *Journal of Forensic Sciences* 53(1): 219–221.

Wang, M., Chu, J., Wang, Y., Li, F., Liao, M., Shi, H., Zhang, Y., Hu, G. & Wang, J. (2019) Forensic entomology application in China: four case reports. *Journal of Forensic and Legal Medicine* 63: 40–47.

Williams, K.A., Wallman, J.F., Lessard, B.D., Kavazos, C.R., Mazungula, D.N. & Villet, M.H. (2017) Nocturnal oviposition behavior of blowflies (Diptera: Calliphoridae) in the southern hemisphere (South Africa and Australia) and its forensic implications. *Forensic Science, Medicine, and Pathology* 13(2): 123–134.

Supplemental reading

Al-Mesbah, H., Al-Osaimi, Z. & El-Azazy, O.M. (2011) Forensic entomology in Kuwait: the first case report. *Forensic Science International* 206(1–3): e25–e26.

Benecke, M. (1998) Six forensic entomology cases: description and commentary. *Journal of Forensic Science* 43(4): 797–805.

Benecke, M. & Lessig, R. (2001) Child neglect and forensic entomology. *Forensic Science International* 120(1–2): 155–159.

Beyer, J.C., Enos, W.F. & Stajić, M. (1980) Drug identification through analysis of maggots. *Journal of Forensic Science* 25(2): 411–412.

Bugelli, V., Papi, L., Fornaro, S., Stefanelli, F., Chericoni, S., Giusiani, M., Vanin, S. & Campobasso, C.P. (2017) Entomotoxicology in burnt bodies: a case of maternal filicide-suicide by fire. *International Journal of Legal Medicine* 131(5): 1299–1306.

Campobasso, C.P., Henry, R., Disney, L. & Introna, F. (2004) A case of *Megaselia scalaris* (Loew)(Dipt., Phoridae) breeding in a human corpse. *Anil Aggrawal's Internet Journal of Forensic Medicine and Toxicology* 5(1): 3–5.

Charabidze, D., Colard, T., Vincent, B., Pasquerault, T. & Hedouin, V. (2014) Involvement of larder beetles (Coleoptera: Dermestidae) on human cadavers: a review of 81 forensic cases. *International Journal of Legal Medicine* 128(6): 1021–1030.

Goff, M.L. (1992) Problems in estimation of postmortem interval resulting from wrapping of the corpse: a case study from Hawaii. *Journal of Agricultural Entomology* 9(4): 237–243.

Huchet, J.B. & Greenberg, B. (2010) Flies, Mochicas and burial practices: a case study from Huaca de la Luna, Peru. *Journal of Archaeological Science* 37(11): 2846–2856.

Levine, B., Golle, M. & Smialek, J.E. (2000) An unusual drug death involving maggots. *The American Journal of Forensic Medicine and Pathology* 21(1): 59–61.

Lindgren, N.K., Sisson, M.S., Archambeault, A.D., Rahlwes, B.C., Willett, J.R. & Bucheli, S.R. (2015) Four forensic entomology case studies: records and behavioral observations on seldom reported cadaver fauna with notes on relevant previous occurrences and ecology. *Journal of Medical Entomology* 52(2): 143–150.

Nolte, K.B., Pinder, R.D. & Lord, W.D. (1992) Insect larvae used to detect cocaine poisoning in a decomposed body. *Journal of Forensic Science* 37(4): 1179–1185.

Pujol-Luz, J.R., Marques, H., Ururahy-Rodrigues, A., Rafael, J.A., Santana, F.H., Arantes, L.C. & Constantino, R. (2006) A forensic entomology case from the Amazon rain forest of Brazil. *Journal of Forensic Sciences* 51(5): 1151–1153.

Salleh, A.F.M., Marwi, M.A., Jeffery, J., Hamid, N.A.A., Zuha, R.M. & Omar, B. (2007). Review of forensic entomology cases from Kuala Lumpur hospital and hospital universiti kebangsaan Malaysia, 2002. *Journal of Tropical Medicine and Parasitology* 30: 51–54.

Samnol, A., Kumar, R., Mahipal Singh, S. & Parihar, K. (2020) Study of insect larva used to detect toxic substance through decomposed bodies. *International Journal of Forensic Science & Pathology* 7(1): 420–422.

Turchetto, M., Lafisca, S. & Costantini, G. (2001). Postmortem interval (PMI) determined by study sarcophagous biocenoses: three cases from the province of Venice (Italy). *Forensic Science International* 120(1–2): 28–31.

Turchetto, M. & Vanin, S. (2004) Forensic evaluations on a crime case with monospecific necrophagous fly population infected by two parasitoid species. *Anil Aggrawal's Internet Journal of Forensic Medicine and Toxicology* 5(1): 12–18.

Vanin, S., Bonizzoli, M., Migliaccio, M.L., Buoninsegni, L.T., Bugelli, V., Pinchi, V. & Focardi, M. (2017) A case of insect colonization before the death. *Journal of Forensic Sciences* 62(6): 1665–1667.

Wallace, J.R. (2019) Computer-aided image analysis of crayfish bitemarks—reinterpreting evidence: A case report. *Forensic Science International* 299: 203–207.

Additional resources

American Board of Forensic Entomology: https://forensicentomologist.org/

Blow flies solve murder in Vegas: https://www.inverse.com/article/51681-absence-of-blow-flies-overturned-kirstin-blaise-lobato-murder-conviction

Case studies: http://aboutforensics.co.uk/case-studies/

Case studies of seven serial killers in America: http://faculty.fortlewis.edu/burke_b/Forensic/Class%20Readings/Murder.pdf

Crime Scene Creatures: Fly Justice: https://video.search.yahoo.com/search/video;_ylt=AwrEwhGILwJgyosADBlXNyoA;_ylu=Y29sbwNiNiZjJEEcG9zAzEEdnRpZAMEc2VjA3BpdnM-?p=forensic+entomology+case+studies&fr2=piv-web&fr=yfp-t-s#action=view&id=40&vid=f160b4fdd6f3f2fa77f0c54cffc46d56

Forensic entomology: https://aboutforensics.co.uk/forensic-entomology/

Forensic Entomology: Effective use of insects in death investigations: https://forensicyard.com/forensic-entomology/

Forensic Entomology is more than just flies and beetles: https://entomologytoday.org/2015/01/22/forensic-entomology-is-more-than-just-blow-flies-and-beetles/

Forensic Files: Insect clues: https://video.search.yahoo.com/search/video;_ylt=AwrEwhGILwJgyosADBlXNyoA;_ylu=Y29sbwNiNiZjJEEcG9zAzEEdnRpZAMEc2VjA3BpdnM-?p=forensic+entomology+case+studies&fr2=piv-web&fr=yfp-t-s#action=view&id=45&vid=250025eba82ac707d987a66bb65130d1

Appendix I

Collection and preservation of calyptrate Diptera

Collecting adult flies

The calyptrate Diptera includes the Calliphoridae (blow flies), Sarcophagidae (flesh flies), and several other families of forensic interest. Well-preserved specimens will be much easier to identify (or get identified by others) than poorly preserved specimens. This preservation process begins when a specimen is first collected. In many ways, insects are pretty tough and can withstand some rough treatment without destruction of key characters needed for identification. Live specimens can generally withstand more manipulation than dead specimens. However, once an insect dies, changes begin in their bodies, making them more susceptible to damage.

A good summary of recommended procedures for collecting specimens at a potential crime scene are provided by Sanford *et al.* (2020). This reference provides a range of details that must be considered and acted upon when specimens are being collected that may be important for a legal matter. We will not repeat this information here. However, there is one procedure often suggested by forensic entomology specialists that consistently causes problems in the identification process: the placement of adult flies directly into ethanol after they have been killed. Many of the characters that need to be seen for morphological identification of blow flies and flesh flies become obscured by this common

procedure. While it is possible to identify specimens that have been preserved in alcohol, it is much more time-consuming. Extra effort during the collection and preservation process will result in superior-looking specimens that are much easier and quicker to identify. This appendix is intended to provide information and tips for collection, preservation, and identification based on Dr Dahlem's 30 plus years of experience collecting and identifying calyptrate Diptera.

The basic collecting equipment that needs to be assembled before heading out into the field includes:

- A fine mesh insect net.
- A minimum of five medium-sized killing jars, charged with killing agent.
- Fine forceps.
- Mechanical pencil and small notebook.
- A GPS unit (or smartphone with GPS capability).
- A camera or smartphone
- 20 or more small (4–6 dram or 15–22 mL) glass or plastic vials with tight lids.
- A backpack or collecting (photographer's) vest with lots of pockets.
- A bottle of water (to drink).
- A pack of tissues or couple of napkins.

Entomologists studying different taxonomic groups of insects prefer different kinds of equipment and will use different techniques to find and capture

The Science of Forensic Entomology, Second Edition. David B. Rivers and Gregory A. Dahlem.
© 2023 John Wiley & Sons Ltd. Published 2023 by John Wiley & Sons Ltd.
Companian website: www.wiley.com/go/forensicentomology2

specimens. The following recommendations are meant for those whose main focus is flies, particularly the species of higher Diptera generally involved in forensic studies. Additional information dealing with techniques for collecting and preserving flies can be found in chapter 2 of Chandler (2010).

For a general insect net, a professional series insect net with 15-inch (38 cm) net ring diameter, 3-inch (7.6 cm) handle, and soft aerial white net bag is recommended. Killing jars are available for purchase, but it is very simple to make your own. Spice jars make excellent killing jars, especially those from Spice Islands or Penzeys. They are made of heavy glass, which resist breakage, and have metal or hard plastic screw-top lids. Mix up and pour about 3 cm of plaster of Paris into the bottom of a clean jar and allow it to dry thoroughly. Be sure to prepare and attach a label stating "Danger: Poison" and the name of the killing agent you are using for each jar.

For general fly collecting, ethyl acetate is recommended as a killing agent. It kills flies very rapidly, but does not work as well on beetles, which may require much more time to die. This organic solvent is usually fairly easy to obtain, but you can only rely on this poison if you can personally transport the killing jars and/or killing agent to your collecting location (e.g., driving with it in a car). Ethyl acetate should not pose much of a poisoning risk to humans (when used properly) but it is very flammable and cannot be taken on a plane or be sent in the normal mail. In a pinch, many types of nail polish remover have ethyl acetate as the main ingredient, and this can be purchased at a grocery or drug store. Note that using nail polish remover is only mentioned as an emergency solution for a killing agent. While it will work to kill specimens, the other ingredients can cause "greasing" of fly specimens that come into contact with the chemical. "Greasing" melts the surface lipids, blackens the specimen, and makes it look wet, generally making specimens much more difficult to identify and much less visually appealing. When you are traveling by plane, be sure to follow regulations involving banned substances and see if someone at your destination can supply you with needed killing agents.

Adding liquid poison to your killing jar should always be done outside or in a well-ventilated location. To "charge" the killing jar, pour ethyl acetate onto the dry plaster and give it 5 minutes or so for the plaster to absorb the killing agent. Pour off any extra liquid and leave the cap open for a minute or so until the interior of the jar is fully dry. Add a piece of lightly wadded tissue to the jar (if using nail polish remover, add several centimeters of paper towel material firmly at the bottom so that there is no way a fly can contact the plaster with killing agent). The tissue will absorb moisture in the jar, allow the flies to perch and die on material away from the plaster bottom with killing agent, and decreases the chances of greasing of specimens.

Document when and where you are collecting specimens. The main location (state, county, and name of location or nearest city) and date should be entered into your field book to start your collection notes. With each individual collection, note the time, GPS coordinates, and biological or behavioral observations dealing with the specimens. A good working procedure is to divide up collecting events into half-hour intervals. A few photos of the place where you caught the fly, or flies, are always good to compare with your field notes later. Use a different killing jar for different locations and number them so you can keep track of which jar holds which specimens.

For normal-sized blow flies and flesh flies, I would suggest that they stay in a kill jar at least 15 minutes but no longer than 1 hour before moving the specimens to a separate storage jar or container (to make sure they are fully dead). If they are still moving, they are not dead yet. Do not expose the kill jars to any more direct sunlight than possible, as this will lead to rapid internal heating of the jar and may cause condensation on the inside walls of the jar which can cause greasing or other types of destruction of your specimens. Simply keeping the jars in your pocket or in a cloth backpack when you are not adding specimens will stop this problem. If you are collecting for a while at one location (and are catching a good number of flies), alternate catches between two kill jars. This lowers the chance that a resilient fly will escape as you attempt to add a freshly caught specimen.

Once the flies are fully dead, you should transfer them out of the killing jar and into a temporary storage container. Small glass or plastic vials with tight lids work well for this. In the past, many fly collectors used Fuji film containers with snap-top lids, but these are getting very hard to find. Each vial should be individually numbered for easy association between your field notes and the specimens. The temporary storage vials should be lined with absorbent paper. Using one-quarter of a napkin (the thin ones from fast food restaurants work especially well) is a good choice for this. Take the one-quarter of napkin and fold it several times into a strip the same height as the storage vial. Wrap the strip around your finger and slide your finger into the vial. When you let loose, the napkin strip will uncurl to form a barrier around the outside wall of the vial. Press a small folded-up leaf to the bottom of the vial. A "drier" leaf, like a redbud or oak leaf will work better than a "moister" leaflike dandelion. The leaf will maintain moisture in the vial without causing condensation and will keep your specimens fresh until you get back to the lab.

When you transfer flies from one container to another, be sure to do this out of the wind and over a smooth background (one solution is to do this over your net bag on the ground in the field) so that specimens are not lost during the transfer. Use fine forceps for handling and transferring the flies from one container to the other. Good forceps are essential for handling specimens in the field and lab, so be sure to get something like the Swiss style forceps with fine or superfine tips. Picking flies up by their wings will eliminate damage to key characters on their bodies. You can gently move them by grasping a leg, but legs come off much more easily than a wing. Be sure to unfold the tissue in the killing jar to remove specimens that might have crawled inside folds of the tissue as they died. Do not stuff the storage vial if you have a lot of flies; they should be "loose-packed" at most. Be sure to note time, GPS coordinates, and other biological information in your field book for each group of specimens. Keep storage vials cool and away from direct sunlight. Put them into the refrigerator when you get home from your collecting trip (or freezer if you will not have the time to process the specimens in the next 24–48 hours).

Collecting fly larvae

Again, the point of this appendix is not to provide detailed information on what specimens to collect or where to collect them from, but to give some practical information on collecting specimens. Use a resource like Sanford et al. (2020) as your guide for what, where, and how to document larval collections for a legal investigation. Live maggots are pretty tough and can survive rougher handling than you might imagine, especially in the third instar, but they are not immune to damage or death. If you are collecting or manipulating live larvae, do *not* use the Swiss style forceps described above. The tips are too fine to handle live larvae, making it too easy to

accidentally puncture a larva when trying to pick it up. Use flexible featherweight forceps with a narrow squared point to pick up or manipulate live larvae and puparia.

Larvae should *not* be dropped alive into a vial of alcohol. They will die in the alcohol, but they will also constrict themselves, which can affect estimations of postmortem interval based on larval length. It is recommended that larvae be killed by dropping them into hot (near boiling) water for a minute or so before transferring them to 80% ethanol. This expands the larva to full size, kills it very quickly, cleans off much of the gooey residue, accentuates the posterior spiracular plates, and maintains a clean white coloration for a long time. This is a good procedure (for more detailed information on methods for killing and preserving larvae and the effects of several different techniques, see Adams & Hall, 2003). But how do you get access to hot water when you are out at an investigation site? Note that killing in alcohol and then putting larvae into hot water when you get back to the lab does not work – it has to be done at the time you kill the larvae. One technique that solves this problem is to fill a coffee thermos with boiling water just before you leave for the field. Bring a small wide-mouthed container and small kitchen strainer. Then you can collect larvae in the field, pour hot water from the thermos into the container, add larvae, strain out larvae after about 90 seconds, and place "blanched" larvae into ethanol for long-term preservation and storage. Be sure to put only larvae into the alcohol and do not transfer tissue, rocks, or other materials that were associated with the larvae at the time of collection. Note that systematists have been known to refuse to identify specimens if bits of human tissue are in the vial with the larvae.

Mounting and preserving specimens (adult flies)

You can immediately mount your catch when you get back to the lab from the field, but you will find that the specimens are easier to manipulate if you leave them in a refrigerator for 10–24 hours before trying to pin them. Lightly pinch the specimen between your thumb and forefinger so that the dorsal surface of the fly is exposed. Insert an insect pin just below the dorsal suture of the thorax, slightly to the right of the midline. Push slowly and adjust angle of pin when it exits the body so that it does not remove a leg. This takes a bit of practice to reach the point that you can feel when a leg might be catching the tip of the pin. Insect pins come in a variety of sizes. For most blow flies and flesh flies, pins of size 1 or 0 are recommended. You can probably use up to a size 2 or down to 00, but the idea is to use a pin that will cause insignificant damage to the external characters while providing the strength to not bend when pushed into storage containers.

After placing a pin into a specimen, you should look at it under a dissecting microscope and gently manipulate the legs so that they are fully extended below the specimen. A variety of useful morphological characters used for identification are found on the legs, but legs usually need to be extended to see these. If you are collecting tissue from the flies for possible mitochondrial DNA-based identification, this is a good time to do this. If the fly has all six legs, remove the three legs from one side. You will get the maximum amount of muscle tissue if you remove each leg by pinching it with your Swiss style forceps at the coxa, or junction of the coxa and thorax, and pulling the leg away from the body. Sometimes they come off easily; sometimes it can be a little more difficult to remove the legs without damaging the pinned specimen. Place the legs into a microcentrifuge tube filled with 95–100% ethanol for storage. Each tube should have a unique identifier code (e.g., CD209, CD210, and CD211) and that unique code should also be placed on a small label that goes on the pin below the specimen, so that tissue and specimen can always be associated with one another. Three legs work well as each leg contains enough cells and associated DNA to run a genetic test and it leaves a complete set of legs on the pinned specimen for morphological identification. This procedure also allows two "back-ups" in case any problem arises with the initial genetic test. Ideally, the tubes containing tissue (legs) should be placed into an ultra-low temperature freezer for storage.

At this point you are nearly finished with the females. Make sure they look attractive with the head on straight, remaining legs extended downward, and wings extended upward (so you see the ventral side of wing in lateral view). You can often get the wings in an "up" position (if they are not already there) by placing the tips of your forceps on either side of the fly under the "shoulder" area and lifting until the wings snap into the "up" position. In general, fly wings will naturally flip into an "up" or "down" position after the fly dies, and "up" is preferable for seeing morphological characters over "down." Now you are finished with the mounting process for your females and males of many metallic green species of blow flies.

For male flesh flies and some male blow flies (especially those with a metallic blue abdomen) you will want to expose the male genitalia. This is a fairly simple process that gets much easier the more you practice. Take the fresh pinned specimen and insert the pin at an acute angle into a small slice of Styrofoam ($5 \times 9 \times 1$ cm blocks should work well), so that the specimen is lightly resting on its lateral side. You will need a pack of "minutens" for spreading the genitalia. Minutens are very tiny pins. Insert one minuten pin on top of the fifth abdominal tergite, just before the genital capsule. Under the dissecting microscope, use fine forceps and a second minuten to spread the genitalia by placing the minuten on the anterior surface of the cerci and gently pulling back until the cerci are parallel with the main pin. The phallus should extend down at this point and be clearly visible (see figure 1 in Dahlem & Naczi, 2006). Allow the specimen to dry for several days before removing the minutens. The end result is a properly spread specimen that can usually be identified to species without dissection. For most flesh flies and blow flies, the male genitalia show a species-specific diagnostic shape. Be sure to include critical locality or field note information with the pinned specimens so that proper association can be made. Always keep locality data with specimens: do not rely on your memory!

A locality label should be placed on each specimen. The locality labels for dry specimens can be printed in black ink on a laserjet or inkjet printer. Ariel 4 point font works well for locality labels and are easy to construct with a word processing program such as Microsoft Word. The drop-down menu for font size will not show "4" but if you click on the font size it will highlight, then just type in the size you want. You may want to change the "View" on your screen to 200%, so that it is easier to see what you are typing in. If you are making many labels, you may want to change your margins to "Narrow" or "0.5" on all sides, insert a table with seven or eight columns, and type the individual locality label information into individual cells of the table. This makes it very easy to make up multiple labels for the same location by copying the information from one cell to subsequent cells. If you use a table format, be sure to select the table and use the drop-down menu for border lines to

USA: KENTUCKY: Kenton Co.
Fort Wright, suburban backyard
N 32° 50.84′, W 108° 17.99′
Collected on dog dung with hand net
July 4, 2021; Coll. G. A. Dahlem

select "No Border" before you print the labels. A typical label should have country, state, and county on the first line, location name or nearest city on the next line, GPS coordinates on the third line, collecting method or environmental note on the fourth line, date and the name of the collector on the sixth line (see Box AI.1).

References cited

Adams, Z.J.O. & Hall, M.J.R. (2003) Methods used for the killing and preservation of blowfly larvae, and their effect on post-mortem larval length. *Forensic Science International* 138: 50–61.

Chandler, P.J. (ed.) (2010) *A Dipterists Handbook*, 2nd edn. Amateur Entomologists' Society, London.

Dahlem, G.A. & Naczi, R.F.C. (2006) Flesh flies (Diptera: Sarcophagidae) associated with North American pitcher plants (Sarraceniaceae), with descriptions of three new species. *Annals of the Entomological Society of America* 99: 218–240.

Sanford, M.R., Byrd, J.H., Tomberlin, J.K. & Wallace, J.R. (2020) Entomological evidence collection methods. In: J.H. Byrd & J.K. Tomberlin (eds) *Forensic Entomology: The Utility of Arthropods in Legal Investigations*, 3rd edn, pp. 63–85. CRC Press, Boca Raton, FL.

Resources and links

See the article "How to make your tachinids stand out in a crowd" by J.E. O'Hara in Issue 34 of Tachinid Times (2021) for useful information on making your pinned specimens look nice and professional. See: http://www.nadsdiptera.org/Tach/WorldTachs/TTimes/Tach34.html

Appendix II
Getting specimens identified

Morphological identification of specimens on your own

If you wish to try to identify specimens on your own, you should start by identifying to order and family level. A good resource for order- and family-level identifications of adult insects is *Borror and DeLong's Introduction to the Study of Insects* by Johnson and Triplehorn (2004). Another possibility is *American Insects. A Handbook of the Insects of America North of Mexico* by Arnett (2000). For immature insects, a good general reference for family-level identifications is the *Immature Insects* series edited by Stehr (1987, 1991).

If you know what order your insect belongs to, you may be able to find an identification resource for that particular order. For example, if you know that your insect belongs to the order Coleoptera, a good resource to start with is the two-volume *American Beetles* books by Arnett and Thomas (2000) and Arnett *et al.* (2002). Arnett and Thomas (2000) and Arnett *et al.* (2002) includes keys to family and genus. If you know you have a member of the Diptera, you can start with the branching, pictorial, family-level key in Marshall (2012) which includes nice photographic illustrations of key morphological characters. If you do not have access to Marshall's book, another good family key can be found in Volume 1 of the three-volume *Manual of Nearctic Diptera* by McAlpine *et al.* (1983, 1987), which can be freely downloaded online. Volumes 1 and 2 of McAlpine *et al.* (1983, 1987) include keys to family and genus. Once you have a family-level identification, you may wish to search for available systematic publications on that particular family for generic and species level keys.

The most commonly encountered flies of forensic interest are the blow flies (Family Calliphoridae) and flesh flies (Family Sarcophagidae). For identifying adult most forensically important Calliphoridae from North America. This is a photograph-based pictorial key to the subfamilies, genera, and many species. Whitworth (2020) (which is an updated and revised version of Whitworth, 2006) is another extremely useful key to calliphorid species. There is no single reference that can be used for adult Sarcophagidae of North America. Identification of flesh flies is usually accomplished by matching the male genitalia of a specimen with a published figure, rather than by using a dichotomous morphological key. The two essential revisionary works to begin with for North America are Aldrich (1916) and Roback (1954). Note that many name changes (see Chapter 5) have occurred since these revisions were written and you will need to check for the current name by using Pape (1996). A guide to current names for the species included in Aldrich (1916) is provided in Appendix IV. Some other references that include species of high forensic importance are Dodge

The Science of Forensic Entomology, Second Edition. David B. Rivers and Gregory A. Dahlem.
© 2023 John Wiley & Sons Ltd. Published 2023 by John Wiley & Sons Ltd.
Companion website: www.wiley.com/go/forensicentomology2

(1966), Giroux and Wheeler (2009), and Parker (1919, 1923). Note that the key to genera of Sarcophagidae in the *Manual of Nearctic Diptera* is difficult to use, reflects a "splitter's" point of view in handling genera, and is not recommended for use by non-specialists.

Access to a reference collection of identified specimens can be extremely helpful. Look for specimens with a separate identification label attached to the pin. When systematic experts identify specimens, they normally attach a label to each specimen, or the first specimen in a series, that provides the species name and the expert's name. These are the best specimens to use for comparison to confirm your personal identification. Anyone who expects to do many identifications on their own should start assembling their own personal reference collection of specimens.

Identification of specimens (by systematic expert)

The most important thing to keep in mind when contacting a systematic expert for help with identification is the time that will be required by you and the expert. In general, payment should be offered to the expert. The inclusion of identification costs in a research grant proposal is fully justifiable and is not normally questioned by funding agencies. This is for consideration of the time required to handle and put a name on that specimen, and for the specialist's expertise. This is especially true if the identification will be used in court. If you think of the time and expense involved with obtaining a DNA-based identification of a single specimen, it can help to guide your offer to a systematic expert for morphologically derived identifications. Generally, a payment of approximately $55 per identification (or $100/hour) is a good starting point. Things that will affect the price are often the condition of the specimens (pinned nicely with genitalia exposed, or preserved in alcohol and require pinning and dissection), the number of specimens (single specimens can often take just about the same amount of time as a series of 10 when handling time is taken into account), the sex of the specimens (males are often easier to identify than females), and how critical an identification is to a particular case (highly important specimens need to be handled and identified with a higher degree of scrutiny). If the systematic expert is employed in a public service position that requires identification as part of their job description (e.g., research entomologists at the National Museum of Natural History), the expert may not be able to accept payment for identification services.

Many identifications do not require the help of PhD-level systematic entomologists, the top resource for species determinations. A scientist trained in forensic entomology may be perfectly competent to identify most specimens of forensic interest. It is when unusual taxa, especially difficult species groups or poorly preserved and/or damaged specimens, need identification that a higher level of systematic expertise is called for. Finding a specialist in a particular taxonomic group can be difficult; there is no single website or resource that provides contact information for available systematic expertise. You will need to contact people and network with them to find the expert you need or to see if an expert is even available (many important taxonomic groups do not have an active expert working on them right now).

Most of the major orders of insects have separate organizations devoted to the study of their group. For coleopterists, there is the Coleopterists Society (https://www.coleopsoc.org/), for hymenopterists there is the International Society of Hymenopterists (www.hymenopterists.org), and for dipterists there is the North American Dipterists' Society (https://dipterists.org/). These are "order-level" societies. If you are hunting for an expert in a particular group of insects, these society web pages can provide great leads on who is working on what, and how to contact a particular expert. You may need to join the group to be able to use their search functions. Other taxonomic-based web resources can be very useful when searching for a systematic expert in a particular group of insects or confirming an identification of your own.

In some cases, a detailed photograph of diagnostic features of an insect can allow an expert to provide an identification without the time and cost involved with physically shipping the specimen. A good SLR camera attached on a C-mount to a dissecting microscope can provide a photo good enough to allow identification. If you do not have access to that sort of photographic resources, you may be able to find a good mount for your smartphone to take photos through a microscope lens. Really good images can be obtained by taking photos at multiple focus planes and using software to "stack" or combine the multiple photos into one in-focus composite image. Whenever possible, sending a photo along with your email request for identification help may result in a quicker and more positive response from the specialist you are trying to get help from. Note that identifications from photos cannot replace an identification derived from observation of the physical specimen. If an identification is going to be presented in court as coming from a particular expert, the expert should physically see the specimen.

Everyone seems to think they know how to ship fragile objects like insect specimens in the mail, but few actually do. With all the time and effort involved in shipping specimens, and the potential danger of the loss of the physical specimens in the shipping process, it is wise to use extra care when boxing specimens for shipment through the mail. It is highly recommended that an expert is consulted for advice on packaging fragile insects. People who ship specimens often include entomology museum curators and systematic entomologists. For shipment within the United States, all the major carriers seem to work equally well (U.S. Postal Service, UPS, FedEx). If you are shipping specimens outside the country (and this includes Canada), a whole new set of concerns comes into play. To legally send biological specimens to foreign countries you must obtain authorization from the United States Fish and Wildlife Service (FWS). There is a form for the export of specimens (US FWS 3-177) that must be completed and submitted to the regional enforcement office along with some additional documentation. A link to the FWS website that gives instructions on this form is provided in the resources and links section. If you do not follow proper procedure when it comes to international shipments you can end up with hefty fines, legal action, and possible loss and/ or destruction of the specimens at the border. Always realize that when specimens are shipped in the mail there is a possibility that the package will be lost or damaged. So, use extra care and precaution especially when sending valuable specimens.

References cited

Aldrich, J.M. (1916) Sarcophaga *and Allies in North America*, Vol. 1. Entomological Society of America, Thomas Say Foundation, Lafayette, IN. Available as a free pdf download from the Biodiversity Heritage Library at http://www.biodiversitylibrary.org/item/35357.

Arnett, R.H. Jr & Thomas, M.C. (eds) (2000) *American Beetles, Volume I: Archostemata, Myxophaga, Adephaga, Polyphaga: Staphyliniformia*. CRC Press, Boca Raton, FL.

Arnett, R.H. Jr, Thomas, M.C., Skelley, P.E. & Frank, J.H. (eds) (2002) *American Beetles, Volume II: Polyphaga: Scarabaeoidea through Curculionoidea*. CRC Press, Boca Raton, FL.

Arnett, R.H. Jr. (2000) *American Insects. A Handbook of the Insects of America North of Mexico*, pp. 1003. CRC Press.

Dodge, H.R. (1966) Sarcophaga utilis Aldrich and allies (Diptera, Sarcophagidae). *Entomological News* 77: 85–97.

Giroux, M. & Wheeler, T.A. (2009) Systematics and phylogeny of the subgenus Sarcophaga (Neobellieria) (Diptera: Sarcophagidae). *Annals of the Entomological Society of America* 102: 567–587.

Johnson, N.F. & Triplehorn, C.A. (2004) *Borror and DeLong's Introduction to the Study of Insects*. Thompson Brooks/Cole, Belmont, CA.

Marshall, S. A. (2012) *Flies. The Natural History and Diversity of Diptera*, pp. 616. Firefly Books Ltd.

McAlpine, J.F., Peterson, B.V., Shewell, G.E., Teskey, H.J., Vockeroth, J.R. & Wood, D.M. (eds) (1983) *Manual of Nearctic Diptera*, Vol. 1. Biosystematic Research Institute Research Monograph 28. Available for free download as pdf file from the Entomological Society of Canada at http://www.esc-sec.ca/aafcmono.php.

McAlpine, J.F., Peterson, B.V., Shewell, G.E., Teskey, H.J., Vockeroth, J.R. & Wood, D.M. (eds) (1987) *Manual of Nearctic Diptera*, Vol. 2. Biosystematic Research Institute Research Monograph 28. Available for free download as pdf file from the Entomological Society of Canada at http://www.esc-sec.ca/aafcmono.php.

Pape, T. (1996) *Catalogue of the Sarcophagidae of the World (Insecta: Diptera)*. Memoirs on Entomology International, Vol. 8. American Entomological Society, Gainesville, FL.

Parker, R.R. (1919) Concerning the subspecies of Sarcophaga dux Thomson. *Bulletin of the Brooklyn Entomological Society* 14: 41–46.

Parker, R.R. (1923) New Sarcophagidae from Asia, with data relating to the dux group. *The Annals and Magazine of Natural History* 11: 123–129.

Roback, S.S. (1954) *The Evolution and Taxonomy of the Sarcophaginae*. Illinois Biological Monograph, Vol. 23. University of Illinois Press, Urbana.

Stehr, F.W. (ed.) (1987) *Immature Insects*, Vol. 1. Kendall Hunt, Dubuque, IA.

Stehr, F.W. (ed.) (1991) *Immature Insects*, Vol. 2. Kendall Hunt, Dubuque, IA.

Whitworth, T. (2006) Keys to the genera and species of blow flies (Diptera: Calliphoridae) of America north of Mexico. *Proceedings of the Entomological Society of Washington* 108: 689–725.

Whitworth, T. (2020) Keys to the genera and species of blow flies (Diptera: Calliphoridae) of America, North of Mexico. In: J.H. Byrd & J. K. Tomberlin (eds) *Forensic Entomology: The Utility of Arthropods in Legal Investigations*, 3rd edn, pp. 413–443. CRC Press, Boca Raton, FL.

Resources and links

Websites dealing with Diptera

North American Dipterists Society (NADS), with the Dipterist's Directory: https://dipterists.org/
Diptera info site: http://www.diptera.info/news.php
Flesh flies on the "my species" website. This site contains many photos of Sarcophagidae, all authoritatively identified. https://sarcophagidae.myspecies.info/gallery

Examples of focus stacking software

Helicon Focus: http://www.heliconsoft.com/helicon
focus.html
Zerene Stacker: http://zerenesystems.com/stacker/

Websites dealing with mailing issues and international requirements

U.S. Postal Service: www.usps.com
How to safely box and ship pinned specimens: http://www.nku.edu/~dahlem/Shipping%20Specimens/HOW%20TO%20SAFELY%20SHIP%20PINNED%20INSECTS.html
U.S. Fish and Wildlife Service, Office of Law Enforcement: https://www.fws.gov/program/office-of-law-enforcement/information-importers-exporters

Appendix III

Necrophagous fly and beetle life table references

The following references contain information on necrophagous fly and beetle development at different rearing temperatures. This is not meant to be an exhaustive list, and other fly data can be found through internet searches using such sites as Google Scholar, Yahoo, and PubMed, as well as by contacting individual investigators. A good starting point for the latter is by contacting members of the American Board of Forensic Entomology, North American Forensic Entomology Association, the European Forensic Entomology Association, or the American Academy of Forensic Sciences.

Abd Algalil, F.M. & Zambare, S.P. (2015) Effect of temperature on the development of Calliphorid Fly of forensic importance, *Chrysomya rufifacies* (Macquart, 1842). *International Journal of Advanced Research* 3(3): 1099–1103.

Abou Zied, E.M., Gabre, R.M. & Chi, H. (2003) Life table of the Australian sheep blow fly *Lucilia cuprina* (Wiedemann) (Diptera: Calliphoridae). *Egyptian Journal of Zoology* 41: 29–45.

Ames, C. & Turner, B. (2003) Low temperature episodes in development of blowflies: implications for postmortem interval estimation. *Medical and Veterinary Entomology* 17: 178–186.

Amoudi, M.A., Diab, F.M. & Abou-Fannah, S.S. (1994) Development rate and mortality of immature *Parasarcophaga* (*Liopygia*) *ruficornis* (Diptera: Sarcophagidae) at constant laboratory temperatures. *Journal of Medical Entomology* 31(1): 168–170.

Anderson, G.S. (2000) Minimum and maximum development rates of some forensically important Calliphoridae (Diptera). *Journal of Forensic Sciences* 45: 824–832.

Aruna Devi, A., Abu Hassan, A., Kumara, T.K. & Che Salmah, M.R. (2011) Life table of *Synthesiomyia nudiseta* (Van der Wulp) (Diptera: Muscidae) under uncontrolled laboratory environments – a preliminary study. *Tropical Biomedicine* 28(3): 524–530.

Aubernon, C., Charabidzé, D., Devigne, C., Delanno, Y. & Gosset, D. (2015) Experimental study of *Lucilia sericata* (Diptera Calliphoridae) larval development on rat cadavers: effects of climate and chemical contamination. *Forensic Science International* 253: 125–130.

Bambaradeniya, Y.T.B., Karunaratne, W.I.P., Tomberlin, J.K., Goonerathne, I., Kotakadeniya, R.B. & Magni, P.A. (2019) Effect of temperature and tissue type on the development of the forensic fly *Chrysomya megacephala* (Diptera: Calliphoridae). *Journal of Medical Entomology* 56(6): 1571–1581.

Barros-Cordeiro, K.B., Pujol-Luz, J.R., Name, K.P.O. & Báo, S.N. (2016) Intra-puparial development of the *Cochliomyia macellaria* and *Lucilia cuprina* (Diptera, Calliphoridae). *Revista Brasileira de Entomologia* 60(4): 334–340.

Barros-Souza, A.S., Ferreira-Keppler, R.L. & de Brito Agra, D. (2012) Development period of forensic importance Calliphoridae (Diptera: Brachycera) in urban area under natural conditions in Manaus, Amazonas, Brazil. *EntomoBrasilis* 5(2): 99–105.

Bauer, A., Bauer, A.M. & Tomberlin, J.K. (2020) Impact of diet moisture on the development of the forensically important blow fly *Cochliomyia macellaria* (Fabricius) (Diptera: Calliphoridae). *Forensic Science International* 312: 110333.

The Science of Forensic Entomology, Second Edition. David B. Rivers and Gregory A. Dahlem.
© 2023 John Wiley & Sons Ltd. Published 2023 by John Wiley & Sons Ltd.
Companion website: www.wiley.com/go/forensicentomology2

Bernhardt, V., Schomerus, C., Verhoff, M.A. & Amendt, J. (2017) Of pigs and men – comparing the development of *Calliphora vicina* (Diptera: Calliphoridae) on human and porcine tissue. *International Journal of Legal Medicine* 131(3): 847–853.

Beuter, L. & Mendes, J. (2013) Development of *Chrysomya albiceps* (Wiedemann) (Diptera: Calliphoridae) in different pig tissues. *Neotropical Entomology* 42(4): 426–430.

Bharti, M., Singh, D. & Sharma, Y.P. (2007) Effect of temperature on the development of forensically important blowfly, *Chrysomya megacephala* (Fabricius) (Diptera: Calliphoridae). *Entomon-Trivandrum* 32(2): 149.

Bhosale, D. (2020) Determine the time duration in life cycle stages of family Sarcophagidae species of *Sarcophaga bullata* & *Sarcophaga carnaria* during winter season in Poladpurtehsil. *International Journal of Current Research in Life Sciences* 9(5): 3276–3278.

Boatright, S.A. & Tomberlin, J.K. (2010) Effects of temperature and tissue type on the development of *Cochliomyia macellaria* (Diptera: Calliphoridae). *Journal of Medical Entomology* 47: 917–923.

Braet, Y., Bourguignon, L., Vanpoucke, S., Drome, V. & Hubrecht, F. (2015) Preliminary data on pupal development, lifespan and fertility of *Cynomya mortuorum* (L., 1761) in Belgium (Diptera: Calliphoridae). *Biodiversity Data Journal* 3: 1–12.

Braet, Y., Bourguignon, L., Vanpoucke, S., Drome, V. & Hubrecht, F. (2015) New developmental data for *Cynomya mortuorum* (L., 1761) in Belgium (Diptera: Calliphoridae). *Forensic Science International* 252: 29–32.

Brown, K., Thorne, A. & Harvey, M. (2015) *Calliphora vicina* (Diptera: Calliphoridae) pupae: a timeline of external morphological development and a new age and PMI estimation tool. *International Journal of Legal Medicine* 129(4): 835–850.

Byrd, J.H. & Allen, J.C. (2001) The development of the black blow fly, *Phormia regina* (Meigen). *Forensic Science International* 120: 79–88.

Byrd, J.H. & Butler, J.F. (1996) Effects of temperature on *Cochliomyia macellaria* (Diptera: Calliphoridae) development. *Journal of Medical Entomology* 33: 901–905.

Byrd, J.H. & Butler, J.F. (1997) Effects of temperature on *Chrysomya rufifacies* (Diptera: Calliphoridae) development. *Journal of Medical Entomology* 34: 353–358.

Byrd, J.H. & Butler, J.F. (1998) Effects of temperature on *Sarcophaga haemorrhoidalis* (Diptera: Sarcophagidae) development. *Journal of Medical Entomology* 35: 694–698.

Campobasso, C.P., Di Vella, G. & Introna, F. (2001) Factors affecting decomposition and Diptera colonization. *Forensic Science International* 120: 18–27.

Caneparo, M.F.C., Fischer, M.L. & Almeida, L.M. (2017) Effect of temperature on the life cycle of *Euspilotus azureus* (Coleoptera: Histeridae), a predator of forensic importance. *Florida Entomologist* 100(4): 795–801.

Chen, W., Yang, L., Ren, L., Shang, Y., Wang, S. & Guo, Y. (2019) Impact of constant versus fluctuating temperatures on the development and life history parameters of *Aldrichina grahami* (Diptera: Calliphoridae). *Insects* 10(7): 184.

Clark, K., Evans, L. & Wall, R. (2006) Growth rates of the blowfly, *Lucilia sericata*, on different body tissues. *Forensic Science International* 156: 145–149.

Clarkson, C.A., Hobischak, N.R. & Anderson, G.S. (2004) A comparison of the developmental rate of *Protophormia terraenovae* (Robineau-Desvoidy) raised under constant and fluctuating temperature regimes. *Canadian Society of Forensic Sciences* 37: 95–101.

Dallwitz, R. (1984) The influence of constant and fluctuating temperatures on development and survival rate of pupae of the Australian sheep blowfly, *Lucilia cuprina*. *Entomologia Experimentalis et Applicata* 36: 89–95.

Davies, L. & Ratcliffe, G.G. (1994) Developmental rates of some pre-adult stages in blowflies with reference to low temperatures. *Medical and Veterinary Entomology* 8: 245–254.

Day, D.M. & Wallman, J.F. (2006) A comparison of frozen/thawed and fresh food substrates in development of *Calliphora augur* (Diptera Calliphoridae) larvae. *International Journal of Legal Medicine* 120: 391–394.

Donovan, S.E., Hall, M.J.R., Turner, B.D. & Moncrieff, C.B. (2006) Larval growth rates of the blowfly, *Calliphora vicina*, over a range of temperatures. *Medical and Veterinary Entomology* 20: 106–114.

El-Hefnawy, A.A., Abul Dahab, F.F., Ibrahim, A.A., Salama, E.M., Mahmoud, S.H., Sanford, M.R., Kovar, S.J. & Tarone, A.M. (2020) Developmental plasticity of the flesh fly *Blaesoxipha plinthopyga* (Diptera: Sarcophagidae) on different substrates. *Journal of Medical Entomology* 57(6): 1686–1693.

Faris, A.M., West, W.R., Tomberlin, J.K. & Tarone, A.M. (2020) Field validation of a development data set for *Cochliomyia macellaria* (Diptera: Calliphoridae): estimating insect age based on development stage. *Journal of Medical Entomology* 57(1): 39–49.

Flissak, J.C. & Moura, M.O. (2018) Intrapuparial development of *Sarconesia chlorogaster* (Diptera: Calliphoridae) for postmortem interval estimation (PMI). *Journal of Medical Entomology* 55(2): 277–284.

Flores, M., Longnecker, M. & Tomberlin, J.K. (2014) Effects of temperature and tissue type on *Chrysomya rufifacies* (Diptera: Calliphoridae) (Macquart) development. *Forensic Science International* 245: 24–29.

Gabre, R.M., Adham, F.K. & Chi, H. (2005) Life table of *Chrysomya megacephala* (Fabricius) (Diptera: Calliphoridae). *Acta Oecologica* 27: 179–183.

Gallagher, M.B., Sandu, S. & Kimsey, R. (2010) Variation in developmental time for geographically distinct populations of the common green bottle fly, *Lucilia sericata* (Meigen). *Journal of Forensic Sciences* 55: 438–442.

Garcia, D. A., Pérez-Hérazo, A. & Amat, E. (2017) Life history of *Cochliomyia macellaria* (Fabricius, 1775) (Diptera, Calliphoridae), a blowfly of medical and forensic importance. *Neotropical Entomology* 46(6): 606–612.

Goodbrod, J.R. & Goff, M.L. (1990) Effects of larval populations density on rates of development and interactions between two species of *Chrysomya* (Diptera: Calliphoridae) in laboratory culture. *Journal of Medical Entomology* 27: 338–343.

Grassberger, M. & Reiter, C. (2001) Effect of temperature on *Lucilia sericata* (Diptera: Calliphoridae) development with special reference to the isomegalen- and isomorphen-diagram. *Forensic Science International* 120: 32–36.

Grassberger, M. & Reiter, C. (2002a) Effect of temperature on development of *Liopygia* (=*Sarcophaga*) *argyrostoma* (Robineau-Desvoidy) (Diptera: Sarcophagidae) and its forensic implications. *Journal of Forensic Sciences* 47: 1332–1336.

Grassberger, M. & Reiter, C. (2002b) Effect of temperature on development of the forensically important Holarctic blow fly *Protophormia terraenovae* (Robineau-Desvoidy) (Diptera: Calliphoridae). *Forensic Science International* 128: 177–182.

Grassberger, M., Friedrich, E. & Reiter, C. (2003) The blowfly *Chrysomya albiceps* (Weidemann) (Diptera: Calliphoridae) as a new forensic indicator in Central Europe. *International Journal of Legal Medicine* 117: 75–81.

Greenberg, B. (1991) Flies as forensic indicators. *Journal of Medical Entomology* 28: 565–577.

Greenberg, B. & Kunich, J.C. (2002) *Entomology and the Law*. Cambridge University Press, Cambridge, UK.

Greenberg, B. & Tantawi, T.I. (1993) Different developmental strategies in two boreal blow flies (Diptera: Calliphoridae). *Journal of Medical Entomology* 30: 481–483.

Greenberg, B. & Wells, J. D. (1998) Forensic use of *Megaselia abdita* and *M. scalaris* (Phoridae: Diptera): case studies, development rates, and egg structure. *Journal of Medical Entomology* 35(3): 205–209.

Gruner, S.V., Slone, D.H., Capinera, J.L. & Turco, M.P. (2017) Development of the oriental latrine fly, *Chrysomya megacephala* (Diptera: Calliphoridae), at five constant temperatures. *Journal of Medical Entomology* 54(2): 290–298.

Grzywacz, A. (2019) Thermal requirements for the development of immature stages of *Fannia canicularis* (Linnaeus) (Diptera: Fanniidae). *Forensic Science International* 297: 16–26.

Hadura, A.H., Sundharavalli, R., Azulia, N.Z., Zairi, J. & Hamdan, A. (2018) Life table of forensically important blow fly, *Chrysomya rufifacies* (Macquart) (Diptera: Calliphoridae). *Tropical Biomedicine* 35: 413–422.

Hanski, I. (1976) Assimilation by *Lucilia illustris* (Diptera) larvae in constant and changing temperatures. *Oikos* 27: 288–299.

Hanski, I. (1977) An interpolation model of assimilation by larvae of the blowfly, *Lucilia illustris* (Calliphoridae) in changing temperatures. *Oikos* 28: 187–195.

Hu, G., Wang, Y., Sun, Y., Zhang, Y., Wang, M. & Wang, J. (2019) Development of *Chrysomya rufifacies* (Diptera: Calliphoridae) at constant temperatures within its colony range in Yangtze River Delta region of China. *Journal of Medical Entomology* 56(5): 1215–1224.

Hwang, C.C. & Turner, B.D. (2009) Small-scaled geographical variation in life-history traits of the blowfly *Calliphora vicina* between rural and urban populations. *Entomologia Experimentalis et Applicata* 132: 218–224.

Ireland, S. & Turner, B. (2006) The effects of larval crowding and food type on the size and development of the blowfly, *Calliphora vomitoria*. *Forensic Science International* 159: 175–181.

Johl, H.K. & Anderson, G.S. (1996) Effects of refrigeration on development of the blow fly, *Calliphora vicina* (Diptera: Calliphoridae) and their relationship to time of death. *Journal of the Entomological Society of British Columbia* 93: 93–98.

Kamal, A.S. (1958) Comparative study of thirteen species of sarcosaprophagous Calliphorida and Sarcophagidae (Diptera). I. Bionomics. *Annals of the Entomological Society of America* 51: 261–270.

Kaneshrajah, G. & Turner, B. (2004) *Calliphora vicina* larvae grow at different rates on different body tissues. *International Journal of Legal Medicine* 118: 242–244.

Karabey, T. & Sert, O. (2018) The analysis of pupal development period in *Lucilia sericata* (Diptera: Calliphoridae) forensically important insect. *International Journal of Legal Medicine* 132(4): 1185–1196.

Kotzé, Z., Villet, M.H. & Weldon, C.W. (2015) Effect of temperature on development of the blowfly, *Lucilia cuprina* (Wiedemann) (Diptera: Calliphoridae). *International Journal of Legal Medicine* 129(5): 1155–1162.

Lecheta, M.C. & Moura, M.O. (2019) Estimating the age of forensically useful blowfly, *Sarconesia chlorogaster* (Diptera: Calliphoridae), using larval length and weight. *Journal of Medical Entomology* 56(4): 915–920.

Lecheta, M.C., Thyssen, P.J. & Moura, M.O. (2015) The effect of temperature on development of *Sarconesia chlorogaster*, a blowfly of forensic importance. *Forensic Science, Medicine, and Pathology* 11(4): 538–543.

Levot, G.W., Brown, K.R. & Shipp, E. (1979) Larval growth of some calliphorid and sarcophagid Diptera. *Bulletin of Entomological Research* 69: 469–475.

Madubunyi, L.C. (1986) Laboratory life history parameters of the red-tailed fleshfly, *Sarcophaga haemorrhoidalis* (Fallen) (Diptera: Sarcophagidae). *International Journal of Tropical Insect Science* 7(5): 617–621.

Mai, M. & Amendt, J. (2012) Effect of different post-feeding intervals on the total time of development of the blowfly *Lucilia sericata* (Diptera: Calliphoridae). *Forensic Science International* 221(1–3): 65–69.

Marchenko, M.I. (2001) Medicolegal relevance of cadaver entomo-fauna for the determination of the time since death. *Forensic Science International* 120: 89–109.

Meyer, J.A. & Mullens, B.A. (1988) Development of immature *Fannia* spp. (Diptera: Muscidae) at constant laboratory temperatures. *Journal of Medical Entomology* 25(3): 165–171.

Miranda, C.D., Cammack, J.A. & Tomberlin, J.K. (2020) Life-history traits of house fly, *Musca domestica* L. (Diptera: Muscidae), reared on three manure types. *Journal of Insects as Food and Feed* 6(1): 81–90.

Mohr, R.M. & Tomberlin, J.K. (2015) Development and validation of a new technique for estimating a minimum postmortem interval using adult blow fly (Diptera: Calliphoridae) carcass attendance. *International Journal of Legal Medicine* 129(4): 851–859.

Myskowiak, J.B. & Doums, C. (2002) Effects of refrigeration on the biometry and development of *Protophormia terraenovae* (Robineau-Desvoidy) (Diptera: Calliphoridae) and its consequences in estimating post-mortem interval in forensic investigations. *Forensic Science International* 125(2–3): 254–261.

Nabity, P.D., Higley, L.G. & Heng-Moss, T.M. (2006) Effects of temperature on development of *Phormia regina* (Diptera: Calliphoridae) and use of developmental data in determining time intervals in forensic entomology. *Journal of Medical Entomology* 43: 1276–1286.

Nabity, P.D., Higley, L.G. & Heng-Moss, T.M. (2007) Light-induced variability in the development of the forensically important blow fly, *Phormia regina* (Meigen) (Diptera: Calliphoridae). *Journal of Medical Entomology* 44: 351–358.

Nassu, M.P., Thyssen, P.J. & Linhares, A.X. (2014) Developmental rate of immatures of two fly species of forensic importance: *Sarcophaga (Liopygia) ruficornis* and *Microcerella halli* (Diptera: Sarcophagidae). *Parasitology Research* 113(1): 217–222.

Niederegger, S., Pastuschek, J. & Mall, G. (2010) Preliminary studies of the influence of fluctuating temperatures on the development of various forensically relevant flies. *Forensic Science International* 199(1–3): 72–78.

Niederegger, S., Wartenberg, N., Spiess, R. & Mall, G. (2013) Influence of food substrates on the development of the blowflies *Calliphora vicina* and *Calliphora vomitoria* (Diptera, Calliphoridae). *Parasitology Research* 112(8): 2847–2853.

Norris, K.R. (1965) The bionomics of blowflies. *Annual Review of Entomology* 10: 47–68.

Núñez-Vázquez, C., Tomberlin, J.K., Cantú-Sifuentes, M. & García-Martínez, O. (2013) Laboratory development and field validation of *Phormia regina* (Diptera: Calliphoridae). *Journal of Medical Entomology* 50(2): 252–260.

Owings, C.G., Spiegelman, C., Tarone, A.M. & Tomberlin, J.K. (2014) Developmental variation among *Cochliomyia macellaria* Fabricius (Diptera: Calliphoridae) populations from three ecoregions of Texas, USA. *International Journal of Legal Medicine* 128(4): 709–717.

Proença, B., Ribeiro, A.C., Luz, R.T., Aguiar, V.M., Maia, V.C. & Couri, M.S. (2014) Intrapuparial development of *Chrysomya putoria* (Diptera: Calliphoridae). *Journal of Medical Entomology* 51(5): 908–914.

Pujol-Luz, J.R. & Barros-Cordeiro, K.B. (2012) Intra-puparial development of the females of *Chrysomya albiceps* (Wiedemann) (Diptera, Calliphoridae). *Revista Brasileira de Entomologia* 56(3): 269–272.

Putnam, R.J. (1977) Dynamics of the blowfly, *Calliphora erythrocephala*, within carrion. *Journal of Animal Ecology* 46: 853–866.

Rabêlo, K.C., Thyssen, P.J., Salgado, R.L., Araújo, M.S. & Vasconcelos, S.D. (2011) Bionomics of two forensically important blowfly species *Chrysomya megacephala* and *Chrysomya putoria* (Diptera: Calliphoridae) reared on four types of diet. *Forensic Science International* 210(1–3): 257–262.

Ramos-Pastrana, Y., Londoño, C.A. & Wolff, M. (2017) Intra-puparial development of *Lucilia eximia* (Diptera, Calliphoridae). *Acta Amazonica* 47(1): 63–70.

Reibe, S., Doetinchem, P.V. & Madea, B. (2010) A new simulation-based model for calculating post-mortem intervals using developmental data for *Lucilia sericata* (Dipt.: Calliphoridae). *Parasitology Research* 107(1): 9–16.

Richards, C.S., Paterson, I.D. & Villet, M.H. (2008) Estimating the age of immature *Chrysomya albiceps* (Diptera: Calliphoridae), correcting for temperature and geographical latitude. *International Journal of Legal Medicine* 122: 271–279.

Richards, C.S., Crous, K.L. & Villet, M.H. (2009) Models of development for blowfly sister species *Chrysomya chloropyga* and *Chrysomya putoria*. *Medical and Veterinary Entomology* 23: 56–61.

Rivers, D.B., Ciarlo, T., Spelman, M. & Brogan, R. (2010) Changes in development and heat shock protein expression in two species of flies (*Sarcophaga bullata* [Diptera: Sarcophagidae] and *Protophormia terraenovae* [Diptera: Calliphoridae]) reared in different sized maggot masses. *Journal of Medical Entomology* 47: 677–689.

Roe, A. & Higley, L.G. (2015) Development modeling of *Lucilia sericata* (Diptera: Calliphoridae). *PeerJ* 3: e803.

Rueda, L.C., Ortega, L.G., Segura, N.A., Acero, V.M. & Bello, F. (2010) *Lucilia sericata* strain from Colombia: experimental colonization, life tables and evaluation of two artificial diets of the blowfly *Lucilia sericata* (Meigen) (Diptera: Calliphoridae), Bogotá, Colombia Strain. *Biological Research* 43(2): 197–203.

Russo, A., Cocuzza, G.E., Vasta, M.C., Simola, M. & Virone, G. (2006) Life fertility tables of *Piophila casei* L. (Diptera: Piophilidae) reared at five different temperatures. *Environmental Entomology* 35(2): 194–200.

Salazar-Souza, M., Couri, M.S. & Aguiar, V.M. (2018) Chronology of the intrapuparial development of the blowfly *Chrysomya albiceps* (Diptera: Calliphoridae): application in forensic entomology. *Journal of Medical Entomology* 55(4): 825–832.

Saleh, V., Soltani, A., Dabaghmanesh, T., Alipour, H., Azizi, K. & Moemenbellah-Fard, M.D. (2014) Mass rearing and life table attributes of two Cyclorrhaphan Flies, *Lucilia sericata* Meigen (Diptera: Calliphoridae) and *Musca domestica* L. (Diptera: Muscidae) under laboratory conditions. *Journal of Entomology* 11(5): 291–298.

Salimi, M., Rassi, Y., Oshaghi, M., Chatrabgoun, O., Limoee, M. & Rafizadeh, S. (2018) Temperature requirements for the growth of immature stages of blowflies species, *Chrysomya albiceps* and *Calliphora vicina*, (Diptera: Calliphoridae) under laboratory conditions. *Egyptian Journal of Forensic Sciences* 8(1): 1–6.

Sanei-Dehkordi, A., Khamesipour, A., Akbarzadeh, K., Akhavan, A.A., Rassi, Y., Oshaghi, M.A., Miramin-Mohammadi, A., Eskandari, S.E. & Rafinejad, J. (2014) Experimental colonization and life table of the *Calliphora vicina* (Robineau-Desvoidy) (Diptera: Calliphoridae). *Journal of Entomological and Zoological Studies* 2: 45–48.

Saunders, D.S. (1998) Under-sized larvae from short-day adults of the blow fly, *Calliphora vicina*, side-step the diapause programme. *Physiological Entomology* 22: 249–255.

Shiao, S.F. & Yeh, T.C. (2008) Larval competition of *Chrysomya megacephala* and *Chrysomya rufifacies* (Diptera: Calliphoridae): behavior and ecological studies of two blow fly species of forensic significance. *Journal of Medical Entomology* 45(4): 785–799.

da Silva, S.M. & Moura, M.O. (2019) Intrapuparial development of *Hemilucilia semidiaphana* (Diptera: Calliphoridae) and its use in forensic entomology. *Journal of Medical Entomology* 56(6): 1623–1635.

Sim, L.X. & Zuha, R.M. (2019) *Chrysomya megacephala* (Fabricius, 1794) (Diptera: Calliphoridae) development by landmark-based geometric morphometrics of cephalopharyngeal skeleton: a preliminary assessment for forensic entomology application. *Egyptian Journal of Forensic Sciences* 9(1): 1–9.

Smith, K.G.V. (1986) *A Manual of Forensic Entomology*. British Museum (Natural History), London.

Smith, K.E. & Wall, R. (1998) Estimates of population density and dispersal in the blowfly, *Lucilia sericata* (Diptera: Calliphoridae). *Bulletin of Entomological Research* 88: 65–73.

So, P.M. & Dudgeon, D. (1989) Life-history responses of larviparous *Boettcherisca formosensis* (Diptera: Sarcophagidae) to larval competition for food,

including comparisons with oviparous *Hemipyrellia ligurriens* (Calliphoridae). *Ecological Entomology* 14(3): 349–356.

Sukontason, K., Piangjai, S., Siriwattanarungsee, S. & Sukontason, K.L. (2008) Morphology and development rate of blowflies *Chrysomya megacephala* and *Chrysomya rufifacies* in Thailand: application in forensic entomology. *Parasitology Research* 102: 1207–1216.

Tarone, A.M., Picard, C.J., Spiegelman, C. & Foran, D.R. (2011) Population and temperature effects on *Lucilia sericata* (Diptera: Calliphoridae) body size and minimum development time. *Journal of Medical Entomology* 48: 1062–1068.

Thomas, J.K., Sanford, M.R., Longnecker, M. & Tomberlin, J.K. (2016) Effects of temperature and tissue type on the development of *Megaselia scalaris* (Diptera: Phoridae). *Journal of Medical Entomology* 53(3): 519–525.

Thyssen, P.J., de Souza, C.M., Shimamoto, P.M., de Britto Salewski, T. & Moretti, T.C. (2014) Rates of development of immatures of three species of *Chrysomya* (Diptera: Calliphoridae) reared in different types of animal tissues: implications for estimating the postmortem interval. *Parasitology Research* 113(9): 3373–3380.

Tomberlin, J.K., Adler, P.H. & Myers, H.M. (2009) Development of the black soldier fly (Diptera: Stratiomyidae) in relation to temperature. *Environmental Entomology* 38: 930–934.

Vélez, M.C. & Wolff, M. (2008) Rearing five species of Diptera (Calliphoridae) of forensic importance in Columbia in semicontrolled field conditions. *Papeis Avulsos de Zoologia (São Paulo)* 48(6). Available at http://dx.doi.org/10.1590/S0031-10492008000600001.

Verma, K. & Paul, R. (2016) *Lucilia sericata* (Meigen) and *Chrysomya megacephala* (Fabricius) (Diptera: Calliphoridae) development rate and its implications for forensic entomology. *Journal of Forensic Science and Medicine* 2(3): 146.

Villet, M.H., MacKenzie, B. & Muller, W.J. (2006) Larval development of the carrion-breeding flesh fly, *Sarcophaga* (*Liosarcophaga*) *tibialis* Macquart (Diptera: Sarcophagidae), at constant temperatures. *African Entomology* 14(2): 357–366.

Voss, S.C., Cook, D.F., Hung, W.F. & Dadour, I.R. (2014) Survival and development of the forensically important blow fly, *Calliphora varifrons* (Diptera: Calliphoridae) at constant temperatures. *Forensic Science, Medicine, and Pathology* 10(3): 314–321.

Wall, R., French, N. & Morgan, K.L. (1992) Effects of temperature on the development and abundance of the sheep blowfly *Lucilia sericata* (Diptera: Calliphoridae). *Bulletin of Entomological Research* 82(1): 125–131.

Wang, Y., Wang, J.F., Zhang, Y.N., Tao, L.Y. & Wang, M. (2017) Forensically important *Boettcherisca peregrina*

(Diptera: Sarcophagidae) in China: development pattern and significance for estimating postmortem interval. *Journal of Medical Entomology* 54(6): 1491–1497.

Wang, Y., Yang, J.B., Wang, J.F., Li, L.L., Wang, M., Yang, L.J., Tao, L.Y., Chu, J. & Hou, Y.D. (2017) Development of the forensically important beetle *Creophilus maxillosus* (Coleoptera: Staphylinidae) at constant temperatures. *Journal of Medical Entomology* 54(2): 281–289.

Wang, Y., Yang, L., Zhang, Y., Tao, L. & Wang, J. (2018) Development of *Musca domestica* at constant temperatures and the first case report of its application for estimating the minimum postmortem interval. *Forensic Science International* 285: 172–180.

Wang, Y., Zhang, Y.N., Liu, C., Hu, G.L., Wang, M., Yang, L.J., Chu, J. & Wang, J.F. (2018) Development of *Aldrichina grahami* (Diptera: Calliphoridae) at constant temperatures. *Journal of Medical Entomology* 55(6): 1402–1409.

Warren, J.A. & Anderson, G.S. (2013) Effect of fluctuating temperatures on the development of a forensically important blow fly, *Protophormia terraenovae* (Diptera: Calliphoridae). *Environmental Entomology* 42(1): 167–172.

Weidner, L.M., Tomberlin, J.K. & Hamilton, G.C. (2014) Development of *Lucilia coeruleiviridis*

(Diptera: Calliphoridae) in New Jersey, USA. *Florida Entomologist* 97(2): 849–851.

Wells, J.D. & Kurahashi, H. (1994) *Chrysomya megacephala* (Fabricius) (Diptera: Calliphoridae) development: rate, variation and the implications for forensic entomology. *Medical Entomology and Zoology* 45(4): 303–309.

Williams, H. & Richardson, A.M.M. (1983) Life history responses to larval food shortages in four species of necrophagous flies (Diptera: Calliphoridae). *Australian Journal of Ecology* 8: 257–263.

Williams, H. & Richardson, A.M.M. (1984) Growth energetics in relation to temperature for larvae of four species of necrophagous flies (Diptera: Calliphoridae). *Australian Journal of Ecology* 9: 141–152.

Yang, Y.Q., Lyu, Z., Li, X.B., Li, K., Yao, L. & Wan, L.H. (2015) Development of *Hemipyrellia ligurriens* (Wiedemann) (Diptera: Calliphoridae) at constant temperatures: applications in estimating postmortem interval. *Forensic Science International* 253: 48–54.

Zuha, R.M. & Omar, B. (2014) Developmental rate, size, and sexual dimorphism of *Megaselia scalaris* (Loew) (Diptera: Phoridae): its possible implications in forensic entomology. *Parasitology Research* 113(6): 2285–2294.

Appendix IV

Current names for species in Aldrich's *Sarcophaga* and allies

Current names for Sarcophagidae species illustrated in Aldrich, J.M. (1916) *Sarcophaga* and allies in North America. Vol. 1 of the Thomas Say Foundation of the Entomological Society of America, LaFayette, IN.

Aldrich genus	Aldrich species	Figure	Current genus	Current subgenus	Current species
Phrissopodia	*praeceps* (Wiedemann)	1	*Peckia*	Peckia	*praeceps* (Wiedemann)
Wohlfahrtia	*meigenii* (Schiner)	2	*Wohlfahrtia*		*vigil* (Walker)
Wohlfahrtia	*vigil* (Walker)	3	*Wohlfahrtia*		*vigil* (Walker)
Johnsonia	*elegans* (Coquillett)	4	*Lepidodexia*	Johnsonia	*elegans* (Coquillett)
Johnsonia	*setosa* (Aldrich)	5	*Lepidodexia*	Johnsonia	*setosa* (Aldrich)
Camptops	*unicolor* (Aldrich)	6	*Lepidodexia*	Camptops	*unicolor* (Aldrich)
Sarothromyia	*femoralis* (Schiner)	7	*Tricharea*	Sarothromyia	*femoralis* (Schiner)
Sarothromyia	*femoralis* var. *simplex* (Aldrich)	8	*Tricharea*	Sarothromyia	*simplex* (Aldrich)
Sarcophagula	*occidua* (Fabricius)	9	*Tricharea*	Sarcophagula	*occidua* (Fabricius)
Camptopyga	*aristata* (Aldrich)	10	*Microcerella*		*hypopygialis* (Townsend)
Agria	*affinis* (Fallén)	11	*Agria*		*housei* (Shewell)
Harbeckia	*tessellata* (Aldrich)	12	*Lepidodexia*	Neophyto	*sheldoni* (Coquillett)
Hypopelta	*scrofa* (Aldrich)	13	*Microcerella*		*scrofa* (Aldrich)
Notochaeta	*subpolita* (Aldrich)	14	*Lepidodexia*	Notochaeta	*subpolita* (Aldrich)
Notochaeta	*plumigera* (Wulp)	15	*Lepidodexia*	Dexomyophora	*plumigera* (Wulp)
Emblemasoma	*erro* (Aldrich)	16	*Emblemasoma*		*erro* (Aldrich)
Emblemasoma	*faciale* (Aldrich)	17	*Emblemasoma*		*faciale* (Aldrich)
Sthenopyga	*globosa* (Aldrich)	18	*Lepidodexia*	Johnsonia	*rufitibia* (Wulp)
Harpagopyga	*diversipes* (Coquillett)	19	*Lepidodexia*	Harpagopyga	*diversipes* (Coquillett)
Thelodiscus	*indivisus* (Aldrich)	20	*Angiometopa*		*ravinia* (Parker)
Sarcophaga	*sinuata* (Meigen)	21	*Sarcophaga*	Sarcotachinella	*sinuata* (Meigen)
Sarcophaga	*cockerellae* (Aldrich)	22	*Sarcophaga*	Mehria	*cockerellae* (Aldrich)

The Science of Forensic Entomology, Second Edition. David B. Rivers and Gregory A. Dahlem.
© 2023 John Wiley & Sons Ltd. Published 2023 by John Wiley & Sons Ltd.
Companian website: www.wiley.com/go/forensicentomology2

Aldrich genus	Aldrich species	Figure	Current genus	Current subgenus	Current species
Sarcophaga	*hinei* (Aldrich)	23	*Sarcophaga*	Mehria	*hinei* (Aldrich)
Sarcophaga	*pulla* (Aldrich)	24	*Sarcophaga*	Bercaeopsis	*pulla* (Aldrich)
Sarcophaga	*tarsata* (Aldrich)	25	*Sarcophaga*	Bercaeopsis	*tarsata* (Aldrich)
Sarcophaga	*latisterna* (Parker)	26	*Boettcheria*		*latisterna* (Parker)
Sarcophaga	*parkeri* (Aldrich)	27	*Boettcheria*		*parkeri* (Aldrich)
Sarcophaga	*cimbicis* (Townsend)	28	*Boettcheria*		*cimbicis* (Townsend)
Sarcophaga	*bisetosa* (Parker)	29	*Boettcheria*		*bisetosa* (Parker)
Sarcophaga	*taurus* (Aldrich)	30	*Boettcheria*		*taurus* (Aldrich)
Sarcophaga	*cessator* (Aldrich)	31	*Blaesoxipha*	Gigantotheca	*cessator* (Aldrich)
Sarcophaga	*sarraceniae* (Riley)	32	*Sarcophaga*	Bercaeopsis	*sarraceniae* (Riley)
Sarcophaga	*tetra* (Aldrich)	33	*Sarcophaga*	Bercaeopsis	*tetra* (Aldrich)
Sarcophaga	*idonea* (Aldrich)	34	*Sarcophaga*	Bercaeopsis	*idonea* (Aldrich)
Sarcophaga	*sima* (Aldrich)	35	*Sarcophaga*	Bercaeopsis	*sima* (Aldrich)
Sarcophaga	*pervillosa* (Aldrich)	36	*Tulaeopoda*		*pervillosa* (Aldrich)
Sarcophaga	*pervillosa* var. *inchoata* (Aldrich)	37	*Tulaeopoda*		*inchoata* (Aldrich)
Sarcophaga	*fletcheri* (Aldrich)	38	*Fletcherimyia*		*fletcheri* (Aldrich)
Sarcophaga	*davidsoni* (Coquillett)	39	*Sarcophaga*	Mehria	*davidsonii* (Coquillett)
Sarcophaga	*atlanis* (Aldrich)	40	*Blaesoxipha*	Blaesoxipha	*atlanis* (Aldrich)
Sarcophaga	*hunteri* (Hough)	41	*Blaesoxipha*	Tephromyia	*hunteri* (Hough)
Sarcophaga	*spatulata* (Aldrich)	42	*Blaesoxipha*	Tephromyia	*spatulata* (Aldrich)
Sarcophaga	*melampyga* (Aldrich)	43	*Titanogrypa*	Titanogrypa	*melampyga* (Aldrich)
Sarcophaga	*melampyga* var. alata (Aldrich)	44	*Titanogrypa*	Titanogrypa	*alata* (Aldrich)
Sarcophaga	*chaetopygialis* (Williston)	45	*Oxysarcodexia*		*chaetopygialis* (Williston)
Sarcophaga	*albisignum* (Aldrich)	46	*? Unplaced*		?
Sarcophaga	*magna* (Aldrich)	47	*Blaesoxipha*	Acanthodotheca	*magna* (Aldrich)
Sarcophaga	*opifera* (Coquillett)	48	*Blaesoxipha*	Blaesoxipha	*opifera* (Coquillett)
Sarcophaga	*peniculata* (Parker)	49	*Ravinia*		*acerba* (Walker)
Sarcophaga	*parallela* (Aldrich)	50	*Sarcophaga*	Bercaeopsis	*parallela* (Aldrich)
Sarcophaga	*rudis* (Aldrich)	51	*Blaesoxipha*	Acanthodotheca	*rudis* (Aldrich)
Sarcophaga	*excisa* (Aldrich)	52	*Blaesoxipha*	Acanthodotheca	*excisa* (Aldrich)
Sarcophaga	*eleodis* (Aldrich)	53	*Blaesoxipha*	Acanthodotheca	*eleodis* (Aldrich)
Sarcophaga	*masculina* (Aldrich)	54	*Blaesoxipha*	Acanthodotheca	*masculina* (Aldrich)
Sarcophaga	*alcedo* (Aldrich)	55	*Blaesoxipha*	Acanthodotheca	*alcedo* (Aldrich)
Sarcophaga	*prohibita* (Aldrich)	56	*Blaesoxipha*	Acanthodotheca	*prohibita* (Aldrich)
Sarcophaga	*reversa* (Aldrich)	57	*Blaesoxipha*	Acridiophaga	*reversa* (Aldrich)
Sarcophaga	*marginata* (Aldrich)	58	*Blaesoxipha*	Servaisia	*uncata* (Wulp)
Sarcophaga	*falciformis* (Aldrich)	59	*Blaesoxipha*	Servaisia	*falciformis* (Aldrich)
Sarcophaga	*setigera* (Aldrich)	60	*Blaesoxipha*	Tephromyia	*setigera* (Aldrich)
Sarcophaga	*coloradensis* (Aldrich)	61	*Blaesoxipha*	Servaisia	*coloradensis* (Aldrich)
Sarcophaga	*websteri* (Aldrich)	62	*Blaesoxipha*	Servaisia	*websteri* (Aldrich)
Sarcophaga	*angustifrons* (Aldrich)	63	*Blaesoxipha*	Acridiophaga	*angustifrons* (Aldrich)
Sarcophaga	*aculeata* (Aldrich)	64	*Blaesoxipha*	Acridiophaga	*aculeata* (Aldrich)
Sarcophaga	*aculeata* var. gavia (Aldrich)	65	*Blaesoxipha*	Acridiophaga	*caridei* (Bréthes)

Aldrich genus	Aldrich species	Figure	Current genus	Current subgenus	Current species
Sarcophaga	*aculeata* var. taediosa (Aldrich)	66	*Blaesoxipha*	Acridiophaga	*taediosa* (Aldrich)
Sarcophaga	*flavipes* (Aldrich)	67	*Blaesoxipha*	Tephromyia	*flavipes* (Aldrich)
Sarcophaga	*larga* (Aldrich)	68	*Tripanurga*		*importuna* (Walker)
Sarcophaga	*rufiventris* (Wiedemann)	69	*Rafaelia*		*rufiventris* (Townsend)
Sarcophaga	*ampulla* (Aldrich)	70	*Rafaelia*		*ampulla* (Aldrich)
Sarcophaga	*biseriata* (Aldrich)	71	*Sarcodexiopsis*		*biseriata* (Aldrich)
Sarcophaga	*surrubea* (Wulp)	72	*Helicobia*		*surrubea* (Wulp)
Sarcophaga	*hospes* (Aldrich)	73	*Blaesoxipha*	Tephromyia	*blandita* (Brethes)
Sarcophaga	*helicis* (Townsend)	74	*Helicobia*		*rapax* (Walker)
Sarcophaga	*johnsoni* (Aldrich)	75	*Sarcophaga*	Wohlfahrtiopsis	*johnsoni* (Aldrich)
Sarcophaga	*uliginosa* (Kramer)	76	*Sarcophaga*	Varirosellea	*uliginosa* (Kramer)
Sarcophaga	*aldrichi* (Parker)	77	*Sarcophaga*		*aldrichi* (Parker)
Sarcophaga	*houghi* (Aldrich)	78	*Sarcophaga*	Mehria	*houghi* (Aldrich)
Sarcophaga	*tuberosa* var. *harpax* (Pandellé)	79	*Sarcophaga*	Liosarcophaga	*harpax* (Pandelle)
Sarcophaga	*villipes* (Wulp)	80	*Tripanurga*		*villipes* (Wulp)
Sarcophaga	*cabensis* (Aldrich)	81	*Udamopyga*		*cabensis* (Aldrich)
Sarcophaga	*pedata* (Aldrich)	82	*Blaesoxipha*	Aldrichisca	*pedata* (Aldrich)
Sarcophaga	*fulvipes* (Macquart)	83	*Sarcophaga*	Robackina	*triplasia* (Wulp)
Sarcophaga	*singularis* (Aldrich)	84	*Spirobolomyia*		*singularis* (Aldrich)
Sarcophaga	*deceptiva* (Aldrich)	85	*Spirobolomyia*		*basalis* (Walker)
Sarcophaga	*cotyledonea* (Aldrich)	86	*Peckia*	Pattonella	*intermutans* (Walker)
Sarcophaga	*haemorrhoidalis* (Fallén)	87	*Sarcophaga*	Bercaea	*africa* (Wiedemann)
Sarcophaga	*wiedemanni* (Aldrich)	88	*Peckia*	Peckia	*gulo* (Fabricius)
Sarcophaga	*thatuna* (Aldrich)	89	*Sarcophaga*	Neosarcophaga	*thatuna* (Aldrich)
Sarcophaga	*elongata* (Aldrich)	90	*Sarcophaga*	Neosarcophaga	*elongata* (Aldrich)
Sarcophaga	*occidentalis* (Aldrich)	91	*Sarcophaga*	Neosarcophaga	*occidentalis* (Aldrich)
Sarcophaga	*juliaetta* (Aldrich)	92	*Sarcophaga*	Neosarcophaga	*juliaetta* (Aldrich)
Sarcophaga	*perspicax* (Aldrich)	93	*Sarcophaga*	Neosarcophaga	*perspicax* (Aldrich)
Sarcophaga	*gracilis* (Aldrich)	94	*Sarcophaga*	Neosarcophaga	*gracilis* (Aldrich)
Sarcophaga	*securifera* (Villeneuve)	95	*Sarcophaga*	Liopygia	*crassipalpis* (Macquart)
Sarcophaga	*falculata* (Pandellé)	96	*Sarcophaga*	Liopygia	*argyrostoma* (Robineau-Desvoidy)
Sarcophaga	*amoena* (Aldrich)	97	*Peckia*	Peckia	*amoena* (Aldrich)
Sarcophaga	*capitata* (Aldrich)	98	*Peckia*	Peckia	*capitata* (Aldrich)
Sarcophaga	*hillifera* (Aldrich)	99	*Peckia*	Peckia	*hillifera* (Aldrich)
Sarcophaga	*spectabilis* (Aldrich)	100	*Peckia*	Peckia	*spectabilis* (Aldrich)
Sarcophaga	*otiosa* (Williston)	101	*Peckia*	Peckia	*concinnata* (Williston)
Sarcophaga	*scoparia* (Pandellé)	102	*Sarcophaga*	Robineauella	*caerulescens* (Zetterstedt)
Sarcophaga	*peltata* (Aldrich)	103	*Oxysarcodexia*		*peltata* (Aldrich)
Sarcophaga	*subdiscalis* (Aldrich)	104	*Sarcophaga*	Bulbostyla	*subdiscalis* (Aldrich)
Sarcophaga	*impar* (Aldrich)	105	*Blaesoxipha*	Gigantotheca	*impar* (Aldrich)
Sarcophaga	*sulculata* (Aldrich)	106	*Tripanurga*		*sulculata* (Aldrich)
Sarcophaga	*cooleyi* (Parker)	107	*Sarcophaga*	Neobellieria	*cooleyi* (Parker)

Aldrich genus	Aldrich species	Figure	Current genus	Current subgenus	Current species
Sarcophaga	tuberosa var. sarracenioides (Aldrich)	108	Sarcophaga	Liosarcophaga	sarracenioides (Aldrich)
Sarcophaga	tuberosa var. exuberans (Pandellé)	109	Sarcophaga	Liosarcophaga	dux (Thomson)
Sarcophaga	bullata (Parker)	110	Sarcophaga	Neobellieria	bullata (Parker)
Sarcophaga	libera (Aldrich)	111	Sarcophaga	Neobellieria	libera (Aldrich)
Sarcophaga	salva (Aldrich)	112	Mecynocorpus		salvum (Aldrich)
Sarcophaga	rileyi (Aldrich)	113	Fletcherimyia		rileyi (Aldrich)
Sarcophaga	jonesi (Aldrich)	114	Fletcherimyia		jonesi (Aldrich)
Sarcophaga	celarata (Aldrich)	115	Fletcherimyia		celarata (Aldrich)
Sarcophaga	servilis (Aldrich)	116	Sarcodexiopsis		servilis (Aldrich)
Sarcophaga	floridensis (Aldrich)	117	Ravinia		floridensis (Aldrich)
Sarcophaga	planifrons (Aldrich)	118	Ravinia		planifrons (Aldrich)
Sarcophaga	pectinata (Aldrich)	119	Ravinia		pectinata (Aldrich)
Sarcophaga	communis (Parker)	120	Ravinia		querula (Walker)
Sarcophaga	communis var. ochracea (Aldrich)	121	Ravinia		lherminieri (Robineau-Desvoidy)
Sarcophaga	flavipalpis (Aldrich)	122	Spirobolomyia		flavipalpis (Aldrich)
Sarcophaga	bishoppi (Aldrich)	123	Sarcophaga	Wohlfahrtiopsis	bishoppi Aldrich
Sarcophaga	incurva (Aldrich)	124	Tripanurga		incurva (Aldrich)
Sarcophaga	insurgens (Aldrich)	125	Tripanurga		incurva (Aldrich)
Sarcophaga	kellyi (Aldrich)	126	Blaesoxipha	Kellymyia	kellyi (Aldrich)
Sarcophaga	sternodontis (Townsend)	127	Peckia	Sarcodexia	lambens (Wiedemann)
Sarcophaga	robusta (Aldrich)	128	Blaesoxipha	Gigantotheca	plinthopyga (Wiedemann)
Sarcophaga	bakeri (Aldrich)	129	Oxysarcodexia		bakeri (Aldrich)
Sarcophaga	hamata (Aldrich)	130	Blaesoxipha	Acanthodotheca	hamata (Aldrich)
Sarcophaga	xanthosoma (Aldrich)	131	Oxysarcodexia		xanthosoma (Aldrich)
Sarcophaga	utilis (Aldrich)	132	Sarcophaga	Wohlfahrtiopsis	utilis (Aldrich)
Sarcophaga	cistudinis (Aldrich)	133	Cistudinomyia		cistudinis (Aldrich)
Sarcophaga	galeata (Aldrich)	134	Oxysarcodexia		galeata (Aldrich)
Sarcophaga	australis (Aldrich)	135	Oxysarcodexia		conclausa (Walker)
Sarcophaga	timida (Aldrich)	136	Oxysarcodexia		timida (Aldrich)
Sarcophaga	assidua (Walker)	137	Oxysarcodexia		ventricosa (Wulp)
Sarcophaga	cingarus (Aldrich)	138	Oxysarcodexia		cingarus (Aldrich)
Sarcophaga	culminata (Aldrich)	139	Oxysarcodexia		culminata (Aldrich)
Sarcophaga	fissa (Aldrich)	140	Argoravinia		rufiventris (Wiedemann)
Sarcophaga	fimbriata (Aldrich)	141	Titanogrypa	Sarconeiva	fimbriata (Aldrich)
Sarcophaga	texana (Aldrrich)	142	Rafaelia		texana (Aldrich)
Sarcophaga	quadrisetosa (Coquillett)	143	Ravinia		derelicta (Walker)
Sarcophaga	latisetosa (Parker)	144	Ravinia		stimulans (Walker)
Sarcophaga	globulus (Aldrich)	145	Ravinia		globulus (Aldrich)

Glossary

abiogenesis (spontaneous generation)   Idea that life arose from non-living or inorganic matter.

abiotic   Non-living, usually in reference to factors in the environment such as temperature, humidity, photoperiod, etc.

acclimatization/acclimation   Readjustment or adapting the range of tolerance to a particular environmental feature such as temperature over a period of time in response to changes in the environment. Acclimation is used to describe similar changes in a laboratory setting.

accumulated degree day (hour)   Statistical measure that usually refers to the time (in hours or days) taken by an insect to develop to the stage collected from a crime scene.

adipocere   Thick waxy layer derived from neutral lipids that can form a thin protective layer around all or part of a corpse.

administrative law   Branch of public law that governs the creation and operation of government agencies in the United States.

aedeagus   Phallus or penis of adult male insect.

aesthetic-injury level (AIL)   Arbitrary threshold that reflect consumer's desire to have no insects present, usually in an urban environment, as opposed to an actual injury or economic damage limit.

aestivation   Period of quiescence or diapause during hot or dry conditions.

agonal period   The moment of death.

agroterrorism   Deliberate introduction of animal or plant pathogens or pests that directly attack cropping systems or livestock, with the purpose of instilling fear, causing economic losses, or undermining social stability.

algor mortis   Body temperature of deceased gradually cools to ambient conditions.

Allee effect theory   Benefits for individuals arising from interactions with conspecifics.

allelochemical   Chemical signals released from exocrine glands used in interspecific communication.

allomone   Chemical signal that produces a deleterious response in the receiver but not releaser.

allospecific   Of or pertaining to individuals of different species.

allothetic pathway   Control system involved in the detection and processing of external stimuli.

anaphylaxis   Hypersensitivity to allergens that leads to an acute systemic allergic reaction, which could result in death.

anautogeny   Female that requires a protein meal as an adult to provision eggs.

anemotaxis   Innate behavior in which longitudinal axis of body is oriented toward (positive) or away from (negative) air currents in response to external stimulus.

The Science of Forensic Entomology, Second Edition. David B. Rivers and Gregory A. Dahlem.
© 2023 John Wiley & Sons Ltd. Published 2023 by John Wiley & Sons Ltd.
Companian website: www.wiley.com/go/forensicentomology2

angle of impact   Angle at which a blood drop strikes a surface, yielding a characteristic morphology to the resulting stain.

antemortem   Before or prior to death.

antifreeze proteins   Function to induce thermal hysteresis by binding to small ice crystals to prevent expansion.

apneumone   Chemical signal originating from a non-living object, often in the form of carrion.

apodeme   Point of muscle attachment associated with the exoskeleton.

apolysis   Separation of the exoskeleton from the epidermis at the onset of molting.

apoptosis   A form of programmed cell death in which stimuli or death signals trigger the activation of signal transduction pathways associated with several classes of caspases, ultimately killing the cell.

aposematic   Warning, generally in the form of colors, markings, or specific behaviors, used to ward off potential attack by another animal.

Arboweapons   Use of arthropods or their products as biological weapons.

archaeoentomology   Discipline that uses insect remains from archaeological excavations to examine questions related to the environment, insects, and past civilizations.

archaeology   The study of human activity over time through the recovery and analysis of individual and cultural materials and environmental data that are left behind.

area of convergence   Area on a two-dimensional plane that approximates the origin of blood producing blood spatter.

area of origin   Area that represents the location in a three-dimensional space that the blood was projected to yield blood spatter.

autocatalysis   Initiation of chemical reactions by one of the products or the activation of an enzyme by the enzyme itself.

autogeny   Female that does not require a protein meal as an adult to provision first clutch of eggs.

autointoxication   Self-poisoning through production of defensive or other toxic compounds.

autolysis   Destruction of a cell due to the action of enzymes found within that cell.

base temperature   Temperature threshold below which development does not occur. *See also* developmental limit.

biofact   A biological object like a plant seed or animal remains that carry archaeological significance and which has previously been unhandled by humans.

biogeoclimatic zone   A geographic area characterized by a relatively uniform macroclimate with vegetation, soil type, moisture levels, and zoological life reflective of climatic conditions.

bioinfestation   Term applied to an outbreak of a pest species that causes economic damage and it often involves an invasive foreign pest.

biological control program   Insect pest management efforts to reduce the pest status of a given species through the release (inoculative or inundative) of natural enemies to the pest species.

biological terrorism (bioterrorism)   Use of living organisms or their products as weapons of war by militant groups or countries toward another group of people.

biotic   Living components of the environment.

bloodstain   Essentially synonymous with the term "blood spatter" and reflects the pattern of blood that results when the fluid strikes a surface.

cadaver decomposition island (CDI)   Animal carcass and the soil environment that has become saturated with expelled body remains.

cadaveric spasm   A condition that resembles localized rigor mortis in which specific regions of the body, usually hands and fingers, are locked in muscle contraction at the onset of the agonal period.

calyptrate   Dipterans that posses calypters, membranous flaps below the hind wings that typically cover the halteres.

cantharidin   Toxic allomone synthesized by some blister beetles (Family Meloidae) that causes skin lesions and which can be lethal if ingested or it enters body fluids.

capital breeder   Female that does not need to feed as an adult to produce first clutch of eggs as all the nutrients and energy for egg provisioning were acquired during larval stages.

carbamates   Are a group of insecticides derived from carbamic acids.

carrion   A dead animal carcass.

chain of custody   Collection of individuals responsible for maintaining continuity of evidence.

chelae   The pincers located on the pedipalps of scorpions and related chelicerates.

chelicerae   Mouthparts of chelicerates.

chemoreception   Detection of chemical stimuli in the internal or external environment, although generally refers to outside the body.

chemotaxis   Innate behavior in which longitudinal axis of body is oriented toward (positive) or away from (negative) a chemical functioning as an external stimulus.

chill-coma   Cessation of movement that results when temperatures are at or below the critical thermal minimum.

chilling injury   Damage is incurred following a short (direct chilling injury) or prolonged (indirect chilling injury) exposure to low temperatures above freezing.

chorion   Outer shell of an insect egg, lying just outside the vitelline envelope.

civil law   Branch of law that deals with disputes between individuals and organizations over private matters; attempts to right a wrong.

clade   A group consisting of a single species and all its decedents.

classical (Pavlovian) conditioning   Form of associative learning in which the organism is trained to respond to a stimulus (conditioned stimulus) because a reward or punishment (unconditioned stimulus) is associated with the conditioned stimulus.

climatic climax vegetation   Vegetative fauna that has achieved a steady-state condition through ecological succession in a given geographic area.

cohesion   Attraction of like molecules to each other and held together by an intramolecular force throughout the mass of an object.

cold hardiness   Acclimatization to low temperatures as a means to avoid chilling or freezing injury.

cold shock   Rapid unexpected decline in ambient temperatures that may or may not drop below zero and which constitutes a low-temperature stress.

collectors   Aquatic invertebrates that are detritivores, filter feeders or suspension feeders of fine particulate organic matter.

common oviduct   Duct in female reproductive system that connects lateral oviducts to the vagina or bursa copulatrix.

compatible solute   Solutes in body fluids that are inert and which do not interfere with metabolic processes.

conduction   Transference of heat between two stationary objects or bodies.

conspecific   Of or pertaining to individuals of the same species.

continuity of evidence   Physical evidence of a crime is accounted for at all times from the moment it is collected at a crime scene until presented in court.

control variable   Independent variable that could potentially alter what is to be measured and which must be maintained at a constant or static state to decrease its influence on an experimental outcome.

convection   Transfer of heat from liquid or gas by circulation from one region to another.

convergent evolution   Independent evolution of similar features in species of different periods or epochs in time, which yields analogous structures that have similar form or function but were not present in the last common ancestor of those groups.

coprophagous   Insects that breed/feed in dung of other animals.

corpora allata (corpus allatum)   Endocrine glands that produces juvenile hormones. In Diptera, these glands are fused with the corpora cardiaca to form a ring gland.

corpus delicti   The elements or facts needed to "prove" a case.

crime   Any act that violates public law.

criminal intent   Mental state of an individual when committing an overt criminal act.

criminalistics   Branch of forensic science focused on the specific activities conducted by a crime or forensic laboratory.

criminal law   Branch of law that deals with failure to abide by public law; law that deals with crime.

critical thermal maximum   Upper temperature conditions that inhibit aerobic metabolism and denature proteins in cells.

critical thermal minimum   Lower temperature conditions that lead to cold temperature injury and death if exposure is long enough.

Crow–Glassman Scale (CGS)   Subjective rating system to characterize levels of burn injury to a human body.

cryoprotectants   Osmolytes that function to prevent damage during low temperatures commonly by depressing the supercooling point of fluids. Cryoprotectants are typically classified as colligative or non-colligative depending on whether protection is gained through concentration or unique chemical properties.

cryoprotective dehydration   Low-temperature strategy that depends on water loss to the environment and accumulation of non-colligative cryoprotectants to maintain vapor pressure equilibrium with environmental ice.

cryptic species   Species that are morphologically indistinguishable.

CSI effect   Distorted public view of what forensic science can do based on crime shows on television versus the realities of forensic investigation.

cuticle   Non-cellular, outer layer of the exoskeleton produced by the epidermis.

cyanosis   Blue-red coloration of skin that can be the result of exposure to low temperatures, including in water.

cyclorrhaphous   Of the suborder Cyclorrhapha in the order Diptera. All members have circular seams in terms of sutures associated with the aperture used during eclosion from puparia.

cytokines   Small proteins that function in cell–cell communication, modulating functions of immune cells in response to antigens, and which can display cytotoxicity to foreign cells/organisms.

defecatory stain   Type of insect artifact in which spots or specks form from the feces of an insect.

defect action level (DAF)   Amount of naturally occurring or non-preventable defects in food that present no health hazard to humans if consumed. Levels are established by the United States Food and Drug Administration.

delusional parasitosis (or infestation)   Clinical term referencing a broad range of psychopathological conditions characterized by individuals' fixation that they are infested with small pathogens or animals (e.g., insects, "worms") on the skin or that are internal.

dependent variable   Factor or feature to be measured as an outcome of an experiment as influenced by other factors or independent variables.

developmental limit   Temperature threshold below which development does not occur. *See also* base temperature.

diapause   Dynamic physiological state of dormancy commonly associated with winter; consequently, enhanced cold hardiness occurs as part of the diapause program.

disarticulation   Separation of body parts or bones from the skeleton, often as the result of animal activity, blunt force trauma, or fluvial transport.

DNA alignment   DNA sequences of multiple individuals from corresponding regions in the genome that are aligned so that all sequences have the same starting and ending points.

DNA barcoding   A molecular technique utilized by taxonomists where short specific portions (markers) of a specimen's DNA sequence is used to identify it as a member of a particular species.

DNA profiling   Use of DNA for person's identification, especially with regard to deceased individuals associated with suspicious deaths or homicides.

ecdysis   Process of shedding the old exoskeleton during molting.

eclosion   Refers to either a neonate larva hatching from an egg or an imago emerging from pupal casing.

ecological model   Representation of the dynamics and conceptualization of a living ecosystem.

economic-injury level (EIL)   Lowest number of insects that will evoke damage to a crop or product of interest.

ectatomin   Low-molecular-weight peptide in the venom of *Ectatomma tuberculatum* that is lethal to humans.

ectotherm   Animal that does not regulate internal body temperature.

edema   Swelling of tissues or cells by interstitial fluid.

embalm   Practice of preserving a body after death to temporarily delay decomposition.

entomological terrorism   Use of insects as biological weapons by a militant group or country toward another group of people or nation.

entomotoxicology (or forensic entomotoxicology)   Subfield of forensic entomology that deals with analysis of chemical compounds present in arthropods that fed on a human corpse.

envenomation   Process of injecting venom through a bite or sting into potential prey or host or as part of a defensive response.

environmental token   Feature of the environment such as photoperiod, humidity, and temperature that signifies aspects about climatic conditions.

ephemeral   Lasting a brief time, or transitory, such as a non-predictable resource.

epinecroticmicrobiome   Microbial community found on external surfaces. The external surfaces include superficial epithelial tissues, mucosal membranes of the buccal cavity, and the lining of the digestive tract.

evaporative cooling   Loss of heat through evaporation of water from a body surface.

exophilic   Insects associated with archaeological sites independent of human structures.

exploitive competition   Occurs when an organisms uses a limited resource that diminishes amount available for others.

extra-oral digestion   Initiation of chemical digestion outside the digestive tract of an animal.

extrication   Behaviors associated with an adult fly breaking out of a puparium and removing obstacles in its path.

false negative   When a xenobiotic is not detected in insect tissues during toxicological testing but there is some type of drug, toxicant or metabolite present.

false positive   When a xenobiotic is detected in insect tissues during toxicological testing but no drug, toxicant or metabolite is present.

faunal succession   Colonization of carrion by insects based on stage of decomposition.

felony   Most serious violation of public law (crime) and includes such acts as murder, rape, and armed robbery.

fibrinolysis   Digestion or destruction of whole blood clots and fibrin plaques.

fitness   Propensity to survive and reproduce, either as an individual or as population.

fluvial transport   Movement of human or animal remains by water currents from the original point of entry into an aquatic system.

fly spot (speck)   Type of insect artifact in which either regurgitate or feces is deposited in a location that potentially confounds bloodstain evidence.

food assimilation   Conversion of food nutrients into cells and tissues of body; in same cases also includes use of nutrients as cellular energy.

forcipules   Appendages of centipedes which represent modified legs that form a pair of pincers with the venom glands extending to the tips.

forensic archaeology   The field that applies archaeological principles and techniques to matters of legal interest.

forensic microbiology   The subdiscipline of microbiology focused on the role of microorganisms in medicolegal and criminal investigations.

forensic profiling   Process of characterizing the patterns or other features of the modus operandi of a criminal in an effort to apprehend the individual before committing another crime.

forensic science   Application of science to help resolve civil and criminal matters through a judicial system.

frass   Digestive and/or metabolic waste materials passed from digestive tract via anus to external environment.

freeze avoidance   Cold strategy allowing an animal to supercool body fluids without ice formation even at subzero temperatures.

freeze tolerance   Cold strategy that relies on freezing or formation of ice in extracellular fluids as means to deal with low temperatures.

freezing injury   Damage that occurs when ambient temperatures drop below zero and ice forms in body fluids.

frenetic movement   Locomotory activity of fly larvae in feeding aggregations.

froth cone   Small bubbles formed around the nasal and mouth openings, commonly associated with drowning or asphyxiation.

funerary archaeoentomology   Field of research examining insect association with ancient taphonomy and mortuary practices.

Goldilocks effect (phenomenon)   A phrase used to indicate that conditions must be "just right" for an event to occur. This term is used in association with adipocere formation, referencing the conditions necessary for saponification to occur in soft tissues of a corpse.

gravid   Condition in female insects in which mature oocytes have been produced and are ready for oviposition.

gregarious   Term frequently used in reference to clutch sizes greater than one individual (solitary), usually implying that "many" eggs or larvae per brood are deposited in the same location.

guild   Group of organisms (insects) that exploit the same resource (carrion) in a similar way.

gustation   Sense of taste.

halteres   Modification of the hind wings of Diptera into gyroscope-like structures.

heat shock   Rapid unexpected rise in ambient temperatures that constitutes a thermal stress.

heat shock proteins (Hsp)   Array of stress proteins synthesized in response to several types of environmental stressors including temperature, desiccation, anoxia, and overcrowding.

heat stupor   Cessation of movement that results from exposure to temperatures at or above the critical thermal maximum.

hematophagous   Blood-feeding insects.

hemimetabolous   Type of development displayed by insects whereby the immature stages gradually metamorphose in adults through series of growth and molting events.

heraldry   Practice of using symbols, typically in the form of animals, to depict an army, kingdom, tribe, etc.

heterogeneous   Non-uniform in reference to composition, such as in multiple species being present.

heterothermy   Process of generating internal heat by an ectotherm.

holometabolous   Type of development displayed by insects in which immatures metamorphose into an adult through a transition stage known as a pupa.

homeothermy   Maintenance of a stable internal body temperature regardless of external environmental conditions.

homogeneous   Uniform in composition, as in of the same kind.

hypervariable loci   Loci that are likely to be divergent between individuals, or populations, of the same species.

hypogean   Insects that are subterranean.

hypothesis   Explanation to account for observation of an event or natural phenomenon.

hysteretic freezing point   Lower temperature required for ice formation due to the action of antifreeze proteins.

ice nucleating agent (ICN)   Object that can serve as a site for condensation of water molecules before ice crystallization.

imaginal discs   Undifferentiated tissue found in holometabolous insects that gives rise to many adult structures during pupal–pharate adult development.

imago   Adult insect.

imbibe   Act of drinking or absorbing a liquid.

immunoglobulins   Proteins that function in adaptive immune responses including as antibodies, in modulating immune cell function, and in antigen presentation.

income breeder   Female that requires food as an adult to acquire nutrients and/or energy to provision eggs.

independent variable   Factor in an experiment that can potentially affect a desired outcome or feature to be measured.

inoculative release   Release of small numbers of individuals during a seasonably favorable period that permits establishment of the species in that area.

insect artifact   Spots or specks derived from the insect digestive tract; represent either regurgitate or feces. The term is also used to describe blood spots, stains or streaks created when an insect walks through wet blood and distorts the existing stain or produces a new one.

insect stain   Bloodstain resulting from insect activity.

insect (faunal) succession   Progressive arrival or colonization of insects in an ecological community, such as carrion, that augments, competes, or replaces a prior insect community.

insecticide toxicology   Discipline that has a primary focus on any organism affected by insecticides, not just insects.

integument   Outer covering of an insect composed of the epidermis and cuticle.

interspecific   Occurring between individuals of different species, as in allelochemicals used as chemical signals between allospecifics.

intraspecific   Within the same species, such as occurs with chemical signals like pheromones used to convey messages to conspecifics.

inundative release   Release of masses of individuals to essentially overwhelm an area and produce an immediate response.

isothermal   Temperature of two objects or organisms is equal to one another.

kairomone   Chemical signal that benefits the receiver but generally harms the originator of the message.

larval-mass effect   The concept that depending on the number of individual fly larvae in a feeding aggregation and environmental temperatures, larvae release heat that has the potential to increase the local temperature of the mass and surrounding habitat.

latent heat of vaporization   Heat required to evaporate a liquid.

latrotoxin   Neurotoxic peptide produced by black widow spiders.

livor mortis   Non-moving fluids settle to lower portions of body due to gravity.

Locard's exchange principle   Any interaction between two individuals leads to transference of materials that can serve as trace evidence.

loxoscelism   Tissue necrosis resulting from venom of the brown recluse spider.

maggot mass   Feeding aggregation generally formed by some species of necrophagous flies during larval stages.

maggot therapy   Treatment of wounds or injuries with necrophagous fly larvae as a means to remove necrotic tissue (débride) and clean site of microorganisms.

mandaratoxin   Lethal neurotoxin found in venom of large East Asian *Vespa* species.

marbeling   Mosaic pattern of skin discoloration resulting from formation of sulfhemoglobin in capillaries near skin surface.

mastication   Physical manipulation or breakdown of food materials either outside the body or once in the digestive system.

mastoparan   Most abundant peptide in venom from vespid wasps. Produces a wide range of cellular changes in several cell types.

mechanoreception   Sense of touch.

meconium   Excretory products that accumulate within the digestive tract during pupal development that are purged at emergence.

medicocriminal (medicolegal) entomology   Branch of forensic entomology focused on the use of insects or related arthropods to help solve crimes, particularly those that involve violence (i.e., homicides).

microbiome (microbiota)   A community of microorganisms found on or in a living or dead organism.

midden   A trash heap or garbage refuge.

misdemeanor   Less severe acts of crime.

modus operandi   Habits or characteristics of a criminal evident from repeated crimes.

molluscicides   Are pesticides that are often used in agriculture to control unwanted snails and slugs.

molt   Complex series of physiological events that leads to shedding of the old "skin" (*see* ecdysis) and synthesis of a new one.

monophagous herbivore   Animal that feeds on one or a few members of a single plant genus.

monophyletic group   Taxa that form a group consisting of a single species and all its descendants.

monovoltine   Production of only one generation per year.

mouth hooks   Modified mandibles of larval Diptera used for food acquisition by piercing and scraping food substrate.

mummification   Process of rapidly drying soft tissues under high heat and low moisture to yield shrunken dry soft tissues.

myiasis   Invasion or infestation of living or necrotic tissue of a host by flies during the larval stages.

necrobiome   A community composed of prokaryotic and eukaryotic organisms united in the attraction and utilization of a decomposing heterotrophic mass, i.e., animal remains.

necrophagous   Feeding on dead tissue, typically as carrion or a corpse.

necrophilous   Behavioral attraction to carrion.

necrosis   Physiological changes that occur in cells or tissues after death.

negative detection   When no xenobiotic is found in insect tissues following feeding on a human corpse.

next-generation sequencing (massive parallel sequencing)   A broad term that encompasses several modern sequencing techniques. All share the characteristics of permitting sequencing of nucleic acids much more rapidly and cheaply than the previously used Sanger sequencing.

niche   How an individual or population responds to resources in an ecosystem through utilization or modification of the resources.

non-Newtonian fluid   Fluid whose viscosity depends on the shear rate or shear history.

olfaction   Sense of smell.

oligolaraviparity   Condition in which adult female fly contains more than one precocious egg or larva in the oviduct.

oligophagous herbivore   Animal that feeds on select genera of plants within a single family.

oncosis   Form of cell death that can be induced through mechanical injury to a cell, pore or lesion formation, or possibly by stimuli triggering cellular pathways that lead to death. In older literature, the term "necrosis" was used in place of oncosis.

oocyte   Gametes of female insects prior to maturation.

oogenesis   Female gametogenesis or the formation of an egg cell.

osmotic dehydration   Water loss due to accumulation of osmolytes.

ovariole   Portion of ovary responsible for producing and maturing eggs.

overt criminal act   Willful act of violating public law.

oviparity   Mechanism of progeny deposition in which eggs with a chorion are passed from the mother's body into the environment.

oviposition   Act or process of laying eggs (oviparity).

ovisac   Capsule or pouch containing an egg or ovum.

ovoviviparity   Form of vivpary in which eggs with a chorion hatch inside mother before larvae are deposited into environment.

paleoentomology   Study of fossilized insects in an effort to learn more about the environment, biology, and activity of insects long ago.

parasitoid   Specialized insect parasite that always (usually) kills its host as a result of the association.

parasitoidism   A relation existing between various insect larvae and their hosts in which the larva feeds upon the living host tissues in an orderly sequence such that the host is not killed until the larval development is complete.

parturition   Process of giving birth, as occurs with insects that display true viviparity.

pedipalps   Second pair of appendages located anteriorly on the prosoma or cephalothorax of members of the subphylum Chelicerata.

pelage   Fur on a vertebrate animal.

phanerocephalic stage   Time of development between pupa–adult apolysis after head evagination.

pheromone   Chemical signals released from exocrine glands for intraspecific communication.

phylum (pleural phyla)   Major taxonomic category in the Linnaean hierarchy that ranks below kingdom.

physical evidence   Any part of all of a material object used to establish a fact in criminal or civil case.

piperidines   Alkaloid compounds that serve as the active toxic ingredients of fire ant venoms.

pleurite   Sclerite located on lateral surface of insect.

poikilothermic   Body temperature reflects ambient conditions and is not maintained via endothermic heat production.

polymerase chain reaction (PCR)   Biochemical technique used to amplify small amounts of DNA to millions of copies of a particular sequence using thermal cycling.

polyphagous herbivore   Animal that feeds on a wide variety of plant species.

poneratoxin   Neurotoxic peptide produced by the ant *Paraponera clavata* that evokes excruciating pain.

positive detection   When a drug or toxicant has been detected in insect tissues through toxicological testing.

positive identification   Process of individualization that leads to correct identification of an object or person.

postcolonization interval   Phase of postmortem interval which starts with physical contact and extends through consumption of the remains by arthropods and their offspring.

postmortem   After or following death.

Postmortem *Clostridium* effect (PCE)   Essentially a term recognizing the omnipresence of *Clostridium* spp. on a decaying corpse.

postmortem interval (PMI)   Time elapsed since death.

postmortem microbiome   Community that is specifically associated with a human corpse or vertebrate carrion.

postmortem submersion interval (PMSI)   Time elapsed since body enters a water environment until when the decedent is discovered.

precocious egg development   Retention of eggs by oviparous insects until after egg hatch.

precolonization interval   First phase of postmortem interval extending from time of death until the body is detected by arthropods.

proteolysis   Decay of proteins during autolysis.

proteotaxic stress   Thermally unfavorable conditions for an organism, usually in reference to high temperatures.

pseudogenes   Nuclear DNA sequences similar to a different gene that has lost its functionality.

pseudopustule   Capsule-like structure that forms from human skin filled with interstitial fluid following envenomation by fire ants in the genus *Solenopsis*.

ptilinum   Eversible pouch located on the head above the antennae of cyclorrhaphous Diptera to break open the puparium and push soil and other objects away during adult eclosion.

pulvillus (plural pulvilli)   Soft cushion-like pad located between terminal claws in Diptera.

pupa   Stage of development associated with holometabolous insects following completion of larval development, in which transformation to the adult takes place.

pupariate   Development event unique to Diptera in which last stage larvae initiates formation of puparium to serve as outer protective covering of the pupal stage.

puparium (plural puparia)   Sclerotized exoskeleton (exuvium) of last stage larva of some dipterans that surrounds the pupa.

putrefaction   Chemical degradation of soft tissues due to the action of microbes, principally bacteria.

qualitative analysis   Type of investigation designed to classify or group an object or organism.

quantitative analysis   Type of investigation designed to determine the amount of an object or substance.

Quaternary entomology   Essentially a subfield of paleoentomology focused on fossilized insects from the Quaternary period.

quiescence   Physiological state in which metabolic activity is lowered during unfavorable conditions and which returns to normal immediately on return to favorable conditions.

quinone   Type of organic compound derived from an aromatic precursor such as benzene and used in chemical defense by several species of arthropods.

quorum sensing   A mechanism used by bacteria to regulate gene expression based on population density using signal molecules.

rapid cold hardening (RCH)   Cold tolerance acquired from a brief exposure to non-damaging low temperature prior to encountering a lower, potentially lethal temperature.

regurgitate   Expelled food material from foregut of flies in the form of liquid, usually associated with bubbling behavior.

resource partitioning   Concept that when two or more organisms are competing for the same limited resources, they coexist by using the resource differently.

restriction enzymes   Are proteins that bind to the DNA and then digest it at a specific site (location), based upon a match between the DNA sequence and the specific restriction enzyme.

retroinvasion   Infestation of the anus by fly larvae after they have passed through the alimentary canal.

rigor mortis   Stiffening of muscles due to unregulated muscle contractions following death.

riparian buffer   Strip of land that is usually forested near a stream or creek that functions to protect water from other land uses, such as agriculture or urbanization.

rodenticides   Are pesticides that are commonly used to control rat and mice pest problems.

sanguinivorous   Blood feeding.

saprophagous   Feeding on decaying plant and/or animal material.

satiety   Condition of fullness or meeting a specific nutritional need that in turn inhibits hunger drive.

scallop (or spine)   Edge morphology of a bloodstain. Generally used in reference to several sharp edges rather than a single long tail associated with a bloodstain.

scene impression   Markings of an object like a tire or shoe print made in material like dirt or mud that leave a transference image.

scent gland   Sex pheromone-producing glands found in some Lepidoptera.

scientific method   Method of testing using an approach centered on formulating hypotheses, making observations from carefully designed experiments, refining questions, and narrowing possible explanations for observed phenomena.

sclerite   Hard rigid plates of exoskeleton.

scrapers   Aquatic invertebrates that are herbivores and grazing scrapers of mineral and organic surfaces.

semiochemical   Chemicals released into the environment to modify the behavior and/or physiology of the receiver.

sensillum (plural sensilla)   Chemical receptors used by insects to detect chemical stimuli in the external environment.

short tandem repeats   repeating sequences of DNA, typically 2–6 base pairs in length

shredders   Aquatic invertebrates that are detritivores and chewers of particulate organic matter.

spatial aggregation   Separation or formation of distance between masses of organisms.

spatial partitioning   Separation between organisms or groups on the same resource to decrease competition.

spermatheca   Modified accessory gland in adult females that stores sperm following insemination.

spermatophore   Protective sac produced by male insects to deliver spermatozoa to females.

spiracles   External openings of the ventilatory system.

statute of limitations   Code or enactment in a common law legal system that establishes a maximum time period after an event in which legal recourse may be pursued based on that event.

sternite   Sclerite located on the ventral surface of an insect.

stored product entomology   Branch of forensic entomology concerned with legal proceedings stemming from insect or arthropod presence in food and food products.

successional interval   An estimate the amount of time that human remains were present at a specific location based on the pattern of insect colonization.

sulci   Internal cuticular invaginations that serve to increase the rigidity of the exoskeleton.

supercooling   Lowering of the temperature of body fluids well below 0 °C without ice formation.

supercooling point (SCP)   Temperature at which spontaneous ice crystals form in body fluids. The term is used interchangeably with nucleation temperature and temperature of crystallization.

supine   Lying in a flat position.

surface tension   Property of the surface of a liquid to resist an external force.

suture   Internal invagination of the exoskeleton.

synanthropic   In close association with humans, such as insect species that frequent domiciles or other artificial structures or which depend on refuge for food.

synomone   Chemical signal that produces a response in the receiver and which is beneficial to both the recipient and emitter.

Tache noire (Tache noire de la sclerotique)   French term for "black spot of the sclera." It occurs when the eyes are not completely closed at the moment of death, causing the sclera to dry out.

Tardieu spots   Often occur with lividity or livor mortis, in which capillaries burst due to pooling of blood.

tagmosis   Evolutionary process that involves the modification of body segments into functional units, like the head, thorax, and abdomen.

taphonomy   Study of decomposing organisms over time, including the processes leading to fossilized remains.

tarsal tracks   Bloodied impressions of footpads or tarsi are analogous to "insect footprints."

tergite   Sclerite located on the dorsal surface of an insect.

terrestrial ecozone   Classification system that divides the Earth's land surfaces based on the distribution patterns of terrestrial organisms.

terrorism   Use of terror, usually through acts of violence, in the name of religion, politics, or some other ideological purpose, with no regard for non-combatants.

test impression   Markings of an object like a tire or shoe print made in a laboratory in an attempt to match it to impressions found at a crime scene.

tetany   Condition in which muscles are prevented from relaxing.

thanatomicrobiome   The microbial community located within the corpse including internal organs and cavities enclosed within the body.

thermal hysteresis   Protection from freezing that results from antifreeze proteins binding to ice lattices, which in turn lowers the temperature that allows ice to grow.

thermal tolerance range   *See* zone of tolerance.

thigmotaxis   Type of innate behavior in which body orientation or locomotion of an organism is in response to physical contact with a solid object.

tort   Dispute between private individuals or organizations involving accidents, neglect, or libel.

total body score   Method of assigning numerical values to physical decomposition of human remains as a means of assessment and estimating the postmortem interval.

trace evidence   Minute amounts of physical evidence that can be used to make a connection between two individuals, typically a victim and a suspect, or suspect and crime scene.

transference (translocation)   Deposition of stains from blood-covered parts (in the case of insects) to a non-bloodied surface.

tropism   Preferential movement of the compound and metabolites to specific tissues within the body.

urban entomology   Branch of forensic entomology dealing with legal actions associated with nuisance or destructive activity of insects or arthropods in and around human structures.

urticating hairs   Stinging hairs or setae associated with some insects, namely certain caterpillars.

ventriculus   Midgut region of the digestive tract.

vesicant   A blistering agent.

viscosity   Property of a fluid that describes its resistance to flow; usually thick and sticky.

vitellogenesis   Formation of yolk in oocytes by deposition of nutrients into the cytoplasm.

viviparity   Form of vivipary in which eggs without a chorion hatch in mother and parent provides nutrients beyond yolk to larvae.

vivipary   Live birth.

voucher specimen   Representative example of a particular animal, often an insect, used to confirm the identity of an individual collected in the field or as part of a forensic examination.

wandering   Phase of fly development in which third-stage larvae that have completed feeding crawl away from food source seeking a location to initiate pupariation.

wicking   Flow of a liquid by capillary action in narrow spaces, without the need for an external force like gravity, to facilitate the movement.

xenobiotic   A substance that is foreign or exogenous to the body or ecosystem.

yolk   Nutrients located in mature oocyte or egg.

zone of tolerance   Range of temperatures over which an organism can survive indefinitely.

Index

Note: Page numbers in *italics* refer to Figures; those in **bold** to Tables

---

The Science of Forensic Entomology, Second Edition. David B. Rivers and Gregory A. Dahlem.
© 2023 John Wiley & Sons Ltd. Published 2023 by John Wiley & Sons Ltd.
Companian website: www.wiley.com/go/forensicentomology2

Printed and bound by CPI Group (UK) Ltd, Croydon, CR0 4YY

27/10/2024

14580166-0001